T0093985

Advanced
Particle Physics
Volume II

Advanced
Particle Physics
Volume II

The Standard Model
and Beyond

O. M. Boyarkin

CRC Press
Taylor & Francis Group
Boca Raton London New York

CRC Press is an imprint of the
Taylor & Francis Group, an **informa** business

CRC Press
Taylor & Francis
6000 Broken Sound Parkway NW, Suite 300
Boca Raton, FL 33487-2742

First issued in paperback 2018

© 2011 by Taylor and Francis Group, LLC
CRC Press is an imprint of Taylor & Francis Group, an Informa business

No claim to original U.S. Government works

ISBN-13: 978-1-4398-0416-2 (hbk)
ISBN-13: 978-1-138-37411-9 (pbk)

This book contains information obtained from authentic and highly regarded sources. Reasonable efforts have been made to publish reliable data and information, but the author and publisher cannot assume responsibility for the validity of all materials or the consequences of their use. The authors and publishers have attempted to trace the copyright holders of all material reproduced in this publication and apologize to copyright holders if permission to publish in this form has not been obtained. If any copyright material has not been acknowledged please write and let us know so we may rectify in any future reprint.

Except as permitted under U.S. Copyright Law, no part of this book may be reprinted, reproduced, transmitted, or utilized in any form by any electronic, mechanical, or other means, now known or hereafter invented, including photocopying, microfilming, and recording, or in any information storage or retrieval system, without written permission from the publishers.

For permission to photocopy or use material electronically from this work, please access www.copyright.com (http://www.copyright.com/) or contact the Copyright Clearance Center, Inc. (CCC), 222 Rosewood Drive, Danvers, MA 01923, 978-750-8400. CCC is a not-for-profit organization that provides licenses and registration for a variety of users. For organizations that have been granted a photocopy license by the CCC, a separate system of payment has been arranged.

Trademark Notice: Product or corporate names may be trademarks or registered trademarks, and are used only for identification and explanation without intent to infringe.

Visit the Taylor & Francis Web site at
http://www.taylorandfrancis.com

and the CRC Press Web site at
http://www.crcpress.com

To all
  those whom
I sweetly
  was deceived by.

Contents

Part I

Quantum chromodynamics

1

Canonical quantization

> *Thank God, that he has created the world*
> *such, that everything, that is important in it,*
> *is simple, and that is unimportant is difficult.*
> G. Skovoroda, wandering Ukrainian
> philosopher of Eighteenth Century

1.1 Fundamental relations of chromodynamics

In Section 10 we have already gained some insight of the modern theory of the strong interaction. Before proceeding to the quantization procedure, it is worthwhile to systematize and expand our knowledge relative to this theory. This chapter is devoted to this task. In so doing it will be assumed that we deal with classical field theory.

The chromodynamics symmetry group $SU(3)_c$ is described by generators T^a to obey the commutation relations

$$[T^a, T^b] = i f^{abc} T^c, \tag{1.1}$$

where f^{abc} are structural constants ($a, b, c = 1, \ldots, 8$). Generators in the associate representation (dimensionality equals eight) are expressed in terms of the structural constants

$$\left. \begin{array}{c} \left(T^a_{adj} \right)^{bc} = -i f^{abc}, \\[2mm] T^a_{adj} T^a_{adj} = 3I, \quad \mathrm{Sp}\left(T^a_{adj} T^b_{adj} \right) = 3\delta^{ab}. \end{array} \right\} \tag{1.2}$$

In its turn, in the fundamental representation (dimensionality equals three) generators are given by the set of the Gell–Mann matrices λ

$$T^a_{fund} = \frac{1}{2}\lambda^a, \quad T^a_{fund} T^a_{fund} = \frac{4}{3}I, \quad \mathrm{Sp}\left(T^a_{fund} T^b_{fund} \right) = \frac{1}{2}\delta^{ab}. \tag{1.3}$$

The quark field q_j belongs to the fundamental representation of the $SU_c(3)$, that is, it has three components in a color space ($j = 1, 2, 3$). The gluon field falls into the associate representation ($a = 1, \ldots, 8$). Gluon field strengths $G^a_{\mu\nu}$ are expressed by potentials G^a_μ in the following way

$$G^a_{\mu\nu} = \partial_\mu G^a_\nu - \partial_\nu G^a_\mu + g_s f^{abc} G^b_\mu G^c_\nu. \tag{1.4}$$

Sometimes it is more convenient to use the matrix writing down for quantities entering into the theory rather than the componentwise one. Thus, for example, we have for the gluon fields potentials and strengths:

$$G_\mu = T^a_{adj} G^a_\mu, \quad G_{\mu\nu} = T^a_{adj} G^a_{\mu\nu}. \tag{1.5}$$

By virtue of the commutation relation (1.1), we have

$$G_{\mu\nu} = \partial_\mu G_\nu - \partial_\nu G_\mu - ig_s[G_\mu, G_\nu]. \tag{1.6}$$

Acting the covariant derivative D_μ on the gluon fields G_ν is defined as follows:

$$D_\mu G_\nu = \partial_\mu G_\nu - ig_s[G_\mu, G_\nu],$$

or in the componentwise written as

$$D_\mu G_\nu^a = [\partial_\mu \delta_{ac} + g_s f^{abc} G_\mu^b] G_\nu^c. \tag{1.7}$$

On the other hand, for the quark field q_j we have

$$D_\mu q_j = [\partial_\mu \delta_{jk} - ig_s (T_{fund}^a)_{jk} G_\mu^a] q_k, \tag{1.8}$$

where j, k are color indices. Under the covariant differentiation Jacobi's identity is fulfilled

$$[[D_\mu, D_\nu], D_\rho] + [[D_\nu, D_\rho], D_\mu] + [[D_\rho, D_\mu], D_\nu] = 0.$$

Reasoning from the covariant derivative definition it is easy to check that

$$[D_\mu, D_\nu] G_\rho = -ig_s[G_{\mu\nu}, G_\rho].$$

For the gluon field strengths Bianchi's identity takes place

$$D_\rho G_{\mu\nu} + D_\mu G_{\nu\rho} + D_\nu G_{\rho\mu} = 0. \tag{1.9}$$

At the gauge transformation with parameters $\alpha^a(x)$ the gluon field tensor and potentials are transformed according to the formulae

$$\left. \begin{array}{c} \mathbf{G}_{\mu\nu} \to e^{-i\alpha} G_{\mu\nu} e^{i\alpha}, \\ G_\mu \to e^{-i\alpha} G_\mu e^{i\alpha} - ig_s^{-1} \left(\partial_\mu e^{-i\alpha} \right) e^{i\alpha}, \end{array} \right\} \tag{1.10}$$

where $\alpha = \alpha^a T_{adj}^a$. In the case of the infinitesimal transformations Eqs. (1.10) take the form

$$\left. \begin{array}{c} G_{\mu\nu} \to G_{\mu\nu} - i[\delta\alpha, G_{\mu\nu}], \\ G_\mu \to G_\mu - i[\delta\alpha, G_\mu] - g_s^{-1} \partial_\mu \delta\alpha = G_\mu - g_s^{-1} D_\mu \delta\alpha. \end{array} \right\} \tag{1.11}$$

In the componentwise writing down the infinitesimal gauge transformations will look like:

$$G_\mu^a \to G_\mu^a + f^{abc} \delta\alpha^b G_\mu^c - g_s^{-1} \partial_\mu \delta\alpha^a, \tag{1.12}$$

for the gluon field and

$$q_j \to [\delta_{jk} + i(T_{fund}^a)_{jk} \delta\alpha^a] q_k, \tag{1.13}$$

for the quark field. The total Lagrangian describing the gluon and quarks fields is given by the expression

$$\mathcal{L}_{QCD} = \frac{i}{2} \{ \overline{q}(x)\gamma^\mu [\partial_\mu q(x)] - [\partial_\mu \overline{q}(x)]\gamma^\mu q(x) \} + g_s \overline{q}(x)\gamma^\mu T_{fund}^a G_\mu^a(x) q(x) -$$

$$-m\overline{q}(x)q(x) - \frac{1}{4} G_{\mu\nu}^a(x) G^{a\mu\nu}(x). \tag{1.14}$$

Using the least action principle, we obtain the following equations of motion

$$
\left.\begin{array}{l}
i\gamma_\mu\partial^\mu q(x) + g_s\gamma_\mu G^\mu(x)q(x) - mq(x) = 0, \\
-i[\partial^\mu\overline{q}(x)]\gamma_\mu + g_s\overline{q}(x)\gamma_\mu G^\mu(x) - m\overline{q}(x) = 0.
\end{array}\right\}
\tag{1.15}
$$

for the quark field and

$$
D_\nu(x)G^{\mu\nu}(x) = g_s J^\mu(x),
\tag{1.16}
$$

where

$$
J_\mu^a(x) = \overline{q}(x)\gamma_\mu T_{fund}^a q(x)
\tag{1.17}
$$

is a fermion current, for the gluon field. Taking into account Eq. (1.16) one may show that the current J_μ^a satisfies the so-called compatibility condition

$$
D_\mu(x)J^\mu(x) \equiv \partial_\mu J^\mu(x) - ig_s[G_\mu(x), J^\mu(x)] = 0.
\tag{1.18}
$$

From Eq. (1.16) follows that $\partial_\mu J^\mu(x)$ is determined by the expression

$$
\partial_\mu J^\mu(x) = -i\partial_\mu[G_\nu(x), G^{\mu\nu}(x)].
\tag{1.19}
$$

Now we can define the conserved total color current

$$
J_c^\mu(x) \equiv J^\mu(x) + i[G_\nu(x), G^{\mu\nu}(x)]
\tag{1.20}
$$

and the corresponding charge

$$
Q_c = g_s \int d^3x J_c^0(x) = \int d^3x \partial_i G^{0i}(x) = \oint_S dS_i G^{0i}(x).
\tag{1.21}
$$

Note that Q_c is not changed under the gauge transformations to vanish at infinity.

If we wish to consider external sources J_a^μ of the Yang–Mills fields being given in the space-time (for example, heavy quarks), then we cannot set their color components arbitrarily. The compatibility condition demands orientations of J_a^μ in the color space to be considered as dynamical observables connected with the field G_a^μ.

By analogy with electrodynamics we introduce chromoelectric and chromomagnetic fields

$$
E_{ai} = G_{a0i} = G_a^{i0}, \quad H_{ai} = -\frac{1}{2}\varepsilon^{ijk}G_a^{jk},
\tag{1.22}
$$

that is,

$$
G^{\mu\nu} = \begin{pmatrix}
0 & -E_1 & -E_2 & -E_3 \\
E_1 & 0 & -H_3 & H_2 \\
E_2 & H_3 & 0 & -H_1 \\
E_3 & -H_2 & H_1 & 0
\end{pmatrix}.
\tag{1.23}
$$

The electric and magnetic fields are expressed in terms of the potentials in a nonlinear way:

$$
\left.\begin{array}{l}
\mathbf{E}_a = -\partial_0\mathbf{G}_a - \nabla G_a^0 + g_s f^{abc}\mathbf{G}_b G_c^0, \\
\mathbf{H}_a = [\nabla \times \mathbf{G}_a] - \frac{1}{2}g_s f^{abc}[\mathbf{G}_b \times \mathbf{G}_c].
\end{array}\right\}
\tag{1.24}
$$

Let us rewrite the motion equations for the gluon field in terms of chromoelectric and chromomagnetic fields

$$
\left.\begin{array}{l}
-\nabla \cdot \mathbf{E}_a + g_s f^{abc}\mathbf{G}_b \cdot \mathbf{E}_c = g_s J_a^0, \\
\partial_0\mathbf{E}_a - [\nabla \times \mathbf{H}_a] + g_s f^{abc}G_b^0\mathbf{E}_c + g_s f^{abc}[\mathbf{G}_b \times \mathbf{H}_c] = g_s\mathbf{J}_a.
\end{array}\right\}
\tag{1.25}
$$

These equations are analogous to the pair of the Maxwell ones to involve sources. One more equations pair corresponds to the Bianchi identity (1.9)

$$\left.\begin{array}{l} \nabla \cdot \mathbf{H}_a - g_s f^{abc} \mathbf{G}_b \cdot \mathbf{H}_c = 0, \\ \partial_0 \mathbf{H}_a + [\nabla \times \mathbf{E}_a] + g_s f^{abc} G_b^0 \mathbf{H}_c - g_s f^{abc} [\mathbf{G}_b \times \mathbf{E}_c] = 0. \end{array}\right\} \qquad (1.26)$$

There is a fundamental distinction between the chromodynamics and Maxwell equations. First, the chromodynamics equations are nonlinear. It leads to a self-action of the gluon field, that is, different field components are interacting with each other. Second, the chromodynamics equations being gauge invariant nevertheless involve, along with the field tensor, the potentials G_μ^a, which depend on a particular gauge choice. Thus, in chromodynamics not only does the field tensor has physical meaning, but the potentials being the field equation solutions are physical observables as well.

In theory a gluon field pseudotensor $\tilde{G}^{\mu\nu}$ that is dual to the tensor $G^{\mu\nu}$:

$$\tilde{G}^{\mu\nu} = \frac{1}{2} \varepsilon^{\mu\nu\rho\sigma} G_{\rho\sigma}$$

is also used. It is easy to verify that components of $\tilde{G}^{\mu\nu}$ are expressed in terms of the strengths $\mathbf{E}$ and $\mathbf{H}$ by the following way:

$$\tilde{G}^{i0} = \tilde{G}_{0i} = H_i, \quad \tilde{G}^{ij} = \varepsilon^{ijk} E_k, \quad E_i = \frac{1}{2} \varepsilon^{ijk} \tilde{G}^{jk},$$

that is, just as in the case of electrodynamics they follow from the gluon field tensor components by means of the dual transformation

$$\mathbf{E} \to \mathbf{H}, \qquad \mathbf{H} \to -\mathbf{E}.$$

Multiplying the Bianchi identity by $\varepsilon^{\sigma\mu\nu\rho}$ and contracting the umbral indices, we arrive at

$$D_\mu \tilde{G}^{\mu\nu} = 0.$$

With the help of $\tilde{G}^{\mu\nu}$ one may define an important type of gauge fields. The gauge field is called *self-dual (antiself-dual)* if the components of its tensor and the ones of the dual tensor are connected by the condition

$$\tilde{G}_a^{\mu\nu} = i G_a^{\mu\nu} \qquad (\tilde{G}_a^{\mu\nu} = -i G_a^{\mu\nu}). \qquad (1.27)$$

The validity of Eqs. (1.27) follows from the obvious relation

$$\tilde{\tilde{G}}_{\mu\nu} = -G_{\mu\nu}.$$

When we address the evident writing down of the tensors $G_a^{\mu\nu}$ and $\tilde{G}_a^{\mu\nu}$, the condition (1.27) is reduced to the demand

$$\mathbf{H}_a = i \mathbf{E}_a \qquad (\mathbf{H}_a = -i \mathbf{E}_a). \qquad (1.28)$$

Appearance of the imaginary unit is caused by using the pseudo-Euclidean metric of the Minkowski space. That, by the way, explains why self-dual fields are not usually considered in electrodynamics where all fields are real.

As is well known, electrodynamic fields are classified by means of values of their invariants $F^{\mu\nu} F_{\mu\nu}$ and $\tilde{F}^{\mu\nu} F_{\mu\nu}$. In non-Abelian $SU(3)_c$-theory quantities

$$G_a^{\mu\nu} G_{a\mu\nu} = 2(\mathbf{H}_a^2 - \mathbf{E}_a^2), \quad G_a^{\mu\nu} \tilde{G}_{a\mu\nu} = -4(\mathbf{E}_a \cdot \mathbf{H}_a) \qquad (1.29)$$

are their analogues. It should be stressed that in parallel with (1.29) there is a reach variety of other invariants in the chromodynamics. As a result, the total classification of fields is very complicated.

The energy-momentum tensor is a dynamical characteristic of classical fields

$$T^{\mu\nu} = \frac{\partial \mathcal{L}}{\partial (\partial_\mu Q_i)} \partial\nu Q_i - \delta^{\mu\nu}\mathcal{L}, \tag{1.30}$$

where $Q_i = G_\mu^a(x), q(x), \overline{q}(x)$. Substituting (1.30) into the Lagrangian (1.14), we get

$$T^{\mu\nu} = -G_\rho^{a\mu}\partial^\nu G^{a\rho} + \frac{1}{4}g^{\mu\nu}G_{\rho\sigma}^a G^{a\rho\sigma} + \frac{i}{2}[\overline{q}\gamma^\mu(\partial^\nu q) - (\partial^\nu\overline{q})\gamma^\mu q]. \tag{1.31}$$

Since the spins of the gluon and quark fields do not equal zero, the obtained canonical energy-momentum tensor appears to be asymmetric. However, the energy-momentum tensor is defined ambiguously. We may always add to it a total divergence of the kind $\partial_\lambda f_\mu^{[\nu\lambda]}$, where $f_\mu^{[\nu\lambda]}$ is the third-rank tensor antisymmetric with regard to indices ν and λ. To make the expression (1.30) symmetric and obtain the right relation between the angular momentum and energy-momentum tensors, we should define the quantity $f_\mu^{[\nu\lambda]}$ with the help of the spin-moment tensor (see Section 4.3 of *Advanced Particle Physics Volume 1*). Carrying out corresponding computations, we arrive at the following expression for the symmetrized (metric) energy-momentum tensor (Belinfante's form):

$$T_{\mu\nu}^{(metr)} = T_{\mu\nu} + \frac{1}{2}\partial^\lambda[\Pi_\lambda^{(j)}\Sigma_{\mu\nu}^{(j)}Q_j - \Pi_\mu^{(j)}\Sigma_{\lambda\nu}^{(j)}Q_j + \Pi_\nu^{(j)}\Sigma_{\mu\lambda}^{(j)}Q_j] \tag{1.32},$$

where summing over all the fields $Q_j = G_\mu^a, q, \overline{q}$ is implied and

$$\Pi_\lambda^{(j)} = \frac{\partial \mathcal{L}}{\partial (\partial_\lambda Q_j)}, \qquad \left(\Sigma_{\mu\nu}^{(G)}\right)_{\alpha\beta} = g_{\mu\alpha}g_{\nu\beta} - g_{\mu\beta}g_{\nu\alpha}, \qquad \Sigma_{\mu\nu}^{(q)} = \frac{1}{4}[\gamma_\mu, \gamma_\nu].$$

By means of the motion equations the symmetrized energy-momentum tensor $T_{\mu\nu}^{(metr)}$ may be represented in the form

$$T_{\mu\nu}^{(metr)} = T_{\mu\nu}^{(metr)G} + T_{\mu\nu}^{(metr)q}, \tag{1.33}$$

where

$$T_{\mu\nu}^{(metr)G} = -G_{\nu\rho}^a G_\mu^{a\rho} + \frac{1}{4}g_{\mu\nu}G_{\rho\sigma}^a G^{a\rho\sigma},$$

$$T_{\mu\nu}^{(metr)q} = \frac{i}{4}\left[\overline{q}\gamma_\mu(\partial_\nu q) - (\partial_\nu\overline{q})\gamma_\mu q + \overline{q}\gamma_\nu(\partial_\mu q) - (\partial_\mu\overline{q})\gamma_\nu q\right] + \frac{g_s}{2}\left(\overline{q}\gamma_\mu G_\nu q + \overline{q}\gamma_\nu G_\mu q\right).$$

The energy-momentum tensor of the gluon field $T_{\mu\nu}^{(metr)G}$ has the track equal to zero. Having expressed its components in terms of the electric and magnetic strengths, we obtain the expressions for the gluon energy density:

$$T_{00}^{(metr)G} = \frac{1}{2}\left(\mathbf{E}_a^2 + \mathbf{H}_a^2\right), \tag{1.34}$$

for the Poynting vector components:

$$T_{0i}^{(metr)G} = [\mathbf{E}^a \times \mathbf{H}^a]_i, \tag{1.35}$$

and for the stress tensor:

$$T_{ij}^{(metr)G} = -\left(H_i^a H_j^a + E_i^a E_j^a\right) + \frac{1}{2}\delta_{ij}\left(\mathbf{E}_a^2 + \mathbf{H}_a^2\right). \tag{1.36}$$

We see that the expressions obtained are completely analogous to the corresponding ones of the electrodynamics.

The quark part of the energy-momentum tensor is manifestly gauge invariant because it is written in terms of the fields $q, \bar{q}$, and the covariant derivatives to act on these fields

$$T_{\mu\nu}^{(metr)q} = \frac{i}{4} \left[\bar{q}\gamma_\mu(D_\nu q) - (D_\nu \bar{q})\gamma_\mu q + \bar{q}\gamma_\nu(D_\mu q) - (D_\mu \bar{q})\gamma_\nu q \right], \tag{1.37}$$

Using the motion equation, one could show that the track of $T_{\mu\nu}^{(metr)q}$ is proportional to a quark mass

$$(T^{(metr)q})_\mu^\mu = m\bar{q}q. \tag{1.38}$$

Hereon, we complete the formulae summary of chromodynamics and proceed to the quantization of this theory.

1.2   Gauge invariance in QCD

All quantum field theories to successfully describe the world surrounding us are non-Abelian gauge theories. Those, in turn, are based on gauge-invariance principles generalizing the ordinary gauge $U(1)$-invariance of quantum electrodynamics (QED). The significance of these principles lies in the fact that the main properties and even the gauge field existence itself can be deduced from the invariance of a theory with respect to gauge transformations. Thus, for example, in quantum chromodynamics (QCD) the gauge invariance (GI) guarantees the following: First, as follows from the covariant derivative definition, the GI demands universality of a coupling constant, that is, one and the same coupling constant g_s governs the interaction of quarks with gluons and self-interaction gluons. Second, as was shown in Ref. [1], the GI ensures the renormalizability. Third, only non-Abelian gauge theory can possesses the property of the asymptotic freedom [2].

However, under quantization, the GI gives rise to considerable troubles. The GI of a theory accounts for the fact that couplings are imposed on dynamical variables of a system under consideration, that is, among these variables are those not to be associated with true dynamical degrees of freedom. Before proceeding to consideration of QCD, it is useful to address the simpler construction, namely, to QED where troubles connected with the GI are being met in a simple form.

The gauge invariant Lagrangian of the free electromagnetic field has the form

$$\mathcal{L} = \frac{1}{2} F^{\mu\nu} \left(\frac{1}{2} F_{\mu\nu} - \partial_\mu A_\nu + \partial_\nu A_\mu \right). \tag{1.39}$$

Varying $\mathcal{L}$ over A_μ and $F_{\mu\nu}$ leads to the equations

$$\partial_\sigma \frac{\partial \mathcal{L}}{\partial(\partial_\sigma A_\mu)} - \frac{\partial \mathcal{L}}{\partial A_\mu} = 0. \tag{1.40}$$

$$\partial_\sigma \frac{\partial \mathcal{L}}{\partial(\partial_\sigma F_{\mu\nu})} - \frac{\partial \mathcal{L}}{\partial(F_{\mu\nu})} = 0, \tag{1.41}$$

From (1.39) – (1.41) it follows the Maxwell equations and definition of the electromagnetic field tensor. Within the canonical formalism the canonical variables $A_\mu(x)$ and the conjugated momenta

$$\pi^\mu(x) = \frac{\partial \mathcal{L}}{\partial[\partial_0 A_\mu(x)]}$$

are considered as operators to obey the commutation relations

$$[A^\mu(\mathbf{r}, t), \pi^\nu(\mathbf{r}', t)] = ig^{\mu\nu}\delta^{(3)}(\mathbf{r} - \mathbf{r}'). \tag{1.42}$$

Subjecting the electromagnetic field potentials $A_\mu(x)$ to the gauge transformation

$$A_\mu(x) \to A'_\mu(x) = A_\mu(x) + \partial_\mu\chi(x),$$

where $\chi(x)$ is an arbitrary function, we obtain already other commutation relations. It happens because photons possessing zero mass have only two degree of freedom while $A_\mu(x)$ is a four-dimensional vector. And what is more, it is impossible to fulfill the relations (1.42) for all components of the potential $A_\mu(x)$ since there exist several constraints in the theory. Really, using (1.39), we find

$$\pi_0(x) = \frac{\partial \mathcal{L}}{\partial[\partial_0 A_0(x)]} = F_{00}(x) = 0. \tag{1.43}$$

This restriction is called a *primary constraint* inasmuch as it follows directly from the Lagrangian structure. At $\mu = 0$, Eq. (1.40) gives the second constraint for that quantity to be fixed by the primary constraint:

$$\partial_\sigma F^{\sigma 0}(x) = -\partial_i \pi^i(x) = \frac{\partial \mathcal{L}}{\partial A_0} = 0. \tag{1.44}$$

Now the quantities $\pi_0(x)$ and $\nabla \cdot \boldsymbol{\pi}(x)$ vanish. As a result, the operators $A_0(x)$ and $\nabla \cdot \mathbf{A}(x)$ commute with all canonical operators, that is, they are actually *c*-numbers. Therefore, the canonical commutation relations for transverse fields $\mathbf{A}_\perp(x)$ and $\boldsymbol{\pi}_\perp(x)$, to be in line with two independent freedom degrees, must be determined so that they will be compatible with the aforementioned constraints. Fictitious degrees of freedom that are present due to the gauge invariance must be removed by means of an appropriate choice of a condition fixing a gauge. And at the same time, two possibilities exist. The first consists in a choice of such a gauge to ensure the absence of the nonphysical degrees of freedom. Since, in so doing, the Lorentz invariance is manifestly violated, these gauges are said to be *noncovariant*. Among these are

$$\nabla \cdot \mathbf{A}(x) = 0 \qquad \text{Coulomb, or radiation gauge,}$$

$$A_3(x) = 0 \qquad \text{axial gauge,}$$

$$A_0(x) = 0 \qquad \text{temporal, or timelike gauge.}$$

The second possibility resides in the fact that all the field components $A_\mu(x)$ are considered on an equal footing (*covariant gauges*). Since nonphysical degrees of freedom are retained, then the necessity of introducing an indefinite metric space appears.

Needles to say the canonical quantization method is not unique in quantum field theory. A quantization formalism with the help of functional integrals is very convenient for gauge theories. We shall become acquainted with it in Chapter 22 and now we consider using the covariant gauges in QED.

Quantizing a free electromagnetic field showed that there is no way to impose the Lorentz gauge

$$\partial_\mu A^\mu(x) = 0 \tag{1.45}$$

and at the same time to preserve the commutation relations. A consistent theory will be obtained only in the event that a physical state Ψ satisfies the condition

$$< \Psi|\partial_\mu A^\mu(x)|\Psi >= 0 \qquad (1.46).$$

In Section 17.3 of *Advanced Particle Physics Volume 1* in order to take into account gauge ambiguity of potentials, we introduced the gauge-fixing term to the electromagnetic field Lagrangian

$$\mathcal{L}_{GF} = -\frac{1}{2\xi}\partial_\mu A^\mu \partial_\nu A^\nu. \qquad (1.47)$$

Such a modification does not lead to any physical consequences since on the strength of the condition (1.46) matrix elements of (1.47) over the physical vectors Ψ vanish. Now all the momenta to be canonical conjugated to the fields $A_\mu(x)$ do not equal to zero and the relations (1.42) are fulfilled for all values of μ and ν. In this case the photon propagator is defined by the expression

$$D^c_{\mu\nu}(k) = -\left[g_{\mu\nu} + (\xi - 1)\frac{k_\mu k_\nu}{k^2}\right]\frac{1}{k^2}. \qquad (1.48)$$

($\xi = 1$ is Feynman gauge, $\xi = 0$ is transverse gauge or Landau one).

Let us investigate consequences of introducing the term $\mathcal{L}_{GF}$. We consider an infinitesimal gauge transformation for the potential A_μ

$$A_\mu \to A_\mu - e^{-1}(\partial_\mu \delta\alpha). \qquad (1.49)$$

At that the action variation is given by the expression

$$\delta S = \delta \int (\mathcal{L} + \mathcal{L}_{GF})d^4x = \delta \int \mathcal{L}_{GF}d^4x = -\frac{1}{2\xi}\int(\partial_\mu A^\mu - e^{-1}\partial_\mu\partial^\mu\delta\alpha)\times$$

$$\times(\partial_\nu A^\nu - e^{-1}\partial_\nu\partial^\nu\delta\alpha)d^4x - \frac{1}{2\xi}\int \partial_\mu A^\mu \partial_\nu A^\nu d^4x = \frac{1}{\xi e}\int(\partial_\mu A^\mu \partial_\nu\partial^\nu\delta\alpha)d^4x =$$

$$= \frac{1}{\xi e}\int \delta\alpha(\partial_\nu\partial^\nu\partial_\mu A^\mu)d^4x = 0. \qquad (1.50)$$

From (1.50) follows the equation

$$\partial_\nu\partial^\nu\eta(x) = 0, \qquad (1.51)$$

where $\eta(x) = \partial_\mu A^\mu(x)$, defining the behavior of massless scalar η-particles which, as is easy to see, do not interact with any other particles. There is nothing to prevent a choice of initial conditions in the view

$$\eta(x)\Big|_{t=0} = 0, \qquad \frac{\partial\eta(x)}{\partial t}\Big|_{t=0} = 0. \qquad (1.52)$$

Then, at any instants of time, we shall always have $\eta(x) = 0$.

However, introducing $\mathcal{L}_{GF}$ violates manifest gauge invariance that is very undesirable. To restore this important property of the theory one should append the term

$$\mathcal{L}^\omega = -\frac{1}{2}\partial_\mu\omega(x)\partial^\mu\omega(x)$$

corresponding to a free neutral scalar massless field $\omega(x)$, to the total QED Lagrangian

$$\mathcal{L}^\xi_{QED} = \overline{q}(x)(i\hat{D} - m)q(x) + \frac{1}{2}F^{\mu\nu}(x)\left[\frac{1}{2}F_{\mu\nu}(x) - \partial_\mu A_\nu(x) + \partial_\nu A_\mu(x)\right] -$$

$$-\frac{1}{2\xi}\partial_\mu A^\mu(x)\partial_\nu A^\nu(x). \tag{1.53}$$

The field $\omega(x)$ is nonphysical and the particles connected with it will be called *ghosts*. Let us generalize the gauge transformations in such a way as to include the ghost fields. Defining infinitesimal transformation parameters in the form $\delta\alpha(x) = \epsilon\omega(x)$, we obtain for $q(x)$ and $A_\mu(x)$

$$q(x) \to q(x) + ie\epsilon\omega(x)q(x), \qquad A_\mu(x) \to A_\mu(x) - \epsilon\partial_\mu\omega(x). \tag{1.54}$$

If one demands $\omega(x)$ to be transformed by the law

$$\omega(x) \to \omega(x) - \frac{\epsilon}{\xi}\partial_\mu A^\mu(x), \tag{1.55}$$

the total Lagrangian

$$\mathcal{L}_{QED} = \mathcal{L}_{QED}^\xi + \mathcal{L}^\omega, \tag{1.56}$$

will be invariant relative to the transformations (1.54) and (1.55) with a precision of an inessential four-dimensional divergence. Thanks to the fact that photons are neutral and do not interact with each other, the fields ω may be chosen in the form of free real fields. One could show [3], that the gauge transformations (1.54) and (1.55) generate all the Ward identities. The S-matrix unitarity, that is, the absence of transitions in the Hilbert space between the physical and nonphysical states (states with the opposite metric parity), is one of consequences of these identities. In order to check this circumstance it is enough to prove that the photon propagator is transverse in all orders of the perturbation theory. We shall demonstrate that such is the case.

Invariance of the quantity $< 0|T(A^\mu(x)\omega(0))|0 >$ with respect to the gauge transformations (1.54) and (1.55) leads to the relation

$$\frac{1}{\xi} < 0|T(A^\mu(x)\partial_\nu A^\nu(0))|0 > = < 0|T(\partial^\mu\omega(x)\omega(0))|0 > . \tag{1.57}$$

Passing on to Fourier images in the right and left sides of (1.57), we get

$$\frac{1}{\xi}\int < 0|T(A^\mu(x)\partial_\nu A^\nu(0))|0 > \exp[ikx]d^4x =$$

$$= \frac{ik_\nu}{\xi}\int < 0|T(A^\mu(x)A^\nu(0))|0 > \exp[ikx]d^4x = \frac{ik_\nu}{\xi}D^{c\mu\nu}(k) \tag{1.58}$$

and

$$\int < 0|T(\partial^\mu\omega(x)\omega(0))|0 > \exp[ikx]d^4x = ik^\mu\int < 0|T(\omega(x)\omega(0))|0 > \exp[ikx]d^4x =$$

$$= \frac{k^\mu}{k^2 + i0}, \tag{1.59}$$

where it has been taken into account that $\omega(x)$ describes a scalar field. Equating (1.58) with (1.59), as well as allowing for separating the photon propagator on the transverse and longitudinal parts

$$D_{\mu\nu}^c(k) = \left(-g_{\mu\nu} + \frac{k_\mu k_\nu}{k^2}\right)D^{(tr)}(k^2) + \frac{k_\mu k_\nu}{k^2}D^{(l)}(k^2), \tag{1.60}$$

we obtain

$$D^{(l)}(k^2) = -\frac{i\xi}{k^2 + i0}. \tag{1.61}$$

So, the longitudinal part of the photon propagator has the same view as in the case of a free electromagnetic field. In other words, if one expands the propagator $D^c_{\mu\nu}(k)$ as a power series in the interaction constant

$$D^c_{\mu\nu}(k) = D^{c(0)}_{\mu\nu}(k) + \frac{e^2}{4\pi} D^{c(2)}_{\mu\nu}(k) + ..., \qquad (1.62)$$

then all the quantities $D^{c(n)}_{\mu\nu}(k)$ $(n = 2, 4, ...)$ will satisfy the transversality condition

$$k^\mu D^{c(n)}_{\mu\nu} = 0. \qquad (1.63)$$

Taking into consideration that the photon polarization operator $\Pi^{(n)}_{\nu\sigma}(k)$ is connected with $D^{c(n)}_{\mu\nu}$ by the relation

$$D^{c(n)}_{\nu\sigma}(k) = D^{c(0)}_{\nu\tau}(k)\Pi^{(n)\tau\lambda}(k)D^{c(0)}_{\lambda\sigma}(k), \qquad (1.64)$$

we also find the transversality condition of the electromagnetic field polarization tensor in all orders of the perturbation theory

$$k^\nu \Pi^{(n)}_{\nu\sigma}(k) = 0.$$

Thus, the aforementioned procedure allows us to carry out quantization in the QED and at the same time to conserve the manifest gauge invariance and the S-matrix unitarity.

1.3 Quantization of the free gluon field

Using canonical formalism, we shall quantize a free gluon field. The Lagrangian of the theory under investigation is as follows:

$$\mathcal{L}_G = \frac{1}{2} G_{a\mu\nu}(x) \left[\frac{1}{2} G^{\mu\nu}_a(x) - \partial^\mu G^\nu_a(x) + \partial^\nu G^\mu_a(x) - g_s f_{abc} G^\mu_b(x) G^\nu_c(x) \right]. \qquad (1.65)$$

Simultaneous commutation relations are written in the form

$$[G^\mu_a(\mathbf{r}, t), \pi^\nu_b(\mathbf{r}', t)] = i\delta_{ab} g^{\mu\nu} \delta^{(3)}(\mathbf{r} - \mathbf{r}'), \qquad (1.66)$$

where the momenta $\pi^\nu_b(x)$ canonically conjugated to the fields $G^\nu_b(x)$ are given by the expression

$$\pi^\nu_b(x) = G^{\nu 0}_b(x). \qquad (1.67)$$

Since zero components of the momenta $\pi^0_b(x)$ are identically equal to zero, then we should fix a gauge to remove contradictions. We shall use the covariant gauge. Let us introduce the gauge-fixing term

$$\mathcal{L}_{GF} = -\frac{1}{2\xi} \sum_a (\partial_\mu G^\mu_a)(\partial_\nu G^\nu_a) \qquad (1.68)$$

into the Lagrangian (1.65) and demand the physical state vectors to satisfy the relation

$$< 0|\Psi|\partial_\mu G^\mu_a)(x)|\Psi >= 0. \qquad (1.69)$$

Now the momenta $\pi^\nu_b(x)$ will look like

$$\pi^\nu_b(x) = G^{\nu 0}_b(x) - \frac{1}{\xi} g^{\nu 0} \partial_\mu G^\mu_b(x). \qquad (1.70)$$

Since none of the components $\pi_b^\nu(x)$ turns into zero, the canonical quantization scheme is free from contradictions. However, the Hilbert space of the theory under consideration possesses the indefinite metric. In order to be certain that such is the case, we examine the relation (1.66) when $\mu = 0$:

$$\frac{1}{\xi}[\partial_\mu G_a^\mu(\mathbf{r}, t), G_b^\nu(\mathbf{r}', t)] = i\delta_{ab}g^{0\nu}\delta^{(3)}(\mathbf{r} - \mathbf{r}').\tag{1.71}$$

In the momentum space we introduce canonical tetrads $\epsilon_\mu^{(\sigma)}(k)$ $(k_\mu^2 = 0)$:

$$\left.\begin{array}{ll}\epsilon_0^{(i)} = 0, & (\epsilon^{(i)} \cdot \mathbf{k}) = 0, \qquad i = 1, 2, \\[2mm] \epsilon_\mu^{(3)} = k_\mu(k^0)^{-1} - \delta_{\mu 0}, & \epsilon_\mu^{(i)}\epsilon^{(j)\mu} = -\delta_{ij}, \qquad i, j = 1, 2, 3, \qquad \epsilon_\mu^{(0)} = \delta_{\mu 0}\end{array}\right\}\tag{1.72}$$

and choose the gauge parameter as $\xi = 1$. The components $\epsilon_\mu^{(i)}$ $(i = 1, 2)$ are associated with the physical freedom degrees, $\epsilon_\mu^{(3)}$ represent the longitudinal component, while the component $\epsilon_\mu^{(0)}$ describes the scalar particle. Then, expansion over the production and destruction operators for the gluon field becomes

$$G_b^\mu(x) = \frac{1}{(2\pi)^{3/2}} \int \frac{d^3k}{\sqrt{2k_0}} \sum_\sigma [\epsilon^{(\sigma)\mu}(k)a_\sigma(b, k)\exp(-ikx)+$$

$$+\epsilon^{(\sigma*)\mu}(k)a_\sigma^\dagger(b, k)\exp(ikx)].\tag{1.73}$$

Substituting (1.73) into (1.66), we arrive at the following commutation relations:

$$[a_\mu(b, k), a_\nu^\dagger(b', k')] = -\delta_{bb'}g_{\mu\nu}\delta^{(3)}(\mathbf{k} - \mathbf{k}').\tag{1.74}$$

From (1.74) it follows that the vacuum expectation value of the scalar particles number operator

$$N_0 = a_0(b, k)a_0^\dagger(b, k)$$

in the gauge at hand is negative, that is, we deal with the indefinite metric theory.

Let us pass on to the gluon field potentials into the Lagrangian (1.65) and represent the sum $\mathcal{L}_G + \mathcal{L}_{GF} = \mathcal{L}_t$ in the form of two terms:

$$\mathcal{L}_t = \mathcal{L}_0(G) + \mathcal{L}_{self}(G, g_s).\tag{1.75}$$

The former involves terms that are quadratic in the gluon field and do not include self-interaction

$$\mathcal{L}_0(G) = -\frac{1}{4}(\partial_\mu G_{a\nu} - \partial_\nu G_{a\mu})^2 - (1/2\xi)(\partial_\mu G_a^\mu)^2,\tag{1.76}$$

while the latter contains terms describing self-interaction of gluons

$$\mathcal{L}_{self}(G, g_s) = \mathcal{L}_{self}^{(1)} + \mathcal{L}_{self}^{(2)},\tag{1.77}$$

$$\mathcal{L}_{self}^{(1)} = -\frac{1}{2}g_s f_{abc}(\partial_\mu G_{a\nu} - \partial_\nu G_{a\mu})G_b^\mu G_c^\nu = -g_s f_{abc}(\partial_\mu G_{a\nu})G_b^\mu G_c^\nu,\tag{1.78}$$

$$\mathcal{L}_{self}^{(2)} = -\frac{1}{4}g_s^2 f_{abc}f_{ade}G_{b\mu}G_{c\nu}G_{d\mu}G_{e\nu}.\tag{1.79}$$

Writing the Lagrangian as (1.75) allows us to quantize a free gluon field in the linear approximation $(\mathcal{L}_t \to \mathcal{L}_0(G))$ and then take into consideration the triple and quadruple vertices using the perturbation theory in powers of the coupling constant g_s. As switching on the gluon interaction with matter leads to the appearance of the quark-gluon vertices involving g_s, this procedure is altogether lawful. As a result we have the perturbation theory series in powers of the same coupling constants g_s.

The expression $\mathcal{L}_0(G)$ may be chosen in the form of the componentwise sum:

$$\mathcal{L}_0(G) = \sum_a \mathcal{L}_0(G_a) \tag{1.80}$$

which represents the sum of the quadratic nondegenerate forms to be analogous to the electromagnetic field Lagrangian

$$\mathcal{L} = -\frac{1}{4}(\partial_\mu A_\nu - \partial_\nu A_\mu)^2 - \frac{1}{2\xi}\partial_\mu A^\mu \partial_\nu A^\nu \tag{1.81}$$

Using the same methods as in the case of the electromagnetic field, we get the expression for the gluon propagator

$$D^c_{ab\mu\nu}(k) = -\delta_{ab}\left[g_{\mu\nu} + (\xi - 1)\frac{k_\mu k_\nu}{k^2}\right]\frac{1}{k^2}. \tag{1.82}$$

The quadratic Lagrangian $\mathcal{L}_0(G)$ is often called a *free gluon field Lagrangian*. In this case the fields being described by the Lagrangian (1.75) are named by gluon fields in a vacuum. Further, in order to avoid confusion we choose the following terminology. The field to obey linear equations resulting from the quadratic Lagrangian $\mathcal{L}_0(G)$ will be called a *linear approximation field* or a *linear gluon field*. The gluon field being described by the Lagrangian $\mathcal{L}_t$, which does not include the gluon interaction with matter fields, will be referred to as a *free gluon field*.

Applying the ordinary quantization rules to $\mathcal{L}_t$, one can build up the perturbation theory in powers of the coupling constant g_s. In the momentum representation the Feynman rules for the free gluon field involve the following elements:

1) triple gluon vertex (Fig. 1.1)

$$V_{3G} = g_s f_{abc}[g_{\mu\nu}(k-p)_\sigma + g_{\nu\sigma}(p-q)_\mu + g_{\sigma\mu}(q-k)_\nu](2\pi)^4\delta^{(4)}(p+k+q) \tag{1.83}$$

2) quadruple gluon vertex (Fig. 1.2)

$$V_{4G} = -ig_s^2\left\{f_{abl}f_{cdl}(g_{\mu\rho}g_{\nu\sigma} - g_{\mu\sigma}g_{\nu\rho}) + \begin{pmatrix} b \leftrightarrow c \\ \nu \leftrightarrow \rho \end{pmatrix} + \begin{pmatrix} b \leftrightarrow d \\ \nu \leftrightarrow \sigma \end{pmatrix}\right\}(2\pi)^4\delta^{(4)}(\textstyle\sum k) \tag{1.84}$$

3) motion of virtual gluon with momentum k (Fig. 1.3)

$$P_{ab}^{(G)\mu\nu}(k) = -\frac{i\delta_{ab}}{(2\pi)^4}\left[g_{\mu\nu} + (\xi - 1)\frac{k_\mu k_\nu}{k^2}\right]\frac{1}{k^2}. \tag{1.85}$$

1.4   Feynman rules in QCD with a covariant gauge

Because photons are not subjected to self-interaction, then using covariant gauges does not produce any additional troubles within QED. In the case of QCD, however, gluon self-interaction results in considerable complications.

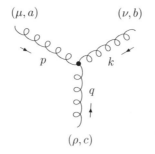

FIGURE 1.1
Triple gluon vertex.

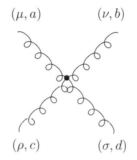

FIGURE 1.2
Quadruple gluon vertex.

FIGURE 1.3
Gluon propagator.

Let us extensively investigate the consequences of introducing the gauge-fixing term into the Lagrangian. To find an equation the scalar field $\eta_a = \partial_\mu G_a^\mu$ obeys, we subject a free gluon field to an infinitesimal gauge transformation

$$G'_{a\mu}(x) = G_{a\mu}(x) + \delta G_{a\mu}(x), \tag{1.86}$$

with

$$\delta G_{a\mu}(x) = -\frac{1}{g_s} D_\mu \delta\alpha_a(x). \tag{1.87}$$

Further, we compute the action variation of the free gluon field being described by the Lagrangian $\mathcal{L}_G + \mathcal{L}_{GF}$. In so doing we take into account that $\delta\mathcal{L}_G = 0$. From the least action principle follows:

$$\delta S = \int d^4x \delta \mathcal{L}_{GF} = -\frac{1}{\xi} \int d^4x (\partial_\mu G_a^\mu)(\partial_\nu \delta G_a^\nu) =$$

$$= -\frac{1}{\xi} \int d^4x \, (\partial_\mu G_a^\mu) \left[\partial_\nu \left(f_{abc}\delta\alpha_b(x)G_c^\nu(x) - \frac{1}{g_s}\partial^\nu \delta\alpha_a(x) \right)\right] =$$

$$= \frac{1}{\xi g_s} \int d^4x [\delta\alpha_a(x)\partial_\nu\partial^\nu(\partial_\mu G_a^\mu) + g_s f_{abc}G_c^\nu \delta\alpha_b(x)(\partial_\nu\partial_\mu G_a^\mu)] = 0. \qquad (1.88)$$

In such a way, for η-particles we get the equation:

$$D_\nu \partial^\nu \eta_a(x) = 0,$$

or in an explicit form

$$[\delta_{ab}\partial_\nu\partial^\nu - g_s f_{abc}G_c^\nu(x)\partial_\nu]\eta_a(x) = 0. \qquad (1.89)$$

Unlike QED, this equation is not free, since it includes interaction with the field G_c^ν. By this reason, the field η_a will be generated by the field G_c^ν even at zero initial conditions for η_a and $\partial\eta_a/\partial t$. In quantum language it means that the field G_c^ν excites quantum fluctuations of the field η_a, that is, generates η-particles that can be both in real and virtual states.

Let us switch on the gluons interaction with the matter field. In this case the QCD Lagrangian is given by the expression

$$\mathcal{L}_{QCD} = \frac{i}{2}[\overline{q_k}(x)\gamma_\mu\partial^\mu q_k(x) - \partial^\mu\overline{q_k}(x)\gamma_\mu q_k(x)] - m\overline{q_k}(x)q_k(x) -$$

$$- g_s[\overline{q_k}(x)\gamma_\mu G_{kj}^\mu(x)q_j(x)] - \frac{1}{4}(G_{\mu\nu}^a(x))^2 - \frac{1}{2\xi}(\partial_\mu G_a^\mu)^2, \qquad (1.90)$$

where

$$G_{kj}^\mu(x) = \frac{1}{2}G_a^\mu(x)\,(\lambda_a)_{kj}$$

and k, j are color indices. It is evident that the Lagrangian (1.90) also generates Eq. (1.89) for η-particles. Inasmuch as the equation (1.89) contains the source giving rise to η-particles, then its solution cannot be expanded into positive and negative frequency parts ($\eta^{(+)}$ and $\eta^{(-)}$) in a relativistically invariant way. Therefore, in QCD we cannot choose physical states $|\Psi>$ in such a way as to

$$\eta^{(+)}|\Psi >= 0, \qquad (1.91)$$

as it was possible in QED. Since η-particles possess the negative metric parity, transitions between the Hilbert space parts having the opposite metric exist, that is, the theory is nonunitarity. To put it differently, the Lagrangian $\mathcal{L}_t$ transfers physical states whose metric is positive into nonphysical states to possess a negative metric. For the first time this circumstance was mentioned in Ref. [4]. A solution of the problem for some particular cases was proposed by Feynman [5] while Faddeev and Popov found a solution in the general case [6]. The idea is as follows. Along with the fields being present at the Lagrangian, additional nonphysical fields, ghosts, are introduced in such a way that they turn nonphysical states produced by the Lagrangian $\mathcal{L}_t$ into zero. A multiplet of scalar fields $\omega_a(x)$ described by the Lagrangian

$$\mathcal{L}_{FPG} = (\partial_\mu\omega_a^\dagger(x))[\delta_{ab}\partial^\mu - g_s f_{abc}G_c^\mu(x)]\omega_b(x) = (\partial_\mu\omega_a^\dagger(x))D^\mu\omega_a(x) \qquad (1.92)$$

is used as ghosts. From $\mathcal{L}_{FPG}$, in its turn, follows the equations

$$\left.\begin{array}{l} [\delta_{ab}\partial_\nu\partial^\nu - g_s f_{abc}G_c^\nu(x)\partial_\nu]\omega_a(x) = 0, \\[2mm] [\delta_{ab}\partial_\nu\partial^\nu - g_s f_{abc}G_c^\nu(x)\partial_\nu]\omega_a^\dagger(x) = 0, \end{array}\right\} \qquad (1.93)$$

that is, in the linear approximation $\omega_a(x)$ and $\omega_a^\dagger(x)$ obey the Klein–Gordon free equation for the massless particle. Let us introduce a new quantum number n_{gh}, the number of ghosts. It is equal: to +1 for the ghost field ω_a, to -1 for the antighost field and to zero for all other fields. Then, from the Lagrangian $\mathcal{L}_{FPG}$ follows the ghosts number conservation law. Ghost lines enter into diagrams only in the form of loops. In parallel with every diagram involving a closed loop of a gauge field there exists a topologically equivalent diagram with a closed ghost line in the same place. Ghost fields are subjected to the Fermi–Dirac statistics.* This leads to the fact that a closed loop of the $\omega_a(x)$ field has a redundant minus sign as compared with a loop of the η-particles. Since the multiplet of the $\omega_a(x)$ ghosts satisfies the same equation as does the field $\eta_a(x)$, then loops connected with η- and ω-fields are mutually cancelled. Thus, the aforementioned trick results in mutual cancellation of contributions coming from nonphysical states η and ω and allows us not to include these particles in real states.

The term $\mathcal{L}_{FPG}$ causes the appearance of the following Feynman diagrams:

4) motion of virtual ghost with momentum p (Fig. 1.4)

$$P_{ab}^{(\omega)}(p) = \frac{i}{(2\pi)^4} \frac{\delta_{ab}}{p^2},\tag{1.94}$$

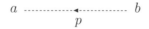

$$a \ \text{-----------}\blacktriangleleft\text{------------} \ b$$
$$p$$

FIGURE 1.4
The line associated with the ghost as well as the fermion line, has a direction, that is, the ghost differs from its antiparticle.

5) ghost-ghost-gluon vertex (Fig. 1.5)

$$V_{\omega\omega G} = g_s f_{abc} p_\mu (2\pi)^4 \delta\left(\sum k\right).\tag{1.95}$$

In the theory the following diagrams associated with quarks are also present:

6) propagation of virtual quark with momentum p (Fig. 1.6)

$$P^{(q)}(p) = \frac{1}{(2\pi)^4} \frac{i}{\hat{p} - m} = \frac{1}{(2\pi)^4} \frac{i(\hat{p} + m)}{p^2 - m^2}\tag{1.96}$$

7) quark-gluon vertex (Fig. 1.7)

$$V_{qqG} = \frac{ig_s}{2}(2\pi)^4 \gamma_\mu \, (\lambda_a)_{\overline{k}j}\, \delta\left(\sum k\right).\tag{1.97}$$

To obtain the total set of Feynman rules in the QCD, we have to supplement (1.83)–(1.85), (1.94)–(1.97) with rules describing particles in the initial and final states. For quarks and

---

*Since the fields $\omega_a(x)$ do not appear in both the initial and final states, the inconsistency between their spin and statistics must not give concern.

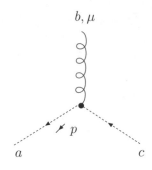

FIGURE 1.5

Ghost-ghost-gluon vertex.

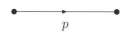

FIGURE 1.6

Quark propagator.

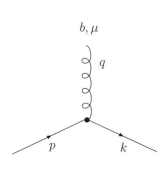

FIGURE 1.7

Quark-gluon vertex.

gluons these rules follow from the corresponding QED ones under the replacements

$$leptons \rightarrow quarks, \qquad photons \rightarrow gluons.$$

In the Feynman gauge, the summation over gluon polarizations is fulfilled using the relation

$$\sum_{\lambda} \epsilon^{(\lambda)\mu}(k)\epsilon^{(\lambda)\nu}(k) = -g^{\mu\nu}. \tag{1.98}$$

Note, that only the gauge field propagator depends on the gauge parameter ξ.

    Introducing the ghosts into the QCD Lagrangian that is written in a covariant gauge was stimulated by the demand of the S-matrix unitarity reconstruction. However, in view of the gauge invariance of the theory, the S-matrix unitarity property must be carried out in any gauge. It is evident that the given violation is connected with introducing the gauge-fixing term that does not possess the gauge invariance property. In Section 1.3, we showed that in the QED case introducing the ghosts may be interpreted as a manner of the gauge invariance reconstruction of the Lagrangian. Let us demonstrate that the analogous property takes place in the QCD as well.

    For a non-Abelian theory Becchi–Rouet–Stora (BRS) transformations [7] are generalizations of the gauge transformations (1.45) and (1.55). Here, the ghost fields, as well as all other fields, are subjected to gauge transformations and, as a result, $\mathcal{L}_{QCD}$ becomes gauge invariant with a precision of a four-dimensional divergence. The BRS transformations result in Slavnov[8]–Taylor[16] identities that represent the analogue of the Ward identities in QED.

It is more convenient to use real Grassmann fields ρ and σ to be defined as

$$\omega_a = \frac{1}{\sqrt{2}}(\rho_a + i\sigma_a), \qquad \omega_a^\dagger = \frac{1}{\sqrt{2}}(\rho_a - i\sigma_a)$$

rather than the complex ghost fields ω_a and $\omega_a^\dagger$. Making use of the anticommutativity property of Grassmann fields

$$\rho_a^2 = \sigma_a^2 = 0, \qquad \rho_a\sigma_b = -\sigma_b\rho_a,$$

we obtain

$$\mathcal{L}_{FPG} = -i\sum(\partial_\mu\rho_a(x))[\delta_{ab}\partial^\mu - g_s f_{abc}G_c^\mu(x)]\sigma_b(x) = -i\sum(\partial_\mu\rho_a(x))D^\mu\sigma_a(x). \quad (1.99)$$

Infinitesimal BRS transformations have the form

$$\left.\begin{array}{c} G_a^\mu \to G_a^\mu + \epsilon(\delta_{ab}\partial^\mu - g_s f_{abc}G_c^\mu)\sigma_b, \\[6pt] q \to q + i\epsilon g_s(\lambda^a/2)\sigma_a q, \qquad \sigma_a \to \sigma_a - (\epsilon/2)g_s f_{abc}\sigma_b\sigma_c, \\[6pt] \rho_a \to \rho_a - (\imath\epsilon/\xi)\partial_\mu G_a^\mu, \end{array}\right\} \quad (1.100)$$

where an infinitesimal BRS transformation parameter ϵ is an anticommuting Grassmann variable not dependent on x. It should be noted that the production $\epsilon\sigma_a$ is an ordinary number. Thus, we have

$$\epsilon^2 = 0, \qquad \epsilon\sigma_a = -\sigma_a\epsilon, \qquad q\epsilon = -\epsilon q, \qquad \epsilon G_a^\mu = G_a^\mu\epsilon. \quad (1.101)$$

Using these transformation in just the same way as it was done in the QED, one may show that the longitudinal propagator part of a gluon is defined by the expression

$$D_{ab}^{(l)\mu\nu}(k) = -\frac{i\delta_{ab}\xi k^\mu k^\nu/k^2}{k^2}, \quad (1.102)$$

that is, has the same form as in the case of a free gluon field. Then, expanding $D_{ab}^{c\mu\nu}(k)$ into series in square powers of the coupling constant g_s

$$D_{ab}^{c\mu\nu}(k) = D_{ab}^{c(0)\mu\nu}(k) + \left(\frac{g_s^2}{4\pi}\right)D_{ab}^{c(2)\mu\nu}(k) + \ldots, \quad (1.103)$$

we get the transversality condition of a gluon field in all orders of the perturbation theory

$$k_\mu D_{ab}^{c(n)\mu\nu}(k) = 0, \qquad n = 2, 4, \ldots \quad (1.104)$$

Since (1.104) is the direct consequence of the gauge invariance, we convince ourselves that introducing the Faddeev–Popov ghosts really restores the gauge invariance of the theory.

The gauge invariance of the total QCD Lagrangian could be demonstrated by the example of the action invariance with respect to the BRS transformation. Because

$$\delta G_a^\mu = \epsilon(\partial^\mu\sigma_a - g_s f_{abc}\sigma_b G_c^\mu),$$

then the BRS transformation is actually a gauge transformation associated with a particular choice of a gauge function

$$\delta\alpha_a(x) = -\epsilon g_s\sigma_a(x).$$

In that way, the action

$$S_{QCD} = \int d^4 x \mathcal{L}_{QCD}$$

is not changed under this transformation, that is, $\delta S_{QCD} = 0$ and proving the total action invariance

$$S_{eff} = S_{QCD} + S_{gf} + S_{gh} \tag{1.105}$$

is reduced to proving the relation

$$\delta S' = \delta \int d^4 x [\mathcal{L}_{GF} + \mathcal{L}_{FPG}] = 0.$$

We have for $\delta S'$

$$\delta S' = -\frac{1}{\xi} \int d^4 x (\partial_\nu G_a^\nu)(\partial_\mu \delta G_a^\mu) - i \int d^4 x (\partial_\nu \delta \rho_a)(D^\nu \sigma_a) - i \int d^4 x (\partial_\nu \rho_a) \delta (D^\nu \sigma_a). \tag{1.106}$$

First, we find variation of the covariant derivative of σ_b:

$$-\delta(\partial^\nu \delta_{ab} - g_s f_{abc} G_c^\nu)\sigma_b = \frac{g_s}{2} \epsilon f_{abc} \partial^\nu (\sigma_b \sigma_c) + g_s f_{abc} \sigma_b [\epsilon(\partial^\nu \sigma_c - g_s f_{cde} \sigma_d G_e^\nu) +$$

$$-g_s f_{abc} G_c^\nu \left(\frac{g_s}{2} \epsilon f_{bde} \sigma_d \sigma_e \right). \tag{1.107}$$

Because σ_a is a Grassmann quantity, then the derivative in the first term is

$$f_{abc} \partial^\nu (\sigma_b \sigma_c) = f_{abc} [(\partial^\nu \sigma_b)\sigma_c + \sigma_b (\partial^\nu \sigma_c)] = f_{abc} [(\partial^\nu \sigma_b)\sigma_c - (\partial^\nu \sigma_c)\sigma_b] =$$

$$= 2 f_{abc} (\partial^\nu \sigma_b)\sigma_c. \tag{1.108}$$

Allowing for the structural constants f_{abc} to obey the Jacobi identity

$$f_{acb} f_{bde} + f_{adb} f_{bec} + f_{aeb} f_{bcd} = 0, \tag{1.109}$$

we can rewrite the last term in Eq. (1.107) as

$$\frac{g_s^2 \epsilon}{2} f_{acb} f_{bde} \sigma_d \sigma_e G_c^\nu = -\frac{g_s^2 \epsilon}{2} [f_{adb} f_{bec} \sigma_d \sigma_e G_c^\nu + f_{aeb} f_{bcd} \sigma_d \sigma_e G_c^\nu] =$$

$$= -g_s^2 \epsilon f_{adb} f_{bec} \sigma_d \sigma_e G_c^\nu. \tag{1.110}$$

Bringing together the results obtained, we arrive at

$$\delta(\partial^\nu \sigma_a - g_s f_{abc} G_c^\nu \sigma_b) = 0. \tag{1.111}$$

With allowance made for Eqs. (1.100) and (1.111), the final result is as follows

$$\delta S' = -\frac{\epsilon}{\xi} \int d^4 x (\partial_\nu G_a^\nu)(\partial_\mu D^\mu \sigma_a) - \frac{\epsilon}{\xi} \int d^4 x (\partial_\nu \partial_\mu G_a^\mu)(D^\nu \sigma_a) =$$

$$= -\frac{\epsilon}{\xi} \int d^4 x \partial_\mu [(\partial_\nu G_a^\nu)(D^\mu \sigma_a)] = 0. \tag{1.112}$$

So, we have proved that S_{eff} is invariant with regard to the BRS transformations.

2

Formalism of functional integration

Way down below you'll never meet
This magic beauty—a tenth of it!
V. Vysotsky

2.1   Functional integral in quantum mechanics

In this chapter we shall again obtain the same set of the QCD Feynman rules as in the previous chapter. But now, our approach will be based on a formalism of a functional (or continual) integration. The paths integral idea introduced by R. Feynman [10] in quantum mechanics lies at the heart of this formalism (detailed presentation of this method could be found in the book [11]). Later on this method received wide recognition in statistical physics from which it was extended to quantum field theory. The functional integration method has a whole series of advantages over other quantization methods. It traces the linkage with classical dynamics. Besides, under such an approach we are operating with ordinary functions. This permits carrying out of any nonlinear transformations of fundamental dynamical variables in a more simple style and to directly see effects appearing under these transformations. The functional integration method is a general and self-contained quantization method that works for a wide class of systems including those with couplings. It easily deduces the Feynman rules and, at the same time, gives an opportunity for calculating nonperturbative Green functions. The advantage of the method is that it establishes the commonness of quantum field theory (QFT) with quantum statistical physics. This, in turn, permits the use of the quantum statistics technique in QFT. Formulation compactness of fundamental relations of QFT enables us to easily trace the structure of one or another model and its gauge invariance in particular.

An analogue with Gauss integrals lies at the heart of the functional integration. In a theory of the function of a single variable a Gauss integral has the form

$$\int_{-\infty}^{\infty} \frac{dy}{\sqrt{\pi}} \exp\left(-ay^2\right) = \frac{1}{\sqrt{a}}, \tag{2.1}$$

where $a > 0$. Let us consider a quadratic form of k variables

$$(\mathbf{y}, A\mathbf{y}) \equiv \sum_{i,j=1}^{k} y_i A_{ij} y_j \geq 0. \tag{2.2}$$

Diagonalizing this expression with the help of the orthogonal linear transformation, we can easily ascertain the validity of the following value of the k-dimensional Gauss integral:

$$\int_{-\infty}^{\infty} dy_1 \ldots \int_{-\infty}^{\infty} \frac{dy_k}{(\pi)^{k/2}} \exp\left[-(\mathbf{y}, A\mathbf{y})\right] = [\det A]^{-1/2}. \tag{2.3}$$

Further, we carry out generalization of the finite-dimensional case on infinite-dimensional functional space. At that the index i is replaced with a continuous variable x ($i \to x$, $-\infty < x < \infty$) and $y_i \to y(x)$. Then the sum over i turns into an integral and the quadratic form (2.2) passes on to a double integral whose kernel is composed of an operator A with matrix elements $A(x, x')$:

$$(\mathbf{y}, A\mathbf{y}) \to (y, Ay) = \int dx \int dx' y(x) A(x, x') y(x') \geq 0. \tag{2.4}$$

Under such a limiting transition the finite-dimensional integral (2.3) is transformed into an infinite-dimensional integral, a functional integral

$$\int D[y(x)] \exp\left[-(y, Ay)\right] = [\det A]^{-1/2},$$

where $D[y(x)]$ will be called the *measure*. In the right side of this equality, as well as in (2.3), there is $\det A$ that is understood as the product of all eigenvalues of the operator A

$$A\xi_i = \lambda_i \xi_i, \qquad \det A = \prod \lambda_i,$$

where ξ_i is the eigenvector of A. Every so often it is convenient to use the notion

$$\det A = \exp(\mathrm{Sp}\ln A). \tag{2.5}$$

By way of example we examine nonrelativistic quantum mechanics (NQM) in one-dimensional space. In this case the Hamiltonian H, representing a function of generalized momenta P and coordinates Q

$$H = \frac{P^2}{2m} + V(Q). \tag{2.6}$$

is an evolution operator of a quantum system. The matrix element of the transition between the initial and final states

$$< q'; t' | q; t > = < q' | \exp\left[-iH(t - t')\right] | q >, \tag{2.7}$$

where states $|q>$ obey the condition $Q|q> = q|q>$, symbols $|q; t>$ label states in the Heisenberg picture, and $|q; t> = \exp(iHt)|q>$, is a fundamental quantity in the NQM. There is need to bear in mind that states in the Heisenberg picture do not depend on time and our designations mean the states $|q; t>$ and $|q'; t'>$ coincide with two Schrödinger states $|q(t)>$ and $|q'(t')>$ in the time moments t and t', respectively. We divide the interval (t', t) into a great number of small periods

$$t - t_{n-1}, t_{n-1} - t_{n-2}, \ldots, t_j - t_{j-1}, \ldots, t_1 - t'.$$

Then the transition amplitude may be represented as

$$< q' | \exp\left[-iH(t - t')\right] | q > = \int dq_1 ... dq_{n-1} < q' | \exp\left[-iH(t - t_{n-1})\right] | q_{n-1} > \times$$

$$\times < q_{n-1} | \exp\left[-iH(t_{n-1} - t_{n-2})\right] | q_{n-2} > ... < q_1 | \exp\left[-iH(t_1 - t')\right] | q >, \tag{2.8}$$

where we have substituted total eigenstate sets of the operator Q^S in the Schrödinger picture. For arbitrarily small $\delta t_j = t_j - t_{j-1}$ we have

$$< q_j | \exp\left[-iH(t_j - t_{j-1})\right] | q_{j-1} > = < q_j | [1 - iH(P, Q)\delta t_j] | q_{j-1} > + O[(\delta t_j)^2]. \tag{2.9}$$

In Eq. (2.9) we replace the operator Q by its eigenvalue at a time t_j and pass on to the momentum representation to get rid of the operator P. Assuming the symmetric ordering of operators in $V(Q)$ as well, we get

$$< q_j | \exp\left[-iH\delta t_j\right] | q_{j-1} > \approx < q_j | q_{j-1} > -i\delta t_j \left[< q_j | \frac{P^2}{2m} | q_{j-1} > +V\left(\frac{q_j + q_{j-1}}{2}\right) \times \right.$$

$$\times \delta(q_j - q_{j-1}) \Bigg] = \int \frac{dp_j}{2\pi} \exp\left[ip_j(q_j - q_{j-1})\right] \left\{ 1 - i\delta t_j \left[\frac{p_j^2}{2m} + V\left(\frac{q_j + q_{j-1}}{2}\right)\right]\right\} \approx$$

$$\approx \int \frac{dp_j}{2\pi} \exp\left[ip_j(q_j - q_{j-1})\right] \exp\left[-i\delta t_j H(p_j, (q_j + q_{j-1})/2)\right] \tag{2.10}$$

where $H(p_j, (q_j + q_{j-1})/2)$ is by now the classical Hamiltonian. Substitution of (2.10) into (2.8) gets

$$< q' | \exp\left[-iH(t' - t)\right] | q > \approx \int \frac{dp_1}{2\pi} ... \frac{dp_n}{2\pi} \int dq_1 ... dq_{n-1} \times$$

$$\times \exp\left\{ i \sum_{j=1}^{n} \left[p_j(q_j - q_{j-1}) - \delta t_j H\left(p_j, \frac{q_j + q_{j-1}}{2}\right)\right]\right\}. \tag{2.11}$$

With an allowance made for the obtained expressions, the transition amplitude can be represented in the view

$$< q' | \exp\left[-iH(t' - t)\right] | q > = \lim_{n \to \infty} \int \left(\frac{dp_1}{2\pi}\right) ... \left(\frac{dp_n}{2\pi}\right) \int dq_1 ... dq_{n-1} \times$$

$$\times \exp\left\{ i \sum_{j=1}^{n} \delta t_j \left[p_j \left(\frac{q_j - q_{j-1}}{\delta t_j}\right) - H\left(p_j, \frac{q_j + q_{j-1}}{2}\right)\right]\right\} \equiv$$

$$\equiv \int \left[\frac{dpdq}{2\pi}\right] \exp\left\{ i \int_{t}^{t'} dt[p\dot{q} - H(p,q)]\right\}, \tag{2.12}$$

where

$$\left[\frac{dpdq}{2\pi}\right] = \prod_{i=1}^{n} \left(\frac{dp_i}{2\pi}\right) \prod_{j=1}^{n-1} dq_j. \tag{2.13}$$

In Eq. (2.12) we shall fulfill the momentum integration over

$$\left[\frac{dp}{2\pi}\right] = \prod_{i=1}^{n} \frac{dp_i}{2\pi}.$$

Since the integrand is an oscillating function, it may be analytically continued to the Euclidean space. Then, considering $(i\delta t_j)$ to be a real quantity, we find

$$\int \frac{dp_j}{2\pi} \exp\left[\frac{-i\delta t_j}{2m} p_j^2 + ip_j(q_j - q_{j-1})\right] = \left(\frac{m}{2\pi i\delta t}\right)^{1/2} \exp\left[\frac{im(q_j - q_{j-1})^2}{2\delta t}\right], \tag{2.14}$$

where we have taken into account (2.1) and set all δt_j to be equal to each other ($\delta t_j = \delta t$). Gathering the obtained results, we get the final expression for the transition amplitude (2.7) in the functional integration formalism

$$< q'; t' | q; t > = \lim_{n \to \infty} \left(\frac{m}{2\pi i\delta t}\right)^{n/2} \int \prod_{i}^{n-1} dq_i \exp\left\{ i \sum_{j=1}^{n} \delta t \left[\frac{m}{2} \left(\frac{q_j - q_{j-1}}{\delta t}\right)^2 - V(q)\right]\right\} =$$

$$= N \int [dq] \exp \left\{ i \int_t^{t'} d\tau \left[\frac{m\dot{q}^2}{2} - V(q) \right] \right\} = N \int [dq] \exp \left[i \int_t^{t'} d\tau \mathcal{L}(q, \dot{q}) \right], \qquad (2.15)$$

where N is a normalization factor and $[dq]$ is the measure on the functional space of trajectories $q(t)$. Restoring the Plank constant, we rewrite this expression in the form

$$< q'; t' | q; t >= N \int [dq] \exp \left[\frac{i}{\hbar} S(t', t) \right], \qquad (2.16)$$

where S is the action. So, we have written the matrix element of the transition between the initial and final states in the form of the functional integral that represents the sum of contributions over all paths connecting points $(q; t)$ and $(q'; t')$. These contributions are taken with weights which are equal to the exponentials of the action multiplied by the imaginary unit. The functional integration trick actually consists of the following. At sufficiently small divisions of time, we get rid of operator quantities and can deal only with classical variables, but at the price of introducing infinite large number of intermediate time values t and allowing for all possible values of coordinates q_j at every instant of time t_j. The superposition principle reflects a fundamental property of continual integrals. It lies in the fact that, when t belongs to the interval t_0, t_f, then the equality

$$\int [dq] \exp \left[\frac{i}{\hbar} S(t_f, t_0) \right] = \int dq(t) \int [dq] \exp \left[\frac{i}{\hbar} S(t_f, t) \right] \int [dq] \exp \left[\frac{i}{\hbar} S(t, t_0) \right] \qquad (2.17)$$

must be fulfilled.

The formula (2.16) is of a mathematical expression of the Huygens principle and with its help one can clearly understand the difference between classical and quantum mechanics or, what is the same, between geometrical and wave optics. In classical mechanics, in general, the action is very large as compared with $\hbar$. For this reason, $\exp(iS/\hbar)$ extremely oscillates when we pass on from trajectory to trajectory; in so doing nearly a total cancellation of all contributions takes place except the contribution coming from the region in the immediate vicinity to that trajectory on which the action S reaches an extreme value (minimum more often than not). Therefore, in classical mechanics, a trajectory associated with minimum action is the only important one. In contrast, in quantum mechanics, the action S may be comparable or less than $\hbar$, so that many various trajectories give more or less identical contributions. Therefore, the quantum-mechanical description may be interpreted as the inclusion of fluctuations close to a classical trajectory.

Paths integration formalism allows us to calculate not only transition amplitudes like $< q'; t' | q; t >$, but matrix elements between states $< q'; t' |$ and $| q; t >$ for products of arbitrary operators $A(P(t), Q(t))$ as well. In so doing, we bear in mind one subtlety. Because the subdivision points t_j in (2.8) correspond to an increase of t_j from right to left ($\delta t_j = t_j - t_{j-1} > 0$), then the functional integral actually represents a time-ordering product of operators (T product). In quantum field theory (QFT) one usually studies the T-products of vacuum expectation values of field operators that are called *n-point Green functions*

$$G(t_1, t_2, ..., t_n) =< 0 | T(\psi(t_1)\psi(t_2)...\psi(t_n)) | 0 > .$$

With their help, n-partial amplitudes of various processes are obtained. As in QFT, wave functions play a part of coordinates, averages over the ground system state $| 0 >$ of the T-product of coordinate operators taken at the time moments $t_1, \ldots, t_n$

$$\beta(t_1, \ldots, t_n) =< 0 | T(Q(t_1) \ldots Q(t_n)) | 0 >$$

are analogues of the Green functions. So then, our following task is to produce $\beta(t_1, \ldots, t_n)$ in the form of functional integrals.

From the beginning we consider a quantity

$$\beta(t_1, t_2) = < 0|T(Q^H(t_1)Q^H(t_2)|0 > . \qquad (2.18)$$

Substituting the total state sets into (2.18), we get

$$\beta(t_1, t_2) = \int dq \int dq' < 0|q'; t' >< q'; t'|T(Q^H(t_1)Q^H(t_2))|q; t >< q; t|0 >=$$

$$= \int dq \int dq' \Phi_0(q', t') \Phi_0^*(q, t) < q'; t'|T(Q^H(t_1)Q^H(t_2))|q; t >, \qquad (2.19)$$

where we have taken into account the definition of the wave functions of the ground state

$$< 0|q; t >= \Phi_0(q, t), \qquad < q; t|0 >= \Phi_0^*(q, t).$$

Further, assuming that $t_1 > t_2$ (i.e., $t' > t_1 > t_2 > t$) and allowing for (2.12), we find

$$< q'; t'|T(Q^H(t_1)Q^H(t_2))|q; t >=< q'|\exp[-iH(t' - t_1)]Q^S \exp[-iH(t_1 - t_2)]Q^S \times$$

$$\times \exp[-iH(t_2 - t)]|q >= \int dq_1 \int dq_2 < q'|\exp[-iH(t' - t_1)]|q_1 >< q_1|Q^S \times$$

$$\times \exp[-iH(t_1 - t_2)]|q_2 >< q_2|Q^S \exp[-iH(t_2 - t)]|q >=< q'|q_1(t)q_2(t) \times$$

$$\times \exp[-iH(t' - t)]|q >= \int \left[\frac{dpdq}{2\pi}\right] q_1(t_1)q_2(t_2) \exp\left\{i \int_t^{t'} d\tau[p\dot{q} - H(p, q)]\right\}. \qquad (2.20)$$

It is evident, that the same formula holds true for the case $t_2 > t_1$ (i.e., $t' > t_2 > t_1 > t$) as well. Substituting (2.20) into (2.19) results in

$$\beta(t_1, t_2) = \int dq \int dq' \Phi_0(q', t') \Phi_0^*(q, t) \int \left[\frac{dpdq}{2\pi}\right] q_1(t_1)q_2(t_2) \exp\left\{i \int_t^{t'} d\tau[p\dot{q} - H(p, q)]\right\}. \qquad (2.21)$$

The presence of the wave functions of the ground states is the disadvantage of Eq. (2.21) and our next task is to get rid of them. Examine a matrix element

$$< q'; t'|T(Q^H(t_1)Q^H(t_2))|q; t >= \int dQ \int dQ' < q'; t'|Q'; \tau' > \times$$

$$\times < Q'; \tau'|T(Q^H(t_1)Q^H(t_2))|Q; \tau >< Q; \tau|q; t >, \qquad (2.22)$$

$t' \geq \tau' \geq (t_1, t_2) \geq \tau \geq t$. We write the first factor in the right side of (2.22) as

$$< q'; t'|Q'; \tau' >=< q'|\exp(-iH(t' - \tau')|Q' >= \sum_n < q'|n >< n|\exp(-iH(t' - \tau')|Q' >=$$

$$= \sum_n \Phi_n^*(q')\Phi_n(Q') \exp(-iE_n(t' - \tau')), \qquad (2.23)$$

where we have accepted

$$H|n >= E_n|n >, \qquad \Phi_n^*(q) =< q|n > \qquad \Phi_n(Q') =< n|Q' > . \qquad (2.24)$$

Later we allow for

$$E_n > E_0 \qquad n \neq 0$$

and pass to the limit $t' \to -i\infty$. Then, the expression (2.23) becomes

$$\lim_{t' \to -i\infty} < q'; t'|Q'; \tau' > = \Phi_0^*(q')\Phi_0(Q') \exp\left[-E_0|t'| + iE_0\tau'\right]. \tag{2.25}$$

Similar calculations give

$$\lim_{t \to i\infty} < Q; \tau|q; t > = \Phi_0(q)\Phi_0^*(Q) \exp\left(-E_0|t| - iE_0\tau\right). \tag{2.26}$$

To substitute (2.25) and (2.26) into (2.22) results in

$$\lim_{\substack{t' \to -i\infty \\ t \to i\infty}} < q'; t'|T(Q^H(t_1)Q^H(t_2))|q; t > = \int dQ \int dQ' \Phi_0^*(q')\Phi_0(Q') \times$$

$$\times < Q'; \tau'|T(Q^H(t_1)Q^H(t_2))|Q; \tau > \Phi_0^*(Q)\Phi_0(q) \exp\left(-E_0|t'| + iE_0\tau' - E_0|t| - iE_0\tau\right) =$$

$$= \Phi_0^*(q')\Phi_0(q) \exp\left(-E_0|t'| - E_0|t|\right)\beta(t_1, t_2), \tag{2.27}$$

where at the final stage we have taken into consideration Eq. (2.19). In turn, from Eqs. (2.25) and (2.26) follows

$$\lim_{\substack{t' \to -i\infty \\ t \to i\infty}} < q'; t'|q; t > = \Phi_0^*(q')\Phi_0(q) \exp\left(-E_0|t'| - E_0|t|\right). \tag{2.28}$$

Making use of (2.27) and (2.28), we get the sought formula for $\beta(t_1, t_2)$

$$\beta(t_1, t_2) = \lim_{\substack{t' \to -i\infty \\ t \to i\infty}} \left[\frac{< q'; t'|T(Q^H(t_1)Q^H(t_2))q; t >}{< q'; t'|q; t >}\right] =$$

$$= \lim_{\substack{t' \to -i\infty \\ t \to i\infty}} \frac{1}{< q'; t'|q; t >} \int \left[\frac{dq\,dp}{2\pi}\right] q(t_1)q(t_2) \exp\left\{i \int_t^{t'} d\tau[p\dot{q} - H(p, q)]\right\}, \tag{2.29}$$

where the factor $< q'; t'|q; t >$ being in the denominator is defined by the expression (2.15). The evident generalization of (2.29) for the case of n-products of chronologically ordered coordinate operators has the view

$$\beta(t_1, \ldots t_n) = < 0|T(Q^H(t_1), \ldots Q^H(t_n))|0 > =$$

$$= \lim_{\substack{t' \to -i\infty \\ t \to i\infty}} \frac{1}{< q'; t'|q; t >} \int \left[\frac{dq\,dp}{2\pi}\right] q(t_1) \ldots q(t_n) \exp\left\{i \int_t^{t'} d\tau[p\dot{q} - H(p, q)]\right\}. \tag{2.30}$$

To introduce the idea of a generating functional, we make a small mathematical digression. In the case of finite numbers of variables y_j $(j = 1, 2, \ldots, k)$ it is defined as follows:

$$F[y] \equiv F(y_1, \ldots, y_k) = \sum_{n=0}^{\infty} \sum_{j_1=1}^{k} \ldots \sum_{j_n=1}^{k} \frac{1}{n!} F_n(j_1, j_2, \ldots, j_n)y_{j_1} \ldots y_{j_n}, \tag{2.31}$$

where $F_n(j_1, j_2, \ldots, j_n)$ is a coefficient symmetrical over indices $j_1, j_2, \ldots, j_n$. Thus, for example, at $k = 2$ we have the Taylor series expansion:

$$F(y_1, y_2) = F(0) + y_1 \left(\frac{\partial F}{\partial y_1}\right)_0 + y_2 \left(\frac{\partial F}{\partial y_2}\right)_0 + \frac{1}{2} y_1 y_1 \left(\frac{\partial^2 F}{\partial y_1^2}\right)_0 +$$

$$+ \frac{1}{2} y_1 y_2 \left(\frac{\partial^2 F}{\partial y_1 \partial y_2} + \frac{\partial^2 F}{\partial y_2 \partial y_1}\right)_0 + \frac{1}{2} y_2 y_2 \left(\frac{\partial^2 F}{\partial y_2^2}\right)_0 + \ldots =$$

$$= F(0) + \sum_{j=1}^{2} y_j \left(\frac{\partial F}{\partial y_j}\right)_0 + \sum_{j_1=1}^{2} \sum_{j_2=1}^{2} y_{j_1} y_{j_2} \left(\frac{\partial^2 F}{\partial y_{j_1} \partial y_{j_2}}\right)_0 + \ldots =$$

$$= F(0) + \sum_{j=1}^{2} y_j F_1(j) + \sum_{j_1=1}^{2} \sum_{j_2=1}^{2} y_{j_1} y_{j_2} F_2(j_1, j_2) + \ldots$$

It is clear, that

$$F_1(1) = \left(\frac{\partial F(y_1, y_2)}{\partial y_1}\right)_0, \qquad F_2(1, 1) = \left(\frac{\partial^2 F(y_1, y_2)}{\partial y_1^2}\right)_0$$

and so on, that is, $F(y_1, y_2)$ is the generating functional (in the given case, it is a function) allowing us to calculate all the expansion coefficients $F_n(j_1, j_2, \ldots, j_n)$.

Let us generalize the situation on a functional space. We introduce the functional

$$F[y(x)] = \sum_{n=0}^{\infty} \frac{1}{n!} \int dx_1 \ldots \int dx_n F_n(x_1, \ldots, x_n) y(x_1) \ldots y(x_n). \qquad (2.32)$$

Now we have to use functional derivatives $\delta/\delta y(x_j)$ rather than the partial ones. For $\delta/\delta y(x_j)$ the following relation takes place

$$\frac{\delta}{\delta y(x_j)} y(x_k) = \delta(x_j - x_k).$$

Then, the coefficient functions in the functional (2.32) may be written in terms of the functional derivatives of $F[y(x)]$:

$$F_n(x_1, \ldots, x_n) = \frac{\delta}{\delta y(x_1)} \cdots \frac{\delta}{\delta y(x_n)} F[y(x)] \Bigg|_{y=0}. \qquad (2.33)$$

By this reason, $F[y(x)]$ is referred to as the generating functional for the functions $F_n(x_1, \ldots, x_n)$.

Let us return to the question at hand. We shall consider that during a finite time interval a force $J(t)$ (source) acts on a system. The corresponding Lagrangian

$$\mathcal{L}_{int} = J(t)q(t)$$

must be supplemented to the total one. Then, using the generating functional $W[J]$

$$W[J] = \lim_{\substack{t' \to -i\infty \\ t \to i\infty}} \frac{1}{<q'; t'|q; t>} \int \left[\frac{dq\,dp}{2\pi}\right] \exp\left\{i \int_t^{t'} d\tau [p\dot{q} - H(p, q) + J(\tau)q(\tau)]\right\} =$$

$$= N \lim_{\substack{t' \to -i\infty \\ t \to i\infty}} \frac{1}{<q'; t'|q; t>} \int [dq] \exp\left\{i \int_t^{t'} d\tau [\mathcal{L}(q, \dot{q}) + J(\tau)q(\tau)]\right\}, \qquad (2.34)$$

where N is a constant factor, one can obtain all $\beta(t_1, \ldots, t_n)$ with the help of the following formula

$$\beta(t_1, \ldots t_n) = (-i)^n \frac{\delta^n}{\delta J(t_1) \ldots \delta J(t_n)} W[J] \Big|_{J=0}. \tag{2.35}$$

We shall demonstrate later that in QFT the factors $< q; t | q'; t' >$ and N, which do not depend on $J(\tau)$ are inessential and can be omitted under the calculation of connected Green functions. We may choose the following normalization condition for the generating functional:

$$W[0] = 1. \tag{2.36}$$

To compare (2.35) with (2.30) leads to the key conclusion, the generating functional $W[J]$ is in line with the transition amplitude between the ground state at a time t and that at a time t' that is computed under the addition of a source $J(\tau)$

$$W[J] = < 0 | 0 >_J. \tag{2.37}$$

All the aforesaid is directly extended on the case of n degrees of freedom. In doing so, it is enough to understand $q_1(t) \ldots q_n(t)$ as a totality of generalized coordinates $q_1(t) \ldots q_n(t)$. In this case the generating functional is given by the expression

$$W[J_1, \ldots J_n] =$$

$$= N \lim_{\substack{t' \to -i\infty \\ t \to i\infty}} \frac{1}{< q'; t' | q; t >} \int \prod_{j=1}^{n} [dq_j] \exp \left\{ i \int_t^{t'} d\tau \sum_j^n [\mathcal{L}(q_j, \dot{q}_j) + J_j(\tau) q_j(\tau)] \right\}. \tag{2.38}$$

It is well to bear in mind that functional integrals were defined as a limiting case of Gauss integrals (converging!) when dimensions tend to infinity. By this reason the functional integrals entering into (2.29) and (2.30) would be well defined if they have the Gaussian form. It is needed for that to pass on from the Minkowski space with the pseudo-Euclidean metric to the Euclid space, that is, to carry out the Wick rotation

$$(x_0, x_1, x_2, x_3) \to (x_4 = ix_0, x_1, x_2, x_3).$$

Just this condition is reflected by the presence of nonphysical limits $t' \to -i\infty$, $t \to i\infty$ in Eq. (2.38).

2.2 Boson fields quantization by means of the functional integrals method

The functional approach is immediately generalized on QFT to be considered as a quantum-mechanical system with infinite numbers of degrees of freedom. Now, rather than divide a time interval by small periods ϵ, we divide a space-time on four-dimensional cells having a volume ϵ^4. Besides, the previous trajectory definition such that a value x_j is confronted with every time moment t_j should be changed too. We shall define a trajectory giving a value of a field $\psi(x) = \psi(\mathbf{r}, t)$, which may be multicomponents, in every cell. The paths integral presently involves a summation over all possible field values in every cell. We confront objects that were introduced in NQM with the following ones:

$$\prod_{j=1}^{n} [dq_j, dp_j] \to [d\psi(x), d\pi(x)],$$

$$\mathcal{L}(q_j, \dot{q}_j), H(q_j, p_j) \to \int d^3x \mathcal{L}(\psi(x), \partial_\mu \psi(x)), \quad \int d^3x H(\psi(x), \pi(x)),$$

where $\pi(x)$ is a momentum canonically conjugate to a field.

In QFT a physical process in question is defined by the numbers of initial and final quanta with given values of spin and momentum. Therefore, our goal is as follows. Within formalism under consideration we are going to build up such a mechanism that automatically produces initial and final particles from a vacuum independently on a specific shape of interaction. This is realized by the introduction of terms with an arbitrary source into the Lagrangian. Because in QFT a vacuum state is a ground state, then the generating functional $W[J]$ is nothing else than a vacuum-vacuum transition amplitude in the presence of an external source $J(x)$. Then the evident generalization of the formula (2.38) has the form

$$W[J] = N \int [d\psi] \exp \left\{ i \int d^4x [\mathcal{L}(\psi(x), \partial_\mu \psi(x)) + J(x)\psi(x)] \right\}, \tag{2.39}$$

where N is a normalization factor. In the expression obtained we have to do the Wick's rotation

$$t \to x_0 = it, \qquad W[J] \to W_E[J] \tag{2.40}$$

and, after that, $W_E[J]$ may be used to calculate Green functions. When formulating the Feynman rules, we have to take into account that disconnected diagrams — diagrams describing the propagation of groups of particles not connected with each other—are beyond our interest. Scattering amplitudes can be obtained by connected diagrams only. Therefore, we are interested in merely those Green functions that are associated with connected diagrams. To remove a disconnected part of the Green function we have to divide the expression (2.34) into the vacuum-vacuum transition amplitude, $W_E[J]$. Thus, in the Euclidean space, connected Green functions are defined in the following way:

$$G_E^{(n)}(x_1, \ldots x_n) = \left[\frac{1}{W_E[J]} \frac{\delta^n W_E[J]}{\delta J(x_1) \ldots \delta J(x_n)} \right]_{J=0} = \left[\frac{\delta^n \ln W_E[J]}{\delta J(x_1) \ldots \delta J(x_n)} \right]_{J=0}. \tag{2.41}$$

Such a definition has an important practical consequence, namely, the normalization factor of the quantity $W[J]$ being independent on J proves to be inessential under the subsequent calculations of Green functions. All this could be demonstrated by the example of a real scalar field ϕ with the Lagrangian

$$\mathcal{L} = \mathcal{L}_0 + \mathcal{L}_{int}, \tag{2.42}$$

where

$$\mathcal{L}_0 = \frac{1}{2} \left(\partial_\mu \phi \partial^\mu \phi - m^2 \phi^2 \right), \qquad \mathcal{L}_{int} = -\frac{\lambda}{4!} \phi^4.$$

We set $\mathcal{L}_{int} = 0$ and compute the Euclidean generating functional for the noninteracting field $W_E^0[J]$. After integration by parts, we obtain

$$W_E^0[J] = \int [d\phi] \exp \left\{ \int d^4x \left[-\frac{1}{2} \phi(x)(\partial_\mu \partial^\mu + m^2)\phi(x) + J(x)\phi(x) \right] \right\}, \tag{2.43}$$

where

$$\partial_\mu \partial^\mu = -\frac{\partial^2}{\partial x_0^2} - \nabla^2.$$

The integral in the exponential is the limit of the four-dimensional cells sum. Having denoted the field in a cell α as ϕ_α, the one in a next cell β as ϕ_β, and so on, one may consider $1/2(\partial_\mu \partial^\mu + m^2)$ to be a limit of symmetric matrix $A_{\alpha\beta}$ connecting neighbor cells.

Further, allowing for symmetry of the matrix A and using Eq. (2.1), one may show the validity of the following formula:

$$\int_{-\infty}^{\infty} \prod_{i=1}^{N} d^4 x_i \exp\left(-x_i A_{ij} x_j + 2S_k x_k\right) = \pi^{N/2}[\det A]^{-1/2} \exp\left(S_i A_{ij}^{-1} S_j\right). \qquad (2.44)$$

Taking into account this result, setting $S_k = J(x)/2$ and throwing away the inessential infinite factor being independent on $J(x)$, we arrive at the following expression for $W_E^0[J]$

$$W_E^0[J] = \exp\left\{\frac{1}{2}\int d^4 x d^4 y J(x)[\partial_\mu \partial^\mu + m^2]^{-1} J(y)\right\} = \exp\left\{\frac{1}{2}\int d^4 x d^4 y J(x)\Delta(x-y)J(y)\right\}.$$
$$(2.45)$$

The quantity $\Delta(x-y) = (\partial_\mu \partial^\mu + m^2)^{-1}$ may be represented by way of the Fourier integral

$$\Delta(x-y) = \frac{1}{(2\pi)^4}\int \frac{d^4 k}{k^2 + m^2}\exp\left[-ik(x-y)\right],$$

where $k = (ik_0, \mathbf{k})$ is a four-dimensional Euclidean momentum vector. Carrying out analytical continuation into the Minkowski space, we lead to a well-known result—to the Feynman propagator of a scalar particle with a mass m.

$$-\Delta(x-y) \to i\Delta^c(x-y) = \frac{i}{(2\pi)^4}\int \frac{d^4 p}{p^2 - m^2 + i\epsilon}\exp\left[-ip(x-y)\right] \qquad (2.46)$$

($\Delta^c(x-y)$ is a causal Green function of a free scalar field). The term $-i\epsilon$ ensures the Feynman prescription of the propagator poles detour. In the Lagrangian, it corresponds to the additional term $+i\epsilon\phi^2/2$, which ensures convergence of the functional integral (2.43) in the Minkowski space.

The factor in the exponential of the right side of (2.45) has the following meaning: in the point y the source $J(y)$ produces a free scalar particle that propagates from y to x and the source $J(x)$ annihilates this particle in the point x. Expanding the exponential gives a series of terms corresponding to $0, 1, 2, \ldots$ free scalar particles.

Now we switch on interaction ($\mathcal{L}_{int} \neq 0$). To obtain a series of the perturbation theory for $W_E[J]$, we have to expand $\exp\left\{\int d^4 x \mathcal{L}_{int}\right\}$ as a power series. Then, we get the sum of terms, each representing the functional integral of the Gauss function

$$\exp\left\{\int d^4 x[\mathcal{L}_0 + J(x)\phi(x)]\right\},$$

multiplied by the expression of the following view

$$\frac{1}{n!}\left[\int d^4 x \mathcal{L}_{int}\right]^n.$$

Such integrals are calculated when $\mathcal{L}_{int} \sim \phi^n$. It is easy to see that every power of $\phi(x)$ can be obtained by differentiating the exponential in

$$W_E[J] = \int [d\phi]\exp\left\{\int d^4 x \left[\mathcal{L}_0 + \mathcal{L}_{int} + J(x)\phi(x)\right]\right\} = \exp\left\{\int d^4 x \mathcal{L}_{int}\right\}W_E^0[J] \qquad (2.47)$$

with respect to the source $J(x)$, that is, by the change $\phi(x) \to \delta/\delta J(x)$,

$$W_E[J] = \sum_{n=0}^{\infty}\frac{1}{n!}\left[\int d^4 z \mathcal{L}_{int}\left(\frac{\delta}{\delta J(z)}\right)\right]^n \int [d\phi]\exp\left\{\int d^4 x[\mathcal{L}_0 + J(x)\phi(x)]\right\} =$$

$$= \exp\left\{ \int d^4 z \mathcal{L}_{int}\left(\frac{\delta}{\delta J(z)}\right)\right\} \int [d\phi] \exp\left\{ \int d^4 x [\mathcal{L}_0 + J(x)\phi(x)]\right\} =$$

$$= \sum_{n=0}^{\infty} \frac{1}{n!}\left\{ \int d^4 z\left[-\frac{\lambda}{4!}\left(\frac{\delta}{\delta J(z)}\right)^4\right]\right\}^n \exp\left\{\frac{1}{2}\int d^4 x \int d^4 y J(x)\Delta(x-y)J(y)\right\}. \quad (2.48)$$

In order to deeper realize the generating functional structure, we display $W_E[J]$ in the form of graphs, Feynman diagrams. First and foremost we note that, in the absence of an interaction, $W_E^0[J]$ is represented as the diagrams sum shown in Fig. 2.1. Here, the solid line indicates $\Delta(x-y)$, the cross does $J(x)$ and integrating over x_j, y_j is meant. When

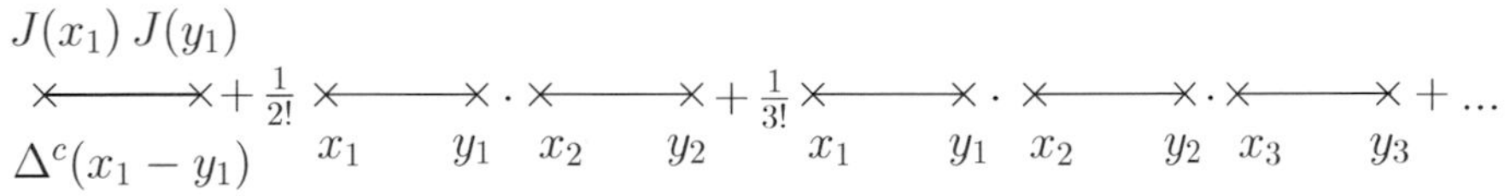

$$J(x_1)\,J(y_1)$$

$$\underset{\Delta^c(x_1-y_1)}{\times\!\!\!-\!\!\!\!-\!\!\!\!-\!\!\!\times} + \frac{1}{2!}\underset{x_1 \quad y_1 \quad x_2}{\times\!\!\!-\!\!\!\!-\!\!\!\times\cdot\times} \underset{y_2}{-\!\!\!\!-\!\!\!\times} + \frac{1}{3!}\underset{x_1 \quad y_1 \quad x_2}{\times\!\!\!-\!\!\!\times\cdot\times} \underset{y_2 \quad x_3}{-\!\!\!\times\cdot\times} \underset{y_3}{-\!\!\!\times} + \ldots$$

FIGURE 2.1

The Feynman diagrams for $W_E^0[J]$.

differentiating, every term of interaction (2.48) destroys four sources pinching four points x_j into one point z and ascribing the factor λ to this point. As a result, from the graphs of Fig. 2.1 follows additional set of graphs (Fig. 2.2) associated with the second and third terms in $W_E[J]$:

$$W_E[J] = W_E^0[J]\{1 + \lambda w_1[J] + \lambda^2 w_2[J] + \ldots\}, \quad (2.49)$$

where

$$w_1[J] = -\frac{1}{4!}(W_E^0[J])^{-1}\left\{\int d^4 x\left[\frac{\delta}{\delta J(x)}\right]^4\right\}W_E^0[J] = -\frac{1}{4!}[\Delta(x-y_1)\Delta(x-y_2)\times$$

$$\times\Delta(x-y_3)\Delta(x-y_4)J(y_1)J(y_2)J(y_3)J(y_4) + 3!\Delta(x-y_1)\Delta(x-y_2)\Delta(x-x)J(y_1)J(y_2)], \quad (2.50)$$

$$w_2[J] = -\frac{1}{2(4!)^2}(W_E^0[J])^{-1}\left\{\int d^4 x\left[\frac{\delta}{\delta J(x)}\right]^4\right\}W_E^0[J] =$$

$$= -\frac{1}{2(4!)}W_0^{-1}[J]\left\{\int d^4 x\left[\frac{\delta}{\delta J(x)}\right]^4\right\}w_1[J] = \frac{1}{2}w_1^2[J] + \frac{1}{2(3!)^2}\Delta(x_1-y_1)\Delta(x_1-y_2)\Delta(x_1-y_3)\times$$

$$\times\Delta(x_1-x_2)\Delta(x_2-y_4)\Delta(x_2-y_5)\Delta(x_2-y_6)J(y_1)J(y_2)J(y_3)J(y_4)J(y_5)J(y_6)+$$

$$+\frac{3}{2(4!)}\Delta(x_1-y_1)\Delta(x_1-y_2)[\Delta(x_1-x_2)]^2\Delta(x_2-y_3)\Delta(x_2-y_4)J(y_1)J(y_2)J(y_3)J(y_4)+$$

$$+\frac{2}{4!}\Delta(x_1-y_1)\Delta(x_1-x_1)\Delta(x_1-x_2)\Delta(x_2-y_2)\Delta(x_2-y_3)\Delta(x_2-y_4)J(y_1)J(y_2)J(y_3)J(y_4)+$$

$$+\frac{1}{8}\Delta(x_1-y_1)\Delta(x_1-x_1)\Delta(x_1-x_2)\Delta(x_2-x_2)\Delta(x_2-y_2)J(y_1)J(y_2)+\frac{1}{8}\Delta(x_1-y_1)[\Delta(x_1-x_2)]^2\times$$

$$\times\Delta(x_2-x_2)\Delta(x_1-y_2)J(y_1)J(y_2)+\frac{1}{12}\Delta(x_1-y_1)[\Delta(x_1-x_2)]^3\Delta(x_2-y_2)J(y_1)J(y_2). \quad (2.51)$$

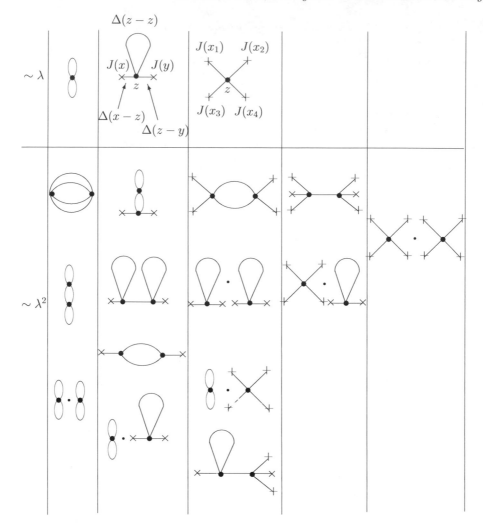

FIGURE 2.2
The Feynman diagrams for $W_E[J]$.

In Eqs. (2.50) and (2.51) terms independent on J are omitted and integration over location points of vertices and crosses is again meant.

When $J = 0$, the quantity $W[0]$ gives the diagrams set of the first column in Fig. 2.2 that describes vacuum-vacuum transitions. In QFT, when finding the Green functions, these diagrams are excluded at the expense of division operation by the quantity $W[0]$, which is equivalent to redefining a vacuum.

In Eq. (2.51), the term $w_1^2[J]/2$ governs disconnected diagrams. Restricting ourselves to the second order of the perturbation theory, we demonstrate this term not to actually contribute to connected Green functions. Using Eq. (2.47) with

$$\mathcal{L}_{int}(\phi) \to \mathcal{L}_{int}\left(\frac{\delta}{\delta J(x)}\right),$$

we get

$$\ln W_E[J] = \ln W_E^0[J] + \ln\{1 + (W_E^0[J])^{-1}(W_E[J] - W_E^0[J])\} = \ln W_E^0[J]+$$

$$+ \ln\{1 + (W_E^0[J])^{-1}(e^{S_{int}} - 1)W_E^0[J]\}, \tag{2.52}$$

where

$$\int d^4x \mathcal{L}_{int} = S_{int}.$$

Since at small values of λ the quantity $(W_E^0[J])^{-1}(e^{S_{int}} - 1)W_E^0[J]$ is small, we can expend the logarithm in Eq. (2.52). The inclusion of (2.49) leads to the result

$$\ln W_E[J] = \ln W_E^0[J] + (\lambda w_1[J] + \lambda^2 w_2[J] + \ldots) - \frac{1}{2}(\lambda w_1[J] + \lambda^2 w_2[J] + \ldots)^2 + \ldots =$$

$$= \ln W_E^0[J] + \lambda w_1[J] + \lambda^2\left(w_2[J] - \frac{1}{2}w_1^2\right) + \ldots \tag{2.53}$$

to confirm cancellation of disconnected diagrams.

Thus, in the Minkowski space the recipe of calculating the cross sections for n interacting particles by means of the generating functional $W[J]$ is as follows. The connected Green functions $G^{(n)}(x_1, \ldots, x_n)$ are determined by the expression:

$$G^{(n)}(x_1, \ldots, x_n) = (-i)^n \left[\frac{1}{W[J]}\frac{\delta^n}{\delta J(x_1)\ldots\delta J(x_n)}W[J]\right]\Bigg|_{J=0}, \tag{2.54}$$

which is to say that

$$\ln W[J] = \sum_n \frac{i^n}{n!}\int dx_1 \ldots dx_n G^{(n)}(x_1, \ldots x_n)J(x_1)\ldots J(x_n). \tag{2.55}$$

From (2.54) it follows that every differentiation with respect to $J(x_j)$ removes a cross in the diagrams associated with $W[J]$ and ascribes a coordinate x_j to a free end of the field ϕ. In such a way the two-particle Green function $G^{(2)}(x_1, x_2)$ is derived from the second column diagrams (see Fig. 2.2), the four-particle Green function $G^{(4)}(x_1, x_2, x_3, x_4)$ from the third column diagrams and so on. Now, so as to pass on to the real process amplitude, it is necessary to take the Green function $G^{(n)}(x_1 \ldots, x_n)$ with n external lines, cut off n external lines (i.e., divide by n free Green functions corresponding to these lines*), and multiply by n wave functions of initial and final states. After that, activity is reduced to computing a required totality of Feynman diagrams. In what follows, unless otherwise stated, we shall work with amputated one-particle irreducible Green functions $\Gamma^{(n_G, n_q)}(k_i; p_j)$, the Green functions that have no external lines and do not break up into disconnected parts under splitting one of the diagram lines.

2.3 Fermion fields quantization by means of the functional integrals method

Fields subjected to the Fermi–Dirac statistics could be described by the Feynman path integral language as well. But now, unlike the Bose-system, one has to introduce new kind of functionals to produce the Grassmann algebra. In this section we shall give the necessary knowledge about this topic.

---

*This operation is called *amputation*.

n-Dimensional Grassmann algebra G_n is specified by n generatrices $y_1, y_2 \ldots y_n$ to obey the conditions

$$\{y_j, y_i\} = 0, \qquad j, i = 1, 2, \ldots n. \tag{2.56}$$

The arbitrary Grassmann algebra element $F \in G_n$ can be represented in the forms of the finite series

$$F(y) = F_0 + F_{i_1}^{(1)} y_{i_1} + F_{i_1 i_2}^{(2)} y_{i_1} y_{i_2} + \ldots + F_{i_1 \ldots i_n}^{(n)} y_{i_1} \cdots y_{i_n}, \tag{2.57}$$

where every summation index takes the values from 1 to n. Thanks to the conditions (2.56) this expansion is broken off. As an example, we consider one-dimensional Grassmann algebra G_1

$$\{y, y\} = 0, \qquad\qquad y^2 = 0.$$

For any element $F(y) \in G_1$, we have

$$F(y) = F_0 + y F_1. \tag{2.58}$$

To be specific, let us hold $F(y)$ to be an ordinary number. Then F_0 and F_1 are ordinary and Grassmann numbers, respectively. We introduce differentiation and integration operations into G_1. Two kind of derivatives, left and right ones, exist in the Grassmann algebra owing to the anticommutation relations. The following designations for them are used:

$$\frac{\overrightarrow{d}}{dy} \equiv \frac{d^L}{dy} \qquad\qquad \frac{\overleftarrow{d}}{dy} \equiv \frac{d^R}{dy}.$$

We have for the element $F(y)$

$$\frac{\overrightarrow{d}}{dy} F(y) = F_1 \qquad\qquad \frac{\overleftarrow{d}}{dy} F(y) = -F_1, \tag{2.59}$$

where the anticommutation of d/dy with F_1 has been taken into account. The relation

$$\left\{ \frac{d}{dy}, y \right\} = 1 \tag{2.60}$$

takes place for any derivative as well. The integration operation is usually defined as a operation being inverse with respect to the differentiation one. In the Grassmann algebra, however, there is the relation

$$\frac{\overrightarrow{d}^2}{dy^2} F(y) = \frac{\overleftarrow{d}^2}{dy^2} F(y) = 0 \tag{2.61}$$

which hampers the ordinary definition of the integration operation. As a consequence, the integration definition is rather artificial. The simplest way of defining integration rules rests on the assumption that the integration operation acts on a function in the same manner as the differentiation one. This avoids the presented dilemma. We demand the integration to lead to the same result as does the left differentiation operation, that is,

$$\int dy F(y) = \frac{d}{dy} F(y) = F_1. \tag{2.62}$$

Hence, it follows that

$$\int dy = 0, \qquad \int y \, dy = 1. \tag{2.63}$$

These two integrals prove to be sufficient for finding all integrals in the Grassmann algebra.

Now we investigate the problem of the integration variable replacement $y \to y' = a + by$, where a and b are Grassmann and ordinary numbers, respectively. Allowing for the relation (2.62), we get

$$\int dy' F(y') = \frac{d}{dy'} F(y') = F_1. \tag{2.64}$$

On the other hand, we have

$$\int dy F(y') = \int dy F_1 by = F_1 b \tag{2.65}$$

to finally give

$$\int dy' F(y') = \int dy \left(\frac{dy'}{dy}\right)^{-1} F(y'(y)). \tag{2.66}$$

Thus, the formula for the replacement of anticommuting variables involves the quantity that is inverse to Jacobian rather than Jacobian, as it is in the case of ordinary variables.

Let us generalize the formulae obtained on a case of n-dimensional Grassmann algebra. The formulae for the left and right derivatives have the view

$$\frac{\partial^L}{\partial y_j}(y_1 y_2 \ldots y_n) = \delta_{j1} y_2 \ldots y_n - \delta_{j2} y_1 y_3 \ldots y_n + \ldots + (-1)^{n-1} \delta_{jn} y_1 y_2 \ldots y_{n-1} \tag{2.67}$$

$$(y_1 y_2 \ldots y_n)\frac{\partial^R}{\partial y_j} = \delta_{jn} y_1 y_2 \ldots y_{n-1} - \ldots + (-1)^{n-1} \delta_{j1} y_2 \ldots y_n. \tag{2.68}$$

Integration rules are assigned by the relations

$$\int dy_j = 0, \qquad \int y_k dy_j = \delta_{jk}, \tag{2.69}$$

where $\{dy_j, dy_k\} = 0$. Multiple integrals are defined in the following way:

$$\int y_{k_1} \ldots y_{k_n} dy_n \ldots dy_1 - \epsilon_{k_1 \ldots k_n}, \tag{2.70}$$

where $\epsilon_{k_1 \ldots k_n}$ is a n-dimensional Levi–Civita tensor.

Further we find generalization of the formula (2.66). To do this, we fulfill the replacement of variables $y_j \to y'_j = a_{jk} y_k$ and compare the expressions

$$\int dy'_n \ldots dy'_1 F(y') \qquad \int dy_n \ldots dy_1 F(y'(y)).$$

In $F(y')$ only the terms containing n factors y'_j contribute to these integrals

$$y'_1 \ldots y'_n = a_{1j_1} \ldots a_{nj_n} y_{j_1} \ldots y_{j_n} = a_{1k_1} \ldots a_{nk_n} \epsilon_{k_1 \ldots k_n} y_1 \ldots y_n = (\det a) y_1 \ldots y_n.$$

Then, for the relations (2.69) to be fulfilled it needs to have

$$\int dy'_n \ldots dy'_1 F(y') = \int dy_n \ldots dy_1 \left[\det\left(\frac{dy'}{dy}\right)\right]^{-1} F(y'(y)), \tag{2.71}$$

as distinguished from the ordinary rule of variables replacement.

The integral of the Gauss type

$$f(A) = \int dy_n \ldots dy_1 \exp\left[\frac{1}{2}(y, Ay)\right], \tag{2.72}$$

where A is an antisymmetric matrix and $(y, Ay) = y_j A_{jk} y_k$, is a fundamental integral to be dealt with in the Grassmann variables theory. In the simplest case $n = 2$, the matrix A is given by the expression

$$A = \begin{pmatrix} 0 & A_{12} \\ -A_{12} & 0 \end{pmatrix}$$

and the integral (2.72) is taken without difficulty

$$f(A) = \int dy_2 dy_1 \exp\left[y_1 y_2 A_{12}\right] = \int dy_2 dy_1 \left[1 + y_1 y_2 A_{12}\right] = A_{12} = \sqrt{\det A}. \qquad (2.73)$$

In the general case the matrix A represents an antisymmetric $n \times n$ matrix with a even value n (if n is odd, the integral (2.72) vanishes). With the help of a unitarity transformation $UAU^\dagger$, the matrix A is reduced to the standard view

$$UAU^\dagger = A_s = \begin{bmatrix} a\begin{pmatrix} 0 & 1 \\ -1 & 0 \end{pmatrix} & & \\ & b\begin{pmatrix} 0 & 1 \\ -1 & 0 \end{pmatrix} & \\ & & \ddots \end{bmatrix}, \qquad (2.74)$$

where all elements outside the main diagonal are equal to zero. To find the matrix U, we proceed as follows. Since the matrix iA is Hermitian, there exists a unitary transformation V that reduces it to the diagonal form

$$A_d = V(iA)V^\dagger, \qquad (2.75)$$

where the diagonal elements of the matrix A is found from the secular equation

$$\det |iA - \lambda I| = 0. \qquad (2.76)$$

On the other hand,

$$\det |iA - \lambda I|^T = \det | -iA - \lambda I| = 0. \qquad (2.77)$$

If λ is a solution of Eq. (2.76), then $-\lambda$ will be its solution too. Therefore, the matrix A_d has the form

$$A_d = \begin{bmatrix} \lambda_1 & & & & & \\ & -\lambda_1 & & & & \\ & & \lambda_2 & & & \\ & & & -\lambda_2 & & \\ & & & & \ddots & \\ & & & & & \ddots \end{bmatrix}, \qquad (2.78)$$

where $\lambda_1 = a$, $\lambda_2 = b \ldots$. Let us introduce two more $n \times n$-matrices, D and C

$$D = \begin{pmatrix} B & & & \\ & B & & \\ & & \ddots & \\ & & & \ddots \end{pmatrix}, \qquad C = \begin{pmatrix} \lambda_1^{-1/2} & & & & \\ & \lambda_1^{-1/2} & & & \\ & & \lambda_2^{-1/2} & & \\ & & & \lambda_2^{-1/2} & \\ & & & & \ddots \\ & & & & & \ddots \end{pmatrix}, \qquad (2.79)$$

where

$$B = \frac{1}{\sqrt{2}} \begin{pmatrix} i & 1 \\ 1 & i \end{pmatrix}$$

and

$$\det(C^{-1}) = \sqrt{\det A}.$$

Taking into account the matrix B to obey the relation

$$B(-i) \begin{pmatrix} 1 & 0 \\ 0 & -1 \end{pmatrix} B^\dagger = \begin{pmatrix} 0 & 1 \\ -1 & 0 \end{pmatrix},$$

we find

$$A_s = D(-iA_d)D^\dagger = (DV)A(DV)^\dagger,$$

that is, $U = DV$. Collecting together the results obtained, we arrive at

$$C(UAU^\dagger)C = CA_sC \equiv \overline{A}_s = \begin{bmatrix} \begin{pmatrix} 0 & 1 \\ -1 & 0 \end{pmatrix} & & \\ & \begin{pmatrix} 0 & 1 \\ -1 & 0 \end{pmatrix} & \\ & & \ddots \end{bmatrix}. \tag{2.80}$$

Now, the Gauss integral (2.72) could be reported in the form

$$f(A) - \int dy_n \dots dy_1 \exp\left[\frac{1}{2}(y, U^\dagger C^{-1}\overline{A}_sC^{-1}Uy)\right]. \tag{2.81}$$

Passing on to new integration variables $y' = (C^{-1}U)y$ and allowing for Eq. (2.71), we get

$$f(A) - \int dy'_n \dots dy'_1 \exp\left[\frac{1}{2}(y', \overline{A}_sy')\right] \left[\det\left(\frac{dy'}{dy}\right)\right] = \det\left(\frac{dy'}{dy}\right) -$$

$$= \det(C^{-1}U) = \det(C^{-1}) = \sqrt{\det A}. \tag{2.82}$$

This formula is essentially distinguished from that for the Gauss integral in the case of commuting real variables (see Eq. (2.3)). It is easy to show that in the case of complex Grassmann variables the Gauss integral is given by

$$\int dy_n dy_n^* \dots dy_1 dy_1^* \exp\left[(y^*, Ay)\right] = \det A, \tag{2.83}$$

where y_j and y_j^* are independent generatrices of the Grassmann algebra.

To describe fermion fields, we have to pass on to infinite-dimensional Grassmann algebra. Transition from discrete variables

$$y_1, \dots, y_j, \dots, y_n$$

to continuous ones $y(x)$ is realized by $n \to \infty$, and the replacement $j \to x$ on a constrained interval of changing a variable y. As a consequence, the Grassmann function $F(y)$ (2.57) turns to the Grassmann functional defined by the expression

$$F[y] = F_0 + \int dx_1 F^{(1)}(x_1)y(x_1) + \int dx_1 dx_2 F^{(2)}(x_1, x_2)y(x_1)y(x_2) + \dots +$$

$$+ \int dx_1 \ldots dx_n F^{(n)}(x_1, \ldots x_n) y(x_1) \ldots y(x_n), \qquad (2.84)$$

where the generators $y(x)$ are subjected to the relations

$$\{y(x_i), y(x_j)\} = 0.$$

Differentiation of Grassmann functionals is defined as a functional generalization of the rules (2.67) and (2.68)

$$\frac{\delta^L}{\delta y(x)} y(x_1) \ldots y(x_n) = \delta(x - x_1) y(x_2) \ldots y(x_n) - \delta(x - x_2) y(x_1) y(x_3) \ldots y(x_n) + \ldots \quad (2.85)$$

The same is true for $\delta^R / \delta y(x)$. Generalization of the integrals over Grassmann variables (2.69) and (2.70) is realized by means of the transition to functional integrals. In the case of complex Grassmann variables the fundamental integral used in the fermion fields theory will look like:

$$\int [dy(x)] \int [dy^*(x)] \exp\{\int dx \int dx' y^*(x) A(x, x') y(x)\} = \det A, \qquad (2.86)$$

where A is a function being antisymmetric with respect to variables x and x'.

Now we shall consider classical fermion fields to be elements of the infinite-dimensional Grassmann algebra. The generating functional of fermion fields is given by the expression:

$$W[\eta, \overline{\eta}] = \int [d\overline{\psi}(x)][d\psi(x)] \exp\left\{ i \int d^4x [\mathcal{L}(\psi(x), \overline{\psi}) + \overline{\eta}\psi(x) + \overline{\psi}\eta] \right\}, \qquad (2.87)$$

where η and $\overline{\eta}$ are auxiliary Grassmann sources to obey the conditions

$$\{\eta(x), \eta(x')\} = 0, \qquad \{\eta(x), \overline{\eta}(x') = 0$$

and so on. The Green functions are obtained by the previous recipe:

$$G^{(n)}(x_1, \ldots, x_n) = <0|T(\overline{\psi}(y_1) \ldots \psi(x_1) \ldots)|0> =$$

$$= \left(i\frac{\delta}{\delta\eta(y_1)} \right) \ldots \left(-i\frac{\delta}{\delta\overline{\eta}(x_1)} \right) \ldots \ln W[\eta, \overline{\eta}] \bigg|_{\eta = \overline{\eta} = 0}. \qquad (2.88)$$

In doing so, it has to take into consideration that the quantities $\overline{\psi}, \psi, \eta, \overline{\eta}, \delta/\delta\eta, \delta/\delta\overline{\eta}$ anti-commute. Since fermion fields usually enter into the Lagrangian quadratically, $\mathcal{L} = \overline{\psi}A\psi$, we must calculate the following integral

$$W[\eta, \overline{\eta}] = \int [d\overline{\psi}(x)][d\psi(x)] \exp\left\{ i \int d^4x [\overline{\psi}(x) A\psi(x) + \overline{\eta}(x)\psi(x) + \overline{\psi}(x)\eta(x)] \right\}. \qquad (2.89)$$

After the replacement of variables

$$\psi(x) \to \psi(x) + \beta(x), \qquad \overline{\psi}(x) \to \overline{\psi}(x) + \overline{\beta}(x),$$

where

$$\beta(x) = -A^{-1}\eta(x), \qquad \overline{\beta}(x) = -\overline{\eta}(x)A^{-1}$$

the integrand in the exponential takes the view

$$\overline{\psi}A\psi + \overline{\eta}\psi + \overline{\psi}\eta + \overline{\psi}A\beta + \overline{\beta}A\psi + \overline{\beta}A\beta + \overline{\eta}\beta + \overline{\beta}\eta = \overline{\psi}A\psi - \overline{\eta}A^{-1}\eta. \qquad (2.90)$$

With allowance made for (2.83), we finally have

$$W[\eta,\overline{\eta}] = \int [d\overline{\psi}(x)][d\psi(x)] \exp\left\{i\int d^4x[\overline{\psi}(x)A\psi(x) + \overline{\eta}\psi(x) + \overline{\psi}\eta]\right\} =$$

$$= \det A \exp\left[-i\int d^4x\overline{\eta}(A^{-1})\eta\right]. \tag{2.91}$$

When the terms with the external sources are omitted in Eq. (2.91), then $W[\eta = \overline{\eta} = 0] \equiv W[0]$ would describe vacuum-vacuum transitions. Connected Feynman diagrams caused by $\ln W[0]$ represent the set of graphs with a single fermion closed loop. When switching from the functional integral over commuting complex variables

$$W_b[J = 0] = \int \left[\frac{d\phi(x)}{\sqrt{\pi}}\right] \left[\frac{d\phi^*(x)}{\sqrt{\pi}}\right] \exp\left\{i\int d^4x\phi^*(x)A\phi(x)\right\} = (\det A)^{-1}$$

to that over anticommuting variables (2.89), $(\det A)^{-1}$ is replaced by $\det A$ to lead to changing the sign of $\ln W[0]$. The well-known rule that the factor (-1) is ascribed to every fermion closed loop follows from this.

In the case of a free field of particles with the spin $1/2$, the Lagrangian is given by the expression

$$\mathcal{L}_0 = \overline{\psi}(x)[i\gamma^\mu\partial_\mu - m]\psi(x), \tag{2.92}$$

to lead to the following form of the generating functional

$$W^0[\eta,\overline{\eta}] = N\exp\left\{-i\int d^4xd^4y\overline{\eta}(x)S^c(x-y)\eta(y)\right\}, \tag{2.93}$$

where $S^c(x-y)$ is the free Green function of the Dirac field

$$S^c(x-y) = \frac{1}{(2\pi)^4}\int d^4p\frac{\gamma_\mu p^\mu + m}{p^2 - m^2 + i\epsilon}\exp\left[-ip(x-y)\right] \tag{2.94}$$

and N is an inessential normalization factor.

When the interaction is switched on

$$\mathcal{L}_0 \to \mathcal{L}_0 + \mathcal{L}_{int},$$

we act the following way. First, in the integrand of the formula (2.87) we fulfill the expansion in powers of $\mathcal{L}_{int}$ as in the case of boson fields. Next, we calculate Gauss integrals and build up the perturbation theory series. If boson and fermion fields, $\psi(x)$ and $\phi(x)$, take part in the interaction, that is, $\mathcal{L}_{int} = \mathcal{L}_{int}(\psi(x),\overline{\psi}(x),\phi(x))$, then the expression

$$W[J,\eta,\overline{\eta}] = \exp\left\{i\int d^4x\mathcal{L}_{int}\left(-i\frac{\delta}{\delta\overline{\eta}(x)}, i\frac{\delta}{\delta\eta(x)}, -i\frac{\delta}{\delta J(x)}\right)\right\}\times$$

$$\times \exp\left\{-i\int d^4xd^4y\left[\overline{\eta}(x)S^c(x-y)\eta(y) + \frac{1}{2}J(x)\Delta^c(x-y)J(y)\right]\right\}, \tag{2.95}$$

where the normalization factor is missed, is the evident generalization of the formula (2.48).

2.4   Functional formulation of QCD

Functional formulation of gauge theories, like electrodynamics and especially chromodynamics, has a certain specificity. In these theories, fields are described by vector potentials $A_\mu(x)$ and $G_\mu^a(x)$, respectively, the potentials being determined with an accuracy of gauge transformations. When quantizing gauge theories, the main source of troubles is this very circumstance. It could be illustrated by the example of a free electromagnetic field. In this case the formal expression of the generating functional must have the following view

$$W[\mathbf{J}] = \int [dA_\mu(x)] \exp\left\{ i \int d^4x \left[\mathcal{L}_{em} + J_\lambda(x)A^\lambda(x) \right] \right\} =$$

$$= \int [dA_\mu(x)] \exp\left\{ i \int d^4x \left[\frac{1}{2}F^{\mu\nu}(x) \left(\frac{1}{2}F_{\mu\nu}(x) - \partial_\mu A_\nu(x) + \partial_\nu A_\mu(x) \right) + J_\lambda(x)A^\lambda(x) \right] \right\}.$$
(2.96)

Integrating by parts, we get

$$W[\mathbf{J}] = \int [dA_\mu(x)] \exp\left\{ i \int d^4x \left[\frac{1}{2}A_\mu(x)(g^{\mu\nu}\partial_\lambda\partial^\lambda - \partial^\mu\partial^\nu)A_\nu(x) + J_\lambda(x)A^\lambda(x) \right] \right\}.$$
(2.97)

In the integrand the matrix $\Lambda_{\mu\nu} = g^{\mu\nu}\partial_\lambda\partial^\lambda - \partial^\mu\partial^\nu$ being contained in the plates of A_μ and A_ν does not have an inverse one, on account of

$$\det \Lambda = 0.$$

As was shown in Section 17.3 of *Advanced Particle Physics Volume 1*, it is connected with the fact that $\Lambda_{\mu\nu}$ has zero modes to be in the form of a pure gauge

$$A_\mu(x) = \partial_\mu\alpha(x),$$
(2.98)

where $\alpha(x)$ is a gauge function. It is evident that

$$(g^{\mu\nu}\partial_\lambda\partial^\lambda - \partial^\mu\partial^\nu)\partial_\nu\alpha(x) = 0.$$

Consequently, the integration of the Gauss integral (2.97) over the fields $A_\mu(x)$ gives a divergence of the form $(\det \Lambda)^{-1/2} \to \infty$. Infinity is collected at the expense of integration over zero modes, that is, at the expense of integration over pure gauge variables having the form (2.98), to contribute

$$\int [d \text{ (zero modes)}] \exp(-0) \to \infty.$$

Thus, the singular character of the functional integral is associated with the gauge invariance of the theory. In the expression (2.97) we summarize overall possible field configurations, including those connected with each other by gauge transformations.* The inclusion of these redundant configurations (zero modes) are responsible for a functional integral divergency. Therefore, it is necessary to find the method of singling out the infinite functional volume of the orbit. For a gauge theory to be quantized it is needed to fix a gauge. In doing so, redundant η-particles associated with longitudinal field components appear in a theory.

---

*These field configurations constitute *orbits* of a gauge group.

In QED they are fortunately removed because no particles interact with them. In QCD, however, the η-particles interact with the G_μ field itself owing to nonlinearity of the non-Abelian gauge field. As it was discussed earlier, eliminating contribution of nonphysical η fields is achieved at the expense of introducing the Faddeev–Popov ghost fields. We shall show in what way this is realized in functional integral formalism.

By way of a preparatory exercise, we pick out the volume factor by an example of a two-dimensional integral

$$I = \int d\mathbf{r}\exp\left[iF(\mathbf{r})\right], \tag{2.99}$$

where a function $F(\mathbf{r})$ is constant on an orbit being a circle, that is,

$$F(\mathbf{r}) = F(\mathbf{r}_\phi) \tag{2.100}$$

at

$$\mathbf{r} = (r, \varphi) \rightarrow \mathbf{r}_\phi = (r, \varphi + \phi) \tag{2.101}$$

If we are going to take into account the contribution into the integral coming only from nonequivalent values of $F(\mathbf{r})$, then a *volume factor* associated with the integration over the angle variable φ should be estimated. With this aim in view, we realize the following. We substitute the delta-function definition

$$\int d\phi\delta(\varphi - \phi) = 1$$

into (2.99)

$$I = \int d\phi \int d\mathbf{r}\exp\left[iF(\mathbf{r}|\delta(\varphi - \phi) = \int d\phi I_\phi, \tag{2.102}\right.$$

where I_ϕ has been calculated at the definite value of the angle ϕ. Hence, at first we find the quantity I at the fixed value of the angle $\varphi = \phi$ and then integrate over contributions corresponding to all values of ϕ. Since $I_\phi = I_{\phi'}$ in accordance with (2.100), the orbit *Advanced Particle Physics Volume 1* singled out as a factor

$$I = \int d\phi I_\psi = I_\psi \int d\phi = 2\pi I_\phi.$$

Let us complicate the task. In place of $\varphi = \phi$ we choose the more complicated linkage to be represented by the equation of the curve

$$g(\mathbf{r}) = 0 \tag{2.103}$$

which intersects every orbit only once. It means that the equation $g(\mathbf{r}_\phi) = 0$ must have the single solution ϕ under the given value of $\mathbf{r}$. We consider the quantity

$$\Delta_g^{-1}(\mathbf{r}) = \int d\phi\delta[g(\mathbf{r}_\phi)]. \tag{2.104}$$

It is clear that this quantity is invariant with respect to two-dimensional rotations (2.101) as

$$\Delta_g^{-1}(\mathbf{r}_{\phi'}) = \int d\phi\delta[g(\mathbf{r}_{\phi+\phi'})] = \int d\tilde{\phi}\delta[g(\mathbf{r}_{\tilde{\phi}})] = \Delta_g^{-1}(\mathbf{r}).$$

And so, we may write down

$$1 = \int d\phi\Delta_g^{-1}(\mathbf{r})\delta[g(\mathbf{r}_\phi)] = \int d\phi\Delta_g^{-1}(\mathbf{r}_\phi)\delta[g(\mathbf{r}_\phi)]. \tag{2.105}$$

Since the quantity I_ϕ is invariant under rotations:

$$I_{\phi'} = \int d\mathbf{r} \exp\left[iF(\mathbf{r})\right]\Delta_g(\mathbf{r})\delta[g(\mathbf{r}_{\phi'})] = \int d\mathbf{r}' \exp\left[iF(\mathbf{r}')\right]\Delta_g(\mathbf{r}')\delta[g(\mathbf{r}'_\phi)] = I_\phi, \qquad (2.106)$$

where $\mathbf{r}' = (r, \phi')$ and we have taken into consideration the integration measure invariance under rotations $d\mathbf{r}$, then singling out the volume factor is again reduced to computing the integral

$$I = \int d\phi I_\phi, \qquad (2.107)$$

$$I_\phi = \int d\mathbf{r} \exp\left[iF(\mathbf{r})\right]\Delta_g(\mathbf{r})\delta[g(\mathbf{r}_\phi)].$$

For the factor $\Delta_g(\mathbf{r})$ to be found the equality

$$\int dx_1 \ldots dx_k \prod_{i=1}^{k} \delta[f_i(x_1, \ldots, x_k)] = \left\{\det \frac{\partial f_i}{\partial x_j}\right\}^{-1} \qquad (2.108)$$

must be used. As a result, we get

$$\Delta_g(\mathbf{r}) = \left.\frac{\partial g(\mathbf{r})}{\partial \varphi}\right|_{g=0}.$$

Now we are proceeding to QCD. Let $U(\theta)$ be an infinitesimal gauge transformation with parameters $\theta(x)$, and G^θ be fields that originate from the G ones under this gauge transformation:

$$G_a^{\theta\mu}(x) = G_a^\mu(x) + f_{abc}\theta_b(x)G_c^\mu(x) - g_s^{-1}\partial^\mu\theta_a(x). \qquad (2.109)$$

The QCD action is constant on the orbit of the gauge group consisting of all $G_a^{\theta\mu}$, which are obtained by transformation $U(\theta)$, running over all elements of the $SU(3)_c$ group, from some fixed configuration. At the right quantization the functional integration must be fulfilled merely on the hypersurface intersecting every orbit only once. Thus, if we write the hypersurface equation in the view

$$K_a[G_\mu] = 0, \qquad (2.110)$$

where K is a gauge-fixing functional, the equation

$$K_a[G_\mu^\theta] = 0 \qquad (2.111)$$

has to possess the single solution θ at the given field configuration G^μ. The quantity

$$\Delta_K^{-1}[G_\mu] = \int [d\theta(x)] \prod_a \delta(K_a[G_\mu^\theta(x)]), \qquad (2.112)$$

where

$$[d\theta] = \prod_a d\theta_a,$$

is invariant with respect to the gauge transformations (2.109). For this statement to be proved we rewrite Eq. (2.112) in the following form

$$\Delta_K^{-1}[G_\mu] = \int [d\theta'(x)] \prod_a \delta(K_a[G_\mu^{\theta'}(x)]).$$

Then we have

$$\Delta_K^{-1}[G_\mu^\theta] = \int [d\theta'(x)] \prod_a \delta(K_a[G_\mu^{\theta\theta'}(x)]) = \int [d(\theta(x)\theta'(x))] \prod_a \delta(K_a[G_\mu^{\theta\theta'}(x)]) =$$

$$= \int [d\theta''(x)] \prod_a \delta(K_a[G_\mu^{\theta''}(x)]) = \Delta_K^{-1}[G_\mu], \qquad (2.113)$$

where we have allowed for the invariance of the integration measure over the group space

$$[d(\theta\theta')] = [d\theta'] \qquad (2.114)$$

(in the case of infinitesimal transformations $U(\theta)U(\theta') = U(\theta\theta') = U(\theta + \theta')$ this property is evident). With allowance made for this circumstance, we get

$$\int [d\theta(x)] \Delta_K[G_\mu] \prod_a \delta(K_a[G_\mu^\theta(x)]) = 1. \qquad (2.115)$$

Further, from Eq. (2.108) it follows that the quantity $\Delta_K[G_\mu]$ represents the determinant of the infinite-dimensional matrix M_K

$$\Delta_K[G_\mu] = \det M_K, \qquad (2.116)$$

where

$$(M_K)_{ab} = \frac{\partial K_a}{\partial \theta_b}.$$

For the time being, we will forget about the existence of quarks, a role of that is completely evident under gauge transformations. Then, using (2.117), we obtain for the vacuum-vacuum transition amplitude the following expression:

$$\int [dG_\mu(x)] \exp\left\{i \int d^4x \mathcal{L}_G(x)\right\} = \int [d\theta(x)][dG_\mu(x)]\Delta_K[G_\mu] \prod_a \delta(K_a[G_\mu^\theta(x)]) \times$$

$$\times \exp\left\{i \int d^4x \mathcal{L}_G(x)\right\} = \int [d\theta(x)][dG_\mu(x)]\Delta_K[G_\mu] \prod_a \delta(K_a[G_\mu(x)]) \exp\left\{i \int d^4x \mathcal{L}_G(x)\right\},$$
$$(2.117)$$

where $\mathcal{L}_G(x)$ is the gluon field Lagrangian (see Eq. (1.65)). Thus, we have achieved the posed goal—in (2.117) we have obtained the integrand independent on $\theta(x)$. Now, to integrate over $[d\theta(x)]$ gives the infinite orbit volume we aimed to single out. With allowance made for the results obtained, the generating functional of the gluon field takes the form

$$W[\mathbf{J}] = \int [dG_\mu(x)](\det M_K) \prod_a \delta(K_a[G_\mu(x)]) \exp\left\{i \int d^4x [\mathcal{L}_G(x) + J_{a\mu}G_a^\mu]\right\} \qquad (2.118)$$

(Faddeev–Popov ansatz [6]). So, introducing the factor

$$\det M_K \prod_a \delta(K_a[G_\mu(x)])$$

into the functional measure, we can remove all redundant integrations under the procedure of functional quantization. Taking into account the definition of the connected Green functions of the gluon field

$$D_G^{(n)}(x_1, \ldots x_n) = (-i)^n \left[\frac{1}{W[\mathbf{J}]} \frac{\delta^n}{\delta J(x_1) \ldots \delta J(x_n)} W[\mathbf{J}]\right]\Bigg|_{J=0} \qquad (2.119)$$

we convince ourselves that divergencies connected with the zero modes are cancelled, that is, they may be included into an inessential normalization factor.

Thus, the main problem has been removed and we could proceed to deducing the QCD generating functional in the form to be convenient for practical calculations. Since the Feynman rules are more simple in covariant (Lorentz) gauges, we carry out detail discussion of this very case. First of all, we get rid of the δ function in the generating functional of the gluon field. To this end we generalize the gauge-fixing condition, that is, fulfill the transition

$$K_a[G_\mu] = 0 \rightarrow K_a[G_\mu] = \Phi_a(x), \tag{2.120}$$

where $\Phi_a(x)$ is an arbitrary function independent on the gauge field. At that

$$\prod_a \delta(K_a[G_\mu]) \rightarrow \prod_a \delta(K_a[G_\mu] - \Phi_a(x)).$$

Let us integrate the expression (2.118) over Φ with the weight

$$\exp\left\{-\frac{i}{2\xi} \int d^4x [\Phi_a(x)]^2\right\}.$$

In the left side we get the generating functional $W[\mathbf{J}]$ multiplied by the factor

$$\int [d\Phi] \exp\left\{-\frac{i}{2\xi} \int d^4x [\Phi_a(x)]^2\right\},$$

which does not depend on the fields G and, as a result, could be included into the normalization factor again. The integration of the right side is trivially realized with the help of the δ function. Thus, we obtain the generating functional of the gluon field in the form

$$W[\mathbf{J}] = \int [dG_\mu(x)](\det M_K) \exp\left\{i \int d^4x [\mathcal{L}_G(x) - \frac{1}{2\xi}\{K_a[G_\mu]\}^2 + J_{a\mu}G_a^\mu\right\}. \tag{2.121}$$

The next task is to calculate the factor $\Delta_K[G_\mu]$. We shall work in the Lorentz gauge

$$K_a[G_\mu] = \partial_\mu G_a^\mu = 0. \tag{2.122}$$

Carrying out the infinitesimal transformations (2.109), we have

$$K_a[G_\mu^\theta] = K_a[G_\mu] - \partial_\mu \left[\frac{1}{g_s}\partial^\mu\theta_a(x) - f_{abc}\theta_b(x)G_c^\mu(x)\right] = K_a[G_\mu] + \int d^4y [M_K(x,y)]_{ab}\theta_b(y), \tag{2.123}$$

where

$$[M_K(x,y)]_{ab} = -g_s^{-1}\partial_\mu [\delta_{ab}\partial^\mu - g_s f_{abc}G_c^\mu(x)] \delta^4(x - y). \tag{2.124}$$

So, we have found the explicit form of the matrix M_K (2.116) whose determinant defines the quantity $\Delta_K[G_\mu]$. Notice that the operator standing in the square bracket of the expression (1.126) appears in Eq. (1.89) for the η-particles. The determinant of the matrix M_K entering into Eq. (2.121) could be also written as the generating functional. With that end in view, we introduce an octet of complex scalar fictitious fields, Faddeev–Popov ghosts, to be represented by anticommuting c-number functions ω and $\omega^\dagger$. Making use of the formula for the Gauss integral over fermion fields, we get the generating functional describing these fields and their interactions

$$\det M_K \sim \int [d\omega^\dagger(x)][d\omega(x)] \exp\left\{i \int d^4x d^4y \sum_{a,b} \omega_a^\dagger(x)[M_f(x,y)]_{ab}\omega_b(y)\right\} \tag{2.125}$$

(when switching on an interaction with gluons, we have the massless scalar field Lagrangian in the exponential). Consequently, det M_K governs the amplitudes of the vacuum-vacuum transitions that are caused by particles subjected to Fermi statistics. If one introduces external sources η and $\bar\eta$ into the formula for det M_K, then the connected Feynman diagrams produced by ln det M_K will form graphs set with a single closed-fermion loop. Carrying out integration by parts in (2.125) and substituting the result obtained into (2.121), we get the final expression for the generating functional of the gluon field

$$W[\mathbf{J},\boldsymbol{\eta},\boldsymbol{\eta}^\dagger] = \int [dG_\mu, d\omega, d\omega^\dagger] \exp\left\{i\int d^4x [\mathcal{L}_G(x) + \mathcal{L}_{GF}(x) + \mathcal{L}_{FPG}(x) + J_a^\mu G_{a\mu} + \right.$$

$$\left. + \eta_a^\dagger \omega_a + \eta_a \omega_a^\dagger]\right\} = \int [dG_\mu, d\omega, d\omega^\dagger] \exp[iS_{eff}], \tag{2.126}$$

where

$$S_{eff} = \int d^4x [\mathcal{L}_G(x) + \mathcal{L}_{GF}(x) + \mathcal{L}_{FPG}(x) + J_a^\mu G_{a\mu} + \eta_a^\dagger \omega_a + \eta_a \omega_a^\dagger],$$

$\mathcal{L}_{GF}(x), \mathcal{L}_{FPG}(x)$ are defined by Eqs. (1.68) and (1.92), respectively, and the factor g_s^{-1} has been included into the redefinition of the fields ω_a and $\omega_a^\dagger$.

We illustrate the calculations technique of the perturbation theory by the example of a free gluon field. Again, as in the case of canonical formalism, we shall work in the linear approximation, that is, report S_{eff} in the form

$$S_{eff} = S_0 + S_{int}, \tag{2.127}$$

where the free action S_0 is quadratic in the fields:

$$S_0 = \int d^4x \left[-\frac{1}{4}(\partial_\mu G_{a\nu} - \partial_\nu G_{a\mu})^2 - \frac{1}{2\xi}(\partial_\mu G_a^\mu)^2 + \partial_\mu \omega_a^\dagger \partial^\mu \omega_a + J_a^\mu G_{a\mu} + \eta_a^\dagger \omega_a + \eta_a \omega_a^\dagger \right\}, \tag{2.128}$$

S_{int} is the term that describes self-interaction of the gluon field and interaction between the gluon and ghost fields:

$$S_{int} = \int d^4x [\mathcal{L}_{self} - ig_s(\partial_\mu \omega_a^\dagger) f_{abc} G_c^\mu \omega_b], \tag{2.129}$$

and $\mathcal{L}_{self}$ is determined by Eq. (1.77). Replacing the functional integration by differentiation, we derive for $W[\mathbf{J},\boldsymbol{\eta},\boldsymbol{\eta}^\dagger]$ the following expression

$$W[\mathbf{J},\boldsymbol{\eta},\boldsymbol{\eta}^\dagger] = \exp\left[iS_{int}\left(-i\frac{\delta}{\delta \mathbf{J}_\mu}, -i\frac{\delta}{\delta\boldsymbol{\eta}}, -i\frac{\delta}{\delta\boldsymbol{\eta}^\dagger}\right)\right] W_G^0[\mathbf{J}] W_\omega^0[\boldsymbol{\eta},\boldsymbol{\eta}^\dagger], \tag{2.130}$$

where

$$W_G^0[\mathbf{J}] = \int [dG_\mu] \exp\left\{i\int d^4x \left[-\frac{1}{4}(\partial_\mu G_{a\nu} - \partial_\nu G_{a\mu})^2 - \frac{1}{2\xi}(\partial_\mu G_a^\mu)^2 + J_a^\mu G_{a\mu} \right]\right\}, \tag{2.131}$$

$$W_\omega^0[\boldsymbol{\eta},\boldsymbol{\eta}^\dagger] = \int [d\omega][d\omega^\dagger] \exp\left\{i\int d^4x \left[-\omega_a^\dagger \partial_\mu \partial^\mu \omega_a + \eta_a^\dagger \omega_a + \eta_a \omega_a^\dagger \right]\right\}. \tag{2.132}$$

After integration by parts, $W_G^0[\mathbf{J}]$ becomes the form

$$W_G^0[\mathbf{J}] = \int [dG_\mu] \exp\left\{i\int d^4x \left[\frac{1}{2}G_{a\mu}\left(g^{\mu\nu}\partial_\lambda\partial^\lambda - \frac{\xi-1}{\xi}\partial^\nu\partial^\mu \right)\delta_{ab}G_{b\nu} + J_a^\mu G_{a\mu} \right]\right\} =$$

$$= \int [dG_\mu] \exp \left\{ i \int d^4x \left[\frac{1}{2} G_{a\mu} K_{ab}^{\mu\nu} G_{b\nu} + J_a^\mu G_{a\mu} \right] \right\}, \tag{2.133}$$

where

$$K_{ab}^{\mu\nu} = \left[g^{\mu\nu} \partial_\lambda \partial^\lambda - \left(1 - \frac{1}{\xi} \right) \partial^\nu \partial^\mu \right] \delta_{ab}. \tag{2.134}$$

To integrate (2.133) over $[dG_\mu]$ could be easily performed by means of Eq. (2.44) to give

$$W_G^0[\mathbf{J}] = (\det K)^{-1/2} \exp \left[-\frac{i}{2} \int d^4x d^4y J_{a\mu}(x) \Delta_{ab}^{\mu\nu}(x-y) J_{b\nu}(y) \right], \tag{2.135}$$

where $\Delta_{ab}^{\mu\nu}(x)$ is the causal Green function for a free gluon field

$$\Delta_{ab}^{\mu\nu}(x-y) = \frac{\delta_{ab}}{(2\pi)^4} \int d^4p \left[-\left(g^{\mu\nu} - \frac{p^\mu p^\nu}{p^2} \right) - \xi \frac{p^\mu p^\nu}{p^2} \right] \frac{\exp\left[-ip(x-y) \right]}{p^2 + i\epsilon}. \tag{2.136}$$

It is easy to check that the relation

$$\int d^4y K_{ab\mu\nu}(x-y) G_{bd}^{\nu\lambda}(y-z) = g_\mu^\lambda \delta_{ad} \delta^4(x-z) \tag{2.137}$$

takes place. Employing the formula (2.44) to $W_\omega^0[\boldsymbol{\eta}, \boldsymbol{\eta}^\dagger]$ gives the result

$$W_\omega^0[\boldsymbol{\eta}, \boldsymbol{\eta}^\dagger] = \exp \left\{ -i \int d^4x d^4y \eta_a^\dagger(x) G_{ab}(x-y) \eta_b(y) \right\}, \tag{2.138}$$

where $G_{ab}(x)$ is the causal Green function of the Faddeev–Popov ghosts

$$G_{ab}(x) = -\frac{\delta_{ab}}{(2\pi)^4} \int d^4p \frac{\exp\left(-ipx \right)}{p^2 + i\epsilon}. \tag{2.139}$$

Thus, the gluon and Faddeev—Popov ghost propagators in the momentum space are given by the expressions

$$P_{ab}^{(G)m\mu\nu}(p) = -i\delta_{ab} \left[g^{\mu\nu} - (1-\xi) \frac{p^\mu p^\nu}{p^2} \right] \frac{1}{p^2 + i\epsilon}, \tag{2.140}$$

$$P_{ab}^{(\omega)}(p) = -i\delta_{ab} \frac{1}{p^2 + i\epsilon}. \tag{2.141}$$

From the expression (2.129) the Feynman rules for the vertex functions V_{3G}, V_{4G}, and $V_{\omega\omega G}$ follow. Using the procedure stated for the theory $\mathcal{L}_{int} = -\lambda\phi^4/(4!)$, we arrive at the expressions (1.83), (1.84), and (1.94), respectively.

Inserting quarks into the theory does not lead to any troubles. It is enough to add the term

$$\mathcal{L}_q(x) = \frac{i}{2} [\overline{q_k}(x)\gamma_\mu D_{kj}^\mu(x)q_j(x) - D_{kj}^{\mu\dagger}\overline{q_k}(x)\gamma_\mu q_j(x)] - m\overline{q_k}(x)q_k(x)$$

to $\mathcal{L}_{eff}$ and introduce for quarks the anticommuting sources $\gamma_j(x)$ and $\overline{\gamma}_j(x)$. Thus, the final expression for the QCD generating functional has the form:

$$W[\mathbf{J}, \boldsymbol{\eta}, \boldsymbol{\eta}^\dagger, \gamma, \overline{\gamma}] = \int [dG_\mu, d\omega, d\omega^\dagger, dq, d\overline{q}] \exp \left\{ i \int d^4x [\mathcal{L}_{eff}(x) + J_{a\mu} G_a^\mu + \right.$$

$$\left. + \eta_a^\dagger \omega_a + \omega_a^\dagger \eta_a + \overline{\gamma}_j q_j + \overline{q}_j \gamma_j] \right\}, \tag{2.142}$$

where

$$\mathcal{L}_{eff}(x) = \mathcal{L}_G(x) + \mathcal{L}_{GF}(x) + \mathcal{L}_{FPG}(x) + \mathcal{L}_q(x). \tag{2.143}$$

In what follows, we shall omit the color indices both at quarks and at their sources for the sake of simplicity.

We briefly discuss the case of noncovariant gauges. The appearance of ghosts is caused by the fact that the Hilbert space of the fields involves the indefinite metric. Choosing the gauge in which all the gluon states correspond to the physical ones could be a way out of the existing situation. Such gauges, however, violate the explicit gauge invariance of the theory. The Coulomb gauge $\nabla \cdot \mathbf{G}_a = 0$ is one example of noncovariant gauges, but it brings about additional complications into the theory and is not free from the ghosts [12] too. In the case of the axial and light-like gauges, the necessity of introducing the ghosts disappears. In functional integration formalism it must denote as follows. The matrix M_K does not depend on the gluon field and, as a result, the factor $\det M_K$ has to be omitted in the generating functional. Let us show that such is the case. The above-listed gauges may be represented by the single equation

$$n^\mu G_\mu^a(x) = 0, \qquad n_\mu n^\mu \le 0. \tag{2.144}$$

When a vector n_μ is space-like ($n_\mu^2 < 0$), then we deal with the axial gauge [13]. In the case of a light-like vector n_μ ($n_\mu^2 = 0$) we have a light-like gauge [14]. Since the vector n_μ is outside with respect to the task, its introduction violates the explicit Lorentz-invariance of intermediate calculations. It is self-evident that, thanks to the gauge invariance, final results are independent of this vector and, therefore, are Lorentz-invariant.

Under the infinitesimal gauge transformations (2.109) Eq. (2.123) will look like:

$$K_a[G_\mu^\theta] = n^\mu G_{a\mu}(x) + f_{abc}\theta_b(x)n^\mu G_{c\mu}(x) - g_s^{-1}n^\mu\partial_\mu\theta_a(x) = -g_s^{-1}n^\mu\partial_\mu\theta_a(x), \tag{2.145}$$

where Eq. (2.144) has been allowed for, that is, the necessity of introducing the ghosts is really absent. In spite of this attenuating circumstance, however, using noncovariant gauges greatly complicates the Feynman rules. Thus, for example, in the axial gauge the gluon propagator is given by the expression

$$P_{ab}^{(G)\mu\nu} = i\delta_{ab}\left[\frac{-g^{\mu\nu} - n_\gamma n^\gamma p^\mu p^\nu/(n_\sigma p^\sigma)^2 + (p^\nu n^\mu + p^\mu n^\nu)/(n_\sigma p^\sigma)}{p^2 - i\epsilon}\right]. \tag{2.146}$$

3

Renormalization and unitarity

By three methods we may learn wisdom:
First, by reflection, which is noblest;
Second, by limitation, which is easiest;
and third by experience, which is the bitterest.
Confucius

3.1 Primitive divergent diagrams

When studying the perturbation theory in quantum field theory (QFT), one cannot restrict oneself to tree diagrams only. Including loops does not change a process type but could lead to divergency of matrix elements. These divergencies arise from the fact that integration over a loop momentum is always fulfilled from zero to infinity. To put it differently, in QFT a natural momentum cut-off is absent. The renormalization theory gives the recipe that allows to systematically extract and eliminate such divergencies from observable quantities. Here, we discuss both general peculiarities of divergent diagrams and possibilities of their regularization. From quantum electrodynamics (QED) it is well known (Section 22.1 of *Advanced Particle Physics Volume 1*), there is no need to consider all divergent diagrams. It is quite enough to investigate primitive divergent integrals only. To enumerate primitive divergent diagrams corresponds to the fundamental divergencies list, all remaining divergencies being available with the aid of corresponding insertions.

To analyze the structure of a primitive divergent diagram, by analogy with QED we use the notion of the integral dimensionality D as difference between the total power of momenta in the numerator and denominator of integrand. This quantity is sometimes referred to as the *superficial degree of divergence*, or as the *overall divergence of the diagram*. The sufficient condition of the convergency has the form

$$D < 0, \tag{3.1}$$

and the divergencies existence condition is written as

$$D \geq 0.$$

($D = 0$ is associated with the logarithmic divergency). The analysis based on the definition of D is referred to as *naive counting* of the divergence degree (or *seeming divergence degree*). We introduce the following notations, to characterize a typical diagram: $Q_e =$ number of external quark lines; $Q_i =$ number of internal quark lines; $G_e =$ number of external gluon lines; $G_i =$ number of internal gluon lines; $g_i =$ number of internal ghost lines; $V_{4G} =$ number of four-gluon vertices; $V_{3G} =$ number of three-gluon vertices; $V_{qG} =$ number of quark-gluon vertices; and $V_{Gg} =$ number of gluon-ghost vertices.

To calculate D we take into account that every internal quark line gives one momentum power in the denominator while every internal gluon (or ghost) line does the second momentum power in the denominator. The volume element brings momentum power $4l$ (l is the loops number) to the numerator and the ghost-gluon (or three-gluon) vertex gives one momentum power in the numerator. As a result we get

$$D = 4l - Q_i - 2G_i - 2g_i + V_{3G} + V_{Gg}. \tag{3.2}$$

For the loops number to be found, the number of external lines, over which one may fulfill the only one sequential transition from the first to last vertices $V_{4G} + V_{3G} + V_{qG} + V_{Gg} - 1$, must be subtracted from the total number of internal lines $Q_i + G_i + g_i$. This gives

$$l = G_i + Q_i + g_i - (V_{4G} + V_{3G} + V_{qG} + V_{Gg} - 1). \tag{3.3}$$

Setting the number and the external lines type concretizes a process while setting the numbers Q_i, G_i, and g_i defines the approximation in which a process is calculated. Therefore, we need to express D in terms of the external lines number that is fixed for a process in question. To this end we take into account the following combinatorial reasons. Since the quark lines are continuous, we get for them

$$Q_e + 2Q_i = 2V_{qG}. \tag{3.4}$$

While the number of the gluon and ghost lines is defined by

$$\left. \begin{aligned} G_e + 2G_i &= 4V_{4G} + 3V_{3G} + V_{Gg} + V_{qG} \\ 2g_i &= 2V_{Gg}, \end{aligned} \right\} \tag{3.5}$$

respectively. Using (3.3), we obtain

$$D = 4(G_i + Q_i + g_i - V_{4G} - V_{3G} - V_{qG} - V_{Gg} + 1) - Q_i - 2G_i - 2g_i + V_{3G} + V_{Gg}. \tag{3.6}$$

Excluding from here the internal lines number with the help of Eqs. (3.4) and (3.5), we arrive at

$$D = 4 - G_e - \frac{3}{2}Q_e. \tag{3.7}$$

The obtained equation exactly coincides with the corresponding one for D in QED. Consequently, in just the same way as QED, QCD proves to be a renormalizable theory. In other words, in QCD the number of the primitive divergent diagrams is finite.

It is convenient to introduce the notion of a divergence index. To this end, we rewrite Eq. (3.3) in the following form

$$D = 4l - Q_i - 2G_i - 2g_i + nn_D,$$

where n_D is the number of derivatives at the vertex and n is the number of vertices (for an nth order Feynman diagram). The formula for l could be represented as

$$l = G_i + Q_i + g_i - (n - 1).$$

Further on, enter the number of quarks fields (Q), the number of gluon fields (G), and the number of ghost fields (Gh) that interact at each point in space-time. Then, the superficial divergence in terms of the number of external lines (which depends only on the physical process under consideration), Q, G, and Gh (which depend only on the theory), and the diagram order n is given by the expression

$$D = nI + 4 - G_i - \frac{3}{2}Q_i,$$

where

$$I = G + Gh + \frac{3}{2}Q - 4 + n_D \tag{3.8}$$

is the index of the divergence. Evaluating Eq. (3.10) for each vertex of QCD gives the following computations:

$$
\left.
\begin{array}{ll}
V_{3G}: & 3 + 1(0) + \dfrac{3}{2}(0) - 4 + 1 = 0, \\[2mm]
V_{4G}: & 4 + 1(0) + \dfrac{3}{2}(0) - 4 + 0 = 0, \\[2mm]
V_{Gq}: & 1 + 1(0) + \dfrac{3}{2}(2) - 4 + 0 = 0, \\[2mm]
V_{Gg}: & 1 + 2 + \dfrac{3}{2}(0) - 4 + 1 = 0.
\end{array}
\right\} \tag{3.9}
$$

In every case the divergence index is zero. The diagrams that correspond to these vertices should be supplemented by those associated with the gluon, ghost, and quark self-energies. Thus, in QCD, the number of primitive divergences are seven in number (see, Fig. 3.1).

3.2 Slavnov–Taylor identities

Absorbtion of all the QCD divergencies by restricted numbers of counter-terms is ensured by the fact that there are relations between different divergent Green functions. These relations are called Slavnov [15]–Taylor [16] identities (STIs). The STIs represent the generalization of the well-known Ward identities on non-Abelian gauge theories. They reflect the gauge invariance of QCD, constrain the number of independent ultraviolet-divergent matrix elements, and guarantee the absence of gauge noninvariant counter-terms. Thus, for example, in QED, where three renormalization constants $Z_{1,2,3}$ exist, the Ward identity gives

$$Z_1 = Z_2, \tag{3.10}$$

where Z_1 defines the vertex function renormalization and Z_2 is associated with the wave function renormalization of a Dirac particle. This, in its turn, should be interpreted as the cancellation of the vertex divergence with the wave function renormalization divergence in all orders of the perturbation theory. In QCD, there are considerably more renormalization constants, some of them being dependent because of the STIs. Thus, for example $Z_G = Z_\xi$. One more problem connected with introducing nonphysical particles, Faddeev–Popov ghosts, into the theory exists. One needs to prove that their presence in the theory does not violate the S-matrix unitarity. Here once again, the STIs appear in the role of the protagonist.

The STIs reflect the symmetry associated with the invariance of the effective action with respect to the BRS-transformation. To obtain these relations between the Green function, we consider the QCD generating functional in which the complex ghost fields ω_a and $\omega_a^\dagger$ have been changed by real Grassmann fields ρ and σ (see Eq. (1.92)). To prevent the appearance of complicated nonlinear terms it is convenient to introduce additional sources connected with the change of fields under the BRS-transformations, that is, to add the sources $\kappa_a^\mu, \nu_a, \beta$, and $\overline{\beta}$ for the composite operators $D^\mu \sigma_a, f_{abc}\sigma_b \sigma_c/2, T_a \sigma_a q$, and $\overline{q} T_a \sigma_a$, which enter into Eq. (1.100), to the ordinary sources set $J_a^\mu, \eta_a, \eta_a^\dagger, \gamma$, and $\overline{\gamma}$. Hereafter,

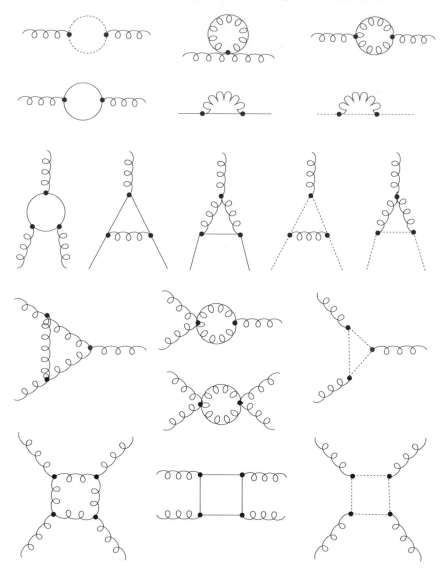

FIGURE 3.1
The primitive divergent diagrams of QCD.

generators T_a mean the fundamental representation generators, $\lambda_a/2$. It is clear that the sources ν_a are commuting variables while κ^μ, β, and $\overline{\beta}$ are anticommuting ones. The generating functional takes the form

$$W[\mathbf{J}, \boldsymbol{\eta}', \boldsymbol{\eta}'', \gamma, \overline{\gamma}, \boldsymbol{\kappa}, \boldsymbol{\nu}, \beta, \overline{\beta}] = \int [dG_\mu][d\rho][d\sigma][dq][d\overline{q}] \exp\left\{i \int d^4x [\mathcal{L}_{eff}(x) + \Sigma(x)]\right\},$$
(3.11)

where the term with the sources is given by the expression

$$\Sigma = J_{a\mu}G_a^\mu + \eta_a'\rho_a + \eta_a''\sigma_a + \overline{\gamma}q + \overline{q}\gamma + \kappa_a^\mu D_\mu\sigma_a - \frac{1}{2}f_{abc}\nu_a\sigma_b\sigma_c + \overline{\beta}T_a\sigma_a q + \overline{q}T_a\sigma_a\beta. \quad (3.12)$$

Subject $W[\mathbf{J}, \ldots \overline{\beta}]$ to the infinitesimal BRS-transformation (23.100)

$$\delta W[\mathbf{J}, \boldsymbol{\eta}', \boldsymbol{\eta}'', \gamma, \overline{\gamma}, \boldsymbol{\kappa}, \boldsymbol{\nu}, \beta, \overline{\beta}] = \int \Delta(x, y)[dG_\mu][d\rho][d\sigma][dq][d\overline{q}](\delta\Sigma) \times$$

$$\times \exp\left\{i \int d^4y[\mathcal{L}_{eff}(y) + \Sigma(y)]\right\}, \tag{3.13}$$

where $\Delta(x, y)$ is the Jacobian,

$$\delta\Sigma = J_{a\mu}\delta G_a^\mu + \eta_a'\delta\rho_a + \eta_a''\delta\sigma_a + \overline{\gamma}\delta q + \delta\overline{q}\gamma + \kappa_a^\mu\delta(D_\mu\sigma_a) + \nu_a\delta\left(-\frac{1}{2}f_{abc}\sigma_b\sigma_c\right) +$$

$$+\overline{\beta}\delta(T_a\sigma_a q) + \delta(\overline{q}T_a\sigma_a)\beta, \tag{3.14}$$

$$\left.\begin{array}{ll}
\delta G_a^\mu = \epsilon D^\mu\sigma_a, & \delta q = i\epsilon g_s T_a\sigma_a q, \\
\delta\sigma_a = -(\epsilon/2)g_s f_{abc}\sigma_b\sigma_c, & \delta\rho_a = -(i\epsilon/\xi)\partial_\mu G_a^\mu
\end{array}\right\} \tag{3.15}$$

and we have taken into account the invariance of the effective action with respect to the BRS-transformation. First, demonstrate that the coefficients attached to the new sources are invariant. Invariance of the coefficient at κ_a^μ was shown by Eq. (1.11). Now we shall prove the coefficients at ν_a, β and $\overline{\beta}$ possess this property as well. Using (3.15), we get

$$\delta(f_{abc}\sigma_b\sigma_c) - f_{abc}[\delta(\sigma_b)\sigma_c + \sigma_b\delta(\sigma_c)] - -\frac{\epsilon g_s}{2}\left(f_{abc}f_{bmn}\sigma_m\sigma_n\sigma_c + f_{abc}f_{cmn}\sigma_b\sigma_m\sigma_n\right) -$$

$$= -\epsilon g_s f_{abc}f_{cmn}\sigma_m\sigma_n\sigma_b = -\frac{\epsilon g_s}{3}f_{abc}f_{cmn}\left(\sigma_m\sigma_n\sigma_b + \sigma_b\sigma_m\sigma_n + \sigma_n\sigma_b\sigma_m\right) =$$

$$= -\frac{\epsilon g_s}{3}\left(f_{abc}f_{cmn} + f_{anc}f_{cbm} + f_{amc}f_{cnb}\right)\sigma_m\sigma_n\sigma_b = 0. \tag{3.16}$$

In passing we note that, as may be inferred from (23.100), the change of σ_b is also nilpotent:

$$\delta^2(\delta\sigma_b) = 0. \tag{3.17}$$

Further, we have

$$\delta(T_a\sigma_a q) = T_a(\delta\sigma_a)q + T_a\sigma_a\delta q = T_a\left(-\frac{g_s}{2}\epsilon f_{abc}\sigma_b\sigma_c\right)q - ig_s\epsilon T_a\sigma_a T_b\sigma_b q =$$

$$= T_a\left(-\frac{g_s}{2}\epsilon f_{abc}\sigma_b\sigma_c\right)q - \frac{ig_s\epsilon}{2}\left(T_aT_b - T_bT_a\right)\sigma_a\sigma_b =$$

$$= T_a\left(-\frac{g_s}{2}\epsilon f_{abc}\sigma_b\sigma_c\right)q + \frac{g_s\epsilon}{2}f_{abc}T_c\sigma_a\sigma_b = 0. \tag{3.18}$$

Our next task is to find Jacobian $\Delta(x, y)$. By definition we have

$$\Delta(x, y) =$$

$$= \partial\left(\frac{G_a^\mu(x) + \delta G_a^\mu(x), \sigma_a(x) + \delta\sigma_a(x), \rho_a(x) + \delta\rho_a(x), q(x) + \delta q(x), \overline{q}(x) + \delta\overline{q}(x)}{G_b^\nu(y), \sigma_b(y), \rho_b(y), q(y), \overline{q}(y)}\right). \tag{3.19}$$

In this determinant only the following terms are nonzero

$$\frac{\delta[G_a^\mu(x) + \delta G_a^\mu(x)]}{\delta G_b^\nu(y)} = \delta_\nu^\mu\delta^4(x - y)(\delta_{ab} + \epsilon g_s f_{abc}\sigma_c), \tag{3.20}$$

$$\frac{\delta[\sigma_a(x) + \delta\sigma_a(x)]}{\delta\sigma_b(y)} = \delta^4(x - y)\left[\delta_{ab} - \frac{\epsilon g_s}{2}\frac{\delta}{\delta\sigma_b}(f_{adc}\sigma_d\sigma_c)\right] =$$

$$= \delta^4(x-y)\left[\delta_{ab} - \frac{\epsilon g_s}{2}(f_{abc}\sigma_c - f_{adb}\sigma_d)\right] = \delta^4(x-y)[\delta_{ab} + \epsilon g_s f_{abc}\sigma_c], \qquad (3.21)$$

$$\frac{\delta[q(x) + \delta q(x)]}{\delta q(y)} = \delta^4(x-y)[1 + i\epsilon g_s T_a \sigma_a], \qquad (3.22)$$

$$\frac{\delta[\rho_a(x) + \delta\rho_a(x)]}{\delta\rho_b(y)} = \delta^4(x-y)\delta_{ab}, \qquad (3.23)$$

$$\frac{\delta[G_a^\mu(x) + \delta G_a^\mu(x)]}{\delta\sigma_b(y)} = -\delta^4(x-y)[\epsilon g_s f_{abc}G_c^\mu]. \qquad (3.24)$$

Thus, the Jacobian is schematically represented in the form

$$\Delta(x,y) = \delta_\nu^\mu[\delta^4(x-y)]^5 \begin{vmatrix} 1+\epsilon g_s f\sigma & -\epsilon g_s fG & 0 & 0 & 0 \\ 0 & 1+\epsilon g_s f\sigma & 0 & 0 & 0 \\ 0 & 0 & 1 & 0 & 0 \\ 0 & 0 & 0 & 1+i\epsilon g_s T\sigma & 0 \\ 0 & 0 & 0 & 0 & 1-i\epsilon g_s T\sigma \end{vmatrix} =$$

$$= \delta_\nu^\mu[\delta^4(x-y)]^5. \qquad (3.25)$$

Therefore, it is equal to 1. So, since the Lagrangian $\mathcal{L}_{eff}(x)$ is invariant under the BRS-transformations, then from invariance of the generating functional $W[\mathbf{J}, \ldots \overline{\beta}]$ it follows

$$W[\mathbf{J}, \boldsymbol{\eta}', \boldsymbol{\eta}'', \gamma, \overline{\gamma}, \boldsymbol{\kappa}, \boldsymbol{\nu}, \beta, \overline{\beta}] =$$

$$= \int [dG_\mu][d\rho][d\sigma][dq][d\overline{q}] \exp\left\{i\left[S_t + \int d^4x(J_{a\mu}\delta G_a^\mu + \eta_a'\delta\rho_a + \eta_a''\delta\sigma_a + \overline{\gamma}\delta q + \delta\overline{q}\gamma)\right]\right\} =$$

$$= \int [dG_\mu][d\rho][d\sigma][dq][d\overline{q}]\left[1 + \int d^4x(J_{a\mu}\delta G_a^\mu + \eta_a'\delta\rho_a + \eta_a''\delta\sigma_a + \overline{\gamma}\delta q + \delta\overline{q}\gamma)\right]\exp(iS_t),$$

where $S_t = \int d^4x[\mathcal{L}_{eff}(x) + \Sigma(x)]$. As a result, we get

$$\int [dG_\mu][d\rho][d\sigma][dq][d\overline{q}]\int d^4x(J_{a\mu}\delta G_a^\mu + \eta_a'\delta\rho_a + \eta_a''\delta\sigma_a + \overline{\gamma}\delta q + \delta\overline{q}\gamma)\exp(iS_t) =$$

$$= \epsilon\int d^4x\left[J_{a\mu}D^\mu\sigma_a - \frac{i\eta_a'}{\xi}\partial_\mu G_a^\mu - \frac{\eta_a''g_s}{2}f_{abc}\sigma_b\sigma_c + i\overline{\gamma}g_s(T_a\sigma_a)q - \right.$$

$$\left. -i\gamma g_s(T_a\sigma_a)\overline{q}\right]W[\mathbf{J}, \boldsymbol{\eta}', \boldsymbol{\eta}'', \gamma, \overline{\gamma}, \boldsymbol{\kappa}, \boldsymbol{\nu}, \beta, \overline{\beta}] = 0. \qquad (3.26)$$

From Eqs. (3.12) and (3.15) it is seen that δG_a^μ and $\delta\sigma_a$ are coefficients at the sources κ_a^μ and ν_a, respectively. The quantity $\delta\rho_a$ is proportional to $\partial_\mu G_a^\mu$, where G_a^μ, in its turn, may be considered as the coefficient at J_a^μ. As for the quantities δq and $\delta\overline{q}$, they are proportional to the coefficients at the sources $\overline{\beta}$ and β. Therefore, the expression (3.26) may be rewritten in the following symbolic form

$$\int d^4x\left\{J_a^\mu(x)\frac{\delta}{\delta\kappa_a^\mu(x)} - \frac{i\eta_a'}{\xi}\left[\partial^\mu\frac{\delta}{\delta J_a^\mu(x)}\right] + \eta_a''\frac{\delta}{\delta\nu_a(x)} + ig_s\overline{\gamma}\frac{\delta}{\delta\overline{\beta}(x)} - \right.$$

$$\left. -ig_s\gamma\frac{\delta}{\delta\beta(x)}\right\}W[\mathbf{J}, \boldsymbol{\eta}', \boldsymbol{\eta}'', \gamma, \overline{\gamma}, \boldsymbol{\kappa}, \boldsymbol{\nu}, \beta, \overline{\beta}] = 0. \qquad (3.27)$$

As we see, Eq. (3.27) involves the derivatives only of the first order, what is a consequence of introducing the sources κ_a^μ and ν_a for the nonlinear terms δG_a^μ and $\delta\sigma_a$. The expressions (3.26) and (3.27) represent the generalized Ward identities for QCD (the STIs), which connect different Green functions. To get the relations for some specific collection of the Green functions one needs to differentiate the expression (3.26) with respect to the sources $\mathbf{J}, \boldsymbol{\eta}', \boldsymbol{\eta}'', \gamma, \overline{\gamma}$ and set all the sources equal to zero. When proving the theory renormalizability, the identities obtained are extremely essential. They decrease the number of independent ultra-violet divergences removing them by means of the redefinition of initial parameters. By way of illustration let us find some relations for the Green functions. Taking the functional derivative

$$\frac{\delta^2}{\delta J_a^\mu(x_1)\delta J_b^\nu(x_2)}$$

of (3.26), we arrive at

$$<0|T(D^\mu\sigma_a(x_1)G_b^\nu(x_2))|0> + <0|T(G_a^\mu(x_1)D^\nu\sigma_b(x_2))|0> = 0. \qquad (3.28)$$

By analogy, acting

$$\frac{\delta^4}{\delta\eta_a'(x_1)\delta J_b^\nu(x_2)\delta\gamma(x_3)\delta\overline{\gamma}(x_4)}$$

on (3.26) results in

$$\frac{1}{\xi}<0|T(\partial_\mu G_a^\mu(x_1)G_b^\nu(x_2)\overline{q}(x_3)q(x_4))|0> + <0|T(\rho_a(x_1)\partial^\nu\sigma_b(x_2)\overline{q}(x_3)q(x_4))|0> = 0.$$
$$(3.29)$$

3.3   Ghosts and S-matrix unitarity

So, we have been earlier convinced that for the gauge invariance of the theory to be ensured ghosts are needed. However, the ghosts prove to perform one more significant duty—they ensure the theory unitarity [17]. The presence of particles with the indefinite metric makes the unitarity problem nontrivial. In this section we consider the particular reaction

$$q\overline{q} \to q\overline{q} \qquad (3.30)$$

and use it to illustrate the theory unitarity in the fourth order of the perturbation theory. In doing so, one will be shown that cancellation of nonphysical contributions in the S-matrix is caused by the existence of the generalized Ward identities.

First of all we find out what information may be extracted from the S-matrix unitarity condition. The reaction amplitude $\mathcal{A}_{ab}$ is connected with the S-matrix elements by the relation

$$S_{ab} = \delta_{ab} + i(2\pi)^4\delta^4(p_a - p_b)\mathcal{A}_{ab}. \qquad (3.31)$$

Then, from the unitarity condition it follows

$$\text{Im}\mathcal{A}_{ab} = \frac{1}{2}\sum_c \mathcal{A}_{ac}\mathcal{A}_{cb}^\dagger. \qquad (3.32)$$

Consider initial and final states (a and b) to be two-particle states ($|p_1; p_2 >$ and $|p_3; p_4 >$). To calculate the expression (3.32) insert in its right side the total set of intermediate states

$|n>$ including particles with momenta $k_1, k_2, \ldots, k_n$. Then Eq. (3.32) takes the form

$$\text{Im} < p_3; p_4|\mathcal{A}|p_1; p_2 > = \frac{1}{2(2\pi)^4} \sum_n \int \frac{d^3 k_1}{\sqrt{2E_1}} \cdots \frac{d^3 k_n}{\sqrt{2E_n}} \times$$

$$\times \delta^4(p_1 + p_2 - k_1 - k_2 - \ldots - k_n) < p_3; p_4|\mathcal{A}|k_1; \ldots k_n >< p_1; p_2|\mathcal{A}|k_1; \ldots k_n >^\dagger, \quad (3.33)$$

where summation is realized over all real (rather than virtual) intermediate states that obey the 4-momentum conservation law. So, from the S-matrix unitarity demand it follows that the imaginary part of the scattering amplitude $\mathcal{A}_{ab}$ is expressed in terms of the products sum of the matrix elements to connect initial and final states with all physical states having the same energy and momentum. The obtained relation presents the relativistic analogue of the optical theorem. In Fig. 3.2 the graphical presentation of Eq. (3.33) is shown. So far,

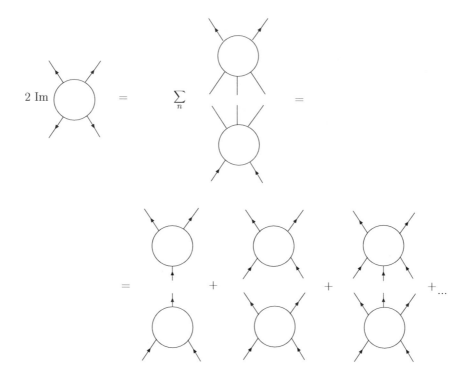

FIGURE 3.2
The graphical presentation of Eq. (3.33)

the initial and final states have been supposed to be two-particle ones. However, we could equally well use one- and many-particle asymptotical states. It should be stressed that the unitarity condition obtained may be utilized in every order of the perturbation theory. In the fourth order of the perturbation theory the unitarity condition for the process (3.30) is graphically represented in Fig. 3.3.

In order to calculate the imaginary part of the scattering amplitude in the left side of Eq. (3.33) one should replace the propagators in intermediate states by their imaginary parts to multiply them by the scattering amplitudes $\mathcal{A}(q\bar{q} \to NN)$ and $\mathcal{A}(NN \to q\bar{q})^\dagger$, which are taken at the mass shell [18]. In the Feynman gauge the gluon propagator is given by the

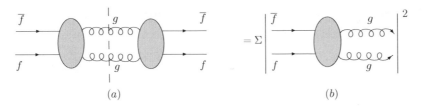

FIGURE 3.3
The unitarity condition for the process (3.30) in (a) and (b). Here gluons and Faddeev–Popov ghosts feature as intermediate particles.

expression

$$P_{ab}^{(G)\mu\nu} = -\frac{\delta_{ab}g^{\mu\nu}}{k^2 + i\epsilon}. \tag{3.34}$$

Making use of the formula

$$\frac{1}{x \pm i\epsilon} = P\frac{1}{x} \mp i\pi\delta(x), \tag{3.35}$$

we get the imaginary part of the gluon propagator in the form

$$\pi\delta_{ab}g^{\mu\nu}\delta(k^2)\theta(\omega), \tag{3.36}$$

where $\omega = |k_0|$ and the factor $\theta(\omega)$ ensures the energy positivity of the gluon field. By the analogy, the imaginary part of the ghost propagator is as follows

$$\pi\delta_{ab}\delta(k^2)\theta(\omega).$$

So, in the left side of Eq. (3.33) the propagators are replaced by the functions that guarantee gluon and ghost momenta to be on the mass shell. Therefore, these particles become real but with a tinge of surrealism. Really, since the gluon propagator is proportional to $g^{\mu\nu}$ and

$$-g_{\mu\nu} = \sum_{\lambda=0}^{3} \epsilon_\mu^\lambda(p)\epsilon_\nu^\lambda(p), \tag{3.37}$$

in the imaginary part of the amplitude the gluons are not entirely real owing to the fact that, whilst $p^2 = 0$ takes place for them, they could possess both the physical and nonphysical polarization states. As regards the ghosts, they are absent in the real states by definition. In the right side of Eq. (3.33) the situation with initial and final particles is already quite different. Here, all the particles are real, that is, the ghosts are absent and the summation over the gluon polarizations involves only the transverse physical states due to using the physical gauge. Earlier we were discussing the possibility of defining the axial and light-like gauges with the help of the four-dimensional vector n_μ. In the case of space-like vector n_μ we deal with the axial gauge and the summation rule over the gluon polarizations is given by the expression

$$\Lambda_{\mu\nu}(p) = \sum_{\lambda=1,2} \epsilon_\mu^\lambda(p)\epsilon_\nu^\lambda(p) = -g_{\mu\nu} - \frac{(nn)p_\mu p_\nu - (np)(p_\nu n_\mu + p_\mu n_\nu)}{(np)^2}. \tag{3.38}$$

In the event of the light-like gauge the vector n_μ is a zero one ($n_\mu n^\mu = 0$) and the summation rule will look like:

$$\Lambda_{\mu\nu}(p) = \sum_{\lambda=1,2} \epsilon_\mu^\lambda(p)\epsilon_\nu^\lambda(p) = -g_{\mu\nu} + \frac{p_\nu n_\mu + p_\mu n_\nu}{(np)}. \tag{3.39}$$

After all these preparations, it presents no problem to write down the unitarity condition
for the process (3.30) in the fourth order of the perturbation theory

$$\frac{1}{2} V_{ab}^{\mu\nu} V_{ab}^{\mu'\nu'\,*} g_{\mu\mu'} g_{\nu\nu'} - S_{ab} S_{ab}^{*} = \frac{1}{2} V_{ab}^{\mu\nu} V_{ab}^{\mu'\nu'\,*} \Lambda_{\mu\mu'}(p_1) \Lambda_{\nu\nu'}(p_2), \qquad (3.40)$$

where $V_{ab}^{\mu\nu}$ and S_{ab} are the transition amplitudes $q\bar{q} \to G_a^{\mu} G_b^{\nu}$ and $q\bar{q} \to \omega_a^{\dagger}\omega_b$, respectively.
The imaginary part of the amplitude $q\bar{q} \to q\bar{q}$ is given by cut diagrams* shown in Fig. 3.4.

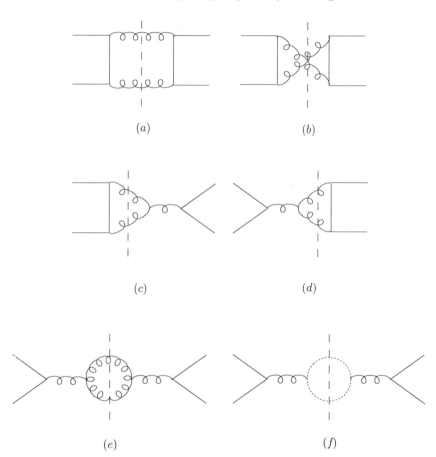

(a)   (b)

(c)   (d)

(e)   (f)

FIGURE 3.4
The cut diagrams (a)–(f) for the process $q\bar{q} \to q\bar{q}$.

It represents the squares sum of the amplitudes $q\bar{q} \to G_a^{\mu} G_b^{\nu}$ (Fig. 3.5) and $q\bar{q} \to \omega_a^{\dagger}\omega_b$ (Fig. 3.6).

   The factor 1/2 being in the left side of Eq. (3.40) is caused by the fact that, under rasing
to the second power of the amplitudes shown in Fig. 3.5, nine diagrams appear, eight of
them correspond to the ones of Fig. 3.4(a)–(d) repeated twice and the nine is associated
with the one displayed in Fig. 3.4(e). The last-named contains the closed-gluon loop and
must be multiplied by the symmetrization factor 1/2. Since the ghost field behaves as a
fermion for which $\omega \neq \omega^{\dagger}$, in front of $S_{ab}S_{ab}^{*}$, the minus sign stands and the symmetrization
factor is absent.

---

*In the literature such a cutting is frequently called *Cutkosky's cutting* [18].

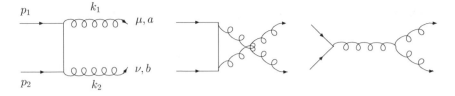

FIGURE 3.5

The Feynman diagrams are associated with $q\bar{q} \to G_a^\mu G_b^\nu$.

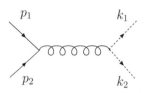

FIGURE 3.6

The Feynman diagram for $q\bar{q} \to \omega_a^\dagger \omega_b$.

Using the Feynman rules leads to the following expressions for $V_{ab}^{\mu\nu}$ and S_{ab}, respectively:

$$V_{ab}^{\mu\nu} = -ig_s^2\bar{v}(p_2)\left[\frac{\lambda_b}{2}\gamma^\nu\frac{\hat{p}_1 - \hat{k}_1 + m_q}{(p_1 - k_1)^2 - m_q^2}\frac{\lambda_a}{2}\gamma^\mu + \frac{\lambda_a}{2}\gamma^\mu\frac{\hat{k}_1 - \hat{p}_2 + m_q}{(k_1 - p_2)^2 - m_q^2}\frac{\lambda_b}{2}\gamma^\nu\right]u(p_1)-$$

$$-g_s^2 f_{abc}\bar{v}(p_2)\left[\frac{(k_1 - k_2)^\sigma g^{\mu\nu} + (k_1 + 2k_2)^\mu g^{\nu\sigma} - (2k_1 + k_2)^\nu g^{\mu\sigma}}{(k_1 + k_2)^2}\right]\frac{\lambda_c}{2}\gamma_\sigma u(p_1), \qquad (3.41)$$

$$S_{ab} = -ig_s^2 f_{abc}\bar{v}(p_2)\left[\frac{1}{(k_1 + k_2)^2}\right]\frac{\lambda_c}{2}\hat{k}_1 u(p_1). \qquad (3.42)$$

We shall work in the light-like gauge. Substitute Eq. (3.39) into the right side of Eq. (3.40) and calculate the terms of the type $k_{i\mu}V_{ab}^{\mu\nu}$, $k_{i\nu}V_{ab}^{\mu\nu}$ ($i = 1, 2$) to appear in the process. We find for $k_{1\mu}V_{ab}^{\mu\nu}$

$$k_{1\mu}V_{ab}^{\mu\nu} = -ig_s^2\bar{v}(p_2)\left[\frac{\lambda_b}{2}\frac{\lambda_a}{2}\gamma^\nu\frac{\hat{p}_1 - \hat{k}_1 + m_q}{(p_1 - k_1)^2 - m_q^2}\hat{k}_1 + \frac{\lambda_a}{2}\frac{\lambda_b}{2}\hat{k}_1\frac{\hat{k}_1 - \hat{p}_2 + m_q}{(k_1 - p_2)^2 - m_q^2}\gamma^\nu\right]u(p_1)-$$

$$-g_s^2 f_{abc}\bar{v}(p_2)\left[\frac{(k_1 - k_2)^\sigma k_1^\nu + 2(k_1 k_2)g^{\nu\sigma} - (2k_1 + k_2)^\nu k_1^\sigma}{(k_1 + k_2)^2}\right]\frac{\lambda_c}{2}\gamma_\sigma u(p_1) =$$

$$= g_s^2 f_{abc}\bar{v}(p_2)\frac{\lambda_c}{2}\gamma^\nu u(p_1) - g_s^2 f_{abc}\bar{v}(p_2)\left[\frac{k_1^\nu}{(k_1 + k_2)^2}\frac{\lambda_c}{2}(\hat{p}_1 + \hat{p}_2) + \frac{\lambda_c}{2}\gamma_\nu + \right.$$

$$\left. + \frac{k_2^\nu}{(k_1 + k_2)^2}\frac{\lambda_c}{2}\hat{k}_1\right]u(p_1) = -g_s^2 f_{abc}\bar{v}(p_2)\frac{k_2^\nu}{(k_1 + k_2)^2}\frac{\lambda_c}{2}\hat{k}_1 u(p_1), \qquad (3.43)$$

where we have taken into account the Dirac equation and the 4-momentum conservation law

$$p_1^\mu + p_2^\mu = k_1^\mu + k_2^\mu.$$

Comparing the obtained expression with Eq. (3.42) results in

$$k_{1\mu}V_{ab}^{\mu\nu} = -iS_{ab}k_2^\nu. \qquad (3.44)$$

The analogous computations give

$$V_{ab}^{\mu\nu} k_{2\nu} = -iS_{ab}k_1^{\mu}. \qquad (3.45)$$

The relations found are nothing else than the example of the generalized Ward identities in QCD. They were obtained by us earlier (see Eq. (3.29)).

Now, it presents no special problems to verify the unitarity condition fulfillment for the process in question. Using the relations (3.44), (3.45), we get (indices a and b are missed for the time being)

$$\frac{1}{2}V_{\mu\nu}V_{\mu'\nu'}^{*}\left[-g^{\mu\mu'} + \frac{k_1^{\mu'}n_1^{\mu} + k_1^{\mu}n_1^{\mu'}}{(n_1k_1)}\right]\left[-g^{\nu\nu'} + \frac{k_2^{\nu'}n_2^{\nu} + k_2^{\nu}n_2^{\nu'}}{(n_2k_2)}\right] = \frac{1}{2}\{VV^{*}gg+$$

$$+\left[\frac{(k_1Vn_2)(n_1V^{*}k_2) + (n_1Vk_2)(k_1V^{*}n_2)}{(k_1n_1)(k_2n_2)}\right] - \left[\frac{(k_1V)\cdot(n_1V^{*}) + (n_1V)\cdot(k_1V^{*})}{(k_1n_1)}\right] -$$

$$-\left[\frac{(Vk_2)\cdot(V^{*}n_2) + (Vn_2)\cdot(V^{*}k_2)}{(k_2n_2)}\right]\} = \frac{1}{2}\{VV^{*}gg + 2SS^{*} - 2SS^{*} - 2SS^{*}\} =$$

$$= \frac{1}{2}[VV^{*}gg - 2SS^{*}]. \qquad (3.46)$$

Thus, we have proved the S-matrix unitarity for the particular process in the fourth order of the perturbation theory. It may be shown that this condition is fulfilled in all the orders as well. When considering diagrams with loops (and not just tree diagrams), the unitarity condition demands taking into consideration diagrams with closed ghost lines. Consequently, the presence of ghosts simultaneously solves the problems of both the unitarity and gauge invariance. It may be possible to state that the unitarity is provided with the generalized Ward identities. Both of these statements are equivalent.

3.4 Renormalization of one-loop diagrams

QCD is a theory with the inviolate gauge symmetry $SU(3)_c$. This circumstance determines its renormalizability. Therefore, when treating diverging expressions, one cannot lose the gauge invariance. On the other hand, for divergent terms to be operated, a definite meaning must be assigned to them. This operation is called *regularization*. A simple cut-off of momenta is accompanied by the inclusion of parameters with mass dimensionality, resulting in gauge invariance violation. The dimensional regularization method based on the transition into a space-time of lower dimensionality d ($d < 4$) (see Section 21.2 of *Advanced Particle Physics Volume 1*) preserves the gauge invariance. After completion of all mathematical operations with regularized expressions, the dimensionality of a space-time tends to four and divergences appear again. The procedure of removing these divergences, which is connected with a transition from *naked* parameters to renormalized (physical) ones, is the essence of the multiplicative renormalization scheme. It was already used under the proof of QED renormalizability. We shall follow this scheme considering the QCD as well.

The proof of renormalizability of non-Abelian gauge theories, QCD in particular, was first developed by t'Hooft [1]. Subsequently, similar proofs were presented in Refs. [16, 19]. When proving renormalizability, the use of the definite gauges class is very essential. Recall the different behavior of the gluon propagator under $p \to \infty$ in the cases of the Feynman and transverse gauges. In 1972 t'Hooft and Veltman [20] gave the general proof of the

S-matrix independence on the gauge choice. As a result, the proof of the gauge invariance of the renormalized S-matrix has been obtained. In what follows we shall be constrained by renormalization of one-loop diagrams only. Detailed discussion of a general renormalization procedure is beyond the scope of this book and we refer the reader who is interested in this problem to an excellent monograph by Collins [21].

So, our task is to calculate all divergent QCD expressions in the one-loop approximation and demonstrate the way in which appearing divergences are removed by the multiplicative renormalization scheme. As was shown, the primitively divergent diagrams are involved in the definitions of propagators, three-point and four-point vertices. By way of the first example we investigate the two-point Green function of the gluon $G_{ab}^{\mu\nu}$, that is, the gluon propagator. According to the general rules (2.54), we have

$$G_{ab}^{\mu\nu}(x_1, x_4) = iD_{ab}^{\mu\nu}(x_1 - x_4) - D_{ad}^{\mu\rho}(x_1 - x_2)\Pi_{dc\rho\sigma}(x_2 - x_3)D_{cb}^{\sigma\nu}(x_3 - x_4),$$

where $\Pi_{dc\rho\sigma}(x_2 - x_3)$ is the polarization operator of the gluon field. Carrying out into the momentum space, we get

$$G_{ab}^{\mu\nu}(p) = iD_{ab}^{\mu\nu}(p) - D_{ad}^{\mu\rho}(p)\Pi_{dc\rho\sigma}(p)D_{cb}^{\sigma\nu}(p). \tag{3.47}$$

Further, we shall use for n-point Green functions the following designations: $\Gamma^{(n_G, n_q)}(p_i; k_j)$, where $i = 1, \dots, n_G; j = 1, \dots, n_q; n_G$ (n_q) is the number of gluon (quark) variables and $n = n_G + n_q$. Therefore,

$$G_{ab\mu\nu}(p) \equiv \Gamma_{ab\mu\nu}^{(2,0)}(p).$$

For the polarization tensor $\Pi_{dc\rho\sigma}(p)$ to be calculated, it is necessary to address the gluon self-energy diagrams. In the second order of the perturbation theory these diagrams are represented in Fig. 3.7. The diagram of Fig. 3.7(a) describes the polarization of the quark

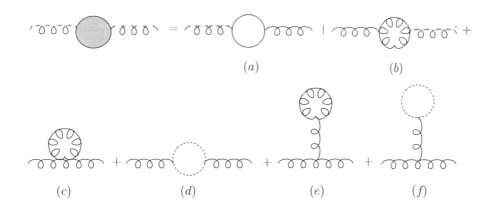

(a) (b)

(c) (d) (e) (f)

FIGURE 3.7
The one-loop corrections (a)–(f) to the vacuum polarization.

vacuum (production of virtual quark-antiquark pairs), the diagram of Fig. 3.7(b) describes the polarization of the gluon vacuum (production of virtual gluon pairs), the diagrams of Fig. 3.7(c),(e) describe the direct gluon self-interaction, and the diagrams of Fig. 3.7(d),(f) describe the gauge corrections (production of virtual ghost pairs). During the process of calculations we shall use the Feynman gauge for the gluon propagator. Begin with the diagram shown in Fig. 3.7(a). The polarization operator of the quark vacuum has the form (remember about the minus sign for a fermion loop)

$$\Pi_{ab}^{(q)\mu\nu}(p) = -g_s^2 \int \frac{d^4k}{(2\pi)^4} \mathrm{Sp} \left\{ \gamma^\mu \frac{1}{[\hat{k} + \hat{p} - m]} \gamma^\nu \frac{1}{[\hat{k} - m]} \right\} \frac{1}{4} (\lambda_a)_{dc} (\lambda_b)_{cd}. \tag{3.48}$$

Note, that the structure of this integral is the same as in the QED case apart from the group-theoretical factors. We obtain for these factors

$$\frac{1}{4} (\lambda_a)_{dc} (\lambda_b)_{cd} = \frac{1}{4} \mathrm{Sp}(\lambda_a \lambda_b) = \frac{1}{2} \delta_{ab}. \tag{3.49}$$

If only quarks belonging to one representation of the $SU(3)_c$ group contribute to vacuum polarization, i.e., if only one quark flavor exists, this result would be substituted into Eq. (3.48). When the number of quark flavors is equal to N_f, then instead of Eq. (3.49) we have

$$\frac{1}{4} \mathrm{Sp}(\lambda_a \lambda_b) = \frac{N_f}{2} \delta_{ab}. \tag{3.50}$$

In accordance with the naive counting of the momenta power, the integral (3.48) is quadratically divergent. However, as we shall see later, the divergence proves to be logarithmic only. This circumstance is caused by the gauge invariance of QCD. To operate with the integral (3.48) we have to regularize it. In doing so, we must trouble about conservation of the gauge invariance of the theory. The dimensional regularization is the very thing we need. Pass on from the four-dimensional Minkowski space to the d-dimensional Minkowski space ($d < 4$) with one-time-dimension and $d - 1$-space-dimension. In this space-time the metric tensor will look like:

$$g_{\mu\nu} = \mathrm{diag} \, (+, -, -, \ldots)$$

($g_\mu^\mu = d$). At that we have

$$\int \frac{d^4k}{(2\pi)^4} \to \int \frac{d^dk}{(2\pi)^d}.$$

It should be taken into account that the coupling constant g_s becomes dimensional in the d space ($d \neq 4$). Really, as long as

$$\left. \begin{array}{c} [\int \overline{q}(x)q(x)d^dx] = m^{-1}, \qquad [\int G_a^{\mu\nu}(x)G_{\mu\nu a}(x)d^dx] = 1, \\ [g_s \int \overline{q}(x)G_a^\mu(x)q(x)d^dx] = 1, \end{array} \right\} \tag{3.51}$$

then

$$[q(x)] = m^{(d-1)/2}, \qquad [G_a^\mu(x)] = m^{-1+d/2}$$

and we get

$$[g_s] = m^{(4-d)/2}. \tag{3.52}$$

For the dimension of g_s^2 to be taken into consideration, in front of the integral (3.48) we introduce the factor $\overline{\mu}^{4-d}$, where $\overline{\mu}$ is a parameter with the mass dimension. Since we shall use Feynman's parametrization, the integrals of interest are as follows:

$$I_d^{(l)}(p) = \int d^dk \frac{F_l(k)}{(k^2 + 2kp - m^2)^n}, \tag{3.53}$$

where $k = (k_0, \mathbf{k})$, $n = 0, 1, \ldots$, $F_l(k)$ is a polynomial of power l with respect to d-momenta. Before computing these integrals, we make one remark. After the dimensional regularization the integral is convergent. Then, singling out the perfect square $k^2 + 2kp = (k + p)^2 - p^2$ in the denominator and carrying out the replacement $k + p \to k$, we arrive at the integral

$$I_d^{(l)}(p) = \int d^d k \frac{f_l(k)}{[k^2 - \Delta^2]^n}, \tag{3.54}$$

where $\Delta^2 = p^2 + m^2$. Evidently, that now we may be constrained by the case when $f_l(k)$ represents a certain scalar function of k^2. Really, if $f_1(k) = k_\mu F_1(k^2)$, then $I_{d=4}^{(1)} = 0$; if $f_2(k) = k_\mu k_\nu F_1(k^2)$, then

$$\int d^d k \frac{k_\mu k_\nu F_1(k^2)}{[k^2 - \Delta^2]^n} = \frac{g_{\mu\nu}}{d} \int d^d k \frac{k^2 F_1(k^2)}{[k^2 - \Delta^2]^n}$$

and so on. This, in turn, allows us to neglect the integrals of the kind (3.54) that have odd powers of k in the numerator.

It turns out that all necessary integrals can be obtained from the basic integral

$$I_d(p) = \int \frac{d^d k}{(k^2 + 2kp - m^2)^n}, \tag{3.55}$$

by means of the differentiation with respect to p. To take it, we pass on to the polar coordinates

$$k = (k_0, r, \varphi, \theta_1, \theta_2, \dots \theta_{d-3}).$$

Then, in d-dimensional space the volume element becomes the view

$$d^d k = dk_0 r^{d-2} dr d\varphi \prod_{k=1}^{d-3} \sin^k \theta_k d\theta_k$$

$$-\infty < k_0 < \infty, \qquad 0 < r < \infty, \qquad 0 < \varphi < 2\pi, \qquad 0 < \theta_k < \pi.$$

At present, the starting integral is defined by the expression

$$I_d(p) = 2\pi \int_0^\infty dk_0 \int_0^\infty r^{d-2} dr \int_0^\pi \prod_{l=1}^{d-3} \sin^l \theta_l d\theta_l \frac{1}{(k^2 + 2kp - m^2)^n}. \tag{3.56}$$

Making use of the formula

$$\int_0^{\pi/2} (\sin \theta)^{2n-1} (\cos \theta)^{2m-1} d\theta = \frac{\Gamma(n)\Gamma(m)}{2\Gamma(n+m)}$$

and taking into consideration that

$$\Gamma\left(\frac{1}{2}\right) = \sqrt{\pi},$$

we rewrite $I_d(p)$ in the following form

$$I_d(p) = \frac{2\pi^{(d-1)/2}}{\Gamma\left(\dfrac{d-1}{2}\right)} \int_{-\infty}^\infty dk_0 \int_0^\infty \frac{r^{d-2} dr}{(k_0^2 - r^2 + 2kp - m^2)^n}. \tag{3.57}$$

Since the given integral is Lorentz-invariant, we may turn to the rest frame in which $p_\mu = (\mu, \mathbf{0})$. Introducing the variable $k'_\nu = k_\nu + p_\nu$, which obeys the relation $k'^2_0 - p^2 = k^2_0 + 2\mu k_0$, we get

$$I_d(p) = \frac{2\pi^{(d-1)/2}}{\Gamma\left(\frac{d-1}{2}\right)} \int_{-\infty}^{\infty} dk'_0 \int_0^{\infty} \frac{r^{d-2} dr}{[k'^2_0 - r^2 - (p^2 + m^2)]^n}. \tag{3.58}$$

In what follows we need the definition of the beta-function $B(x, y)$

$$B(x, y) = \frac{\Gamma(x)\Gamma(y)}{\Gamma(x+y)} = 2 \int_0^{\infty} dt(1 + t^2)^{-x-y} t^{2x-1} \tag{3.59}$$

to be valid at Re $x > 0$ and Re $y > 0$. Then, inserting new variables

$$x = \frac{1+\gamma}{2}, \qquad y = n - \frac{1+\gamma}{2}, \qquad t = \frac{s}{M}$$

and taking into consideration (3.59), we have

$$\int_0^{\infty} ds \frac{s^\gamma}{(s^2 + M^2)^n} = \frac{\Gamma\left(\frac{1+\gamma}{2}\right)\Gamma\left(n - \frac{1+\gamma}{2}\right)}{2(M^2)^{n-(1+\gamma)/2}\Gamma(n)}. \tag{3.60}$$

Now the starting integral $I_d(p)$ can be taken by means of the double application of Eq. (3.60) at $\gamma = d - 2$ and $M^2 = p^2 + m^2 - k'^2_0$

$$I_d(p) = (-1)^n \pi^{(d-1)/2} \frac{\Gamma\left(n - \frac{d-1}{2}\right)}{\Gamma(n)} \int_{-\infty}^{\infty} \frac{dk'_0}{(p^2 + m^2 - k'^2_0)^{n-(d-1)/2}} =$$

$$= (-1)^{(d-1)/2} \pi^{(d-1)/2} \frac{\Gamma\left(n - \frac{d-1}{2}\right)}{\Gamma(n)} \int_{-\infty}^{\infty} \frac{dk'_0}{(k'^2_0 - p^2 - m^2)^{n-(d-1)/2}} =$$

$$= i\pi^{d/2} \frac{\Gamma\left(n - \frac{d}{2}\right)}{\Gamma(n)} \frac{1}{(-p^2 - m^2)^{n-d/2}}.$$

So then, we have shown

$$\int \frac{d^d k}{(k^2 + 2kp - m^2)^n} = i\pi^{d/2} \frac{\Gamma\left(n - \frac{d}{2}\right)}{\Gamma(n)} \frac{1}{(-p^2 - m^2)^{n-d/2}}. \tag{3.61}$$

The obtained result may be represented in an easy-to-use form

$$\int \frac{d^d k}{(2\pi)^d} \frac{1}{(k^2 - \Delta^2)^n} = \frac{(-1)^n i}{(4\pi)^{d/2}} \frac{\Gamma\left(n - \frac{d}{2}\right)}{\Gamma(n)} \left(\frac{1}{\Delta^2}\right)^{n-d/2}. \tag{3.62}$$

Employing the operation of the double and fourfold differentiation with respect to p to both sides of Eq. (3.61), we easily find the values of the following integrals

$$\int \frac{d^d k}{(2\pi)^d} \frac{k_\mu k_\nu}{(k^2 - \Delta^2)^n} = g_{\mu\nu} \frac{(-1)^{n-1} i}{2(4\pi)^{d/2}} \frac{\Gamma\left(n - \frac{d+2}{2}\right)}{\Gamma(n)} \left(\frac{1}{\Delta^2}\right)^{n-(d+2)/2}, \tag{3.63}$$

$$\int \frac{d^d k}{(2\pi)^d} \frac{k_\mu k_\nu k_\rho k_\sigma}{(k^2 - \Delta^2)^n} = \frac{(-1)^n i}{4(4\pi)^{d/2}} \frac{\Gamma\left(n - \frac{d+4}{2}\right)}{\Gamma(n)} \left(\frac{1}{\Delta^2}\right)^{n-(d+4)/2} \times$$

$$\times [g_{\mu\nu} g_{\rho\sigma} + g_{\mu\rho} g_{\nu\sigma} + g_{\mu\sigma} g_{\nu\rho}]. \tag{3.64}$$

Note, that the expression (3.62) converges only at the condition $d < 2n$. In Ref. [20] it was shown that, using an analytical continuation, the expression (3.62) can be transformed in such a way that the convergence takes place at $d < 6$.

Now we are entirely ready to fulfillment of integral operations in d space-time. Introducing the Feynman's parameter and assuming $k' = k - pz$, we get the following expression for $\Pi_{\mu\nu}^{(q)}$

$$\Pi_{\mu\nu}^{(q)}(p) = -g_s^2 \bar{\mu}^{4-d} \frac{N_f}{2} \int_0^1 dz \int \frac{d^d k'}{(2\pi)^d} \frac{\text{Sp}\left\{\gamma_\mu \left[\hat{k}' + \hat{p}z + m\right] \gamma_\nu \left[\hat{k}' - \hat{p}(1-z) + m\right]\right\}}{[k'^2 - m^2 + p^2 z(1-z)]^2}. \tag{3.65}$$

To take the γ-matrices spur in (3.65), we have to consider the definition of the Dirac matrices algebra in d-dimensions. From the fundamental relation of the Clifford algebra

$$\{\gamma_\mu, \gamma_\nu\} = 2g_{\mu\nu}$$

the rules follow

$$\left.\begin{array}{ll} \text{Sp } I = f(d), & \text{Sp } \gamma_\mu \gamma_\nu = f(d) g_{\mu\nu}, \\ \text{Sp } \gamma_\mu \gamma_\nu \gamma_\lambda \gamma_\sigma = f(d)(g_{\mu\nu} g_{\lambda\sigma} - g_{\mu\lambda} g_{\nu\sigma} + g_{\mu\sigma} g_{\nu\lambda}), \end{array}\right\} \tag{3.66}$$

where $f(d)$ is an arbitrary function possessing the good behavior and submitting to the condition $f(4) = 4$. It is clear that its explicit form is inessential for our subsequent calculations. It is clear that the spur of odd number of the γ-matrices vanishes in d space-time as well. Note, it is not possible to define an analogue of γ_5 in d space-time. This difficulty is connected with a problem of the so-called chiral anomalies.

When taking the spur in the numerator of Eq. (3.65), we allow for the terms being odd on k' not to contribute to the integral. With allowance made for (3.66), we get

$$\text{Sp}\left\{\gamma_\mu \left[\hat{k}' + \hat{p}z + m\right] \gamma_\nu \left[\hat{k}' - \hat{p}(1-z) + m\right]\right\} = f(d)\{2k'_\mu k'_\nu - 2z(1-z)(p_\mu p_\nu - p^2 g_{\mu\nu}) -$$

$$- g_{\mu\nu}[k'^2 - m^2 + p^2 z(1-z)]\}, \tag{3.67}$$

where the quantity $g_{\mu\nu} p^2 z(1-z)$ has been added and subtracted. The expression for $\Pi_{\mu\nu}^{(q)}(p)$ will look like:

$$\Pi_{\mu\nu}^{(q)}(p) = -g_s^2 \bar{\mu}^{4-d} f(d) \frac{N_f}{2} \int_0^1 dz \int \frac{d^d k}{(2\pi)^d} \left[2k_\mu k_\nu - 2z(1-z)(p_\mu p_\nu - g_{\mu\nu} p^2) -\right.$$

$$\left. - g_{\mu\nu}[k^2 - m^2 + p^2 z(1-z)]\right] \frac{1}{[k^2 - m^2 + p^2 z(1-z)]^2}, \tag{3.68}$$

where we have removed the prime at the integration variable ($k' \to k$).

From Eqs. (3.62) and (3.63) it follows that the contributions from the first and third terms in the integrand cancel each other. Then, taking into account Eq. (3.62), we arrive at

$$\Pi_{\mu\nu}^{(q)}(p) = -\frac{ig_s^2 (-1)^{(d-4)/2}}{(4\pi)^{d/2}} \frac{N_f}{2} \Gamma\left(2 - \frac{d}{2}\right) f(d)(g_{\mu\nu} p^2 - p_\mu p_\nu) \times$$

$$\times \int_0^1 dz\, 2z(1-z) \left[\frac{p^2 z(1-z) - m^2}{4\pi\overline{\mu}^2} \right]^{(d-4)/2}, \tag{3.69}$$

where the relation $\Gamma(2) = 1$ has been used. The gamma-function $\Gamma(2 - d/2)$ has poles at $d = 4, 6, \ldots$. The expansion of $\Gamma(x)$ in the vicinity of the pole $x = 0$ has the view

$$\Gamma(x) = \frac{1}{x} - \gamma + O(x), \tag{3.70}$$

where γ is the Euler–Mascheroni constant, $\gamma \approx 0.5772$. Setting $\nu = 4 - d \to 0$, we can expand the quantity $a^{-\nu/2} = \{[p^2 z(1-z) - m^2]/4\pi\overline{\mu}^2\}^{-\nu/2}$ as a power series in ν

$$a^{-\nu/2} = \exp\left[-\frac{\nu}{2} \ln a \right] = 1 - \frac{\nu}{2} \ln a + \ldots. \tag{3.71}$$

In the high-energy approximation one may also set $m \to 0$. Based on (3.70) and (3.71), the integral (3.69) becomes the form

$$\Pi_{\mu\nu}^{(q)}(p) = -\frac{ig_s^2}{16\pi^2} \frac{4}{3} \frac{N_f}{2} (g_{\mu\nu} p^2 - p_\mu p_\nu) \left(\frac{2}{\nu} - \gamma \right) \left[1 - \frac{\nu}{2} \left(\ln \frac{p^2}{4\pi\overline{\mu}^2} - \frac{5}{9} \right) \right]. \tag{3.72}$$

Let us proceed to the diagrams describing the gluon vacuum polarization (Fig. 3.7(b)). Remembering the symmetry factor, we obtain for the corresponding matrix element

$$\Pi_{\mu\nu}^{(G)ab}(p) = \frac{1}{2} g_s^2 \overline{\mu}^{4-d} f_{acd} f_{bdc} \int \frac{d^d k}{(2\pi)^d} \frac{A_{\mu\nu}(k,p)}{k^2 (k+p)^2}, \tag{3.73}$$

$$A_{\mu\nu}(k,p) = [(2k+p)_\mu g_{\sigma\rho} - (2p+k)_\sigma g_{\mu\rho} + (p-k)_\rho g_{\mu\sigma}][(2k+p)_\nu g^{\sigma\rho} - (2p+k)^\sigma g_\nu^\rho + (p-k)^\rho g_\nu^\sigma] =$$

$$= p_\mu p_\nu (d-6) + (p_\mu k_\nu + k_\mu p_\nu)(2d-3) + k_\mu k_\nu (4d-6) + g_{\mu\nu}[(2p+k)^2 + (p-k)^2],$$

where the relation $g_\mu^\mu = d$ has been allowed for. Introducing the Feynman parameter and carrying out the replacement of the integration variable in the loop, we rewrite the integral (3.73) in the form

$$\Pi_{\mu\nu}^{(G)ab}(p) = \frac{1}{2} g_s^2 \overline{\mu}^{4-d} f_{acd} f_{bdc} \int_0^1 dz \int \frac{d^d k}{(2\pi)^d} \frac{A_{\mu\nu}(k-pz,p)}{[k^2 + p^2 z(1-z)]^2}, \tag{3.74}$$

where $A_{\mu\nu}(k-pz,p)$ is given by the expression

$$A_{\mu\nu}(k-pz,p) = (4d-6)k_\mu k_\nu + [(4d-6)z(z-1) + d - 6]p_\mu p_\nu + \{2k^2 + p^2[2z(z-1)+5]\}g_{\mu\nu}$$

and we have neglected the terms linear on k because they give zero under the integration (see discussion after the formula (3.54)). Carrying out the integration over the loop variable results in

$$\Pi_{\mu\nu}^{(G)ab}(p) = \frac{g_s^2 i \overline{\mu}^{4-d}}{2(4\pi)^{d/2}} f_{acd} f_{bdc} \int_0^1 dz \left\{ \frac{(3d-3)g_{\mu\nu}\Gamma(1-d/2)}{[p^2 z(1-z)]^{1-d/2}} + \right.$$

$$\left. + \frac{\Gamma(2-d/2)}{[p^2 z(1-z)]^{2-d/2}} \left[g_{\mu\nu} p^2 (5 - 2z(1-z)) + p_\mu p_\nu (d - 6 - z(1-z)(4d-6)) \right] \right\}. \tag{3.75}$$

Before one fulfills the integration over z and passes on to the limit $\nu \to 0$ it is worthwhile to find the contribution coming from the ghost loop and add it to the expression (3.75). Recalling the minus sign of the ghost loop, we obtain

$$\Pi_{\mu\nu}^{(gh)ab}(p) = g_s^2 \overline{\mu}^{4-d} f_{dca} f_{cdb} \int \frac{d^d k}{(2\pi)^d} \frac{(k+p)_\mu k_\nu}{k^2 (k+p)^2} =$$

$$= -g_s^2 \overline{\mu}^{4-d} f_{acd} f_{bdc} \int_0^1 dz \int \frac{d^d k}{(2\pi)^d} \frac{(k-pz)_\nu [k+p(1-z)]_\mu}{[k^2 + p^2 z(1-z)]^2} =$$

$$= -\frac{g_s^2 i \overline{\mu}^{4-d}}{(4\pi)^{d/2}} f_{acd} f_{bdc} \int_0^1 dz \left\{ \frac{g_{\mu\nu} \Gamma(1-d/2)}{2[p^2 z(1-z)]^{1-d/2}} - \frac{p_\mu p_\nu z(1-z)\Gamma(2-d/2)}{[p^2 z(1-z)]^{2-d/2}} \right\}. \tag{3.76}$$

The contribution caused by the diagram shown in Fig. 3.7(c) is proportional to

$$\int \frac{d^d k}{k^2}.$$

At the dimensional regularization in accordance with the formula (3.62) such an integral is equal to zero. For the same reason the contributions coming from the two other tadpole-diagrams (Figs. 3.7(e),(f)) vanish. This is the quite expected fact. Really, since the dependence on the momentum does not appear in the tadpole-diagram vertices, then these diagrams can give a correction to the gauge particle mass only. But owing to the gauge invariance the mass term is absent in the gluon field Lagrangian. Note, if we used any regularization procedure violating the gauge invariance, then these diagrams would lead to a nonzero mass term.

To sum up (3.75) and (3.76) results in

$$\Pi_{\mu\nu}^{(YM)ab}(p) = \Pi_{\mu\nu}^{(G)ab}(p) + \Pi_{\mu\nu}^{(gh)ab}(p) = \frac{g_s^2 i \overline{\mu}^{4-d}}{2(4\pi)^{d/2}} f_{acd} f_{bdc} \int_0^1 dz \left\{ \frac{(3d-2)g_{\mu\nu}\Gamma(1-d/2)}{[p^2 z(1-z)]^{1-d/2}} + \right.$$

$$\left. + \frac{\Gamma(2-d/2)}{[p^2 z(1-z)]^{2-d/2}} \left[g_{\mu\nu} p^2 (5 - 2z(1-z)) + p_\mu p_\nu (d - 6 - z(1-z)(4d-4)) \right] \right\}. \tag{3.77}$$

Reforming the first term with the help of the formula

$$x\Gamma(x) = \Gamma(x+1), \tag{3.78}$$

singling out pole terms ($\nu \to 0$) and doing the integration over z, we obtain the following expression for the purely Yang–Mills contribution to the vacuum polarization

$$\Pi_{\mu\nu}^{(YM)ab}(p) = -\frac{g_s^2 i}{16\pi^2} c_V (p_\mu p_\nu - g_{\mu\nu} p^2) \left[\frac{5}{3} \left(\frac{2}{\nu} - \gamma - \ln \frac{p^2}{4\pi\overline{\mu}^2} \right) + \frac{31}{9} \right] \delta_{ab}. \tag{3.79}$$

In Eq. (3.79) it has been taken into account that

$$f_{acd} f_{bdc} = -c_V \delta_{ab}, \tag{3.80}$$

where c_V is equal to N for the group $SU(N)$. The sum of (3.72) and (3.79) gives the final expression for the gluon field polarization tensor in the second order of the perturbation theory

$$\Pi_{\mu\nu}^{(2)ab}(p) = \frac{i g_s^2}{16\pi^2} (g_{\mu\nu} p^2 - p_\mu p_\nu) \left(\frac{5}{3} c_V - \frac{2N_f}{3} \right) \left(\frac{2}{\nu} - \gamma - \ln \frac{p^2}{4\pi\overline{\mu}^2} \right) \delta_{ab}. \tag{3.81}$$

Given a gauge with an arbitrary longitudinal part ξ, the additional term appears in the expression (3.81)

$$\frac{5}{3} c_V \to \left(\frac{13}{6} - \frac{\xi}{2} \right) c_V. \tag{3.82}$$

The fact, that the gluon self-energy depends on the gauge choice, does not contradict the general theorem to read the S-matrix elements to be independent on ξ. In a gauge theory the total set of one-loop corrections to the S-matrix elements always involves the collection of radiative corrections to vertices and propagators, dependence on the gauge being cancelled between these various terms in a complicated manner.

An interesting property of the relations (3.72) and (3.79) consists in that the contribution of purely Yang–Mills terms and the one of quarks into the quantity $\Pi_{\mu\nu}^{(2)ab}(p)$ are of different signs. As a result, at a not-too-large fermions number, the vacuum polarization tensor has an opposite sign as compared with the QED case.

Considering Eq. (3.47), we write the expression for the gluon Green function in the transverse gauge

$$\Gamma_{ab\mu\nu}^{(2,0)}(p) = i\left[\frac{-g_{\mu\nu} + p_\mu p_\nu/p^2}{p^2}\right]\left[1 + \frac{g_s^2}{16\pi^2}\left(\frac{13}{6}c_V - \frac{2N_f}{3}\right)\left(\frac{2}{\nu} - \gamma - \ln\frac{p^2}{4\pi\overline{\mu}^2}\right)\right]\delta_{ab}. \tag{3.83}$$

Inasmuch as the parameter ξ equals zero in this gauge, then the problem of its renormalization does not appear.

In QCD there is a particular arbitrariness of the renormalization scheme choice. The existence of the renormalization invariance presents a reflection of this circumstance. A minimal substraction scheme [22, 23] denoted as MS is one of the more obvious renormalization schemes. At the dimensional regularization, while divergences appear as poles of expressions for $d \to 4$, it is especially convenient. The MS scheme consists of discarding the poles on the parameter $2/\nu$, that is, in supplementing the Lagrangian with counter-terms to cancel infinities. Emphasize that counter-terms have no finite parts. Since counter-terms do not depend both on an arbitrary parameter κ and on a mass particle m, the MS scheme is also called the *massless one*, that is, this renormalization procedure is independent on a mass. Truly, as counter-terms have no finite parts, they involve only that *bare skeletal* structure that is needed to cancel divergences at small distances, but there, given infinity momenta, all masses may be neglected (if amplitudes have good behavior at $p \to \infty$). As it follows from Eq. (3.72) and (3.79), the poles $2/\nu$ always appear in the combination

$$N_\nu = \frac{2}{\nu} - \gamma + \ln 4\pi.$$

Therefore, if one discard the term $2/\nu$ only, then the transcendental terms $\gamma, \ln 4\pi$ would be left. It is quite natural to subtract these quantities as well. This demand results in the modified minimal subtraction scheme ($\overline{\text{MS}}$-scheme) in which the factor N_ν is fully excluded. The renormalization scheme $\overline{\text{MS}}$ involves one-loop corrections in a particularly simple form. However, one should pay for this simplicity. After regularization, an arbitrary parameter $\overline{\mu}$ having the mass dimension appears in all expressions. So the task is to banish it from observable quantities. For the Green functions independent on the parameter $\overline{\mu}$ to be obtained, we introduce a parameter $\kappa^2 > 0$ (normalization constant)

$$\ln\frac{p^2}{\overline{\mu}^2} = \ln\frac{p^2}{\kappa^2} + \ln\frac{\kappa^2}{\overline{\mu}^2}.$$

Further, including the term $\ln(\kappa^2/\overline{\mu}^2)$ in the divergent factor $Z_3(\nu)$, we rewrite $\Gamma_{ab\mu\nu}^{(2,0)}$ in the form

$$\Gamma_{ab\mu\nu}^{(2,0)}(p) = Z_3(\nu)\Gamma_{\mu\nu}^{(2,0)R}(p)\delta_{ab}, \tag{3.84}$$

where

$$Z_3(\nu) = 1 + \frac{g_s^2}{16\pi^2}\left[\frac{13}{6}c_V - \frac{2N_f}{3}\right]N_\nu^{\overline{\mu}}, \tag{3.85}$$

$$\Gamma_{\mu\nu}^{(2,0)R}(p) = i\left[\frac{-g_{\mu\nu} + p_\mu p_\nu/p^2}{p^2}\right]\left\{1 - \frac{g_s^2}{16\pi^2}\left[\frac{13}{6}c_V - \frac{2N_f}{3}\right]\ln\frac{p^2}{\kappa^2}\right\}, \tag{3.86}$$

$$N_\nu^{\bar\mu} = \frac{2}{\nu} - \gamma + \ln 4\pi + \ln \frac{\bar\mu^2}{\kappa^2}. \qquad (3.87)$$

Apparently $Z_3(\nu)$ possesses the pole divergence on the dimension $\nu = 4 - d$. It corresponds to the logarithmic divergence on the momentum L

$$N_\nu^{\bar\mu} \to \ln\left(\frac{L^2}{p^2}\right) \qquad (3.88)$$

resulting in the momentum cut-off in the integral over q. The renormalization constant choice in the form $\kappa^2 = Q^2$ ensures the natural result

$$\Gamma_{\mu\nu}^{(2,0)R}(p)\Big|_{p^2=\kappa^2} = i\left[\frac{-g_{\mu\nu} + p_\mu p_\nu/p^2}{p^2}\right]. \qquad (3.89)$$

It is clear that the divergent part associated with the gluon self-energy may be removed from the theory by means of introducing the following counter-term into the initial Lagrangian:

$$\delta\mathcal{L}_{G_s} = (Z_3 - 1)\left[\frac{1}{2}\left(\partial^\mu G_a^\nu - \partial^\nu G_a^\mu\right)^2\right], \qquad (3.90)$$

where in the case of an arbitrary gauge Z_3 is given as

$$Z_3 = 1 + \frac{g_s^2}{16\pi^2}\left[\left(\frac{13}{6} - \frac{\xi}{2}\right)c_V - \frac{2N_f}{3}\right]N_\nu^{\bar\mu}, \qquad (3.91)$$

and the argument ν is omitted. There is no need to introduce the counter-terms of the kind $(G_a^\mu)^2$ and $(\partial_\mu G_a^\mu)^2$ into the Lagrangian because of transversality of the gluon field polarization tensor.

Now we proceed to the calculation of the two-point quark Green function. In the momentum space we have

$$\Gamma^{(0,2)}(p) - iS(p) - S(p)\Sigma(p)S(p), \qquad (3.92)$$

where $\Sigma(p)$ is an operator of the quark self-energy (mass operator of a quark). In the one-loop approximation the quark self-energy diagram is represented in Fig. 3.8. In the

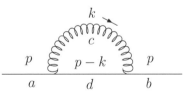

FIGURE 3.8
The diagram of the quark self-energy.

Feynman gauge $\Sigma(p)$ is defined by the expression

$$\Sigma_{ab}(p) = -g_s^2 \int \frac{d^4k}{(2\pi)^4} \gamma_\mu \frac{1}{\hat{p} - \hat{k} - m} \gamma_\nu \frac{g^{\mu\nu}}{k^2}\left(\frac{\lambda_c}{2}\right)_{ad}\left(\frac{\lambda_c}{2}\right)_{db}. \qquad (3.93)$$

It is evident that it is equal to the quantity $(\lambda_c)_{ad}(\lambda_c)_{db}/4$ multiplied by the corresponding expression for the fermion self-energy to be taken from QED:

$$\Sigma_{ab}(p) = \frac{1}{4}(\lambda_c\lambda_c)_{ab}\Sigma(QED). \qquad (3.94)$$

Making use of the expression for Σ found in Chapter 21.3 of *Advanced Particle Physics Volume 1* (see Eq. (21.74)) and neglecting the quark mass, we arrive at

$$\Sigma(QED) = \frac{ig_s^2}{16\pi^2}\hat{p}\left(\frac{2}{\nu} - \gamma - \ln\frac{p^2}{4\pi\overline{\mu}^2}\right), \tag{3.95}$$

where we have taken into account Eq. (3.88). Find the group-theoretical factor $c_F = (\lambda_c)_{ad}(\lambda_c)_{db}/4$ (summation over c is meant). We have

$$\frac{1}{4}(\lambda_c\lambda_c)_{ab} = \frac{4}{3}\delta_{ab}$$

for the $SU(3)$-group and

$$c_F = \frac{N^2 - 1}{2N}\delta_{ab} \tag{3.96}$$

for the $SU(N)$-group. So, the quark mass operator will look like:

$$\Sigma_{ab}(p) = \frac{ig_s^2}{16\pi^2}c_F\hat{p}\left(\frac{2}{\nu} - \gamma - \ln\frac{p^2}{4\pi\overline{\mu}^2}\right)\delta_{ab}. \tag{3.97}$$

As calculations show, in the case of an arbitrary gauge the expression (3.97) should be multiplied by ξ. Now we can write the two-point quark Green function. Given an arbitrary gauge, it takes the form:

$$\Gamma^{(0,2)}(p) = i\frac{\hat{p}}{p^2}\left[1 - \xi c_F\frac{g_s^2}{16\pi^2}\left(\frac{2}{\nu} - \gamma - \ln\frac{p^2}{4\pi\overline{\mu}^2}\right)\right]\delta_{ab}. \tag{3.98}$$

Hence, the renormalized quark propagator is

$$\Gamma^{(0,2)R}(p) = i\frac{\hat{p}}{p^2}\left(1 + \xi c_F\frac{g_s^2}{16\pi^2}\ln\frac{p^2}{\kappa^2}\right)\delta_{ab}. \tag{3.99}$$

And, at the same time, for the appeared divergences to be compensated introducing the counter-term

$$\delta\mathcal{L}_{\overline{q}q}^0 = (Z_2 - 1)i\overline{q}(x)\partial_\mu\gamma^\mu q(x), \tag{3.100}$$

where

$$Z_2 = 1 - \xi c_F\frac{g_s^2}{16\pi^2}N_\nu^{\overline{\mu}} \tag{3.101}$$

into the Lagrangian is needed. Taking into account the quark mass leads to the replacement

$$\delta\mathcal{L}_{\overline{q}q}^0 \to \delta\mathcal{L}_{\overline{q}q} = \delta\mathcal{L}_{\overline{q}q}^0 - (Z_m - 1)m\overline{q}(x)q(x), \tag{3.102}$$

where

$$Z_m = 1 - \frac{c_F g_s^2}{16\pi^2}(3 + \xi)N_\nu^{\overline{\mu}}. \tag{3.103}$$

Proceed to the calculation of the three-point Green function, the quark-gluon vertex function $\Gamma_\mu^{(1,2)}(p, p'; q)$. If one neglects the tadpole diagrams, then the one-loop correction to this quantity consists of two diagrams displayed in Fig. 3.9. The contribution of the diagram shown in Fig. 3.9(a) into the vertex function is equal to

$$\Lambda_\mu(p, p', q)(1) = -ig_s^2\int\frac{d^4k}{(2\pi)^4}\gamma_\nu\frac{1}{\hat{p}' - \hat{k} - m}\gamma_\mu\frac{1}{\hat{p} - \hat{k} - m}\frac{g^{\nu\rho}}{k^2}\frac{1}{8}(\lambda_d)_{cj}(\lambda_a)_{ji}(\lambda_d)_{ib}. \tag{3.104}$$

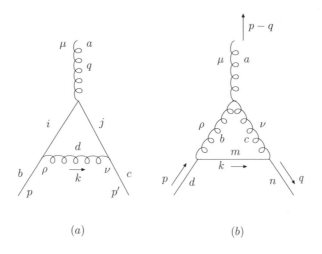

FIGURE 3.9

The corrections (a) and (b) to the quark-gluon vertex in the one-loop approximation in (a) and (b).

Comparing the obtained expression with the corresponding one of QED (see Eq. (21.110) of *Advanced Particle Physics Volume 1*), we see that they are distinguished by the group-theoretical factor only, that is

$$\Lambda_\mu(p, p', q)(1) = \frac{1}{8}(\lambda_d \lambda_a \lambda_d)\Lambda_\mu(p, p', q)(QED). \tag{3.105}$$

This factor is calculated with no trouble

$$\frac{1}{8}(\lambda_d \lambda_a \lambda_d) = \frac{1}{8}\left(\lambda_d[\lambda_a, \lambda_d] + \lambda_d \lambda_d \lambda_a\right) = \left(\frac{i}{4}f_{adc}\lambda_d\lambda_c + \frac{1}{2}c_F\lambda_a\right) -$$

$$= -\frac{1}{4}f_{adc}f_{dcb}\lambda_b + \frac{1}{2}c_F\lambda_a = \left(-\frac{1}{2}c_V + c_F\right)\frac{\lambda_a}{2}. \tag{3.106}$$

With allowance made for $\Lambda_\mu(p, p', q)(QED)$ (see, Eq. (21.130) of *Advanced Particle Physics Volume 1*) and Eq. (3.106), we arrive at the following result

$$\Lambda_\mu(p, p', q)(1) = \frac{g_s^2}{16\pi^2}\left(-\frac{1}{2}c_V + c_F\right)\left(\frac{2}{\nu} - \gamma - \ln\frac{p^2}{4\pi\overline{\mu}^2}\right)\gamma_\mu\frac{\lambda_a}{2}. \tag{3.107}$$

Now we shall find the contribution of the vertex shown in Fig. 3.9(b). After the dimensional regularization, it becomes the view

$$\Lambda_\mu(p, p', q)(2) = g_s^2(\overline{\mu}^{4-d})\int\frac{d^d k}{(2\pi)^d}\gamma_\rho\left(\frac{\lambda_b}{2}\right)_{dm}\frac{1}{(k-p)^2}f_{abc}[(2p - q - k)_\nu g_{\mu\rho}+$$

$$+(2k - p - q)_\mu g_{\rho\nu} + (2q - k - p)_\rho g_{\nu\mu}]\frac{1}{(q-k)^2}\frac{1}{\hat{k} - m}\gamma_\nu\left(\frac{\lambda_c}{2}\right)_{mn} \tag{3.108}$$

or

$$\Lambda_\mu(p, p', q)(2) = \frac{g_s^2\overline{\mu}^{4-d}}{(2\pi)^d}\frac{1}{4}f_{abc}\lambda_b\lambda_c J_\mu, \tag{3.109}$$

where

$$J_\mu = \int d^d k \frac{\gamma_\rho(\hat{k} + m)\gamma_\nu[(2p - q - k)_\nu g_{\mu\rho} + (2k - p - q)_\mu g_{\rho\nu} + (2q - k - p)_\rho g_{\nu\mu}]}{(k - p)^2(k^2 - m^2)(q - k)^2}.$$

(3.110)

To compute J_μ it is convenient to use the two-parameter Feynman formula

$$\frac{1}{abc} = 2\int_0^1 dx \int_0^{1-x} dy \frac{1}{[a(1 - x - y) + bx + cy]^3}.$$

(3.111)

Using Eq. (3.111) and carrying out the substitution $k' = k - px - qy$, we rewrite J_μ in the following form

$$J_\mu = 2\int_0^1 dx \int_0^{1-x} dy \int d^d k' \gamma^\rho(\hat{k}' + \hat{p}x + \hat{q}y + m)\gamma^\nu \{[(2 - x)p - (1 + y)q - k']_\nu g_{\mu\rho}+$$

$$+[2k' + (2x - 1)p + (2y - 1)q]_\mu g_{\rho\nu} + [(2 - y)q - (1 + x)p - k']_\rho g_{\nu\mu}\}\times$$

$$\times \frac{1}{[k'^2 - m^2(1 - x - y) + p^2 x + q^2 y - (px + qy)^2]^3}.$$

(3.112)

When integrating, the terms that are linear in k' give zero. Under $d \to 4$ the terms not containing k' in the numerator are finite and they may be neglected. The divergent part appears because of the term in the numerator that is quadratic in k'

$$Q_\mu = \gamma^\rho \hat{k}' \gamma^\nu(-k'_\nu g_{\mu\rho} + 2k'_\mu g_{\rho\nu} - k'_\rho g_{\nu\mu}) = -2\gamma_\mu k'^2 + (4 - 2d)k'_\mu \hat{k}'.$$

(3.113)

Therefore, the divergences are involved in the term

$$J'_\mu = 2\int_0^1 dx \int_0^{1-x} dy \int d^d k \frac{(4 - 2d)k_\mu \hat{k} - 2\gamma_\mu k^2}{[k^2 - m^2(1 - x - y) + p^2 x + q^2 y - (px + qy)^2]^3}.$$

(3.114)

Using Eq. (3.63) and turning $\nu \to 0$, we get the following expression for the pole part of J'_μ

$$J^d_\mu = -3i\pi^2 \gamma_\mu \left(\frac{2}{\nu} - \gamma - \ln\frac{p^2}{4\pi\overline{\mu}^2}\right).$$

(3.115)

Calculating the group-theoretical factor results in

$$\frac{1}{4}f_{abc}\lambda_b\lambda_c = \frac{i}{8}f_{abc}f_{bcd}\lambda_d = \frac{i}{2}c_V\frac{\lambda_a}{2}.$$

(3.116)

Consequently, the divergent part of $\Lambda_\mu(p, p', q)(2)$ should then be

$$\Lambda^d_\mu(p, p', q)(2) = \frac{3g_s^2}{32\pi^2}c_V\frac{\lambda_a}{2}\left(\frac{2}{\nu} - \gamma - \ln\frac{p^2}{4\pi\overline{\mu}^2}\right).$$

(3.117)

To sum the contributions coming from the vertices (3.107) and (3.117) gives the following expression for the divergent part of the three-point Green function

$$\Lambda^d_\mu(p) = \frac{g_s^2}{16\pi^2}[c_V + c_F]\gamma_\mu\frac{\lambda_a}{2}\left(\frac{2}{\nu} - \gamma - \ln\frac{p^2}{4\pi\overline{\mu}^2}\right).$$

(3.118)

In the case of an arbitrary gauge, the expression (3.118) is replaced by

$$\Lambda^d_\mu(p) = \frac{g_s^2}{16\pi^2}\left[\left(1 - \frac{1-\xi}{4}\right)c_V + \xi c_F\right]\left(\frac{2}{\nu} - \gamma - \ln\frac{p^2}{4\pi\overline{\mu}^2}\right).$$

(3.119)

Further, allowing for the definition of the three-point quark-gluon Green function in the one-loop approximation

$$\Gamma^{(1,2)}_{a\mu}(p) = -ig_s\gamma_\mu\frac{\lambda_a}{2} + \Lambda_{a\mu}, \tag{3.120}$$

we get the renormalized function $\Gamma^{(1,2)R}_{a\mu}(p)$ in the view

$$\Gamma^{(1,2)R}_{a\mu}(p) = Z_{1F}\Gamma^{(1,2)}_{a\mu}(p), \tag{3.121}$$

where

$$Z_{1F} = 1 - \frac{g_s^2}{16\pi^2}\left[\left(1 - \frac{1-\xi}{4}\right)c_V + \xi c_F\right]N_\nu^{\bar\mu}, \tag{3.122}$$

$$\Gamma^{(1,2)R}(p) = -ig_s\gamma_\mu\frac{\lambda_a}{2}\left\{1 + \frac{g_s^2}{16\pi^2}\left[\left(1 - \frac{1-\xi}{4}\right)c_V + \xi c_F\right]\ln\frac{p^2}{\kappa^2}\right\}. \tag{3.123}$$

For this divergence to be compensated, the counter-term

$$\delta\mathcal{L}_{\bar qGq} = (Z_{1F} - 1)\left(-ig_s\bar q(x)G_a^\mu(x)\gamma_\mu\frac{\lambda_a}{2}q(x)\right) \tag{3.124}$$

is needed. By analogy radiative corrections of the order g_s^2 to other propagators and vertices can be calculated. The corrections to the ghost propagator is defined by the diagram pictured in Fig. 3.10 (ghost self-energy). Simple calculations leads to the result

FIGURE 3.10
The Feynman diagram describing the ghost self-energy.

$$G^{(gh)}(p) = G_0^{(gh)}(p)\left[1 + \frac{g_s^2}{16\pi^2}c_V\left(\frac{3}{4} - \frac{\xi}{4}\right)\left(\frac{2}{\nu} - \gamma - \ln\frac{p^2}{4\pi\bar\mu^2}\right)\right], \tag{3.125}$$

where $G_0^{(gh)}(p)$ is the free ghost propagator. Singling out the divergent factor in Eq. (3.125), we obtain

$$G^{(gh)}(p) = \tilde Z_3 G^{(gh)R}(p), \tag{3.126}$$

where

$$\tilde Z_3 = 1 + \frac{g_s^2}{16\pi^2}c_V\left(\frac{3}{4} - \frac{\xi}{4}\right)N_\nu^{\bar\mu} \tag{3.127}$$

and the expression for the renormalized propagator is written in the form

$$G^{(gh)R}(p) = G_0^{(gh)}(p)\left[1 - \frac{g_s^2}{16\pi^2}c_V\left(\frac{3}{4} - \frac{\xi}{4}\right)\ln\frac{p^2}{\kappa^2}\right], \tag{3.128}$$

The Feynman rules are such that the momentum of a coming ghost line may be always factorized. As a consequence, the divergences connected with the ghost self-energy do not demand introducing any mass counter-term and we simply have

$$\delta\mathcal{L}_{\omega^\dagger\omega} = (\tilde Z_3 - 1)[-\omega_a^\dagger(x)\partial^2\omega_a(x)]. \tag{3.129}$$

The list of divergent irreducible diagrams is completed by two three-point Green functions: three-gluon and four-gluon vertex functions ($\Gamma^{(3,0)}(p)$ and $\Gamma^{(4,0)}(p)$). The radiative corrections to the vertex with three gluon lines are given by the diagrams set shown in Fig. 3.11. To compute these diagrams results in

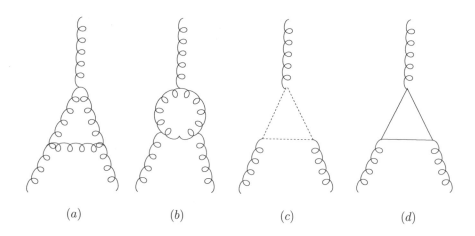

$$(a) \qquad\qquad (b) \qquad\qquad (c) \qquad\qquad (d)$$

FIGURE 3.11

The Feynman diagrams (a)–(d) describing the radiative corrections to the three-gluon vertex.

$$\Gamma^{(3,0)}(p) = \Gamma_0^{3,0}(p)\left\{1 - \frac{g_s^2}{16\pi^2}\left[\left(\frac{17}{12} - \frac{3\xi}{4}\right)c_V - \frac{4}{3}\frac{N_f}{2}\right]\left(\frac{2}{\nu} - \gamma - \ln\frac{p^2}{4\pi\overline{\mu}^2}\right)\right\}, \quad (3.130)$$

where $\Gamma_0^{3,0}(p)$ is the three-point Green function in the first order of g_s. As usual we obtain

$$\Gamma^{(3,0)}(p) = Z_1\Gamma^{(3,0)R}(p), \tag{3.131}$$

where

$$Z_1 = 1 + \frac{g_s^2}{16\pi^2}\left[\left(\frac{17}{12} - \frac{3\xi}{4}\right)c_V - \frac{4}{3}\frac{N_f}{2}\right]N_\nu^{\overline{\mu}}, \tag{3.132}$$

$$\Gamma^{(3,0)R}(p) = 1 - \frac{g_s^2}{16\pi^2}\left[\left(\frac{17}{12} - \frac{3\xi}{4}\right)c_V - \frac{4}{3}\frac{N_f}{2}\right]\ln\frac{p^2}{\kappa^2}. \tag{3.133}$$

The counter-term to compensate this divergence is

$$\delta\mathcal{L}_{3G} = (Z_1 - 1)\{-g_s[\partial^\mu G_a^\nu(x) - \partial^\nu G_a^\mu(x)]f_{abc}G_{b\mu}(x)G_{c\nu}(x)\}. \tag{3.134}$$

Finally, the counter-term caused by the existence of the four-point vertex functions (four last diagrams displayed in Fig. 3.1) is given by the expression

$$\delta\mathcal{L}_{4G} = (Z_4 - 1)\left[\frac{g_s^2}{2}f_{abc}G_b^\nu(x)G_c^\mu(x)f_{adn}G_{d\nu}(x)G_{n\mu}(x)\right], \tag{3.135}$$

where

$$Z_4 = 1 + \frac{g_s^2}{16\pi^2}\left[\left(\frac{2}{3} - \xi\right)c_V - \frac{4N_f}{3}\right]N_\nu^{\overline{\mu}}. \tag{3.136}$$

$$\Gamma^{(4,0)R}(p) = 1 - \frac{g_s^2}{16\pi^2}\left[\left(\frac{2}{3} - \xi\right)c_V - \frac{4N_f}{3}\right]\ln\frac{p^2}{\kappa^2}. \tag{3.137}$$

Thus, we have calculated all the divergent diagrams of QCD in the one-loop approximation. It turned out, therefore, that the theory contains only the logarithmic divergences while the naive counting of momenta power predicted both the linear and quadratic divergences. Such a cancellation of divergences takes place in QED too. In QCD, just as in QED, the reason of this phenomenon is the gauge invariance of the theory. To remove the logarithmic divergences the counter-terms were introduced into the initial Lagrangian. In so doing, the structure of the counter-terms and the monomials of the initial QCD Lagrangian is the same. However, this fact does not yet guarantee the renormalized and bare Lagrangians to have the same symmetry. Check whether it is the case or not by the example of a free gluon field. In this situation the renormalized Lagrangian will look like:

$$\mathcal{L} + \delta\mathcal{L} = -\frac{1}{4}Z_3(\partial_\mu G_{a\nu} - \partial_\nu G_{a\mu})(\partial^\mu G_a^\nu - \partial^\nu G_a^\mu) - \frac{1}{2\xi}(\partial_\mu G_a^\mu)^2 + g_s Z_1(\partial_\mu G_{a\nu} - \partial_\nu G_{a\mu})\times$$

$$\times(f_{abc}G_b^\mu G_c^\nu) + \frac{g_s^2}{2}Z_4 f_{abc}G_{b\mu}G_{c\nu}f_{adn}G_d^\mu G_n^\nu + \tilde{Z}_3\partial_\mu\omega_a^\dagger\partial^\mu\omega_a + g_s\tilde{Z}_1(f_{abc}\partial_\mu\omega_b^\dagger G_a^\mu\omega_c), \tag{3.138}$$

where in Z_1, Z_3, and Z_4 one should set N_f to be equal to zero. Define the bare (unrenormalized) fields and bare parameters in the following way:

$$\left.\begin{array}{ll} G_0(x) = Z_3^{1/2}G(x), & \omega_0(x) = \tilde{Z}_3^{1/2}\omega(x), \qquad \omega_0^\dagger(x) = \tilde{Z}_3^{1/2}\omega^\dagger(x), \\[2mm] g_{s0} = Z_1 Z_3^{-3/2}g_s, & \xi_0 = Z_3\xi. \end{array}\right\} \tag{3.139}$$

In doing so $\mathcal{L} + \delta\mathcal{L}$ may be considered as the initial Lagrangian $\mathcal{L}(G_0, \omega_0, \omega_0^\dagger; g_{s0}, \xi_0)$ expressed in terms of bare quantities provided the following identities are valid:

$$\frac{Z_4}{Z_1} = \frac{Z_1}{Z_3} = \frac{\tilde{Z}_1}{\tilde{Z}_3}. \tag{3.140}$$

At a single glance on Eqs. (3.91), (3.127), (3.132), and (3.136) it is evident that they are satisfied in the one-loop approximation. Note that the equality of the renormalization constants being at $(\partial_\mu G_{a\nu} - \partial_\nu G_{a\mu})^2$ and ξ is predicted by the Slavnov–Taylor identities. The expressions (3.140) are analogous to the expression $Z_1 = Z_2$ in QED. They reflect the fact that the renormalization of the coupling constants both in the vertices of the third and fourth orders in the field, and in the vertices $\omega^\dagger G\omega$ coincide. To put this another way, the interaction universality is conserved under the renormalization.

By switching on massless quarks into consideration, in place of Eq. (3.140) we shall have

$$\frac{Z_4}{Z_1} = \frac{Z_1}{Z_3} = \frac{\tilde{Z}_1}{\tilde{Z}_3} = \frac{Z_{1F}}{Z_2}. \tag{3.141}$$

Comparing the obtained expressions for the renormalization constants, we are easily convinced that the relations (3.141) are valid. In that way we have practically shown that the matrix elements are equally expressed in terms of unrenormalized and renormalized parameters of the theory. The proof of QCD renormalizability in the one-loop approximation is concluded by the establishment of this fact.

In conclusion we find the renormalized charge in QCD. Substituting the values of Z_3 and Z_1 (see Eqs. (3.91) and (3.132)) and being limited by the third order in g_s^2, we get

$$g_{s0} = Z_g g_s = Z_1 Z_3^{-3/2} g_s = \left[1 - \frac{g_{s0}^2}{16\pi^2}\left(\frac{11}{6}c_V - \frac{N_f}{3}\right)\right]\ln\left(\frac{L^2}{\kappa^2}\right)g_s, \tag{3.142}$$

where $N_\nu^{\bar\mu}$ has been replaced by

$$\ln\left(\frac{L^2}{\kappa^2}\right).$$

Note that Z_g does not depend on the gauge. As far as the cut-off momentum L is concerned, it can be interpreted as a limiting momentum to the point of which the theory is valid. This is equivalent to the statement that the theory is true up to distances $r_{\min} \sim 1/L$. With all this going on, $g_s = Z_g^{-1}g_{s0}$ may be considered as the expression for the effective charge with a precision of g_s^2 at the distance $r \sim 1/\kappa$:

$$g_s^2(r) = \frac{g_s^2(r_0)}{Z_g^{-2}} = \frac{g_s^2(r_0)}{\left[1 - \frac{g_s^2(r_0)}{16\pi^2}\left(\frac{11}{6}c_V - \frac{N_f}{3}\right)\ln\left(\frac{r^2}{r_0^2}\right)\right]}, \qquad (3.143)$$

where $g_s(r_0)$ is the initial charge at the distance r_0 and we have taken into account that $Z_g Z_g^{-1} \approx 1$. Practically the factor Z_g^{-1} plays the same role as a medium permittivity ϵ in macroscopic electrodynamics where an electric field density $\mathbf{E}$ is determined by an effective charge density ρ_{eff}:

$$\mathrm{div}\mathbf{E} = \rho_{eff}, \qquad \rho_{eff} = \frac{\rho_0}{\epsilon}, \qquad (3.144)$$

where ρ_0 is a bare charge density.

A significant distinction of a non-Abelian gauge theory from QED consists in the fact that in the former $g_s(r) > g_s(r_0)$, that is, the charge antishielding takes place, the antishielding effect being caused by the vector field self-interaction (the term $11c_V/6$ in Eq. (3.143)). It is clear that the antishielding occurs at non-too-large fermions number that gives the charge shielding (the term $-N_f/3$ in Eq. (3.143)) in exactly the same manner as for QED. The charge property described above leads to the asymptotical freedom in a non-Abelian theory. Later this phenomenon will be comprehensively examined on the base of the differential equations for the renormalization group we dealt with studying QED (see Section 22.6 of *Advanced Particle Physics Volume 1*).

4

Asymptotical freedom

> *Our flowers are merely—flowers,*
> *And the shadow of thy perfect bliss*
> *Is the sunshine of ours.*
> Edgar Allan Poe

4.1 Callan–Symanzik equation

At high energy the running coupling constant of QCD approaches to zero. Theories possessing such a property are called *asymptotically free*. First, this property of non-Abelian theories was ascertained in works by Gross and Wilczek [24, 25] and Politzer [26, 27] as well. It should be stressed, besides non-Abelian theories of gauge fields, none of the renormalizable field theory can be asymptotically free [2, 28]. Thus, the distinguishing feature of the Yang–Mills theories from all other field theories consists in the fact that the coupling constant falls with a decrease of distance. In the well-known QED the effective coupling constant decrease may be intuitively understood reasoning from the fact that, when increasing a distance between charges, a dielectric shielding of charges is strengthened by a cloud of virtual electron-positron pairs. Therefore, our task is to establish the reason of the origin of the shielding effect in non-Abelian theories.

At first we investigate Green function properties at large transferred momenta in a space-like region. These properties prove to be useful both in calculating radiative corrections to processes at high energies and in studying the effective coupling constant evolution. It is evident that in the large momenta region the mass terms in the Lagrangian may be omitted so as to take them into account as corrections on the following stages of calculations. This is why we are constrained by the case of the massless QCD.

In the previous section we found the renormalized Green functions in the one-loop approximation. It is worth noting that these functions must be defined by way of the renormalized coupling constant g_s rather than the initial one g_{s0}. Thus, the renormalized Green functions become dependent both on g_s and on the normalization point κ. In QED the normalization point is fixed by the condition that a photon has zero momentum and an electron is on the mass shell ($\hat{p} = m$). The advantage of such a scheme follows from Thirring's theorem [29] which reads: *at zero photon energy the Compton scattering amplitude in all orders of constant α is exactly given by the classical formula.* Hence, for the fundamental theory parameters α and m_e to be determined, classical expressions may be used. In QCD such a chosen scheme based on physical reasons is absent. Moreover, in QCD, as in any non-Abelian gauge theory, there are badly controllable infrared singularities and so the normalization point is taken under space-like momenta where vertices have no singularities. At that the numerical value of the color charge must be determined by comparing the theoretical and experimental values of the cross sections for the process, in which the vertex with external momenta at the normalization point enters as the subprocess. However, according to the contemporary conceptions free quarks and gluons must not exist. Therefore, we should discuss the problem concerning changes to appear in observables under variation of the normalization point.

Since the initial Lagrangian involves only one parameter, the initial coupling constant g_{s0}, there must exist a dependence between g_s and κ, which reveals oneself in the existence of the invariant charge

$$\overline{g}_s = \overline{g}_s(p^2/\kappa^2, g_s(\kappa)), \tag{4.1}$$

to remain constant at a variation of κ and $g_s(\kappa)$.

Derive a differential equation defining connection between κ and $g_s(\kappa)$. In accordance with the multiplicative renormalization program considered, we introduce renormalization constants $Z_G(\nu)$ and $Z_q(\nu)$ in such a way that the finite limit exists

$$\lim_{\nu \to 0} Z_G^{n_G}(\nu) Z_q^{n_q}(\nu) \Gamma^{(n_G, n_q)}(k_i; p_j, \nu).$$

This, in turn, allows us to write the renormalized finite Green function as

$$\Gamma^{(n_G, n_q)R}(k_i; p_j) = \lim_{\nu \to 0} Z_G^{n_G}(\nu) Z_q^{n_q}(\nu) \Gamma^{(n_G, n_q)}(k_i; p_j, \nu). \tag{4.2}$$

The functions Γ^R, along with the renormalized constants Z, must depend on the normalization point κ. Earlier we imposed the demand: *at $p_j^2 = -\kappa^2$ or $k_i^2 = -\kappa^2$ the renormalized Green functions turn to the corresponding expression of the tree approximation.* Thus, for example, the renormalization condition for the quark-gluon vertex defines the renormalized coupling constant

$$\Gamma^{(1,2)R}(p, p) \Big|_{p^2 = -\kappa^2} = \lim_{\nu \to 0} Z_G(\nu) Z_q^2(\nu) \Gamma^{(1,2)}(p, p, \nu) \Big|_{p^2 = -\kappa^2} = g_s \frac{\lambda_a}{2} \gamma_\mu. \tag{4.3}$$

Unrenormalized functions depend on unrenormalized (bare) coupling constant g_{s0}. Under the dimension regularization g_{s0} appears in the Green functions as $g_{s0} \overline{\mu}^{\nu/2}$. The condition (4.3) expresses the combination $g_{s0} \overline{\mu}^{\nu/2}$ in terms of the renormalized coupling constant g_s:

$$g_s = g_s(g_{s0} \overline{\mu}^{\nu/2}, \kappa). \tag{4.4}$$

Thus, according to (4.2) the renormalized Green functions prove to depend both on the renormalized charge g_s and on the renormalization point κ:

$$\Gamma^{(n_G, n_q)R} = \Gamma^{(n_G, n_q)R}(k_i; p_j, g_s, \kappa). \tag{4.5}$$

In Section 22.5 of *Advanced Particle Physics Volume 1* we have already dealt with the renormalization group in the QED case. Remember, that this equation is due to the fact that any observable must not depend on a choice of the normalization point κ. The reason of the renormalization invariance is evident: the initial Lagrangian and the unrenormalized Green function along with it, did not involve the parameter κ. Independence of $\Gamma^{(n_G, n_q)}(k_i; p_j, g_s, \kappa)$ on the normalization point κ could be written by means of the differential equation

$$\kappa \frac{d}{d\kappa} \Gamma^{(n_G, n_q)}(k_i; p_j, g_s, \kappa) = 0 \tag{4.6}$$

or

$$\left(\kappa \frac{\partial}{\partial \kappa} + \kappa \lim_{\nu \to 0} \frac{\partial g_s}{\partial \kappa} \frac{\partial}{\partial g_s} \right) Z_G^{-n_G} Z_q^{-n_q} \Gamma^{(n_G, n_q)R}(k_i; p_j, g_s, \kappa) = 0, \tag{4.7}$$

where Eq. (4.2) has been taken into account. When taking into consideration quark masses, in Eq. (4.7) one should include the term

$$\kappa \frac{\partial m}{\partial \kappa} \frac{\partial}{\partial m}.$$

Fulfilling the differentiation and multiplying the both side of Eq. (4.7) by $Z_G^{n_G} Z_q^{n_q}$, we arrive at

$$\left[\kappa \frac{\partial}{\partial \kappa} + \beta(g_s) \frac{\partial}{\partial g_s} + n_G \gamma_G(g_s) + n_q \gamma_q(g_s) \right] \Gamma^{(n_G, n_q)^R}(k_i; p_j, g_s, \kappa) = 0, \qquad (4.8)$$

where the following designations

$$\beta(g_s) = \lim_{\nu \to 0} \left[\kappa \frac{\partial}{\partial \kappa} g_s(g_{s0} \overline{\mu}^{\nu/2}, \nu, \kappa) \right], \qquad \gamma_i(g_s) = \lim_{\nu \to 0} \left[-\kappa \frac{1}{Z_i} \frac{\partial}{\partial \kappa} Z_i(g_{s0} \overline{\mu}^{\nu/2}, \nu, \kappa) \right] \quad (4.9)$$

have been introduced and $i = G, q$. The equation obtained is called the *renormalization group (RG) equation*. Its physical meaning is as follows. Any small change of the normalization point κ is accompanied by the corresponding changes of g_s and Z_i at which any physical quantity Γ^R remains invariable. Such transformations form a renormalization group. The existence of such a group was established in Ref. [30]. At first the RG method was employed under studying the asymptotic behavior of the Green functions in QED [31]. The RG role in quantum field theory was comprehensively investigated in works by Bogoliubov and Schirkov [32].

The RG equation describes amplitudes change under fixed momenta p_j, k_i when varying the normalization point κ. By definition the quantity κ determines the momenta scale in a theory. Therefore, Eq. (4.8) may be also used to describe an amplitudes change when moving along the momenta scale at fixed κ. Modify the momenta scale:

$$p_j \to \lambda p_j, \qquad k_i \to \lambda k_i. \qquad (4.10)$$

The function $\Gamma^{(n_G, n_q)R}$ has the mass dimension D_c, that is

$$[\Gamma^{(n_G, n_q)R}] = m^{D_c}.$$

The quantity D_c is given by the expression

$$D = 4 - n_G - \frac{3}{2} n_q \qquad (4.11)$$

and named the kinematic or canonical dimension of Green functions. Hence, from the consideration of the dimensions one may write

$$\Gamma^{(n_G, n_q)R}(\lambda p, g_s, \kappa) = \kappa^{D_c} f\left(\frac{\lambda p}{\kappa}, g_s \right).$$

Since the dependence on κ is either manifest (κ^D) or has the form $\lambda p / \kappa$, then one may switch from derivatives with respect to κ to those with respect to λ:

$$\kappa \frac{\partial}{\partial \kappa} \Gamma^{(n_G, n_q)R} = D_c \kappa^{D_c} f\left(\frac{\lambda p}{\kappa}, g_s \right) - \kappa^{D_c} \left[\lambda \frac{\partial}{\partial \lambda} f\left(\frac{\lambda p}{\kappa}, g_s \right) \right] =$$

$$= \left(-\lambda \frac{\partial}{\partial \lambda} + D_c \right) \Gamma^{(n_G, n_q)R}(\lambda p, g_s, \kappa). \qquad (4.12)$$

Substituting (4.12) into the RG equation, we get the Callan [33]–Symanzik [34] equation for the case of the massless QCD:

$$\left[\lambda\frac{\partial}{\partial\lambda} - \beta(g_s)\frac{\partial}{\partial g_s} - n_G\gamma_G(g_s) - n_q\gamma_q(g_s) - D_c\right]\Gamma^{(n_G,n_q)R}(\lambda k_i; \lambda p_j, g_s, \kappa) = 0. \quad (4.13)$$

This equation describes the Green functions change under the momentum scale transformations. In real experiments, for example, in deep inelastic scattering, the parameter λ is confronted with a momentum to take a large value. Note, if the relation

$$\beta(g_s) = \gamma_G(g_s) = \gamma_q(g_s) = 0$$

takes place, then the result of changing the momenta scale would be simply determined by the canonical dimension as was to be expected from the naive scale analysis. The necessity of considering the renormalizations, and consequently, nonzero functions $\beta(g_s), \gamma_G(g_s), \gamma_q(g_s)$ and deviation from a purely scaling behavior of Green functions is the result of switching on an interaction. Note, in particular, in the case of a massless theory the Lagrangian has the scale invariance, but the Green functions are not invariant because the functions $\beta(g_s)$ and $\gamma_i(g_s)$ are nonzero. They contribute to the so-called anomalous dimensions. In such a way, the reason for appearing anomalous dimensions lies deep. The renormalization inevitably introduces a scale either in the form of the mass κ at the dimensional regularization, or in the form of the momenta cut-off parameter L at the cut-off regularization. Therefore, even a scale-invariant classical theory does not lead to a scale-invariant quantum theory.

Now we try to solve Eq. (4.13). For a start, exclude terms not involving the derivatives. It is achieved by the transformation:

$$\Gamma^{(n_G,n_q)R}(\lambda k_i; \lambda p_j, g_s, \kappa) = \lambda^D \exp\left\{\int_0^{g_s} dg_s'\left[\frac{n_G\gamma_G(g_s') + n_q\gamma_q(g_s')}{\beta(g_s')}\right]\right\}\Gamma_0(\lambda k_i; \lambda p_j, g_s, \kappa).$$
$$(4.14)$$

Then

$$\left[\lambda\frac{\partial}{\partial\lambda} - \beta(g_s)\frac{\partial}{\partial g_s}\right]\Gamma_0(\lambda k_i; \lambda p_j, g_s, \kappa) = 0. \quad (4.15)$$

Let us pass on to the new variable $t = \ln\lambda$. To solve the equation

$$\left[\frac{\partial}{\partial t} - \beta(g_s)\frac{\partial}{\partial g_s}\right]\Gamma_0(e^t k_i; e^t p_j, g_s, \kappa) = 0, \quad (4.16)$$

we introduce the effective or running coupling constant $\bar{g}_s(t, g_s)$ depending on the momentum scale λ, which is the solution of the equation

$$\frac{\partial\bar{g}_s(t, g_s)}{\partial t} = \beta(\bar{g}_s) \quad (4.17)$$

with the boundary condition $\bar{g}_s(t = 0, g_s) = g_s$. Eq. (4.17) may be written in other form. To this end we integrate it over t:

$$t = \int_{g_s}^{\bar{g}_s(t,g_s)} \frac{dx}{\beta(x)}, \quad (4.18)$$

and then differentiate both sides with respect to g_s:

$$0 = \frac{1}{\beta(\bar{g}_s)}\frac{d\bar{g}_s}{dg_s} - \frac{1}{\beta(g_s)},$$

or

$$\left[\frac{\partial}{\partial t} - \beta(g_s)\frac{\partial}{\partial g_s}\right]\overline{g}_s(t, g_s) = 0. \tag{4.19}$$

Therefore, if the function Γ_0 satisfies (4.16), it depends on t and g_s as $\overline{g}_s(t, g_s)$. Then $\Gamma^{(n_G, n_q)R}(\lambda k_i; \lambda p_j, g_s, \kappa)$ is defined by the expression

$$\Gamma^{(n_G, n_q)R}(\lambda k_i; \lambda p_j, g_s, \kappa) = \lambda^{D_c}\exp\left\{\int_0^{g_s} dg'_s\left[\frac{n_G\gamma_G(g'_s) + n_q\gamma_q(g'_s)}{\beta(g'_s)}\right]\right\}\Gamma_0(k_i; p_j, \overline{g}_s(t, g_s), \kappa). \tag{4.20}$$

Transform the exponential in the equation obtained

$$\exp\left\{\int_0^{g_s} dg'_s\left[\frac{n_G\gamma_G(g'_s) + n_q\gamma_q(g'_s)}{\beta(g'_s)}\right]\right\} = \exp\left\{\int_0^{\overline{g}_s} dg'_s\left[\frac{n_G\gamma_G(g'_s) + n_q\gamma_q(g'_s)}{\beta(g'_s)}\right] +\right.$$

$$\left. + \int_{\overline{g}_s}^{g_s} dg'_s\left[\frac{n_G\gamma_G(g'_s) + n_q\gamma_q(g'_s)}{\beta(g'_s)}\right]\right\} =$$

$$= f(\overline{g}_s)\exp\left\{-\int_0^t dt'\left[n_G\gamma_G(\overline{g}_s(t', g_s)) + n_q\gamma_q(\overline{g}_s(t', g_s))\right]\right\}, \tag{4.21}$$

where

$$f(\overline{g}_s) = \exp\left\{\int_0^{\overline{g}_s} dg'_s\left[\frac{n_G\gamma_G(g'_s) + n_q\gamma_q(g'_s)}{\beta(g'_s)}\right]\right\}$$

and, under the variable replacement, we have allowed for Eq.(4.17) in the last line. So, we have

$$\Gamma^{(n_G, n_q)R}(\lambda k_i; \lambda p_j, g_s, \kappa) =$$

$$= \exp\left\{D_c t - \int_0^t dt'\left[n_G\gamma_G(\overline{g}_s(t', g_s)) + n_q\gamma_q(\overline{g}_s(t', g_s))\right]\right\}f(\overline{g}_s)\Gamma_0(k_i; p_j, \overline{g}_s(t, g_s), \kappa). \tag{4.22}$$

Setting $t = 0$ into (4.22), we arrive at the conclusion

$$f(\overline{g}_s)\Gamma_0(k_i; p_j, \overline{g}_s(t, g_s), \kappa) = \Gamma^{(n_G, n_q)R}(k_i; p_j, \overline{g}_s(\lambda, g_s), \kappa).$$

Then, the Callan–Symanzik equation solution will look like:

$$\Gamma^{(n_G, n_q)R}(\lambda k_i; \lambda p_j, g_s, \kappa) =$$

$$= \exp\left\{D_c t - \int_0^t dt'\left[n_G\gamma_G(\overline{g}_s(t', g_s)) + n_q\gamma_q(\overline{g}_s(t', g_s))\right]\right\}\Gamma^{(n_G, n_q)R}(k_i; p_j, \overline{g}_s(\lambda, g_s), \kappa). \tag{4.23}$$

The found solution describes the Green function transformation in changing the momenta scale. Evidently this transformation

$$\Gamma^{(n_G, n_q)R}(k_i; p_j, g_s, \kappa) \to \Gamma^{(n_G, n_q)R}(\lambda k_i; \lambda p_j, g_s, \kappa)$$

is attained at the expense of changing both the coupling constant

$$g_s \to \overline{g}_s(1, g_s) \to \overline{g}_s(\lambda, g_s)$$

and the multiplicative factor. Note that this factor is distinguished by the presence of the quantities γ_i from the factor λ^{D_c} caused by the canonical dimension only. Then it becomes evident why γ_i are called the *anomalous dimensions*. At the momenta scale transformation they are summed with the *ordinary dimension*. In other words, if one does not speak of momentum dependence of the effective charge $\overline{g}_s$, at large momenta the vertex $\Gamma^{(n_G,n_q)R}$ behaves as $p^{D_c+\gamma_G+\gamma_q}$, where the quantity $D_c+\gamma_G+\gamma_q$ is named a *dynamical dimension*. In that way, at large λ the behavior of scattering amplitudes is defined by the effective coupling constant and the anomalous dimensions being in the exponential of Eq. (4.23). This exponential taking into account all leading logarithms in all orders of g_s, is unambiguously determined by the quantities β and γ_i to be calculated in the one-loop approximation assuming that $\overline{g}_s < 1$.

In conclusion in this case, we address to Eq. (4.17). Rewrite it in terms of the variable λ

$$\lambda\frac{\partial\overline{g}_s(\lambda,g_s)}{\partial\lambda} = \beta(\overline{g}_s). \tag{4.24}$$

and examine some examples of the behavior of $\overline{g}_s(\lambda,g_s)$ when $\lambda \to \infty$, that is, at large momenta. In the general case the following situations are possible.

(i) The function $\beta(\overline{g}_s)$ is everywhere positive at $\overline{g}_s > 0$ (Fig. 4.1(a)). Then, with growing

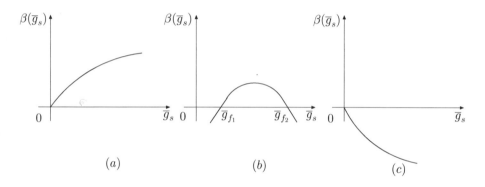

FIGURE 4.1
The plots of $\beta(\overline{g}_s)$ (a) and (b) versus $\overline{g}_s$ (c).

λ, $\overline{g}_s^2$ is increased beyond all bounds. This case is realized in QED, where the behavior of the fine structure constant is given by the formula

$$\overline{\alpha}(\lambda,\alpha) = \frac{\alpha}{1+b\alpha\ln\lambda^2},$$

with $b = -1/(3\pi)$ and $\alpha = \overline{\alpha}(0,\alpha)$.

(ii) The function $\beta(\overline{g}_s)$ has zero at some value of $\overline{g}_s = g_f > 0$. We agree to name zeros of the function $\beta(\overline{g}_s)$ as *fixed points*. If the coupling constant is situated in the fixed point at $\lambda = 1$, then it is left in this point at all momentum values. It is needed to distinguish two kinds of fixed points. Expand the right side of Eq. (4.24) in the vicinity of g_f

$$\lambda\frac{\partial(\overline{g}_s - g_f)}{\partial\lambda} \approx \beta'(g_f)(\overline{g}_s - g_f)\dots. \tag{4.25}$$

Integrating the obtained expression results in

$$\overline{g}_s \approx g_f + \lambda^{\beta'(g_f)}c_0 + \dots. \tag{4.26}$$

The behavior of $\overline{g}_s = \overline{g}_s(\lambda)$ in the vicinity of the fixed point g_f is governed by the derivative sign of the function β. From the solution (4.26) follows

$$\overline{g}_s(\lambda) \to g_f \quad \text{at} \quad \begin{cases} \lambda \to \infty, & \text{when} \quad \beta'(g_f) < 0; \\ \lambda \to 0, & \text{when} \quad \beta'(g_f) > 0. \end{cases} \tag{4.27}$$

In accordance with (4.27), g_f is called an *infrared-stable fixed point* when $\beta' > 0$ and an *ultraviolet-stable fixed point* when $\beta' < 0$.

(iii) If $\beta(\overline{g}_s)$ has two fixed points $0 < g_{f_1} < g_{f_2} < \infty$, then the coupling constant $\overline{g}_s$, being within the interval $[g_{f_1}, g_{f_2}]$ at some value of λ, will be left there at all other values of λ (Fig. 4.1(b)).

(iv) The function $\beta(\overline{g}_s)$ is negative at $\overline{g}_s > 0$ and $\beta(0) = 0$ (Fig. 4.1(c)). In this case with growing λ, the charge $\overline{g}_s$ tends to zero. It means that with increasing a momentum, the condition of the perturbation theory applicability is improved because the value of the coupling constant becomes less and less and $\overline{g}_s^2 = 0$ at $|p^2| \to \infty$. This phenomenon, the decrease of the coupling constant up to zero with growing a momenta (or with decreasing a distance), is called an *asymptotical freedom*. In QCD the coupling constant displays just this behavior. To the contrary, if $|p^2|$ is decreased, the coupling constant is increased, that is, with increasing distances, the interaction becomes more and more stronger. Such a behavior, in turn, is evidence of the confinement of quarks.

From the aforesaid becomes evident that finding the β function zeros represents a fundamental problem. It should be stressed, this problem is sufficiently complicated because its solution demands the way out beyond the scope of the perturbation theory. However, one may use the simplifying fact, namely, the function β has the trivial zero at the origin of coordinates $g_s = 0$. In this point the anomalous dimension turns to zero as well. What is more, in some cases the point $g_s = 0$ represent the special interest from the phenomenological point of view. If the point $g_s = 0$ is the infrared-stable fixed one, then the corresponding theory is asymptotical free. Thus, in order to define whether one or another theory possesses this property, one must calculate the β function and check whether the inequality $\beta(g_s) < 0$ is fulfilled at $g_s \geq 0$. To put it differently, we are interested in the β-function plot slope at the origin of coordinates because the sign of this slope defines whether the point g_s would be the ultraviolet or infrared fixed one. It is important to notice that at small values of the coupling constant we have the right to calculate the β-function using the perturbation theory.

4.2 Evolution of the effective coupling constant in QCD

To find out the dependence of the effective coupling constant $\overline{g}_s$ on a momentum we have to define the β function in the one-loop approximation and then solve the equation

$$\lambda \frac{\partial}{\partial \lambda} \overline{g}_s(\lambda, g_s) = \beta(\overline{g}_s). \tag{4.28}$$

In doing so, there are several possibilities. One may simply use the β function definition (4.9). There exists other way connected with using the Callan–Symanzik (CS) equation. To realize this idea one should write the CS equation for several Green functions whose manifest forms are known. Further, one should find the anomalous dimensions γ_G, γ_q from these equations to define β. Let us go in this way.

For auxiliary Green functions we take the gluon and quark propagators and the quark-gluon vertex as well, which have been calculated in the one-loop approximation (see Eqs. (3.88), (3.99) and (3.121)). Let us use the transversal gauge. Fulfill the scale transformation of momenta $p \to \lambda p$ into (25.88) and substitute the expression obtained into the CS equation (4.13), where D is defined by Eq. (4.11). From Eqs. (3.141) and (4.17) it is evident that β has the third order of g_s and, as a result, its contribution may be neglected. Having implemented all operations, we find the following expression for γ_G in the second order of g_s:

$$\gamma_G = \frac{g_s^2}{16\pi^2}\left(\frac{13}{6}c_V - \frac{2N_f}{3}\right) + O(g_s^4). \tag{4.29}$$

The analogous operations with $\Gamma^{(0,2)R}(p, g_s, \kappa)$ result in

$$\gamma_q = 0 + O(g_s^4). \tag{4.30}$$

Further, write the equation for $\Gamma^{(1,2)R}(p, g_s, \kappa)$:

$$\left\{\lambda\frac{\partial}{\partial\lambda} - \beta(g_s)\frac{\partial}{\partial g_s} + [1 - \gamma_G(g_s)] + 2\left[\frac{3}{2} - \gamma_q(g_s)\right] - 4\right\}\Gamma^{(1,2)R}(\lambda p, g_s, \kappa) = 0. \tag{4.31}$$

Substituting Eq. (3.121) into the obtained equation, we get with a precision of g_s^3

$$\lambda\frac{\partial}{\partial\lambda}\left[g_s\left(\frac{3}{4}c_V\frac{g_s^2}{16\pi^2}\ln\frac{\lambda^2 p^2}{\kappa^2}\right)\right] - \beta(g_s)\frac{\partial}{\partial g_s}g_s - [2\gamma_q(g_s) + \gamma_G(g_s)]g_s = 0. \tag{4.32}$$

From Eq. (4.32) immediately follows

$$\beta(g_s) = -\left[\gamma_G + 2\gamma_q + \frac{3}{2}c_V\frac{g_s^2}{16\pi^2}\right]g_s = -\frac{g_s^2}{16\pi^2}\left(\frac{11}{3}c_V - \frac{2N_f}{3}\right) + O(g_s^5), \tag{4.33}$$

where at the end we have taken into account the obtained values of γ_i (Eqs. (4.29) and (4.30)). So, in the main approximation ($\sim g_s^3$) the β function is given by the expression

$$\beta(g_s) = -\frac{b}{16\pi^2}g_s^3, \tag{4.34}$$

where $b = 11 - 2N_f/3$. The essential, that follows from this result, is the negative sign of the β function at a sufficiently small g_s and at $N_f < 16$. The negativity of β guarantees the asymptotical freedom of the theory. As it follows from calculations of one-loop corrections to the QCD Green functions, this phenomenon is entirely connected with a presence of three-gluon vertices resulting in the vacuum polarization by gluon pairs.

Carry out the replacement $g_s \to \overline{g}_s$ in (4.34) and substitute the expression obtained into Eq. (4.28):

$$\lambda\frac{\partial}{\partial\lambda}\overline{g}_s(\lambda, g_s) = -\frac{b}{16\pi^2}\overline{g}_s^3. \tag{4.35}$$

Integrating Eq. (4.35) and allowing for the initial condition at $\lambda = 1$

$$\overline{g}_s(1, g_s) = g_s,$$

we arrive at

$$\overline{g}_s^2(\lambda, g_s) = \frac{16\pi^2 g_s^2}{16\pi^2 + bg_s^2\ln\lambda^2}. \tag{4.36}$$

At positive b and $\lambda \to \infty$ from this formula it follows

$$\overline{g}_s^2(\lambda, g_s) \to 0. \tag{4.37}$$

Consequently, when increasing momenta, the effective QCD coupling constant is really decreased and turns into zero in the limit of infinitely large momenta. The origin of coordinates $g_s = 0$ represents the ultraviolet-stable fixed point.

The solution (4.36) may be rewritten in a more convenient form choosing the scale λ to be equal to

$$\lambda = \sqrt{\frac{Q^2}{\kappa^2}}, \qquad Q^2 = -p^2. \tag{4.38}$$

Such a choice is convenient when processes with large transferred momentum, as for instance, deep inelastic scattering of leptons by hadrons, are investigated. With allowance made for Eq. (4.38), we obtain for the running coupling constant $\overline{g}_s^2 = \overline{g}_s^2(Q^2)$

$$\overline{g}_s^2(Q^2) = \frac{16\pi^2 g_s^2}{16\pi^2 + b g_s^2 \ln(Q^2/\kappa^2)}, \tag{4.39}$$

where $g_s = \overline{g}_s(\kappa^2)$. Under the transition from momenta to distances, the equation obtained exactly coincides with Eq. (3.142).

Pass on from κ^2 to the new parameter

$$\Lambda_{QCD}^2 = \kappa^2 \exp\left(-\frac{16\pi^2}{b g_s^2}\right)$$

and, on the analogy of QED, introduce the fine structure constant of the strong interaction

$$\alpha_s(Q^2) \equiv \frac{\overline{g}_s^2(Q^2)}{4\pi}. \tag{4.40}$$

Then, instead of (4.39), we get

$$\alpha_s(Q^2) = \frac{12\pi}{(33 - 2N_f)\ln(Q^2/\Lambda_{QCD}^2)}. \tag{4.41}$$

The quantity Λ_{QCD}, the integration constant, to appear at the solution of the RG equation, is a single dimension free parameter of the theory. Its role consists in the fact that it fixes the momenta scale at which the interaction becomes strong and the perturbation theory ceases to be true.

With allowance made for three-loop diagrams, the RG equation for α_s takes the form

$$Q\frac{\partial \alpha_s}{\partial Q} = -\frac{\beta_0}{2\pi}\alpha_s^2 - \frac{\beta_1}{4\pi^2}\alpha_s^3 - \frac{\beta_2}{64\pi^3}\alpha_s^4, \tag{4.42}$$

where

$$\beta_0 = b = 11 - \frac{2}{3}N_f, \qquad \beta_1 = 51 - \frac{19}{3}N_f, \qquad \beta_2 = 2857 - \frac{5033}{9}N_f + \frac{325}{27}N_f^2. \tag{4.43}$$

Note that the values of the coefficients β_0 and β_1 do not depend on the renormalization scheme, while the expression for β_2 has been found for the scheme $\overline{\text{MS}}$. To solve Eq. (4.42) results in the following expression for the running coupling constant [35, 36]

$$\alpha_s(Q^2) = \frac{4\pi}{\beta_0 \ln(Q^2/\Lambda_{QCD}^2)}\left\{1 - \frac{2\beta_1}{\beta_0^2}\frac{\ln(\ln(Q^2/\Lambda_{QCD}^2))}{\ln(Q^2/\Lambda_{QCD}^2)} + \frac{4\beta_1^2}{\beta_0^4 \ln^2(Q^2/\Lambda_{QCD}^2)} \times \right.$$

$$\left. \times \left[\left(\ln(\ln(Q^2/\Lambda_{QCD}^2)) - \frac{1}{2}\right)^2 + \frac{\beta_2\beta_0}{8\beta_1^2} - \frac{5}{4}\right]\right\}. \tag{4.44}$$

It is well to bear in mind that N_f, entering into the formulae cited above, represents the flavors number of quarks whose masses are less than typical considered energies Q. Every one of the intervals between successive masses of quarks is associated both with the definite value of N_f and with the definite value of the integration constant Λ_{QCD} to be defined by the continuity condition over $g_s(Q)$. In particular, the experiments of deep inelastic electrons scattering, as a rule, are carried out under energies to exceed the masses of only four quarks (u, d, s, and c). On the other hand, experiments on electron-positron colliders are usually realized under energies that are greatly in excess of the fifth quark mass. It should be stressed that the parameter Λ_{QCD} assigned the scale of all hadron physics can be found from experiments only. It may be defined from different independent experiments, the comparison of obtained values Λ_{QCD} being the significant testing of QCD. At present, joint efforts of experimentalists and theorists lead to the conclusion that Λ_{QCD} lies in the vicinity of 200 MeV.

What are the physical reasons of the asymptotical freedom in QCD? Two different effects contribute to the vacuum polarization to cause the renormalization of the coupling constant g_s. The former consists in the fact that the bare charge g_{s0} polarizes a quark vacuum owing to the production of virtual quark-antiquark pairs. As a consequence, around g_{s0} a shielding cloud of virtual quarks with an opposite color charge is generated. This process is completely analogous to the corresponding shielding of the electric charge in QED. The latter appears thanks to a boson vacuum polarization (production of virtual gluons) as a result of which g_{s0} proves to be surrounded with the same sign charges. This effect is called an *antishielding*. Since the total color charge has to remain constant, with decreasing of the distance to a charge, its effective value tends to zero, that is, the asymptotical freedom appears.

Below we shall present considerations roughly estimating variation bounds of Λ_{QCD}. At

$$Q^2 \approx < r_p^2 >^{-1} \approx (0.8 \text{ Fermi})^{-2} \approx (0.3 \text{ GeV})^2,$$

where $< r_p^2 >$ is a mean square proton radius, the strong interaction indeed becomes strong. This value corresponds to the momentum value of a quark to be bound inside a hadron. On the other hand, under $Q^2 \sim 2 \text{ GeV}^2$, an experiment demonstrates the scale invariance in deep inelastic scattering of leptons by nucleons to a good approximation to form the inequality

$$\frac{\alpha_s(2 \text{ GeV}^2)}{\pi} \ll 1.$$

Then, using Eq. (4.41) gives the following bounds for the scale parameter

$$0.1 \text{ GeV} \leq \Lambda_{QCD} \leq 0.7 \text{ GeV}. \tag{4.45}$$

Examine an other example, namely, e^-e^+-annihilation into hadrons. The majority of hadrons appearing in a final state is defined as fragments of quark and antiquark to be produced in the subprocess

$$e^- + e^+ \to \gamma^* \to q + \bar{q}, \tag{4.46}$$

whose Feynman diagram is displayed in Fig. 4.2. Later we shall prove this when calculating the cross section of the higher order

$$e^- + e^+ \to q + \bar{q} + g. \tag{4.47}$$

The cross section of the electrodynamic subprocess (4.46) may be easily obtained from that of the process $e^-e^+ \to \tau^-\tau^+$:

$$\sigma_{e^-e^+ \to \tau^-\tau^+} = \frac{4\pi\alpha^2}{3s^2}. \tag{4.48}$$

FIGURE 4.2
The Feynman diagram describing the process $e^- + e^+ \to q + \overline{q}$.

Making use of Eq. (20.113) of *Advanced Particle Physics Volume 1* gives

$$\sigma_{e^- e^+ \to q\overline{q}} = 3e_{q_f}^2 \sigma_{e^- e^+ \to \tau^- \tau^+}, \tag{4.49}$$

where e_{q_f} is a charge of f quark in units of the electron charge. The additional factor 3 appears because we have a single diagram for every quark flavor and the corresponding cross sections must be added up. In Eq. (4.49) it was also assumed that the energy in the center-of-mass frame ($\sqrt{s}$) is so much higher that the quark masses may be neglected. For the production cross sections of all possible hadrons to be found, summation over all quark flavors must be fulfilled as well. So, in the second order of the electromagnetic coupling constant (in the leading order) the cross section of $e^- e^+$-annihilation into hadrons is given by the expression

$$\sigma_{e^- e^+ \to hadrons} = \frac{4\pi\alpha^2}{s^2} \sum_f e_{q_f}^2. \tag{4.50}$$

The basic corrections to $e^- e^+$-annihilation probability that are caused by gluon exchange and gluon emitting, are shown in Fig. 4.3. Calculations of these diagrams will be carried

FIGURE 4.3
The basic corrections to $e^- e^+$-annihilation.

out later and here we only present the final result

$$\sigma_{tot} = \frac{4\pi\alpha^2}{s^2} \sum_f e_{q_f}^2 \left[1 + \frac{3c_F}{4\pi}\alpha_s(s^2) + O(\alpha_s^2)\right], \tag{4.51}$$

where α_s has been defined in a time-like region. Now, using the experimental value of σ_{tot}, we can find $\alpha_s(s^2)$ to form

$$\Lambda_{QCD} \approx 0.7 \text{ GeV}.$$

Comparing the obtained value with that given by (4.45), we see that the analysis of deep inelastic scattering of leptons by hadrons results in smaller values of Λ_{QCD}.

It should be stressed that all considerations above are valid only in a space-like region. In massless non-Abelian gauge theory infrared singularities exist and, as a consequence, to subtract divergent quantities is fulfilled not at $p = 0$ but at a space-like point $p^2 =$

$-Q^2 < 0$ where amplitude singularities are absent. In this connection, strictly speaking, the statement about the asymptotical freedom is also correct only in a space-like region. A region of space-like momenta could be covered only by the analytical continuation of results under assumption of their *uncritical smoothness*.

Having known the momentum dependence of the effective coupling constant, we can easily compute the momentum dependence of scattering amplitudes that is predicted by the CS equation. In this solution, as it follows from Eq. (4.23), the nontrivial contribution is caused by the exponential of the anomalous dimensions

$$N_i = \exp \int_0^{\ln \lambda} d(\ln \lambda') n_i \gamma_i(\overline{g}_s(\lambda')). \tag{4.52}$$

Because

$$\frac{\partial \overline{g}_s}{d \ln \lambda} = \beta(\overline{g}_s),$$

then

$$N_i = \exp \int_{\lambda'=1}^{\lambda'=\lambda} \frac{d\overline{g}_s}{\beta(\overline{g}_s)} n_i \gamma_i(\overline{g}_s). \tag{4.53}$$

Further, substitute the earlier obtained values of β and γ_i into (4.53)

$$\beta(\overline{g}_s) = -b\frac{\overline{g}_s^3}{16\pi^2}, \qquad n_G \gamma_G(\overline{g}_s) = 2c\frac{\overline{g}_s^2}{16\pi^2}, \tag{4.54}$$

where $b = 11 - 2N_f/3$, and, in the transverse gauge, c and γ_q are

$$c = \left(\frac{13c_V}{6} - \frac{2N_f}{3}\right), \qquad \gamma_q = 0.$$

The result is as follows

$$N_G = \exp\left[-\frac{2c}{b}\int_{\lambda'=1}^{\lambda'=\lambda} \frac{d\overline{g}_s}{\overline{g}_s}\right] = \exp\left\{-\frac{2c}{b}[\ln \overline{g}_s(\lambda) - \ln g_s]\right\} = \left[\frac{g_s}{\overline{g}_s(\lambda)}\right]^{2c/b}. \tag{4.55}$$

So, the CS equation solution will look like:

$$\Gamma^R(\lambda p, g_s) = \lambda^D \left[\frac{g_s}{\overline{g}_s(\lambda)}\right]^{2c/b} \Gamma^R(p, \overline{g}_s(\lambda)). \tag{4.56}$$

On the other hand, from Eq. (4.42) follows

$$\left[\frac{g_s}{\overline{g}_s(\lambda)}\right]^{2c/b} \sim \left[\ln \frac{Q^2}{\Lambda_{QCD}^2}\right]^{2c/b}. \tag{4.57}$$

Therefore, at the scale transformations the Green functions are varied as momenta logarithm functions. Reasoning from this fact, we can establish the momentum dependence of an amplitude for any QCD process.

Up to now, examining the CS equation, we disregarded the quark masses m_{q_i} and the gauge parameter ξ. However, from the viewpoint of the Lagrangian formalism, it is convenient sometimes to consider m_{q_i} and ξ as some coupling constants to be analogous to α_s. Let us take a look what modifications of the theory will be caused by the inclusion of these corrections.

Since the quark masses are small, they may be neglected at high energies. However, it is not always right even for the light quarks u, d, and s. In the case of the heavy quarks c, b, and t their masses could lead to significant effects. Alternatively, as quarks have not been observed in free states up to the present, precise values of their masses are not known. In what follows, we shall consider the quark mass as a parameter of the QCD perturbation theory that is defined by the renormalization condition at some renormalization scale μ. As long as the quark mass is determined by the same way as the coupling constant, one should await this parameter to be changed in accordance with the RG evolution. And so different values of the mass parameter are needed for different processes. This recipe leads to an effective quark mass depending on a momentum scale at which this mass is calculated. In a similar manner, the gauge parameter also displays dependence on a transferred momentum. Hence, both quark masses and gauge parameter represent running parameters of the theory. Their inclusion leads to the appearance of the following terms in Eq. (4.8):

$$\sum_i m_{q_i} \gamma_{m_{q_i}}(\alpha_s)\frac{\partial}{\partial m_{q_i}} + \delta(\alpha_s)\frac{\partial}{\partial \xi}, \tag{4.58}$$

where

$$\gamma_{m_{q_i}}(\alpha_s) = -\frac{1}{Z_{m_{q_i}}(\kappa)}\kappa\frac{\partial Z_{m_{q_i}}(\kappa)}{\partial \kappa}, \qquad \delta(\alpha_s) = -\frac{1}{Z_3(\kappa)}\kappa\frac{\partial Z_3(\kappa)}{\partial \kappa}.$$

To solve the obtained equation, one introduces effective or running parameters

$$Q^2\frac{\partial m(Q^2)}{\partial Q^2} = \gamma_{m_{q_i}}(\alpha_s)m(Q^2), \qquad Q^2\frac{\partial \zeta(Q^2)}{\partial Q^2} = \delta(\alpha_s)\xi(Q^2). \tag{4.59}$$

In the MS-renormalized scheme the running (effective) mass is defined by the expression [??]

$$m(Q^2) = \frac{m_0}{[\frac{1}{2}\ln(Q^2/\Lambda_{QCD}^2)]^{d_m}}, \tag{4.60}$$

where

$$d_m - \frac{12}{33 - 2N_f}$$

and m_0 is an integration constant. In the two-loop approximation the expression for the effective mass $m(Q^2)$ was obtained in Ref. [37]. When calculating mass corrections to

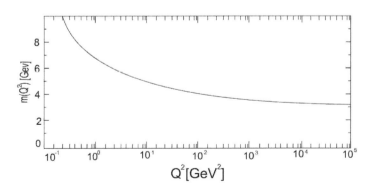

FIGURE 4.4
Running of the b-quark mass.

quark production processes or scattering processes with the transferred momentum Q, the effective mass $m(Q^2)$ must be used. In Fig. 4.4 the dependence of the effective b-quark mass on the square-transferred momentum is shown. Analogous calculations lead to the following expression for the running gauge parameter [38]

$$\xi(Q^2) = \frac{\xi_0}{[\frac{1}{2}\ln(Q^2/\Lambda_{QCD}^2)]^{d_\xi}} \left\{ 1 + \frac{9}{(39 - 4N_f)} \frac{\xi_0}{[\ln(Q^2/\Lambda_{QCD}^2)]^{d_\xi}} \right\}^{-1}, \qquad (4.61)$$

where

$$d_\xi = \frac{39 - 4N_f}{2(33 - 2N_f)}.$$

5

Chiral symmetries

Life is tricky with enchanting pathos.
That is why it is so powerful, and
It composes its pernicious letters
With its outrageous, rugged hand.
S. Esenin

5.1 Currents algebra

If one neglects quark masses, then fields of the u, d, s quarks would not be distinguished and, as a consequence, the QCD Lagrangian would be invariant with respect to rotations in the quark flavors space. And at the same time, owing to the vector character of interaction between quarks and gluons, one may independently rotate left and right constituents of quark fields (q_L and q_R). Transformations of this kind are described by eight independent parameters ξ_L^a for the left-handed particles and eight ones ξ_R^a for the right-handed particles:

$$q_{L,R} \rightarrow \left(1 + i \sum_a \xi_{L,R}^a \lambda_a\right) q_{L,R}, \qquad (5.1)$$

where λ_a operates in the flavor space of u, d, and s quarks. If $\xi_R^a = \xi_L^a$, the transformations (5.1) conserve the parity. Invariance with respect to these transformations takes place when the quark masses are nonzero but are equal to each other (originally just this possibility has been discussed). At current data on the masses of the u, d, and s quarks, there are no reasons to believe the approximation of equal quark masses to be better than that of zero quark masses. In the latter case the Lagrangian is also invariant with respect to the transformations with $\xi_R^a = -\xi_L^a$ that do not conserve the parity (such transformations will be called *chiral ones*). From the mathematical point of view, the invariance with respect to the transformations (5.1) denotes the chiral $SU(3)_L \times SU(3)_R$-symmetry of the strong interaction Lagrangian. Note, that the approximate $SU(3)_L \times SU(3)_R$-symmetry had been discovered before QCD was formulated.

As long as we do not observe any particles degeneration that may be associated with the chiral symmetry, then this symmetry has to be violated. The existence of bosons set with masses being close to zero, three π-mesons and the whole octet of 0^--mesons, is a physical manifestation of such a realization of the symmetry. In this chapter we will show that in definite kinematic limits matrix elements of these light pseudoscalar mesons may be calculated with the help of commutation relations and conservation laws for currents corresponding to the chiral symmetry.

In accordance with the Noether theorem, exact symmetry results in conserved currents. As this takes place, corresponding charges generate a group algebra associated with a given symmetry, that is, these charges are group generators. Thus, for example, in the case of the isospin symmetry (see Chapter 4 of *Advanced Particle Physics Volume 1*), the conserved charges $S_\gamma^{(is)}$ ($\gamma = 1, 2, 3$) satisfy the commutation relations (CRs)

$$[S_\gamma^{(is)}, S_\beta^{(is)}] = i\epsilon_{\gamma\beta\alpha}S_\alpha^{(is)}. \tag{5.2}$$

Significance of (5.2) is caused by the fact that with their help we can classify particle states. It turns out that the CRs (5.2) remain valid even if the Lagrangian involves the terms to violate the symmetry. Let us examine the Lagrangian of the form

$$\mathcal{L} = \mathcal{L}_0 + \mathcal{L}_I, \tag{5.3}$$

where $\mathcal{L}_0$ is invariant under the group of the global symmetry G while $\mathcal{L}_I$ is not invariant with respect to G. Subject a system to the infinitesimal transformation

$$\psi_k'(x) = \psi_k(x) + \delta\psi_k(x), \qquad \delta\psi_k(x) = i\,(T^\gamma)_k^j\,\psi_j\delta\theta^\gamma, \tag{5.4}$$

where $\delta\theta^\gamma$ are transformation parameters independent on x, $(T_\gamma)_k^j$ is a set of Hermitian matrices that obey the CRs of Lie algebra of the group G:

$$[T^a, T^b] = iC_{abc}T^c, \tag{5.5}$$

C_{abc} are structural constants of the group G. As before, we can define the Noether's current $J_\mu^a(x)$ by the expression

$$J_\mu^a(x) = -i\frac{\partial\mathcal{L}(x)}{\partial\psi_{i;\mu}(x)}\,(T^a)_i^j\,\psi_j(x). \tag{5.6}$$

But now, this current is not conserved and the charges determined as

$$Q^a(t) = \int J_0^a(x)d\mathbf{r} = -i\int \frac{\partial\mathcal{L}(x)}{\partial\psi_{i;0}(x)}(T^a)_i^j\psi_j(x)d\mathbf{r}, \tag{5.7}$$

depend on time. Nevertheless, despite the presence of the term $\mathcal{L}_I$, the quantity $\partial\mathcal{L}/(\partial\psi_{i;0})$ will continue in use for the momentum canonically conjugate to ψ_i:

$$\pi_i(x) = \frac{\partial\mathcal{L}(x)}{\partial\psi_{i;0}(x)}$$

and satisfy the canonical CRs at equal times:

$$[\pi_i(\mathbf{r}, t), \psi_j(\mathbf{r}', t)] = -i\delta^3(\mathbf{r} - \mathbf{r}')\delta_{ij}. \tag{5.8}$$

Using these relations and the identity

$$[AB, CD] = A[B, C]D - C[D, A]B,$$

which is true at $[A, C] = [D, B] = 0$, we can compute the charges commutator at equal times for arbitrary $\mathcal{L}_I$:

$$[Q^a(t), Q^b(t)] = -\int d\mathbf{r}\int d\mathbf{r}'[\pi_i(\mathbf{r}, t)(T^a)_i^j\psi_j(\mathbf{r}, t), \pi_k(\mathbf{r}', t)(T^b)_k^l\psi_l(\mathbf{r}', t)] =$$

$$= -\int d\mathbf{r}\int d\mathbf{r}'\{\pi_i(\mathbf{r}, t)(T^a)_i^j[\psi_j(\mathbf{r}, t), \pi_k(\mathbf{r}', t)](T^b)_k^l\psi_l(\mathbf{r}', t)+$$

$$+\pi_k(\mathbf{r}', t)(T^b)_k^l[\pi_i(\mathbf{r}, t), \psi_l(\mathbf{r}', t)](T^a)_i^j\psi_j(\mathbf{r}, t)\} = -i\int d\mathbf{r}\{\pi_k(\mathbf{r}, t)[T^a, T^b]_k^j\psi_j(\mathbf{r}, t). \tag{5.9}$$

Taking into consideration (5.5), we get

$$[Q^a(t), Q^b(t)] = iC^{abc}Q^c(t). \tag{5.10}$$

So, we have shown, that even if the quantities $Q^a(t)$ vary with time, then the CRs of the group algebra are fulfilled at any given instants of time t. These relations are usually called *charges algebra*.

Now we proceed to the object of our interest. Consider a free system of u, d, s quarks possessing the $SU(3)$ flavor symmetry. The quark fields constitute the triplet representation

$$q(x) = \begin{pmatrix} u(x) \\ d(x) \\ s(x) \end{pmatrix} \tag{5.11}$$

with the transformation properties

$$q_k(x) \rightarrow q_k(x) + i\left(\frac{\lambda^a}{2}\right)_{kj} q_j(x)\delta\theta^a. \tag{5.12}$$

The Lagrangian of the theory has the form

$$\mathcal{L} = \mathcal{L}_0 + \mathcal{L}_I,$$

where the Lagrangian $\mathcal{L}_0$ is invariant with respect to the $SU(3)$-transformations while $\mathcal{L}_I$ does not

$$\mathcal{L}_0 = i\overline{q}(x)\gamma^\mu\partial_\mu q(x), \qquad \mathcal{L}_I = m_u\overline{u}(x)u(x) + m_d\overline{d}(x)d(x) + m_s\overline{s}(x)s(x). \tag{5.13}$$

The currents and corresponding charges associated with the $SU(3)$-transformations are given by the expressions

$$V_\mu^a(x) - q(x)\gamma_\mu\left(\frac{\lambda^a}{2}\right)q(x), \tag{5.14}$$

$$Q^a(t) = \int V_0^a(x)d\mathbf{r}. \tag{5.15}$$

It easy to verify that, in consequence of the canonical CRs

$$\{q_{\alpha k}(\mathbf{r}, t), q_{\beta j}^\dagger(\mathbf{r}', t)\} = \delta_{kj}\delta_{\alpha\beta}\delta^3(\mathbf{r} - \mathbf{r}'), \tag{5.16}$$

where k, j are flavor indices and α, β are spinor indices, the charges obey the CRs of the $SU(3)$ group algebra

$$[Q^a(t), Q^b(t)] = if^{abc}Q^c(t). \tag{5.17}$$

In order to be the symmetry exact the fulfillment of the equalities

$$m_u = m_d = m_s \tag{5.18}$$

are needed. In the limit $\mathcal{L}_I = 0$, the Lagrangian $\mathcal{L}_0$ is invariant with respect to a group that is wider than $SU(3)$. Aside from the transformation (5.12), $\mathcal{L}_0$ remains invariant under the axial infinitesimal transformation

$$q_k(x) \rightarrow q_k(x) + i\left(\frac{\lambda^a}{2}\right)_{kj} \gamma_5 q_j(x)\delta\alpha^a. \tag{5.19}$$

The corresponding Noether current

$$A_\mu^a(x) = \bar{q}(x)\gamma_\mu\gamma_5\left(\frac{\lambda^a}{2}\right)q(x) \tag{5.20}$$

is named an *axial-vector one*. Thus, even at the presence of the term $\mathcal{L}_I$ violating the symmetry, we can define the axial charge Q^{5a}

$$Q^{5a}(t) = \int A_0^a(x)d\mathbf{r}. \tag{5.21}$$

The vector and axial charges generate the CRs totality

$$[Q^a(t), Q^{5b}(t)] = if^{abc}Q^{5c}(t), \tag{5.22}$$

$$[Q^{5a}(t), Q^{5b}(t)] = if^{abc}Q^c(t), \tag{5.23}$$

that, combined with (5.17), define the chiral $SU(2)_L \times SU(2)_R$ algebra. To make sure of this, we determine the left and right charges using the relations

$$Q_{L,R}^a = \frac{1}{2}\left(Q^a \mp Q^{5a}\right). \tag{5.24}$$

Then, the relations (5.17), (5.22), and (5.23) may be written in the view

$$\left.\begin{aligned}
[Q_L^a(t), Q_L^b(t)] &= if^{abc}Q_L^c(t), \\
[Q_R^a(t), Q_R^b(t)] &= if^{abc}Q_R^c(t), \\
[Q_L^a(t), Q_R^b(t)] &= 0.
\end{aligned}\right\} \tag{5.25}$$

Now, it is evident that the quantities Q_L^c generate the $SU(3)_L$ algebra while $Q_R^c(t)$—$SU(3)_R$ algebra.

Considering the commutators of charges and corresponding currents at equal times, one may expand charges algebra (5.10). Making use of the canonical CRs (5.8), one could show the validity of the following relation

$$[Q^a(t), J_0^b(\mathbf{r}, t)] = iC^{abc}J_0^c(\mathbf{r}, t). \tag{5.26}$$

Further, employing the Lorentz-invariance, one may also include other current components:

$$[Q^a(t), J_\mu^b(\mathbf{r}, t)] = iC^{abc}J_\mu^c(\mathbf{r}, t). \tag{5.27}$$

One cannot be constrained by this to consider the relation

$$[J_0^a(\mathbf{r}, t), J_0^b(\mathbf{r}', t)] = iC^{abc}J_0^c(\mathbf{r}, t)\delta^3(\mathbf{r} - \mathbf{r}') \tag{5.28}$$

in conjunction with (5.26). These relations and the analogous extension of the relations (5.17), (5.22), and (5.23) are called *currents algebra*. It is clear, by analogy we can define currents algebra for any symmetry group.

If one tries to include space components in currents algebra (5.28), then additional terms disappearing under space integration emerge. Thus, for example,

$$[J_0^a(\mathbf{r}, t), J_j^b(\mathbf{r}', t)] = iC^{abc}J_j^c(\mathbf{r}, t)\delta^3(\mathbf{r} - \mathbf{r}') + S_{jk}^{ab}(\mathbf{r})\frac{\partial}{\partial x_k'}\delta^3(\mathbf{r} - \mathbf{r}'), \tag{5.29}$$

where $S_{jk}^{ab}(\mathbf{r})$ is an operator depending on the explicit form of the quantity $J_j^c(\mathbf{r}, t)$. As this term vanishes under the space integration, it does not change currents algebra. The terms of such a type are named *Schwinger's terms* [39].

It should be noted as well that two more $U(1)$ symmetries exist in the case of a free quark model. The former is associated with the $\mathcal{L}$ invariance with respect to a variation of a general phase for all quark fields

$$q_j(x) \rightarrow \exp[i\beta]q_j(x).$$

As a result, the baryon current

$$J_\mu^B = \beta \overline{q}_j(x)\gamma_\mu q_j(x) \tag{5.30}$$

is conserved. The latter $U(1)$ symmetry is appropriate to the $\mathcal{L}_0$ invariance under the transformations

$$q_j(x) \rightarrow \exp[i\alpha\gamma_5]q_j(x).$$

The current

$$J_\mu^A = \alpha \overline{q}_j(x)\gamma_\mu\gamma_5 q_j(x), \tag{5.31}$$

which is connected with an axial baryon charge, is partially conserved.

Emphasize, that the currents associated with the symmetries are exactly the same ones that enter into the electroweak and strong interactions, that is, the Noether's currents themselves or some of their linear combinations enter into the interaction Lagrangian.

5.2 Weak and electromagnetic currents of hadrons

The weak interaction leads to the necessity of detail investigating a structure and properties of hadron currents. In the phenomenological aspect this interaction is well described by the effective Lagrangian having the form *current×current*:

$$\mathcal{L}_{eff} = -\frac{G_F}{\sqrt{2}}j^\lambda(x)j_\lambda^\dagger(x), \tag{5.32}$$

where the total current j_λ is the sum of the lepton (J_λ^l) and hadron (J_λ^h) currents:

$$j_\lambda(x) = J_\lambda^l(x) + J_\lambda^h(x). \tag{5.33}$$

Each of these currents represents the superposition of the vector (V) and axial (A) parts. At present the structure of the lepton current may be considered to be determined

$$J_\lambda^l(x) = \overline{\nu}_e(x)\gamma_\lambda(1 - \gamma_5)e(x) + \overline{\nu}_\mu(x)\gamma_\lambda(1 - \gamma_5)\mu(x) + \overline{\nu}_\tau(x)\gamma_\lambda(1 - \gamma_5)\tau(x). \tag{5.34}$$

However, the situation connected with the hadron current is appreciably more complex. To find out its structure and properties we address a typical semilepton hadron decay, namely, a β-decay of a neutron

$$n(p_n) \rightarrow p(p_p) + e^-(p_e) + \overline{\nu}_e(p_{\overline{\nu}_e}), \tag{5.35}$$

where four-dimensional particle momenta are indicated in brackets. Let us constrain ourselves by the first order in the constant G_F. In the momentum representation the amplitude will look like:

$$\mathcal{A} = \frac{G_F}{\sqrt{2}}(V_\lambda^h - A_\lambda^h)\sqrt{\frac{m_e}{E_e}}\sqrt{\frac{m_{\overline{\nu}_e}}{E_{\overline{\nu}_e}}}[\overline{u}_e(p_e)\gamma^\lambda(1 - \gamma_5)u_{\overline{\nu}_e}(p_{\overline{\nu}_e})], \tag{5.36}$$

where $J_\lambda^h = V_\lambda^h - A_\lambda^h$ represents the part of the hadron weak current to conserve the strangeness. To determine its explicit form, we must build a four-dimensional vector and axial vector using wave functions, four-dimensional momenta of particles involved in the reaction and γ-matrices as well. In doing so, we take into consideration that the matrix element can depend on the sum of the nucleon momenta

$$P_\lambda = (p_p + p_n)_\lambda$$

and on that of the lepton momenta

$$q_\lambda = (p_n - p_p)_\lambda = (p_e + p_{\bar{\nu}_e})_\lambda,$$

rather than on every momentum separately. The second demand is caused by the fact that, in accordance with (5.32), the lepton pair is emitted at a single point. From the quantities above mentioned one may form three independent vectors:

$$\bar{u}\gamma_\lambda u, \qquad \bar{u}\sigma_{\lambda\beta}q^\beta u, \qquad \bar{u}q_\lambda u,$$

and three pseudovectors:

$$\bar{u}\gamma_5\gamma_\lambda u, \qquad \bar{u}\gamma_5 P_\lambda u, \qquad \bar{u}\gamma_5 q_\lambda u.$$

Thus, the vector and axial parts of the hadron current is defined by the expressions:

$$V_\lambda^h = \bar{u}(p_p)[f_1(q^2)\gamma_\lambda + f_2(q^2)\sigma_{\lambda\beta}q^\beta + f_3(q^2)q_\lambda]u(p_n), \qquad (5.37)$$

$$A_\lambda^h = \bar{u}(p_p)[g_1(q^2)\gamma_\lambda + g_2(q^2)P_\lambda + g_3(q^2)q_\lambda]\gamma_5 u(p_n), \qquad (5.38)$$

where f_i and g_i are nucleon form-factors. From hermicity and CP-invariance it follows that they have to be real functions.

When calculating the neutron decay probability, one may neglect with a quite good accuracy the terms to be proportional to q and q^2. Then the total Lagrangian becomes the form

$$J_\lambda^h = c_V^n \bar{u}(p_p)\gamma_\lambda(1 - \alpha_n\gamma_5)u(p_n), \qquad (5.39)$$

where $c_V^n \equiv f_1(0)$, $c_A^n \equiv g_1(0)$ $\alpha_n = c_A^n/c_V^n$. Since at the decay, the modest energy is released ($\Delta E \approx m_n - m_p \approx 1.3$ MeV), then in the laboratory reference frame the neutron and produced proton may be considered to be at rest ($\mathbf{p}_p = \mathbf{p}_n = 0$). In accordance with Eq. (15.86) of *Advanced Particle Physics Volume 1* their wave functions have the view

$$u(p_p) \equiv u_p = \begin{pmatrix} w_p \\ 0 \end{pmatrix}, \qquad u(p_n) \equiv u_n = \begin{pmatrix} w_n \\ 0 \end{pmatrix}, \qquad (5.40)$$

where $w_{p,n}$ are nonrelativistic (two-component) spinors normalized on unit. Using the explicit form of the γ-matrices (see Eq. (15.7) of *Advanced Particle Physics Volume 1*), we obtain for the weak hadron current

$$J_\lambda^h = c_V^n(w_p^\dagger w_n, -\alpha_n w_p^\dagger \boldsymbol{\sigma} w_n). \qquad (5.41)$$

Its product with the lepton current gives the decay amplitude

$$\mathcal{A} = \frac{G_F c_V^n}{\sqrt{2}}\sqrt{\frac{m_e}{E_e}}\sqrt{\frac{m_{\bar{\nu}_e}}{E_{\bar{\nu}_e}}}\{w_p^\dagger w_n[\bar{u}_e\gamma_0(1 - \gamma_5)u_{\bar{\nu}_e}] + \alpha_n(w_p^\dagger \boldsymbol{\sigma} w_n)[\bar{u}_e\boldsymbol{\gamma}(1 - \gamma_5)u_{\bar{\nu}_e}]\}. \qquad (5.42)$$

Then the amplitude square is

$$|\mathcal{A}|^2 = \frac{G_F^2(c_V^n)^2 m_e m_{\bar{\nu}_e}}{E_e E_{\bar{\nu}_e}}\{\mathrm{Sp}(\rho_n\rho_p)\mathrm{Sp}[\rho_e\gamma_0\rho_{\bar{\nu}_e}\gamma_0(1 - \gamma_5)] + \alpha_n^2\mathrm{Sp}(\rho_n\sigma_k\rho_p\sigma_i)\mathrm{Sp}[\rho_e\gamma^i\rho_{\bar{\nu}_e}\times$$

$$\times \gamma^k (1 - \gamma_5)] + \alpha_n \mathrm{Sp}(\rho_n \boldsymbol{\sigma} \rho_p) \mathrm{Sp}[\rho_e \gamma_0 \rho_{\overline{\nu}_e} \boldsymbol{\gamma} (1 - \gamma_5)] + \alpha_n \mathrm{Sp}(\rho_n \rho_p \boldsymbol{\sigma}) \mathrm{Sp}[\rho_e \boldsymbol{\gamma} \rho_{\overline{\nu}_e} \gamma_0 (1 - \gamma_5)]\},$$

$$(5.43)$$

where ρ_j ($j = e, \overline{\nu}_e, n, p$) is the polarization density matrix of the j particle. Because the electron and electron antineutrino are relativistic objects, then ρ_e and $\rho_{\overline{\nu}_e}$ will be described by the expression (see Eq. (15.134) of *Advanced Particle Physics Volume 1*):

$$\rho_e = \frac{1}{2} \left[\frac{\hat{p} + m_e}{2m_e} (1 + \gamma_5 \hat{s}) \right], \qquad \rho_{\overline{\nu}_e} = \frac{\hat{p}_{\overline{\nu}_e}}{2m_{\overline{\nu}_e}}.$$

It is evident that the nonrelativistic polarization density matrices should be used for ρ_n, ρ_p (see Eq. (15.125) of *Advanced Particle Physics Volume 1*)

$$\rho_n = \frac{1}{2}[1 + (\boldsymbol{\xi}^n \cdot \boldsymbol{\sigma})], \qquad \rho_p = \frac{1}{2}[1 + (\boldsymbol{\xi}^p \cdot \boldsymbol{\sigma})],$$

where $\boldsymbol{\xi}^{n,p}$ are the neutron and proton polarization vectors, respectively. To sum up over the electron and proton polarizations results in

$$\sum_{e^-, p} |\mathcal{A}|^2 = \frac{G_F^2 (c_V^n)^2}{4 E_e E_{\overline{\nu}_e}} [\mathrm{Sp}(\hat{p}_e \gamma^0 \hat{p}_{\overline{\nu}_e} \gamma^0) + \alpha_n^2 \mathrm{Sp}(\hat{p}_e \boldsymbol{\gamma} \hat{p}_{\overline{\nu}_e} \boldsymbol{\gamma}) +$$

$$+ i \alpha_n^2 (\epsilon_{lki} \xi_l^p) \mathrm{Sp}(\hat{p}_e \gamma^i \hat{p}_{\overline{\nu}_e} \gamma^k \gamma_5) + 2\alpha_n \mathrm{Sp}(\hat{p}_e \gamma^0 \hat{p}_{\overline{\nu}_e} (\boldsymbol{\xi}^n \cdot \boldsymbol{\gamma}))] =$$

$$= G_F^2 (c_V^n)^2 [1 + 3\alpha_n^2 - (\alpha_n^2 - 1)(\mathbf{v}_e \cdot \mathbf{v}_{\overline{\nu}_e}) - 2\alpha_n (\alpha_n - 1)(\mathbf{v}_e \cdot \boldsymbol{\xi}^n) + 2\alpha_n (\alpha_n + 1)(\mathbf{v}_{\overline{\nu}_e} \cdot \boldsymbol{\xi}^n)], \quad (5.44)$$

where $\mathbf{v}_e = \mathbf{p}_e / E_e$, $\mathbf{v}_{\overline{\nu}_e} = \mathbf{p}_{\overline{\nu}_e} / E_{\overline{\nu}_e}$ are the velocities of e^- and $\overline{\nu}_e$ ($|\mathbf{v}_{\overline{\nu}_e}| \approx 1$).

The expression in the square brackets defines the angle distributions of the electrons and antineutrinos. Having done averaging over the directions of $\mathbf{v}_e$ and $\mathbf{v}_{\overline{\nu}_e}$, we get

$$\sum_{e^-, p} |\mathcal{A}|^2 = G_F^2 (c_V^n)^2 (1 + 3\alpha_n^2). \qquad (5.45)$$

Substitute (5.45) into the formula for the differential decay probability

$$d\Gamma = (2\pi)^4 \delta^{(4)}(k - \sum_{j=1}^{n_f} p_j) |\sum \mathcal{A}_{i \to f}|^2 \prod_{j=1}^{n_f} \frac{d^3 p_j}{(2\pi)^3}, \qquad (5.46)$$

and integrate the expression obtained over $d^3 p_p d^3 p_{\overline{\nu}_e}$ and over the directions of $\mathbf{p}_e$. As a result, we arrive at the following expression for the energy electron distribution

$$d\Gamma = \frac{G_F^2 (c_V^n)^2}{2\pi^3} (1 + 3\alpha_n^2) \sqrt{(E_e^2 - m_e^2)} (\Delta E - E_e)^2 E_e dE_e, \qquad (5.47)$$

where ΔE is the energy transferred to the lepton pair. Now, using the observed electrons spectrum, one can graphically present the dependence

$$\frac{1}{E_e} \left(\frac{d\Gamma}{dE_e} \right)^{1/2}$$

on E_e. This is a so-called *Curie graph*. It could be employed to find the electron antineutrino mass $m_{\overline{\nu}_e}$. If $m_{\overline{\nu}_e} = 0$, then the graph will be presented by a straight line to be ended at the point ΔE. When the electron antineutrino mass is nonzero, the linear dependence will be violated especially at E_e close to ΔE. To integrate over the interval $m_e \leq E_e \leq \Delta E$ gives the total decay probability

$$\Gamma = C \frac{G_F^2 (c_V^n)^2 (\Delta E)^5}{60\pi^3}(1 + 3\alpha_n^2), \qquad (5.48)$$

where

$$C = \frac{1}{60}\left[2\frac{(\Delta E)^4}{m_e^4} - 9\frac{(\Delta E)^2}{m_e^2} - 8\right] + \frac{(\Delta E)}{4m_e}\text{Arch}\,\frac{\Delta E}{m_e} \approx 0.47.$$

The most accurate value of the quantity α_n is obtained by comparing the theoretical β-decay probability with the experimental data concerning the neutron lifetime ($\tau = 1/\Gamma$). The current value is

$$\alpha_n = 1.267. \qquad (5.49)$$

If the quantity α_n has been determined from other experiments (for example, from investigations of nuclei β-decay), then the neutron lifetime data can be used for the determination of the quantity $G_F c_V^n$. On the other hand, studying the pure lepton decays, such as muon decay[*]

$$\mu^- \rightarrow e^- + \nu_\mu + \overline{\nu}_e \qquad (5.50)$$

one can measure the quantity $G_F c_V^\mu$. Measurements show that the constants c_V^μ and c_V^n are equal to each other with a precision of a few percents, that is, $c_V^n \approx 1$. This fact is not trivial. Really, in the case of nuclei β-decay the strong interaction, along with the weak one, is present. And for the μ-decay the strong interaction is absent. Coincidence of the constants means that the strong interaction does not influence the value of the vector coupling constant c_V^n, as the saying goes, c_V^n is not renormalized by the strong interaction. In physics such a situation is not new. The electron and proton electromagnetic interaction constants are also equal to each other despite the fact that for the proton the strong interaction, along with the electromagnetic one, exists. To put this another way, the electric charge is not renormalized by the strong interaction as well. The reason lies in the fact that the electromagnetic vector current is conserved to give the electric charge conservation law. The electric charge is renormalized only because of radiative corrections to the photon Green function. However, these corrections have to be such that the electric charge is conserved. As the particle structure or the charge distribution in a cloud surrounding a bare particle may be, the electric charge always remains the same.

Assume, that, by analogy with electrodynamics, the weak hadron vector current responsible for the neutron β-decay is also conserved

$$\frac{\partial}{\partial x_\mu} V_\mu^h = 0. \qquad (5.51)$$

This assumption, in particular, leads to equality of the vector constants for the β and μ decays, that is, to nonrenormalizability of the vector constant of the β decay by means of the strong interaction. To visually elucidate the nonrenormalizability of the vector constant of the β-decay is as follows. When calculating the dressed neutron decay, one needs to take into account three types of virtual strong interactions: (i) neutron interaction before decay (Fig. 5.1(a)), (ii) proton interaction after decay (Fig. 5.1(b)), and (iii) interaction between proton and neutron by means of hadrons exchange (Fig. 5.1(c)). However, since the neutron and proton belong to one and the same isotopic doublet, their strong interactions are equal. This leads to the fact that contributions of all three diagrams are canceled and the dressed vector coupling constant proves to be equal to the bare one.

---

[*]Note, in this case we have $c_V^\mu = c_A^\mu = 1$.

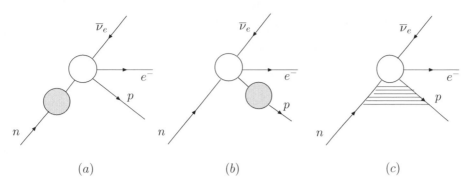

$$(a) \qquad\qquad (b) \qquad\qquad (c)$$

FIGURE 5.1

The graphical interpretations of nonrenormalizability of the vector β-decay constant in (a)–(c).

As the electromagnetic current J_μ^{em} and weak hadron vector current V_μ^h are conserved, one may assume these currents to be unified into one isotopic vector, a triplet. The isovector part of the electromagnetic current will be the third component of this vector, while the weak vector hadron current and the Hermitian conjugate one are two other components. Then, the conservation law follows

$$\frac{\partial}{\partial x_\mu} J_\mu^\alpha = 0, \tag{5.52}$$

where α is the isotopic index ($\alpha = 1, 2, 3$). Of course, we are dealing with an approximate conservation of the currents. They are conserved in the exact isospin symmetry approximation. The given scheme is commonly called the *hypothesis of conservation of the vector current* (CVC) [40]. In such a way, knowing matrix elements of the electromagnetic current (they are called *form-factors*) and using isospin rotations, we can calculate the form-factor values that are associated with the weak vector current to conserve the strangeness.

Reasons of insignificant differences between c_V^μ and c_V^n are as follows. The weak hadron current represents the sum of two currents, one of them conserves the strangeness $J_\lambda^{\Delta S=0}$ while the second does not $J_\lambda^{|\Delta S|=1}$. At that, in accordance with Cabibbo's hypothesis [41], the total weak hadron current is given by the expression

$$J_\lambda^h(x) = J_\lambda^{\Delta S=0}(x) c_{\theta_c} + J_\lambda^{|\Delta S|=1}(x) s_{\theta_c}, \tag{5.53}$$

where $c_{\theta_c} = \cos\theta_c, s_{\theta_c} = \sin\theta_c$ and θ_c is the Cabibbo angle ($c_{\theta_c} \approx 0.975$). Consequently, in the formulae for the neutron β-decay probability one should fulfill the replacement

$$c_V^n \rightarrow c_V^n c_{\theta_c} \tag{5.54}$$

to ensure $c_V^\mu = c_V^n c_{\theta_c}$.

Similar to the lepton current, the weak hadron current may be written in terms of fundamental quark fields

$$J_\lambda^h(x) = \overline{u}(x)\gamma_\lambda(1 - \gamma_5)d(x) c_{\theta_c} + \overline{u}(x)\gamma_\lambda(1 - \gamma_5)s(x) s_{\theta_c}. \tag{5.55}$$

Making use of this circumstance, we show how the weak hadron currents may be placed to one and the same the $SU(3)$ octet. In Section 9.3 of *Advanced Particle Physics Volume 1* we have obtained the $SU(3)$ irreducible eight-component spinor (see Eq. (9.151) of *Advanced Particle Physics Volume 1*)

$$\Phi_l^k - \frac{1}{3}\delta_l^k \Phi_n^n =$$

$$= \begin{pmatrix} \frac{2}{3}\Phi_1^1 - \frac{1}{3}(\Phi_2^2 + \Phi_3^3) & \Phi_1^2 & \Phi_1^3 \\ \Phi_2^1 & \frac{2}{3}\Phi_2^2 - \frac{1}{3}(\Phi_1^1 + \Phi_3^3) & \Phi_2^3 \\ \Phi_3^1 & \Phi_3^2 & \frac{2}{3}\Phi_3^3 - \frac{1}{3}(\Phi_1^1 + \Phi_2^2) \end{pmatrix}. \tag{5.56}$$

Setting

$$\Phi_j^i = \overline{q}_i q_j,$$

we get the quark filling of the pseudoscalar mesons octet

$$(P_j^i) = \begin{pmatrix} \frac{2}{3}\overline{u}u - \frac{1}{3}(\overline{d}d + \overline{s}s) & \overline{d}u & \overline{s}u \\ \overline{u}d & \frac{2}{3}\overline{d}d - \frac{1}{3}(\overline{u}u + \overline{s}s) & \overline{s}d \\ \overline{u}s & \overline{d}s & \frac{2}{3}\overline{s}s - \frac{1}{3}(\overline{u}u + \overline{d}d) \end{pmatrix}. \tag{5.57}$$

Modify this expression not changing its transformation properties with respect to the $SU(3)$ group. In the first place, replace every quark by the corresponding operator, for example, $u \to U$ and $\overline{u} \to \overline{U}$, where U and $\overline{U}$ is the Dirac u quark field and Dirac-conjugate u quark field, respectively. Next, insert the matrices γ_μ. As a result, we get the octet of the vector currents V_μ:

$$\begin{pmatrix} \frac{2}{3}\overline{U}\gamma_\mu U - \frac{1}{3}(\overline{D}\gamma_\mu D + \overline{S}\gamma_\mu S) & \overline{D}\gamma_\mu U & \overline{S}\gamma_\mu U \\ \overline{U}\gamma_\mu D & \frac{2}{3}\overline{D}\gamma_\mu D - \frac{1}{3}(\overline{U}\gamma_\mu U + \overline{S}\gamma_\mu S) & \overline{S}\gamma_\mu D \\ \overline{U}\gamma_\mu S & \overline{D}\gamma_\mu S & \frac{2}{3}\overline{S}\gamma_\mu S - \frac{1}{3}(\overline{U}\gamma_\mu U + \overline{D}\gamma_\mu D) \end{pmatrix}. \tag{5.58}$$

This octet may be written in the convenient designations:

$$V_\mu^a(x) = \overline{q}(x)\gamma_\mu \frac{\lambda^a}{2} q(x) = \begin{pmatrix} V^3 + V^8/\sqrt{3} & V^1 - iV^2 & V^4 - iV^5 \\ V^1 + iV^2 & -V^3 + V^8/\sqrt{3} & V^6 - iV^7 \\ V^4 + iV^5 & V^6 + iV^7 & -V^8\sqrt{2/3} \end{pmatrix}. \tag{5.59}$$

Obviously, the first element being positioned on the central diagonal represents the electromagnetic current with a precision of the factor $|e|$. Therefore, one may write

$$V_\mu^{em} = J_\mu^1 = |e|\left(V_\mu^3 + \frac{1}{\sqrt{3}}V_\mu^8\right), \tag{5.60}$$

which represents the linear combination of the third component of the isospin current V^3 and isoscalar current V^8. $J^{2,3} = V^1 \pm iV^2$ are the other components of the isospin current. It is evident, that they correspond to the weak charged hadron vector current with $\Delta S = 0$ and its Hermitian-conjugate counterpart. At last, $V^4 \pm iV^5$ are associated with the weak charged hadron vector current with $|\Delta S| = 1$ and its Hermitian-conjugate counterpart.

By analogy, one may define the axial currents octet

$$A_\mu^a = \overline{q}(x)\gamma_\mu \gamma_5 \frac{\lambda^a}{2} q(x). \tag{5.61}$$

It is clear, that the components $A^1 \pm iA^2$ and $A^4 \pm iA^5$ are connected with the axial parts of the weak charged hadron currents with $\Delta S = 0$ and $|\Delta S| = 1$, respectively. However, the component $A^3 + A^8/\sqrt{3}$ does not have any direct physical meaning.

Unlike its vector partner, the weak axial hadron current conserving the strangeness is not conserved. This immediately follows from the fact that the axial constants of the β and μ decays are not equal to each other. In other words, the axial constant c_A^n is renormalized by the strong interaction. However, the difference of c_A^n from c_A^μ is nothing but 20%. As we shall see later, this circumstance is caused by an anomalously small π-meson mass (m_π) as compared with masses of other hadrons. In the hypothetical limit $m_\pi \to 0$, the axial current conservation becomes exact and the chiral symmetry is realized. In so doing, a pion appears as a Goldstone boson under spontaneous symmetry breaking. This approximation is called the *partial conservation of the axial current* (PCAC).[*]

In closing of the theme, we give the relations of the $SU(3)_L \times SU(3)_R$ algebra to which the vector and axial weak hadron currents must obey [42]

$$[V_0^a(\mathbf{r}, t), V_0^b(\mathbf{r}', t)] = if^{abc}V_0^c(\mathbf{r}, t)\delta^3(\mathbf{r} - \mathbf{r}'), \tag{5.62}$$

$$[V_0^a(\mathbf{r}, t), A_0^b(\mathbf{r}', t)] = if^{abc}A_0^c(\mathbf{r}, t)\delta^3(\mathbf{r} - \mathbf{r}'), \tag{5.63}$$

$$[A_0^a(\mathbf{r}, t), A_0^b(\mathbf{r}', t)] = if^{abc}V_0^c(\mathbf{r}, t)\delta^3(\mathbf{r} - \mathbf{r}'). \tag{5.64}$$

All this allows us to use the interaction Lagrangian that has the form *current-current* (5.32). So, the currents algebra to reproduce the strong interaction symmetries could be checked in the electroweak interactions of hadrons. Applications of the currents algebra are comprehensively considered in the book [43].

5.3 Adler's sum rules for neutrino reactions

Sum rules (SRs) are theoretical relations to fix some sum (integral) value of matrix elements characterizing transitions between states of a system under examination. Wide application of these rules is connected with the fact that in many cases, from theoretical considerations it manages to find only some sum of physical matrix elements while every individual term of this sum could not be calculated. However, all individual terms may be measured in experiments. As a consequence, we have the possibility to check theoretical principles underlying a theory. The SRs to have received wide acceptance in hadron physics may be obtained on the basis both of dispersion relations and of currents algebra (current SRs). Here we discuss some examples of getting the current SRs. Since, from the point of view of the quantum field theory, observed hadrons are described by a complicated (and unknown) wave function being in line with the quark-gluon system, we need to address the auxiliary objects, currents. As for currents, on the one hand, they are simple bilinear combinations of fundamental quark fields while, on the other hand, their matrix elements could be measured in weak and electromagnetic transitions between hadrons. In particular, investigating the permutation relations between components of the electromagnetic hadron current results in the Drell–Hearn sum rule [44]:

$$\int_0^\infty \frac{d\nu}{\nu}[\sigma_P(\nu) - \sigma_A(\nu)] = \frac{2\pi^2\alpha}{M_p^2}\mu_p^2 \approx 205 \ \mu b, \tag{5.65}$$

where σ_P (σ_A) is the total cross section of the interaction of the photon with a polarized proton when the photon spin is parallel (P) and antiparallel (A) to the proton one, ν is the photon energy, M_p is the proton mass, and μ_p is the anomalous magnetic proton moment in units of the nuclear magneton. There exists the analogous rule for the magnetic neutron moment that contains corresponding neutron characteristics.

Let us show how to work the currents algebra by the example of reactions with neutrinos participation. Consider scattering of high-energy neutrinos by a nucleon target (see Fig. 5.2)

$$\nu_l + N \to l^- + X_n, \tag{5.66}$$

---

[*]PCAC hypothesis.

where X_n is some hadron state (n particles with common momentum p_n). Introduce the

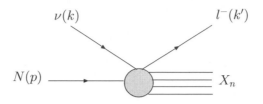

FIGURE 5.2
The graphical interpretation of nonrenormalizability of the vector β-decay constant.

variables

$$q = k - k', \qquad \nu = \frac{(pq)}{M}, \tag{5.67}$$

where M is a nucleon mass. In the laboratory system of coordinates

$$p_\mu = (M, 0, 0, 0), \qquad k_\mu = (E, \mathbf{k}), \qquad k'_\mu = (E', \mathbf{k}'),$$

we have

$$q^2 = -4EE' \sin^2 \frac{\theta}{2}, \qquad \nu = E - E', \tag{5.68}$$

where θ is the angle between $\mathbf{k}$ and $\mathbf{k}'$ while the energy is so high that we may neglect the charged lepton mass. With allowance made for Eqs. (5.32) and (5.34), the process amplitude takes the form

$$\mathcal{A}_n^{(\nu)} = \frac{G_F}{2} \sqrt{\frac{m_l}{E'}} \sqrt{\frac{m_{\nu_l}}{E}} \overline{u}_l(k', \lambda') \gamma^\rho (1 - \gamma_5) u_\nu(k, \lambda) < n |J_\rho^h| p, \sigma >, \tag{5.69}$$

where σ, λ, and λ' are the spin indices of the initial and final leptons, respectively. In the case of the unpolarized particles the differential cross section is given by the expression

$$d\sigma_n^{(\nu)} = \frac{d^3 k'}{16(2\pi)^3 M E E' |\mathbf{v}|} \prod_{i=1}^n \left[\frac{d^3 p_i}{2(2\pi)^3 p_{i0}} \right] \sum_{\sigma, \lambda, \lambda'} |\mathcal{A}_n^{(\nu)}|^2 (2\pi)^4 \delta^4(k + p - k' - p_n), \tag{5.70}$$

where $\mathbf{v}$ is the incoming neutrino velocity and

$$p_n = \sum_{i=1}^n p_i.$$

Summing up over all possible hadron states, we get the inclusive cross section

$$\frac{d^2 \sigma_n^{(\nu)}}{d|q^2| d\nu} = \frac{G_F^2}{32\pi E^2} L^{\alpha\beta} W_{\alpha\beta}^{(\nu)}, \tag{5.71}$$

where the lepton tensor $L^{\alpha\beta}$ and the hadron one $W_{\alpha\beta}^{(\nu)}$ will look like:

$$L^{\alpha\beta} = \mathrm{Sp}[\hat{k}'\gamma_\alpha(1 - \gamma_5)\hat{k}\gamma_\beta(1 - \gamma_5)] = 8[k_\alpha k'_\beta + k_\beta k'_\alpha - (kk')g_{\alpha\beta} + i\epsilon_{\alpha\beta\gamma\delta}k'^\gamma k^\delta], \tag{5.72}$$

$$W_{\alpha\beta}^{(\nu)} = \frac{1}{4M} \sum_\sigma \sum_n \int \prod_{i=1}^n \left[\frac{d^3 p_i}{2(2\pi)^3 p_{i0}} \right] < p, \sigma |J_\beta^{h\dagger}(0)| n >< n |J_\alpha^h(0)| p, \sigma > \times$$

$$\times (2\pi)^3 \delta^4(p_n - p - q) = \frac{1}{4M} \sum_\sigma \int \frac{d^4x}{2\pi} < p, \sigma |J_\beta^{h\dagger}(x) J_\alpha^h(0)|p, \sigma > e^{iqx}. \tag{5.73}$$

Note, that the hadron tensor is nonzero only at $q_0 = E - E' > 0$. Express the hadron tensor in terms of Lorentz-invariant structure nucleon functions as it was done under investigation of electron-nucleon collisions (see Section 10.2 of *Advanced Particle Physics Volume 1*). In the case in question, the number of structure functions will be bigger because parity violation effects come into play (this is reflected by the presence of the γ_5-matrix in the lepton tensor). To find the lepton tensor we use the relativistic invariance demand and take into account that two four-dimensional vectors p_μ, q_μ, metric tensor $g_{\mu\nu}$, and antisymmetric unit four-rank tensor $\epsilon_{\mu\nu\lambda\sigma}$ are at our disposal. Then, the result is as follows:

$$W_{\alpha\beta}^{(\nu)} = -W_1^{(\nu)}(q^2, \nu)g_{\alpha\beta} + W_2^{(\nu)}(q^2, \nu)\frac{p_\alpha p_\beta}{M^2} - iW_3^{(\nu)}(q^2, \nu)\epsilon_{\alpha\beta\gamma\delta}\frac{p^\gamma q^\delta}{M^2} + W_4^{(\nu)}(q^2, \nu)\frac{q_\alpha q_\beta}{M^2} +$$

$$+ W_5^{(\nu)}(q^2, \nu)\left(\frac{p_\alpha q_\beta + q_\alpha p_\beta}{M^2}\right) + iW_6^{(\nu)}(q^2, \nu)\left(\frac{p_\alpha q_\beta - q_\alpha p_\beta}{M^2}\right), \tag{5.74}$$

where $W_j^{(\nu)}(q^2, \nu)$ are nucleon structure functions. Multiplying the lepton and hadron tensors and neglecting the charged lepton mass, we arrive at

$$\frac{d^2\sigma_n^{(\nu)}}{d|q^2|d\nu} = \frac{G_F^2}{2\pi}\left(\frac{E'}{E}\right)\left[2W_1^{(\nu)}\sin^2\frac{\theta}{2} + W_2^{(\nu)}\cos^2\frac{\theta}{2} - \frac{E + E'}{M}W_3^{(\nu)}\sin^2\frac{\theta}{2}\right]. \tag{5.75}$$

Now we consider scattering of the antineutrino by a nucleon target

$$\overline{\nu}_l + N \to l^+ + X. \tag{5.76}$$

In this case the hadron tensor has the view

$$W_{\alpha\beta}^{(\overline{\nu})} = \frac{1}{4M} \sum_\sigma \int \frac{d^4x}{2\pi} e^{iqx} < p, \sigma |J_\beta^h(x) J_\alpha^{h\dagger}(0)|p, \sigma > . \tag{5.77}$$

As in the previous case, it is nonzero only at $q_0 = E - E' > 0$. Making use of the translation invariance

$$J_\beta^h(x) = e^{-ipx} J_\beta^h(0) e^{ipx},$$

we rewrite the expression (5.77) in the form

$$W_{\alpha\beta}^{(\overline{\nu})} = \frac{1}{4M} \sum_\sigma \int \frac{d^4x}{2\pi} e^{iqx} < p, \sigma |J_\beta^h(0) J_\alpha^{h\dagger}(-x)|p, \sigma >=$$

$$= \frac{1}{4M} \sum_\sigma \int \frac{d^4x}{2\pi} e^{-iqx} < p, \sigma |J_\beta^h(0) J_\alpha^{h\dagger}(x)|p, \sigma > . \tag{5.78}$$

To represent the hadron tensor in terms of structure functions is analogous to the expression (5.74), that is

$$W_{\alpha\beta}^{(\overline{\nu})} = W_{\alpha\beta}^{(\nu)}(\nu \to \overline{\nu}). \tag{5.79}$$

Calculations lead to the following result for the differential cross section of the process (5.76)

$$\frac{d^2\sigma_n^{(\overline{\nu})}}{d|q^2|d\nu} = \frac{G_F^2}{2\pi}\left(\frac{E'}{E}\right)\left[2W_1^{(\overline{\nu})}\sin^2\frac{\theta}{2} + W_2^{(\overline{\nu})}\cos^2\frac{\theta}{2} + \frac{E + E'}{M}W_3^{(\overline{\nu})}\sin^2\frac{\theta}{2}\right]. \tag{5.80}$$

So, investigations of the reactions (5.66) and (5.76) give us the possibility to define the nucleon structure functions $W_j^{(\nu)}, W_j^{(\bar{\nu})}$ $(j = 1, 2, 3)$. At present, our task is as follows. Employing the currents algebra, find the theoretical values of these structure functions and compare them with experimental data.

Introduce the following tensor

$$W_{\alpha\beta}(p,q) = \frac{1}{4M} \sum_\sigma \int \frac{d^4x}{2\pi} e^{iqx} < p, \sigma |[J_\beta^{h\dagger}(x), J_\alpha^h(0)]| p, \sigma >. \qquad (5.81)$$

Comparing it with (5.73) and (5.78), we come to the conclusion

$$\left.\begin{array}{ll} W_{\alpha\beta} = W_{\alpha\beta}^{(\nu)}(p,q) & q_0 > 0, \\ W_{\alpha\beta} = -W_{\alpha\beta}^{(\bar{\nu})}(p,-q) & q_0 < 0. \end{array}\right\} \qquad (5.82)$$

Further, integrate the quantity W_{00} over q_0:

$$\int_{-\infty}^{\infty} W_{00}(p,q)dq_0 = \int_0^\infty [W_{00}^{(\nu)}(p,q) - W_{00}^{(\bar{\nu})}(p,q)]dq_0 =$$

$$= \frac{1}{4M} \sum_\sigma \int d^3x e^{-i\mathbf{qr}} < p, \sigma |[J_0^{h\dagger}(\mathbf{r},0), J_0^h(0)]| p, \sigma >. \qquad (5.83)$$

The commutator could be computed with the help of the currents algebra. The most simplest way consists in using the fact that the relations (5.62)–(5.64) are also fulfilled in a free quark model where the current is given by the expression (5.55). Using the canonical anticommutation relations

$$\{q_{\alpha j}^\dagger(\mathbf{r},0), q_{\beta k}(0)\} = \delta^3(\mathbf{r})\delta_{\alpha\beta}\delta_{jk}, \qquad (5.84)$$

where α, β are spinor indices and j, k are flavor indices, we get

$$[J_0^{h\dagger}(\mathbf{r},0), J_0^h(0)] = [(d^\dagger(\mathbf{r},0)c_{\theta_c} + s^\dagger(\mathbf{r},0)s_{\theta_c})(1-\gamma_5)u(\mathbf{r},0), \ u^\dagger(0)(1-\gamma_5)(d(0)c_{\theta_c}+$$

$$+s(0)s_{\theta_c})] = 2(d^\dagger(\mathbf{r},0)c_{\theta_c} + s^\dagger(\mathbf{r},0)s_{\theta_c})(1-\gamma_5)\{u(\mathbf{r},0), u^\dagger(0)\}(d(0)c_{\theta_c} + s(0)s_{\theta_c})) -$$

$$-2u^\dagger(0)(1-\gamma_5)(c_{\theta_c}^2\{d^\dagger(\mathbf{r},0), d(0)\} + s_{\theta_c}^2\{s^\dagger(\mathbf{r},0), s(0)\})u(\mathbf{r},0) =$$

$$= 2\delta^3(\mathbf{r})(c_{\theta_c}^2 d^\dagger(0)(1-\gamma_5)d(0) + s_{\theta_c}^2 s^\dagger(0)(1-\gamma_5)s(0) + s_{\theta_c}c_{\theta_c}(d^\dagger(0)(1-\gamma_5)s(0)+$$

$$+s^\dagger(0)(1-\gamma_5)d(0) - u^\dagger(0)(1-\gamma_5)u(0)) = -\delta^3(\mathbf{r})[4c_{\theta_c}^2(V_0^3(0) - A_0^3(0))+$$

$$+s_{\theta_c}^2(3(V_0^Y(0) - A_0^Y(0)) + 2(V_0^3(0) - A_0^3(0))) + 4s_{\theta_c}c_{\theta_c}(V_0^6(0) - A_0^6(0))]. \qquad (5.85)$$

In the expression obtained we have used (5.59)

$$V_0^3 = \frac{1}{2}\left(u^\dagger u - d^\dagger d\right), \qquad A_0^3 = \frac{1}{2}\left(u^\dagger\gamma_5 u - d^\dagger\gamma_5 d\right),$$

$$V_0^6 = \frac{1}{2}\left(d^\dagger s + s^\dagger d\right), \qquad A_0^6 = \frac{1}{2}\left(d^\dagger\gamma_5 s + s^\dagger\gamma_5 d\right),$$

$$-\sqrt{2/3}V_0^8 = \frac{1}{3}\left(u^\dagger u + d^\dagger d - 2s^\dagger s\right), \qquad -\sqrt{2/3}A_0^8 = \frac{1}{2}\left(u^\dagger\gamma_5 u + d^\dagger\gamma_5 d - 2s^\dagger\gamma_5 s\right)$$

and introduced the designations

$$V_0^Y \equiv -\sqrt{2/3}V_0^8, \qquad A_0^Y \equiv -\sqrt{2/3}A_0^8. \qquad (5.86)$$

Owing to the fact that the operators V_0^6 and A_0^6 change the strangeness, their matrix elements between the nucleon states are equal to zero. Since the γ_5-matrix enter into A_0^j, then averaging over the nucleon spin leads to

$$\frac{1}{2} \sum_\sigma < p, \sigma | A_0^j | p, \sigma > = 0 \qquad \text{at any } j. \tag{5.87}$$

As a result, the relation (5.83) takes the form

$$\int_0^\infty [W_{00}^{(\nu)}(p, q) - W_{00}^{(\bar\nu)}(p, q)] dq_0 = -\frac{p_0}{M} [4I_3 c_{\theta_c}^2 + (3Y + 2I_3) s_{\theta_c}^2], \tag{5.88}$$

where I_3 and Y denote the isospin projection on the third axis and the hypercharge of a nucleon state, respectively:

$$\frac{1}{2} \sum_\sigma < p, \sigma | V_0^3 | p, \sigma > = 2I_3 p_0, \tag{5.89}$$

$$\frac{1}{2} \sum_\sigma < p, \sigma | V_0^Y | p, \sigma > = 2Y p_0. \tag{5.90}$$

Next, making use of Eqs. (5.74) and (5.79), we express $W_{00}^{(\nu)}, W_{00}^{(\bar\nu)}$ in terms of the structure functions. We have for $W_{00}^{(\nu)}$

$$W_{00}^{(\nu)} = -W_1^{(\nu)} + W_2^{(\nu)} \left(\frac{p_0}{M}\right)^2 + W_4^{(\nu)} \left(\frac{q_0}{M}\right)^2 + W_5^{(\nu)} \frac{p_0 q_0}{M^2}. \tag{5.91}$$

Choosing an appropriate system of coordinates, this expression could be simplified. Instead of the rest frame of the nucleon, we consider an infinite momentum frame [45] in which the nucleon has an infinite momentum being orthogonal to $\mathbf{q}$. In this frame we have

$$\left.\begin{array}{l} p_0 = (\mathbf{p}^2 + M^2)^{1/2} \approx |\mathbf{p}| \to \infty, \\ \nu = (\mathbf{p} \cdot \mathbf{q})/M - p_0 q_0/M, \\ q^2 = q_0^2 - \mathbf{q}^2 = (\nu M/p_0)^2 - \mathbf{q}^2 \to -\mathbf{q}^2. \end{array}\right\} \tag{5.92}$$

Employing both (5.91) and the analogous expression for $W_{00}^{(\bar\nu)}$ and tending $\mathbf{p}$ to infinity, we rewrite the left side of the relation (5.88) in the following form

$$\lim_{|\mathbf{p}| \to \infty} \int_0^\infty [W_{00}^{(\nu)}(p, q) - W_{00}^{(\bar\nu)}(p, q)] dq_0 = \frac{p_0}{M} \int_0^\infty [W_2^{(\nu)}(q^2, \nu) - W_2^{(\bar\nu)}(q^2, \nu)] d\nu, \tag{5.93}$$

where we have assumed, that the operations sequence of integration and transition to the limit may be changed, and also allowed for, that $W_2^{(\nu)}$ and $W_2^{(\bar\nu)}$ give the greatest contribution to $W_{00}^{(\nu)}$ and $W_{00}^{(\bar\nu)}$, respectively. As a result, the Adler sum rule for a neutrino becomes the view [46]

$$\int_0^\infty [W_2^{(\bar\nu)}(q^2, \nu) - W_2^{(\nu)}(q^2, \nu)] d\nu = 4I_3 c_{\theta_c}^2 + (3Y + 2I_3) s_{\theta_c}^2 =$$

$$= \begin{cases} 2c_{\theta_c}^2 + 4s_{\theta_c}^2 & \text{for proton target} \\ -2c_{\theta_c}^2 + 2s_{\theta_c}^2 & \text{for neutron target} \end{cases} \tag{5.94}$$

Let us fix the values of q^2 and ν and consider the case when the energy E (and E' as well) is increased. Then, the scattering angle θ tends to zero and the cross sections (5.75) and (5.80) may be approximated only by the contribution of W_2. Therefore,

$$\lim_{E \to \infty} \frac{d\sigma^{(\nu,\bar{\nu})}}{d|q^2|} = \frac{G_F^2}{2\pi} \int_{-q^2/2M}^{\infty} W_2^{(\nu,\bar{\nu})}(q^2,\nu)d\nu. \tag{5.95}$$

In this case, the Adler sum rule will look like:

$$\lim_{E \to \infty} \left(\frac{d\sigma^{(\nu)}}{d|q^2|} - \frac{d\sigma^{(\bar{\nu})}}{d|q^2|} \right) = \frac{G_F^2}{2\pi} \left[4I_3 c_{\theta_c}^2 + (3Y + 2I_3)s_{\theta_c}^2 \right]. \tag{5.96}$$

Despite the fact that the Adler sum rule was derived in the infinite momentum frame, the final result expressed in terms of the Lorentz invariants is valid in an arbitrary reference frame. A marvellous property of this rule consists in the fact that it does not involve the dependence on q^2. The Adler sum rule could be used for any hadron target with corresponding values of I_3 and Y.

5.4 Goldstone theorem

The $SU(3)_L \times SU(3)_R$ algebra to be generated by different currents suggests that the strong interaction Lagrangian has the form

$$\mathcal{L} = \mathcal{L}_0 + \epsilon \mathcal{L}_I,$$

where $\mathcal{L}_0$ and $\mathcal{L}_I$ are given by Eq. (5.13). Then, in the limit $\epsilon \to 0$, all chiral algebra generators present the conserved quantities. As this takes place, we would expect particles to form generated multiplets corresponding to irreducible representations of the $SU(3)_L \times SU(3)_R$ group. For example, the pseudoscalar meson octet would be accompanied by an scalar meson octet while the $(1/2)^+$-baryons would have partners with the opposite parities. However, in reality, there is no evidence that such a broader multiplet structure exists. This leads us to the conclusion that the $SU(3)_L \times SU(3)_R$ symmetry is spontaneously broken, that is, this symmetry of the Lagrangian $\mathcal{L}_0$ is not realized in particles spectrum. The question arises: *What is the condition of this symmetry violation?* To answer that we will consider a theory with the Hamiltonian H_0 possessing a symmetry with respect to transformations group with matrices U

$$U H_0 U^\dagger = H_0. \tag{5.97}$$

Assume, there is the relation

$$U|A> = |B>, \tag{5.98}$$

where $|A>$ and $|B>$ are states constituting an irreducible group representation. From (5.97) and (5.98) it follows

$$E_A = <A|H_0|A> = <B|H_0|B> = E_B. \tag{5.99}$$

Thus, the symmetry of the Hamiltonian H_0 is displayed in the degeneracy of eigenstates of the energy operator that are associated with irreducible symmetry group representations. However, the relations (5.98) and hence (5.99) are indirectly based on the assumption of a ground state invariance. Indeed, since the states $|A>$ and $|B>$ must be connected with the ground state $|0>$ by means of corresponding production operators

$$|A>= \varphi_A|0>, \qquad |B>= \varphi_B|0>, \tag{5.100}$$

where

$$U\varphi_A U^\dagger = \varphi_B, \tag{5.101}$$

then the relation (5.98) is true only when

$$U|0>= |0>. \tag{5.102}$$

If the condition (5.102) fails, then the relation (5.99) is violated, and, as a result, the conclusion about the symmetry of degenerated energy levels becomes invalid. Such a state is called *spontaneous symmetry breaking* (SSB). There exists a theorem, the Goldstone theorem [47], which states that for every spontaneously broken continuous symmetry, the theory must contain massless particles with zero spin. The massless fields with the spin 0 that arise through the SSB are called *Goldstone bosons* (GBs). Many light bosons seen in physics, such as the pions, may be interpreted (at least approximately) as GBs. When the SSB is realized in the theory with massless gauge fields, the Goldstone theorem may fail. We already faced a similar situation in Section 11.2 of *Advanced Particle Physics Volume 1* considering the SSB in the theory with the $SU(2)_L \times U(1)_Y$ gauge group. In that case might-have-been GBs turn into additional degrees of freedom of gauge bosons. In that process the massless gauge bosons became massive. The spontaneous breaking of discrete symmetries does not also lead to the appearance of Goldstone particles.

At the SSB the necessity of the Goldstone particles appearance may be obviously explained by the example of an isotropic ferromagnetic. In the ground state a magnetization vector **M** is nonzero and arbitrarily oriented in space. Every direction of **M** is associated with its own vacuum (ground state). The vacuum corresponding to the given **M** is invariant with respect to rotations around the axis directed along **M** and is not invariant with respect to any other rotations. To rotate the magnetization vector in a volume $\sim R^3$ one must rotate the spin magnetic moments of $\sim R^3$ particles or excite $\sim R^3$ spin waves (magnons). Let us take into account that range of action of forces between spin magnetic moments (a) is finite. Then, in order to make such a rotation it is necessary to spend an energy only in the surface layer of the volume $R^2 a$ because inside this volume the state is a vacuum. So, at $R \to \infty$ the energy that is accounted for by one magnon ($R^2 a/R^3$) is arbitrarily small and the magnon mass equals zero, that is, the magnon is the Goldstone particle. The assumption concerning finite range of action of forces is crucial. If long-range interaction exists, the conclusion fails. It is precisely this reason that may lead to the violation of the Goldstone theorem for theories with massless gauge fields.

At first we shall deal with the classical fields. As a case in point, we consider a neutral field $\varphi^{(j)}$ ($j = 1, 2, 3$), which is scalar with respect to the Lorentz transformations and isovector with respect to the $SU(2)$ transformations. The Lagrangian of the theory has the form

$$\mathcal{L} = \frac{1}{2}\partial_\mu\varphi^{(j)}\partial^\mu\varphi^{(j)} - V(\varphi^{(j)}), \tag{5.103}$$

where

$$V(\varphi^{(j)}) = \frac{m^2}{2}\varphi^{(j)}\varphi^{(j)} + \lambda(\varphi^{(j)}\varphi^{(j)})^2 \tag{5.104}$$

and we hold $m^2 < 0$. The Lagrangian (5.103) is invariant under isospin rotations

$$\varphi^{(j)} \to \varphi^{(j)\prime} = \left\{\exp[iS_k^{(is)}\theta_k]\right\}_{jl}\varphi^{(l)}, \tag{5.105}$$

where θ_k are rotation angles in the isospin space and $S_k^{(is)}$ are defined by the expressions (9.69) of *Advanced Particle Physics Volume 1*. The minimum of the potential $V(\varphi^{(j)})$ is

$$|\varphi_0| = \sqrt{\varphi^{(1)2} + \varphi^{(2)2} + \varphi^{(3)2}} = \sqrt{\frac{-m^2}{4\lambda}} \equiv a. \qquad (5.106)$$

The potential minimum points lie on the sphere having the radius a that forms a set of degenerate vacua. These vacua are connected with each other by transformations of the symmetry put into the theory. Now, at switching on interaction, physical fields to represent excitations above a vacuum are realized nearly values $\varphi_0 = a$ rather than nearly $\varphi_0 = 0$. To choose a definite vacuum demands settling of definite field values. As a result, a vacuum becomes noninvariant with respect to symmetry transformations. Let us direct the vacuum field value along the third axis in the isospin space

$$\varphi_0 = a\mathbf{e}_3, \qquad (5.107)$$

that is

$$\varphi_0^{(j)} = (0, 0, a).$$

(see Fig. 5.3). Thus, the isovector φ_0 is not invariant under the rotations around the first

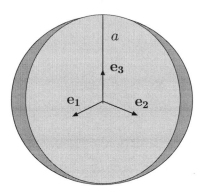

FIGURE 5.3
The vacuum expectation value of the field φ.

and second axes in the isotopic space. However, this vector is invariant with respect to the $U(1)$ subgroup of the $SU(2)$ group that consists of the rotations around the third axis:

$$\varphi_0' = \exp[iS_3^{(is)}\theta_3]\varphi_0 = \varphi_0. \qquad (5.108)$$

At the same time, the potential V is invariant with respect to the whole $SU(2)$ group and just this circumstance results in the appearance of massless particles, GBs. To find them we, as it usually is, set

$$\varphi^{(3)} = \kappa + a. \qquad (5.109)$$

Then, $\varphi^{(1)}, \varphi^{(2)}$, and κ will be physical fields because their vacuum expectation values are equal to zero. Express these fields in terms of the potential

$$V = \frac{m^2}{2}[\varphi^{(1)2} + \varphi^{(2)2} + (\kappa + a)^2] + \lambda[\varphi^{(1)2} + \varphi^{(2)2} + (\kappa + a)^2]^2 =$$

$$= 4a^2\lambda\kappa^2 + 4a\lambda\kappa(\varphi^{(1)2} + \varphi^{(2)2} + \kappa^2) + \lambda(\varphi^{(1)2} + \varphi^{(2)2} + \kappa^2)^2 - \lambda a^4 =$$

$$= \lambda[(\varphi^{(i)}\varphi^{(i)} - a^2)^2 - a^4], \tag{5.110}$$

where the relation (5.106) has been taken into account. In this expression the quadratic term and hence the mass correspond to the field κ only:

$$m_\kappa^2 = 8a^2\lambda, \qquad m_{\varphi^{(1)}}^2 = m_{\varphi^{(2)}}^2 = 0. \tag{5.111}$$

Therefore, after the SSB we have two GBs and one massive scalar field. Note, that the number of the GBs is equal to the number of spontaneously broken symmetry generators. It is not difficult to show that the mass matrix of the theory is given by the expression

$$M_{jk}^2 = \frac{\partial^2 V}{\partial\varphi_0^{(j)}\partial\varphi_0^{(k)}}. \tag{5.112}$$

Now, we can interpret this situation in the very general case. There is a scalar field $\varphi^{(j)}$ with the Lagrangian

$$\mathcal{L} = \frac{1}{2}\partial_\mu\varphi^{(j)}\partial^\mu\varphi^{(j)} - V(\varphi^{(j)}), \tag{5.113}$$

where $\varphi^{(j)}$ $(j = 1, 2, \ldots)$ belong to some representation of the G group of internal symmetry with a dimension dim $G = N$ and V is a function of $\varphi^{(j)}$ being symmetric with respect to the transformations of the G group. This means

$$\delta V - \frac{\partial V}{\partial\varphi^{(j)}}\delta\varphi^{(j)} - i\frac{\partial V}{\partial\varphi^{(j)}}T_{jk}^a\varphi^{(k)}\delta\theta^a = 0, \tag{5.114}$$

where T_{jk}^a is a matrix of a group generator in φ representation, θ^a are group parameters $(a = 1, 2, \ldots, N)$. From Eq. (5.114) follows the symmetry condition

$$\frac{\partial V}{\partial\varphi^{(j)}}T_{jk}^a\varphi^{(k)} = 0. \tag{5.115}$$

Differentiate (5.115) over $\varphi^{(l)}$

$$\frac{\partial^2 V}{\partial\varphi^{(j)}\partial\varphi^{(l)}}T_{jk}^a\varphi^{(k)} + \frac{\partial V}{\partial\varphi^{(j)}}T_{jl}^a = 0. \tag{5.116}$$

The SSB takes place when $V(\varphi)$ has the minimum at $\varphi = v \neq 0$. The minimum will exist at the following conditions

$$\left.\frac{\partial V}{\partial\varphi^{(j)}}\right|_{\varphi=v} = 0, \qquad \left.\frac{\partial^2 V}{\partial\varphi^{(j)}\partial\varphi^{(l)}}\right|_{\varphi=v} = (M^2)_{jl}, \tag{5.117}$$

where the matrix M^2 must be larger or equal to zero $(\det M^2 \geq 0)$. Next, we expand the potential nearly the minimum $\varphi^{(j)} = v^{(j)}$:

$$V(\varphi) = V_0 + \frac{1}{2}(M^2)_{jk}(\varphi - v)^j(\varphi - v)^k + \ldots. \tag{5.118}$$

Substituting (5.118) into the Lagrangian and taking into consideration that the constant V_0 does not play any role in dynamics, we get

$$\mathcal{L} = \frac{1}{2}\partial_\mu\varphi^{(j)}\partial^\mu\varphi^{(j)} - \frac{1}{2}(M^2)_{jk}(\varphi - v)^j(\varphi - v)^k + \ldots. \tag{5.119}$$

It is evident that the matrix (M^2) makes sense of masses square. In the case of the minimum, Eq. (5.116) may be rewritten in the form

$$(M^2)_{jk} T^a_{jl} v^{(l)} = 0. \tag{5.120}$$

The vector $v = (v^{(1)}, v^{(2)}, \ldots, v^{(\hat{N})})$ may be an invariant of a subgroup H belonging to G ($H \in G$)) with a dimension dim $H = \hat{N} < N$. Then

$$T^{\hat{a}}_{jk} v^{(k)} = 0, \tag{5.121}$$

where $\hat{a} = 1, 2, \ldots, \hat{N}$ and $T^{\hat{a}}$ are H-subgroup generators forming Lie algebra. In this case the matrix $(M^2)_{jk}$ has $N - \hat{N}$ zero eigenvalues

$$||(M^2)_{jk}|| = \operatorname{diag}(\lambda_1, \lambda_2, \ldots \lambda_{\hat{N}}, 0, 0, \ldots, 0), \tag{5.122}$$

while, due to (5.121), the remaining $\hat{N}$-equations are trivially satisfied and have $\hat{N}$ solutions with nonzero eigenvalues λ_j. It means that out of the overall number of freedom degrees N, $N - \hat{N}$ are responsible for massless scalar particles, GBs. The rest of the particles $\hat{N}$ are massive and their masses square are positive eigenvalues of the matrix $(M^2)_{jk}$. So, the spontaneous continuous symmetry breaking results in the appearance of massless particles, GBs, whose number is defined by that of generators of the spontaneously broken symmetry. This statement is referred to as the *Goldstone theorem*.

To stress the commonness of the obtained conclusions concerning the number of the GBs, we note that they still stand in two extreme cases as well. When a symmetry is not spontaneously broken, there is only one vacuum being invariant with respect to the G group itself. Then, the number of broken symmetry generators is equal to zero and the GBs are absent. In other extreme case, a vacuum is done so that there is no subgroup H keeping one of the vacuum states φ_0 to be invariant. Then, the H subgroup coincides with the unit element and the GBs' number is equal to a group order.

Our next goal is to find out the status of the developed classical considerations in the quantum field theory. Here, the Goldstone theorem reads: *if some field operator $\varphi(x)$ possesses a nonzero vacuum average*

$$< 0|\varphi(x)|0 > \neq 0, \tag{5.123}$$

which is not a singlet with respect to transformations of some symmetry group, then massless particles must be present in the states spectrum.

In accordance with the Noether theorem, any continuous symmetry of the Lagrangian results in the existence of a conserved current:

$$\partial^\mu j^{(j)}_\mu(x) = 0. \tag{5.124}$$

At ordinary conditions, from this equation the charges conservation law follows

$$\frac{d}{dt} Q^{(j)}(t) = 0,$$

where

$$Q^{(j)}(t) = \int d^3x j^{(j)}_0(\mathbf{r}, t). \tag{5.125}$$

The charges satisfy the commutation relations of a symmetry group

$$[Q^{(j)}, Q^{(k)}] = iC^{jkl} Q^{(l)},$$

where C^{jkl} are structural constants of Lie algebra. The unitary operator associated with a group transformation has the form

$$U = \exp[iQ^{(j)}\theta^j].$$

If a vacuum is invariant under a group transformation, then

$$U|0> = |0>$$

and hence

$$Q^{(j)}|0> = 0, \tag{5.126}$$

that is, the charges annihilate a vacuum. This situation is typical when a symmetry holds. Under the SSB we have degenerate vacua and

$$Q^{(j)}|0> = |0>' \qquad \text{or} \qquad Q^{(j)}|0> \neq 0.$$

In the case (5.123), the quantity $Q^{(j)}$ is not well defined because the field operator of the integrand (5.125) is insufficiently rapidly decreased at infinity. Even a weak limit corresponding to the matrix element $< 0|Q^{(j)2}(t)|0 >$ does not exists. Really, the translation invariance of a vacuum leads to the expression

$$< 0|Q^{(j)2}(t)|0 > = \int d^3x < 0|j_0^{(j)}(0)Q^{(j)}(0)|0 > .$$

Since the integrand is nonzero and independent on $\mathbf{r}$, this expression is divergent. Therefore, the quantity $Q^{(j)}(t)|0 >$ is not available in the Hilbert space because its norm turns into infinity. However, this circumstance does not matter since only commutators with $Q^{(j)}$ appear in real calculations. Inasmuch as the operator $\varphi(x)$ does not represent a group singlet, there must exist an operator $\varphi'(x)$ such that, at some j, the relation

$$[Q^{(j)}, \varphi'(x)] = \varphi(x) \tag{5.127}$$

takes place. Then, taking into account that $< 0|\varphi(x)|0 > \neq 0$, we get the inequality

$$< 0|[Q^{(j)}, \varphi'(x)]|0 > \neq 0. \tag{5.128}$$

In its turn, Eq. (5.128) means, that the relation (5.126) is inapplicable and hence a symmetry is absent in the customary sense (degenerate multiplets do not exist). Now we demonstrate that Eq. (5.128) predicts the existence of massless particles. Substitute the expression (5.125) into (5.128) and inserting a total system of intermediate states, we obtain

$$\sum_n \int d^3y \left[< 0|j_0^{(k)}(y)|n >< n|\varphi'(x)|0 > - < 0|\varphi'(x)|n >< n|j_0^{(k)}(y)|0 > \right]\Bigg|_{x_0=y_0} \neq 0 \tag{5.129}$$

(recall that the commutator (5.128) is taken at equal times). Next, with allowance made for the translation invariance of the current

$$j_0^{(k)}(y) = e^{-ipy}j_0^{(k)}(0)e^{ipy},$$

we rewrite the relation (5.129) in the following manner

$$\sum_n \int d^3y \left\{ < 0|j_0^{(k)}(0)|n >< n|\varphi'(x)|0 > e^{ip_n y} - < 0|\varphi'(x)|n >< n|j_0^{(k)}(0)|0 > e^{-ip_n y} \right\}\Bigg|_{x_0=y_0} =$$

$$= \sum_n (2\pi)^3 \delta^{(3)}(\mathbf{p}_n) \left\{ <0|j_0^{(k)}(0)|n><n|\varphi'(x)|0> e^{iE_n y_0} - \right.$$

$$\left. - <0|\varphi'(x)|n><n|j_0^{(k)}(0)|0> e^{-iE_n y_0} \right\} \Bigg|_{x_0=y_0} =$$

$$= \sum_n (2\pi)^3 \delta^{(3)}(\mathbf{p}_n) \left\{ <0|j_0^{(k)}(0)|n><n|\varphi'(x)|0> e^{im_n y_0} - \right.$$

$$\left. - <0|\varphi'(x)|n><n|j_0^{(k)}(0)|0> e^{-im_n y_0} \right\} \Bigg|_{x_0=y_0} \neq 0, \tag{5.130}$$

where m_n is a mass of an intermediate state n. Note that a vacuum state does not contribute to the sum over n. Now, it only remains for us to prove the expression (5.130) to be independent on y_0. If we shall manage to do it, then we make a conclusion that $m_n = 0$, that is, all intermediate states have zero mass that represents the statement of the Goldstone theorem. Moreover, for the expression (5.130) to be nonzero these states should exist. To prove independence (5.130) on y_0 we integrate the current conservation law (5.124) over the three-dimensional volume to give the result

$$\frac{\partial}{\partial y_0} \int d^3 y j_0^{(k)}(y) = \int d^3 y \boldsymbol{\nabla} \cdot \mathbf{j}^{(k)}(y). \tag{5.131}$$

Next, let us act on the expression (5.128) by the operator $\partial/\partial y_0$

$$\frac{\partial}{\partial y_0} <0|[Q^{(k)}, \varphi'(x)]|0> = \frac{\partial}{\partial y_0} \int d^3 y <0|[j_0^{(k)}(y), \varphi'(x)]|0> =$$

$$= \int d^3 y <0|[\boldsymbol{\nabla} \cdot \mathbf{j}^{(k)}(y), \varphi'(x)]|0> = \oint_S (d\mathbf{S} \cdot <0|[\mathbf{j}^{(k)}(y), \varphi'(x)])|0> . \tag{5.132}$$

Since the closed surface S is arbitrary, it can be taken off at infinity. As a result, the surface integral in (5.132) vanishes. Consequently, the intermediate states $|n>$ represent massless ones, that is, the GBs. As follows from (5.130) these particles possess the following properties:

$$<n|\varphi(x)|0> \neq 0, \qquad <0|j_0^{(k)}(0)|n> \neq 0. \tag{5.133}$$

The relations (5.133) mean that the states $|n>$ can be connected with a vacuum by means of the operator $\varphi(x)$ or the current $j_0^{(k)}(0)$. At this point the proof of the Goldstone theorem is completed.

In the examples considered above the Goldstone particles had zero spin. This circumstance is not obvious. Thus, for example, in theories with the SSB the Goldstone particles with the spin 1/2 exist. Note that quantum numbers of the Goldstone particles are determined by generators with regard to which a vacuum is not invariant.

The relations (5.133) are worthy of notice. Let us show that the direct connection between the GBs and the nonconservation of corresponding currents follows from them. By way of the simplest example, we consider an idealized case, namely, an Abelian-theory describing the interaction of two scalar fields, π-meson and σ fields:

$$\mathcal{L} = \frac{1}{2}[(\partial_\nu \sigma)^2 + (\partial_\nu \pi)^2] - V(\sigma^2 + \pi^2), \tag{5.134}$$

where

$$V(\sigma^2 + \pi^2) = -\frac{\mu^2}{2}(\sigma^2 + \pi^2) + \frac{\lambda}{4}(\sigma^2 + \pi^2)^2. \tag{5.135}$$

The Lagrangian (5.134) exhibits the continuous symmetry $U(1)$ or $O(2)$:

$$\begin{pmatrix} \sigma \\ \pi \end{pmatrix} \rightarrow \begin{pmatrix} \sigma' \\ \pi' \end{pmatrix} = \begin{pmatrix} \cos\theta & \sin\theta \\ -\sin\theta & \cos\theta \end{pmatrix} \begin{pmatrix} \sigma \\ \pi \end{pmatrix}. \tag{5.136}$$

The extremum of the potential V is found from the conditions

$$\frac{\delta V}{\delta \sigma} = \sigma[-\mu^2 + \lambda(\sigma^2 + \pi^2)] = 0,$$

$$\frac{\delta V}{\delta \pi} = \pi[-\mu^2 + \lambda(\sigma^2 + \pi^2)] = 0.$$

When $\mu^2 > 0$, we have minimum at

$$\sigma^2 + \pi^2 = \frac{\mu^2}{\lambda} \equiv v^2.$$

In the plane (σ, π) minima are associated with the points being situated on the circle with the radius v. These points are connected with each other by means of $O(2)$ rotations. Therefore, all of them are equivalent and there exists an infinite number of degenerate vacua. For a theory to be built, it is necessary to chose one of them (one, and only one). Let us opt for the point

$$< 0|\sigma|0 >= v, \qquad < 0|\pi|0 >= 0. \tag{5.137}$$

In so doing, the $U(1)$ symmetry is violated by a vacuum state. To find the particles spectrum in the perturbation theory, we examine small oscillations near the genuine minimum and define the shifted field

$$\sigma' = \sigma - v. \tag{5.138}$$

The Lagrangian expressed in terms of the new fields becomes the view

$$\mathcal{L} = \frac{1}{2}[(\partial_\nu \sigma')^2 + (\partial_\nu \pi)^2] - \mu^2 \sigma'^2 - \lambda v \sigma'(\sigma^2 + \pi^2) - \frac{\lambda}{4}(\sigma'^2 + \pi^2)^2. \tag{5.139}$$

Because in (5.139) a term quadratic in the π field is absent, then π represents a massless Goldstone boson. In turn, the σ' field gets the mass $m_{\sigma'} = \sqrt{2}\mu$ as a result of the SSB.

Track the connection of these properties with the relations (5.133). The conserved current associated with the $U(1)$ symmetry is given by the expression

$$j_\nu(x) = [(\partial_\nu \pi(x)]\sigma(x) - [\partial_\nu \sigma(x)]\pi(x). \tag{5.140}$$

According to the formal proof of the Goldstone theorem, from the symmetry violation condition (5.137) the masslessness of the π particle follows. At that, the π field quanta obey the relations

$$< \pi|\pi(x)|0 >\neq 0, \qquad < 0|j_0(0)|\pi >\neq 0. \tag{5.141}$$

To make sure that the inequalities (5.141) are valid, we note that in the case in question the expression (5.130) takes the form

$$\sum_n (2\pi)^3 \delta^{(3)}(\mathbf{p}_n) \left[< 0|j_0(0)|n >< n|\pi(x)|0 > e^{iE_n t} - \right.$$

$$\left. - < 0|\pi(x)|n >< n|j_0(0)|0 > e^{-iE_n t} \right] = -iv. \tag{5.142}$$

In the left side of this equality, the nonzero contribution is given only a massless π state: $|n >= |\pi(p) >$ where $|\pi(p) >$ is a one-pion state with the momentum p. So, we have

$$\int \frac{d^3 p}{2p_0} \delta^{(3)}(\mathbf{p}_n) \left[< 0|j_0(0)|\pi(p) >< \pi(p)|\pi(x)|0 > - \right.$$

$$- < 0|\pi(x)|\pi(p) >< \pi(p)|j_0(0)|0 >] = -iv. \tag{5.143}$$

If one uses the normalization condition

$$< 0|\pi(x)|\pi(p) >= e^{ipx},$$

the equality (5.143) will be fulfilled at

$$< 0|j_0(0)|\pi(p) >= ivp_0. \tag{5.144}$$

Note, from the covariance condition the demand follows

$$< 0|j_\mu(0)|\pi(p) >= ivp_\mu. \tag{5.145}$$

Consequently, the matrix element of the current divergence will look like:

$$< 0|\partial^\mu j_\mu(0)|\pi(p) >= vm_\pi^2, \tag{5.146}$$

and from the current conservation it is inferred that either

$$v =< 0|\sigma(0)|0 >= 0, \tag{5.147}$$

or

$$m_\pi = 0. \tag{5.148}$$

Our next task is to consider a theory with several currents and ascertain connection of the GBs with them. Consider the $SU(2)_L \times SU(2)_R$ invariant theory describing the interaction of the pion isotriplet $\boldsymbol{\pi} = (\pi_1, \pi_2, \pi_3)$ with the isoscalar σ field. The corresponding Lagrangian is

$$\mathcal{L} = \frac{1}{2}[(\partial_\mu \sigma)^2 + (\partial_\mu \boldsymbol{\pi})^2] - V(\sigma^2 + \boldsymbol{\pi}^2), \tag{5.149}$$

where

$$V(\sigma^2 + \boldsymbol{\pi}^2) = -\frac{\mu^2}{2}(\sigma^2 + \boldsymbol{\pi}^2) + \frac{\lambda}{4}(\sigma^2 + \boldsymbol{\pi}^2)^2. \tag{5.150}$$

The Lagrangian is invariant with respect to the infinitesimal transformations of the $SU(2)_L \times SU(2)_R$ group

$$\left.\begin{array}{l} \boldsymbol{\pi}' = \boldsymbol{\pi} + [\boldsymbol{\alpha} \times \boldsymbol{\pi}], \\ \sigma' = \sigma, \end{array}\right\} \tag{5.151}$$

$$\left.\begin{array}{l} \boldsymbol{\pi}' = \boldsymbol{\pi} - \boldsymbol{\beta}\sigma, \\ \sigma' = \sigma + (\boldsymbol{\beta} \cdot \boldsymbol{\pi}). \end{array}\right\} \tag{5.152}$$

Corresponding conserved currents and charges are given by the expressions

$$J_\mu^a(x) = \epsilon^{abc}\pi^b(x)\partial_\mu\pi^c(x), \qquad Q^a = \int J_0^a(x)d^3x \tag{5.153}$$

$$A_\mu^a(x) = [\partial_\mu\sigma(x)]\pi^a(x) - [\partial_\mu\pi^a(x)]\sigma(x), \qquad Q^{5a} = \int A_0^a(x)d^3x. \tag{5.154}$$

The SSB is realized at $\mu^2 > 0$ and the potential minimum is achieved providing

$$\sigma^2 + \boldsymbol{\pi}^2 = \sqrt{\left(\frac{\mu^2}{\lambda}\right)} \equiv v^2. \tag{5.155}$$

Choosing the point

$$< 0|\boldsymbol{\pi}|0 >= 0, \qquad < 0|\sigma|0 >= v \tag{5.156}$$

as the genuine vacuum and defining the shifted field as $\sigma' = \sigma - v$, we lead to the conclusion that the pions are massless Goldstone bosons. And at the same time, the choice (5.156) means that the axial charges Q^{5a} will not annihilate a vacuum

$$< 0|A^a_\mu(0)|\pi^a >\neq 0, \tag{5.157}$$

that is, the GBs are associated with the nonconservation of the axial current. So, the $SU(2)_L \times SU(2)_R$ symmetry is spontaneously broken up to the $SU(2)$ one generated by the charges Q^a to satisfy

$$Q^a|0 >= 0 \qquad \text{at} \qquad a = 1, 2, 3. \tag{5.158}$$

Note, if $\mu^2 < 0$, the symmetry would not be hidden. As a result, $\boldsymbol{\pi}$ and σ would be degenerate in the mass and form the irreducible representation of the $SU(2)_L \times SU(2)_R$ group.

5.5   Hypothesis of the partially conserved axial current

A Lagrangian symmetry is always reflected by the currents algebra. But at the SSB, particles spectrum realizes only that symmetry part, which inheres a ground state as well. Thus, for example, in the $SU(3)_L \times SU(3)_R$ algebra of the electromagnetic and weak currents at the SSB, a vacuum remains only $SU(3)$-invariant. Then, in accordance with the Goldstone theorem, it follows that eight massless pseudoscalar mesons associated with spontaneously broken axial charges Q^{5a} ($a = 1, \ldots, 8$) should exist. However, in reality, such massless particles are not observed and there are eight relatively light mesons $\pi^+, \pi^-, \pi^0, K^+, K^0, \overline{K}^-, \overline{K}^0$, and η^0. Therefore, the flavor $SU(3)_L \times SU(3)_R$ symmetry should be manifestly violated as well, this violation of the chiral symmetry affecting masses of 0^- mesons. It means that the total Lagrangian is represented by the sum of two terms

$$\mathcal{L} = \mathcal{L}_0 + \epsilon\mathcal{L}_I, \tag{5.159}$$

where $\mathcal{L}_0$ is invariant under chiral transformations while $\mathcal{L}_I$ does not possess such an invariance. The π-meson isotriplet is considerably lighter than K and η^0 mesons. To explain this circumstance one should divide the Lagrangian to break the symmetry into two parts:

$$\epsilon\mathcal{L}_I = \epsilon_1\mathcal{L}_{I1} + \epsilon_2\mathcal{L}_{I2}, \tag{5.160}$$

where the Lagrangian $\mathcal{L}_{I1}$ ($\mathcal{L}_{I2}$) is invariant (noninvariant) with respect to the $SU(2)_L \times SU(2)_R$ transformations and $\epsilon_1 \gg \epsilon_2$. In such a way, the $SU(2)_L \times SU(2)_R$ symmetry is performed better than the $SU(3)_L \times SU(3)_R$ one. Really, in the free quark model the chiral symmetry is explicitly broken by the quark mass term (5.13) at $m_s \gg m_u, m_d$ allowing us to suppose $\epsilon_1\mathcal{L}_{I1} = m_s\bar{s}s$ and $\epsilon_2\mathcal{L}_{I2} = m_u\bar{u}u + m_d\bar{d}d$. Thus, we think that the masses of the π mesons are proportional to the strangeless quark masses while the masses of the kaons and the η^0 meson are proportional to the strange quark mass.

Let us investigate the physical consequences of dividing the Lagrangian (5.160). Since, at the chiral symmetry breaking, the axial currents are not conserved, the following relation

$$< 0|A^\alpha_\mu(0)|\pi^\beta(p) >= if^{\alpha\beta}p_\mu, \qquad \alpha, \beta = 1, 2, 3, \tag{5.161}$$

where $f^{\alpha\beta}$ is a constant quantity ($f^{\alpha\beta} \neq 0$), takes place. If one assumes that the isospin symmetry is not broken, then $f^{\alpha\beta}$ can be represented in the form

$$f^{\alpha\beta} = f_\pi \delta^{\alpha\beta}, \tag{5.162}$$

where f_π is the π-meson decay constant to be measured in the process

$$\pi^+ \to l^+ + \nu_l$$

($f_\pi \approx 93$ MeV). Having taken the divergence of (5.161), we get

$$< 0|\partial^\mu A_\mu^\alpha(0)|\pi^\beta(p) > = \delta^{\alpha\beta} f_\pi m_\pi^2. \tag{5.163}$$

So, if one sets the parameter ϵ_2 to equal zero in the expression (5.160), the Lagrangian will be symmetric with respect to the $SU(2)_L \times SU(2)_R$ transformations and

$$\partial^\mu A_\mu^\alpha(0) = 0. \tag{5.164}$$

From (5.164) immediately follows that $m_\pi^2 = 0$ in (5.163), which is in agreement with the Goldstone theorem. But in the case, when the symmetry is explicitly broken ($\epsilon_2 \neq 0$), we can rewrite (5.163) in the view

$$< 0|\partial^\mu A_\mu^\alpha(0)|\pi^\beta(p) > = f_\pi m_\pi^2 < 0|\varphi^\alpha(0)|\pi^\beta(p) >, \tag{5.165}$$

where φ^α is the π-meson field operator satisfying the normalization condition

$$< 0|\varphi^\alpha(0)|\pi^\beta(p) > = \delta^{\alpha\beta}.$$

An assumption, that the more general operator relation

$$\partial^\mu A_\mu^\alpha(x) = f_\pi m_\pi^2 \varphi^\alpha(x) \tag{5.166}$$

is true, is known as the hypothesis of a partially conserved axial current (PCAC) [48, 49]. Generalization on the $SU(3)_L \times SU(3)_R$ symmetry is realized at once. In the case of the axial-vector currents octet, the PCAC relation will look like:

$$\partial^\mu A_\mu^a(x) = f_a m_a^2 \varphi^a(x), \qquad a = 1, 2, \dots, 8, \tag{5.167}$$

where φ^a are field operators of the pseudoscalar meson octet.

Consequences of (5.166) have been checked up in series of processes with participation of low-energy π mesons. Since at the derivation of (5.166) the total energy of the π meson has been neglected, then predictions bear the approximate character. Having defined the axial current by the expression

$$A_\mu^\alpha(x) = \overline{u}(x)\gamma_\mu \frac{\tau^\alpha}{2}\gamma_5 d(x) \tag{5.168}$$

and taken the divergence of it, we obtain

$$\partial^\mu A_\mu^\alpha(x) = (m_u + m_d)\overline{u}(x)\frac{\tau^\alpha}{2}\gamma_5 d(x), \tag{5.169}$$

where m_u and m_d are the current masses of the u- and d-quarks, respectively. The last quantity has the quantum numbers of the π^+ meson. Consequently, it can be used as a compound pion field and, as a result, we arrive at Eq. (5.166). The relation (5.169) is employed to estimate current quark masses. Other well-known consequences of the PCAC hypothesis are: (i) calculations of amplitudes of pion-hadron scattering [50, 51], (ii) relations between matrix elements of weak decays of the K mesons [52], and so on. The most well-known result of the PCAC is the Goldberger–Treiman relation [53], the derivation of which we now proceed.

Consider the neutron β-decay. The matrix element of the weak axial current not violating the strangeness is

$$< p(k'|A_1^\mu(x) + iA_2^\mu(x)|n(k) >\equiv< p(k'|A_{ud}^\mu(x)|n(k) >,$$

where $< p(k'$ and $|n(k) >$ are nucleon states. For the reasons of the relativistic covariance this quantity should be given by the expression

$$< p(k'|A_{ud}^\mu(x)|n(k) >= -\exp(-iqx)\overline{\psi}_p(k')[\gamma^\mu\gamma_5 g_A(q^2) + q^\mu\gamma_5 h_A(q^2)]\psi_n(k), \qquad (5.170)$$

where $q = k - k'$ and g_A, h_A are nucleon form factors. When analyzing the different contributions to (5.170), we shall act in the following way. First, we forget about quarks and assume that the neutron is a naked particle, and then add various corrections caused by the strong interaction. The bare neutron decay is associated with the diagram pictured in Fig. 5.3(a), which contributes $\sim \overline{\psi}_p(k')\gamma_\mu\gamma_5\psi_n(k)$ into (5.170). Pion corrections to the

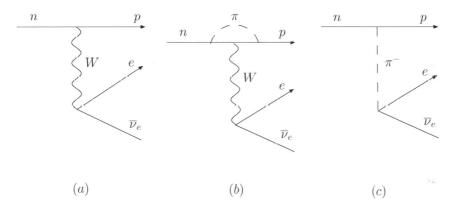

(a) (b) (c)

FIGURE 5.4
The bare neutron decay in (a)–(c).

vertex should be added to this contribution (Fig. 5.3.(b)). Of course, no one knows how to calculate a total contribution of a correction connected with the diagram shown in Fig. 5.3(b) and analogous corrections of higher orders. However, we undoubtedly are aware that all such corrections contribute to $g_A(q^2)$. In spite of the considered diagrams, there exists a one-pion exchange diagram (Fig. 5.3(c)) (and higher order diagrams associated with it too). In accordance with the common rules, the amplitude associated with the diagram shown in Fig. 5.3(c) is

$$\mathcal{A}^c = (\text{vertex } n \to p\pi^-) \times (\text{propagator } \pi^-) \times (\text{vertex } \pi^- \to e^-\overline{\nu}_e). \qquad (5.171)$$

However,

$$(\text{vertex } n \to p\pi^-) = ig_{\pi NN}\sqrt{2}\overline{\psi}_p(k')\gamma_5\psi_n(k), \qquad (5.172)$$

where $g_{\pi NN}$ is the pion-nucleon coupling constant,

$$(\text{vertex } \pi^- \to e^-\overline{\nu}_e) = i\frac{G_F}{\sqrt{2}}f_\pi q_\mu L^\mu, \qquad (5.173)$$

and $L^{\mu} = \overline{\psi}_e \gamma^{\mu}(1 - \gamma_5)\psi_{\nu_e}$. To substitute (5.172) and (5.173) into (5.171) results in

$$\mathcal{A}^c = -\frac{G_F}{\sqrt{2}} f_{\pi} g_{\pi NN} \sqrt{2} \frac{1}{q^2 - m_{\pi}^2} \overline{\psi}_p(k') q_{\mu} \gamma_5 \psi_n(k) L^{\mu}. \tag{5.174}$$

It is clear that this expression contributes to the term $h_A(q^2)$ of Eq. (5.170), and if other corrections would be absent, then

$$h_A(q^2) = -f_{\pi} g_{\pi NN} \frac{\sqrt{2}}{q^2 - m_{\pi}^2}. \tag{5.175}$$

As a mater of fact, taking into account of the vertex corrections gives

$$h_A(q^2) = -f_{\pi} g_{\pi NN} F(q^2) \frac{\sqrt{2}}{q^2 - m_{\pi}^2}, \tag{5.176}$$

where $F(q^2)$ is a smooth function of q^2 satisfying

$$F(q^2)\bigg|_{q^2 = m_{\pi}^2} = 1.$$

Diagrams of many-pions exchange would give additional corrections to (5.176), but we neglect them. Thus, (5.170) takes the form

$$< p(k'|A_{ud}^{\mu}(x)|n(k) >= -\exp(-iqx)\overline{\psi}_p(k')[\gamma^{\mu} g_A(q^2) - \frac{f_{\pi} g_{\pi NN} F(q^2)\sqrt{2}}{q^2 - m_{\pi}^2} q^{\mu}]\gamma_5 \psi_n(k). \tag{5.177}$$

Having taken the divergence of both sides of the relation (5.177) and used the PCAC hypothesis

$$\lim_{m_{\pi} \to 0} \partial_{\mu} A_{ud}^{\mu}(x) = 0, \tag{5.178}$$

we get

$$0 = \lim_{m_{\pi} \to 0} \overline{\psi}_p(k')[g_A(q^2)(m_p + m_n) - \frac{f_{\pi} g_{\pi NN} F(q^2)\sqrt{2}}{q^2 - m_{\pi}^2} q^2]\gamma_5 \psi_n(k). \tag{5.179}$$

If one also assumes $F(q^2)$ to be a slowly changing function, then it can be extrapolated from point $q^2 = m_{\pi}^2$ to point $q^2 = 0$, where

$$F(0) = 1.$$

Then, Eq. (5.179) leads to the Goldberger–Treiman relation

$$g_A(0) \approx \frac{f_{\pi} g_{\pi NN}}{\sqrt{2} m_p}, \tag{5.180}$$

that coincides with experimental results to an accuracy of 5%.

For kaons the PCAC hypothesis reads

$$\partial_{\mu} A_{us}^{\mu}(x) = f_K m_K^2 \varphi_K(x), \tag{5.181}$$

where f_K is a constant fixing the intensity of the decay $K^- \to \mu^- + \overline{\nu}_{\mu}$ and $\varphi_K(x)$ is a wave kaon function. The experimental value of $f_K = 110$ MeV is consistent with that of f_{π} to an accuracy of 20%. This would be expected in reality, since in the limit $m_{\alpha} \to 0$ ($\alpha = u, d, s$)

the difference between pions and kaons is absent and $f_K = f_\pi$. The fact, that in the real world the values of f_π and f_K prove to be so close, is a weighty argument in favor of chiral symmetry.

Thus, in order to obtain relations connecting physical quantities from the PCAC hypothesis, an additional assumption concerning smoothness of the function $F(q^2)$ is needed. In particular, we have to extrapolate the pion field from the point, being situated beyond the mass shell $q^2 = 0$, to the mass shell $q^2 = m_\pi^2$. We believe that this extrapolation makes a modest mistake because the pion mass m_π^2 is small compared with masses of other hadrons. The Goldberger–Treiman relations (5.180) serve an accuracy scale that is typical for these types of extrapolations. It means, if one extends the PCAC hypothesis to other pseudoscalar mesons, that is, K- and η^0-mesons, then the extrapolation must be fulfilled in a much wider kinematic range (from 0 to m_K^2 or m_η^2). For this reason, relations, to follow from the PCAC for kaons and the η-meson, are not so good as those for pions.

5.6 $U(1)$-problem

The $SU(3) \times U(1)$ vector transformations represent explicit symmetries of the strong interaction and the corresponding currents lead to the conservation laws. The axial transformations, which are orthogonal to them, do not correspond to any obvious symmetries of the strong interaction. In 1961 Nambu and Jona-Lasinio [54] put forward the hypothesis that axial transformations are exact spontaneously broken symmetries of the strong interaction. This idea results in correct and surprisingly detailed description of strong interaction properties at low energies.

The QCD Lagrangian $\mathcal{L}_{QCD}$ undoubtedly possesses all the strong interaction symmetries. It conserves the charge and space parities. Since gluons do not depend on the flavor, $\mathcal{L}_{QCD}$ conserves the strangeness, charm, beauty, and so on. Obviously, $\mathcal{L}_{QCD}$ has all the flavor symmetries of the free quark model. In the limit $m_u = m_d = m_s = 0$, the QCD Lagrangian is invariant with respect to the chiral transformations just as it does in the free quark model. The diagonal subgroups $SU(3)$ and $U(1)$ are realized in the ordinary way, that is, a vacuum is invariant under these transformations. Hadrons form degenerate $SU(3)$-multiplets and the baryon number is conserved. Other symmetries, axial $SU(3)_A$ and $U(1)_A$ symmetries, are not associated with explicit particles degeneration. As any elementary scalar fields are not used in the theory, then one should assume that these axial symmetries are broken by the QCD vacuum. It is easy to understand the reason for spontaneous chiral symmetry breaking. In superconductivity theory a small electron-electron attraction leads to the appearance of electron pairs condensate in a ground metal state. In QCD, quarks and antiquarks exhibit a strong attracting force and when these quarks are massless, then the energy of producing additional quark-antiquark pairs is small. Therefore, one would expect that the QCD vacuum contains quark-antiquark pairs condensate. These fermion pairs should possess zero values both of total momentum and of angular momentum. For this reason, at the pairing of left-handed quarks with right-handed antiquarks, the pairs must have a nonzero chiral charge (Fig. 5.5). A vacuum state with quark pairs condensate is characterized by a nonzero vacuum expectation value of a scalar operator

$$< 0|\bar{q}q|0 >=< 0|(\bar{q}_L q_R + \bar{q}_R q_L)|0 >\neq 0. \qquad (5.182)$$

Such a nonzero vacuum expectation value means that a vacuum mixes two quark helicities. This allows $u-$, $d-$, and s-quarks to acquire effective masses during a motion through a

FIGURE 5.5

The quark-antiquark pair with zero total momentum and angular momentum.

vacuum. Inside quark-antiquark bound states, $u-, d-$, and s-quarks will be moving as though they have finite effective masses even if they had zero masses in the initial QCD Lagrangian. The quark condensate (5.182) may be used as an order parameter. When the chiral symmetry has been broken, then

$$< 0|\bar{q}q|0 > \neq 0,$$

while the chiral symmetry has been restored, then

$$< 0|\bar{q}q|0 > = 0.$$

In accordance with the Goldstone theorem, at the spontaneous breaking of the $SU(3)_A$ symmetry, eight massless pseudoscalar mesons must be present in the hadron sector. The given chiral symmetry is also broken in an explicit way because three mass terms associated with light quarks were introduced into the Lagrangian. As a result of such a violation (it is called *soft violation*), the GBs are not strictly massless. Now, they may be identified with three π mesons, four K mesons, and a single η^0 meson.

$U(1)_A$ symmetry is exhibited by the theory invariance under the transformations

$$q_k \rightarrow q'_k = \exp(i\gamma_5\theta)q_k, \tag{5.183}$$

where θ takes on the same value for all k, that is, u_L, d_L, s_L are multiplied by the common phase factor $\exp(-i\theta)$, and u_R, d_R, s_R—by $\exp(i\theta)$. Because we do not observe the doubling of hadrons by the parity, then this approximate symmetry is not realized by the ordinary way. On the other hand, the SSB must lead to the existence of the ninth pseudoscalar meson. As calculations show [55], the mass of this meson m' is comparable with the pion mass

$$m' \leq \sqrt{3}m_\pi.$$

However, such a strong interacting particle is not observed. This circumstance is the essence of a so-called $U(1)_A$ problem. As will be shown later, this problem can be solved by means of taking into account the nonperturbative effects that violate the $U(1)_A$ symmetry.

6

Anomalies

> In the distorted world all dogmas are arbitrary,
> including the dogma about the arbitrariness of dogmas too.
> Robert Sheckley

6.1 Adler–Bell–Jackiw anomaly (perturbative approach)

Some conservation laws, to be valid in classical theory, cease to be fulfilled in quantum field theory (QFT) at the right inclusion of quantum effects. These phenomena are called *anomalies*. Their origin is connected with ultraviolet divergences in QFT. Even in renormalizable theories, the presence of divergences causes us to apply regularization or cut-off in real calculations. In some cases, regularization is impossibly carried out so that all symmetry demands of the initial classical field theory will be simultaneously satisfied. Moreover, the regularization used may violate the initial symmetry and even after it will be removed in the end of calculations, violation tracks may be left. As a case in point, we consider QED. Here, we have the vector current conservation law:

$$\frac{\partial}{\partial x_\mu} J_\mu(x) = 0, \tag{6.1}$$

where $J_\mu(x) = e\overline{\psi}(x)\gamma_\mu\psi(x)$. Along with this current, one may examine the axial-vector current: $J_{\mu 5}(x) = e\overline{\psi}(x)\gamma_\mu\gamma_5\psi(x)$. Thanks to the Dirac equation, the divergence of the axial-vector current is given by the expression

$$\frac{\partial}{\partial x_\mu} J_{\mu 5}(x) = 2iem\overline{\psi}(x)\gamma_5\psi(x) \equiv 2mJ_5(x), \tag{6.2}$$

where a quantity J_5 is called a *chiral density*. From this equation it follows that, in the limit of a zero electron mass, the axial-vector current is conserved. This represents the consequence of chiral symmetry. However, the more accurate consideration demonstrates that this conclusion fails. Really, in the definition of the axial-vector current, the product of anticommuting operators $\overline{\psi}(x)$ and $\psi(x)$ taken in one and the same point x is contained. Such a product needs extension of the definition (regularization). If the regularization is carried out so that the vector current conservation law is not violated,* then it turns out that the right expression for the divergence of the axial-vector current becomes the view:

$$\frac{\partial}{\partial x_\mu} J_{\mu 5}(x) = 2mJ_5(x) - \frac{e^2}{16\pi^2}\epsilon_{\mu\nu\lambda\sigma}F^{\mu\nu}(x)F^{\lambda\sigma}(x), \tag{6.3}$$

---

*Inviolability of this law is connected with the fact that it guarantees the charge conservation law.

where $F^{\mu\nu}$ is the electromagnetic field tensor. Thus, the axial-vector current is not conserved even in the limit of a massless electron. This phenomenon is named an *axial anomaly*. It was discovered by Schwinger in 1951 [56] under analysis of the decay $\pi^0 \to \gamma\gamma$. In 1969 Bell and Jackiw [57] found out that violation of the chiral symmetry is a source of this anomaly. The violation, in its turn, is caused by introducing an regularization that is needed to obtain consequences resulting from the neutral axial-vector current conservation for one-loop Feynman diagrams. In gauge theories Ward identities are associated with current conservation laws. Therefore, in theories with fermions, axial Ward identities are not automatically true. In QED chiral anomalies were discovered by Adler [58] under the investigation of axial-vector Ward identities. Afterwards, these anomalies are called *Adler–Bell–Jackiw (ABJ) ones*. The appearance of the ABJ anomalies happen because particular one-loop diagrams give anomalous terms to prevent the reproduction of Ward identities in higher orders of the perturbation theory. In QED and QCD, the vector interaction of the type

$$g_V J_\mu W^\mu,$$

where W^μ is a gauge field, $J_\mu \sim \overline{\psi}\gamma_\mu\psi$ is a vector current of Fermi matter fields and internal symmetry indices have been omitted, is the unique type of interaction between matter and gauge fields. Just due to the vector current conservation law, the Ward identity for a vertex function is fulfilled. Thus, for example, the graph shown in Fig. 6.1 is associated with the amplitude

$$\mathcal{A} = < p'|g_V J_\mu W^\mu|p >$$

and we safely get

$$(p' - p)^\mu J_\mu = q^\mu J_\mu = 0.$$

However, in the presence of an axial-vector coupling between matter and gauge fields, as

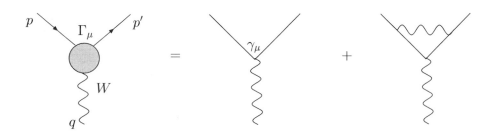

FIGURE 6.1
The expansion of the vector coupling between the gauge field and matter fields with an accuracy of e^3.

this takes place in the electroweak interaction theory, the situation becomes complicated. In Fig. 6.2 we display the expansion of the axial-vector vertex $\Gamma_{\mu 5}$ to an accuracy of g^5. In every diagram entering into this expansion, the low vertex equals $\gamma_\mu\gamma_5$, that is, it is the axial coupling, while other vertices represent the vector couplings. Calculations show that the last graph containing the triangular closed loop of the fermion fields does not satisfy the axial Ward identities resulting in the axial (chiral or triangular) anomaly. Great importance of this anomaly is connected with the fact that the Ward identities represent the essential element under proving the gauge theories renormalizability. The only way for renormalizability rescue is to ensure the total contribution of the triangle graphs to be equal

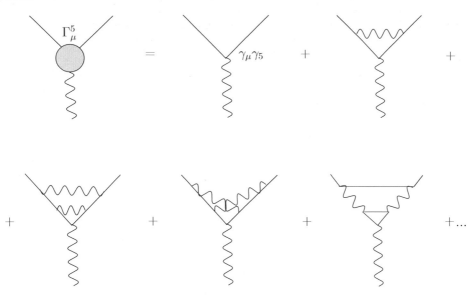

FIGURE 6.2

The expansion of the axial-vector coupling between the gauge field and matter fields with an accuracy of e^5.

to zero, that is, to secure mutual anomalies cancellation. This, in its turn, puts particular demands on the fermion theory contents.

Two fermion triangles AVV containing the axial anomaly ($\gamma_\mu \gamma_5$ corresponds to the A-vertex and γ_μ—to V-vertex) are displayed in Fig. 6.3. Aside from the graphs pictured in

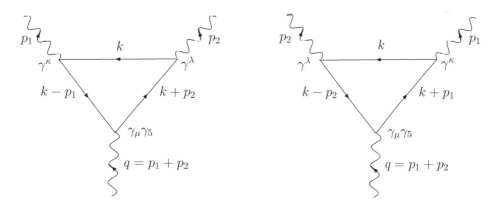

FIGURE 6.3

The triangle diagrams.

Fig. 6.3, the AAA triangle, along with quadratic and pentagonal configurations that enter into diagrams of higher orders, generates the axial anomalies. However, if the anomaly is cancelled in the AVV graphs, it disappears in all the indicated graphs [59]. Calculations also show that the inclusion of radiative corrections does not affect the anomalies [60]. Thus, in order to obtain the axial anomaly within the perturbation theory, we need to choose the particular process and constrain ourselves by considering the diagrams displayed in Fig. 6.3.

By way of example let us consider the following situation. In the vertices 2 and 3 two photons with momenta p_1, p_2 are produced, while in the vertex 1 a pseudoscalar particle A is annihilated. An effective Lagrangian can be represented as the product of an axial-vector current $J_{\mu 5}(x)$ by combination of a wave function of a particle A with some four-dimensional vector. If one amputates the external line associated with the particle A, the obtained diagram will describe the transition from the vacuum state to the two-photon state. The corresponding matrix element is

$$\int \exp(-iqx) < p_1, p_2 | J_{\mu 5}(x) | 0 >=< p_1, p_2 | J_{\mu 5}(q) | 0 > .$$

On the other hand, in accordance with the Feynman rules, we have

$$< p_1, p_2 | J_{\mu 5}(q) | 0 >= T_{\kappa\lambda\mu}(p_1, p_2) e^{*\kappa}(\mathbf{p}_1) e^{*\lambda}(\mathbf{p}_2), \tag{6.4}$$

where the fermion contribution to the amplitude is given by the expression:

$$T_{\kappa\lambda\mu}(p_1, p_2) = N_{\kappa\lambda\mu}(p_1, p_2) + N_{\lambda\kappa\mu}(p_2, p_1), \tag{6.5}$$

$$N_{\kappa\lambda\mu}(p_1, p_2) = -(-ie)^2 \int \frac{d^4k}{(2\pi)^4} \mathrm{Sp} \left(\gamma_\kappa \frac{i}{\hat{k} - \hat{p}_1 - m} \gamma_\mu \gamma_5 \frac{i}{\hat{k} + \hat{p}_2 - m} \gamma_\lambda \frac{i}{\hat{k} - m} \right), \tag{6.6}$$

while the expression for $N_{\lambda\kappa\mu}(p_2, p_1)$ follows from (6.6) under the replacement $\kappa \leftrightarrow \lambda$, $p_1 \leftrightarrow p_2$. To avoid complications we investigate the case of massless fermions. Then, the expression (6.6) is reduced to

$$N_{\kappa\lambda\mu}(p_1, p_2) = -ie^2 \int \frac{d^4k}{(2\pi)^4} \frac{\mathrm{Sp} \left[\gamma_\kappa (\hat{k} - \hat{p}_1) \gamma_\mu \gamma_5 (\hat{k} + \hat{p}_2) \gamma_\lambda \hat{k} \right]}{(k - p_1)^2 (k + p_2)^2 k^2}. \tag{6.7}$$

The required Ward identities arising from the conservation laws of the axial and vector currents in three vertices have the form

$$(p_1 + p_2)^\mu T_{\kappa\lambda\mu}(p_1, p_2) = 0 \qquad (A), \tag{6.8}$$

$$p_1^\kappa T_{\kappa\lambda\mu}(p_1, p_2) = 0 \qquad (V), \tag{6.9}$$

$$p_2^\lambda T_{\kappa\lambda\mu}(p_1, p_2) = 0 \qquad (V). \tag{6.10}$$

Since the integral (6.6) is symmetric with respect to the replacement $(p_1, \kappa) \leftrightarrow (p_2, \lambda)$, the presence of the crossed graphs simply results in the factor 2. So, the identities (6.8)–(6.10) have to be fulfilled for the quantity $N_{\mu\nu\lambda}$ at once.

At first glance it seems that there are no reasons for the violation of these relations. Let us try to confirm this point of view by calculations. Start with the check of the identity (6.8). Acting on (6.7) by the operator $(p_1 + p_2)^\mu$, we find

$$(p_1 + p_2)^\mu N_{\kappa\lambda\mu}(p_1, p_2) = -ie^2 \int \frac{d^4k}{(2\pi)^4} \frac{\mathrm{Sp} \left[\gamma_\kappa (\hat{k} - \hat{p}_1)(\hat{p}_1 + \hat{p}_2) \gamma_5 (\hat{k} + \hat{p}_2) \gamma_\lambda \hat{k} \right]}{(k - p_1)^2 (k + p_2)^2 k^2}. \tag{6.11}$$

Using the identity

$$(\hat{p}_1 + \hat{p}_2) \gamma_5 = -(\hat{k} - \hat{p}_1) \gamma_5 - \gamma_5 (\hat{k} + \hat{p}_2),$$

we transform the expression obtained in the following way

$$(p_1 + p_2)^\mu N_{\kappa\lambda\mu}(p_1, p_2) = ie^2 \int \frac{d^4k}{(2\pi)^4} \left\{ \frac{\mathrm{Sp}\left[\gamma_\kappa\gamma_5(\hat{k} + \hat{p}_2)\gamma_\lambda\hat{k}\right]}{k^2(k + p_2)^2} + \frac{\mathrm{Sp}\left[\gamma_\kappa(\hat{k} - \hat{p}_1)\gamma_5\gamma_\lambda\hat{k}\right]}{k^2(k - p_1)^2} \right\}.$$

(6.12)

Demonstrate that all of the terms being in the right side of (6.12) turn into zero. Carrying out the spur operation and employing the Feynman parametrization, we get for the first integral

$$\int d^4k \frac{\mathrm{Sp}\left[\gamma_\kappa\gamma_5(\hat{k} + \hat{p}_2)\gamma_\lambda\hat{k}\right]}{k^2(k + p_2)^2} = \int d^4k \frac{-4i\epsilon_{\kappa\nu\lambda\sigma}(k + p_2)^\nu k^\sigma}{k^2(k + p_2)^2} =$$

$$= \int_0^1 dx \int d^4k \frac{-4i\epsilon_{\kappa\nu\lambda\sigma}p_2^\nu k^\sigma}{[k^2 + 2(p_2k)x + p_2^2x]^2}.$$

(6.13)

For the divergent integral to be taken, we use the formula (21.26) of *Advanced Particle Physics Volume 1*

$$\int_0^1 dx \int d^4k \frac{-4i\epsilon_{\kappa\nu\lambda\sigma}p_2^\nu k^\sigma}{[k^2 + 2(p_2k)x + p_2^2x]^2} = 4\pi^2\epsilon_{\kappa\nu\lambda\sigma}p_2^\sigma p_2^\nu \int_0^1 xdx \left\{ \ln\left[\frac{L^2}{p_2^2x(x-1)}\right] - \frac{3}{2} \right\} - 0.$$

(6.14)

Analogous computations give a zero value for the second integral in (6.12) as well. So, the Ward identity for the axial-vector current proves to be valid. However, the happy final is still a long way off. Renormalizability of the theory needs the fulfillment of all the Ward identities.

Let us proceed to an analysis of the relation (6.9). We have

$$p_1^\kappa N_{\kappa\lambda\mu}(p_1, p_2) = -ie^2 \int \frac{d^4k}{(2\pi)^4} \frac{\mathrm{Sp}\left[\hat{p}_1(\hat{k} - \hat{p}_1)\gamma_\mu\gamma_5(\hat{k} + \hat{p}_2)\gamma_\lambda\hat{k}\right]}{(k - p_1)^2(k + p_2)^2k^2}.$$

(6.15)

Carrying out the variables replacement

$$k'_\mu = (k + p_2)_\mu$$

(6.16)

and making use of the cyclicity property under the spur sign, we get

$$p_1^\kappa N_{\kappa\lambda\mu}(p_1, p_2) = -ie^2 \int \frac{d^4k'}{(2\pi)^4} \frac{\mathrm{Sp}\left[(\hat{k}' - \hat{p}_2)\hat{p}_1(\hat{k}' - \hat{p}_1 - \hat{p}_2)\gamma_\mu\gamma_5\hat{k}'\gamma_\lambda\right]}{(k' - p_2)^2(k' - p_1 - p_2)^2k'^2}.$$

(6.17)

With the help of the identity

$$\hat{p}_1 = (\hat{k}' - \hat{p}_2) - (\hat{k}' - \hat{p}_1 - \hat{p}_2)$$

the expression (6.17) could be represented in the form

$$p_1^\kappa N_{\kappa\lambda\mu}(p_1, p_2) = -ie^2 \int \frac{d^4k'}{(2\pi)^4} \left\{ \frac{\mathrm{Sp}\left[(\hat{k}' - \hat{p}_2 - \hat{p}_1)\gamma_\mu\gamma_5\hat{k}'\gamma_\lambda\right]}{(k' - p_1 - p_2)^2k'^2} - \frac{\mathrm{Sp}\left[(\hat{k}' - \hat{p}_2)\gamma_\mu\gamma_5\hat{k}'\gamma_\lambda\right]}{(k' - p_2)^2k'^2} \right\}.$$

(6.18)

Employing the same methods as in the case of analyzing the expression (6.12), we find that (6.18) turns into zero. In a similar manner, one may make sure of the performance of Eq. (6.10). To this end, we fulfill the replacement

$$k''_\mu = (k - p_1)_\mu$$

(6.19)

and use the identity

$$\hat{p}_2 = (\hat{k}'' + \hat{p}_1 + \hat{p}_2) - (\hat{k}'' + \hat{p}_1).$$

It would seem the validity of the Ward identities has been checked and we have no causes for concerns. However, our considerations contained a mistake. The integration variable replacements (6.16) and (6.19) are permissible only for convergent and logarithmic divergent integrals. And we have applied these replacements for linear divergent integrals. Now, in order to be logical, in the considered integrals we should single out linear divergent terms and calculate additions arising under the integration variable replacements (6.16) and (6.19).

The linearly divergent part in the expression $N_{\kappa\lambda\mu}$ has the view

$$N_{\kappa\lambda\mu}^d(p_1, p_2) = -ie^2 \int \frac{d^4k}{(2\pi)^4} \frac{\mathrm{Sp}\left(\gamma_\kappa \hat{k}\gamma_\mu\gamma_5\hat{k}\gamma_\lambda\hat{k}\right)}{k^6}. \tag{6.20}$$

If the variable k is transformed using the shift

$$k'_\mu = (k + a)_\mu, \tag{6.21}$$

then changing the quantity $N_{\kappa\lambda\mu}^d$ is found in the following way:

$$\int d^4k f(k) = \int d^4k' f(k' - a) = \int d^4k' \left[f(k') - \frac{\partial f}{\partial k} a + \dots \right]. \tag{6.22}$$

With the help of the four-dimensional Gauss theorem, the last term in (6.22) can be transformed into the surface integral. Since $f \sim k^{-3}$ and the hypersurface area is $\sim k^3$, then this surface integral is not equal to zero. Accomplishing the replacement (6.21) in (6.20) and taking into account (6.22), we arrive at

$$N_{\kappa\lambda\mu}^{d'} = N_{\kappa\lambda\mu}^d + M_{\kappa\lambda\mu\nu}^d a^\nu, \tag{6.23}$$

where

$$M_{\kappa\lambda\mu\nu}^d = -ie^2 \int \frac{d^4k}{(2\pi)^4} \frac{\partial}{\partial k^\nu} \left[\frac{\mathrm{Sp}\left(\gamma_\kappa \hat{k}\gamma_\mu\gamma_5\hat{k}\gamma_\lambda\hat{k}\right)}{k^6} \right]. \tag{6.24}$$

For the spur of γ-matrices to be found, the formula

$$\mathrm{Sp}\left(\gamma_5\gamma_\rho\gamma_\nu\gamma_\sigma\gamma_\kappa\gamma_\tau\gamma_\mu\right) = 4i[\epsilon_{\kappa\tau\mu\lambda}(\delta_\rho^\lambda g_{\nu\sigma} - \delta_\nu^\lambda g_{\rho\sigma} + \delta_\sigma^\lambda g_{\rho\nu}) -$$

$$-\epsilon_{\rho\nu\sigma\lambda}(\delta_\kappa^\lambda g_{\tau\mu} - \delta_\tau^\lambda g_{\kappa\mu} + \delta_\mu^\lambda g_{\kappa\tau})] \tag{6.25}$$

is needed. With its help the quantity $M_{\kappa\lambda\mu\nu}^d$ takes the form

$$M_{\kappa\lambda\mu\nu}^d = 4e^2 \int \frac{d^4k}{(2\pi)^4} \frac{\partial}{\partial k^\nu} \left[\frac{\epsilon_{\kappa\lambda\mu\rho} k^\rho k^2}{k^6} \right]. \tag{6.26}$$

Carrying out the Wick rotation $(k \to k_E)$, we get

$$M_{\kappa\lambda\mu\nu}^d = 4ie^2 \epsilon_{\kappa\lambda\mu\rho} \int \frac{d^4k_E}{(2\pi)^4} \frac{\partial}{\partial k_E^\nu} \left(\frac{k_E^\rho}{k_E^4} \right) = \frac{ie^2 \delta_{\nu\rho}\epsilon_{\kappa\lambda\mu\rho}}{(2\pi)^4} \int d^4k_E \frac{\partial}{\partial k_E^\tau} \left(\frac{k_E^\tau}{k_E^4} \right) =$$

$$= \frac{ie^2 \epsilon_{\kappa\lambda\mu\nu}}{(2\pi)^4} \oint \left(d^3 S_E \right)_\tau \frac{k_E^\tau}{k_E^4} = \frac{ie^2 \epsilon_{\kappa\lambda\mu\nu}}{(2\pi)^4} \oint \frac{k_E^\tau}{k_E} \left(k_E^3 d\Omega \right) \frac{k_E^\tau}{k_E^4} = \frac{ie^2 \epsilon_{\kappa\lambda\mu\nu}}{8\pi^2}, \tag{6.27}$$

where we have allowed that for a hypersurface the relation

$$\oint d\Omega = 2\pi^2$$

is true. So, under the shift transformation (6.16), the expression $N_{\kappa\lambda\mu}$ acquires the addition

$$\frac{ie^2\epsilon_{\kappa\lambda\mu\sigma}p_2^\sigma}{8\pi^2},$$

that, in turn, results in the following modification of the Ward identity (6.9)

$$p_1^\kappa N_{\kappa\lambda\mu}(p_1, p_2) = \frac{ie^2\epsilon_{\kappa\lambda\mu\sigma}p_1^\kappa p_2^\sigma}{8\pi^2}. \tag{6.28}$$

By an analogy, the relation (6.10) becomes the view

$$p_2^\lambda N_{\kappa\lambda\mu}(p_1, p_2) = -\frac{ie^2\epsilon_{\kappa\lambda\mu\sigma}p_2^\lambda p_1^\sigma}{8\pi^2}. \tag{6.29}$$

To sum up, it is impossible to simultaneously satisfy both the vector and axial-vector Ward identities. Without question, all our efforts must be directed to the fulfillment of the vector Ward identities because just they ensure the charge conservation law and the gauge invariance connected with it. The only way lies in the redefinition of the amplitude (6.5) as

$$T_{\kappa\lambda\mu}(p_1, p_2) \to T'_{\kappa\lambda\mu}(p_1, p_2) = N_{\kappa\lambda\mu}(p_1, p_2) + N_{\lambda\kappa\mu}(p_2, p_1) + \frac{ie^2}{4\pi^2}\epsilon_{\kappa\lambda\mu\sigma}(p_1 - p_2)^\sigma. \tag{6.30}$$

Now, instead of (6.8)–(6.10), we have

$$(p_1 + p_2)^\mu T'_{\kappa\lambda\mu}(p_1, p_2) = \frac{ie^2}{2\pi^2}\epsilon_{\kappa\lambda\mu\sigma}p_2^\mu p_1^\sigma \qquad (A), \tag{6.31}$$

$$p_1^\kappa T'_{\kappa\lambda\mu}(p_1, p_2) = 0 \qquad (V), \tag{6.32}$$

$$p_2^\lambda T'_{\kappa\lambda\mu}(p_1, p_2) = 0 \qquad (V). \tag{6.33}$$

In such a way, the vector Ward identities are satisfied while the axial Ward one contains an anomaly that is impossibly excluded by any regularization method. Thus, for example, in dimension regularization, a generalization of the γ_5-matrix in the case of D-dimensions cannot be built what represents the oblique indication on the anomaly existence.

It is easy to repeat calculations for massive fermions too $(m \neq 0)$. In doing so, the value of the axial anomaly proves to be the same as in the case $m = 0$. On the other hand, in the massless case, the fact that the Ward identity contains the anomaly, means the axial current is not conserved and, as a result, an additional term is present in the relation (6.2). Substituting (6.30) into (6.4) and forming the convolution of the obtained expression with iq^ν, we obtain

$$< p_1, p_2|iq^\nu J_{\nu 5}(q)|0 > = \frac{e^2}{2\pi^2}\epsilon_{\mu\nu\lambda\sigma}(-ip_2^\nu)e^{*\lambda}(\mathbf{p}_2)(-ip_1^\sigma)e^{*\mu}(\mathbf{p}_1) =$$

$$= \frac{e^2}{16\pi^2} < p_1, p_2|\epsilon_{\mu\nu\lambda\sigma}F^{\nu\lambda}(p_2)F^{\sigma\mu}(p_1)|0 > = -\frac{e^2}{16\pi^2} < p_1, p_2|\epsilon_{\nu\lambda\mu\sigma}F^{\nu\lambda}(p_2)F^{\mu\sigma}(p_1)|0 > . \tag{6.34}$$

Pass on to the coordinate representation. Then, at $m \neq 0$, from (6.34) follows

$$\frac{\partial}{\partial x_\mu}J_{\mu 5}(x) = 2mJ_5(x) - \frac{e^2}{16\pi^2}F^{\mu\nu}(x)\tilde{F}_{\mu\nu}(x), \tag{6.35}$$

where the dual tensor $\tilde{F}_{\mu\nu}$ is given by the expression

$$\tilde{F}_{\mu\nu} = \frac{1}{2}\epsilon_{\mu\nu\lambda\sigma}F^{\lambda\sigma}(x). \tag{6.36}$$

Note, that one may define other axial current to which the ordinary Ward identity (of the kind (6.8)) corresponds

$$\tilde{J}_{\mu5}(x) = J_{\mu5}(x) + \frac{e^2}{4\pi^2}\epsilon_{\mu\nu\lambda\sigma}A^{\nu}(x)\partial^{\lambda}A^{\sigma}(x). \tag{6.37}$$

Having rewritten the second term in (6.35) as

$$F^{\mu\nu}(x)\tilde{F}_{\mu\nu}(x) = \frac{1}{2}\epsilon_{\mu\nu\lambda\sigma}\left[\partial^{\mu}A^{\nu}(x) - \partial^{\nu}A^{\mu}(x)\right]\left[\partial^{\lambda}A^{\sigma}(x) - \partial^{\sigma}A^{\lambda}(x)\right] =$$

$$= 2\epsilon_{\mu\nu\lambda\sigma}\partial^{\mu}A^{\nu}(x)\partial^{\lambda}A^{\sigma}(x) = 4\partial^{\mu}[\epsilon_{\mu\nu\lambda\sigma}A^{\nu}(x)\partial^{\lambda}A^{\sigma}(x)], \tag{6.38}$$

one easily checks that the divergence of (6.37) turns into zero in the limit $m \to 0$. However, this current violates the gauge invariance and cannot be viewed as a physical one.

Within the perturbation theory the expression (6.38) can be obtained under investigation of amplitudes already regularized. In the case of the two-dimensional QED just this method was used by Schwinger [61]. The relation (6.36) also can be obtained by an operator approach [60]. The idea is as follows. An axial current represents a compound operator built from fermion fields. Since the product of these local operators is singular, it needs regularization. By these reasons, an axial current is defined in the following way. Two fermion fields are placed at different points separated by a distance ϵ. Then, one accurately proceeds to the limit tending the fields to each other. Thus, it becomes evident that the axial anomaly is not somewhat fortuitous, but it naturally follows from the renormalization procedure and reflects deeper aspects of QFT.

The analysis developed above may be carried out for non-Abelian theories with massless fermions. In this case the vector and axial-vector currents are determined by the expressions

$$J_{\mu}^{\alpha}(x) = g\overline{\psi}(x)T^{\alpha}\gamma_{\mu}\psi(x), \qquad J_{\mu5}^{\alpha}(x) = g\overline{\psi}(x)T^{\alpha}\gamma_{\mu}\gamma_5\psi(x), \tag{6.39}$$

where T^{α} are matrices associated with an internal symmetry. At that, the vertices of the considered diagrams will contain the matrices T^{α}, T^{β}, and T^{γ}, respectively. This leads to the multiplication of the amplitude by the group factor Sp $(T^{\alpha}T^{\beta}T^{\gamma})$. As a result, we arrive at the following relation

$$\frac{\partial}{\partial x_{\mu}}J_{\mu5}^{\alpha}(x) = -\frac{g^2}{4\pi^2}D^{\alpha\beta\gamma}\epsilon^{\mu\nu\lambda\sigma}\partial_{\mu}G_{\nu}^{\beta}(x)\partial_{\lambda}G_{\sigma}^{\gamma}(x), \tag{6.40}$$

where G_{μ}^{α} are the gauge field potentials and $D^{\alpha\beta\gamma}$ is the entirely symmetric quantity:

$$D^{\alpha\beta\gamma} = \frac{1}{2}\text{Sp}\left(\{T^{\alpha}, T^{\beta}\}T^{\gamma}\right) \tag{6.41}$$

(for details, see the monograph [63]).

Consider QCD. Since the group factor for the axial isospin currents is

$$\text{Sp}\,(\tau^{\alpha}T^{b}T^{c}),$$

where τ^{α} is the isospin matrix, T^{b}, T^{c} are the color matrices and the spur is taken over the colors and flavors, we obtain

$$\frac{\partial}{\partial x_{\mu}}J_{\mu5}^{\alpha}(x) = -\frac{g_s^2}{4\pi^2}\epsilon^{\mu\nu\lambda\sigma}\partial_{\mu}G_{\nu}^{b}(x)\partial_{\lambda}G_{\sigma}^{c}(x)\text{Sp}\,(\tau^{\alpha}T^{b}T^{c}). \tag{6.42}$$

In this case we have

$$\text{Sp } (\tau^\alpha T^b T^c) = \text{Sp } (\tau^\alpha)\text{Sp } (T^b T^c) = 0. \qquad (6.43)$$

Therefore, the ABJ anomaly does not influence the conservation of the axial isospin currents. But in the case of the isospin-singlet axial-vector current the matrix τ^α is replaced by the unit flavor matrix resulting in

$$\frac{\partial}{\partial x_\mu} J_{\mu 5}(x) = -\frac{g_s^2}{4\pi^2}\epsilon^{\mu\nu\lambda\sigma}\partial_\mu G_\nu^b(x)\partial_\lambda G_\sigma^c(x)\text{Sp } (T^b T^c) = -\frac{g_s^2 N_f}{32\pi^2}\epsilon^{\mu\nu\lambda\sigma}G_{\mu\nu}^c(x)G_{\lambda\sigma}^c(x), \qquad (6.44)$$

where N_f is the flavors number. So, in QCD the isospin-singlet axial-vector current is not actually conserved. Eq. (6.44) means that QCD has neither isosinglet axial symmetry nor associated Goldstone boson. To put this another way, Eq. (6.44) explains why the strong interaction does not involve a light isosinglet pseudoscalar meson possessing the mass comparable with the pion mass, that is, performs one more solution of the $U(1)$ problem.

Though the axial-vector isospin currents do not have axial anomaly because of QCD interactions, they have an anomaly connected with the electromagnetic interaction of quarks. It is obvious that the electromagnetic anomaly of the axial-vector isospin currents is given by the expression

$$\frac{\partial}{\partial x_\mu} J_{\mu 5}^\alpha(x) = -\frac{e^2}{16\pi^2}\epsilon^{\mu\nu\lambda\sigma}F_{\mu\nu}(x)F_{\lambda\sigma}(x)\text{Sp } (\tau^\alpha Q^2), \qquad (6.45)$$

where Q is the matrix of the electrical quark charges

$$Q = \begin{pmatrix} 2/3 & 0 \\ 0 & -1/3 \end{pmatrix},$$

and the spur is taken over the colors and flavors. As the matrices under the spur sign do not depend on the color, then the sum over colors simply gives the factor 3. Next, taking into account that the spur over the flavors is not equal to zero only at $\alpha - 3$, we get

$$\frac{\partial}{\partial x_\mu} J_{\mu 5}^3(x) = -\frac{e^2}{32\pi^2}\epsilon^{\mu\nu\lambda\sigma}F_{\mu\nu}(x)F_{\lambda\sigma}(x). \qquad (6.46)$$

6.2 Adler–Bell–Jackiw anomaly (nonperturbative approach)

In 1979 Fujikawa [62] ascertained that at the QFT formulation in terms of path integrals the chiral anomaly enters only into a measure used for the definition of fermion path integrals. The existence of the anomaly represents the indication of the fact that the appropriate invariant measure of the integration over fermion fields cannot be determined. Fujikawa's analysis was based both on using the path integrals in the Euclidian space and on expanding the fermion integration variables in terms of the eigenfunctions of the Dirac operator $\hat{D}$ being gauge-invariant and Hermitian in the four-dimensional space. Repeat this derivation for the case of massive fermions.

Consider the QCD functional integral

$$W = \int [d\psi][d\overline{\psi}][dG] \exp\left\{ i \int d^4 x[\overline{\psi}(x)(i\hat{D} - m)\psi(x) - \frac{1}{4}G_{\mu\nu}^a(x)G^{a\mu\nu}(x)] \right\}. \qquad (6.47)$$

Pass on to the Euclidean space. After the Wick rotation $x_0 \to -ix_4$ and $G_0^a \to iG_4^a$, the operator $\hat{D}$ becomes the Hermitian one

$$\hat{D} = i\gamma_0 D_4 + \gamma^k D_k = \gamma^4 D_4 + \gamma^k D_k,$$

and the metric takes the form

$$g_{\mu\nu} = \text{diag}\,(-1,-1,-1,-1).$$

Note that in the Euclidean space the fields ψ and $\overline{\psi}$ must be considered as the independent ones. Introduce the local chiral transformations

$$U(x) = \exp[i\gamma_5 t\alpha(x)], \tag{6.48}$$

where t is an ordinary Hermitian matrix not including γ_5. It is clear that the fermion part of the measure is not invariant with respect to the chiral transformations because Eq. (2.71) results in

$$[d\psi][d\overline{\psi}] \to [d\psi'][d\overline{\psi}'] = (\det\,U)^{-2}[d\psi][d\overline{\psi}]. \tag{6.49}$$

Since a gluon field is not subjected to the transformations (6.48), next we shall work only with the fermion part of the functional integral. In the case of the infinitesimal chiral transformations the fermion fields are transformed in the following way

$$\left.\begin{aligned} \psi(x) \to \psi'(x) &= [1 + i\gamma_5 t\alpha(x)]\psi(x), \\ \overline{\psi}(x) \to \overline{\psi}'(x) &= \overline{\psi}(x)[1 + i\gamma_5 t\alpha(x)]. \end{aligned}\right\} \tag{6.50}$$

In the new variables the fermion integral takes the view

$$W_f' = [d\psi'][d\overline{\psi}']\exp\left[i\int d^4x\,\overline{\psi}'(x)(i\hat{D} - m)\psi'(x)\right] = [\det\,U_{inf}]^{-2}\,[d\psi][d\overline{\psi}]\times$$

$$\times \exp\left\{i\int d^4x[\mathcal{L}_f(x) - \partial_\mu\alpha(x)\overline{\psi}(x)\gamma^\mu\gamma^5 t\psi(x) - 2im\alpha(x)\overline{\psi}(x)\gamma^5 t\psi(x)]\right\} =$$

$$= \Lambda^{-2}[d\psi][d\overline{\psi}]\exp\left\{i\int d^4x[\mathcal{L}_f(x) + \alpha(x)\partial_\mu J_5^\mu(x) - 2mJ_5(x)]\right\}, \tag{6.51}$$

where U_{inf} is the matrix of the infinitesimal transformation (6.50), $\Lambda \equiv \det\,U_{inf}$ and

$$\mathcal{L}_f(x) = \overline{\psi}(x)(i\hat{D} - m)\psi(x), \qquad J_5^\mu(x) = \overline{\psi}(x)\gamma^\mu\gamma^5 t\psi(x), \qquad J_5(x) = i\overline{\psi}(x)\gamma^5 t\psi(x).$$

Introduce the eigenfunctions of the Hermitian operator $\hat{D}$

$$\hat{D}\varphi_m(x) = \lambda_m\varphi_m(x) \tag{6.52}$$

that form the total and orthonormalized set

$$\int d^4x\varphi_m^\dagger(x)\varphi_n(x) = \delta_{mn}. \tag{6.53}$$

At zero value of the gauge field $(G_a^\mu(x) = 0)$, φ_m are Dirac wave functions and their eigenvalues are as follows

$$\lambda_m^2 = p_0^2 + \mathbf{p}^2. \tag{6.54}$$

For the fixed gauge field $G_a^\mu(x)$, λ_m^2 also are an asymptotic form of eigenvalues under large p. Further, we expand the fermion field in terms of the basis φ_n

$$\psi(x) = \sum_n a_n\varphi_n(x) = \sum_n a_n < n|x>, \qquad \overline{\psi}(x) = \sum_n \varphi_n^\dagger(x)\overline{b}_n = \sum_n < x|n > \overline{b}_n, \tag{6.55}$$

where a_n and $\bar{b}_n$ are Grassmann algebra elements. Now, the functional integration measures before and after the transformation (6.50) can be represented in terms of the eigenfunctions of the Hermitian operator $\hat{D}$

$$[d\psi][d\bar{\psi}] = \prod_n d\bar{b}_n \prod_m da_m, \qquad [d\psi'][d\bar{\psi}'] = \prod_n d\bar{b}'_n \prod_m da'_m. \tag{6.56}$$

From (6.50) one may establish the transformation law of the coefficients a_m

$$a'_m = \sum_n \int d^4x \varphi_m^\dagger(x)[1 + i\gamma_5 t\alpha(x)]a_n\varphi_n(x) = \sum_n (\delta_{mn} + C_{mn})a_n. \tag{6.57}$$

Using the identity

$$\det B = \exp[\operatorname{Sp}\ln B]$$

and the relation $\ln(1+x) \to x$ at $x \to 0$, we find the connection between Λ and C

$$\Lambda = \det(1 + C) = \exp[\operatorname{Sp}\ln(1 + C)] = \exp\left(\sum_n C_{nn} + \ldots\right). \tag{6.58}$$

Then

$$\ln \Lambda - i \int d^4x \alpha(x) \sum_n \varphi_n^\dagger(x)\gamma_5 t\varphi_n(x). \tag{6.59}$$

At first glance it does not look as if we could obtain any definite result for $\ln \Lambda$. On the one hand, the coefficient at $\alpha(x)$ looks like $\operatorname{Sp}\gamma_5 = 0$, while, on the other hand, the eigenvalues of φ_n turn into infinity under $p \to \infty$. For the expression (6.59) to be assigned a specific meaning a regularizator must be inserted. And this has to be done by a gauge invariant manner. The simplest way is to choose a regularizator in the form of the exponential $\exp[-\lambda_n^2/M^2]$. Then the sum in the integrand (6.59) will look like:

$$\sum_n \varphi_n^\dagger(x)\gamma_5 t\varphi_n(x) \to \lim_{M\to\infty} \sum_n \varphi_n^\dagger(x)\gamma_5 t \exp[-\lambda_n^2/M^2]\varphi_n(x) =$$

$$= \lim_{M\to\infty} \sum_n <x|n> \{\gamma_5 t \exp[-(\hat{D})^2/M^2]\} <n|x> =$$

$$= \lim_{M\to\infty} <x|\operatorname{Sp}\left[\gamma_5 t \exp[-(\hat{D})^2/M^2]\right]|x> =$$

$$= \lim_{M\to\infty} <x|\operatorname{Sp}\{\gamma_5 t \exp[(-D_\mu D^\mu - \frac{g_s}{4}[\gamma^\mu, \gamma^\nu]G_{\mu\nu}^a T^a)/M^2]\}|x>, \tag{6.60}$$

where $<x|y> = \delta^4(x - y)$ and we have taken into account the relation

$$(\hat{D})^2 = \frac{1}{4}\{D_\mu, D_\nu\}\{\gamma^\mu, \gamma^\nu\} + \frac{1}{4}[D_\mu, D_\nu][\gamma^\mu, \gamma^\nu] = D_\mu D^\mu + \frac{g_s}{4}[\gamma^\mu, \gamma^\nu]G_{\mu\nu}^a T^a.$$

Because the limit $M \to \infty$ is taken, we can concentrate attention on an asymptotic spectrum part where a momentum is large and one may carry out an expansion in gauge field powers. To have a nonzero spur over the Dirac indices with γ_5, one needs to obtain four γ-matrices from the exponential. The main item is given by the exponential expansion up to the order $[\gamma^\mu, \gamma^\nu]G_{\mu\nu}$ and by neglecting the gauge field in all remaining items. This results in

$$\lim_{M\to\infty} <x|\operatorname{Sp}\left\{\gamma_5 t \exp\left[\left(-D^\mu D_\mu - \frac{g_s}{4}[\gamma^\mu, \gamma^\nu]G_{\mu\nu}^a T^a\right)/M^2\right]\right\}|x> =$$

$$= \lim_{M \to \infty} \mathrm{Sp} \left\{ \gamma_5 t \frac{1}{2} \left[\frac{g_s}{4M^2} [\gamma^\mu, \gamma^\nu] G^a_{\mu\nu} T^a \right]^2 \right\} < x| \exp[-\partial^2/M^2]|x > . \qquad (6.61)$$

To calculate the matrix element in (6.61) produces

$$< x| \exp[-\partial^2/M^2]|x > = \lim_{x \to y} \int \frac{d^4 p_E}{(2\pi)^4} \exp[-\partial^2/M^2] \exp[i p_E (x - y)] =$$

$$= \int \frac{d^4 p_E}{(2\pi)^4} \exp(-p_E^2/M^2) = \frac{M^4}{16\pi^2}. \qquad (6.62)$$

Then the relation (6.61) is reduced to

$$\lim_{M \to \infty} \frac{g_s^2 M^4}{128\pi^2} \mathrm{Sp} \left[\gamma_5 \gamma^\mu \gamma^\nu \gamma^\tau \gamma^\lambda \right] \mathrm{Sp} \left[t T^a T^b \right] \frac{1}{M^4} G^a_{\mu\nu}(x) G^b_{\tau\lambda}(x) =$$

$$= \frac{g_s^2}{32\pi^2} \epsilon^{\mu\nu\tau\lambda} G^a_{\mu\nu}(x) G^b_{\tau\lambda}(x) \mathrm{Sp} \left[t T^a T^b \right]. \qquad (6.63)$$

Thus, we have

$$\Lambda = \exp \left\{ i \int d^4 x \alpha(x) \left[\frac{g_s^2}{32\pi^2} \epsilon^{\mu\nu\tau\lambda} G^a_{\mu\nu}(x) G^b_{\tau\lambda}(x) \mathrm{Sp} \left[t T^a T^b \right] \right] \right\}. \qquad (6.64)$$

As a result, we find that the transformed functional integral is

$$W'_f = [d\psi][d\overline{\psi}] \exp \left\{ i \int d^4 x \left[\overline{\psi}(x)(i\hat{D} - m)\psi(x) + \right. \right.$$

$$\left. \left. + \alpha(x) \left(\partial_\mu J^\mu_5(x) - 2m J_5(x) - \frac{g_s^2}{16\pi^2} \epsilon^{\mu\nu\tau\lambda} G^a_{\mu\nu}(x) G^b_{\tau\lambda}(x) \mathrm{Sp} \left[t T^a T^b \right] \right) \right] \right\}. \qquad (6.65)$$

Varying W'_f over $\alpha(x)$, we lead to the following alteration law of the axial current

$$\partial_\mu J^\mu_5(x) = 2m J_5(x) + \frac{g_s^2}{16\pi^2} \epsilon^{\mu\nu\tau\lambda} G^a_{\mu\nu}(x) G^b_{\tau\lambda}(x) \mathrm{Sp} \left[t T^a T^b \right]. \qquad (6.66)$$

In the right side of Eq. (6.66) the sign in front of the second item is opposite relatively to the one we obtained in the Minkowski space (see for example (6.42)). This difference is connected with the fact that in finding this item two factors i were missed as compared with the case of the Minkowski space. One of them is associated with taking the spur in the Euclidean space $\mathrm{Sp}[\gamma_5 \gamma_\mu \gamma_\nu \gamma_\lambda \gamma_\sigma]$, and the second appears due to the replacement $d^4 p \to d^4 p_E$.

6.3 $\pi^0 \to \gamma\gamma$ **decay**

Inasmuch as pions are coupled states of quark and antiquark, then they have negative internal parity. So, they may be produced by the axial isospin currents. Therefore, the axial anomaly has to somehow influence observable properties of these particles. Historically, the axial anomaly appeared as an attempt to understand the decay rate of the neutral pion

$$\pi^0 \to \gamma\gamma.$$

In 1966 Sutherland [64] showed that application of the currents algebra and PCAC to the π^0 decay gives very small value of the decay rate (in reality this value turns into zero at nonphysical limit $m_\pi \to 0$) to contradict experimental data. Later, in the works by Adler [58], Bell and Jackiw [57], it was shown that the main contribution into the decay amplitude $\pi^0 \to \gamma\gamma$ gives the axial anomaly and its inclusion results in the coincidence of the theory with the experiment.

Let us investigate the decay $\pi^0 \to \gamma\gamma$ in the limit of massless u- and d-quarks. In this case the theory is chiral symmetric with an accuracy of anomaly effects. Consider the matrix element of the axial isospin current between a vacuum and two photon state:

$$< p_1, p_2 | J^3_{\mu 5}(q) | 0 >= e^{*\nu}(\mathbf{p}_1) e^{*\lambda}(\mathbf{p}_2) \mathcal{M}_{\mu\nu\lambda}(p_1, p_2), \tag{6.67}$$

where p_1 and p_2 are the photon momenta, ν and λ are their polarization indices and $q = p_1 + p_2$. We shall study the general properties of this matrix element expanding it in terms of formfactors. In general, the amplitude can be expanded in all possible tensor structures with the help of constraints that follow both from the symmetry with respect to the permutations (p_1, ν) and (p_2, λ) and from the Ward identities for the vertices describing the photons emitting

$$p^\nu_1 \mathcal{M}_{\mu\nu\lambda}(p_1, p_2) = p^\lambda_2 \mathcal{M}_{\mu\nu\lambda}(p_1, p_2) = 0. \tag{6.68}$$

The amplitude $\mathcal{M}_{\mu\nu\lambda}$ represents a third-rank pseudotensor. To build it there are the two four dimensional vectors p^μ_1, p^μ_2 and the fourth rank pseudotensor $\epsilon_{\nu\lambda\alpha\beta}$ at our disposal. The analysis shows that the following combination

$$\mathcal{M}_{\mu\nu\lambda}(p_1, p_2) = q_\mu \epsilon_{\nu\lambda\alpha\beta} p^\alpha_1 p^\beta_2 F_1(q^2) + (\epsilon_{\mu\nu\alpha\beta} p_{2\lambda} - \epsilon_{\mu\lambda\alpha\beta} p_{1\nu}) p^\alpha_2 p^\beta_1 F_2(q^2) +$$

$$+ [(\epsilon_{\mu\nu\alpha\beta} p_{1\lambda} - \epsilon_{\mu\lambda\alpha\beta} p_{2\nu}) p^\alpha_2 p^\beta_1 - \epsilon_{\mu\nu\lambda\sigma}(p_1 - p_2)^\sigma (p_1 \cdot p_2)] F_3(q^2), \tag{6.69}$$

where F_i ($i = 1, 2, 3$) are pion formfactors, satisfies the demands listed above. To take the divergence of the axial current, we convolve (6.69) with iq^μ

$$iq^\mu \mathcal{M}_{\mu\nu\lambda}(p_1, p_2) = iq^2 \epsilon_{\nu\lambda\alpha\beta} p^\alpha_1 p^\beta_2 F_1(q^2) - i\epsilon_{\mu\nu\lambda\sigma} q^\mu (p_1 - p_2)^\sigma (p_1 \cdot p_2) F_3(q^2) =$$

$$= iq^2 \epsilon_{\nu\lambda\alpha\beta} p^\alpha_1 p^\beta_2 [F_1(q^2) + F_3(q^2)]. \tag{6.70}$$

Employing the expression for the axial current divergence (6.45), we find

$$< p_1, p_2 | iq^\mu J^3_{\mu 5}(q) | 0 >= -\frac{e^2}{4\pi^2} \epsilon_{\nu\lambda\alpha\beta} p^\alpha_1 p^\beta_2 e^{*\nu}(\mathbf{p}_1) e^{*\lambda}(\mathbf{p}_2) \text{Sp} \, [\sigma_3 Q^2]. \tag{6.71}$$

In the limit $q^2 = m^2_\pi \to 0$ the divergence of the matrix element $iq^\mu \mathcal{M}_{\mu\nu\lambda}$ turns into zero what, however, is in contrast with the relation (6.71). The contradiction may be removed by the assumption that one of the formfactors appearing in (6.70) contains a pole in q^2. The process, in which the current $J^3_{\mu 5}$ emits a pion that is next absorbed in the vertex (Fig. 6.4), can be the pole source. Calculate this pole item. Let us parametrize the pion decay amplitude in the following manner:

$$\mathcal{A}(\pi^0 \to 2\gamma) = iC \epsilon_{\nu\lambda\alpha\beta} p^\alpha_1 p^\beta_2 e^{*\nu}(\mathbf{p}_1) e^{*\lambda}(\mathbf{p}_2), \tag{6.72}$$

where C is a constant being subject to definition. Allowing for that the π^0-meson production amplitude by means of the current $J^3_{\mu 5}$ have the form

$$< 0 | J^3_{\mu 5}(0) | \pi(q) >= iq_\mu f_\pi,$$

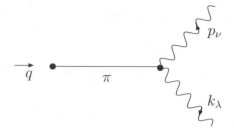

FIGURE 6.4
Contribution resulting in a pole in the formfactor F_1.

we find the contribution of Fig. 6.4 in the current vertex (6.67)

$$(iq_\mu f_\pi)\frac{i}{q^2}(iC\epsilon_{\nu\lambda\alpha\beta}p_1^\alpha p_2^\beta). \tag{6.73}$$

It is obvious that this defines the formfactor F_1

$$F_1(q^2) = -\frac{i}{q^2}Cf_\pi. \tag{6.74}$$

Next, equating (6.70) with (6.71) and discarding the inessential term proportional to q^2, we express the constant C in terms of the anomaly coefficient

$$C = \frac{e^2}{4\pi^2 f_\pi}\mathrm{Sp}\,[\sigma_3 Q^2] = \frac{e^2}{4\pi^2 f_\pi}\frac{n_c}{3}, \tag{6.75}$$

where n_c are the color freedom degrees number. Now, it is a straightforward matter to compute the probability of the decay $\pi^0 \to 2\gamma$. In spite of the fact that we work in the limit of massless π^0, we need to allow for the physically right kinematic to depend on the π^0 mass. Taking into consideration this circumstance, summing over the photon spins and accounting for the photons identity, we arrive at the result

$$\Gamma(\pi^0 \to 2\gamma) = \frac{1}{2m_\pi}\frac{1}{8\pi}\frac{1}{2}\sum |\mathcal{A}(\pi^0 \to 2\gamma)|^2 = \frac{1}{32\pi m_\pi}2[(p_1 \cdot p_2)C]^2 =$$

$$= \frac{\alpha^2}{64\pi^3}\frac{m_\pi^3}{f_\pi^2}\frac{n_c^2}{9}. \tag{6.76}$$

The observable decay probability is in line with the theoretical estimation only at $n_c = 3$, the experimental and theoretical results coinciding with a precision of few percents. The success of the mentioned calculation was one of the first arguments in favor of the existence of three quark colors. The relation (6.76) also allows one to measure the axial anomaly in an experiment at once.

6.4   Chiral anomalies and chiral gauge theories

In QCD the interaction between gauge fields and fermions was introduced by the way being invariant with respect to the parity transformation. This is achieved by replacing the ordinary derivatives by the covariant ones in the Dirac equation. In doing so, the gauge field is connected with the fermion current. But it gives only part of the possible interactions between fermions and gauge fields. It may be a situation when only fermion fields ψ possessing a definite chirality interact with the gauge fields. As an example, we consider a theory with the Lagrangian

$$\mathcal{L} = i\overline{\psi}(x)\gamma^\mu \left[\partial_\mu - igG_\mu^a(x)T^a \left(\frac{1-\gamma_5}{2} \right) \right] \psi(x). \tag{6.77}$$

By direct calculation, one may convinces oneself that (6.77) is invariant under the local infinitesimal gauge transformations

$$\psi(x) \rightarrow \left[1 + ig\alpha^a(x)T^a \left(\frac{1-\gamma_5}{2} \right) \right] \psi(x), \tag{6.78}$$

$$G_\mu^a(x) \rightarrow G_\mu^a(x) + g^{-1}\partial_\mu\alpha^a(x) + f^{abc}G_\mu^b(x)\alpha^c(x). \tag{6.79}$$

Because the right-handed field $\psi_R = (1+\gamma_5)/2$ are free, they can be removed. Then, one may work with the gauge-invariant Lagrangian only for the left-handed fermions $\psi_L = (1-\gamma_5)/2$. The idea of gauge fields interacting only with left-handed fermions plays a principal role in the electroweak interaction theory.

Establish a criteria of the axial anomalies absence in known chiral gauge theories. To realize this program, we consider a theory in which arbitrary gauge fields interact with left-handed massless fermions. Investigate the diagrams pictured in Fig. 6.5 Let external fields

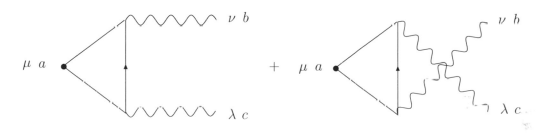

FIGURE 6.5
Diagrams resulting in to an axial vector anomaly.

be non-Abelian gauge bosons and marked vertex represents the current associated with the gauge symmetry:

$$J_\mu^a = \overline{\psi}\gamma_\mu \left(\frac{1-\gamma_5}{2} \right) T^a \psi. \tag{6.80}$$

Gauge boson vertices also contain the projectors $(1-\gamma_5)/2$. These three projectors can be interchanged and unified into a single factor. Regularization of the term involving γ_5 results in the appearance of the axial anomaly, that is, nonconservation of the axial current

$$\frac{\partial}{\partial x_\mu} J_{\mu 5}^\alpha(x) = -\frac{g^2}{4\pi^2} D^{\alpha\beta\gamma} \epsilon^{\mu\nu\lambda\sigma} \partial_\mu G_\nu^\beta(x) \partial_\lambda G_\sigma^\gamma(x). \tag{6.81}$$

However, the gauge invariance demands the absence of anomalies. The only one way to solve this problem is to demand the fulfillment of the relation

$$D^{\alpha\beta\gamma} = 0, \tag{6.82}$$

where

$$D^{\alpha\beta\gamma} = \frac{1}{2}\text{Sp}\ \left(\{T^\alpha, T^\beta\}T^\gamma\right), \tag{6.83}$$

as a fundamental condition of the self-consistency for chiral gauge fields [65]. Gauge fields satisfying these conditions are called *anomaly free*. The condition (6.82) can be satisfied for any gauge group when fermion fields realize an appropriate reducible or irreducible representation of this group. Besides, there exist several gauge groups for which this condition is satisfied in any representation [66].

Let us examine how to work the formula (6.82). As a case in point, we consider the Glashow–Weinberg–Salam theory of the electroweak interaction based on the $SU(2)_L \times U(1)_Y$ gauge group. If two gauge bosons shown in Fig. 6.5 are the $SU(2)$ gauge bosons and the current J_μ^a is the $SU(2)$ gauge current, then $T^a = \sigma^a/2$ and we obtain

$$D^{\alpha\beta\gamma} = \frac{1}{16}\text{Sp}\ \left(\{\sigma^\alpha, \sigma^\beta\}\sigma^\gamma\right) = \frac{1}{16}\text{Sp}\ \left(2\delta^{\alpha\beta}\sigma^\gamma\right) = 0. \tag{6.84}$$

When the electromagnetic current is present in the vertex marked in Fig. 6.5, then $D^{\alpha\beta\gamma} \to D^{\alpha\beta}$ and we have

$$D^{\alpha\beta} = \frac{1}{8}\text{Sp}\ \left(\{\sigma^\alpha, \sigma^\beta\}Q\right) = \frac{1}{4}\text{Sp}(Q)\delta^{\alpha\beta}. \tag{6.85}$$

This factor is proportional to the sum of the fermion electric charges that does not equal to zero both for quarks and for leptons. Summing over the quark colors and over fermion families gives

$$\text{Sp}(Q) = f_c \times n_c \times \left(\frac{2}{3} - \frac{1}{3}\right) + f_l \times (0 - 1), \tag{6.86}$$

where f_c (f_l) is the number of the quark (lepton) families and n_c is the number of the color degrees of freedom. From (6.86) it follows that $D^{\alpha\beta}$ turns into zero at the fulfillment of the conditions

$$n_c = 3, \qquad f_c = f_l \tag{6.87}$$

which are realized in the standard model of the electroweak interaction.

It is obvious that the condition (6.82) is satisfied when left-handed fermion (or antifermion) fields realize the gauge algebra representation to be equivalent to the conjugate representation in the sense that

$$(iT^a)^* = S(iT^a)S^{-1}.$$

With allowance made for the hermicity of T^a, this relation can be rewritten in the form

$$(T^a)^T = -ST^aS^{-1}. \tag{6.88}$$

Clarify what is understood by the conjugate representation. There is a corresponding conjugate representation $\bar{r}$ for every irreducible representation r of a group $\mathcal{U}$. The representation generates the following infinitesimal transformation

$$\varphi \to (1 + i\alpha^a T_r^a)\varphi. \tag{6.89}$$

The complex-conjugate transformation

$$\varphi^* \to [1 - i\alpha^a (T_r^a)^*]\varphi^* \tag{6.90}$$

must also give an infinitesimal element of $\mathcal{U}$ group representation. Thus, the representation conjugate to r is given by the matrices:

$$T_{\bar{r}}^a = -(T_r^a)^* = -(T_r^a)^T. \tag{6.91}$$

Inasmuch as the product $\varphi^*\varphi$ is invariant with respect to unitary transformations, one may form a group invariant from fields that are transformed by r and $\bar{r}$ representations. If there exists a unitary transformation with the property

$$T_{\bar{r}}^a = S T_r^a S^\dagger,$$

then r representation and $\bar{r}$ one are equivalent. At that, the representation r is real. In this case when η and ξ belong to the representation r, there exists a matrix C_{ab} such that the combination $C_{ab}\eta_a\xi_b$ is invariant. Two cases are distinguished: (i) C_{ab} is symmetric, (ii) C_{ab} is antisymmetric. In the former the representation is definitely real, in the latter the one is pseudoreal. Thus, for example, in the $SU(2)$ group one may form an invariant combination $p_a k_a$ from two vectors and, as a result, a vector is transformed under real representation. From two spinors an invariant combination $\epsilon_{ab}\eta_a\xi_b$ may be built, therefore a spinor is transformed under pseudoreal representation.

As T_r^a and $T_{\bar{r}}^a$ are connected by an unitary transformation and (6.83) is invariant with respect to unitary transformations, replacing T_r^a by $T_{\bar{r}}^a$ and taking into account (6.91), we get

$$D^{\alpha\beta\gamma} = \frac{1}{2}\mathrm{Sp}\ \left(\{(-T^\alpha)^T, (-T^\beta)^T\}(-T^\gamma)^T\right) = -\frac{1}{2}\mathrm{Sp}\ \left(T^\gamma\{T^\beta, T^\alpha\}\right) = -D^{\alpha\beta\gamma} = 0. \quad (6.92)$$

Hence, if a representation is real, a gauge theory is automatically free from axial anomalies. In the more general situation, calculating $D^{\alpha\beta\gamma}$ may be simplified noting that this quantity is a gauge group invariant to be completely symmetric over three indices in an associate representation. For some gauge groups a corresponding invariant may not exist. In these cases $D^{\alpha\beta\gamma}$ has to be equal to zero. For example, in $SU(2)$ an associate representation has a spin 1. A symmetric product of two multiplets with the spin 1 gives the spins 0 and 2 and does not involve a component with the spin 1. Consequently, there exists no a symmetric tensor coupling for two indices of the spin 1 that gives the spin 1. By this reason the factor $D^{\alpha\beta\gamma}$ needs to be equal to zero in any $SU(2)$ gauge theory. Earlier we have convinced ourselves of the validity of this statement. Among gauge algebras having only real or pseudoreal representations also are [67]: $SO(2n+1)$, $SO(4n)$ for $n \geq 2$, $USp(2n)$ for $n \geq 3$, G_2, F_4, E_7, E_8 and all their direct sums. In parallel with these listed above there are several algebras to have representations only with $D^{\alpha\beta\gamma} - 0$ despite the fact that some their representations represent neither real nor pseudoreal [66]. These are $SO(4n+1)$ (except $SO(2) \equiv U(1)$ and $SO(6) \equiv SU(4)$) and E_6 as well as their direct sums, both with each other and with algebras mentioned above.

In $SU(n)$ groups with $n \geq 3$, one symmetric invariant of the required kind $d^{\alpha\beta\gamma}$ exists. It appears in the anticommutator of the fundamental representation matrices:

$$\{T_n^\alpha, T_n^\beta\} = \frac{1}{n}\delta^{\alpha\beta} + d^{\alpha\beta\gamma}T_n^\gamma. \quad (6.93)$$

The uniqueness of this invariant means that in any $SU(n)$ gauge theory any spur of the type (6.83) is proportional to $d^{\alpha\beta\gamma}$. And, at the same time, the anomalous coefficient $A(r)$ can be determined for every representation r

$$\mathrm{Sp}\ (\{T_r^\alpha, T_r^\beta\}T^\gamma) = \frac{1}{2}A(r)d^{\alpha\beta\gamma}. \quad (6.94)$$

So, the existence of the axial anomalies is possible only for such algebras that include $SU(n)$ (for $n \geq 3$) or $U(1)$ by way of the factors. As is often the case, such algebras are major ones in modern physics. Thus, for example, the $SU(3)_c \times SU(2)_L \times U(1)_Y$ gauge group underlies the standard model and, here, mutual cancellation of anomalies is ensured only by equality of the families number in the quark and lepton sectors.

6.5   Anomalous breaking of scale invariance

There exists one more major symmetry type that is conserved at the classical level and violated by quantum corrections. This is a classical scale invariance of a massless QFT with a nondimensional coupling constant, that is, the invariance with respect to coordinates dilatation $x_\mu \to \lambda x_\mu$ with simultaneous multiplication of field operators by the factor λ in a power equal to a field dimension. In accordance with the Noether theorem, in classical field theory such an invariance results in the conserved current j_μ^D (dilatation current)

$$j_\mu^D(x) = x^\nu T_{\mu\nu}^{metr}(x), \tag{6.95}$$

where $T_{\mu\nu}^{metr}$ is the symmetric energy-momentum tensor (in what follows the index $metr$ of the energy-momentum tensor will be omitted). Really, by virtue of the motion equations the energy-momentum tensor is conserved

$$\frac{\partial}{\partial x_\mu} T_{\mu\nu}(x) = 0,$$

so that the divergence of the dilatation current is equal to the spur of the energy-momentum tensor

$$\frac{\partial}{\partial x_\mu} j_\mu^D(x) = T_\mu^\mu(x), \tag{6.96}$$

T_μ^μ being equal to zero. However, QFT with a nondimensional coupling constant involves logarithmic ultraviolet divergences that must be regularized and renormalized. As a result, regularized expressions prove to be dependent on a dimension quantity, a normalization momentum (or scale parameter), and the scale invariance is broken. Thus, with allowance made for quantum effects, we have

$$\frac{\partial}{\partial x_\mu} j_\mu^D(x) = T_\mu^\mu(x) \neq 0. \tag{6.97}$$

This relation is known as a *spur anomaly*. As in the case of the axial anomaly, the spur anomaly could be found by different ways. Here, we obtain this anomaly using the charge renormalization.

In the QCD, the symmetrized energy-momentum tensor is given by the expression (1.33). The spur of items containing gluon fields is automatically cancelled. The remaining quark items could be transformed with the help of the Dirac equation to give

$$T_\mu^\mu = m\bar{q}(x)q(x).$$

So, at the classical level the spur of the energy-momentum tensor really turns into zero in the massless limit.

When taking into account quantum corrections, in the massless QCD the scale invariance proves to be broken because the charge is renormalized and becomes dependent on the scale $\bar{g}_s = \bar{g}_s(\lambda, g_s)$. For the sake of convenience we change the definition of potentials and strengths for the time being

$$G_{\mu\nu}^a = \partial_\mu G_\nu^a - \partial_\nu G_\mu^a + f^{abc} G_\mu^b G_\nu^c, \tag{6.98}$$

where

$$G_\mu^a = g_s \overset{o}{G}{}^a{}_\mu, \qquad G_{\mu\nu}^a = g_s \overset{o}{G}{}^a{}_{\mu\nu},$$

$\overset{\scriptscriptstyle 0}{G}{}^a{}_\mu$ and $\overset{\scriptscriptstyle 0}{G}{}^a{}_{\mu\nu}$ are the potentials and strengths in the initial definition. At such a normalization, the gauge constant g_s disappears from the covariant derivative and appears in the coefficient at the kinetic item for the gauge field. The transformation laws of G^a_μ and q do not also depend on g_s:

$$\delta G^a_\mu = \partial_\mu \alpha^a + f^{abc} G^b_\mu \alpha^c, \qquad \delta q = i\alpha^a T^a q.$$

In the new designations the QCD action takes the form:

$$S = \int d^4x \left[-\frac{1}{4g_s^2} G^a_{\mu\nu}(x) G^{a\mu\nu}(x) + i\bar{q}(x)\hat{D}q(x) \right]. \tag{6.99}$$

Quantum fluctuations will lead to changing the action of the initial field $S \to S_{eff}$ and their inclusion results in the replacement of the charge in Eq. (6.99) by its renormalized value:

$$S_{eff} = \int d^4x \left[-\frac{1}{4\bar{g}_s(\lambda, g_s)^2} G^a_{\mu\nu}(x) G^{a\mu\nu}(x) + \dots \right]. \tag{6.100}$$

Varying the effective action (6.100) and allowing for the invariance of the initial action, we get

$$\delta S_{eff} = \frac{\delta \bar{g}_s(\lambda, g_s)}{2\bar{g}_s^3(\lambda, g_s)} \int d^4x\, G^a_{\mu\nu}(x) G^{a\mu\nu}(x). \tag{6.101}$$

Now, we remember that the effective charge $\bar{g}_s(\lambda, g_s)$ satisfies the differential equation

$$\frac{d\bar{g}_s(\lambda, g_s)}{d\ln\lambda} = \beta(\bar{g}_s), \tag{6.102}$$

that is, not far from $\lambda = 1$

$$\delta\bar{g}_s(\lambda, g_s) = \beta(\bar{g}_s)\delta\lambda. \tag{6.103}$$

Substituting this equality into (6.101), we find

$$\delta S_{eff} = \delta\lambda \frac{\beta(\bar{g}_s)}{2\bar{g}_s^3(\lambda, g_s)} \int d^4x\, G^a_{\mu\nu}(x) G^{a\mu\nu}(x). \tag{6.104}$$

On the other hand, according to the Noether theorem we have

$$\delta S_{eff} = \delta\lambda \int d^4x [\partial^\mu j^D_\mu(x)]. \tag{6.105}$$

Therefore, the divergence of the dilatation current is given by the expression [68]

$$\frac{\partial}{\partial x_\mu} j^D_\mu(x) = \frac{\beta(\bar{g}_s)}{2\bar{g}_s^3(\lambda, g_s)} G^a_{\mu\nu}(x) G^{a\mu\nu}(x). \tag{6.106}$$

Taking into account the connection between the dilatation current and the energy-momentum tensor, we arrive at the spur anomaly relation

$$T^\mu_\mu(x) = \frac{\beta(\bar{g}_s)}{2\bar{g}_s^3(\lambda, g_s)} G^a_{\mu\nu}(x) G^{a\mu\nu}(x). \tag{6.107}$$

Substituting the value $\bar{g}_s(\lambda, g_s)$, which has been found in Section 4.2 with an accuracy of g_s^3, into (6.106), we finally obtain

$$\frac{\partial}{\partial x_\mu} j^D_\mu(x) = -\frac{b}{32\pi^2} G^a_{\mu\nu}(x) G^{a\mu\nu}(x) = \frac{1}{32\pi^2} \left(\frac{2}{3} N_f - 11 \right) G^a_{\mu\nu}(x) G^{a\mu\nu}(x). \tag{6.108}$$

From (6.107) it also follows, when the vacuum expectation value of the gauge field tensor square $< 0|(G^a_{\mu\nu})^2|0 >$ does not equal to zero, then that of the energy-momentum tensor spur does not equal to zero as well. Later, we shall show that in QCD the relation

$$< 0|(G^a_{\mu\nu})^2|0 > \neq 0$$

is valid. This is caused by the fact that a true vacuum of gauge fields is different from a vacuum of the perturbation theory because fluctuations not described by the perturbation theory lead to the appearance of a gluon condensate with a nonzero average field square in a system (see for review [69] and references therein).

The spur anomaly equation could be obtained calculating the spur $T_{\mu\nu}$ in the one-loop approximation and using one or an other regularization. For each regularization method there exists its own derivation of the spur anomaly that exploits possible pathology of this concrete regularization. Thus, for example, the derivation of Eq. (6.106) based on the dimensional regularization is given in the book [70].

7

Hard processes in QCD

We're not scared by life's pace,
In response we run in place.
V. Vysotsky

7.1 e^-e^+ annihilation into hadrons

The asymptotical freedom property simplifies calculations of many processes. In order to understand it we, for example, consider a quark being contained into a proton. In a space the effective color quark charge is distributed in the form of a cloud (Fig. 7.1). We shall

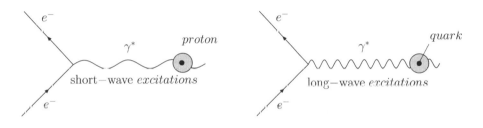

FIGURE 7.1
Probing the proton structure with the help of the $e^-p \to e^-p$ elastic scattering (for the sake of convenience we direct the time axis upwards).

use virtual photons as a probe of a proton structure. Probes with various wavelengths *see* a color cloud of a quark differently. Long-wave photons see this cloud to be large, while short-wave photons do only a small part of the cloud. At sudden transfer of large momentum to the quark, which is realized by a short-wave photon, the quark is released from the color cloud and, as a result, its effective color charge becomes smaller. The outgoing quark will be propagated practically freely as long as a new cloud appears. At that, the electric and weak quark charges, which are independent on color cloud polarization, remain concentrated on the quark. For this reason short-wave electroweak probes see point-like quarks whose strong (color) interaction give small corrections to their propagation. To be more exact, these corrections change an energy-momentum of color currents, those, in itself, must be small in such processes.

In this chapter we shall consider the QCD processes in a high-energies region, that is, in a small distances region. The small distances limit can be easily studied using the Feynman diagrams method. In this limit the asymptotical freedom leads to smallness of the coupling constant. As a consequence, there exists the reasonable diagrammatic perturbation theory where the model of free quarks and gluons can be taken as a zero approximation. It should be noted that one may investigate QCD in the strong coupling condition as well, however, only with the help of nonperturbative methods. Among them we mention Wilson's lattice QCD, the currents algebra, the PCAC hypothesis, and so on. Nevertheless, it should be stressed that the perturbative approach represents a principal computation instrument of QCD.

In Section 10.3 of *Advanced Particle Physics Volume 1* we considered the deep-inelastic e^-p-scattering process. A virtual photon with large Q^2 and small wavelength being prepared in this process probes a proton revealing its constituents and their color interactions. An appearing picture may be easily interpreted, since the short-range character of the quark-gluon interactions gives an opportunity to compare experimental data with calculation results fulfilled by the perturbation theory method. Virtual photons with a large resolution can be also prepared in collisions of electron-positron high-energy beams (Fig. 7.2). Therefore, electron-positron colliders can be used for the experimental check-up both

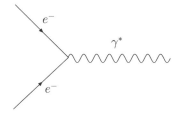

FIGURE 7.2
The probing virtual photons obtained under e^-e^+-collisions.

of the electroweak interaction and of QCD. e^-e^+-annihilation offers major advantages over deep-inelastic e^-p-scattering, since it represents a clean process in the sense that in an initial state we deal with structureless leptons rather than hadrons representing composite formation built up from partons. Note that the above mentioned also remains valid for muon colliders scheduled to be built in the near future.

The simplest reaction with quarks participation is a quark-pair production under e^-e^+-annihilation

$$e^- + e^+ \to \gamma^* \to q + \overline{q}. \tag{7.1}$$

The diagram associated with this process is shown in Fig. 7.3a. Its contribution has the order of $O(\alpha^2)$. In the center-of-mass frame the quark and antiquark have the identical energy and opposite-directed three-dimensional momenta. The quark and antiquark are produced in one point and flown away at the light velocity. When they are separated at a distance comparable with the confinement radius $r_{conf} \sim 1$ fm, a dipole color field of the quark-antiquark pair is compressed into a tube (string). Under further separation, the current tube is increased, and more and more amounts of kinetic energy of the pair is transformed into the field energy. When the energy involved in the current tube E exceeds two quark masses, a new quark-antiquark pair appears that shields the color charge of the initial pair. The string decays into two strings at the ends of which the initial and newly arisen quarks are located. If a relative momentum of the quark-antiquark pair is sufficiently large, the string is again exploded and so on. A parton cascade is developed: more and more number of quarks and gluons with smaller energy appears.

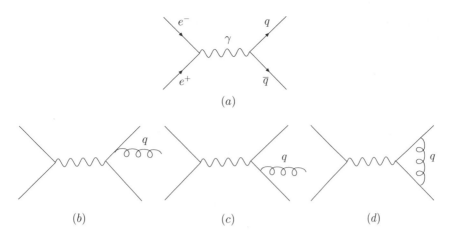

FIGURE 7.3

The diagrams (a)–(d) describing e^-e^+-annihilation.

At high energies ($E \gg M_N$, where M_N is a nucleon mass) the current tube has a large longitudinal momentum ($p_L \sim E$) while a transverse momentum is limited ($p_\perp \sim M_N$). Therefore, in a cascade, partons have large longitudinal momenta and constrained transverse ones. Following a parton cascade evolution, a hadronization process of partons begins, that is, recombination (fragmentation) of partons into color singlet states happens. It is believed that the present transition is not accompanied by a large momentum transfer between partons, that is, it presents a soft process (local parton-hadron duality hypothesis). This mechanism, for example, suppresses a process of unification of quarks belonging to different jets. So, the transition from parton jet to hadron jet does not practically change the transverse and longitudinal momentum distributions. Final particles, hadrons, prove to be collimated in small angular directions. Moreover, the angular distribution of jets with respect to e^-e^+-beams coincides with that of the initial quark and antiquark. It is evident that an energetic quark can emit a gluon at a large angle. In this case we observe a three-jet event.

Assume that the total hadron cross section is equal to the total parton cross section

$$\sigma(e^-e^+ \to \text{hadrons}) = \sigma(e^-e^+ \to \text{partons}) = \sigma(e^-e^+ \to q\bar{q}) + \sigma(e^-e^+ \to q\bar{q}g) + \ldots, \quad (7.2)$$

that is, quarks fragment into hadrons with the probability equal to 1. There are physical reasons for this assumption. The typical time of a parton process (the time of a virtual photon decay) is

$$t_p \sim \frac{1}{E}, \quad (7.3)$$

where E is an energy in the center-of-mass frame. The hadronization time is given as

$$t_h \sim \frac{1}{\Lambda_{QCD}}. \quad (7.4)$$

Since at high-energies $E \gg \Lambda_{QCD}$, then $t_p \ll t_h$ and the hadronization occurs far later than parton processes to guarantee the correctness of (7.2).

Principal corrections to the probability of e^-e^+-annihilation into hadrons are caused by the diagrams pictured in Fig. 7.3(b),(c). In the lowest order of the perturbation theory the cross section of the process (7.1) was found earlier (see Eq. (4.49)). Here, we shall calculate the contribution coming from the diagrams shown in Fig. 29.3b,c, that is, the cross section of the process

$$e^- + e^+ \rightarrow \gamma^* \rightarrow q + \overline{q} + g. \tag{7.5}$$

Let us denote portions of a beam energy E which are transferred by quark, antiquark, and gluon by x_i ($i = q, \overline{q}, g$):

$$x_q = \frac{E_q}{E}, \qquad x_{\overline{q}} = \frac{E_{\overline{q}}}{E}, \qquad x_g = \frac{E_g}{E}, \tag{7.6}$$

where

$$E = \frac{1}{2}\sqrt{s} = \frac{1}{2}Q^2 = \frac{1}{2}(p_e + p_{\overline{e}})^2, \tag{7.7}$$

and $0 \leq x_i \leq 1$. For the sake of simplicity, we shall consider leptons and quarks to be massless. Choose the Feynman gauge for the photon propagator. Then, the amplitude of the process (7.5) will look like:

$$\mathcal{A} = \mathcal{A}_a + \mathcal{A}_b = = -ie^2 e_q g_s \frac{\lambda_a}{2} \sqrt{\frac{1}{2^5 E_{e^-} E_{e^+} E_q E_{\overline{q}} E_g}} \overline{v}(\mathbf{p}_{e^+})\gamma_\mu u(\mathbf{p}_{e^-})\frac{g^{\mu\nu}}{s}\overline{u}'(\mathbf{p}_q)\times$$

$$\times \left(\gamma_\rho \frac{\hat{p}_q + \hat{k}}{(p_q + k)^2}\gamma_\nu - \gamma_\nu \frac{\hat{p}_{\overline{q}} + \hat{k}}{(p_{\overline{q}} + k)^2}\gamma_\rho\right)v'(\mathbf{p}_{\overline{q}})\epsilon_a^\rho(\mathbf{k}), \tag{7.8}$$

where e_q is a quark charge in the electron charge units ($|e|$), $p_q(p_{\overline{q}})$ is a four-dimensional quark (antiquark) momentum, $p_{e^-}(p_{e^+})$ is a four-dimensional electron (positron) momentum, and k is a four-dimensional gluon momentum. Note, since all particles are considered to be massless, then the normalization factors are taken in the form $1/\sqrt{2E}$ and, instead of the factor

$$\frac{\hat{p} + m}{2m},$$

the factor $\hat{p}$ appears under summing and averaging over fermion spins. We shall consider all particles to be unpolarized. Then, the amplitude square should be summed over polarizations of final particles and averaged over those of initial particles, that is, we should find

$$\frac{1}{4}\sum_{pol}|\mathcal{A}|^2.$$

Summation over gluon polarizations is fulfilled using the formula

$$\sum_{pol}\epsilon_a^\rho(\mathbf{k})\epsilon_b^\sigma(\mathbf{k}) = \left[-g^{\rho\sigma} + \frac{k^\rho k^\sigma}{k^2}\right]\delta_{ab}. \tag{7.9}$$

Direct calculations show that the term $k^\rho k^\sigma/k^2$, which may be a source of additional operations, namely, introducing a gluon mass m_g and tending m_g to zero at the final stage, does not contribute. Taking into account the relative velocity of the e^-e^+-pair in the center-of-mass frame to be $v_{rel} = (p_{e^-} \cdot p_{e^+})/(E_{e^-}E_{e^+}) = 2$, we get

$$d\sigma = \frac{1}{2^3\pi^5}\sum\int|\mathcal{A}|^2\delta^{(4)}(Q - p_q - p_{\overline{q}} - k)d^3p_q d^3p_{\overline{q}}d^3k =$$

$$= -\frac{4c_F^2 e^4 e_q^2 g_s^2}{2^{13}\pi^5 s^3}L_{\mu\nu}(p_{e^-}, p_{e^+})T^{\mu\nu}(Q), \tag{7.10}$$

where $Q_\mu = (p_{e^-} + p_{e^+})_\mu$, the lepton and hadron tensors have the view

$$L_{\mu\nu}(p_{e^-}, p_{e^+}) = \mathrm{Sp}\ [\hat{p}_{e^-}\gamma_\mu\hat{p}_{e^+}\gamma_\nu],$$

$$T^{\mu\nu}(Q) = \int \frac{d^3p_q}{E_q}\frac{d^3p_{\bar{q}}}{E_{\bar{q}}}\frac{d^3k}{E_g}\delta^4(p_q + p_{\bar{q}} + k - Q)H^{\mu\nu}(p_q, p_{\bar{q}}, k),$$

$$H^{\mu\nu}(p_q, p_{\bar{q}}, k) = \mathrm{Sp}\ \left[\hat{p}_q\left(\gamma_\rho\frac{\hat{p}_q + \hat{k}}{(p_q + k)^2}\gamma_\mu - \gamma_\mu\frac{\hat{p}_{\bar{q}} + \hat{k}}{(p_{\bar{q}} + k)^2}\gamma_\rho\right)\hat{p}_{\bar{q}}\left(\gamma_\nu\frac{\hat{p}_q + \hat{k}}{(p_q + k)^2}\gamma_\rho - \right.\right.$$
$$\left.\left. -\gamma_\rho\frac{\hat{p}_{\bar{q}} + \hat{k}}{(p_{\bar{q}} + k)^2}\gamma_\nu\right)\right].$$

These tensors are symmetric and transverse relative to Q_μ as a result of the electromagnetic current conservation

$$Q^\mu L_{\mu\nu} = 0, \qquad Q_\mu T^{\mu\nu} = 0. \tag{7.11}$$

Therefore, the tensor $T^{\mu\nu}$ can be parametrized in the following way

$$T^{\mu\nu}(Q) = \left(g^{\mu\nu} - \frac{Q^\mu Q^\nu}{Q^2}\right)T(Q^2), \tag{7.12}$$

where

$$T(Q^2) = \frac{1}{3}g^{\mu\nu}T_{\mu\nu}(Q). \tag{7.13}$$

Then, the convolution of the tensors $L_{\mu\nu}T^{\mu\nu}$ becomes the form

$$T^{\mu\nu}L_{\mu\nu} = \left(g^{\mu\nu} - \frac{Q^\mu Q^\nu}{Q^2}\right)T(Q^2)L_{\mu\nu} = g^{\mu\nu}T(Q^2)L_{\mu\nu} = \frac{1}{3}g_{\lambda\sigma}T^{\lambda\sigma}g^{\mu\nu}L_{\mu\nu} =$$

$$= \frac{1}{3}g_{\lambda\sigma}T^{\lambda\sigma}\mathrm{Sp}\ [\hat{p}_{e^-}\gamma^\mu\hat{p}_{e^+}\gamma_\mu] = -\frac{4}{3}Q^2 g_{\lambda\sigma}T^{\lambda\sigma}. \tag{7.14}$$

In such a way, calculations are reduced to finding the quantity $g_{\lambda\sigma}H^{\lambda\sigma}(p_q, p_{\bar{q}}, k)$. Allowing for the relations

$$p_{\bar{q}} + k = Q - p_q, \qquad p_q + k = Q - p_{\bar{q}}, \tag{7.15}$$

we get

$$g_{\lambda\sigma}H^{\lambda\sigma}(p_q, p_{\bar{q}}, k) = \frac{a_{11}}{(Q - p_{\bar{q}})^4} + \frac{a_{12} + a_{21}}{(Q - p_{\bar{q}})^2(Q - p_q)^2} + \frac{a_{22}}{(Q - p_q)^4}, \tag{7.16}$$

where

$$a_{11} = \mathrm{Sp}\ [\hat{p}_q\gamma_\rho(\hat{Q} - \hat{p}_{\bar{q}})\gamma_\mu\hat{p}_{\bar{q}}\gamma^\mu(\hat{Q} - \hat{p}_{\bar{q}})\gamma^\rho] = 16[2(p_q \cdot Q)(p_{\bar{q}} \cdot Q) - Q^2(p_q \cdot p_{\bar{q}})], \tag{7.17}$$

$$a_{12} = -\mathrm{Sp}\ [\hat{p}_q\gamma_\rho(\hat{Q} - \hat{p}_{\bar{q}})\gamma_\mu\hat{p}_{\bar{q}}\gamma^\rho(\hat{Q} - \hat{p}_q)\gamma^\mu] = 32(p_q \cdot p_{\bar{q}})[(p_{\bar{q}} - Q)^\mu(p_q - Q)_\mu], \tag{7.18}$$

$$a_{21} = -\mathrm{Sp}\ [\hat{p}_q\gamma_\mu(\hat{Q} - \hat{p}_{\bar{q}})\gamma_\rho\hat{p}_{\bar{q}}\gamma^\mu(\hat{Q} - \hat{p}_{\bar{q}})\gamma^\rho] = a_{12} \tag{7.19}$$

$$a_{22} = \mathrm{Sp}\ [\hat{p}_q\gamma_\mu(\hat{Q} - \hat{p}_q)\gamma_\rho\hat{p}_{\bar{q}}\gamma^\rho(\hat{Q} - \hat{p}_q)\gamma^\mu] = a_{11}. \tag{7.20}$$

Substituting the obtained values of a_{ij} into (7.16), we arrive at

$$g_{\lambda\sigma}H^{\lambda\sigma} = 16\left[2(p_q \cdot Q)(p_{\bar{q}} \cdot Q) - Q^2(p_q \cdot p_{\bar{q}})\right]\left[\frac{1}{(Q - p_{\bar{q}})^4} + \frac{1}{(Q - p_q)^4}\right] +$$

$$+\frac{64(p_q \cdot p_{\bar{q}})[(p_{\bar{q}} - Q)^\mu(p_q - Q)_\mu]}{(Q - p_{\bar{q}})^2(Q - p_q)^2}. \tag{7.21}$$

   The most evident experimental indication of emitting a gluon is the fact that by this time q and $\bar{q}$ are born not in the opposite directions. The antiquark is produced with some portion of the transverse momentum (with respect to the quark direction). By this reason, the corresponding observable is $d\sigma/(dx_q dx_{\bar{q}})$. Hence, we must integrate both over gluon momenta and over angular variables of the quark-antiquark pair. Integration over a three-dimensional gluon momentum in the center-of-mass frame removes $\delta^3(\mathbf{p}_q + \mathbf{p}_{\bar{q}} + \mathbf{k})$ and ensures

$$E_g = |\mathbf{k}| = |\mathbf{p}_q + \mathbf{p}_{\bar{q}}| = \sqrt{E_q^2 + E_{\bar{q}}^2 + 2E_q E_{\bar{q}} \cos\theta_{q\bar{q}}}. \tag{7.22}$$

Next, we choose the reference frame in which the vector $\mathbf{p}_q$ is directed along the axis z. In this case we have

$$d^3 p_{\bar{q}} = E_{\bar{q}}^2 dE_{\bar{q}} d\cos\theta' d\varphi',$$

where $\cos\theta_{q\bar{q}} \equiv \cos\theta'$. Now with allowance made for the formula

$$\left[\frac{\partial E_g}{\partial\cos\theta'}\right]^{-1} = \frac{E_g}{E_q E_{\bar{q}}}, \tag{7.23}$$

we can integrate over the angular variables of the $q\bar{q}$-pair

$$\int \frac{E_q dE_q d\Omega E_{\bar{q}} dE_{\bar{q}} d\cos\theta' d\varphi'}{E_g} \delta\left(Q_0 - E_q - E_{\bar{q}} - E_g\right) g_{\mu\nu} H^{\mu\nu} = \int dE_q dE_{\bar{q}} d\Omega d\varphi' g_{\mu\nu} H^{\mu\nu} =$$

$$= 8\pi^2 dE_q dE_{\bar{q}} g_{\mu\nu} H^{\mu\nu} = 2\pi^2 s dx_q dx_{\bar{q}} g_{\mu\nu} T^{\mu\nu}, \tag{7.24}$$

where we have taken into account the definition (7.6). Further, the hadron tensor should be expressed in terms of the variables x_q and $x_{\bar{q}}$. Using the formulae

$$(p_q \cdot Q) = \frac{x_q Q^2}{2}, \qquad (p_{\bar{q}} \cdot Q) = \frac{x_{\bar{q}} Q^2}{2}, \qquad (p_{\bar{q}} \cdot p_q) = \frac{(x_q + x_{\bar{q}} - 1)Q^2}{2}, \tag{7.25}$$

we find

$$g_{\mu\nu} H^{\mu\nu} = 8\left[\frac{1-x_q}{1-x_{\bar{q}}} + \frac{1-x_{\bar{q}}}{1-x_q} + \frac{2(x_q + x_{\bar{q}} - 1)}{(1-x_q)(1-x_{\bar{q}})}\right] = 8\frac{x_q^2 + x_{\bar{q}}^2}{(1-x_q)(1-x_{\bar{q}})}. \tag{7.26}$$

Now we can write the final result

$$\frac{d^2\sigma}{dx_q dx_{\bar{q}}} = \frac{2c_F \alpha^2 e_q^2 \alpha_s}{3s} \frac{x_q^2 + x_{\bar{q}}^2}{(1-x_q)(1-x_{\bar{q}})} dx_q dx_{\bar{q}}. \tag{7.27}$$

   For the total cross section to be obtained the expression (7.27) must be integrated over x_q and $x_{\bar{q}}$ from 0 to 1. In so doing we face a problem that is usual under calculations by the QCD perturbation theory. When x_q or $x_{\bar{q}}$ tends to 1, the integrand of (7.27) is divergent. In order to track the origin of this problem, we consider, for example, the factor $1 - x_q$ in the denominator. From the relation

$$1 - x_q = 1 - \frac{2(p_q \cdot Q)}{Q^2} = \frac{2(p_{\bar{q}} \cdot k)}{Q^2} = \frac{2}{Q^2} E_{\bar{q}} E_g (1 - \cos\theta_{\bar{q}g}) \tag{7.28}$$

it follows that $1 - x_q$ turns into zero when a gluon becomes soft or when the gluon and antiquark becomes collinear ($\cos\theta_{\bar{q}g} \to 1$). The first type of the divergence is called *infrared one*, while the second type is *collinear one*. If the mass of the quark or gluon is not equal to zero, then the value $\cos\theta_{\bar{q}g} = 1$ is kinematically unachievable. For this reason the collinear divergence is called the *mass singularity* as well. In QCD leptons are massive particles and, as a result, the mass singularity is absent.

For subsequent progress we have to regularize the infrared and collinear divergences. The method of excluding the infrared divergences (ID) is already known to us from QED. We have to assign an effective mass to a gluon and repeat the calculation of the Feynman diagrams displayed in Fig. 7.3. Not complicated but cumbersome calculations result in

$$\sigma(e^-e^+ \to q\bar{q}g) = \int dx_q dx_{\bar{q}} \frac{d^2\sigma(e^-e^+ \to q\bar{q}g)}{dx_q dx_{\bar{q}}} = \sigma_0 \frac{2\alpha_s}{3\pi} \left[\ln^2\left(\frac{m_g}{Q}\right) + \right.$$

$$\left. +3\ln\left(\frac{m_g}{Q}\right) - \frac{\pi^2}{3} + 5 \right]. \tag{7.29}$$

where σ_0 denotes the cross section of the reaction $e^-e^+ \to q\bar{q}$ in the tree approximation

$$\sigma_0 = \frac{4\pi\alpha^2 e_q^2}{s} \tag{7.30}$$

and it is has been taken into account that $c_F = 4/3$ for the $SU(3)$ group.

As one would expect, the expression (7.29) diverges at $m_g \to 0$. Obviously, it can not be a final answer because it depends on a fictitious mass m_g whereas this must not be the case. The reason consists in the fact that we did not allow for all contributions into the cross section of the process $e^-e^+ \to$ hadrons. There are additional contributions of the order of $\alpha^2\alpha_s$, which arise from interference of the diagrams, involving virtual gluon loops (Fig. 7.3c), with the diagram of the tree approximation (Fig. 7.3a). Their contribution into the cross section has the form (see for example [71, 72])

$$\sigma(e^-e^+ \to q\bar{q}) - \sigma_0 \frac{2\alpha_s}{3\pi} \left[-\ln^2\left(\frac{m_g}{Q}\right) - 3\ln\left(\frac{m_g}{Q}\right) + \frac{\pi^2}{3} - \frac{7}{2} \right]. \tag{7.31}$$

The summation of (7.29)–(7.31) gives the total cross section of e^-e^+-annihilation into hadrons. At that, one may improve the result replacing the fixed coupling constant α_s by the running coupling constant $\alpha_s(Q^2)$ (see Eq. (4.42)). Thus, the final answer is

$$\sigma(e^-e^+ \to \text{hadrons}) = \frac{4\pi\alpha^2}{Q^2} \sum_f e_{q_f}^2 \left[1 + \frac{\alpha_s(Q^2)}{\pi} + O(\alpha_s^2) \right]. \tag{7.32}$$

As we see, the obtained cross section is finite and independent on m_g. Cancellations of singularities into the contributions of emitting the real and virtual gluons is specific not only for this process. It takes place in many cases and is the manifestation of a general theorem by Kinoshita [73], Lee and Nauenberg [74]. The cancellation of infrared divergences is easy to understand. The primary process $e^-e^+ \to q\bar{q}$ happens much too quickly. As a virtual photon is beyond the mass shell at a distance s, quarks are produced during the time of the order of $1/\sqrt{s}$. However, emitting collinear gluons and corrections, connected with soft gluon exchanges, happen for a much larger time interval. On the diagrams with emitting gluons, deviation of a virtual quark or antiquark from a position on the mass shell is of the order of $p_{g\perp}^2$, where $p_{g\perp}$ is a transverse gluon momentum in respect of the $q\bar{q}$-pair. And so this virtual state lives during the time of the order of $1/p_{g\perp}$. Such a slow process cannot influence the probability of the $q\bar{q}$ pair production. It can influence only properties of the final state the $q\bar{q}$ system passes on to. In line with this logic, sole corrections, which can influence the total cross section, are those for which $p_{g\perp} \sim \sqrt{s}$. Alternatively, this may be said, that after cancellations of the contributions caused by the infrared regions, only the contributions coming from regions with large momentum values of real and virtual gluons remain.

If one gives a glance at the problem of soft and collinear gluons from the experimental view point, then the problem is absent at all. This is connected with the resolution of elementary particle detectors. Since detectors have a finite angular resolution $\delta > 0$, then it is impossible to distinguish isolated quark (antiquark) from quark (antiquark) accompanied by a gluon that is emitted at the angle $\theta < \delta$. Analogously, because of a finite energy resolution $\epsilon > 0$ the separation of an isolated quark (antiquark) from quark (antiquark) accompanied by a gluon with the energy $E < \epsilon$ is impossible. To put it differently, if a gluon is much too soft or/and collinear, then the $q\bar{q}g$ state is detected as the $q\bar{q}$ state:

$$|q\bar{q}g> \xrightarrow{E<\epsilon,\; \theta<\delta} |q\bar{q}>_{phys}.$$

The finite energy and angular resolutions of detectors constrain the integration region in Eq. (7.27), so that singular regions are excluded. As this takes place, the observable cross section proves to depend on the parameters ϵ and δ.

7.2 Deep inelastic scattering of leptons by a proton

Let us proceed to the investigation of the scattering of leptons by a proton. We start with the case when a charged lepton (electron or muon) is used as a probe

$$l^- + p \to l^- + X, \tag{7.33}$$

where X denotes any hadron state. To be definite, we shall say about an electron ($l^- = e^-$). At the most elementary level, this reaction may be considered as the scattering of an electron by a quark being inside a proton. In Fig. 7.4 the diagram associated with the process (7.33), where P is a proton momentum, p is an initial momentum of a quark, k and k' are the initial and final momenta of an electron, and $q = k - k'$ is a momentum transferred to a hadron system (q is space-like), is displayed. If the transferred momentum square $-q^2 = Q^2$

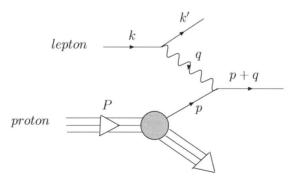

FIGURE 7.4
The diagram presents the momentum flow when a high-energy lepton scatters from a quark taken from the wave-function of the proton.

is large, the quark is emitted so that subsequent soft processes cannot balance this reaction. Soft processes, however, will lead to the production of gluons and quark-antiquark pairs, which finally neutralize the color and convert the knocked (being interacted or active) quark into hadron jets flying along the momentum transferred by the electron. Since the active quark carries a large momentum in respect to the rest of quarks, quark-spectators, the total mass square of a final hadron system is large as usual. In this case they say, the process represents a deep inelastic scattering.

Consider this reaction in the center-of-mass frame. We shall assume the energy to be large in this reference frame and, as a result, when considering kinematic, the proton mass may be neglected. Then the proton has almost a light-like momentum along the collision axis. Quarks constituting the proton also possess light-like momenta to be nearly collinear to the proton momentum. This is the case, since the quark could acquire a large transverse momentum only by means of exchange of a hard gluon, but this exchange is suppressed by the parameter α_s, which is small at large energy scales. Then, in the leading order of the QCD perturbation theory, we have for the constituent quark:

$$p = xP.$$

In the leading order in α_s one may disregard emitting or exchanging gluons during a collision process. As a result, to find the cross section of the electron-proton scattering we must take the cross section of the electron-quark scattering at a specified value of a longitudinal portion of a quark momentum x, multiply it by the probability that the proton contains the quark with such a value x, and integrate the expression obtained over x.

However, the probability that the proton contains the quark with some value of the momentum cannot be calculated within the QCD perturbation theory in so far as the probability depends on soft processes determining the proton structure as a bound state of quarks and gluons. For this reason, we consider this probability as an unknown function that has to be defined by experiments. In the end, we have to use such a probabilistic function for every type of parton that could be found inside the proton. The probability that a parton of a type i with a longitudinal portion value x is situated inside the proton has the form $f_i(x)dx$, where $f_i(x)$ is the parton distribution function. Then, in the leading order in α_s, the cross section of deep inelastic scattering of the electron by the proton is as follows:

$$\sigma\left(e^-(k) + p(P) \to e^-(k') + X\right) = \int_0^1 dx \sum_i f_i(x)\sigma\left(e^-(k) + q_i(xP) \to e^-(k') + q_i(p')\right),$$

(7.34)

where we have written out the particle momenta in the explicit form, and the sum contains the contributions of constituent quarks and antiquarks. It is important to keep in mind that the formula (7.34) represents not the exact prediction of QCD but merely the first term of the expansion in α_s (this approximation is called the *parton model*). High-order corrections to Eq. (7.34) will include the change both of the cross section of e^-p scattering and of the parton distribution functions.

In the same way all the reactions with protons, in which a large momentum transfer is in existence, are described by the parton model. In QCD all the cross sections of these reactions are calculated from scattering amplitudes of quarks and gluons. It should be stressed that in all cases an initial parton motion is determined by one and the same parton distribution function $f_i(x)$.

So, for the cross section of the process (7.34) to be calculated, the cross section of the subprocess

$$e^- + q \to e^- + q \tag{7.35}$$

must be found before. The Feynman diagram of this subprocess is shown in Fig. 7.5 and

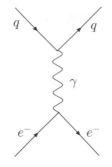

FIGURE 7.5
The Feynman diagram corresponding to the subprocess $e^- + q \to e^- + q$.

the corresponding amplitude is given by the expression:

$$\mathcal{A} = -e^2 e_{q_i} \sqrt{\frac{m_e m_e}{E_e E'_e}} \sqrt{\frac{m_{q_i} m_{q_i}}{E_{q_i} E'_{q_i}}} \overline{u}(k')\gamma^\mu u(k) \frac{1}{q^2} \overline{u}(p')\gamma_\mu u(p), \qquad (7.36)$$

where e_{q_i} is the electric charge of the i-quark in the units of $|e|$ and $q = k - k'$. Obviously, calculation of the subprocess (7.35) is completely analogous to that of the process

$$e^- + \mu^+ \to e^- + \mu^+,$$

which has been fulfilled in Section 10.2 of *Advanced Particle Physics Volume 1*. We shall follow the same scheme. We shall not be interested in the particle polarizations in the initial and final states, that is, we find the quantity

$$\frac{1}{4} \sum_{pol} |\mathcal{A}|^2.$$

Report it in terms of the product of the lepton and quark tensors

$$\frac{1}{4} \sum_{pol} |\mathcal{A}|^2 = \frac{e^4 e_{q_i}^2}{q^4} L_e^{\mu\nu} H_{\mu\nu}^q, \qquad (7.37)$$

where

$$L_e^{\mu\nu} = \frac{1}{2} \frac{m_e m_e}{E_e E'_e} \sum_{pol} [\overline{u}(k')\gamma^\mu u(k)][\overline{u}(k')\gamma^\nu u(k)]^* \qquad (7.38)$$

and for $H_{\mu\nu}^q$ the similar expression is valid (since the second square bracket in the expression (7.38) is the 1×1 matrix, the operations of the complex and Hermitian conjugations give the same result). Using the rules of calculating the γ matrix spurs, we find the electron and quark tensors

$$L_e^{\mu\nu} = \frac{1}{8E_e E'_e} \left[\text{Sp } (\hat{k}'\gamma^\mu \hat{k}\gamma^\nu) + m_e^2 \text{Sp } (\gamma^\mu\gamma^\nu) \right] =$$

$$= \frac{1}{2E_e E'_e} \left\{ k'^\mu k^\nu + k'^\nu k^\mu - g^{\mu\nu}[(k' \cdot k) - m_e^2] \right\}, \qquad (7.39)$$

$$H_{\mu\nu}^q = \frac{1}{2E_{q_i} E'_{q_i}} \left\{ p'^\mu p^\nu + p'^\nu p^\mu - g^{\mu\nu}[(p' \cdot p) - m_{q_i}^2] \right\}. \qquad (7.40)$$

Constituting the product of the expressions (7.39) and (7.40), we get the amplitude square of the process (7.35)

$$\frac{1}{4} \sum_{pol} |\mathcal{A}|^2 = \frac{1}{E_e E'_e E_{q_i} E'_{q_i}} \frac{e^2 e_i^2}{2q^4} \left[(k' \cdot p')(k \cdot p) + (k' \cdot p)(k \cdot p') - \right.$$

$$\left. - m_e^2(p' \cdot p) - m_{q_i}^2(k' \cdot k) + 2m_e^2 m_{q_i}^2 \right]. \tag{7.41}$$

At energies under consideration one may neglect the lepton and quark masses. In this case the Mandelstam variables take the view

$$s = (k + p)^2 \approx 2(k \cdot p) = 2(k' \cdot p'), \tag{7.42}$$

$$t = (k - k')^2 \approx -2(k \cdot k') = -2(p \cdot p'), \tag{7.43}$$

$$u = (k - p')^2 \approx -2(k \cdot p') = -2(k' \cdot p). \tag{7.44}$$

Now we can write the differential cross section of the subprocess (7.35) in terms of the Mandelstam variables

$$\frac{d\sigma}{d\hat{t}} (e^- q_i \to e^- q_i) = \frac{2\pi \alpha^2 e_{q_i}^2}{\hat{s}^2} \left(\frac{\hat{s}^2 + \hat{u}^2}{\hat{t}^2} \right), \tag{7.45}$$

where hereafter we use the symbols $\hat{s}, \hat{u}$, and $\hat{t}$ for the designation of the Mandelstam variables in the cases of two-particle processes at the parton level. These variables are connected with observable characteristics of the electron and a hadron system by the obvious relations:

$$\hat{t} = -Q^2, \qquad \hat{s} = 2(p \cdot k) = 2x(P \cdot k) = xs. \tag{7.46}$$

Using (7.46) as well as

$$\hat{s} + \hat{u} + \hat{t} = 0,$$

we can represent the differential cross section of the process (7.33) in the following form:

$$\frac{d\sigma}{dQ^2} (e^- p \to e^- X) = 2\pi \alpha^2 \int_0^1 dx \sum_i f_i(x) \frac{e_{q_i}^2}{Q^4} \left[1 + \left(1 - \frac{Q^2}{xs} \right)^2 \right] \theta(xs - Q^2), \tag{7.47}$$

where $\theta(xs - Q^2)$ expresses the kinematic constraint $\hat{s} \geq |\hat{t}|$. The expression (7.47) is the first approximation to the cross section of the deep inelastic $e^- p$-scattering at large Q^2. In this case corrections to (7.47) that are caused by emitting and exchanging hard gluons are of the order of $\alpha_s(Q^2)$. It should be noted that x may be defined only by measuring the electron momentum when one assumes the electron to be elastically scattered from a parton. Because the mass of this parton is small as compared with s and Q^2, then

$$0 \approx (p + q)^2 = 2(p \cdot q) + q^2 = 2x(P \cdot q) - Q^2 \tag{7.48}$$

to give

$$x = \frac{Q^2}{(P \cdot q)}. \tag{7.49}$$

It is convenient to write the cross section by means of dimensionless variables. Along with x, we introduce one more dimensionless variable

$$y = \frac{(P \cdot q)}{(P \cdot k)} = \frac{2(P \cdot q)}{s}. \tag{7.50}$$

Since in the rest frame of the proton $y = (E_e - E'_e)/E_e$, then y is an energy portion of the ongoing electron that is transferred to a hadron system ($0 \leq y \leq 1$). From (7.49) and (7.50) the formula

$$xy = \frac{Q^2}{s} \qquad (7.51)$$

follows. Eq. (7.51), in its turn, permits proceeding to new variables

$$dxdQ^2 = \frac{dQ^2}{dy}dxdy = xsdxdy. \qquad (7.52)$$

Then, we can rewrite the differential cross section in the form:

$$\frac{d^2\sigma}{dxdy}\left(e^-p \to e^-X\right) = \left(\sum_i xf_i(x)e_{q_i}^2\right)\frac{2\pi\alpha^2 s}{Q^4}\left[1 + (1-y)^2\right], \qquad (7.53)$$

where we have used the relation connecting y with the parton variables

$$y = \frac{2p_\mu(k-k')^\mu}{2(P \cdot k)} = \frac{\hat{s} + \hat{u}}{\hat{s}}. \qquad (7.54)$$

If one looks aside at the existence of the factor $1/Q^4$ caused by the photon propagator, the dependence on x and y is completely factorized. Each half of this relation contains physical information. The fact, that the parton distribution $f_i(x)$ depends only on x and does not depend on Q^2, is evidence of Bjorken's scaling. It tells us that an initial distribution of partons is independent on hard scattering details. The cross section dependence on y appears because of scattering processes by partons.

Since the sum over the quark flavors is not factorized in (7.53), it is impossible to find an individual parton distribution function $f_i(x)$ only from experiments of the deep inelastic e^-p scattering. A additional information concerning the proton structure can be obtained by investigating a deep inelastic scattering of a neutrino and antineutrino by the proton. In this case, the interaction between a neutrino (antineutrino) and quarks happens owing to the weak interaction. If the interaction is caused by the neutral currents, then the initial neutrino (antineutrino), having emitted a Z boson, turns into the final neutrino (antineutrino). Inasmuch as the neutrino detection is very complicated, this process is of questionable value for us. When the interaction is stipulated for the charge currents, then the neutrino (antineutrino), having emitted a W^+ boson (W^- boson), turns into the corresponding charged lepton. From the viewpoint of experimental measuring of quark-parton distributions, the advantage of these processes lies in the fact that the neutrino and antineutrino interact with different quarks: a neutrino with d and $\bar{u}$ quarks, and an antineutrino with $\bar{d}$ and u ones. Note, as the proton contains small admixture of heavy quarks (s and c) as well, these quarks contribute to the deep inelastic neutrino scattering. However, we shall ignore these corrections. On account of technical reasons, the deep inelastic neutrino scattering is realized most simply with the help of muon neutrino beams. Therefore, the subject of our interest will be the subprocess:

$$\nu_\mu + d \to \mu^- + u, \qquad (7.55)$$

whose Feynman diagram is displayed in Fig. 7.6. Making use of the Lagrangian describing the interaction between quarks and the W boson (see Eq. (11.49) of *Advanced Particle Physics Volume 1*) and setting the Cabibbo angle θ_C to equal zero, we can write the amplitude of the reaction (7.55) in the following view:

$$\mathcal{A} = -\frac{g^2}{8}[\bar{\mu}(p')\gamma^\sigma(1-\gamma_5)\nu_\mu(p)]\frac{g_{\sigma\lambda}}{(p'-p)^2 - m_W^2}[\bar{u}(k')\gamma^\lambda(1-\gamma_5)d(k)], \qquad (7.56)$$

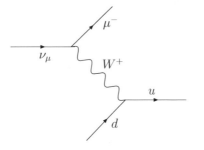

FIGURE 7.6
The Feynman diagrams corresponding to the subprocess $\nu_\mu + d \to \mu^- + u$.

where $p\,(k)$ is the four-dimensional momentum of the muon neutrino (d quark) in the initial state, $p'\,(k')$ is the four-dimensional momentum of the muon (u quark) in the final state, and the factor $g_{\sigma\lambda}/[(p'-p)^2 - m_W^2]$ appears thanks to the W-boson propagator to be taken in the unitary gauge. At energies under consideration, the first term in the W-boson propagator denominator may be neglected and, as a result, we arrive at

$$\mathcal{A} - \frac{G_F}{\sqrt{2}}[\overline{\mu}(p')\gamma^\lambda(1 \quad \gamma_5)\nu_\mu(p)][\overline{u}(k')\gamma_\lambda(1-\gamma_5)d(k)], \tag{7.57}$$

where we have taken into account

$$\frac{G_F}{\sqrt{2}} = \frac{g^2}{8m_W^2}.$$

Considering particles to be massless and fulfilling the summation and the average over the particle polarizations, we get

$$\frac{1}{2}\sum_{pol}|\mathcal{A}|^2 - \frac{G_F^2}{4}\text{Sp}\,[\gamma^\lambda(1-\gamma_5)\hat{k}\gamma^\sigma(1-\gamma_5)\hat{k}'] \times \text{Sp}\,[\gamma_\lambda(1-\gamma_5)\hat{p}\gamma_\sigma(1 \quad \gamma_5)\hat{p}'] =$$

$$= 64G_F^2(p\cdot k)(p'\cdot k') = 16G_F^2\hat{s}^2, \tag{7.58}$$

where $\hat{s} = (k+p)^2$. So, the differential cross section is given by the expression:

$$\frac{d\sigma}{d\Omega}\left(\nu_\mu d \to \mu^- u\right) = \frac{G_F^2\hat{s}}{4\pi^2} \tag{7.59}$$

or

$$\frac{d\sigma}{d\hat{t}}\left(\nu_\mu d \to \mu^- u\right) = \frac{G_F^2}{\pi}. \tag{7.60}$$

Next, we find the cross section of the subprocess

$$\overline{\nu}_\mu + u \to \mu^+ + d \tag{7.61}$$

whose Feynman diagram is pictured in Fig. 7.7. It is a straightforward matter to see this diagram to be connected with that of the reaction (7.55) by means of the crossing operation. Therefore, it is enough to replace $\hat{s}$ by $\hat{t}$, where

$$\hat{t} = (k - k')^2 \approx \frac{\hat{s}}{2}(\cos\theta_{\overline{\nu}\mu} - 1),$$

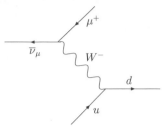

FIGURE 7.7

The Feynman diagram describing the subprocess $\overline{\nu}_\mu + u \to \mu^+ + d$.

in the expression (7.58). As a consequence, we obtain

$$\frac{d\sigma}{d\hat{t}}\left(\overline{\nu}_\mu u \to \mu^+ d\right) = \frac{G_F^2}{\pi}(1-y)^2,. \tag{7.62}$$

where $y = \cos\theta_{\overline{\nu}\mu}$.

In order to transform these scattering cross sections obtained at the parton level, we combine them with the parton distribution functions. Kinematics is the same as in the case of the deep inelastic e^-p scattering. By analogy we get

$$\frac{d^2\sigma}{dxdy}\left(\nu_\mu p \to \mu^- X\right) = \frac{G_F^2 s}{\pi}\left[x f_d(x) + x f_{\overline{u}}(x)(1-y)^2\right], \tag{7.63}$$

$$\frac{d^2\sigma}{dxdy}\left(\overline{\nu}_\mu p \to \mu^+ X\right) = \frac{G_F^2 s}{\pi}\left[x f_u(x)(1-y)^2 + x f_{\overline{d}}(x)\right]. \tag{7.64}$$

In accordance with these relations, the deep inelastic scattering of neutrino and antineutrino by the proton allows us to separate the distribution functions for u, d, $\overline{u}$, and $\overline{d}$ quarks. Within the accuracy, according to which a nucleon may be considered to consist of three constituent quarks with a small admixture of $\overline{q}q$-pairs, the cross section of the deep inelastic $\nu_\mu p$-scattering must not depend on y while the cross section of the $\overline{\nu}_\mu p$-scattering must drop as $(1-y)^2$. Measured dependence on y is shown in Fig. 7.8. From Fig. 7.8 it follows that the qualitative behavior predicted by the parton model is reproduced with sufficient accuracy. Deviations from the exact prediction may be ascribed to a small admixture of sea quarks in the wave proton function.

7.3   Parton distribution functions

Since the parton model predictions for the cross sections of the deep inelastic lepton-nucleon scattering correlate well with experimental data, then these cross sections can be used for the definition of parton distribution functions (detailed discussion of this problem may be found in the review [76]). Based on the conception that a nucleon is built from valence quarks, gluons, and a slow sea of quark-antiquark pairs, one can qualitatively picture the form of a parton distribution function. This form can be guessed by means of successive approximations (Fig. 7.9).

At first we consider a simple model of a nucleon, consisting of three free quarks, for which

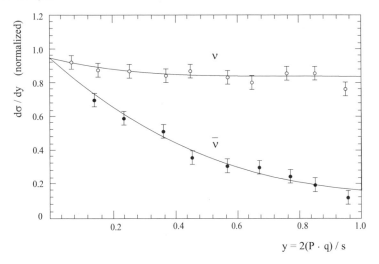

FIGURE 7.8

The distribution in y of neutrino and antineutrino deep inelastic scattering from an iron target [75]. The solid curves are fits to the form $A + B(1 - y^2)$.

a parton distribution function practically represents the delta function centered at point $x = 1/3$, that is, $f(x) \sim \delta(x - 1/3)$ (Fig. 7.9(a)). Switch on the interaction of quarks with gluons. Then the given distribution is smoothed owing to the fact that at the interaction, a momentum is redistributed between quarks and sharply determined momentum values $x = 1/3$ are smeared changing into the momenta distribution with the maximum at $x = 1/3$ (Fig. 7.9(b)). Next, gluons emitted with the probability $\sim dx/x$ can produce $\bar{q}q$-pairs, in the same manner as in the QED case where a virtual photon whose momentum spectrum dk/k is in line with the *bremsstrahlung* can emit c^-c^+-pairs. The processes of the internal conversion and the bremsstrahlung create a qq sea at small values of x resulting in the final distribution form shown in Fig. 7.9(c).

For the proton the distribution functions set satisfying to all experimental data is pictured in Fig. 7.10 [77]. Since all of these functions, especially those corresponding to antiquarks, greatly grow at small values of x, we depict the functions $xf_i(x)$ for each kind of distribution. As noted before, a small deviation from the Bjorken scaling is observed, so these functions are slowly changed versus Q^2. As will be showed later, the scaling violation is caused by higher-order corrections and it allows us to define the parton distribution function for gluons. Passing ahead of events, we also display this function in Fig. 7.10. As can be seen from this figure, a momentum spectrum of sea quarks is softer than that of valence quarks. This qualitatively comes to an agreement with the earlier employed conception that a nucleon is made of a central kern surrounded with a π-mesons cloud (Yukawa hypothesis). From the distribution functions it follows that valence quarks bear about 35% of total proton momentum, sea ones—about 10%, and the rest 55% is connected with gluons. Note, that the gluons distribution is directly measured in processes of heavy quarkoniums production. Since the distribution functions represented in Fig. 7.10 were found without taking into consideration the quark polarizations, they are called *unpolarized distribution functions*. Principles of determining the polarized distribution functions are given in works [79, 80].

Information about the parton distributions follows from the measurement of the structure nucleon functions. For the first time, the connection between the nucleon structure functions $W_{1,2}$ and parton distributions was established for the case of the deep inelastic

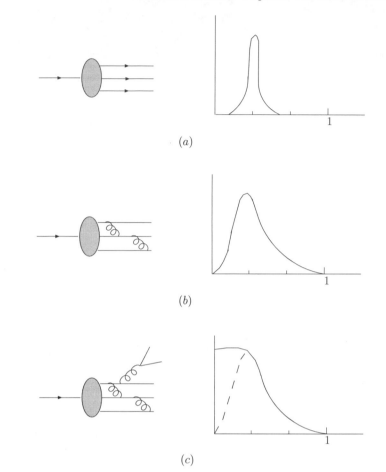

FIGURE 7.9

The parton distribution functions displayed according to different assumptions (a)–(c) concerning the proton structure.

e^-p-scattering in Section 10.3 of *Advanced Particle Physics Volume 1*. Considering the deep inelastic scattering of neutrino and antineutrino by nucleons in Section 5.3, we also found the corresponding nucleon structure functions $W_i^{(\nu,\bar{\nu})}(q^2,\nu)$ ($i = 1, 2, 3$). Obviously, acting just as in the case of the e^-p-scattering, we can define the dimensionless structure functions:

$$\left.\begin{array}{l} m_N W_1^{(\nu,\bar{\nu})}(q^2,\nu) = G_1^{(\nu,\bar{\nu})}(x, q^2/m_N^2), \\ \nu W_2^{(\nu,\bar{\nu})}(q^2,\nu) = G_2^{(\nu,\bar{\nu})}(x, q^2/m_N^2), \\ \nu W_3^{(\nu,\bar{\nu})}(q^2,\nu) = G_3^{(\nu,\bar{\nu})}(x, q^2/m_N^2), \end{array}\right\} . \tag{7.65}$$

At $|q^2| \to \infty$, $\nu \to \infty$ and a fixed value of x, these functions obey the Bjorken scaling

$$G_i^{(\nu,\bar{\nu})}(x, q^2/m_N^2) \to F_i^{(\nu,\bar{\nu})}(x). \tag{7.66}$$

However, it is frequently more convenient to use the nucleon structure functions with a particular helicity. Find them. To be definite we shall discuss the neutrino scattering. In the laboratory frame, we choose the axis z in such a way that the proton momentum P_λ

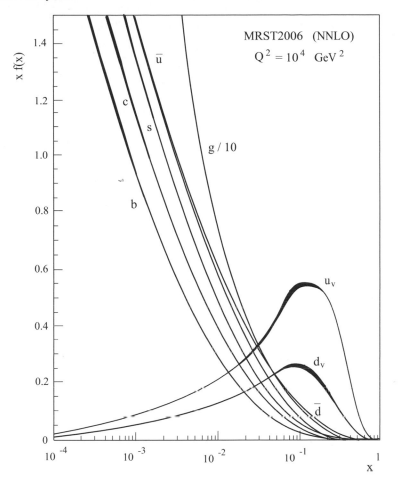

FIGURE 7.10

Distributions of x times the unpolarized parton distributions $f(x)$ ($f = u_v, d_v, \overline{u}, \overline{d}, s, c, b, g$) and their associated uncertainties using the NNLO MRST2006 [78] parametrization at scale $Q^2 = 10^4$ GeV^2.

and momentum $q_\lambda = (p_{\nu_\mu} - p_{\mu^-})_\lambda$ have the form

$$P_\lambda = (m_N, 0, 0, 0), \qquad q_\lambda = (q_0, 0, 0, q_3). \tag{7.67}$$

In this case the vector of the longitudinal polarization of the virtual W-boson is given by the expression

$$\epsilon_\mu^{(S)}(\mathbf{q}) = \frac{1}{\sqrt{-q^2}}(q_3, 0, 0, q_0). \tag{7.68}$$

The corresponding structure function is

$$W_S(q^2, \nu) = \epsilon_\mu^{(S)*}(\mathbf{q})\epsilon_\nu^{(S)}(\mathbf{q})W^{\mu\nu}(q^2, \nu),$$

where $W^{\mu\nu}$ is determined by Eq. (5.74) and we have omitted the index denoting the neutrino (ν). Uncomplicated calculations result in

$$W_S(q^2, \nu) = -W_1(q^2, \nu) - \frac{q_3^2}{q^2}W_2(q^2, \nu) = \left(1 - \frac{\nu^2}{q^2}\right)W_2(q^2, \nu) - W_1(q^2, \nu). \tag{7.69}$$

The vectors of the left and right transverse polarizations of the virtual W-boson have the view:

$$\epsilon_\mu^{(R)}(\mathbf{q}) = \frac{1}{\sqrt{2}}(0, 1, i, 0), \qquad \epsilon_\mu^{(L)}(\mathbf{q}) = \frac{1}{\sqrt{2}}(0, 1, -i, 0) \tag{7.70}$$

and the corresponding structure functions are

$$W_R(q^2, \nu) = W_1(q^2, \nu) + \frac{\sqrt{\nu^2 - q^2}}{2m_N} W_3(q^2, \nu), \tag{7.71}$$

$$W_L(q^2, \nu) = W_1(q^2, \nu) - \frac{\sqrt{\nu^2 - q^2}}{2m_N} W_3(q^2, \nu). \tag{7.72}$$

Note, that these structure functions with definite helicities W_L, W_R, and W_S have to be positive. In the conditions of the scaling realization, the following helical structure functions depend only on one variable x

$$\left.\begin{aligned}
2m_N W_S(q^2, \nu) &\to F_S(x) = -2F_1(x) + F_2(x)/x, \\
m_N W_L(q^2, \nu) &\to F_L(x) = F_1(x) - F_3(x)/2, \\
m_N W_R(q^2, \nu) &\to F_R(x) = F_1(x) + F_3(x)/2.
\end{aligned}\right\} \tag{7.73}$$

Further, one should find out whether data on lepton-nucleon scattering at high energies are in line with those quantum number values that were obtained for quarks from spectroscopic phenomenology. Within the model involving light quarks u, d, and s only, the hadron electromagnetic current is given by the expression:

$$J_\mu^{em} = \frac{2}{3}\bar{u}\gamma_\mu u - \frac{1}{3}\bar{d}\gamma_\mu d - \frac{1}{3}\bar{s}\gamma_\mu s, \tag{7.74}$$

while the hadron charged current has the form

$$J_\mu^h = \bar{u}\gamma_\mu(1 - \gamma_5)(d\cos\theta_c + s\sin\theta_c) \tag{7.75}$$

(in what follows we again fix $\theta_c = 0$). Studying the deep inelastic $e^- p$-scattering in Section 10.3 of *Advanced Particle Physics Volume 1*, we have established the relations:

$$\left.\begin{aligned}
m_N W_1 &= \mathcal{C}_1(x, q^2/m_p^2) = \frac{1}{2}\sum_i e_{q_i}^2 f_i(x), \\
\nu W_2 &= \mathcal{C}_2(x, q^2/m_p^2) = x\sum_i e_{q_i}^2 f_i(x),
\end{aligned}\right\} \tag{7.76}$$

those could be rewritten introducing scaling functions $F_{1,2}(x)$

$$\left.\begin{aligned}
m_N W_1(x, q^2/m_p^2) &\to F_1(x) = \frac{1}{2}\sum_i e_{q_i}^2 f_i(x), \\
\nu W_2(x, q^2/m_p^2) &\to F_2(x) = x\sum_i e_{q_i}^2 f_i(x).
\end{aligned}\right\} \tag{7.77}$$

Then, we have

$$F_T^{ep}(x) \equiv F_L^{ep}(x) + F_R^{ep}(x) = 2F_1^{ep}(x) = \sum_i e_{q_i}^2 f_i(x) = \frac{4}{9}[f_u^p(x) + f_{\bar{u}}^p(x)]+$$

$$+\frac{1}{9}[f_d^p(x) + f_{\bar{d}}^p(x)] + \frac{1}{9}[f_s^p(x) + f_{\bar{s}}^p(x)], \tag{7.78}$$

$$F_T^{en}(x) = 2F_1^{en}(x) = \frac{4}{9}[f_u^n(x) + f_{\bar{u}}^n(x)] + \frac{1}{9}[f_d^n(x) + f_{\bar{d}}^n(x)] + \frac{1}{9}[f_s^n(x) + f_{\bar{s}}^n(x)], \quad (7.79)$$

where $f_q^N(x)$ is the probability of the fact that the N parton, possessing quantum numbers of a quark q and being in a state with a longitudinal momentum portion x, may be revealed in a target nucleon. As long as the parton distributions are defined as the probability of finding different hadron constituents, they have to be normalized taking into account the quantum numbers of the hadron. Thus, for example, the proton is the bound state *uud* with some addition of quark-antiquark pairs. Therefore, it must contain an abundance in the form of two *u*-quarks and one *d*-quark over the corresponding antiquarks. This implies the fulfillment of the conditions:

$$\int_0^1 [f_u^p(x) - f_{\bar{u}}^p(x)]dx = 2, \qquad \int_0^1 [f_d^p(x) - f_{\bar{d}}^p(x)]dx = 1. \quad (7.80)$$

The values of isospin, strangeness, and electric charge of a proton demand the implementation of the following relations:

$$\frac{1}{2}\int_0^1 \{[f_u^p(x) - f_{\bar{u}}^p(x)] - [f_d^p(x) - f_{\bar{d}}^p(x)]\}dx = \frac{1}{2}, \quad (7.81)$$

$$\int_0^1 [f_s^p(x) - f_{\bar{s}}^p(x)]dx = 0, \quad (7.82)$$

$$\int_0^1 \{\frac{2}{3}[f_u^p(x) - f_{\bar{u}}^p(x)] - \frac{1}{3}[f_d^p(x) - f_{\bar{d}}^p(x)] - \frac{1}{3}[f_s^p(x) - f_{\bar{s}}^p(x)]\}dx = 1. \quad (7.83)$$

Making use of the isospin symmetry, that is, the invariance with respect to the replacements $p \leftrightarrow n$ and $u \leftrightarrow d$), we arrive at

$$\left.\begin{array}{l} f_u^p(x) = f_d^n(x) = f_u(x), \\ f_d^p(x) = f_u^n(x) \equiv f_d(x), \\ f_s^p(x) = f_s^n(x) \equiv f_s(x). \end{array}\right\} \quad (7.84)$$

Then, the expressions (7.78) and (7.79) can be written in the view:

$$F_T^{ep}(x) = \frac{4}{9}[f_u(x) + f_{\bar{u}}(x)] + \frac{1}{9}[f_d(x) + f_{\bar{d}}(x)] + \frac{1}{9}[f_s(x) + f_{\bar{s}}(x)], \quad (7.85)$$

$$F_T^{en}(x) = \frac{4}{9}[f_d(x) + f_{\bar{d}}(x)] + \frac{1}{9}[f_u(x) + f_{\bar{u}}(x)] + \frac{1}{9}[f_s(x) + f_{\bar{s}}(x)]. \quad (7.86)$$

The ratio of the proton structure function to the neutron one is

$$\frac{F_T^{ep}(x)}{F_T^{en}(x)} = \frac{4(f_u + f_{\bar{u}}) + (f_d + f_{\bar{d}}) + (f_s + f_{\bar{s}})}{(f_u + f_{\bar{u}}) + 4(f_d + f_{\bar{d}}) + (f_s + f_{\bar{s}})}. \quad (7.87)$$

Since all $f_q(x)$ are positive, then the constraints prevail [81]

$$\frac{1}{4} \leq \frac{F_T^{en}(x)}{F_T^{ep}(x)} \leq 4, \quad (7.88)$$

which are in line with experimental data.

It is worthwhile to subdivide the quark distribution function into two parts corresponding to valence and sea quarks:

$$f_q(x) = f_q^v(x) + f_q^s(x). \quad (7.89)$$

Sea quarks are associated with quark pairs produced by gluons: they are symmetric with respect to $SU(3)_c$ and, as was found out earlier, must be localized in a region of small values of x. In the case of the proton target we have

$$
\left.
\begin{aligned}
& f_u^v(x) = 2f_d^v(x), \\
& f_s^v(x) = f_{\bar{u}}^v(x) = f_{\bar{d}}^v(x) = f_{\bar{s}}^v(x) = 0, \\
& f_u^s(x) = f_{\bar{u}}^s(x) = f_d^s(x) = f_{\bar{d}}^s(x) = f_s^s(x) = f_{\bar{s}}^s(x) \equiv G(x).
\end{aligned}
\right\}
\tag{7.90}
$$

Thus, the expressions (7.85) and (7.86) may be represented in the form

$$
F_T^{ep}(x) = \frac{1}{2}f_u^v(x) + \frac{4}{3}G(x), \qquad F_T^{en}(x) = \frac{1}{3}f_u^v(x) + \frac{4}{3}G(x).
\tag{7.91}
$$

The difference in these quantities gives the valence quarks distribution:

$$
F_T^{ep}(x) - F_T^{en}(x) = \frac{1}{6}f_u^v(x),
\tag{7.92}
$$

which needs to have the peak nearly $x = 1/3$ in accordance with Fig. 7.9. Using Eq. (7.91), one can find the sea quarks distribution function as well

$$
\frac{1}{2}F_T^{en}(x) - \frac{1}{3}F_T^{ep}(x) = \frac{2}{9}G(x).
\tag{7.93}
$$

Furthermore, since we assume the contribution of the function $G(x)$ to be essential in the region $x \to 0$ only, then the following is true

$$
\frac{F_T^{ep}(x)}{F_T^{en}(x)} \to
\begin{cases}
1 & \text{when} \quad x \to 0, \\
3/2 & \text{when} \quad x \to 1.
\end{cases}
\tag{7.94}
$$

These predictions of the quark-parton model are also confirmed by experimental observations.

In the case of the deep inelastic neutrino-nucleon scattering, axial currents come into play too. As a result, the helical structure function $F_3(x)$ becomes nonzero and $F_L(x) \neq F_R(x)$. Below we cite only the basic results for the neutrino-nucleon structure functions

$$
\left.
\begin{aligned}
& F_L^{\nu p}(x) = 2f_d(x), & & F_L^{\nu n}(x) = 2f_u(x), \\
& F_R^{\nu p}(x) = 2f_{\bar{u}}(x), & & F_R^{\nu n}(x) = 2f_{\bar{d}}(x), \\
& F_S^{\nu p}(x) = 0, & & F_S^{\nu n}(x) = 0, \\
& F_L^{\bar{\nu} p}(x) = 2f_u(x), & & F_L^{\bar{\nu} n}(x) = 2f_d(x), \\
& F_R^{\bar{\nu} p}(x) = 2f_{\bar{d}}(x), & & F_R^{\bar{\nu} n}(x) = 2f_{\bar{u}}(x), \\
& F_S^{\bar{\nu} p}(x) = 0, & & F_S^{\bar{\nu} n}(x) = 0.
\end{aligned}
\right\}
\tag{7.95}
$$

The functions being in the right side of these equalities represent the distribution functions for the proton target; the appearance of the factor 2 reflects the presence both of the vector and axial-vector parts of the weak current. Now, we can single out the distribution function of the strange quark. Using (7.95) and the equality

$$
F_2(x) = x[F_L(x) + F_R(x) + F_S(x)],
\tag{7.96}
$$

we get

$$
F_2^{\nu p}(x) + F_2^{\nu n}(x) = 2x[f_u(x) + f_{\bar{u}}(x) + f_d(x) + f_{\bar{d}}(x)].
\tag{7.97}
$$

Employing the equalities $F_2(x) = xF_T(x)$, (7.85) and (7.86) gives

$$F_2^{ep}(x) + F_2^{en}(x) = x \left\{ \frac{5}{9} \left[f_u(x) + f_{\bar{u}}(x) + f_d(x) + f_{\bar{d}}(x) \right] + \frac{2}{9} \left[f_s(x) + f_{\bar{s}}(x) \right] \right\}. \quad (7.98)$$

From the obtained equation it immediately follows that

$$F_2^{ep}(x) + F_2^{en}(x) - \frac{5}{18} \left[F_2^{\nu p}(x) + F_2^{\nu n}(x) \right] = \frac{2x}{9} \left[f_s(x) + f_{\bar{s}}(x) \right]. \quad (7.99)$$

This formula comes to an agreement with experimental data in the case when the right side is equal to zero except for a region of small values of x ($x < 0.2$). This tells us that the content of strange quarks and antiquarks in the proton is extremely small. Besides, the demand of the $SU(3)_c$-symmetry of the sea quark distribution functions guarantees smallness of the functions $f_{\bar{u}}(x)$ and $f_{\bar{d}}(x)$.

From found relations for the nucleon structure functions the sum rules may be obtained. We shall illustrate it by several examples. Earlier, studying the currents algebra, we determined the Adler sum rule (5.94). In the scaling limit, $|q^2| \to \infty, \nu \to \infty$ and x is fixed, this relation takes the form

$$\int_0^1 \frac{dx}{x} \left[F_2^{\bar{\nu} p}(x) - F_2^{\nu p}(x) \right] = 2. \quad (7.100)$$

Reproduce this result within the quark-parton model. Using (7.95) and (7.96), we get

$$\int_0^1 \frac{dx}{x} \left[F_2^{\bar{\nu} p}(x) - F_2^{\nu p}(x) \right] = \int_0^1 \frac{dx}{x} \left\{ 2x \left[f_u(x) - f_{\bar{u}}(x) \right] - 2x \left[f_d(x) - f_{\bar{d}}(x) \right] \right\} =$$

$$= \int_0^1 \frac{dx}{x} [4x I_3(x)], \quad (7.101)$$

where $I_3(x)$ is the third component of the isospin density. Taking into account

$$\int_0^1 dx I_3(x) - \frac{1}{2},$$

we arrive at the Adler relation.

By way of the following example, we consider the sum of the scaling functions

$$F_3^{\nu p}(x) + F_3^{\nu n}(x) = -2[f_u(x) + f_d(x) - f_{\bar{u}}(x) - f_{\bar{d}}(x)]. \quad (7.102)$$

This sum can be written in the form of the combination of density functions associated with the baryon charge and strangeness:

$$F_3^{\nu p}(x) + F_3^{\nu n}(x) = -6 \left[B(x) + \frac{1}{3} S(x) \right],$$

where

$$B(x) = \frac{1}{3} \left[f_u(x) + f_d(x) + f_s(x) - f_{\bar{u}}(x) - f_{\bar{d}}(x) - f_{\bar{s}}(x) \right],$$

$$S(x) = -[f_s(x) - f_{\bar{s}}(x)].$$

Allowing for the values of the quantum proton numbers, we arrive at the Gross–Llewellyn sum rule [82]:

$$\int_0^1 dx [F_3^{\nu p}(x) + F_3^{\nu n}(x)] = -6. \quad (7.103)$$

In Table 7.1 we enumerate basic processes that are used for the definition of the parton distribution functions (PDFs) (see, for example, Ref. [83]):

TABLE 7.1
Basic processes employed for the definition of the PDFs.

Process	Basic subprocess	Measured PDFs
$l^\pm N \to l^\pm X$	$\gamma^* q \to q$	$f_g(x \leq 0.01), f_q, f_{\overline{q}}$
$l^+(l^-)N \to \overline{\nu}(\nu)X$	$W^* q \to q'$	
$\nu(\overline{\nu})N \to l^-(l^+)X$	$W^* q \to q'$	
$\nu N \to \mu^- \mu^+ X$	$W^* s \to c \to \mu^+$	f_s
$lN \to lQX \qquad (Q = c, b)$	$\gamma^* Q \to Q$	f_c, f_b
	$\gamma^* g \to Q\overline{Q}$	$f_g(x \leq 0.01)$
$pp \to \gamma X$	$qg \to \gamma q$	f_g
$pN \to \mu^- \mu^+ X$	$q\overline{q} \to \gamma^*$	$f_{\overline{q}}$
$pp, pn \to \mu^- \mu^+ X$	$u\overline{u}, d\overline{d} \to \gamma^*$	$f_{\overline{u}} - f_{\overline{d}}$
	$u\overline{d}, d\overline{u} \to \gamma^*$	
$ep, en \to e\pi X$	$\gamma^* q \to q$	
$p\overline{p} \to W \to l^\pm X$	$ud \to W$	$f_u, f_d, f_u/f_d$

7.4   Quark-gluon dynamics

So far in studying the deep inelastic scattering processes, we did not take into account the interaction between quarks and gluons. Make up for this deficiency. As a case in point, we consider the diagrams represented in Figs. 7.11 and 7.12.

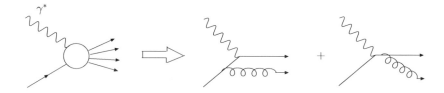

FIGURE 7.11
Contributions of the subprocess $\gamma^* q \to qg$ into the cross section of the process $ep \to eX$.

The inclusion of the diagrams of such a kind explains the scaling violation for the structure functions. Moreover, the diagrams allowing for emitting or absorption of a gluon predict that a final quark, and hence a corresponding jet, are no longer collinear to a virtual photon. Really, in the quark-parton model, a jet of final hadrons has to fly out along the direction of a virtual photon (Fig. 7.13(a)). However, if one takes account of gluons (quark-gluon-parton model), at emitting a gluon (Fig. 7.13(b)) a quark experiences the recoil, and, as a result, two jets (they are marked by arrows) are produced. Each of them has a nonzero momentum transverse relative to a virtual photon. So, we are now being confronted with the problem: *how to allow for the contributions into the deep inelastic cross sections that are caused by the interactions of quarks with gluons.*

Let us address the process $e^- p \to e^- X$ again. Investigation of this process revealed that

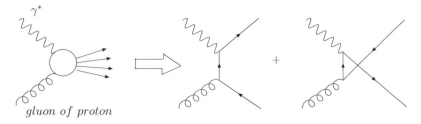

FIGURE 7.12
Contributions of the subprocess $\gamma^* g \to q\bar{q}$ into the cross section of the process $ep \to eX$.

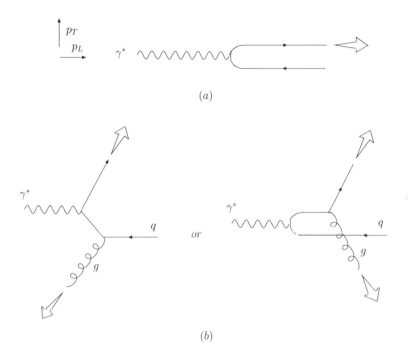

FIGURE 7.13
Diagrams for the subprocesses (a) $\gamma^* q \to$ hadronjet$q$ and (b) $\gamma^* q \to$ hadronjet$q$.

the most important events happen below the dashed line in Fig. 7.4 where a virtual photon interacts with a proton. The role of an electron beam is reduced to the fact that it simply serves a source of virtual photons. Using this circumstance, we can easily write the total cross section for the scattering of a real photon, having the energy q_0 and the polarization $\epsilon_\lambda^\mu(\mathbf{q})$, by a proton with the production of one or more particles in a final state $(\gamma p \to X)$:

$$\sigma(\gamma p \to X) = \frac{4\pi^2 \alpha}{q_0} \sum_\lambda W_{\mu\nu} \epsilon_\lambda^{\mu*}(\mathbf{q}) \epsilon_\lambda^\nu(\mathbf{q}), \qquad (7.104)$$

where the sum is taken over polarizations of a real photon and the hadron tensor was found as far back as the first volume (see Eq. (10.80)). However, to fulfill the total analysis we shall have to know the total cross section in the case of a virtual photon. In so doing we should remember that virtual photon states are not constrained by two transverse polarizations. Moreover, the cross sections of virtual photons cannot be considered as a well-defined quantity. At $q^2 = 0$ initial particles flux J is equal to $4m_p q_0$ while it is arbitrary for virtual photons. The common choice is that at which q_0 satisfies the same demand as in the case of a real photon

$$(P + q)^2 = m_p^2 + 2m_p q_0.$$

In the laboratory frame this gives

$$q_0 = \frac{(P + q)^2 - m_p^2}{2m_p} = \nu + \frac{q^2}{2m_p}. \tag{7.105}$$

Next, we introduce polarization vectors of a virtual photon with the polarization λ

$$\lambda = \pm 1, \qquad \epsilon_\pm = \mp \sqrt{\frac{1}{2}}(0, 1, \pm i, 0), \tag{7.106}$$

$$\lambda = 0, \qquad \epsilon_0 = \frac{1}{\sqrt{-q^2}}(\sqrt{\nu^2 - q^2}, 0, 0, \nu). \tag{7.107}$$

Using Eq. (10.80) of *Advanced Particle Physics Volume 1* and Eq. (7.104), we can connect the total transverse σ_T and total longitudinal σ_L cross sections with the proton structure functions:

$$\sigma_T \equiv \frac{1}{2}(\sigma_+ + \sigma_-) = \frac{4\pi\alpha}{q_0} W_1(\nu, q^2), \tag{7.108}$$

$$\sigma_L \equiv \sigma_0 = \frac{4\pi\alpha}{q_0}\left[\left(1 - \frac{\nu^2}{q^2}\right) W_2(\nu, q^2) - W_1(\nu, q^2)\right]. \tag{7.109}$$

The total cross sections of the absorption of the virtual photon by a proton are expressed in terms of the scaling functions

$$m_p W_1(q^2, \nu) = F_1(x), \qquad \nu W_2(q^2, \nu) = F_2(x),$$

in the following way:

$$\frac{\sigma_T}{\sigma_0} = 2F_1(x), \tag{7.110}$$

$$\frac{\sigma_T + \sigma_L}{\sigma_0} = \frac{1}{x} F_2(x), \tag{7.111}$$

where

$$\sigma_0 = \frac{4\pi^2\alpha}{2m_p q_0} \approx \frac{4\pi^2\alpha}{s} \tag{7.112}$$

and $s = (P + q)^2$.

Now, our task is to find the contributions of subprocesses taking into account the quark-gluon interaction in the process

$$\gamma^* + p \to X. \tag{7.113}$$

Begin with the case of emitting a gluon by a final quark, that is, with the subprocess

$$\gamma^* + q \to q + g. \tag{7.114}$$

In the lowest order of the perturbation theory, the QCD diagram describing contributions of (7.114) into the process (7.113) is displayed in Fig. 7.14. The relations between parameters

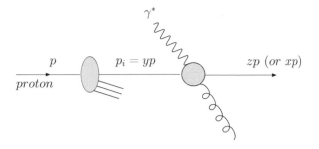

FIGURE 7.14

The diagram associated with the process $\gamma^* + p \to X$.

of two systems has the following form:

system γ^*—proton $\qquad\qquad$ system γ^*—parton

$$P \qquad\qquad \longrightarrow \qquad\qquad p_i = yP,$$

$$x = Q^2/[2(P \cdot q)] \qquad \longrightarrow \qquad z = Q^2/[2(p_i \cdot q)] = x/y.$$

Tie together the cross sections in both systems. Carry out this procedure for the ratio σ_T/σ_0. Since we shall allow for higher corrections in QCD, then the scaling is absent by now, and, as a result, the cross sections in question represent a function of two variables ν and Q^2. Thus, we have

$$\left[\frac{\sigma_T(x, Q^2)}{\sigma_0}\right]_{\gamma^* p} = \sum_i \int_0^1 dz \int_0^1 dy f_i(y) \delta(x - zy) \left[\frac{\hat{\sigma}_T(z, Q^2)}{\hat{\sigma}_0}\right]_{\gamma^* i}, \qquad (7.115)$$

where $f_i(y)$ is the parton distribution function, the coordinate x is fixed and all the quantities associated with subprocesses have been anew furnished with the hat. Fulfilling the integration in (7.115), we finally obtain

$$\left[\frac{\sigma_T(x, Q^2)}{\sigma_0}\right]_{\gamma^* p} = \sum_i \int_x^1 \frac{dy}{y} f_i(y) \left[\frac{\hat{\sigma}_T(x/y, Q^2)}{\hat{\sigma}_0}\right]_{\gamma^* i}. \qquad (7.116)$$

Test the relation (7.116). It is obvious that in the absence of all gluon effects we must obtain all results of the quark-parton model. Check this condition by the example of the relations (7.77). In the considered approximation, $\hat{\sigma}_T$ and $\hat{\sigma}_L$ will describe the reaction

$$\gamma^* + q \to q. \qquad (7.117)$$

The corresponding Feynman diagram (Fig. 7.15) results in the amplitude

$$\mathcal{A}(\gamma^* q \to q) = e_{q_i} e \sqrt{\frac{m_q^2}{2 E_q E_q' q_0}} \overline{u}(p') \gamma^\mu u(p) \epsilon_\mu^\lambda(\mathbf{q}). \qquad (7.118)$$

Averaging over the spins of initial particles and summing over the spins of the final ones, we get

$$\frac{1}{4} \sum |\mathcal{A}_T|^2 = 2 e_{q_i}^2 e^2 (p \cdot q), \qquad (7.119)$$

$$\frac{1}{4} \sum |\mathcal{A}_L|^2 = 0. \qquad (7.120)$$

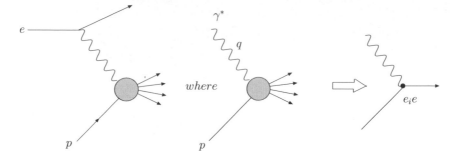

FIGURE 7.15
The process $ep \rightarrow eX$ within the parton model.

The standard computations lead to the relations

$$\frac{\hat{\sigma}_T(z, Q^2)}{\hat{\sigma}_0} = e_{q_i}^2 \,\delta(1 - z), \tag{7.121}$$

$$\hat{\sigma}_L(z, Q^2) = 0. \tag{7.122}$$

Since, in the case under consideration, the variable z will look like:

$$z = \frac{Q^2}{2(p_i \cdot q)} = \frac{m_q^2 + 2(p_i \cdot q) + m_i^2}{2(p_i \cdot q)} \approx 1,$$

the fact that $\hat{\sigma}_T(z, Q^2)$ is proportional to $\delta(1 - z)$ would not be a matter of surprise. Further, for the relations (7.77) to be examined, we substitute the obtained values of the cross sections into (7.116)

$$\frac{F_2(x)}{x} = \sum_i e_{q_i}^2 \int_x^1 \frac{dy}{y} f_i(y) \delta\left(1 - \frac{x}{y}\right) = \sum_i e_{q_i}^2 f_i(x). \tag{7.123}$$

Analogous computations give for the scaling function $F_1(x)$

$$2F_1(x) = \sum_i e_{q_i}^2 f_i(x). \tag{7.124}$$

As we see the quark-parton model results (7.77) are really reproduced, that is, the correspondence principle takes place. Thus, with the help of the working formula (7.116), we can switch on parton subprocesses into the basic process.

By way of the following example, we consider the gluon bremsstrahlung (Fig. 7.16)

$$\gamma^* + q \rightarrow q + g. \tag{7.125}$$

$$\mathcal{A}(\gamma^* q \rightarrow q) = e_{q_i} e \sqrt{\frac{m_q^2}{2E_q E_q' q_0}} \overline{u}(p') \gamma^\mu u(p) \epsilon_\mu^\lambda(\mathbf{q}).$$

The corresponding amplitude is

$$\mathcal{A}(\gamma^* q \rightarrow qg) = \mathcal{A}_1 + \mathcal{A}_2, \tag{7.126}$$

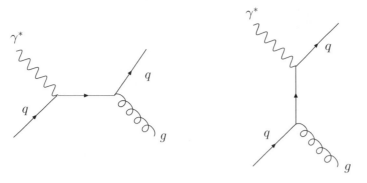

FIGURE 7.16
The gluon bremsstrahlung.

where

$$\mathcal{A}_1 = -ie_{q_i}eg_s\sqrt{\frac{1}{16E_qE'_qq_0E_g}}\frac{1}{(p+q)^2}\overline{u}(p')\gamma^\nu\frac{\lambda_a}{2}(\hat{p}+\hat{q})\gamma^\mu u(p)\epsilon_\mu^{(\lambda)}(\mathbf{q})\epsilon_\nu^{(\lambda')*}(\mathbf{k}), \quad (7.127)$$

$$\mathcal{A}_2 = -ie_{q_i}eg_s\sqrt{\frac{1}{16E_qE'_qq_0E_g}}\frac{1}{(p-k)^2}\overline{u}(p')\gamma^\mu(\hat{p}-\hat{k})\gamma^\nu\frac{\lambda_a}{2}u(p)\epsilon_\mu^{(\lambda)}(\mathbf{q})\epsilon_\nu^{(\lambda')*}(\mathbf{k}), \quad (7.128)$$

q, p (p'), and k are the four-dimensional momenta of the virtual photon, initial (final) quark and gluon, respectively. When writing (7.127) and (7.128), the normalization of the quark functions that is associated with massless particles has been used (see the discussion after Eq. (7.8)) and, for the sake of simplicity, the polarization indices are omitted. We shall not be interested in the polarizations of the initial and final particles. In so doing, we take into account that the gluon and photon polarization sums are

$$\sum_{\lambda'}\epsilon_\nu^{(\lambda')*}(\mathbf{k})\epsilon_\mu^{(\lambda')}(\mathbf{k}) = -g_{\mu\nu}, \qquad \sum_\lambda\epsilon_\nu^{(\lambda)*}(\mathbf{q})\epsilon_\mu^{(\lambda)}(\mathbf{q}) = -g_{\mu\nu}. \quad (7.129)$$

Let us work in the center-of-mass frame. Then, the Mandelstam variables are defined by

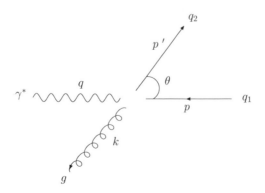

FIGURE 7.17
The center-of-mass frame for the process $\gamma^*q \to qg$.

the expressions (see Fig. 7.17):

$$\hat{s} = (p + q)^2 = (p' + k)^2 = 4|\mathbf{p}'|^2, \tag{7.130}$$

$$\hat{t} = (p' - q)^2 = (p - k)^2 = -2|\mathbf{q}||\mathbf{p}'|(1 - \cos\theta), \tag{7.131}$$

$$\hat{u} = (p' - p)^2 = (q - k)^2 = -2|\mathbf{q}||\mathbf{p}'|(1 + \cos\theta), \tag{7.132}$$

where θ is the angle between four-dimensional momenta of the virtual photon and final quark. The following relation is true

$$\hat{s} + \hat{u} + \hat{t} = q^2. \tag{7.133}$$

In what follows we need the formula

$$\hat{s} - q^2 = -\hat{u} - \hat{t} = 4|\mathbf{q}||\mathbf{p}'|. \tag{7.134}$$

Carrying out calculations, we get

$$\frac{1}{4}\sum|\mathcal{A}_1|^2 = e_i^2 e^2 g_s^2 \frac{1}{16 E_q E_q' q_0 E_g} \frac{1}{(p+q)^2}\text{Sp}\ [\hat{p}'\gamma^\nu(\hat{p}+\hat{q})\gamma^\mu\hat{p}\gamma_\mu(\hat{p}+\hat{q})\gamma_\nu]\text{Sp}\ \left[\frac{\lambda_a}{2}\frac{\lambda_a}{2}\right] =$$

$$= \frac{e_i^2 e^2 g_s^2}{3 E_q E_q' q_0 E_g (p+q)^2}\text{Sp}\ [\hat{p}'(\hat{p}+\hat{q})\hat{p}(\hat{p}+\hat{q})] = \frac{e_i^2 e^2 g_s^2}{3 E_q E_q' q_0 E_g (p+q)^2}\ \{2[p^2 + (p\cdot q)]\times$$

$$\times\text{Sp}\ [\hat{p}'(\hat{p}+\hat{q})] - (p+q)^2\text{Sp}\ [\hat{p}'\hat{p}]\} = \frac{e_i^2 e^2 g_s^2}{3 E_q E_q' q_0 E_g (p+q)^2}\ \{2[p^2 + (p\cdot q)]4[(p'\cdot p)+$$

$$+(p'\cdot q)] - 4(p+q)^2(p'\cdot p)\} = \frac{e_i^2 e^2 g_s^2}{3 E_q E_q' q_0 E_g (p+q)^2}\ \{8(p'\cdot q)(p\cdot q) - 4q^2(p'\cdot p)\} =$$

$$= \frac{e_i^2 e^2 g_s^2}{3 E_q E_q' q_0 E_g \hat{s}^2}\ \{-2(\hat{s}-q^2)(\hat{t}-q^2) + 2q^2\hat{u}\} = \frac{e_i^2 e^2 g_s^2}{3 E_q E_q' q_0 E_g \hat{s}^2}\ \{-2\hat{s}\hat{t}\}\ , \tag{7.135}$$

$$\frac{1}{4}\sum|\mathcal{A}_2|^2 = \frac{e_i^2 e^2 g_s^2}{3 E_q E_q' q_0 E_g \hat{t}^2}\ \{8(p'\cdot k)(p\cdot k) - 4k^2(p'\cdot p)\} = \frac{e_i^2 e^2 g_s^2}{3 E_q E_q' q_0 E_g \hat{t}^2}\ \{-2\hat{s}\hat{t}\}\ , \tag{7.136}$$

$$\frac{1}{4}\sum 2\mathcal{A}_2^*\mathcal{A}_1 = \frac{e_i^2 e^2 g_s^2}{3 E_q E_q' q_0 E_g \hat{s}\hat{t}}\ \{-4\hat{u}q^2\}\ . \tag{7.137}$$

In the case under consideration we are interested in the transverse momentum distribution of final partons. Obviously, a transverse momentum of a final parton $p_T = p'\sin\theta$, where $p' = |\mathbf{p}'|$, is given by the expression

$$p_T^2 = \frac{\hat{s}\hat{u}\hat{t}}{(\hat{s} - q^2)^2}. \tag{7.138}$$

In the limit of small scattering angles $-\hat{t} \ll \hat{s}$

$$p_T^2 = -\frac{\hat{s}\hat{t}}{\hat{s} - q^2},$$

and we have

$$d\Omega = \frac{4\pi}{\hat{s}}dp_T^2. \tag{7.139}$$

Summing (7.135)–(7.137) results in

$$\frac{1}{4}\sum|\mathcal{A}|^2 = \frac{2e_i^2 e^2 g_s^2}{3E_q E_q' q_0 E_g}\left\{-\frac{\hat{t}}{\hat{s}} - \frac{\hat{s}}{\hat{t}} - \frac{2\hat{u}q^2}{\hat{s}\hat{t}}\right\}. \tag{7.140}$$

From (7.140) it follows at once that in the region of high energies the cross section of the subprocess $\gamma^* q \to qg$ has a maximum under $-\hat{t} \to 0$. Therefore, we can approximate the given cross section by its forward peak value, that is, we can put

$$\frac{1}{4}\sum|\mathcal{A}|^2 \approx -\frac{2e_i^2 e^2 g_s^2}{3E_q E_q' q_0 E_g \hat{t}}\left\{\hat{s} - \frac{2(\hat{s} - q^2)q^2}{\hat{s}}\right\}. \tag{7.141}$$

Then, the differential cross section of the forward scattering will look like:

$$\frac{d\hat{\sigma}}{dp_T^2} \approx -\frac{8\pi e_{q_i}^2 \alpha\alpha_s}{3\hat{s}^2\hat{t}}\left[\hat{s} - \frac{2(\hat{s} - q^2)q^2}{\hat{s}}\right]. \tag{7.142}$$

Introduce the dimensionless variable

$$z = -\frac{q^2}{2(p \cdot q)} = \frac{Q^2}{\hat{s} + Q^2}, \tag{7.143}$$

where $Q^2 = -q^2$. Then, the differential cross section (7.142) can be rewritten in the view:

$$\frac{d\hat{\sigma}}{dp_T^2} \approx e_{q_i}^2 \hat{\sigma}_0 \frac{1}{p_T^2}\frac{\alpha_s}{2\pi}P_{qq}(z), \tag{7.144}$$

where

$$\hat{\sigma}_0 = \frac{4\pi^2\alpha}{\hat{s}}$$

and the quantity

$$P_{qq}(z) = \frac{4(1 + z^2)}{3(1 - z)} \tag{7.145}$$

is the probability that a quark emits a gluon and, as a result, decreases its momentum in z times. When $z \to 1$, the singularity is connected with emitting a soft massless gluon (infrared divergence). We have already known how to struggle against it.

The cross section (7.144) is singular at $p_T^2 \to 0$ as well. In the limit $-\hat{t} \ll \hat{s}$ the diagrams Figs. 7.12 and 7.14 either do not contribute (in the quark-parton diagrams the momentum p_T of final partons relative to a virtual photon is equal to zero) or give a contribution that is negligibly small compared with that of the subprocess $\gamma^* q \to qg$. By this reason, in the region $-\hat{t} \ll \hat{s}$ the formula (7.144) gives the entire distribution of p_T^2 for parton jets in a final state.

What experimental consequences have this result? Appeal to Fig. 7.12. Emitting a gluon results in the appearance of quark and gluon jets in a final state, the directions of jets not coinciding with that of a virtual photon. A transverse momentum of such a jet, or bremsstrahlung of hadrons in such a jet, is nonzero and its distribution is defined by Eq. (7.144). This process must be included in the electron-proton interaction using the formula (7.116). In Fig. 7.18 the results of such calculations are compared with the experimental data. Thus, the fact of hadrons production with $p_T \neq 0$ is evidence of gluons emission in a parton subprocess. In the quark-parton model all jets have to be collinear to a virtual photon. Their hadron fragments must be nearly collinear to a virtual photon as well (to be exact they are collinear with p_T dispersion of the order of 300 MeV that is in line with

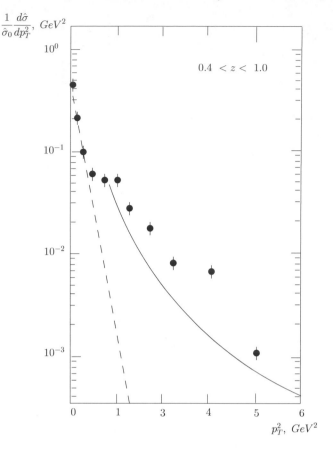

FIGURE 7.18

The p_T^2-distribution of hadrons relative to a virtual photon direction (μN interactions). The solid (dashed) line represent predictions with (without) the inclusion of emitting gluons.

the uncertainty relation for a quark in the confinement state). This prediction is displayed by the dashed line in Fig. 7.18. The experimental data clearly point to the abundance of hadrons with large p_T, which represent fragments of quark and gluon jets kinematically connected with each other.

An experiment with looking for jets with $p_T \neq 0$ may be fulfilled in different variants, for example, in the processes $e^- e^+ \to$ hadrons, $pp \to$ hadrons, and so on. The scheme and calculations technique are analogous to those considered in the given example. Large values of Q^2 assure that we deal with the interaction at small distances where the quantity α_s is small.

Because our final goal is to determine contributions caused by different subprocesses into the parton structure functions that are connected with the total cross sections of subprocesses, then we have to integrate the expression (7.144). Obviously, the maximum transverse momentum p_T^2 may be represented in the form

$$(p_T^2)_{max} = |\mathbf{p}'|^2 = \frac{\hat{s}}{4} = Q^2 \frac{1-z}{4z}. \tag{7.146}$$

To regularize the divergence at $p_T^2 \to 0$ we introduce the cut-off parameter of the transverse momentum μ. So, we have

$$\hat{\sigma}(\gamma^* q \to qg) = \int_{\mu^2}^{\hat{s}/4} dp_T^2 \frac{d\hat{\sigma}}{dp_T^2} \approx \frac{e_{q_i}^2 \alpha_s \hat{\sigma}_0}{2\pi} P_{qq}(z) \int_{\mu^2}^{\hat{s}/4} \frac{dp_T^2}{p_T^2} \approx \frac{e_{q_i}^2 \alpha_s \hat{\sigma}_0}{2\pi} P_{qq}(z) \ln \frac{Q^2}{\mu^2}, \quad (7.147)$$

where we have taken into account

$$\lim_{Q^2 \to \infty} \ln\left(\frac{\hat{s}}{4}\right) = \ln Q^2.$$

Now, adding (7.147) to the cross section of the process $\gamma^* q \to q$ (see Eqs. (7.121) and (7.122)), we find how the processes with gluons emission change the expression (7.123)

$$\frac{F_2(x, Q^2)}{x} = \left[\frac{F_2(x, Q^2)}{x}\right]_{\gamma^* q \to q} + \left[\frac{F_2(x, Q^2)}{x}\right]_{\gamma^* q \to qg} =$$

$$= \sum_i e_{q_i}^2 \int_x^1 \frac{dy}{y} q(y) \left[\delta\left(1 - \frac{x}{y}\right) + \frac{\alpha_s}{2\pi} P_{qq}\left(\frac{x}{y}\right) \ln \frac{Q^2}{\mu^2}\right]. \quad (7.148)$$

So far, we included in deep inelastic $e^- p$ scattering only those processes that are caused by a quark: $\gamma^* q \to q$ and $\gamma^* q \to qg$. In the first order of α_s it is also necessary to allow for the contributions of a quark-antiquark pair production by a gluon being in an initial proton with which a virtual photon interacts next, that is, the process (Fig. 7.19)

$$\gamma^* + g \to q + \bar{q} \quad (7.149)$$

must be taken into account. It is clear that this reaction is connected with the one (7.125)

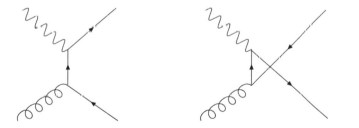

FIGURE 7.19
The diagrams corresponding to the subprocess $\gamma^* + g \to q + \bar{q}$.

by means of the following crossing operation:

$$\hat{u} \to \hat{t}, \qquad \hat{t} \to \hat{s}, \qquad \hat{s} \to \hat{u}. \quad (7.150)$$

Carrying out the replacement (7.150) in the expression (7.140) and taking into consideration changing the color factor ($4/3 \to 1/2$), we find the amplitude module square of the subprocess (7.149) for the unpolarized particles:

$$\frac{1}{4} \sum |\mathcal{A}|^2 = \frac{e_{q_i}^2 e^2 g_s^2}{4 E_q E_{\bar{q}} q_0 E_g} \left\{\frac{\hat{u}}{\hat{t}} - \frac{\hat{t}}{\hat{u}} - \frac{2\hat{s}q^2}{\hat{u}\hat{t}}\right\}. \quad (7.151)$$

Over again, acting the same way as in the case of the process (7.125), that is, approximating the cross section by means of its forward peak value ($u \to 0$), we get

$$\frac{d\hat{\sigma}}{dp_T^2} \approx -\frac{\pi e_{q_i}^2 \alpha \alpha_s}{\hat{s}^2 \hat{u}} \left\{ q^2 - \hat{s} + \frac{2\hat{s}q^2}{q^2 - \hat{s}} \right\}. \tag{7.152}$$

Integrating the obtained expression, we come to recognize that the contribution into the proton structure function (7.148) caused by the subprocess (7.149) will look like:

$$\left[\frac{F_2(x, Q^2)}{x} \right]_{\gamma^* g \to q\bar{q}} = \sum_i \frac{e_{q_i}^2 \alpha_s}{2\pi} \int_x^1 \frac{dy}{y} g(y) P_{qg} \left(\frac{x}{y} \right) \ln \frac{Q^2}{\mu^2}, \tag{7.153}$$

where $g(y)$ is the gluon density in the proton. The function

$$P_{qg}(z) = \frac{1}{2}[z^2 + (1-z)^2]$$

represents the probability that a gluon annihilates into the $q\bar{q}$-pair, a quark taking away the z portion of a gluon momentum.

7.5   Scaling violation. Altarelli–Parisi equation

So, according to QCD, the function F_2 is given by the sum of the expressions (7.148) and (7.153):

$$\frac{F_2(x, Q^2)}{x} = \sum_i e_{q_i}^2 \int_x^1 \frac{dy}{y} \left\{ q_i(y)\delta \left(1 - \frac{x}{y} \right) + \frac{\alpha_s(Q^2)}{2\pi} \ln \frac{Q^2}{\mu^2} \left[q_i(y) P_{qq} \left(\frac{x}{y} \right) + \right. \right.$$

$$\left. \left. + g(y) P_{qg} \left(\frac{x}{y} \right) \right] \right\}, \tag{7.154}$$

where we have accepted $f_{q_i}(x) = q_i(x)$. Because distinction of the quark distributions is caused only by emitting gluons with the transverse momenta of the order of Q^2, then the effective coupling constant $\alpha_s(Q^2)$ is inserted in (7.154). The presence of the factor $\ln Q^2$ means that the scaling predicted by the quark-parton model is broken because, now, F_2 is the function not only of the variable x but the quantity Q^2 as well (although the dependence on Q^2 is only logarithmic). It is obvious that the Bjorken scaling violation is a consequence of the quark-gluon interaction. The expression (7.154) may be considered as the two first terms of the expansion in powers of α_s. Because according to (5.42) at large Q^2 $\alpha_s \sim (\ln Q^2)^{-1}$, then the series expansion in this parameter is completely allowable. All be well in the expression (7.154), if the parameter α_s would not be multiplied by $\ln Q^2$. However, from the evolution equation for the running coupling constant α_s (5.42) it follows that at large Q^2 the product $\alpha_s(Q^2) \ln(Q^2/\mu^2)$ does not turn into zero. Therefore, the formula (7.154), as it exists, cannot be considered a good series expansion. Let us try to improve this state of affairs. Rewrite (7.154) in a parton-like form:

$$\frac{F_2(x, Q^2)}{x} \equiv \sum_i e_{q_i}^2 \int_x^1 \frac{dy}{y} \left[q_i(y) + \Delta q_i(y, Q^2) \right] \delta \left(1 - \frac{x}{y} \right) =$$

$$= \sum_i e_{q_i}^2 \left[q_i(x) + \Delta q_i(x, Q^2) \right], \qquad (7.155)$$

where

$$\Delta q_i(x, Q^2) = \frac{\alpha_s(Q^2)}{2\pi} \ln \frac{Q^2}{\mu^2} \int_x^1 \frac{dy}{y} \left[q_i(y) P_{qq}\left(\frac{x}{y}\right) + g(y) P_{qg}\left(\frac{x}{y}\right) \right]. \qquad (7.156)$$

Now, the quark density depends on Q^2. This dependence is caused by the fact that a photon with large Q^2 probes the wider interval of p_T^2 values inside a proton. The situation is as follows. Under increasing Q^2, say, up to Q_0^2, a photon begins to see point-like quarks inside a proton (Fig. 7.20). If quarks did not interact with each other, at an increase

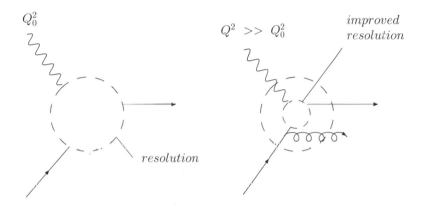

FIGURE 7.20
The parton structure of a proton resolved by a virtual photon.

of Q^2 no additional structure would be resolved and, as a result, the scaling would take place. But, in accordance with QCD, under further increasing of resolution ($Q^2 \gg Q_0^2$), we have to see that every quark itself is surrounded by a partons cloud. We calculated the contributions coming from several diagrams (see Figs. 7.15 and 7.18), but, of course, other diagrams with a larger number of partons exist. The number of observed partons between which a proton momentum is distributed is increased with a growth of Q^2. There is an increasing probability that a quark will be found in a state with a small x. At the same time, since a quark having large momentum loses its momentum thanks to gluons emission, the probability that a quark possesses large values of x is decreasing.

At changing Q^2, the evolution of quark densities is defined by the relation (7.156). Considering the change of the quark density $\Delta q_i(x, Q^2)$ versus the interval $\Delta \ln Q^2$, the formula (7.156) for each quark flavor i can be rewritten in the form of the integro-differential equation:

$$\frac{d}{d \ln Q^2} q_i(x, Q^2) = \frac{\alpha_s(Q^2)}{2\pi} \int_x^1 \frac{dy}{y} \left[q_i(y, Q^2) P_{qq}\left(\frac{x}{y}\right) + g(y, Q^2) P_{qg}\left(\frac{x}{y}\right) \right]. \qquad (7.157)$$

The obtained equation mathematically expresses the fact that a quark with the momentum portion x ($q_i(x, Q^2)$ in the left side) could appear from an initial quark with the large momentum portion y (the first term in the right side $\sim q_i(y, Q^2)$), which has emitted a gluon, or could be the result of the $q\bar{q}$-pair production by an initial gluon with the momentum portion y ($y > x$) (the second term in the right side $\sim g(y, Q^2)$). The probability of the first event is proportional to $\alpha_s P_{qq}(x/y)$, while that of the second event is proportional to $\alpha_s P_{qg}(x/y)$. It is evident that the system of the evolution equations (7.157) is not yet complete. To enclose these equations an evolution law of the gluon distribution must be added to them. This law can be graphically depicted by the diagrams displayed in Fig. 7.21: Repeating the previous considerations (details could be found in Ref. [84]), we get

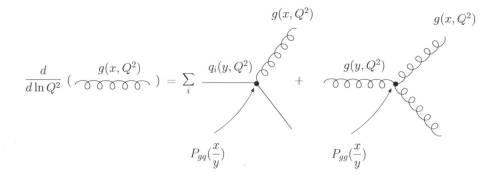

FIGURE 7.21
The graphical interpretation of the gluon distribution evolution law.

$$\frac{d}{d\ln Q^2} g(x, Q^2) = \frac{\alpha_s(Q^2)}{2\pi} \int_x^1 \frac{dy}{y} \left[\sum_i q_i(y, Q^2) P_{gq}\left(\frac{x}{y}\right) + g(y, Q^2) P_{gg}\left(\frac{x}{y}\right) \right], \quad (7.158)$$

where the sum is taken over all quarks and antiquarks of all flavors ($i = 1, 2, \ldots, 2n_f$). For massless quarks the quantity P_{gq} does not depend on the index i. The evolution equations (7.157) and (7.158) are called *Altarelli–Parisi (AP) equations* [85].

Thus, QCD predicts the scaling violation and enables one to exactly calculate the structure functions dependence on Q^2. Knowing the quark structure function $q(x, Q_0^2)$ (hereafter the index i is omitted) or the gluon structure function $g(x, Q_0^2)$, we can compute them with the help of the AP equations for any Q^2. Usually, the structure functions dependence on Q^2 is investigated by means of analyzing the structure function moments

$$M_n^{(q)}(Q^2) = \int_0^1 dx\, x^{n-1} q(x, Q^2), \qquad M_n^{(g)}(Q^2) = \int_0^1 dx\, x^{n-1} g(x, Q^2). \qquad (7.159)$$

Integrating both sides of Eqs. (7.157) and (7.158) with the weight x^{n-1}, rearranging the integrations order in the right sides and passing on to new variables y and $z = x/y$, we find

$$\frac{d}{d\ln Q^2} \begin{bmatrix} M_n^{(q)}(Q^2) \\ M_n^{(g)}(Q^2) \end{bmatrix} = -\frac{\alpha_s(Q^2)}{4\pi} \begin{bmatrix} \gamma_{qq}^n & \gamma_{qg}^n \\ \gamma_{gq}^n & \gamma_{gg}^n \end{bmatrix} \begin{bmatrix} M_n^{(q)}(Q^2) \\ M_n^{(g)}(Q^2) \end{bmatrix}, \qquad (7.160)$$

where the quantity

$$(\gamma^n)_{AB} = -2 \int_0^1 dz\, z^{n-1} P_{AB}(z) \qquad (7.161)$$

($A, B = q, g$) is called an —itanomalous dimensions matrix.

To solve Eqs. (7.160) one needs to substitute the explicit dependence α_s on Q^2

$$\alpha_s(Q^2) = \frac{4\pi}{b} \ln\left(\frac{Q^2}{\Lambda_{QCD}}\right).$$

In the matrix form the solution of (7.160) is given by the expression

$$M_n(Q^2) = \left[\frac{\ln(Q^2/\Lambda_{QCD})}{\ln(Q_0^2/\Lambda_{QCD})}\right]^{-\gamma^n/b} M_n(Q_0^2) = \left[\frac{\alpha_s(Q^2)}{\alpha_s(Q_0^2)}\right]^{\gamma^n/b} M_n(Q_0^2). \qquad (7.162)$$

The experimental data for $F_2(x, Q^2)$ are pictured in Fig. 7.22 [86]. Mark the main features

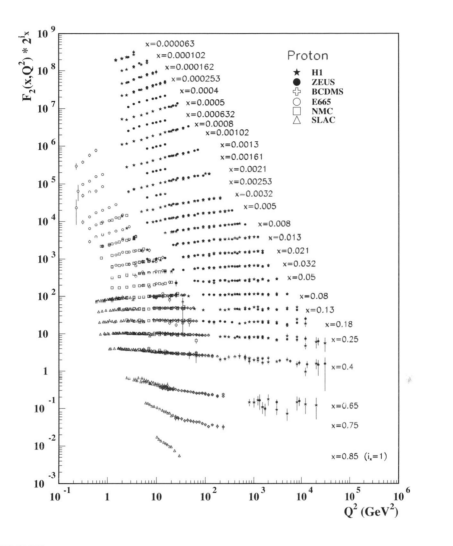

FIGURE 7.22

The proton structure function $F_2(x, Q^2)$ measured in electromagnetic scattering of positrons on protons (collider experiments ZEUS and H1), in the kinematic domain of the HERA data, for $x > 0.00006$, and for electrons (SLAC) and muons (BCDMS, E665, NMC) on a fixed target. For the purpose of plotting, $F_2(x, Q^2)$ has been multiplied by 2^{i_x}, where i_x is the number of the x bin, ranging from $i_x = 1$ ($x = 0.85$) to $i_x = 28$ ($x = 0.000063$).

of the dependence on Q^2. In the vicinity $x = 0.08$ the scaling takes place for the structure functions and the dependence on Q^2 is absent. At $x < 0.08$, the structure functions are increasing with the growth of Q^2 while at $x > 0.08$ they are decreasing. To put it differently, with the growth of Q^2, more and more numbers of soft quarks are resolved. In the result of gluons emission, high-momentum quarks ($x \approx 1$) lose momenta and are shifted towards small momenta ($x \approx 0$). Despite the fact that the parton densities were obtained in the case of the deep inelastic e^-p-scattering, they must characterize a proton-target in a simple style and do not depend on a process in which they are investigated. In other words, the parton densities are universal in the sense that using them, one can connect the structure functions to be found in different processes.

The evolution of the structure functions is determined by the functions P_{qq}, P_{qg}, P_{gq}, and P_{gg}. Make clear their physical meaning. Disregard the gluons existence and rewrite the expression for the quark densities in Eq. (7.155) in the form analogous to (7.115)

$$q(x, Q^2) + \Delta q(x, Q^2) = \int_0^1 dy \int_0^1 dz q(y, Q^2) P_{qq}(z, Q^2) \delta(x - zy), \qquad (7.163)$$

where

$$P_{qq}(z, Q^2) \equiv \delta(1 - z) + \frac{\alpha_s(Q^2)}{2\pi} P_{qq}(z) \ln\left(\frac{Q^2}{\mu^2}\right). \qquad (7.164)$$

It is natural to interpret $P_{qq}(z)$ as a probability density, that a quark will be found in a state with a z portion of an initial quark momentum, which is calculated in the first order in α_s. The term $\delta(1 - z)$ corresponds to the fact that no changes happen in $q(x, Q^2)$. Obviously, this probability will be smaller when all contributions of the order α_s are switched on. Such a physical meaning of the function $P_{qq}(z)$ allows us to call it a *splitting function*. An analogous name is employed for the functions $P_{qg}(z), P_{gq}(z)$, and $P_{gg}(z)$ too.

At $z \to 1$ or at $z \to 0$ the direct calculation of the splitting functions needs consideration of diagrams with virtual gluons. Contributions of these diagrams cancel infinities in the above-mentioned splitting functions that appear due to soft gluons emission. Thus, for example, the diagram of Fig. 7.15 must be supplemented with diagrams displayed in Fig. 7.23. So, there is a contribution of the order $\alpha_s\alpha$ associated with interference between the

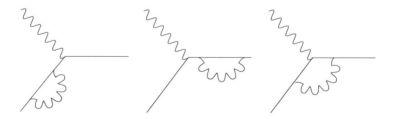

FIGURE 7.23
The radiation corrections caused by a virtual gluon to the process $ep \to eX$.

diagram of the subprocess $\gamma^* q \to q$ and three diagrams involving virtual gluons. These additional interference contributions also are singular under $z \to 1$. It turns out that these singularities are exactly cancelled by the singularity entering into Eq. (7.164), which was obtained by incomplete calculations in the order α_s. Altarelli and Parisi found a method that avoids calculations of similar contributions. This method is based on conservation laws that have to be fulfilled under a momentum redistribution of quarks and gluons.

The first of these laws is a fermion number conservation law (or charge conservation law) under a quark splitting into a quark and a gluon. It means that as a result of a quark splitting the number of quarks is not changed

$$\int_0^1 dz P_{qq}(z) = 0. \qquad (7.165)$$

At $z = 1$ the expression for P_{qq} is singular, therefore, it must be somehow regularized, so that the relation (7.165) is satisfied. To this end, one introduces a function $1/(1-z)_+$ for which the rule of integration with an arbitrary regular function $f(z)$ is

$$\int_0^1 dz \frac{f(z)}{(1-z)_+} \equiv \int_0^1 \frac{f(z) - f(1)}{1-z}. \qquad (7.166)$$

In terms of such a function the expression for $P_{qq}(z)$, meeting (7.165) and coinciding with (7.145) at $z < 1$, can be written in the form

$$P_{qq}(z) = \frac{4(1+z^2)}{3(1-z)_+} + 2\delta(1-z) \qquad (7.167)$$

which completes the definition of $P_{qq}(z)$ when $z = 1$.

At $z = 1$ the functions $P_{qg}(z)$ and $P_{gq}(z)$ are regular and do not demand extension of the definition. As for $P_{gg}(z)$, its extension of the definition is fulfilled with the help of the total momentum conservation law at gluon splitting, that is, in the processes $g \to q+\bar{q}$, $g \to g+g$:

$$\int_0^1 z dz [P_{gg}(z) + 2n_f P_{qg}(z)] = 0, \qquad (7.168)$$

where n_f is the number of quarks with masses $m_f^2 \ll Q^2$ (the factor $2n_f$ in front of the second term appears owing to the inclusion quarks and antiquarks of all considerable flavors). The expression of $P_{gg}(z)$ satisfying this condition will look like:

$$P_{gg}(z) = 6 \left[\frac{z}{(1-z)_+} + \frac{1-z}{z} + z(1-z) \right] + \left(\frac{11}{2} - \frac{n_f}{3} \right) \delta(1-z). \qquad (7.169)$$

Making use of the obtained expressions for $P_{AB}(z)$, we can easily find the values of the anomalous dimensions (7.161):

$$\gamma_{qq}^n = \frac{4}{3} \left[1 - \frac{2}{n(n+1)} + 4 \sum_{j=2}^n \frac{1}{j} \right], \qquad (7.170)$$

$$\gamma_{qg}^n = -\frac{n^2 + n + 2}{n(n+1)(n+2)}, \qquad (7.171)$$

$$\gamma_{gq}^n = -\frac{8(n^2 + n + 2)}{3n(n^2 - 1)}, \qquad (7.172)$$

$$\gamma_{gg}^n = 1 + \frac{2}{3}n_f - \frac{12}{n(n-1)} - \frac{12}{(n+1)(n+2)} + 12 \sum_{j=2}^n \frac{1}{j}. \qquad (7.173)$$

In conclusion in this case, we demonstrate how the Altarelli–Parisi formalism can be used for the investigation of processes with hadrons participation. Go back to the formula (7.144), which defines the probability that a gluon with a momentum portion $1-z$ and a transverse momentum p_T will be produced in the process $\gamma^* q \to qg$. As a matter of fact, this formula should be written in the form of the double-differential cross section

$$\frac{d^2\hat{\sigma}}{dz dp_T^2} = (e_q^2 \hat{\sigma}_0)\Gamma_{qq}(z, p_T^2), \qquad (7.174)$$

where

$$\Gamma_{qq}(z, p_T^2) = \frac{\alpha_s}{2\pi p_T^2} P_{qq}(z). \qquad (7.175)$$

It is useful to display Eq. (7.174) graphically (Fig. 7.24). From Fig. 7.24 follows, in

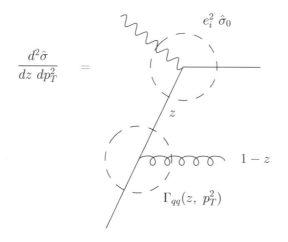

FIGURE 7.24
The graphical representation of Eq. (7.174).

the order $\alpha\alpha_s$, the given cross section is equal to the product of the parton model cross section $e_{q_i}^2 \hat{\sigma}_0$ by the probability Γ_{qq} that a quark emits a gluon with the momentum portion $1 - z$ and the transverse momentum p_T. So, if one retains only the singular part $1/p_T^2$ of the total p_T-distribution, as had been done when deducing the formula (7.144), at not too much large momenta p_T, the process $\gamma^* q \to qg$ can be considered as two sequential processes: $\gamma^* q \to q$ and $q \to qg$. And at the same time, the probability of the processes $e_{q_i}^2 \hat{\sigma}_0$ and Γ_{qq} are calculated independently and then their product gives the cross section of the process $\gamma^* q \to qg$. Such an approach for calculation of the cross sections first employed in QED [87] derives the name *Weizsäcker–Williams method* or *equivalent photons method*. In QCD the similar procedure is named *Altarelli–Parisi method*.

Let us show how this method works by the example of computing the cross section of the process

$$e^- + e^+ \to \gamma^* \to q + \bar{q} + g. \qquad (7.176)$$

Earlier, in Section 7.1, we found the cross section (7.176) using the standard method of the Feynman diagrams. In such a way, we have the criteria to check whether the approximate Altarelli–Parisi method gives the right answer. In Fig. 7.25 we depict the diagram to elucidate the course of our calculation. Begin with the kinematics of the process under consideration. The momentum vectors of particles $q, \bar{q}$ and g that are produced by a resting virtual photon are displayed in Fig. 7.26. The portions of the four-dimensional momenta

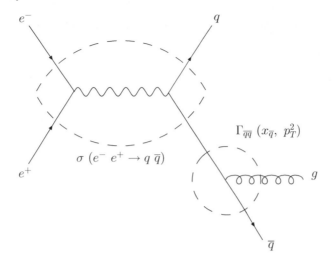

FIGURE 7.25

The Altarelli–Parisi calculation scheme for the process $e^- e^+ \to \gamma^* \to q\bar{q}g$.

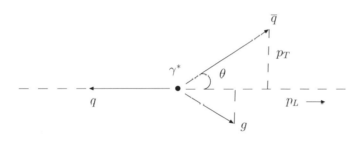

FIGURE 7.26

The process $e^- e^+ \to \gamma^* \to q\bar{q}g$ in the center-of-mass frame.

of quark, antiquark, and gluon are given by the expressions:

$$p_q^0 = (x_q, 0, 0, -x_q), \qquad p_{\bar{q}}^0 = (x_{\bar{q}}, x_T, 0, x_L), \left.\right\}$$
$$p_g^0 = (x_g, -x_T, 0, x_q - x_L), \qquad\qquad\qquad\quad \tag{7.177}$$

$x_i = 2E_i/Q, (i = q, \bar{q}, g)$. These variables are determined with respect to a jet with the greatest energy, for example, to the q-quark jet shown in Fig. 7.26. It is common practice to call this direction by an elongation axis. Quark, antiquark, and gluon are coplanar in the plane $y = 0$. Since $m_{\bar{q}}^2 = m_g^2 = 0$, then from (7.177) it follows that

$$x_{\bar{q}}^2 - x_T^2 - x_L^2 = 0, \qquad x_g^2 - x_T^2 - (x_L - x_q)^2 = 0, \tag{7.178}$$

The conservation laws of the transverse and longitudinal momenta are also contained in the definitions (7.177) while the energy conservation law imposes one more demand:

$$x_q + x_{\bar{q}} + x_g = 2. \tag{7.179}$$

From Eqs. (7.178) and (7.179) we find

$$x_T^2 = \frac{4}{x_q^2}(1 - x_q)(1 - x_{\bar{q}})(1 - x_g). \tag{7.180}$$

Since in the process (7.176) the antiquark emits the softer gluon, then

$$x_q \geq x_{\bar{q}} \geq x_g. \tag{7.181}$$

By analogy with Eq. (7.174), we get

$$\frac{d^2}{dx_{\bar{q}} dp_T^2} \sigma(e^- e^+ \to q\bar{q}g) = \sigma(e^- e^+ \to q\bar{q}) \Gamma_{\bar{q}q}(x_{\bar{q}}, p_T^2), \tag{7.182}$$

where $\Gamma_{\bar{q}q}$ is the probability that the antiquark $\bar{q}$ emits the gluon with the momentum portion $1 - x_q$ and the transverse momentum $|p_T|$. From the relations (4.49) and (7.175) we have

$$\sigma(e^- e^+ \to q\bar{q}) \equiv \sigma_0 = \frac{4\pi\alpha^2}{Q^2} e_q^2, \tag{7.183}$$

$$\Gamma_{\bar{q}q}(x_{\bar{q}}, p_T^2) = \Gamma_{qq}(x_{\bar{q}}, p_T^2) = \frac{\alpha_s}{2\pi p_T^2} P_{qq}(x_{\bar{q}}). \tag{7.184}$$

To substitute (7.183) and (7.184) into (7.182) results in

$$\frac{1}{\sigma_0} \frac{d^2}{dx_{\bar{q}} dx_T^2} \sigma(e^- e^+ \to q\bar{q}g) = \frac{\alpha_s}{2\pi x_T^2} P_{qq}(x_{\bar{q}}), \tag{7.185}$$

where we have taken into account $x_T = 2p_T/Q$.

To compare the obtained expression with (7.27), which was found with the help of the Feynman rules, it is necessary to turn from integration over x_T^2 to integration over x_q. Using the relations (7.179) and (7.180), we arrive at

$$\frac{d^2\sigma}{dx_{\bar{q}} dx_T^2} = \frac{d^2\sigma}{dx_{\bar{q}} dx_q} \frac{dx_q}{dx_T^2} \approx \frac{d^2\sigma}{dx_{\bar{q}} dx_q} \frac{1}{4x_q(1 - x_q)}. \tag{7.186}$$

The expression (7.186) having been obtained, we supposed $x_q \approx 1$. This is due to the fact that under the integration over $x_{\bar{q}}$ the integrand is divergent when $x_{\bar{q}} \to 1$ and, as a consequence, the kinematic situation in which $x_{\bar{q}}$ reaches its maximum value is of chief interest. Then, from (7.181) follows that the most allowable value of $x_{\bar{q}}$ is:

$$x_{\bar{q}} = x_q. \tag{7.187}$$

Supposing $x_q \approx 1$ and taking into consideration that $x_{\bar{q}} \approx 1 - x_g$ in this case, we reduce the relation (7.185) to the form

$$\frac{1}{\sigma_0} \frac{d^2\sigma}{dx_{\bar{q}} dx_q} \approx \frac{2\alpha_s}{3\pi} \frac{x_q^2 + x_{\bar{q}}^2}{(1 - x_q)(1 - x_{\bar{q}})} \tag{7.188}$$

which coincides with (7.27). Integrate the obtained expression over $x_{\bar{q}}$

$$\frac{1}{\sigma_0} \frac{d}{dx_T^2} \sigma(e^- e^+ \to q\bar{q}g) = 2\frac{\alpha_s}{2\pi x_T^2} \int_{(x_{\bar{q}})_{min}}^{(x_{\bar{q}})_{max}} \frac{4(1 + x^2)}{3(1 - x)} dx \tag{7.189}$$

The additional factor 2 appears owing to the inclusion of the diagram with the replacement $q \leftrightarrow \bar{q}$. To approach $x_{\bar{q}}$ to the value $x_{\bar{q}} = x_q$ one needs to make the emitting gluon extremely soft. Since the quantity x_T is fixed, this demand is fulfilled providing

$$x_g = x_T. \tag{7.190}$$

Then, from (7.179) it follows that

$$(x_q)_{min} = (x_{\bar{q}})_{max} \approx 1 - \frac{x_T}{2}. \tag{7.191}$$

With allowance made for (7.191), one can report (7.189) in the form

$$\frac{1}{\sigma_0} \frac{d}{dx_T^2} \sigma(e^- e^+ \to q\bar{q}g) \approx \frac{8\alpha_s}{3\pi x_T^2} \int_{(x_{\bar{q}})_{min}}^{1 - x_T/2} \frac{dx}{1 - x}, \tag{7.192}$$

where we have assumed $1 + x^2 \approx 2$. Integrating the right side of (7.192), we shall omit all terms other than leading logarithmic ones (this approximation is called a *leading logarithmic approximation*). Then, we get

$$\frac{1}{\sigma_0} \frac{d}{dx_T^2} \sigma(e^- e^+ \to q\bar{q}g) \approx \frac{4\alpha_s}{3\pi x_T^2} \ln\left(\frac{1}{x_T^2}\right). \tag{7.193}$$

7.6 Fragmentation functions

Up to this point we considered quantities of the cross sections type that could be completely described by parton distribution functions inside hadrons. However, it is possible to examine more complicated processes where hadrons are detected in a final state. Semi-inclusive processes:

$$e^- + e^+ \to \gamma^* \to h + X, \tag{7.194}$$

$$e^- + e^+ \to \gamma^* \to h_1 + h_2 + X, \tag{7.195}$$

$$\mu^{\pm} + h_1 \to \mu^{\pm} + h_2 + X \tag{7.196}$$

are examples. Consider the process (7.194) to be the simplest of them. In the parton model it develops in two stages: (i) $q\bar{q}$-pair production by a photon; (ii) fragmentation of q or $\bar{q}$ with production of a hadron detected. Until now we did not view the question of how quarks turn into hadrons to be recorded by detectors. We supposed that quarks fragment into hadrons with the probability equal to 1. As a result, we arrived at the formula (7.2). Let us assume that a fast hadron h detected in a final state possesses a momentum portion z of one of the quarks. It is natural to believe that such a hadron is formed from a fast quark (or antiquark) flying along the same direction—a fast quark picks up missing quarks from a sea and produces a fast hadron. To describe fragmentation of quarks into hadrons one employs a formalism that is similar to that used for the description of quarks inside hadrons. In Fig. 7.27 we present the process (7.194) in which a hadron h with the energy E_h is detected. The differential cross section of this process will look like:

$$\frac{d\sigma(e^- e^+ \to \to hX)}{dz} = \sum_q \sigma(e^- e^+ \to \bar{q}q) \left[D_q^h(z) + D_{\bar{q}}^h(z) \right]. \tag{7.197}$$

Here, new quantities—fragmentation functions $D_q^h(z)$ and $D_{\bar{q}}^h(z)$—are introduced. $D_q^h(z)$ is nothing but the probability that at the decay of a bare quark q a hadron h with a momentum portion

$$z = \frac{E_h}{E_q} = \frac{2E_h}{Q}$$

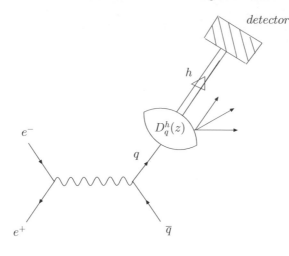

FIGURE 7.27

The process $e^- e^+ \to \gamma^* \to hX$.

will be detected. In a similar way the antiquark fragmentation function $D_{\bar{q}}^h(z)$ is determined. In Eq. (7.197) the sum is taken over all quark flavors because a detector does not know quantum numbers of a quark that produces a hadron h.

It is obvious that the fragmentation functions exist for gluons too. The fragmentation function $D_i^h(z)$ $(i = q, \bar{q}, g)$ describes the transition *parton→hadron* in the same way that the distribution function $f_i(z)$ governs a parton inside a hadron (transition *hadron→parton*). The functions $D_i^h(z)$, just as functions $f_i(z)$, satisfy conditions, following from the conservation laws both of the momentum and of probability. Thus, for example,

$$\sum_h \int_0^1 z D_q^h(z) dz = 1, \qquad \sum_h \int_0^1 z D_{\bar{q}}^h(z) dz = 1, \qquad (7.198)$$

$$\sum_h \int_{z_{min}}^1 [D_q^h(z) + D_{\bar{q}}^h(z)] dz = n_h, \qquad (7.199)$$

where $z_{min} = 2m_h/Q$ corresponds to the threshold energy of a the hadron production with the mass m_h, n_h is then average multiplicity of hadrons of the kind h.* The relations (7.198) simply state the fact that the energies sum of all hadrons is equal to the energy of the quark (or antiquark) that generated these hadrons. The relation (7.199) expresses the circumstance that the number n_h of all hadrons of the kind h is equal to the sum of the probabilities of producing the hadron h by all possible initial quarks q and antiquarks $\bar{q}$ with any flavors.

The mean multiplicity $< n_h >$ is slowly increased with the growth of the energy. Thus, for example, in the process in question, the mean multiplicity of charged hadrons is given by the expression:

$$< n_{ch} >= n_0 + a \exp[b\sqrt{\ln(s/\Lambda_{QCD})}], \qquad (7.200)$$

---

*Multiplicity is the number of secondary hadrons produced in one interaction act of high-energy particles. At particular energy of initial particles multiplicity is vastly varied. For example, at $Sp\bar{p}$ collider (CERN, 1981–1990) under the total energy in the center-of-mass frame $\sqrt{s} = 540$ GeV, events with multiplicity of secondary charged particles (n_{ch}) from 2 to 80 are recorded while average multiplicity $< n_{ch} >= 27$ is much less than the maximum possible number of secondary particles that is allowed by the energy conservation law.

where $n_0 \approx 2, a \approx 0.027, b \approx 1.9$ for the interval $\sqrt{s}$ from 2 to 4 GeV.

Forget the quark existence for a while. Then, for the process $e^- e^+ \to$ hadrons, we have

$$\sigma(e^- e^+ \to \text{hadrons}) = \sum_q \sigma(e^- e^+ \to \overline{q}q). \tag{7.201}$$

Earlier, we established the following connection between the lepton and quark cross sections

$$\sigma(e^- e^+ \to \overline{q}q) = 3e_q^2 \sigma(e^- e^+ \to \mu^- \mu^+). \tag{7.202}$$

To divide the relation (7.197) by (7.201) results in

$$\frac{1}{\sigma_t} \frac{d\sigma(e^- e^+ \to hX)}{dz} = \frac{\sum\limits_q e_q^2 [D_q^h(z) + D_{\overline{q}}^h(z)]}{\sum\limits_q e_q^2} = \mathcal{F}^h(z), \tag{7.203}$$

where $\sigma_t \equiv \sigma(e^- e^+ \to \text{hadrons})$. The quantity $\mathcal{F}^h(z)$ may be called a *total $e^- e^+$ fragmentation function of a hadron h*. From (7.203) it follows that this function has to possess the scaling property. The cross sections $d\sigma/dz$ and σ_t depend on the annihilation energy Q, however, in accordance with (7.203), their ratio is independent on Q. Since we proceeded from the quark-parton model, this is a quite forthcoming result. When including processes of emitting gluons by quarks and antiquarks, in the formula (7.203) terms of the kind $\ln Q^2$ will appear and, as a result, the scaling is broken (weak scaling violation). Their qualitative behavior is the same as in the case of the deep inelastic scattering processes, that is, when increasing Q^2, the function $\mathcal{F}(x, Q^2)$ is increasing at small z and decreasing at z close to 1. In Fig. 7.28 the $e^- e^+$ fragmentation function $\mathcal{F}(z, Q^2)$ for all charged particles is displayed for different center-of-mass energies $\sqrt{s}$ versus z (see [88] and references therein). Pay attention to the fact that now the diagram with the Z-boson exchange is taken into account, that is, the case in point is the process

$$e^- + e^+ \to \gamma^*/Z^* \to h + X. \tag{7.204}$$

At $z \preceq 0.12$ the strong scaling violation that is seen in Fig. 7.28 is caused not only by gluons emitting. The reason has to do with specificity of the $e^- e^+$-annihilation process. In processes of the deep inelastic lepton-nucleon scattering, light quarks (u, d, and s) being in a target in abundance play a dominant role. Charmed quarks are found about in one of ten events while heavier quarks occur even more rarely. It allows us to disregard contributions connected with them. However, in the $e^- e^+$-annihilation the situation is quite different. Above the threshold of heavy quarks production ($Q^2 > 4m_j^2$, $j = c, b, t$) the cross section sharply grows. And at the same time the step to appear in the cross section values gives a signal concerning a new quark. Consider the behavior of the cross section in the vicinity of the charmed quark production. In this case the structure of a final hadron state is greatly distinguished from the typical two-jet events picture that includes the light quarks only. Particles c and $\overline{c}$ are produced almost in rest and, with a big probability, they weakly decay into a rather large number of soft hadrons with a small z. And so increasing the number of events with a small z is associated with a passing through the charm threshold. An analogous state of affairs takes place under passing through the b-quark production threshold and so on. Thus, in the $e^- e^+$-annihilation at $z \preceq 0.12$, the strong scaling violation is connected with heavy quarks production, which leads to events with a big multiplicity and a small z under transition through the heavy quarks production threshold. The given mechanism has a positive aspect too. When increasing the $e^- e^+$-collider energies, the

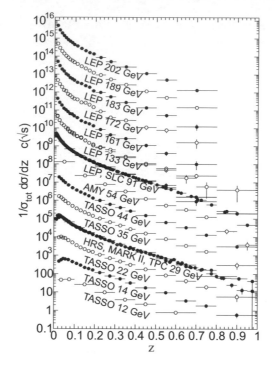

FIGURE 7.28

The $\mathcal{F}(z, Q^2)$ function for all charged particles versus z.

characteristic properties of events with heavy quarks can serve as experimental indications of the heavy quarks existence.

The total $e^- e^+$ fragmentation function $\mathcal{F}^h$ can be represented in terms of the sum over contributions of all partons

$$\mathcal{F}^h(z, Q^2) = \sum_i \int_z^1 \frac{dx}{x} C_i(x, Q^2, \alpha_s) D_i^h(z/x, Q^2), \qquad (7.205)$$

where $i = u, \overline{u}, d, \overline{d}, \ldots g$. In the lowest order in α_s the gluon coefficient function C_g is equal to zero while the quark one is

$$C_i = g_i \delta(1 - x),$$

where g_i is a corresponding electroweak coupling constant. In particular, g_i is proportional to a squared electric charge of a parton i at $Q^2 \ll m_Z^2$ when electroweak interaction effects may be neglected. Fragmentation functions of the deep inelastic lepton-nucleon (lN) scattering and the $e^- e^+$-annihilation supplement each other.

The evolution of the fragmentation function of the i-th parton is also defined by the Altarelli–Parisi equation

$$\frac{d}{d \ln Q^2} D_i(z, Q^2) = \frac{\alpha_s(Q^2)}{2\pi} \sum_j \int_z^1 \frac{dx}{x} P_{ji}(x, \alpha_s) D_j(z/x, Q^2). \qquad (7.206)$$

Now, in the evolution equation, a splitting function P_{ji}, rather than P_{ij}, figures because here D_j represents a fragmentation function of a final parton. Splitting functions can be realized in the form of the series

$$P_{ji}(z, \alpha_s) = P_{ji}^{(0)}(z) + \frac{\alpha_s}{2\pi} P_{ji}^{(1)}(z) + \ldots, \qquad (7.207)$$

where the lowest order functions $P_{ji}^{(0)}(z)$ are the same as in the case of the deep inelastic lN scattering while the higher order quantities $P_{ji}^{(n)}(z)$ $(n \geq 2)$ are different [89, 90].

To determine the gluon fragmentation function $D_g(z)$ the following quantity is used

$$\frac{1}{\sigma_t}\frac{d^2\sigma}{dx d\cos\theta},$$

where θ is the angle between the observed hadron h and the direction of the ingoing electron beam. This quantity can be reported in the view [91]

$$\frac{1}{\sigma_t}\frac{d^2\sigma}{dx d\cos\theta} = \frac{3}{8}F_T(z)(1 + \cos^2\theta) + \frac{3}{4}F_L(z)\sin^2\theta + \frac{3}{4}F_A(z)\cos\theta, \qquad (7.208)$$

where $F_T(z)$, $F_L(z)$, and $F_A(z)$ is the transverse, longitudinal, and asymmetric fragmentation functions for the process $e^-e^+ \to Z^* \to hX$, respectively. $F_T(z)$ $(F_L(z))$ represents contribution of virtual bosons to be transverse (longitudinal) polarized regarding the motion direction of the hadron h. F_A is a P-odd contribution caused by the interference between the vector and axial parts of the neutral current of the electroweak interaction. To integrate (7.208) over the angles gives the total fragmentation function, $F - F_T + F_L$. The procedure of obtaining the gluon fragmentation function D_g from the longitudinal fragmentation function to be defined with the help of Eq. (7.208) is written in Ref. [92]. The gluon fragmentation function could be also deduced from the fragmentation of three-jet events in which a gluon jet is identified, for example, by tagging the other two jets with the aid of heavy quark decays. In the leading order, the measured distributions of $z - E_{had}/E_{jet}$ for particles in gluon jets can be directly identified with the gluon fragmentation functions $D_g(z)$. The reader could find the experimentally measured gluon fragmentation functions in the current issues of the *Review of Particle Physics*.

7.7 Drell–Yan process

If hadrons collide with other hadrons under very high energies, almost all collisions will be described by soft interactions of quarks and gluons to be included in these hadrons. Recall that the process with hadrons participation is called a *soft one* if all detected particles have small transverse momenta (≤ 1 GeV). At high energies of colliding particles this corresponds to a substantial contribution of the big impact parameter region (order of the proton size) into the cross section. The soft interaction cannot be treated using the perturbative QCD, because α_s is large when the momentum transfer is small. However, in some collisions two quarks or gluons will exchange a large momentum $p_\perp$ perpendicular to the collision axis. Then, as in the deep inelastic scattering, the elementary interaction happens very rapidly compared with the internal time scale of the hadron wavefunctions, so the lowest order QCD prediction must correctly describe the process. Anew, we have to find a parton-model formula that is built from a leading order subprocess cross section, integrated with parton distribution functions. For the case of proton-proton scattering, these functions will be the same ones that are measured in the lepton-proton deep inelastic scattering.

As a case in point, we consider a process of a lepton pair production with a large invariant mass $M = \sqrt{P^2}$, where $P = k_1 + k_2$ while k_1 and k_2 are four-dimensional lepton momenta, in a hadron-hadron collision

$$h_1 + h_2 \to l^- + l^+ + X. \qquad (7.209)$$

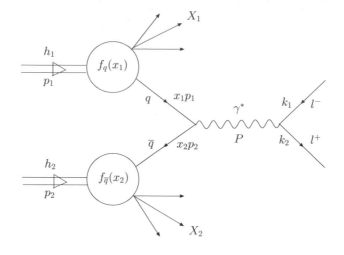

FIGURE 7.29
Lepton pair production in the Drell–Yan model.

This reaction, called the Drell–Yan process [93], is illustrated in Fig. 7.29. At that, just as in the case of deep inelastic scattering, a final hadron state is not detected but only inclusive characteristics of a lepton pair, often called a *dilepton*, are investigated. If one disregards polarization measurements, then the total momentum components of a lepton pair will be these inclusive characteristics. By way of independent variables it is customary to choose the invariant dilepton mass M, the momentum $P_\parallel$ longitudinal with respect to the direction of colliding beams in the center-of-mass frame, and the transverse momentum $P_\perp$. Within the parton model such processes are naturally described as those of the lepton pair production due to annihilation of a quark and antiquark that belong to colliding hadrons, that is,

$$q + \bar{q} \to \gamma^* \to l^- + l^{\cdot} \tag{7.210}$$

As this take place, all remaining hadrons go all the way through the interaction region and form a final hadron state.

Let us find the cross section of a lepton pair production in terms of distributions of partons inside initial hadrons. Denote initial hadron momenta by p_1 and p_2. The total energy square in the center-of-mass frame is $s = (p_1 + p_2)^2 \approx 2(p_1 p_2)$. At dilepton production by partons, the four-dimensional momentum conservation leads to the relation

$$x_1 p_1 + x_2 p_2 = P, \tag{7.211}$$

where $x_1 p_1$ and $x_2 p_2$ are the momenta of q and $\bar{q}$, which, in its turn, results in

$$x_1 x_2 s = M^2. \tag{7.212}$$

Then, the differential cross section of the lepton pair production reaction is given by the expression:

$$\frac{d\sigma}{dM^2}(h_1 h_2 \to l^- l^+ + X) = \left(\frac{1}{3}\right)\left(\frac{1}{3}\right) 3 \sum_i \int_0^1 dx_1 \int_0^1 dx_2 \left[f_{q_i}^{h_1}(x_1) f_{\bar{q}_i}^{h_2}(x_2) + \right.$$

$$\left. + f_{q_i}^{h_2}(x_2) f_{\bar{q}_i}^{h_1}(x_1) \right] \frac{d\sigma}{dM^2}(q_i \bar{q}_i \to l^- l^+), \tag{7.213}$$

where the cross section of the basic parton process (7.210) was found before and with allowance made for (7.212) is equal to

$$\frac{d\sigma}{dM^2}(q_i\bar{q}_i \to l^- l^+) = \frac{4\pi\alpha^2 e_{q_i}^2}{3M^4}\delta(x_1 x_2 s - M^2), \tag{7.214}$$

$f_{q_i}^h(x)$ $(f_{\bar{q}_i}^h(x))$ is the distribution function of i-th quark (antiquark) inside a hadron h and x is the momentum fraction it transfers. Two items in the square bracket take into account that a quark q (antiquark $\bar{q}$) may belong both to hadron h_1 and to hadron h_2. In the expression (7.213) the sum is calculated over all kinds of pairs $q\bar{q}$ that may annihilate into a dilepton. The factors $1/3$ is the consequence of averaging over colors of initial particles q and $\bar{q}$, and the factor 3 appears under the summation over all the possible color combinations of $q\bar{q}$ that annihilate producing a virtual colorless photon. Since the expression (7.213) contains $\delta(x_1 x_2 s - M^2)$, breaking down the data by the invariant mass M one can obtain an additional information about parton distribution functions. When identical hadrons (for example, protons) come into collision, then, with the help of (7.213), one can extract the sea quark distributions that are complexly found under investigating the data concerning the deep inelastic scattering.

If one is interested only in the cross section distribution over the invariant mass M, then the integration over remaining variables should be fulfilled in (7.213). To integrate over x_1 results in:

$$\frac{d}{dM^2}\sigma(h_1 h_2 \to l^- l^+ \mid X) = \frac{4\pi\alpha^2}{9M^4}\sum_i e_{q_i}^2 \tau \int_\tau^1 \frac{dx}{x}\left[f_{q_i}^{h_1}(x)f_{\bar{q}_i}^{h_2}(\tau/x) + f_{q_i}^{h_2}(\tau/x)f_{\bar{q}_i}^{h_1}(x)\right], \tag{7.215}$$

where $\tau = M^2/s$.

Next, it is convenient to pass on to a new variable Y, (longitudinal) rapidity, that is widely used under the analysis of multiple processes. The rapidity is defined as

$$Y = \frac{1}{2}\ln\left(\frac{1 + v_{\parallel}}{1 - v_{\parallel}}\right), \tag{7.216}$$

where $v_{\parallel}$ is a longitudinal (relatively to collision axis) component of a velocity of a particle produced in any collision. For slow particles $(v \ll 1)$ $Y \approx v_{\parallel}$. For high-energy particles $(E \gg m)$ the rapidity is usually expressed in terms of their energy, their momentum p, and the exit angle θ:

$$Y = \frac{1}{2}\ln\left(\frac{E + p_{\parallel}}{E - p_{\parallel}}\right) = \frac{1}{2}\ln\left(\frac{E + p\cos\theta}{E - p\cos\theta}\right). \tag{7.217}$$

The energy and the longitudinal momentum of a particle are expressed by the rapidity, particle mass, and transverse momentum $p_\perp$:

$$E = \sqrt{m^2 + p_\perp}\cosh Y, \qquad p_{\parallel} = \sqrt{m^2 + p_\perp}\sinh Y. \tag{7.218}$$

Obviously, under the Lorentz transformations Y is changed additively. Thanks to this property, at the longitudinal Lorentz transformations, the rapidity particles distribution does not modify its form and only is shifted by the constant value

$$Y_0 = \frac{1}{2}\ln\left(\frac{1 + v_0}{1 - v_0}\right), \tag{7.219}$$

where v_0 is the relative velocity of reference frames.

If both momenta of final leptons are detected, then a total four-dimensional momentum of a virtual photon that is equal to a dilepton momentum P can be reconstructed. In so doing, one can define longitudinal fractions of quark and antiquark momenta as well. In case a transverse momentum of initial partons is small, a transverse momentum of a virtual photon will be small too. As a matter of fact, its longitudinal momentum may be large. We parametrize it using the dilepton rapidity defined by

$$P_0 = M \cosh Y, \tag{7.220}$$

where P_0 is measured in the center-of-mass frame. Express the longitudinal fractions of the quarks and, as a consequence, the Drell–Yan cross section, in terms of the observables M^2 and Y. In the pp center-of-mass frame, the proton momenta are:

$$p_1 = (E, 0, 0, E), \qquad p_2 = (E, 0, 0, -E), \tag{7.221}$$

where $4E^2 = s$. Ignoring their small transverse momenta, we could write the constituent quark and antiquark momenta as $x_1 p_1$ and $x_2 p_2$, respectively, to give

$$P = x_1 p_1 + x_2 p_2 = ((x_1 + x_2)E, 0, 0, (x_1 - x_2)E). \tag{7.222}$$

Comparing (7.220) with (7.222) leads:

$$\cosh Y = \frac{x_1 + x_2}{2\sqrt{x_1 x_2}} = \frac{1}{2} \left(\sqrt{\frac{x_1}{x_2}} + \sqrt{\frac{x_2}{x_1}} \right),$$

which results in

$$e^Y = \sqrt{\frac{x_1}{x_2}}. \tag{7.223}$$

These equations can be inverted to determine x_1 and x_2:

$$x_1 = \frac{M}{\sqrt{s}} e^Y, \qquad x_2 = \frac{M}{\sqrt{s}} e^{-Y}. \tag{7.224}$$

Next, from (7.223) and (7.224) it follows that

$$Y = \ln x_1 + \frac{1}{2} \ln \left(\frac{s}{M^2} \right). \tag{7.225}$$

To substitute (7.225) into (7.215) gives

$$\frac{d\sigma}{dM^2}(h_1 h_2 \to l^- l^+ + X) = \frac{4\pi\alpha^2}{3M^4} \tau \int_{Y_{min}}^{Y_{max}} dY\, g\left(\sqrt{\tau} e^Y, \sqrt{\tau} e^{-Y}, M^2 \right), \tag{7.226}$$

where

$$g(x_1, x_2, M^2) = \frac{1}{3} \sum_i e_{q_i}^2 \left[f_{q_i}^{h_1}(x_1) f_{\bar{q}_i}^{h_2}(x_2) + f_{q_i}^{h_2}(x_2) f_{\bar{q}_i}^{h_1}(x_1) \right] \tag{7.227}$$

while x_1 and x_2 are defined by Eq. (7.224). At a fixed value of the dilepton mass, the limits of changing the dilepton rapidity are represented by the inequalities:

$$-\frac{1}{2} \ln \frac{\sqrt{s} + \sqrt{s - M^2}}{\sqrt{s} - \sqrt{s - M^2}} \le Y \le \frac{1}{2} \ln \frac{\sqrt{s} + \sqrt{s - M^2}}{\sqrt{s} - \sqrt{s - M^2}}. \tag{7.228}$$

The limits (7.228) are found from the relations:

$$Y = \frac{1}{2} \ln \left(\frac{P_0 + P_\parallel}{P_0 - P_\parallel} \right), \qquad P_\parallel = \pm\sqrt{P_0^2 - M^2}, \qquad P_0 = \frac{(x_1 + x_2)\sqrt{s}}{2} \le \sqrt{s}.$$

If one neglects hadron and lepton masses, then the effective dilepton mass will lie in the interval

$$M_0 \leq \sqrt{P^2} \leq \sqrt{s},$$

where M_0 is a minimal value of a dilepton mass at which the quark-parton model may be employed.

In the experiment the double differential cross section over the invariant mass and the rapidity is measured. With allowance made for Eq. (7.226) it will look like:

$$\frac{d^2\sigma}{dM^2 dY}(h_1 h_2 \to l^- l^+ + X) = \frac{4\pi\alpha^2}{3M^4} \tau g\left(\sqrt{\tau}e^Y, \sqrt{\tau}e^{-Y}, M^2\right). \tag{7.229}$$

When increasing the colliding hadron energies it is necessary to include the contributions of the subprocess with the Z-boson exchange

$$q + \bar{q} \to Z^* \to l^+ + l^-. \tag{7.230}$$

Making use of the Lagrangian describing the interaction between the Z-boson and fermions (see Section 11.2 of *Advanced Particle Physics Volume 1*), we obtain the following values for the $\bar{l}Zl$-vertices:

$$V_{\bar{l}Zl} = \frac{g}{4\cos\theta_W}\bar{l}(x)\gamma^\mu\left[L_e(1 - \gamma_5) + R_e(1 - \gamma_5)\right]l(x)Z_\mu(x) \tag{7.231}$$

for charged leptons

$$V_{\bar{\nu}Z\nu} = \frac{g}{4\cos\theta_W}\bar{\nu}(x)\gamma^\mu(1 - \gamma_5)\nu(x)Z_\mu(x) \tag{7.232}$$

for neutrinos

$$V_{\bar{q}_i Z q_i} = \frac{g}{4\cos\theta_W}\bar{q}_i(x)\gamma^\mu\left[L_{q_i}(1 - \gamma_5) + R_{q_i}(1 - \gamma_5)\right]q_i(x)Z_\mu(x), \tag{7.233}$$

and for quarks, where

$$\left.\begin{array}{ll} L_e = 2\sin^2\theta_W - 1, & R_e = 2\sin^2\theta_W, \\ L_{q_i} = 2T_3 - 2e_{q_i}\sin^2\theta_W, & R_{q_i} = -2e_{q_i}\sin^2\theta_W, \end{array}\right\} \tag{7.234}$$

and T_3 is the weak isospin projection of a quark q_i. Employing (7.231)–(7.234), one may show that the inclusion of the subprocess (7.230) results in the replacement

$$e_{q_i}^2 \to e_{q_i}^2 - \frac{M^2(M^2 - m_Z^2)(L_{q_i} + R_{q_i})(L_e + R_e)}{8\sin^2\theta_W\cos^2\theta_W[(M^2 - m_Z^2)^2 + m_Z^2\Gamma_Z^2]}e_{q_i} +$$

$$+ \frac{M^4(L_e^2 + R_e^2)(L_{q_i}^2 + R_{q_i}^2)}{64\sin^4\theta_W\cos^4\theta_W[(M^2 - m_Z^2)^2 + m_Z^2\Gamma_Z^2]} \tag{7.235}$$

in the definition of $g(x_1, x_2, M^2)$ (see Eq. (7.227)). To complete our calculation we need to find the total width of the Z-boson decay. In accordance with Eq. (18.201) of *Advanced Particle Physics Volume 1* the differential decay width $Z \to ab$ is

$$d\Gamma = (2\pi)^4\delta^{(4)}(k - p_a - p_b)\frac{1}{m_Z E_a E_b}|\sum\mathcal{M}'_{Z\to ab}|^2\frac{d^3p_a d^3p_b}{(2\pi)^6}, \tag{7.236}$$

where the prime denotes that the factors m_Z, E_a, and E_b connected with the normalization are separated out from the matrix element square for the sake of convenience. Setting $a = e^-$, $b = e^+$, summing and averaging over particle polarizations, we arrive at

$$|\sum\mathcal{M}'_{Z\to e^+e^-}|^2 = \frac{1}{3}\frac{g^2}{16\cos^2\theta_W}\left(-g^{\alpha\beta} + \frac{k^\alpha k^\beta}{m_Z^2}\right)\frac{1}{8}\mathrm{Sp}\,\{\hat{p}_{e^+}\gamma_\alpha\hat{p}_{e^-}\gamma_\beta[L_e^2(1 - \gamma_5)^2 +$$

$$+R_e^2(1-\gamma_5)^2]\} = \frac{g^2(p_{e^+}\cdot p_{e^-})}{24\cos^2\theta_W}(L_e^2+R_e^2) = \frac{g^2 m_Z^2}{48\cos^2\theta_W}(L_e^2+R_e^2), \qquad (7.237)$$

where the lepton masses have been neglected. Fulfilling the integration over the lepton momenta, we finally obtain

$$\Gamma(Z\to e^-e^+) = \Gamma(Z\to\mu^-\mu^+) = \Gamma(Z\to\tau^-\tau^+) = \frac{G_F m_Z^3}{12\pi\sqrt{2}}\left(1-4\sin^2\theta_W+8\sin^4\theta_W\right]. \qquad (7.238)$$

Analogous computations give

$$\Gamma(Z\to\nu_e\bar\nu_e) = \frac{G_F m_Z^3}{12\pi\sqrt{2}}, \qquad (7.239)$$

$$\Gamma(Z\to u\bar u) = \Gamma(Z\to c\bar c) = 3\left[1-\frac{8}{3}\sin^2\theta_W+\frac{32}{9}\sin^4\theta_W\right]\Gamma(Z\to\nu_e\bar\nu_e), \qquad (7.240)$$

$$\Gamma(Z\to d\bar d) = \Gamma(Z\to s\bar s) = \Gamma(Z\to b\bar b) = 3\left[1-\frac{4}{3}\sin^2\theta_W+\frac{8}{9}\sin^4\theta_W\right]\Gamma(Z\to\nu_e\bar\nu_e). \qquad (7.241)$$

To add up the obtained equations results in

$$\Gamma_Z = 14.5\frac{G_F m_Z^3}{12\pi\sqrt{2}}. \qquad (7.242)$$

In the Drell–Yan process the second-order corrections are important [94, 95]; they include effects of continuation on time-like momenta of photons. Calculations are extremely complicated because of the interrelationship of mass singularities. The indicated corrections change Eqs. (7.226) and (7.229). Particularly, in these equations the factor

$$1+\frac{8\alpha_s(M^2)}{12\pi}\left(1+\frac{4\pi^2}{3}\right) \qquad (7.243)$$

appears. As we see, the corrections are very large (of the order of 1). However, the state of affairs may be improved by the assumption that the total inclusion of radiative corrections leads to the fact that the sum of the terms $\sim\pi^2$ gives the exponent and the factor (7.243) is replaced by the expression:

$$\exp\left[\frac{8\pi\alpha_s(M^2)}{3}\right]\left[1+\frac{8\alpha_s(M^2)}{12\pi}\right], \qquad (7.244)$$

in which the exponential factor is exact in all orders of the perturbation theory. In this case we have the good qualitative agreement with the experimental data.

Instead of the dilepton rapidity variable, on occasion one uses the dimensionless variable

$$x_F = x_1 - x_2 = \frac{2P_\parallel}{\sqrt{s}} = 2\sqrt{\frac{M^2}{s}}\sinh Y. \qquad (7.245)$$

This variable is analogous to the Feynman one that is employed for the scaling description of inclusive hadrons spectrum. The arguments of the quark (antiquark) distribution function x_1 and x_2 are expressed in terms of x_F:

$$x_{1,2} = \frac{1}{2}\left[\pm x_F + (x_F^2 + \frac{4M^2}{s})^{1/2}\right]. \qquad (7.246)$$

The transition from Y to x_F is realized with the help of the relation:

$$dY = dx_F \left[x_F^2 + \frac{4M^2}{s} \right]^{-1/2}$$

and taking into consideration the limits of variation

$$- \left(1 - \frac{4M^2}{s} \right) \leq x_F \leq \left(1 - \frac{4M^2}{s} \right).$$

From the formulae mentioned above follows that the quantity $M^4 d\sigma/dM^2$, as well as its distribution over Y, is a function only of the ratio τ. Really, in the lowest order of the perturbation theory, when the gluons emitting is absent, the scaling needs to be. It is well known, however, QCD interactions lead to the quark distributions evolution. Within the Altarelli–Parise method it is almost evident for the following. To allow for the quark distributions evolution in the leading logarithmic approximation, in the formulae cited above, one should replace the parton distributions by the effective ones measured at $Q^2 = M^2$:

$$f_q^h(x) \to f_q^h(x, M^2), \qquad f_{\bar{q}}^h(x) \to f_{\bar{q}}^h(x, M^2). \tag{7.247}$$

The quark-parton model of the process (7.209) predicts a powerful dependence of a dilepton yield on the type of hadrons. Then, since the number of sea quarks in hadrons is relatively small, the cross section of lepton pairs production with a large effective mass in NN-collisions will be smaller compared with those of $\pi^{\pm} N$-, $K^- N$-, $N\overline{N}$-collisions.

The quark-parton model in its simplest form (7.213) does not predict the character of the $P_\perp$ transverse momentum distribution for dileptons. This distribution can be experimentally measured to give information about the transverse motion of quark-partons in a relativistic hadron. In such an approach the Drell–Yan process offers an advantage as compared to purely hadron processes: the $P_\perp$ distribution of dileptons is determined only by the interaction between quark-partons of colliding hadrons while the transverse momentum distribution of secondary inclusive hadrons is affected by the interaction of quark-partons in a final state in addition. In its turn, the inclusion of quark-gluon interactions predicts the practically linear growth of the average dilepton momentum versus its effective mass in the region $P_\perp > 1$ GeV.

7.8 Gauge bosons production at hadron collisions

Processes of the single production of $W^{\pm}$- and Z-bosons also may be investigated within the Drell–Yan mechanism. At that, the gauge bosons production may happen by the resonance and nonresonance ways. Begin with the former. If produced gauge bosons decay into lepton pairs on the patterns:

$$W^{\pm} \to e^{\pm} + \nu_e(\overline{\nu}_e), \qquad W^{\pm} \to \mu^{\pm} + \nu_\mu(\overline{\nu}_\mu), \tag{7.248}$$

$$Z \to e^- + e^+, \qquad Z \to \mu^+ + \mu^-, \tag{7.249}$$

there is a kinematic analogy between the processes (7.209) and

$$h_1 + h_2 \to W^{\pm} + X, \qquad h_1 + h_2 \to Z + X \tag{7.250}$$

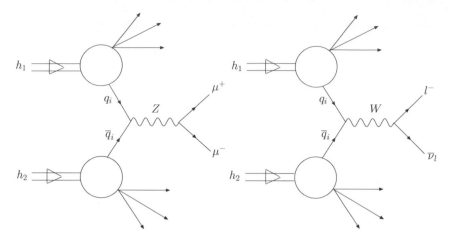

FIGURE 7.30

The diagrams associated with the processes $h_1 + h_2 \to W^{\pm} + X$ and $h_1 + h_2 \to Z + X$.

(See Fig. 7.30). As this takes place, the Drell–Yan formula undergoes the following changes:

(i) In the case of Z-boson production the summation over the quark flavors is carried out in the same manner as for the case of the virtual photon γ^* (see Eq. (7.210)); since the $W^{\pm}$-boson is come into being owing to the annihilation processes of quarks and antiquarks belonging to different flavors, as for instance:

$$u + \overline{d} \to W^+, \qquad u + \overline{s} \to W^+, \tag{7.251}$$

$$d + \overline{u} \to W^-, \qquad s + \overline{u} \to W^-, \tag{7.252}$$

then the summation needs to fulfill over all quark-antiquark pairs with appropriate quantum numbers.

(ii) In the region above the production threshold either we consider the $W^{\pm}$- and Z-boson masses to be fixed ($P^2 = m_{W,Z}^2$) or we deal with the mass spectrum that is described by the Breit–Wigner formula. The latter is more warrantable because the total decay widths of the gauge bosons are quite big

$$\Gamma_W = 2.141 \text{ GeV}, \qquad \Gamma_Z = 2.4952 \text{ GeV}. \tag{7.253}$$

By this reason, in Eq. (7.213) the cross section

$$\frac{d\sigma}{dM^2}(q_i \overline{q}_i \to l^- l^+)$$

should be replaced by the expression

$$\sigma_{W,Z} = \frac{\pi}{3m_{W,Z}^2} \frac{\Gamma_q^{W,Z} \Gamma_l^{W,Z}}{(E - m_{W,Z})^2 + \Gamma_{W,Z}^2/4}, \tag{7.254}$$

where Γ_q^W (Γ_q^Z) is the partial decay width of the W-boson (Z-boson) into a corresponding quark-antiquark pair, Γ_l^W (Γ_l^Z) is the partial decay width of the W-boson (Z-boson) into one of the lepton channels (7.248) ((7.249)), Γ_W (Γ_Z) is the total decay width of the W-boson (Z-boson), and E is the energy of quarks in the center-of-mass frame.

Consider the W^+-boson production process with the subsequent decay into the channel $W^+ \to e^+ \nu_e$ under $p\overline{p}$-annihilation. In so doing, we shall assume that the main contribution to the cross section is caused by the subprocess $u\bar{d} \to W^+$. From the beginning we find the decay width of the W-boson. In accordance with the formulae of Section 11.2 of *Advanced Particle Physics Volume 1*, the $W e \nu_e$ vertex has the view:

$$\frac{g}{2\sqrt{2}} \bar{\nu}_e(x) \gamma^\mu (1 - \gamma_5) e(x) W_\mu^*(x).$$

Again we use Eq. (7.236). Summing and averaging over the particle polarizations as well as neglecting the lepton masses, we get

$$|\sum \mathcal{M}'_{W^+ \to e^+ \nu_e}|^2 = \frac{1}{3} \frac{g^2}{8} \left(-g^{\alpha\beta} + \frac{k^\alpha k^\beta}{m_W^2} \right) \frac{1}{8} \text{Sp} \left[\hat{p}_e \gamma_\alpha \hat{p}_\nu \gamma_\beta (1 - \gamma_5)^2 \right] =$$

$$= \frac{g^2 (p_e \cdot p_\nu)}{12} = \frac{g^2 m_W^2}{24}. \tag{7.255}$$

Fulfilling the integration over the momenta of the decay particles, we finally obtain

$$\Gamma(W^+ \to e^+ \nu_e) = \Gamma(W^+ \to \mu^+ \nu_\mu) = \frac{g^2 m_W}{48\pi} = \frac{G_F m_W^3}{6\sqrt{2}\pi}. \tag{7.256}$$

Analogous computations for quarks give

$$\Gamma(W^+ \to u\bar{d}) = 3\cos^2 \theta_C \Gamma(W^+ \to c^+ \nu_e), \qquad \Gamma(W^+ \to u\bar{s}) = 3\sin^2 \theta_C \Gamma(W^+ \to e^+ \nu_e), \tag{7.257}$$

where we have neglected the quark masses.

A proton and an antiproton represent clusters of quarks and antiquarks whose energies are not fixed but are distributed to form a continuous spectrum. At that, we should remember in order to have an abundance of high-energy quarks, the energy $\sqrt{s}$ of colliding protons and antiprotons needs to exceed in several times the W-boson mass. Therefore, in a real experiment we have $\sqrt{s}_{u\bar{d}} \gg \Gamma_W$. Use this circumstance. Integration of the left side of the expression

$$\sigma(u\bar{d} \to W^+ \to e^+ \nu_e) = \frac{\pi}{3m_W^2} \frac{\Gamma_q^W \Gamma_l^W}{(E - m_W)^2 + \Gamma^2/4}, \tag{7.258}$$

where $\Gamma_q^W = \Gamma(W^+ \to u\bar{d}), \Gamma_l^W = \Gamma(W^+ \to e^+ \nu_e)$ and $\Gamma_W \equiv \Gamma$, over the energy in the vicinity of the resonance (from $-\Gamma$ to $+\Gamma$) would give the result

$$\frac{\pi \Gamma_q^W \Gamma_l^W}{3m_W^2} \frac{2}{\Gamma} \left\{ \arctan \left[\frac{2(\Gamma - E_0)}{\Gamma} \right] + \arctan \left[\frac{2(\Gamma + E_0)}{\Gamma} \right] \right\} \approx \frac{2\pi^2 \Gamma_q^W \Gamma_l^W}{3m_W^2 \Gamma}. \tag{7.259}$$

Then, taking into account that the Breit–Wigner formula defines the cross section in the resonance region, we may represent (7.258) in the form:

$$\sigma(u\bar{d} \to W^+ \to e^+ \nu_e) = \frac{2\pi^2 \Gamma_q^W \Gamma_l^W}{3m_W^2 \Gamma} \delta(E - m_W) = \frac{4\pi^2 \Gamma_q^W \Gamma_l^W}{3m_W \Gamma} \delta(s_{u\bar{d}} - m_W^2) =$$

$$= \frac{4\pi^2 \Gamma_q^W \Gamma_l^W}{3s m_W \Gamma} \delta(x_1 x_2 - \tau), \tag{7.260}$$

where $\tau = m_W^2/s$ and

$$s_{u\bar{d}} = (p_u + p_{\bar{d}})^2 \approx 2(p_u p_{\bar{d}}) = 2x_1 x_2 (p_p p_{\bar{p}}) \approx x_1 x_2 (p_p + p_{\bar{p}})^2 = x_1 x_2 s \tag{7.261}$$

has been allowed for. Making use of the results obtained and neglecting sea quark contributions, we finally get

$$\sigma(p\bar{p} \to W^+ + X) = \frac{4\pi^2}{3} \frac{\Gamma_q^W \Gamma_l^W}{\Gamma m_W^3} \int_0^1 dx_1 \int_0^1 dx_2 f_u^p(x_1) f_{\bar{d}}^{\bar{p}}(x_2) \delta(x_1 x_2 - \tau). \qquad (7.262)$$

If the integrand did not contain the δ-function, then we would have the result

$$\sigma(p\bar{p} \to W^+ + X) \approx 30 , \qquad (7.263)$$

where

$$\int_0^1 f_u^p(x) dx \approx 0.29, \qquad \int_0^1 f_{\bar{d}}^{\bar{p}}(x) dx = \int_0^1 f_d^p(x) dx \approx 0.15 \qquad (7.264)$$

have been employed. The inclusion of the δ-function complicates the situation. Go from the variables x_1 and x_2 to the ones $\beta = x_1 x_2$ and $x_F = x_1 - x_2$. The variable x_F represents the transverse momentum of the W-boson in units of its maximum passible momentum $\sqrt{s}/2$. Based on

$$x_{1,2} = \frac{x_F}{2} \pm \sqrt{\frac{x_F^2}{4} + \beta}, \qquad (7.265)$$

the integral in the expression (7.262) takes the form

$$\int_0^1 dx_1 \int_0^1 dx_2 f_u^p(x_1) f_{\bar{d}}^{\bar{p}}(x_2) \delta(x_1 x_2 - \tau) = \int_{-1+\tau}^{1-\tau} dx_F \frac{f_u^p(x_1) f_{\bar{d}}^{\bar{p}}(x_2)}{\sqrt{x_F^2 + 4\beta}} \qquad (7.266)$$

and could be found only by numerical integration. To fulfill analogous computations for the resonance production of the Z-boson presents no special problems.

When one neglects the transverse motions of the quark and antiquark, then the $W^\pm$- or Z-boson will have the momentum directed parallel to the momentum $\mathbf{p_1}$ of the hadron h_1 in the center-of-mass frame. At these conditions, the $q_\perp$ transverse momentum distribution of the leptons $l^\pm$, which are $W^\pm$ decay products into the channels (7.248), is exemplified by the root singularity at the point $q_\perp = m_W/2$

$$\frac{dw}{dq_\perp} = \frac{2q_\perp}{m_W} \sqrt{\frac{m_W^2}{4} - q_\perp^2} \qquad (7.267)$$

provided $m_W \gg m_l$. Then, measuring events concentration in the vicinity $q_\perp \approx m_W/2$, we can determine the W-boson mass. In the real situation, one should further allow for the finite decay width of the $W^\pm$-boson to lead to a washing out of the root singularity. Then, this singularity will be manifested in the form of the $q_\perp$ distribution maximum of secondary leptons $l^\pm$ nearly the value $q_\perp \approx m_W/2$. An even greater effect of the singularity washing out is expected as a result of the transverse motion of the quark and antiquark that is caused by gluons emitting. And at the same time, however, calculations show that the transverse motion inclusion of the quarks does not practically change the $q_\perp$ distribution maximum location of the observed leptons. In the case of a lepton decay of the Z-boson, both charged leptons could be observed. This allows detection of the Z-boson by the Breit–Wigner peak in the effective mass distribution of the lepton pair $l^+ l^-$.

Let us proceed to the processes of a nonresonance production of the $W^\pm$-bosons. Now, investigating the reaction $h_1 + h_2 \to W^\pm + X$ we shall take into account subprocesses with participation of a strange quark. Since the decay width of the $W^\pm$ is small as compared with their mass, it is enough to consider the production of effectively stable particles multiplying the cross section obtained by a corresponding branching of the $W^\pm$-boson in a final state. Simple computations result in the following expression for the differential cross section:

$$\frac{d\sigma}{dY}(h_1 h_2 \to W^\pm + X) = \frac{G_F \pi \sqrt{2}\tau}{3}\mathcal{F}^\pm\left(\sqrt{\tau}e^Y, \sqrt{\tau}e^{-Y}, m_W^2\right), \qquad (7.268)$$

where $\tau = m_W^2/s$ and in the case of the W^+-boson production

$$\mathcal{F}^+(x_1, x_2, Q^2) = \left[f_u^{h_1}(x_1, Q^2)f_{\bar{d}}^{h_2}(x_2, Q^2) + f_d^{h_1}(x_1, Q^2)f_{\bar{u}}^{h_2}(x_2, Q^2)\right]\cos^2\theta_C +$$

$$\left[f_u^{h_1}(x_1, Q^2)f_{\bar{s}}^{h_2}(x_2, Q^2) + f_s^{h_1}(x_1, Q^2)f_{\bar{u}}^{h_2}(x_2, Q^2)\right]\sin^2\theta_C, \qquad (7.269)$$

while the quarks and antiquarks change places for the function $\mathcal{F}^-$ associated with the W^--boson production.

In Fig. 7.31 the rapidity distributions for the $W^\pm$-boson production in $p\bar{p}$ collisions at $\sqrt{s} = 1.8$ TeV is displayed. Here, we have accepted that $Y > 0$ corresponds to the ingoing proton direction. If we were dealing with neutral particles, then the CP-invariance would

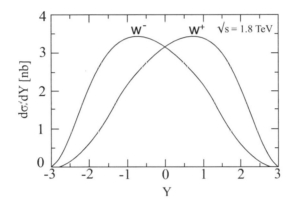

FIGURE 7.31
W^+ and W^- rapidity distribution in $p\bar{p}$ collisions at $\sqrt{s} = 1.8$ TeV.

ensure the distribution symmetry with respect to $Y = 0$. However, for processes of charged particles production such a symmetry must not take place. When one disregards the strange quark existence and neglects the sea quarks distribution, then the W^+-boson production is represented as the annihilation of the valence u-quark, belonging to the proton, with the valence $\bar{d}$-quark, being the member of the antiproton. Inasmuch as, in the proton, the u-quark is moving faster than the d-quark (see Fig. 7.10), the W^+-boson is basically produced in the ingoing proton direction. Accordingly, the preferable direction of the W^--boson production is the antiproton direction. Thus, the rapidity asymmetry

$$A_W(Y) = \frac{d\sigma(W^+)/dY - d\sigma(W^-)/dY}{d\sigma(W^+)/dY + d\sigma(W^-)/dY}, \qquad (7.270)$$

is a sensitive tool for defining the ratio of the distribution functions of the u- and d-quarks [96]. Really, suppressing the dependence on the scale Q and taking into consideration that $f_u^p(x) = f_u^{\bar{p}}(x)$ and so on, we get from (7.268)

$$A_W(Y) \approx \frac{f_u(x_1)f_d(x_2) - f_d(x_1)f_u(x_2)}{f_u(x_1)f_d(x_2) + f_d(x_1)f_u(x_2)} = \frac{R(x_2) - R(x_1)}{R(x_2) + R(x_1)}, \qquad (7.271)$$

where $R(x) = f_d(x)/f_u(x)$. For small Y we have

$$x_{1,2} = \sqrt{\tau} \pm Y\sqrt{\tau} + \ldots, \qquad R(x_{1,2}) = R(\sqrt{\tau}) \pm Y\sqrt{\tau}R'(\sqrt{\tau}) + \ldots,$$

to result in

$$A_W(Y) \approx -Y\sqrt{\tau}\frac{R'(\sqrt{\tau})}{R(\sqrt{\tau})}. \tag{7.272}$$

So, in the case of small values of Y, the asymmetry is a linear function of Y with the coefficient proportional to the slope of the curve $d(x)/u(x)$ at $x = \sqrt{\tau}$ and $Q^2 = m_W^2$. The asymmetry is slightly decreased when we take into account the contributions coming from the s-quark and sea quarks and include $O(\alpha_s)$-corrections. However, the powerful dependence on $f_d(x)/f_u(x)$ remains.

Unfortunately, the $W^\pm$ rapidity is measured in a complicated manner because a neutrino is not detected in a final state. However, the charged lepton asymmetry $A_l(Y)$ connected with this rapidity may be measured with a very high accuracy. True, the inclusion of the lepton decay results in additional complications, namely, owing to the $V - A$ structure of the charged weak current, the lepton is produced in the rest frame of the $W^\pm$-boson nonisotropicly. As a case in point we examine the process:

$$d(p_d) + \overline{u}(p_u) \to e^-(p_e) + \overline{\nu}_e(p_\nu). \tag{7.273}$$

The matrix element of this reaction, being summed and averaged over the particle polarizations, is given by the expression

$$\frac{1}{4}\sum |\mathcal{A}(d\overline{u} \to e^-\overline{\nu}_e)|^2 = 16(\sqrt{2}G_F m_W^2 \cos\theta_C)^2 \frac{(p_u \cdot p_e)^2}{[(p_u + p_d)^2 - m_W^2]^2 + m_W^2\Gamma_W^2}. \tag{7.274}$$

In a similar spirit we obtain the same expression for the charge-conjugate process $u\overline{d} \to e^+\nu_e$, where p_u is the momentum of the ingoing u-quark, and so on. When we define the angle θ^* as the $e^+(e^-)$ polar radiation angle in the $W^-(W^+)$ rest frame that is measured with respect to the direction of the ongoing antiproton (proton), and assume that all the ingoing quarks (antiquarks) belong to p $(\overline{p})$, then we arrive at

$$\frac{1}{4}\sum |\mathcal{A}(d\overline{u} \to e^-\overline{\nu}_e)|^2 = (\sqrt{2}G_F m_W^2 \cos\theta_C)^2 \frac{m_W^4(1 + \cos\theta^*)^2}{[(p_u + p_d)^2 - m_W^2] + m_W^2\Gamma_W^2}. \tag{7.275}$$

So, the cross section displays the maximum when the ongoing electron (positron) is moving along the direction of the ingoing proton (antiproton). This conclusion could be also obtained reasoning from the angular momentum conservation law. Indeed, in the standard model the W-boson is connected with charged fermions having the negative helicity and charged antifermions possessing the positive helicity. Therefore, the angular momentum conservation law demands the ingoing electron to have a preferable direction along the momentum of the ingoing quark which, in its turn, has a momentum along the ingoing proton. It is also evident that the charged lepton asymmetry is smaller than the corresponding quantity for the W-boson. Thus, measuring the charged leptons asymmetry gives additional constraints on the quark distribution functions under global fitting (see, for example, [97]).

To integrate (7.268) leads to the following expression for the total cross section:

$$\sigma(h_1 h_2 \to W^\pm + X) = \frac{G_F \pi \sqrt{2}\tau}{3} \int_\tau^1 \frac{dx}{x} \mathcal{F}^\pm(x, \tau/x, m_W^2). \tag{7.276}$$

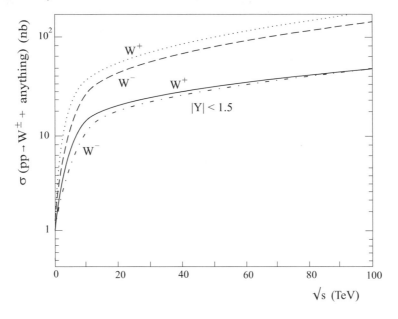

FIGURE 7.32

The cross sections for the W^+-production in pp collisions.

Let us compare the behavior of the total cross sections of the $W^\pm$-bosons production at pp- and $p\bar{p}$-colliders. Employ the quark distribution functions suggested in Ref. [98]. In Figs. 7.32 and 7.33 the total cross sections for the single charged gauge boson production in the proton-proton and proton-antiproton collisions as a function of the energy $\sqrt{s}$ are shown. In the figures the total cross sections of the $W^\pm$-boson productions in the rapidities interval from -1.5 to +1.5 are displayed as well. In pp collisions the total cross section of the W^--bosons production is suppressed by the factor 2 (or so) with regard to that of W^+-bosons production. This is caused by a smaller momentum fraction transferred by down-quarks as compared with up-quarks. It is clear that in the case of $p\bar{p}$ collisions the cross sections of the W^-- and W^+-boson production must be equal to each other. As in the case of dileptons production, the advantage of $p\bar{p}$ proves to be essential only in the region $\tau \geq 0.1$.

The analysis of a single Z-boson production happens in the similar manner. In this case the reaction is developed by way of the subprocesses:

$$u + \bar{u} \to Z, \qquad d + \bar{d} \to Z, \ldots \tag{7.277}$$

The differential cross section is as follows:

$$\frac{d\sigma}{dY}(h_1 h_2 \to Z + X) = \frac{G_F \pi \tau}{3\sqrt{2}} \mathcal{N}(\sqrt{\tau} e^Y, \sqrt{\tau} e^{-Y}, m_Z^2), \tag{7.278}$$

where $\tau = m_Z^2/s$ and

$$\mathcal{N}(x_1, x_2, Q^2) = \sum_i (L_{q_i}^2 + R_{q_i}^2)[f_{q_i}^{h_1}(x_1, Q^2) f_{\bar{q}_i}^{h_2}(x_2, Q^2) + f_{\bar{q}_i}^{h_1}(x_1, Q^2) f_{q_i}^{h_2}(x_2, Q^2)]. \tag{7.279}$$

The integrated cross sections

$$\sigma(h_1 h_2 \to Z + X) = \frac{G_F \pi \tau}{3\sqrt{2}} \int_\tau^1 \frac{dx}{x} \mathcal{N}(x, \tau/x, m_Z^2) \tag{7.280}$$

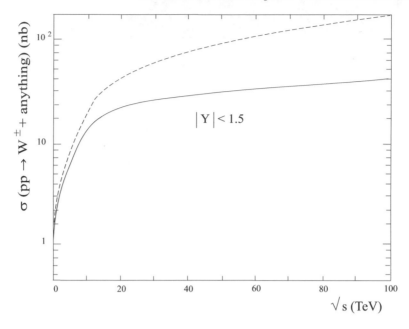

FIGURE 7.33
The cross sections for the $W^\pm$-production in $p\bar{p}$ collisions.

for pp- and $p\bar{p}$-collisions are shown in Fig. 7.34 (the quark distribution functions used are the same as in the case of the $W^\pm$-boson production). As we see, the $p\bar{p}$ cross section is approximately more than the pp cross section by a factor 5 at $\sqrt{s} = 0.54$ TeV. However, the advantage of $p\bar{p}$- over pp-collisions is fast decreased with the energy growth.

The transverse momenta of the $W^\pm$- and Z-bosons produced in the processes under consideration are small. However, processes of gluons emission and gluons absorbtion lead to large values of the transverse momenta of the gauge bosons that are compensated by a hadron jet. In Fig. 7.35 we display the diagrams of the lowest order for the subprocesses:

$$q_i + \bar{q}_j \to W + g, \qquad g + q \to W + q. \tag{7.281}$$

Calculations give the following result for the differential cross section of the W^+-boson production with the rapidity Y [99]

$$\frac{d^2\sigma}{dk_\perp dY}(h_1 h_2 \to W^+ + X) = 2k_\perp \sum_{ij} \int_{x_{min}}^1 dx_1 \frac{f_{q_i}^{h_1}(x_1, Q^2) f_{q_j}^{h_2}(x_2, Q^2) \hat{\sigma}_{ij}(\hat{s}, \hat{t}, \hat{u})}{x_1 s + u - m_W^2}, \tag{7.282}$$

where

$$\left.\begin{array}{l} t = -\sqrt{s}\, m_\perp e^{-Y} + m_W^2, \qquad u = -\sqrt{s}\, m_\perp e^{Y} + m_W^2, \qquad \hat{s} = x_1 x_2 s, \\[2mm] \hat{t} = -\sqrt{s}\, x_1 m_\perp e^{-Y} + m_W^2, \qquad \hat{u} = -\sqrt{s}\, x_2 m_\perp e^{Y} + m_W^2, \\[2mm] m_\perp^2 = k_\perp^2 + m_W^2, \qquad x_2 = \dfrac{-x_1 t - (1 - x_1) m_W^2}{x_1 s + u - m_W^2}, \qquad x_{\min} = -\dfrac{u}{s + t - m_W^2}, \end{array}\right\} \tag{7.283}$$

and the parton cross section is

$$\hat{\sigma}_{ij}(\hat{s}, \hat{t}, \hat{u}) = \frac{2\pi\alpha\alpha_s(Q^2)}{9\sin^2\theta_W} \frac{(\hat{t} - m_W^2)^2 + (\hat{u} - m_W^2)^2}{\hat{s}\hat{t}\hat{u}} \tag{7.284}$$

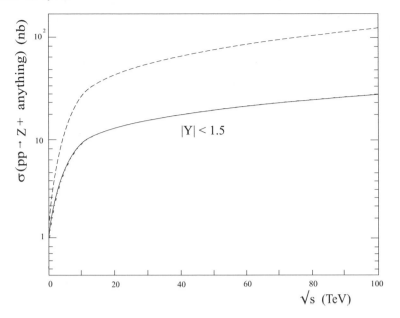

FIGURE 7.34

The cross sections for the Z-production in $p\overline{p}$ (dotted line) and pp (dashed line) collisions. Also shown are the cross sections for Z produced in the rapidity interval $-1.5 < Y < 1.5$: $p\overline{p}$ (solid line); pp (dotted-dashed line).

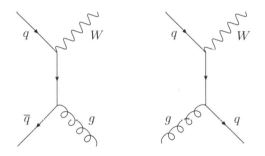

FIGURE 7.35

The Feynman diagrams for the subprocesses $q_i + \overline{q}_j \to W + g$ and $g + q \to W + q$.

for $q + \overline{q} \to W + g$ and

$$\hat{\sigma}_{ij}(\hat{s}, \hat{t}, \hat{u}) = \frac{\pi \alpha \alpha_s(Q^2)}{12 \sin^2 \theta_W} \left[\frac{\hat{s}^2 + \hat{u}^2 + 2m_W^2 \hat{t}}{-\hat{s}^2 \hat{u}} + (\hat{t} \leftrightarrow \hat{u}) \right] \quad (7.285)$$

for $q + g \to W + q$ or $\overline{q} + g \to W + \overline{q}$. In Fig. 7.36 we present the differential cross section $(d^2\sigma/dk_\perp dY)|_{Y=0}$ for the $W^\pm$-bosons production versus the transverse momentum $k_\perp$ for different energy values $\sqrt{s}$.

Now we proceed to the discussion of the pair production of the W-bosons. In the second order of the perturbation theory, the diagrams of the corresponding subprocess

$$q_i + \overline{q}_i \to W^+ + W^- \quad (7.286)$$

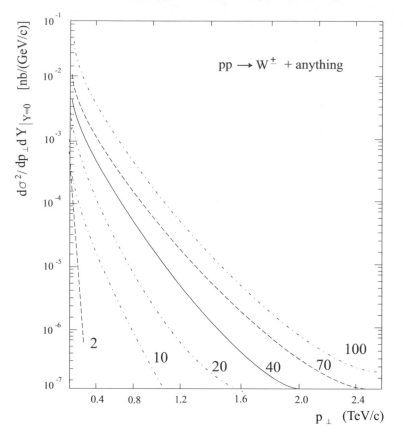

FIGURE 7.36

$\left.\dfrac{d^2\sigma}{dk_\perp \, dY}\right|_{Y=0}$ for the $W^{\pm}$-bosons production versus $k_\perp$.

are pictured in Fig. 7.37. As we see, in the lowest order of the perturbation theory, it is caused by the electroweak interaction of quarks. The subprocess (7.286) is of special interest for the following reason. Everyone of the diagrams represented in Fig. 7.38 leads to the cross section that is increased as a function of s and violates the unitary bound. However, the interference terms reduce to cancellations so that the total cross section reaches its maximum at a finite s and then tends to zero remaining in the bounds that are allowed by the unitarity condition. It should be stressed that this fine cancellation of the terms to violate the unitarity is conditioned by the initial Lagrangian invariance with respect to the $SU(2)_L \times U(1)_Y$ gauge group. Thus, investigating this process provides the experimental verification of the gauge structure of the electroweak interaction. It is evident that the subprocess (7.286) is completely similar to the one

$$e^- + e^+ \rightarrow W^- + W^+, \tag{7.287}$$

which was strongly investigated during the LEP II operation (1999–2000). The calculation of the cross sections of the process (7.287) will be fulfilled in Section 13.2 and here we give only the final results. The differential cross section of the subprocess (7.286) will look like

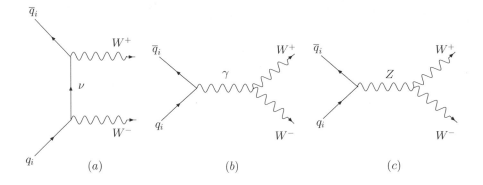

FIGURE 7.37
The Feynman diagrams (a)–(c) for the subprocess $q_i + \bar{q}_i \to W^- + W^+$ in the second order of the perturbation theory.

[100]:

$$\frac{d\sigma(q_i\bar{q}_i \to W^-W^+)}{dz} = \frac{\hat{s}\beta_W}{2}\frac{d\sigma(q_i\bar{q}_i \to W^-W^+)}{d\hat{t}} = \frac{\pi\alpha^2\beta_W}{24x_W^2\,\hat{s}}\left\{\left(\frac{\hat{u}\hat{t}-m_W^4}{\hat{s}^2}\right)\left[3-\right.\right.$$

$$-\left(\frac{\hat{s}-6m_W^2}{\hat{s}-m_Z^2}\right)\frac{L_i}{T_{3i}(1-x_W)}+\left(\frac{\hat{s}}{\hat{s}-m_Z^2}\right)^2\left(\beta_W^2+\frac{12m_W^4}{\hat{s}^2}\right)\left(\frac{L_i^2+R_i^2}{4(1-x_W)^2}\right)\right]-$$

$$-4\left(\frac{m_Z^2}{\hat{s}-m_Z^2}\right)\frac{L_i}{T_{3i}}+\frac{m_Z^2\hat{s}\beta_W^2}{(\hat{s}-m_Z^2)^2}\left(\frac{L_i^2+R_i^2}{1-x_W}\right)+$$

$$+\Theta(-e_i)\left[2\left(1+\frac{m_Z^2}{\hat{s}-m_Z^2}\frac{L_i}{T_{3i}}\right)\left(\frac{\hat{u}\hat{t}-m_W^4}{\hat{t}\hat{s}}-\frac{2m_W^2}{\hat{t}}\right)+\frac{\hat{u}\hat{t}-m_W^4}{\hat{t}^2}\right]+$$

$$+\Theta(e_i)\left[2\left(1+\frac{m_Z^2}{\hat{s}-m_Z^2}\frac{L_i}{T_{3i}}\right)\left(\frac{\hat{u}\hat{t}-m_W^4}{\hat{u}\hat{s}}-\frac{2m_W^2}{\hat{u}}\right)+\frac{\hat{u}\hat{t}-m_W^4}{\hat{u}^2}\right]\right\}, \qquad (7.288)$$

where

$$\beta_W = \sqrt{1-\frac{4m_W^2}{\hat{s}}},$$

$z = \cos\theta^*$, $x_W = \sin^2\theta_W$ and the symbol $*$ is used for the quantities referred to the center-of-mass frame of the colliding partons. To find the cross section in a particular interval of rapidities of both gauge bosons we decompose the rapidity in the hadron center-of-mass (c.m.) frame in terms of the rapidity Y^* in the parton-parton c.m. frame and the motion of the parton-parton system with respect to the overall c.m. frame, as characterized by Y_{boost}:

$$Y = Y_{boost} + Y^*, \qquad (7.289)$$

where the quantity Y_{boost} is related to the parton momentum fractions x_1 and x_2 by

$$Y_{boost} = \frac{1}{2}\ln\left(\frac{x_1}{x_2}\right). \qquad (7.290)$$

In its turn, the rapidity in the parton-parton c.m. frame is simply

$$Y^* = \tanh^{-1}(\beta z), \qquad (7.291)$$

where

$$\beta = \sqrt{1 - \frac{4m_W^2}{s}}.$$

The differential cross section to produce the W^-W^+ pair of invariant mass $M = \sqrt{s\tau}$ such that the both gauge bosons lie in the rapidity interval $(-y, y)$ is then

$$\frac{d\sigma(h_1 h_2 \to W^-W^+ + X)}{dM} = \frac{2M}{s} \sum_i \int_{-y}^{y} dY_{boost} \left[f_{q_i}^{h_1}(x_1, M^2) f_{\bar{q}_i}^{h_2}(x_2, M^2) + \right.$$

$$\left. + f_{\bar{q}_i}^{h_1}(x_1, M^2) f_{q_i}^{h_2}(x_2, M^2) \right] \int_{-z_0}^{z_0} dz \frac{d\sigma(q_i \bar{q}_i \to W^-W^+)}{dz}, \tag{7.292}$$

where

$$x_1 = \sqrt{\tau} e^{Y_{boost}}, \qquad x_2 = \sqrt{\tau} e^{-Y_{boost}}, \qquad z_0 = \min[\beta^{-1} \tanh(y - Y_{boost}), 1] \tag{7.293}$$

and $\tau = x_1 x_2$. To integrate the parton cross section results in

$$\int_{-z_0}^{z_0} dz \frac{d\sigma(q_i \bar{q}_i \to W^-W^+)}{dz} = \frac{\pi \alpha^2}{12 x_W^2 \hat{s}} \left\{ \frac{\beta_W^3 z_0}{4} \left(1 - \frac{z_0^2}{3} \right) \left[3 - \left(\frac{\hat{s} - 6m_W^2}{\hat{s} - m_Z^2} \right) \frac{L_i}{T_{3i}(1 - x_W)} + \right. \right.$$

$$\left. + \left(\frac{\hat{s}}{\hat{s} - m_Z^2} \right)^2 \left(\beta_W^2 + \frac{12 m_W^4}{\hat{s}^2} \right) \left(\frac{L_i^2 + R_i^2}{4(1 - x_W)^2} \right) \right] - \beta_W z_0 \left[4 \left(\frac{m_Z^2}{\hat{s} - m_Z^2} \right) \frac{L_i}{T_{3i}} - \right.$$

$$\left. - \frac{m_Z^2 \hat{s} \beta_W^2}{(\hat{s} - m_Z^2)^2} \left(\frac{L_i^2 + R_i^2}{1 - x_W} \right) \right] + \frac{1}{\hat{s}^2} \left(1 + \frac{m_Z^2}{\hat{s} - m_Z^2} \frac{L_i}{T_{3i}} \right) \left[\hat{t}_+^2 - \hat{t}_-^2 - 2m_W^2(m_W^2 + 2\hat{s}) \ln \left(\frac{\hat{t}_+}{\hat{t}_-} \right) \right] +$$

$$\left. + \left(\frac{\hat{t}_+ + \hat{t}_-}{\hat{s}} \right) \ln \left(\frac{\hat{t}_+}{\hat{t}_-} \right) - \frac{\hat{t}_+ - \hat{t}_-}{\hat{s}} - \frac{m_W^4(\hat{t}_+ - \hat{t}_-)}{\hat{s} \hat{t}_+ \hat{t}_-} \right\}, \tag{7.294}$$

where

$$\hat{t}_\pm = m_W^2 - \frac{1}{2}\hat{s}(1 \mp \beta_W z_0). \tag{7.295}$$

To obtain the total cross section of the W^-W^+ pair production at hadron colliders we have to take the integral

$$\sigma(h_1 h_2 \to W^-W^+ + X) = \int_{2m_W}^{\sqrt{s}} \frac{d\sigma(h_1 h_2 \to W^-W^+ + X)}{dM} dM. \tag{7.296}$$

It can be found only by a numerical integration. In Fig. 7.38 the total cross section of the W^-W^+ pair production in pp collisions is pictured. In this figure the dash-dotted line stands for the total cross section, the dashed (solid) line—for the cross section with rapidity cuts of $|Y| < 1.5$ ($|Y| < 2.5$). The constraint to the central rapidity is needed in order to avoid the collinear singularities in the t-channel as well as to facilitate the detection of decay products of the $W^\pm$-bosons. Ideally, one should like to impose rapidity cuts on $W^\pm$ decay products. However, at energies ~ 1 TeV and more, the gauge bosons are relatively light particles and their decay products have restricted mobility of order ± 1 of the rapidity unit.

In the high-energy region ($\sqrt{s} \geq 20$ TeV) the total cross section of the process $p\bar{p} \to W^-W^+X$ practically coincides with that of the process $pp \to W^-W^+X$. The difference between both the processes is essential at low energies only. Fig. 7.39 shows the W^-W^+ pair production cross sections in pp and $p\bar{p}$ collisions.

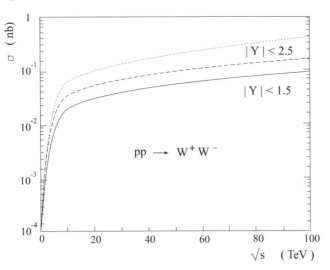

FIGURE 7.38

Yield of W^-W^+ pairs in pp collisions.

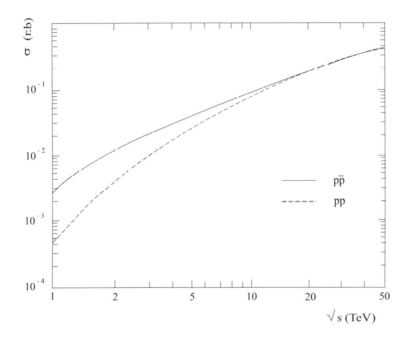

FIGURE 7.39

The W^-W^+ pair production cross sections versus $\sqrt{s}$.

In hadron colliders with luminosity $L = 10^{33}$ cm^{-2}s^{-1}, a flux of the W-boson produced is of the order $10W^-W^+$−pairs/s and lie in the limits $\pm 2^0$ of the beam direction. It means that a detector with the angle coverage down to a few degrees of the beam direction will detect practically all decay products of the $W^\pm$-bosons with $|Y| < 2.5$ (recall that luminosities of TEVATRON and LHC have the values 2.9×10^{32} and 10^{34} cm^{-2}s^{-1}, respectively). From Fig. 7.38, it follows that, in the high-energies region of colliding pp-beams, the yield of the W^-W^+-pair produced is large. Thus, for example, a run with integrated luminosity $\int L dt = 10^{40}$ cm^{-2} will produce about 10^6 pairs.

The mass spectrum of W^-W^+ pairs are of great interest both to the verification of the gauge cancellation and to the background estimation for heavy Higgs-boson decays. In Fig. 7.40 we display $d\sigma(pp \to W^-W^+ + X)/dM$ versus M at $|Y| < 1.5$ for c.m. energies $\sqrt{s} = 2, 10, 20$ TeV. It is evident that the behavior of $d\sigma(pp \to W^-W^+ + X)/dM$ does

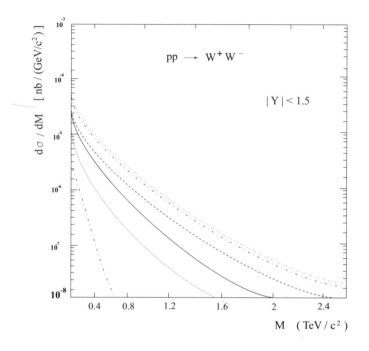

FIGURE 7.40

Mass spectrum of W^-W^+ pairs produced in pp collisions.

not conflict with the unitarity condition and, hence, confirms the gauge cancellation of the terms to violate the unitarity.

In hadron-hadron collisions reactions of $W^{\pm}Z$-pair production takes place as well

$$h_1 + h_2 \to W^{\pm} + Z + X, \qquad (7.297)$$

where the main contribution comes from the subprocess:

$$q_i + \overline{q}_j \to W^{\pm} + Z. \qquad (7.298)$$

In the lowest order of the QCD perturbation theory (tree approximation or leading order), the subprocess (7.298) is the result of the electroweak interaction of quarks. The formulae for the corresponding cross sections could be found in Ref. [101]. Here, we only note that the yield of $W^{\pm}Z$ pairs in pp- or $p\overline{p}$-collisions is approximately a factor of 5 smaller than the W^-W^+ yield.

The reaction

$$h_1 + h_2 \to Z + Z + X \qquad (7.299)$$

was first investigated in Ref. [100]. The main contribution to (7.299) is caused by the subprocess

$$q_i + \overline{q}_i \to Z + Z \qquad (7.300)$$

that, in the leading order (LO), is going due to the electroweak interaction of quarks. The ZZ yield produced in pp- and $p\bar{p}$-collisions is almost one order smaller than the W^-W^+ yield.

The process

$$h_1 + h_2 \rightarrow W^\pm + \gamma + X \tag{7.301}$$

is of primary concern to find out electromagnetic properties of the $W^\pm$ bosons. The elementary process that operates in the reaction (7.301) is

$$q_i + \bar{q}_j \rightarrow W^\pm + \gamma. \tag{7.302}$$

In the lowest order of the perturbation theory its Feynman diagrams are shown in Fig. 7.41. For the first time the differential cross section of the subprocess (7.302) was calculated in

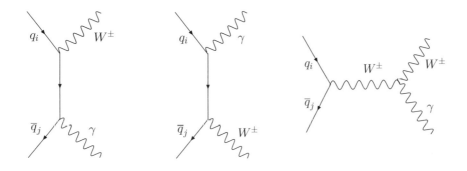

FIGURE 7.41
The process $q_i + \bar{q}_i \rightarrow W^\pm + \gamma$.

Refs. [101, 102]. In the case of the W^--boson production, it has the following view:

$$\frac{d\sigma(q_i\bar{q}_j \rightarrow W^-\gamma)}{d\hat{t}} = \frac{\pi\alpha^2}{6\hat{s}^2} \frac{|\mathcal{M}_{ij}^{CKM}|^2}{x_W} \left[\left(\frac{1}{1+\hat{t}/\hat{u}} - \frac{1}{3}\right)^2 \frac{\hat{t}^2 + \hat{u}^2 + 2m_W^2\hat{s}}{\hat{t}\hat{u}} \right], \tag{7.303}$$

where $\mathcal{M}_{ij}^{CKM}$ are the Cabibbo–Kobayashi–Maskawa matrix elements, $\hat{t} = (p_{q_i} - p_{W^-})^2$ and the average over an initial quark color has been fulfilled. For the $W^+\gamma$ production we obtain the same expression with the only difference that now $\hat{t}$ measures a momentum transferred from q_j to W^+. When

$$k_\gamma - 1 = \lambda_\gamma = \tilde{k}_\gamma = \tilde{\lambda}_\gamma = 0, \tag{7.304}$$

where coefficients k_γ $(\tilde{k}_\gamma)$ λ_γ $(\tilde{\lambda}_\gamma)$ define a magnetic (electric) dipole moment μ_γ $(d_\gamma$, electric (magnetic) quadrupole moment Q_γ $(\tilde{Q}_\gamma)$ of the $W^\pm$-bosons according to the relations

$$\mu_\gamma = \frac{e}{2m_W}[1 + k_\gamma + \lambda_\gamma], \qquad \left(d_\gamma = \frac{e}{2m_W}[\tilde{k}_\gamma + \tilde{\lambda}_\gamma]\right), \tag{7.305}$$

$$Q_\gamma = -\frac{e}{m_W^2}[k_\gamma - \lambda_\gamma], \qquad \left(\tilde{Q}_\gamma = -\frac{e}{m_W^2}[\tilde{k}_\gamma - \tilde{\lambda}_\gamma]\right), \tag{7.306}$$

then the differential cross section (7.303) turns into zero at [103, 104]

$$\hat{t} = 2\hat{u} \tag{7.307}$$

to correspond to $\cos\theta_{c.m.} = -1/3$. Recall that the conditions (7.304) are true only in the standard model and in its extensions. The angle value in the given reference frame depends only on charges of particles but is independent on energies and masses of particles. These radiation zeros (RZs) are also present in the crossing channels of the reaction (7.302). However, if the relations (7.305) become invalid, as it may take place in the case of composite $W^{\pm}$-bosons, the RZs vanish. For the reaction (7.301) the RZs are displayed as sharp slopes in the angular distribution (see Fig. 7.42) The total cross section of the reaction (7.301)

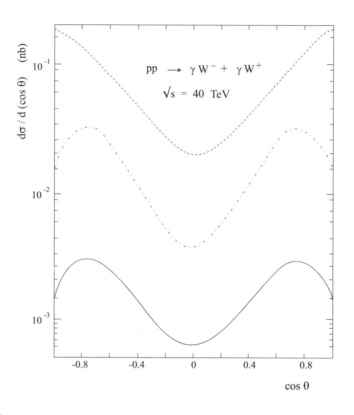

FIGURE 7.42

Distribution in $\cos\theta$ for the process $pp \to W^{\pm}\gamma + $ anything at $\sqrt{s} = 40$ TeV.

depends sensitively on the $W^{\pm}\gamma$ invariant mass and, therefore, on the minimum energy of a photon to be detected in the experiment. Fig. 7.44 shows the total cross section for $pp \to W^{\pm}\gamma$ when the invariant mass of the W and the photon is restricted to be more than 200 GeV. This cut removes the infrared divergence when the photon energy vanishes.

Conclude this section by consideration of the process

$$h_1 + h_2 \to Z + \gamma + X. \tag{7.308}$$

The main contribution is provided with the subprocess

$$q_i + \bar{q}_i \to Z + \gamma, \tag{7.309}$$

whose Feynman diagrams in the tree approximation are pictured in Fig. 7.43. The interest to this process is caused by the fact that it is a sensitive tool in searching for physics beyond the standard model. In Ref. [105] it was shown that in models with a composite Z-boson,

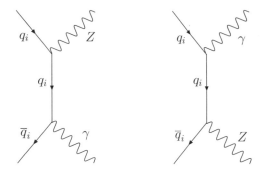

FIGURE 7.43

The process $q_i + \overline{q}_i \to Z + \gamma$.

the process (7.309) may yield large $k_\perp$ photons at a rate substantially greater than predicted by the standard model. In the standard model, the differential cross section for the reaction (7.309) will look like [106]:

$$\frac{d\sigma(q_i \overline{q}_i \to Z\gamma)}{d\hat{t}} = \frac{\pi \alpha^2 (L_i^2 + R_i^2)}{6 x_W (1 - x_W) \hat{s}^2} \left(\frac{\hat{s}^2 + m_Z^4}{2 \hat{u} \hat{t}} - 1 \right), \tag{7.310}$$

where we have averaged over the initial quark colors. Note, that the RZs are absent in the cross section of (7.309). In Fig. 7.44 we represent the total cross section of the process $pp \to Z\gamma X$ for the case when the $Z\gamma$ invariant mass is greater than 200 GeV and the rapidities are constrained by the intervals $|Y| < 1.5$ and $|Y| < 2.5$.

7.9 Higgs bosons production

In the standard model the mass of the physical Higgs boson m_H is not predicted. The Higgs mechanism connects m_H both with the constant λ entering the Yukawa potential (see Section 11.2 of *Advanced Particle Physics Volume 1*) and with the vacuum expectation value $v = 246$ GeV:

$$m_H = \sqrt{2v\lambda}. \tag{7.311}$$

Unfortunately, the quantity λ is unknown and so m_H is not fixed. The bounds on the Higgs boson mass can be obtained on the basis of arguments of the theory self-consistency. So, the unitarity arguments give the upper bound on m_H of the order of 860 GeV while the vacuum stability condition results in the low bound on m_H equal to 132 GeV. Searching for the Higgs boson at LEP II experimentally constrains the Higgs boson mass at the bottom: $m_H > 114.4$ GeV.

Let us consider the Higgs boson production at hadron colliders within the standard model. Begin our analysis with the process:

$$h_1 + h_2 \to H + X, \tag{7.312}$$

which can proceed in the resonance and nonresonance ways. In the former the procedure of finding the cross section is precisely the same as in the case of the resonance production of the $W^\pm$-boson. Therefore, we should start with the definition of all possible decay

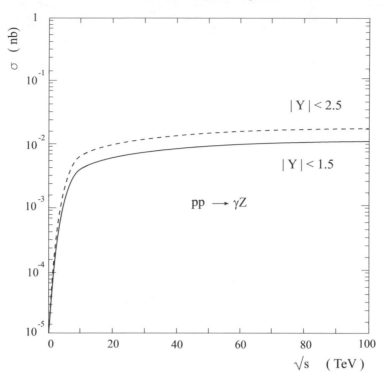

FIGURE 7.44

The total cross section for the process $pp \rightarrow Z\gamma + $ anything as function of $\sqrt{s}$.

widths of the Higgs boson. To calculate these quantities we use the Lagrangian to describe the interaction of the H-boson with fermions and gauge bosons $\mathcal{L}_H$ (see Eq. (11.37) of Volume1). From $\mathcal{L}_H$ it follows that the interaction constant of a particle with the Higgs boson is proportional to a particle mass. Therefore, the heavier particle, the stronger the interaction. If $m_H < 2m_W$, then only the decays through the fermion channels are allowed

$$\Gamma(H \rightarrow f\bar{f}) = \frac{G_F m_f^2 m_H N_c}{4\pi\sqrt{2}} \left(1 - \frac{4m_f^2}{m_H^2}\right)^{3/2}, \tag{7.313}$$

where $N_c = 3$ for quarks and $N_c = 1$ for leptons. In the case of a large Higgs-boson mass, decay channels into gauge boson pairs are opened. These decay widths are given by the expressions [107]

$$\Gamma(H \rightarrow W^- W^+) = \frac{G_F m_H^3}{32\pi\sqrt{2}}(4 - 4a_W + 3a_W^2)\sqrt{1 - a_W}, \tag{7.314}$$

$$\Gamma(H \rightarrow ZZ) = \frac{G_F m_H^3}{64\pi\sqrt{2}}(4 - 4a_Z + 3a_Z^2)\sqrt{1 - a_Z}, \tag{7.315}$$

where

$$a_{W,Z} = \frac{4m_{W,Z}^2}{m_H^2}.$$

In addition to the decay channels (7.313)–(7.315) that may attribute to the leader-order decay channels, certain rare channels appearing beyond the leading-order can be of phenomenological importance. An example is

$$H \to g + g. \tag{7.316}$$

Since the gluon mass is zero, the direct interaction between the H-boson and gluons is absent. However, such an interaction has to appear due to quark loops (Fig. 7.45).

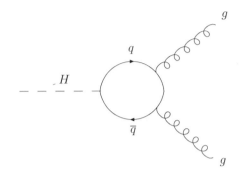

FIGURE 7.45
The diagram corresponding to the decay $H \to g + g$.

Note, the contribution caused by t-quark dominates. To calculate the decay width (7.316) gives [108]

$$\Gamma(H \to gg) = \frac{\alpha_s^2(m_H^2)G_F m_H^3}{36\pi^3\sqrt{2}} \left| \sum_i I\left(\frac{m_{q_i}^2}{m_H^2}\right) \right|^2, \tag{7.317}$$

where

$$I(x) = 3x\left[2 + (4x - 1)F(x)\right], \tag{7.318}$$

$$F(x) = \frac{1}{2}\theta(1 - 4x)\left[\ln\left(\frac{1 + \sqrt{1 - 4x}}{1 - \sqrt{1 - 4x}}\right) + i\pi\right]^2 - 2\theta(4x - 1)\sin^{-2}\left(\frac{1}{2\sqrt{x}}\right). \tag{7.319}$$

For $x < 2 \times 10^{-3}$, $|I(x)|$ is small, therefore, light quark contributions are negligibly small. If $m_H < 2m_t$ and $z > 1/4$, then $I(x) \to 1$ when $x \to \infty$.

By analogy, in the third order of the perturbation theory, the two-photon decay of the Higgs boson becomes possible (Fig. 7.46)

$$H \to \gamma + \gamma. \tag{7.320}$$

The partial width of the decay (7.320) is

$$\Gamma(H \to \gamma\gamma) = \frac{\alpha^2 G_F m_H^3}{128\sqrt{2}\pi^3} \left| \sum_i 3e_{q_i}^2 I_{q_i}\left(\frac{m_{q_i}^2}{m_H^2}\right) + I_W\left(\frac{m_W^2}{m_H^2}\right) \right|^2, \tag{7.321}$$

where

$$I_q(x) = 4x[2 + (4x - 1)F(x)], \qquad I_W(x) = -2[6x + 1 + 6x(2x - 1)F(x)], \tag{7.322}$$

The complete set of phenomenologically relevant Higgs branching ratios as a function of m_H is shown in Fig. 7.47 [109]. Far below the W^-W^+-pair production threshold the decay

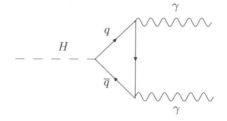

FIGURE 7.46

The Higgs boson decay through the channel $H \rightarrow \gamma + \gamma$.

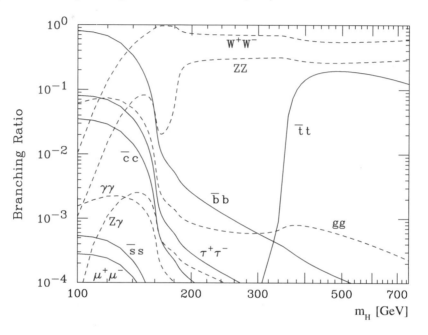

FIGURE 7.47

Branching ratios for the main decays of the SM Higgs boson.

$H \rightarrow b\bar{b}$ is dominant. As the nominal W^-W^+ threshold at $m_H = 2m_W$ is approached from below, the off-shell $H \rightarrow W^-W^+$ and $H \rightarrow ZZ$ decays begin to dominate the decay width. As follows from Fig. 7.47 above the W^-W^+ threshold the Higgs boson decays predominantly into W^-W^+ and ZZ pairs. Away from the thresholds, we have $\mathrm{Br}(H \rightarrow W^-W^+)/\mathrm{Br}(H \rightarrow ZZ) \approx 2$. At $m_H > 2m_t$ the decay channel $H \rightarrow t\bar{t}$ becomes available. For the decay $H \rightarrow t\bar{t}$ the coupling constant of the Higgs boson is proportional to the first power of m_t. Note, the dip in the W^-W^+, ZZ branching ratios is positioned on the $t\bar{t}$ threshold. Of course the $W^\pm$ and Z bosons into which the Higgs boson decays are themselves unstable, and so the actual final states consists of leptons and quark jets. In Table 7.2 we present the branching ratios for the various channels providing $m_H = 500$ GeV. Final states involving hadron jets have large backgrounds. Therefore, a large fraction of the decays are not useful. The much rarer purely leptonic final states yield the main hope for detecting the Higgs boson at hadron colliders.

The total Higgs boson decay width Γ_H grows rapidly with m_H. At $m_H = 600$ GeV, $\Gamma_H \approx 100$ GeV and at $m_H = 1$ TeV, Γ_H becomes comparable with its mass (see Fig. 7.48)

TABLE 7.2
The branching ratios of the Higgs boson decays.

channel	W^-W^+	ZZ	$t\bar{t}$
4jets	24.6	12.6	
2jets+$l\nu$	23.7		
$l^+\nu l^-\bar{\nu}$	5.7		
2jets+l^-l^+		3.6	
2jets+$\nu\bar{\nu}$		7.3	
$l^-l^+l^-l^+$		0.3	
$l^-l^+\nu\bar{\nu}$		1.1	
$\nu\bar{\nu}\nu\bar{\nu}$		1.1	
6jets			9.1
4jets+$l\nu$			8.8
2jets+$l^-\bar{\nu}l^+\nu$			2.1

[109]. It should be stressed, when the Higgs boson decay width becomes large, the Breit–Wigner approximation ceases to work.

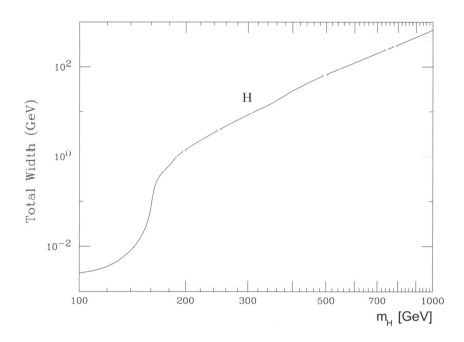

FIGURE 7.48
The total decay width of the SM Higgs boson.

Now we proceed to the investigation of a single Higgs boson production by means of the nonresonance way. Several subprocesses are at the heart of the reaction (7.312). Let us discuss their contributions taken separately. Begin with the simplest subprocess

$$q_i + \bar{q}_i \rightarrow H. \tag{7.323}$$

In the lowest order of the perturbation theory the corresponding diagram is shown in Fig. 7.49. Elementary computations lead to the following expression for the contribution of

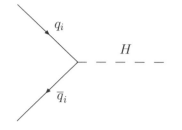

FIGURE 7.49

The Feynman diagram corresponding to the subprocess $q_i + \bar{q}_i \to H$.

(7.323) into the differential cross section of the reaction under consideration

$$\frac{d\sigma_{q\bar{q}}(h_1h_2 \to H + X)}{dY} = \frac{G_F\pi}{3\sqrt{2}} \sum_i \frac{m_{q_i}^2}{m_H^2} \tau[f_{q_i}^{h_1}(x_1, m_H^2)f_{\bar{q}_i}^{h_2}(x_2, m_H^2) +$$

$$+ f_{\bar{q}_i}^{h_1}(x_1, m_H^2)f_{q_i}^{h_2}(x_2, m_H^2)], \tag{7.324}$$

where $\tau = m_H^2/s$, and $x_{1,2}$ are defined by Eqs. (7.224). To integrate the obtained expression gives the total cross section:

$$\sigma_{q\bar{q}}(h_1h_2 \to H + X) = \frac{G_F\pi}{3\sqrt{2}} \sum_i \frac{m_{q_i}^2}{m_H^2} \tau \int_\tau^1 \frac{dx}{x}[f_{q_i}^{h_1}(x, m_H^2)f_{\bar{q}_i}^{h_2}(\tau/x, m_H^2) +$$

$$+ f_{\bar{q}_i}^{h_1}(x, m_H^2)f_{q_i}^{h_2}(\tau/x, m_H^2)]. \tag{7.325}$$

For light quarks the cross section is suppressed by the factor $m_{q_i}^2/m_H^2$, while for heavy quarks it is small as a result of small values of quark distribution functions.

A gluon-gluon fusion

$$g + g \to H \tag{7.326}$$

is a more promising mechanism of a single Higgs boson production. This subprocess is realized via an intermediate top quark loop (Fig. 7.50). In this case the contribution into the differential cross section of the process (7.312) is as follows [110]:

$$\frac{d\sigma_{gg}(h_1h_2 \to H + X)}{dY} = \frac{G_F\pi}{32\sqrt{2}} \left[\frac{\alpha_s(m_H^2)}{\pi}\right]^2 \tau|\eta|^2 f_g^{h_1}(x_1, m_H^2)f_g^{h_2}(x_2, m_H^2), \tag{7.327}$$

where [111]

$$\eta = \sum_i \int_0^1 dx \int_0^{1-x} dy \frac{1 - 4xy}{1 - (xym_H^2/m_{q_i}^2)} = \sum_i \frac{2m_{q_i}^2}{m_H^2}\left[1 + \frac{1}{2}\left(\frac{4m_{q_i}^2}{m_H^2} - 1\right)F\left(\frac{m_{q_i}^2}{m_H^2}\right)\right] \tag{7.328}$$

and the function $F(x)$ is determined by Eq. (7.319). The total cross section is

$$\sigma_{gg}(h_1h_2 \to H + X) = \frac{G_F\pi}{32\sqrt{2}} \left[\frac{\alpha_s(m_H^2)}{\pi}\right]^2 |\eta|^2\tau \int_\tau^1 \frac{dx}{x}f_g^{h_1}(x, m_H^2)f_g^{h_2}(\tau/x, m_H^2), \tag{7.329}$$

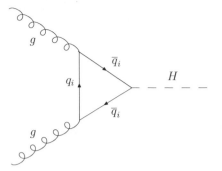

FIGURE 7.50

The Feynman diagram for the Higgs boson production in gluon-gluon fusion

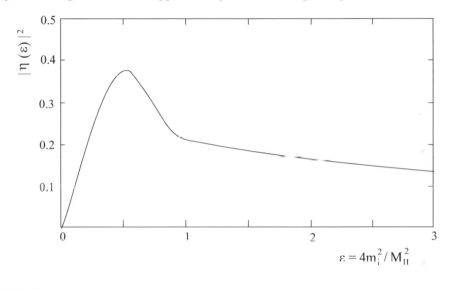

FIGURE 7.51

η versus $\epsilon_i = 4m_{q_i}^2/m_H^2$.

where a quark with $m_{q_i} \geq m_H$ gives $\eta \approx 1/3$. In Fig. 7.51 we display η as a function of $\epsilon_i = 4m_{q_i}^2/m_H^2$. From this figure it follows that the dominant contribution is truly caused by the top quark.

The next-to-leading order corrections to the gluon-gluon fusion subprocess cross section have been calculated in Ref. [112]. Over the relevant Higgs mass range, they enhance the leading-order cross section by a factor 1.5–1.7 depending on m_H. For $m_H^2 < 4m_t^2$ the corrections could be calculated analytically [113].

The gluon fusion mechanism is dominant for the Higgs production at hadron colliders. However, since the coupling constant of the Higgs boson with the gauge bosons is proportional to the gauge boson masses, then, for very heavy Higgs, subprocesses of WW and ZZ fusion

$$q + q \rightarrow W^{-*}W^{+*} \rightarrow H + q + q, \tag{7.330}$$

$$q + q \rightarrow Z^*Z^* \rightarrow H + q + q \tag{7.331}$$

give comparable contributions. The leading-order Feynman diagrams of (7.330) and (7.331) are represented in Fig. 7.52. For the first time the gauge boson fusion mechanism was

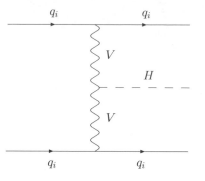

FIGURE 7.52

The gauge-boson fusion mechanism for the Higgs boson production ($V = W, Z$).

considered in Ref. [114]. The corresponding total cross sections are [115, 116]:

$$\sigma_{WW}(h_1 h_2 \to H + X) \approx \frac{\alpha^3}{16 m_W^2 x_W^3} \sum_{i,k} \theta(-e_{q_i} e_{q_k}) \int_\tau^1 \frac{dx}{x} \left[\left(1 + \frac{m_H^2}{\hat{s}}\right) \ln\left(\frac{\hat{s}}{m_H^2}\right) - \right.$$

$$\left. -2 + \frac{2m_H^2}{\hat{s}} \right] f_{q_i}^{h_1}(x, m_H^2) f_{q_k}^{h_2}(\tau/x, m_H^2), \tag{7.332}$$

$$\sigma_{ZZ}(h_1 h_2 \to H + X) \approx \frac{\alpha^3}{64 m_W^2 x_W^3 (1 - x_W)^3} \sum_{i,k} \int_\tau^1 \frac{dx}{x} \left[\left(1 + \frac{m_H^2}{\hat{s}}\right) \ln\left(\frac{\hat{s}}{m_H^2}\right) - \right.$$

$$\left. -2 + \frac{2m_H^2}{\hat{s}} \right] (L_i^2 + R_i^2)(L_k^2 + R_k^2) f_{q_i}^{h_1}(x, m_H^2) f_{q_k}^{h_2}(\tau/x, m_H^2), \tag{7.333}$$

When deriving these expressions, it has been suggested that $\hat{s} \gg m_H^2 \gg m_W^2$ and $W^\pm$-bosons are emitted by quarks at zero angle. The next-to-leading-order QCD corrections to these subprocesses have been computed in Ref. [117]. They give a small positive enhancement to the leading-order cross section.

Now we consider associated production of the Higgs boson and gauge bosons (*Higgsstrahlung*) [118]

$$h_1 + h_2 \to H + V + X, \tag{7.334}$$

where $V = W^\pm, Z$. In the leading-order of the perturbation theory, the corresponding elementary processes

$$q_i + \bar{q}_k \to H + W^\pm \tag{7.335}$$

$$q_i + \bar{q}_i \to H + Z \tag{7.336}$$

are displayed by the diagrams shown in Fig. 7.53. Although the Higgsstrahlung cross sections are small, it may be useful to detect the light Higgs in the decays $H \to b\bar{b}$ and $H \to \gamma\gamma$. The total cross sections, having been averaged over the initial quark colors, are given by the expressions:

$$\sigma(q_i \bar{q}_k \to HW^\pm) = \frac{\pi \alpha^2 |\mathcal{M}_{ik}^{CKM}|^2}{36 x_W^2} \frac{2 p_W (p_W^2 + 3 m_W^2)}{\sqrt{\hat{s}} (\hat{s} - m_W^2)^2}, \tag{7.337}$$

$$\sigma(q_i \bar{q}_i \to HZ) = \frac{\pi \alpha^2 (L_i^2 + R_i^2)}{72 x_W^2 (1 - x_W)^2} \frac{2 p_Z (p_Z^2 + 3 m_Z^2)}{\sqrt{\hat{s}} (\hat{s} - m_Z^2)^2}, \tag{7.338}$$

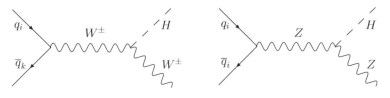

FIGURE 7.53

The Feynman diagrams associated with the subprocesses $q_i + \overline{q}_k \to H + W^{\pm}$ and $q_i + \overline{q}_i \to H + Z$.

where

$$p_V^2 = \frac{1}{4\hat{s}} \left(\hat{s}^2 + m_V^4 + m_H^4 - 2\hat{s}m_V^2 - 2\hat{s}m_H^2 - 2m_V^2 m_H^2 \right). \tag{7.339}$$

The next-to-leading-order QCD corrections are identical to those for the Drell–Yan production of a single, massive off-shell W or Z [119]. They increase the cross sections (7.337) and (7.338) by 20% approximately. For the LHC energies the $W^{\pm}H$ cross section is uniformly about twice as large as that for ZH. Unlike for the dominant $gg \to H$ mechanism, $p\overline{p}$ colliders give larger $q\overline{q} \to W^{\pm}H, ZH$ cross sections than pp colliders.

In the same spirit as the previous mechanism, a Higgs boson can be radiated off a top quark produced in gluon-gluon or quark-antiquark fusion [120]

$$g + g \to t + \overline{t} + H, \qquad q + \overline{q} \to t + \overline{t} + H. \tag{7.340}$$

Two Feynman diagrams of these subprocesses in the leading order of the perturbation theory are shown in Fig. 7.54.

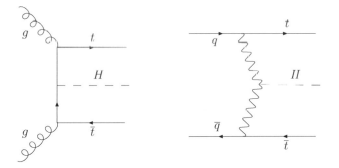

FIGURE 7.54

The Feynman diagrams corresponding to the subprocesses $g + g \to t + \overline{t} + H$ and $q + \overline{q} \to t + \overline{t} + H$.

In Figs. 7.55 we represent the cross sections for the production of the Higgs boson of the standard model for pp collisions at the LHC. Of course, none of the considered production and decay channels of the Higgs boson is free of background. A reader could find detailed discussion of the signal-to-background ratio in the book [121]. The following two channels

$$gg \to H \to \gamma\gamma, \qquad\qquad \text{for } m_H \leq 150 \text{ GeV}, \tag{7.341}$$

$$gg \to H \to Z^* Z^* \to l^+ l^- l^+ l^-, \qquad \text{for } m_H \geq 130 \text{ GeV} \tag{7.342}$$

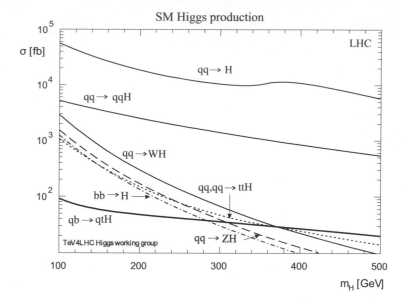

FIGURE 7.55

The SM Higgs production cross sections for pp collisions at 14 TeV.

stand out from the others in that they appear to provide the best chance for the Higgs boson discovery at the LHC. There are other useful channels, as for instance

$$q\bar{q} \to W^+(W^-)H \to l^+\nu_l(l^-\bar{\nu}_l)b\bar{b}. \tag{7.343}$$

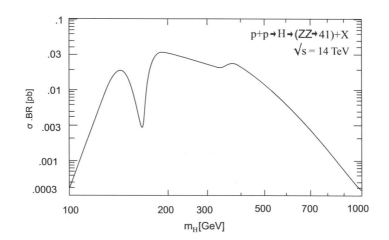

FIGURE 7.56

$\sigma \times \mathrm{BR}$ versus m_H for the process $pp \to H(\to Z^*Z^* \to l^-l^+l^-l^+) + X$.

At sufficiently high luminosity the subprocesses

$$q\bar{q} \to W^-(W^+)H \to l^-\nu_l(l^+\nu_l)b\bar{b}, \tag{7.344}$$

$$gg(q\bar{q}) \to t\bar{t}H \to W^-(W^+)H + X \to l^-\bar{\nu}_l(l^+\nu_l)\gamma\gamma + X \tag{7.345}$$

become important as well. The total Higgs boson production rate could be obtained by multiplying the branching ratio with the cross section. For example, Fig. 7.56 displays the cross section times branching ratio for four-charged-lepton production in the reaction

$$pp \to H(\to Z^*Z^* \to l^-l^+l^-l^+) + X$$

as a function of m_H, at the LHC. This graph reveals an interesting structure. At the low mass end, the rate falls very sharply below about $m_H = 150$ GeV that is connected with the behavior of the corresponding branching ratio. At the high mass end ($m_H > 400$ GeV), the rate falls off slowly reflecting the behavior of the underlying production cross section. The graph dip located at $m_H \approx 170$ GeV is associated with the threshold of the W^-W^+ pair production.

8

Lattice QCD

And you,
Could you perform
A nocturne on a drainpipe flute?
V. Mayakovsky

8.1 Lattice approach

For studying the properties of QCD at large distances it is necessary to regularize the theory in such a way that this regularization is not connected with the usual expansion into a series in terms of the Feynman diagrams, which is admissible only in the case of a small coupling constant. For this purpose Wilson [122] evolved the gauge theory on a lattice, in which the discrete space-time was introduced instead of space-time continuum. Notice that the idea of replacement of the field variable (determined in each point of space-time) with a lattice variable (assigned only in some points of space-time) is probably older than the idea of the field variable itself. In some physical problems, such as the description of electrons in crystals, a lattice spacing a is a physical quantity. The introduction of lattice is a mathematical trick. It is leading to the natural scheme of cut-off because the lengths of waves that are less than twice the spacing do not have sense, that is, the region of changing the momenta is limited by the value π/a. Consequently, the introduction of the lattice provides a cut-off removing the ultraviolet divergences, which are the scourge of the quantum field theory (QFT). Like any regularization, the lattice cut-off must be eliminated after the renormalization. Physical results can be obtained only in a continuous limit, when the lattice spacing is taken to zero. Most often the cubic lattice is used, the points of which (called *sites*) are situated at the vertices of cubes filling the space. The shortest interval between two neighboring cites is called a *link*, the length of a link—a *lattice spacing*.

The simplest example of QFT on a lattice is the scalar field theory, for which only the values on lattice sites are considered, and derivatives appearing in the motion equations are approximated by finite differences. The values of fields in lattice sites are the dynamic variables of a task. As in all practical applications the sites of finite size are considered, the QFT on a lattice changes into the theory with a finite value of degrees of freedom that are determined by the number of sites. For a satisfactory description of continuous field configuration it is necessary for the lattice spacing to be much less than the typical scale of changing the fields (in the case of smooth configurations this can be achieved by a decrease of the lattice spacing). In lattice formulating of a spinor field its values are also ascribed to the lattice sites while the values of a vector field are attributed to the links.

On a lattice the QFT becomes well-defined mathematically and can be studied by different methods. The lattice perturbation theory reproduces all the usual results of other regularization schemes although it meets some technical difficulties. In QCD the discrete space-time is particularly well suited for a strong coupling expansion. It is remarkable that in this limit the confinement (or color confinement) is automatic and the theory reduces to one of quarks on the ends of strings with a finite energy per unit length.

The lattice QCD is very similar to statistical mechanics. In fact along with the momentum cut-off the restriction on the kinetic energy appears, and in the limits of strong coupling it can be considered as a perturbation. It corresponds to a high-temperature expansion in statistical mechanics. The deep ties between these disciplines are manifest in the QFT formalism in the language of the functional integrals. Thus, for example, for QCD in the Euclidean space the average value of an operator $\mathcal{O}$ involving quark and gluon fields has the form

$$< 0|\mathcal{O}|0 > = \frac{\int [dG_\mu][d\bar{\psi}d\psi]\mathcal{O}e^{-S_{QCD}}}{\int [dG_\mu][d\bar{\psi}d\psi]e^{-S_{QCD}}}.$$

In the lattice QCD these integrals are taken using the values of the integrand only in some points of the definition region, namely, at a space-time lattice. So we can see that in the Euclidean space the functional integral is equivalent to the statistical sum for an analogous statistical system. The square of the field theoretical coupling constant corresponds directly to the temperature. Thus, the elementary particle physics receives all the range of methods of the condensed matter theory.

In order for the lattice theory to be an adequate representation of the continuum theory we must satisfy two conditions. The lattice must be large enough that the object we wish to describe fits on it. If this is not the case, the result obtained will be subject to effects from the boundaries called *finite size effects*. In addition, the lattice spacing must be small enough that the physical quantity to be calculated is insensitive to the granularity of the lattice.

The coupling constant on the lattice represents a bare coupling and depends on the lattice spacing. Since non-Abelian gauge theories possess the property of asymptotic freedom, then in their lattice version the bare coupling must be taken to zero as the lattice spacing decreases towards the continuum limit. In statistical physics it corresponds to the transition from the high-temperature regime into a low-temperature domain, which may be followed by a phase transition. In the QFT, in a limit when the lattice spacing tends to zero, the physical consequences of renormalizable field theory do not have to depend on the details of regularization; they do not have to remember the size of the lattice spacing. It means that in this limit the correlation length in the theory must be very large as compared with the lattice spacing. In the statistical physics language the infinity of the correlation length corresponds to the phase transition of the second order (continuous). If in a model there is the phase transition only of the first order, the correlation length never becomes infinite and the required continuous theory does not exist. Besides this, in the lattice gauge theory it is easy to obtain analytically the results in the limits of strong and weak coupling. Along with this it is necessary to find out whether these two limiting cases of the theory are connected by the continuous way, that is, whether phase transitions exist under intermediate values of the coupling constant. For this reason the clarifying of the phase structure of the gauge theory on a lattice is one of the most important problems.

The work by Wilson in which he showed that the lattice QCD (LQCD) has the confinement property in the limit of the strong coupling, resulted in the exploding growth of studies of this theory. At first all these investigations were analytical. The use of the Monte Carlo method for studying statistical systems was already known and in 1979 Creutz, Jacobs, and Rebbi [123] employed these techniques for the investigation of lattice gauge theories. The first computer calculations with a participation of quarks were realized two years later [124, 125]. In the middle of 1980s the LQCD was mostly forming in that form, in which it exists at the present—the mixture of analytical calculations with numerical computations whose size is being constantly increased. With the improvement of the calculating algorithms and growth of computer speed and availability of processors, LQCD has been gradually becoming a more precise and important source of calculations for hadron spectroscopy and matrix elements of QCD.

At present the analytical calculations in QCD are self-consistent only in the ultraviolet region within the perturbation theory, and the most interesting problems, for example, developing the color confinement theory, calculating the spectrum of hadron masses, and other non-perturbative problems do not have the analytical solution, obtained directly from the QCD Lagrangian. Thus, for example, at small values of g_s, for the physical quantity M, which has dimension of mass, the numerical calculations on supercomputers determine the dependence

$$M \propto \exp\left(-\frac{\text{const}}{g_s^2}\right) \tag{8.1}$$

reasoning directly from the QCD Lagrangian. But the relation (8.1) cannot be obtained analytically, nor within the perturbation theory, nor within the nonperturbative methods of QCD (sum rules, instanton model of vacuum, operator expansion, and so on). Any result of this type would mean a big step in developing QCD. Remember that the analytical determination of a glueball mass is among the problems of the millennium.

In such a way a series of the QCD problems can be solved at the present time only numerically within the lattice theory. The problems, theorists have been already investigating and hope to obtain solutions in the nearest future, are as follows: (i) calculating the mass spectrum of hadrons, coupling constants, light quark masses; (ii) predictions of low-energy behavior of different matrix elements taking into account the strong interaction; (iii) determining the masses of exotic states: glueballs, hybrids $(qqg, qq, \overline{qq}, qqqq\ldots)$, and so on; (iv) obtaining the phase diagram of quark-gluon matter in the plane *chemical potential- temperature*; and (v) extraction of the whole information of the color confinement and making the theory of this phenomenon.

There are two widespread methods of introducing the lattice into QCD. In the Euclidean lattice formulation both space and time are discrete [126]. In this case the quantization is carried out in the Euclidean space within the functional integration. This method has the advantage of keeping some tracks of the initial symmetry with respect to the Lorentz transformation. Moreover, as it will be shown later, in the case of a gauge field on a lattice, it is not necessary to introduce terms fixing the gauge. An alternative approach (Hamilton's approach), which was offered in the work by Kogut and Susskind [127], is based on introducing a lattice regularization only in three-dimensional space, while time remains a continuous quantity. Using the Hamilton gauge $G_0 = 0$, they determined the Hamiltonian as a function of space components of a gauge field and canonically conjugate momenta. This theory can be canonically quantized within the usual Hamilton formalism. Its advantage is in the possibility of calculation of some physical quantities (especially the spectrum of masses). In this chapter we will be constrained only by the Euclidean formulation of QCD.

8.2 Scalar field

By way of the simplest example we study a scalar field described by the Lagrangian

$$\mathcal{L} = \frac{1}{2}(\partial_\mu \phi)^2 + V(\phi), \tag{8.2}$$

where

$$V(\phi) = \frac{1}{2}m^2\phi^2 + \frac{\lambda}{4}\phi^4.$$

We assign a four-dimensional hypercubic lattice with the lattice spacing a to reduce the total Lorentz invariance to a hypercubic symmetry. In other words, we restrict our coordinates to the form

$$x_\mu = an_\mu,$$

where n_μ has four integer components. For the purpose of an infrared cut-off, we shall assume the individual components of n to take only a finite number N of possible values:

$$-\frac{N}{2} \leq n_\mu \leq \frac{N}{2}. \tag{8.3}$$

Outside this range we assume that the lattice is periodic for which purpose we identify n with $n + N$. Therefore, our lattice possesses N^4 sites. The scalar field ϕ exists at every lattice site n:

$$\phi(x) \rightarrow \phi_n. \tag{8.4}$$

Now we replace the derivatives of ϕ with nearest-neighbor differences:

$$\partial_\mu\phi(x) \rightarrow \frac{1}{a}(\phi_{n+\hat{\mu}} - \phi_n), \tag{8.5}$$

where $\hat{\mu}$ is a four-dimensional vector of a length a in a direction μ. Then, for the lattice action we get

$$S(\phi) = \sum_n \left[\sum_\mu \frac{a^2}{2}(\phi_{n+\hat{\mu}} - \phi_n)^2 + a^4 \left(\frac{m^2}{2}\phi_n^2 + \frac{\lambda}{4}\phi_n^4 \right) \right] =$$

$$= \sum_{\{m,n\}} \frac{a^2}{2}(\phi_m - \phi_n)^2 + \sum_n a^4 \left(\frac{m^2}{2}\phi_n^2 + \frac{\lambda}{4}\phi_n^4 \right), \tag{8.6}$$

where $\{m,n\}$ denotes the set of all nearest-neighbor pairs of lattice sites. Carrying out the Wick rotation, we go from the Minkowski space to the Euclidean one. In so doing the generating functional is becoming like the statistical sum:

$$W = \int [\prod_n d\phi_n] e^{-S}, \tag{8.7}$$

where the path integration measure is defined as an ordinary integral over each of the lattice fields. It is convenient to represent the action in the following form:

$$S(\phi) = \frac{1}{2}\phi_m M_{mn}\phi_n, \tag{8.8}$$

where M_{mn} is an $N^4 \times N^4$-dimensional square matrix and we have adopted the usual summation convention on repeated indices. Find the matrix M in the free case ($\lambda\phi^4/4 = 0$). With that end in view we introduce the Fourier transformation on the lattice. Define a Fourier image of the function ϕ_n in the following way

$$\tilde{\phi}_k = F_{kn}\phi_n = \sum_n \phi_n e^{2i\pi k_\mu n_\mu/N}. \tag{8.9}$$

Note, since we are working in the Euclidean space, there is no need to use the metric tensor. The index k also has four integer components running the values (8.3). The linear transformation (8.9) could be easily converted with the help of the identity

$$\sum_k e^{-2i\pi(k \cdot n)/N} = N^4 \prod_\nu \delta_{n_\nu,0} \equiv N^4 \delta_{n,0}^{(4)}. \tag{8.10}$$

And so we have

$$(F^{-1})_{nk} = N^{-4} e^{-2i\pi(k \cdot n)/N} = N^{-4} F_{kn}^*,$$

or

$$\phi_n = N^{-4} \sum_k \tilde{\phi}_k e^{-2i\pi(k \cdot n)/N}. \tag{8.11}$$

Employing the obtained expression, we rewrite the kinetic energy term of the action (8.6) in the view:

$$\frac{a^2 N^{-8}}{2} \sum_{\{m,n\}} \sum_k \sum_{k'} e^{-2i\pi(k+k') \cdot n/N} \tilde{\phi}_k \tilde{\phi}_{k'} \left(e^{-2i\pi k_\mu/N} - 1 \right) \left(e^{-2i\pi k'_\mu/N} - 1 \right) =$$

$$= \frac{a^2 N^{-4}}{2} \sum_k \sum_\mu \tilde{\phi}_k \tilde{\phi}_{-k} \left(e^{-2i\pi k_\mu/N} - 1 \right) \left(e^{2i\pi k_\mu/N} - 1 \right) =$$

$$= 2a^2 N^{-4} \sum_k \sum_\mu \tilde{\phi}_k \tilde{\phi}_k^* \sin^2 \left(\frac{\pi k_\mu}{N} \right), \tag{8.12}$$

where $k_\mu = k\hat{\mu}$. Consequently, the free scalar field action will look like:

$$S_0(\phi) = a^4 N^{-4} \frac{1}{2} \sum_k \tilde{M}_k |\tilde{\phi}_k|^2, \tag{8.13}$$

where

$$\tilde{M}_k = m^2 + \frac{4}{a^2} \sum_\mu \sin^2 \left(\frac{\pi k_\mu}{N} \right). \tag{8.14}$$

The Fourier transformation has diagonalized the matrix M:

$$M_{mn} = a^4 N^{-4} \sum_k F_{mk}^* F_{nk} \tilde{M}_k. \tag{8.15}$$

Now we can obtain the generating functional of the free scalar field in the exact form

$$W = \left[\frac{1}{2\pi} \det M \right]^{-1/2} = \prod_k \left[\frac{a^4}{2\pi} \tilde{M}_k \right]^{-1/2}. \tag{8.16}$$

To find Green functions we introduce external sources J_n associated with the field ϕ and situated in the lattice sites as well. Then the free scalar field action takes the form:

$$S(J) = \frac{1}{2} \phi_m M_{mn} \phi_n - J_n \phi_n. \tag{8.17}$$

Next, using the generating functional

$$W[J] = \int \prod_n [d\phi_n] e^{-S(J)}, \tag{8.18}$$

we can define the Green function in the usual fashion

$$G^{(k)}_{n_1,\ldots n_k} = \left[\frac{1}{W[J]} \frac{\delta^k W[J]}{\delta J_{n_1} \ldots \delta J_{n_k}} \right] \Bigg|_{J=0}. \tag{8.19}$$

Completing the square in Eq. (8.17) and shifting the integration in Eq. (8.18), we get the exact expression for the free field generating functional:

$$W[J] = W[0] \exp \left[\frac{1}{2} J_m (M^{-1})_{mn} J_n \right], \tag{8.20}$$

where $W[0]$ is determined by the expression (8.16). From this it follows that the propagator or two-point function is simply the inverse of the matrix M:

$$G^{(2)}_{mn} \equiv\, <0|T(\phi_m \phi_n)|0> = (M^{-1})_{mn}. \tag{8.21}$$

To find the matrix M^{-1} in momentum space presents no difficulties

$$G^{(2)}_{mn} = a^{-4} N^{-4} \sum_k \tilde{M}^{-1}_k e^{2i\pi k \cdot (m-n)/N}. \tag{8.22}$$

To put this expression into a more familiar form, we first take N to infinity and change the momentum sum into an integral with the replacements

$$q_\mu = \frac{2\pi k_\mu}{Na}, \qquad a^{-4} N^{-4} \sum_k \to \int \frac{d^4 q}{(2\pi)^4}. \tag{8.23}$$

The integral over each component of q is taken in the limits

$$-\frac{\pi}{a} \le q_\mu \le \frac{\pi}{a}. \tag{8.24}$$

This explicitly shows the momentum space effect of the lattice cut-off. Now the propagator will look like:

$$G^{(2)}_{mn} = \int_{-\pi/a}^{\pi/a} \frac{d^4 q}{(2\pi)^4} \frac{e^{-iq \cdot x}}{m^2 + 4a^{-2} \sum_\mu \sin^2(aq_\mu/2)}, \tag{8.25}$$

where $x_\mu = a(n_\mu - m_\mu)$. For the continuum limit $a \to 0$ we expand the sine

$$4a^{-2} \sum_\mu \sin^2 \left(\frac{aq_\mu}{2} \right) = q^2 + O(a^2)$$

and get the expression

$$G^{(2)}_{mn} = \int \frac{d^4 q}{(2\pi)^4} \frac{e^{-iq \cdot x}}{m^2 + q^2} + O(a^2) \tag{8.26}$$

that represent the familiar Feynman propagator of the scalar field in the Euclidean space. Having rewritten the expression for $\tilde{M}_k$ as

$$\tilde{M}_k = m^2 + \frac{4}{a^2} \sum_\mu \sin^2 \left(\frac{aq_\mu}{2} \right), \tag{8.27}$$

we obtain the dispersion relation that differs from the standard one, $q^2 + m^2$, but is reduced to it in the limit $q \to 0$ (see Fig. 8.1). If one changes the scales of fields and sources:

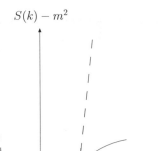

$$S(k) - m^2$$

$$-\pi/\alpha \qquad 0 \qquad \pi/\alpha$$

FIGURE 8.1

The dispersion relation for a free scalar field.

$$\phi'_n = \sqrt{\lambda}\phi_n, \qquad J'_n = \sqrt{\lambda}J_n, \qquad (8.28)$$

the the lattice action takes the view:

$$S(\phi) = \frac{1}{\lambda}S'(\phi'), \qquad (8.29)$$

where

$$S'(\phi') = \sum_{\{m,n\}} \frac{a^2}{2}\left(\phi'_m - \phi'_n\right)^2 + \sum_n \left[a^4\left(\frac{m^2}{2}\phi'^2_n + \frac{1}{4}\phi'^4_n\right) - J'_n\phi'_n\right], \qquad (8.30)$$

that is, the coupling constant λ became the common factor for the whole action. At that, the generating functional is

$$W'[J'] = \int \prod_n [d\phi'_n] \exp\left\{-\frac{1}{\lambda}S'(\phi')\right\}. \qquad (8.31)$$

If one identifies

$$\frac{1}{\lambda} \to \beta \equiv \frac{1}{kT}, \qquad (8.32)$$

where k is the Boltzmann constant, then the expression (8.31) will have the same structure as the partition function in statistical physics. Therefore, the strong coupling expansion, that is, the expansion in powers of λ^{-1}, corresponds to the high-temperature expansion in the statistical physics. In this case, the correlation functions of the statistical variables correspond directly with the Green functions of the corresponding quantum fields. In general, a d-space-time dimensional quantum field theory is equivalent to a d-Euclidean dimensional classical statistical system.

So, in a lattice theory defined in the Euclidean space, the generating functional is reduced to a finite-dimensional integral. The transition from continual integration to integration over Φ_n allows us to calculate quantum averages numerically. The continuous limit is associated with $N \to \infty$ and $a \to 0$, real computations are fulfilled under finite N and a while systematic errors are evaluated by the standard way, namely, by varying the number of lattice sites N^4 and the lattice space a.

An accuracy with which we can derive the physical consequences from the lattice theory certainly depends on the smallness of the lattice spacing. However, the lattice must cover a large enough area of space-time, which is enough for solving a concrete physical problem. It is impossible to study a proton having the size $R \approx 1$ fm if we do not have some lattice sites within its radius, with the general size of the lattice of some fm. Imposing periodical conditions on this basic cell helps but cannot greatly decrease these demands.

8.3   Fermion fields

Consider fermion fields on a lattice. Again, we introduce a four-dimensional hypercubic lattice of N^4 sites. With each site we associate an independent four-component spinor variable ψ_m. In order to keep the lattice action simple, we define the derivative symmetrically:

$$\partial_\mu \psi \rightarrow \frac{1}{2a} \left(\psi_{n+\hat{\mu}} - \psi_{n-\hat{\mu}} \right). \tag{8.33}$$

To sum the Lagrangian over all sites results in the lattice action:

$$S(\psi) = \sum_{\{m,n\}} \overline{\psi}_m M_{mn} \psi_n, \tag{8.34}$$

where

$$M_{mn} = \frac{1}{2} a^3 \sum_\mu \tilde{\gamma}_\mu (\delta^{(4)}_{m+\hat{\mu},n} - \delta^{(4)}_{m-\hat{\mu},n}) + a^4 m \delta^{(4)}_{mn} \tag{8.35}$$

and matrices $\tilde{\gamma}$ are Euclidean, that is,

$$\{\tilde{\gamma}_\mu, \tilde{\gamma}_\nu\} = 2\delta_{\mu\nu}. \tag{8.36}$$

They are connected with the γ^μ-matrices in the Minkowski space by the relations:

$$\tilde{\gamma}^4 = \gamma^0, \qquad \tilde{\gamma}^k = i\gamma^k.$$

As in the case of the scalar field, the matrix M is diagonalized by means of the Fourier transformation

$$\tilde{M}_k = m + ia^{-1} \sum_\mu \tilde{\gamma}_\mu \sin\left(\frac{2\pi k_\mu}{N}\right), \qquad (M^{-1})_{mn} = (aN)^{-4} \sum_k \tilde{M}_k^{-1} e^{2i\pi k \cdot (m-n)/N}. \tag{8.37}$$

Carrying out the replacement $q_\mu = 2\pi k_\mu/(Na)$, turning to a large lattice and replacing the sums over k by integrals, we arrive at the following expression for the fermion propagator:

$$G^{(2)}_{mn} = \int_{-\pi/a}^{\pi/a} \frac{d^4 q}{(2\pi)^4} \frac{e^{-iq \cdot x}}{m + ia^{-1} \sum_\mu \tilde{\gamma}_\mu \sin(aq_\mu)}. \tag{8.38}$$

If we now consider a small lattice spacing and expand

$$\tilde{M}_k = m + ia^{-1} \sum_\mu \tilde{\gamma}_\mu \sin(aq_\mu) \tag{8.39}$$

in powers of a, then we get

$$\tilde{M}_k = m + i\tilde{\gamma}_\mu q_\mu + O(a^2). \tag{8.40}$$

It seems that there are no grounds for trouble and we have really recovered the usual continuum fermion propagator. However it is necessary to analyze more thoroughly the momentum integral (8.38) at the upper limits. So, when q_μ equals π/a, the periodic sine function in (8.38) becomes equal to zero. Here the $O(a^2)$ terms cannot be ignored. Really, in the propagator there is no suppression of momentum values near π/a. Because of this we have to expect fast variations in the fields from site to neighboring site.

Let us show that the formulated theory, the naive lattice fermion formulation, leads to multiplicity of fermion states. To do it we make use of the dispersion relation that may be obtained using (8.39)

$$S(q) = m^2 + \frac{\sin^2(aq_\mu)}{a^2}. \tag{8.41}$$

In Fig. 8.2 we display the graph of the function $S(q)$ in one dimensional space. From

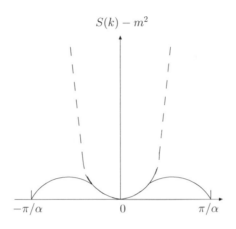

FIGURE 8.2
The dispersion relation for a free fermion field.

Fig. 8.2 we see that there are three minima. One of them is located near the point $q = 0$ and leads to the right continuous limit. Another mode corresponding to the minima in points $q = \pm\pi/a$ possesses infinite momentum at $a \to 0$ and, at finite a, still can be excited. In such a way, in the theory, instead of one mode, 16 modes corresponding to all corners of the Brillouin zone $(-\pi/a \leq q_\mu \leq \pi/a)$ exist. It means that the naive lattice fermion formulation describes 16 independent fermions. This phenomenon is called *fermion doubling* on a lattice (although in reality we are dealing with the 16-fold multiplicity of the fermion spectrum). The main reason for this multiplicity is associated with the fact that the Dirac equation represents the first-order equation. The fermion degeneration must be suppressed with the help of the corresponding modification of the theory on a lattice. If such a modification does not influence the continuous limit, then it is admissible. At first two ways of solving this problem were developed: *Wilson's* fermions [126, 128] and *staggered* fermions of Kogut-Susskind [129]. Later a five-dimensional formulation of *fermions with domain walls* was suggested [130].

A pragmatic approach to the degeneration problem is to add to the naive action new terms that suppress the extra states while vanishing in a continuum limit with the desired fermion species. To keep the local character of the action, we require the new terms to involve nearest-neighbor pairs of lattice sites. The term that is similar to the second derivative

$$S^{add} = \frac{r}{2a} \sum_{n,\mu} \bar{\psi}_n (\psi_{n+\hat{\mu}} - 2\psi_n + \psi_{n-\hat{\mu}}),$$

where a parameter r must lie between zero and one, satisfies such a demand. An addition that accomplishes our needs replaces $\tilde{M}_q$ with

$$\tilde{M}_q = m + ia^{-1} \sum_{\mu} \tilde{\gamma}_\mu \sin(aq_\mu) + ra^{-1} \sum_{\mu} [1 - \cos(aq_\mu)]. \qquad (8.42)$$

Notice, in the case of small momentum the new term has the order of the lattice spacing and could be neglected. However, when a component of q is close to π/a, the addition increases the mass of the unwanted state by $2r/a$. In the continuum limit all the extra states acquire infinite mass and only one species of mass m survives. As a result, we get the required discretizing scheme. Setting r to unity (the case of Wilson's fermion) and going into the coordinate space, we get

$$M_{mn} = \frac{1}{2} a^3 \sum_{\mu} [(1 + \tilde{\gamma}_\mu) \delta^{(4)}_{m+\hat{\mu},n} + (1 - \tilde{\gamma}_\mu) \delta^{(4)}_{m-\hat{\mu},n}] + (a^4 m + 4a^3) \delta^{(4)}_{mn}. \qquad (8.43)$$

Whenever a quark moves from one site to the next, its wave function picks up a factor of $(1 \pm \tilde{\gamma}_\mu)$, rather than the $\tilde{\gamma}_\mu$ from Eq. (8.35). Because $(1 \pm \tilde{\gamma}_\mu)/2$ represent the projector operators

$$\left[\frac{1}{2} (1 \pm \tilde{\gamma}_\mu) \right]^2 = \frac{1}{2} (1 \pm \tilde{\gamma}_\mu), \qquad \frac{1}{2} (1 + \tilde{\gamma}_\mu) + \frac{1}{2} (1 - \tilde{\gamma}_\mu) = 1,$$

then the part of the spinor field no longer propagates. This, in its turn, reduces the degeneracy by a factor of two for each dimension, exactly as needed to remove all the additional states.

However, for n_f flavors, to add S^{add} to the initial action violates the chiral symmetry $SU(n_f)_L \times SU(n_f)_R$, that is, the chiral symmetry proves to be broken even at the absence of mass terms. The value of the symmetry violation is proportional to the lattice spacing, and only in the continuous limit the explicit violation of the chiral symmetry becomes small. With any finite lattice spacing the right representation of the chiral symmetry is provided by the fine-tuning of parameters. In spite of calculation problems, concerning the realization of the chiral symmetry, such a discrete formulation has evident advantages. Thus, for example, Wilson's fermions are the most close to the continuous formulation—in every site of a lattice for every color and flavor of quarks a four-component spinor exists. By this reason, one may apply the general rules for forming currents and states.

The method of *staggered* fermions is often used in studying properties of a quark-gluon plasma. In this method the one-component fermion field, rather than the four-component Dirac's spinor, is employed. The application of the term *staggered* to fermions is connected with the fact that the Dirac's spinors and the quark flavors are built from the combination of corresponding one-component fields in different lattice cites. *Staggered* fermions also violate the chiral symmetry, however, the $U(1) \times U(1)$ symmetry that holds the majority of the physics of the complete chiral symmetry is conserved. Moreover, the manifest chiral symmetry is present under $m_q \to 0$ even for the finite lattice spacing at the condition of the degeneracy of quark masses. On the other hand, the flavor and translation symmetries are mixed with each other resulting in problems because in the real world the flavor symmetry is broken.

As the precise chiral symmetry and the broken flavor symmetry lead to the important physical phenomena influencing the physics of the high-temperature QCD, then the appearance of a third approach, fermions with domain walls, was completely well-grounded.

This formulation assumes the conservation of the precise chiral symmetry even for the finite lattice spacing. Theoretical details of this approach could be found in Ref. [131] while the first computation results for the high-temperature QCD (hot QCD) are contained in Ref. [132].

Since the original Wilson's action is not unique, then the searches of the most precise *improved* action for every important application of LQCD are also continued to date. The improvements can be differently revealed in various regions of the LQCD application because they are connected with the problem of obtaining the physical information from a large quantity of computations.

Thus, it seems that we provided ourselves with all the necessary information for writing the QCD Lagrangian on the lattice. However it is not so, because even in the lattice approach the theory must remain gauge-invariant. To form the theory in such a way we have to comprehend some cobwebs of geometric ground of the conceptions of gauge fields.

8.4   Geometric interpretation of gauge invariance

Let us formulate basic geometric conceptions in the case of curved space. Begin with the parallel transfer operation. To compare vectors (in the general case arbitrary rank tensors) defined in two different space points $A_\mu^a(x)$ and $A_\mu^a(x')$, it is necessary at first to shift (to parallel transfer) $A_\mu^a(x)$ from point x to point x'; to say, before we are going to find the difference between two vectors, we must place them in the same coordinate system. Consequently, the comparison of two vectors is realized in two stages:

$$DA_\mu^a = \delta A_\mu^a + dA_\mu^a, \tag{8.44}$$

where δA_μ^a is the apparent change cased by transferring these two vectors in the same coordinate origin and dA_μ^a is their difference measured in the same coordinate system. The parallel transfer is defined as an operation during which an angle between the transferred vector and the tangent to the trajectory remains fixed. Of course, the parallel transfer is a trivial operation in the flat (Euclidean) space because, in this case, the vector is not changed, $\delta A_\mu^a = 0$, and the covariant differentiation is simply the ordinary one:

$$DA_\mu^a = dA_\mu^a = (\partial_\lambda A_\mu^a)dx^\lambda. \tag{8.45}$$

However, in the curved space a vector is manifestly altered at the parallel transfer as the change of the coordinate axes takes place under the transition from point to point. For points x and x' separated by an infinitesimally small interval dx^λ, one assumes the quantity $\delta A^{a\mu}$ to be linearly depend on dx^λ and A_μ^a:

$$\delta A^{a\mu} = -\Gamma_{\nu\lambda}^\mu A^{a\nu}dx^\lambda, \tag{8.46}$$

while

$$\delta A_\mu^a = \Gamma_{\mu\lambda}^\nu A_\nu^a dx^\lambda, \tag{8.47}$$

where $\Gamma_{\nu\lambda}^\mu$ are *Christoffel symbols* (or *affine connectivity*) and we have taken into account that $\delta(A_\mu^a A^{a\mu}) = 0$. Comparing the vector given in a point x' with the one to be parallel transferred from point x to point x' gives

$$DA^{a\mu} = A^{a\mu}(x') - [A^{a\mu}(x) + \delta A^{a\mu}] = (\partial_\lambda A^{a\mu} + \Gamma_{\nu\lambda}^\mu A^{a\lambda})dx^\lambda, \tag{8.48}$$

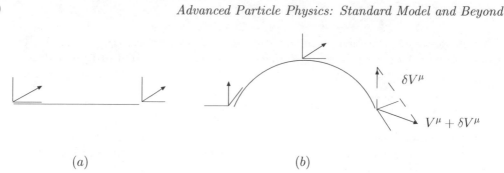

(a) (b)

FIGURE 8.3
The vector changes caused by the parallel transfer in the flat (a) and curved (b) spaces.

where the combination in the square bracket represents the covariant derivative. This discrepancy between the flat and curved spaces is shown in Fig. 8.3. Another important object of non-Euclidean geometry is a curvature tensor. It could be best introduced by analyzing the parallel transfer of a vector along a closed contour (see Fig. 8.4). Consider

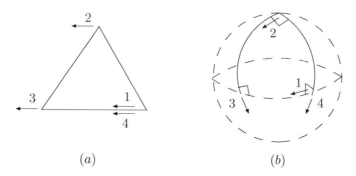

(a) (b)

FIGURE 8.4
The vector changes caused by the parallel transfer: (a)—along the closed contour 1-2-3-4 on the flat and (b)—in the curved space.

changing a vector transferred around a small parallelogram $P_1P_2P_3P_4$ consisting of two vectors b^μ, c^μ and of their parallel shifts (see Fig. 8.5). Let δA_μ^a and $\delta A_\mu^{a\prime}$ be the changes

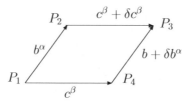

FIGURE 8.5
Changing a vector transferred around a small parallelogram $P_1P_2P_3P_4$.

of the vector at its transference along the paths $P_1 P_2 P_3$ and $P_1 P_4 P_3$, respectively. Then, the total change of the vector at its transference along the whole closed path is

$$\Delta A_\mu^a = \delta A_\mu^a - \delta A_\mu^{a\prime}, \qquad (8.49)$$

where

$$\delta A_\mu^a = (\Gamma_{\mu\alpha}^\nu A_\nu^a)_{P_1} b^\alpha + (\Gamma_{\mu\beta}^\nu A_\nu^a)_{P_2} (c^\beta + \delta c^\beta), \qquad (8.50)$$

$$\delta A_\mu^{a\prime} = (\Gamma_{\mu\beta}^\nu A_\nu^a)_{P_1} c^\beta + (\Gamma_{\mu\alpha}^\nu A_\nu^a)_{P_4} (b^\alpha + \delta b^\alpha). \qquad (8.51)$$

Expanding the quantities taken at the points P_2 and P_4, we can do so that all tensors will be referred to the same point P_1:

$$(\Gamma_{\mu\beta}^\nu A_\nu^a)_{P_2} = (\Gamma_{\mu\beta}^\nu + \partial_\alpha \Gamma_{\mu\beta}^\nu b^\alpha)(A_\nu^a + \Gamma_{\nu\alpha}^\sigma A_\sigma^a b^\alpha), \qquad (8.52)$$

$$(\Gamma_{\mu\alpha}^\nu A_\nu^a)_{P_4} = (\Gamma_{\mu\alpha}^\nu + \partial_\beta \Gamma_{\mu\alpha}^\nu c^\beta)(A_\nu^a + \Gamma_{\nu\beta}^\sigma A_\sigma^a c^\beta). \qquad (8.53)$$

Substituting these expressions into (8.49) and using the relation

$$b^\alpha + \delta b^\alpha = b^\alpha - \Gamma_{\tau\kappa}^\alpha b^\tau c^\kappa$$

as well as the analogous relation for $(c^\beta + \delta c^\beta)$, we get (8.49) in the form:

$$\Delta A_\mu^a = R_{\mu\alpha\beta}^\nu A_\nu^a \sigma^{\alpha\beta}, \qquad (8.54)$$

where $R_{\mu\alpha\beta}^\nu$ is the curvature tensor

$$R_{\mu\alpha\beta}^\nu = \partial_\alpha \Gamma_{\mu\beta}^\nu - \partial_\beta \Gamma_{\mu\alpha}^\nu + \Gamma_{\mu\beta}^\lambda \Gamma_{\lambda\alpha}^\nu - \Gamma_{\mu\alpha}^\lambda \Gamma_{\lambda\beta}^\nu \qquad (8.55)$$

and $\sigma^{\alpha\beta} = b^\alpha c^\beta$ is the area of the contour displayed in Fig. 8.5. In the quite simple case of the spherical triangle every angle of which is equal to 90^0 (Fig. 8.4), the change obviously equals $\pi/2$. Because in the given case the curvature tensor is reduced to the curvature $R = 1/r^2$, where r is the sphere radius, then this result is in line with the relation (8.54). Multiplying R by the triangle area $\pi r^2/2$, we obtain the change $\pi/2$.

To directly compare (8.48) with the gauge covariant derivative

$$D_\mu \psi = (\partial_\mu - igT^a A_\mu^a)\psi, \qquad (8.56)$$

shows that, from the geometric point of view, the quantity $T^a A_\mu^a \equiv A_\mu$ may be interpreted as the connectivity in an internal charge space. By analogy with (8.47) we can find the change of the field $\psi(x)$ under the parallel transfer

$$\psi(x) \to \psi(x + dx) = \psi(x) + \delta\psi(x), \qquad (8.57)$$

where

$$\delta\psi(x) = igA_\mu \psi(x)dx^\mu. \qquad (8.58)$$

At the parallel transfer to a finite distance from point x to point x', Eq. (8.58) should be integrated and we arrive at the following expression for the transfer operator

$$P(x, x') = R_s \left\{ \exp\left[ig \int_x^{x'} T^a A_\mu^a(y)dy^\mu \right] \right\}, \qquad (8.59)$$

where the integral is taken along the trajectory s connecting the points x and x' while R_s represents a path-ordering instruction for noncommuting matrices A_μ. In a power series expansion of the exponential, the matrices are to be ordered as encountered along the path, the largest values of the parameter s being to the left. Therefore, every trajectory may be associated with a group element, that is, the matrix $P(x, x')$ belongs to the gauge group.

The statement proved above gives one more definition of a gauge theory that is based on the conception of the phase [133, 134]. In accordance with this definition, the interaction between particle and gauge field is described by a phase factor associated with any possible world line the particle could intersect. Thus, for example, when a charged particle intersects a space-time contour C, its wave function acquires a phase factor caused by the electromagnetic interaction:

$$\psi \to \exp\left[ie \int_C A_\mu dx^\mu\right]\psi = P_{em}(C)\psi. \tag{8.60}$$

If a particle is at rest, this factor has an especially simple view:

$$P_{em}(C) = \exp[ieA_0 t], \tag{8.61}$$

where t is a contour length in the time direction. From (8.61) it follows that a charged particle is subjected to additional time oscillations whose frequency is proportional to its charge and the scalar potential. As a result, the particle energy is increased on the value eA_0. In the case of non-Abelian theory, the wave function phase caused by its interaction with the gauge field is given by the transfer operator (8.59).

It is useful to demonstrate that confronting a trajectory in an internal space with a group element is consistent with transformation properties of gauge fields. Consider an infinitesimally small parallel transfer under which a field function is transformed by the law (8.57). In the points x and x' we perform the gauge transformation—the axes rotation at every point

$$\psi(x) \to \psi'(x) = U(T^a\theta^a(x))\psi(x) \equiv U(\theta_x)\psi(x), \tag{8.62}$$

$$\psi(x') \to \psi'(x') = U(\theta_{x'})\psi(x'). \tag{8.63}$$

For the product $\overline{\psi}(x+dx)P(x+dx, x)\psi(x)$ to be invariant, the connectivity of the parallel transfer has to be transformed in accordance with the expression

$$1 + igT^a A_\mu^{a'} dx^\mu = U(\theta_{x+dx})[1 + igT^a A_\mu^a dx^\mu]U^{-1}(\theta_x) =$$

$$= [U(\theta_x) + \partial_\nu U(\theta_x)dx^\nu][1 + igT^a A_\mu^a dx^\mu]U^{-1}(\theta_x). \tag{8.64}$$

From (8.64) we obtain

$$T^a A_\mu^{a'} = U(\theta_x)T^a A_\mu^a U^{-1}(\theta_x) - \frac{i}{g}[\partial_\mu U(\theta_x)]U^{-1}(\theta_x), \tag{8.65}$$

that represents the well-known transformation law for the gauge fields. Obviously, this result will take place in the case when the points are separated by a finite interval. Therefore, under the gauge transformations, the parallel transfer operator (8.59) is changed by the following way:

$$P(x', x) \to P'(x', x) = U(\theta_{x'})P(x', x)U^{-1}(\theta_x). \tag{8.66}$$

The direct comparison of the quantity $T^a F_{\mu\nu}^a$, where $F_{\mu\nu}^a$ is the non-Abelian gauge field tensor, with the curvature tensor $R_{\mu\alpha\beta}^\nu$ (8.55) shows that $T^a F_{\mu\nu}^a$ may be interpreted as the curvature in the internal charge space. We can easily check it by considering the parallel transfer along a closed path L. For the sake of simplicity, for L we choose the parallelogram with the vertex in the point x_μ and two generatrices dx_μ and δx_μ:

$$P_\Box = P(x, x+dx)P(x+dx, x+dx+\delta x)P(x+dx+\delta x, x+\delta x)P(x+\delta x, x). \tag{8.67}$$

Using the matrix identity

$$\exp[\lambda A]\exp[\lambda B] = \exp\left\{\lambda(A+B) + \frac{\lambda^2}{2}[A,B]\right\} + O(\lambda^3), \qquad (8.68)$$

we get

$$P(x, x+dx)P(x+dx, x+dx+\delta x) = \exp[igA_\mu(x)dx^\mu]\exp[igA_\nu(x+dx)\delta x^\nu] =$$

$$= \exp\left\{ig(A_\mu(x)dx^\mu + A_\nu(x)\delta x^\nu + \partial_\mu A_\nu(x)dx^\mu\delta x^\nu) - \frac{g^2}{2}[A_\mu(x), A_\nu(x)]dx^\mu\delta x^\nu\right\} \quad (8.69)$$

and

$$P(x+dx+\delta x, x+\delta x)P(x+\delta x, x) = \exp[-igA_\mu(x+\delta x)dx^\mu]\exp[-igA_\nu(x)\delta x^\nu] =$$

$$= \exp\left\{-ig(A_\mu(x)dx^\mu + A_\nu(x)\delta x^\nu + \partial_\nu A_\mu(x)dx^\mu\delta x^\nu) - \frac{g^2}{2}[A_\mu(x), A_\nu(x)]dx^\mu\delta x^\nu\right\}. \tag{8.70}$$

With allowances made for (8.69) and (8.70), the expression (8.67) takes the form

$$P_\square = \exp\left\{ig(\partial_\mu A_\nu(x) - \partial_\nu A_\mu(x) - ig[A_\mu(x), A_\nu(x)])dx^\mu\delta x^\nu\right\}. \qquad (8.71)$$

From the obtained expression it follows that the quantity $T^a F^a_{\mu\nu}$ may be really identified with the curvature tensor in an internal space.

Thus, in any (physical or abstract) space, where coordinates depend on point choice, the comparison of any two tensors of the first and higher rank taken in different points has no physical meaning. The standard method of solving this problem consists in the introduction of the parallel transfer concept or the affine connectivity. In the case of curved physical space-time, the Christoffel symbols are entered, while in the case of abstract space, the gauge fields are introduced. They compensate the change of the local coordinate system at every space-time point.

8.5 QCD action

So, considerable freedom exist in LQCD. We may add to the Lagrangian various terms that will not contribute in the continuum limit. Make use of this freedom of action. Let us consider the gauge transformation of the field functions

$$\psi(x) \to \Phi(\theta_x)\psi(x), \qquad \overline{\psi}(x) \to \overline{\psi}(x)\Phi^\dagger(\theta_x), \qquad (8.72)$$

where

$$\Phi(\theta_x) = \exp\left[iT^a\theta^a(x)\right]. \qquad (8.73)$$

As this takes place, the matrix of the parallel transfer from point x to point x' along the contour C

$$P(x', x) = \exp\left\{ig\int_C T^a A^a_\mu(y)dy^\mu\right\} \qquad (8.74)$$

(hereafter we omit the path-ordering operator) is transformed as

$$P(x', x) \to \Phi(\theta_{x'})P(x', x)\Phi^\dagger(\theta_x). \qquad (8.75)$$

In the lattice version, the transformations (8.72) and (8.75) become the form

$$\psi_n \to \Phi_n \psi_n, \qquad \overline{\psi}_n \to \overline{\psi}_n \Phi_n^\dagger, \tag{8.76}$$

$$U(n + \hat{\mu}, n) \to \Phi_{n+\hat{\mu}} U(n + \hat{\mu}, n) \Phi_n^\dagger, \tag{8.77}$$

where we have renamed the parallel transfer operator by the symbol U. In the case of gauge symmetry with the $SU(3)$-group

$$\Phi_n = \exp\left[i\frac{\lambda^\alpha}{2}\theta_n^\alpha\right], \tag{8.78}$$

$(\alpha = 1, 2 \ldots 8)$, the matrix of the parallel transfer between a nearest-neighbor pair of lattice sites $(n + \hat{\mu}, n)$ is

$$U(n + \hat{\mu}, n) \equiv U_{n,\hat{\mu}} = \exp\left[iag_s\frac{\lambda^\alpha}{2}G_{n,\mu}^\alpha\right], \tag{8.79}$$

where $G_{n,\mu}^\alpha$ is the gluon field potential in a site n. It is common practice to call this matrix by a *link variable*. From the relations (8.76) and (8.77) it follows that the combination

$$\overline{\psi}_n U(n, n + \hat{\mu})\psi_{n+\hat{\mu}}$$

is gauge-invariant. Then, the way of building the quark part of the QCD action in the $SU(3)$ gauge-invariant form becomes obvious. The expression obtained by us earlier

$$S_q' = \overline{\psi}_m M_{mn} \psi_n, \tag{8.80}$$

where M_{mn} is given by Eq. (8.43), must be replaced by

$$S_q = \sum_n \left\{ \frac{a^3}{2} \sum_\mu \overline{\psi}_n \left[(1 + \tilde{\gamma}_\mu)U(n, n + \hat{\mu})\psi_{n+\hat{\mu}} + (1 - \tilde{\gamma}_\mu)U(n, n - \hat{\mu})\psi_{n-\hat{\mu}}\right] + \right.$$

$$\left. + \overline{\psi}_n(a^4 m + 4a^3)\psi_n \right\}. \tag{8.81}$$

Usually, in Wilson's formulation of QCD, it is generally said not about a quark mass but about a so-called *hopping parameter*

$$K = \frac{1}{2(ma + 4)}. \tag{8.82}$$

Introducing the change of fermion fields scale

$$\psi \to \sqrt{2K}\psi,$$

we can rewrite the action (8.81) in terms of new variables

$$S_q = a^3 \sum_n \left\{ \overline{\psi}_n \psi_n + K \sum_\mu \overline{\psi}_n \left[(1 + \tilde{\gamma}_\mu)U(n, n + \hat{\mu})\psi_{n+\hat{\mu}} + (1 - \tilde{\gamma}_\mu)U(n, n - \hat{\mu})\psi_{n-\hat{\mu}}\right] \right\}, \tag{8.83}$$

We have suppressed the internal indices of the quark fields. However, we note that the matrices U describing the transition of a quark between nearest-neighbor sites act on these indices as matrices.

Now we proceed to the definition of the lattice action for the gluon field. We shall follow Wilson's formulation proposed in Ref. [122]. The idea of his approach is heavily motivated by the concept of a gauge field as a path-dependent phase factor. The basic degrees of freedom are group elements associated with lattice links (bonds or straight-line paths connecting nearest neighbor pairs of lattice sites). The group element to correspond to an arbitrary path connecting a sequence of neighboring sites is the group product of the link variables. The main advantage of this formulation consists in the fact that the theory is local gauge-invariant. Thus, the vector potential is introduced with the help of the formula:

$$U_{ij} = \exp\left[ig_s a G_\mu\right], \tag{8.84}$$

where we have passed on to more economical designations

$$U(n + \hat{\mu}, n) \equiv U_{ij},$$

(i, j) is nearest-neighbor pairs of lattice sites and the Lorentz index μ corresponds to the direction of the link (i, j). The spatial coordinate x_μ, that is, the argument of G_μ, should be in the vicinity of the link in question. For the sake of convenience, we take it to lie half way along the link

$$x_\mu = \frac{a}{2}\left(i_\mu + j_\mu\right). \tag{8.85}$$

In the continuum limit, this choice is not contrary to the fact that U_{ij} should be path-ordered along the link according to Eq. (8.73). Next, we should introduce the analog of the tensor $G_{\mu\nu}$ that represents a generalized rotor of the vector potential. This suggests using integrals of G_μ around small closed contours. Then, it becomes obvious that the gluon field action should be a sum over all elementary squares of the lattice:

$$S_G = \sum_\square S_\square. \tag{8.86}$$

The action on each of these squares or elementary edges, *plaquettes* (Fig. 8.6), is the trace of the product of the group elements surrounding the plaquette

$$S_\square = \beta\left[1 - \frac{1}{3}\text{Re Sp}\left(U_{ij}U_{jk}U_{kl}U_{li}\right)\right] = \beta\left[1 - \frac{1}{3}\text{Re Sp } U_\square\right], \tag{8.87}$$

where i, j, k, and l circulate about the square in question, the factor $1/3$ establishes normalization for the matrices of the $SU(3)$-group and the factor β will be later defined under transition to the continuum limit. It should be recorded that the expressions for the gluon

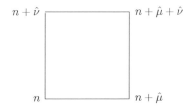

FIGURE 8.6
The lattice plaquette.

field action cited in the literature may sometimes be distinguished from each other by a constant. It is caused by the fact that adding a constant to the action does not result in any physical consequences. In the case under consideration, the constant in the expression (8.87) is chosen in such a way that the action turns into zero for unit group elements.

It is easy to verify that the action (8.86) leads to the ordinary Yang–Mills theory in the continuum limit. To this end, from the beginning we should use the expression (8.84) in order to pass on from U_{ij} to the vector potential. Consider, for example, the plaquette with the center in the point x_μ which is oriented in the plane $(\mu, \nu) = (1, 2)$. Making use of the expression (8.87), we get

$$S_\square = \beta - \frac{\beta}{3} \mathrm{Re}\, \mathrm{Sp}\left\{\exp[ig_s G_1(x_\mu - a\delta_{\mu 2}/2)] \exp[ig_s G_2(x_\mu + a\delta_{\mu 1}/2)] \times \right.$$

$$\left. \exp[-ig_s G_1(x_\mu + a\delta_{\mu 2}/2)] \exp[-ig_s G_2(x_\mu - a\delta_{\mu 1}/2)]\right\}. \tag{8.88}$$

Let us assume the gluon field potential will be a smooth function. Then, its expansion into the Taylor series in the vicinity of the point x will look like:

$$S_\square = \beta - \frac{\beta}{3} \mathrm{Re}\, \mathrm{Sp}\left\{\exp[ig_s a^2 G_{12}(x) + O(a^4)]\right\}. \tag{8.89}$$

Expanding the exponential in (8.89), we arrive at

$$S_\square = \frac{\beta g_s^2}{6} a^4 \mathrm{Sp}\,[G_{12}^2(x)] + O(a^6), \tag{8.90}$$

where we have taken into account the λ-matrix properties. Now, replacing the sum over the plaquettes by the space-time integral, we obtain

$$S_G = \frac{\beta g_s^2}{6} \int d^4x \frac{1}{2} \mathrm{Sp}\,[G_{\mu\nu}(x) G_{\mu\nu}(x)] + O(a^6). \tag{8.91}$$

The factor $1/2$ in the integrand is connected with the symmetry with respect to the permutation of the indices μ and ν. Therefore, we shall get the ordinary gluon field action providing

$$\beta = \frac{6}{g_s^2}. \tag{8.92}$$

The terms to be associated with higher powers of the lattice spacing in the expression (8.91) vanish in the classic continuum limit. The presence of divergences in quantum field theory means that these terms may result in the finite renormalization of the coupling constant.

Note that we have not fixed the gauge. In usual continuum formulations, gauge-fixing terms eliminate divergences that appear when integrating over all possible gauges. Here, however, all the integration variables are elements of a compact group. As a consequence, all the gauge orbits are themselves compact as well. For gauge-invariant observables, it is harmless to include an integral over all gauges. Note, however, when formulating the perturbation theory or using the transfer matrix to find a Hamiltonian formalism, we should introduce the concept of gauge fixing.

So, we have our variables and the Lagrangian. To proceed to the quantum field theory, we insert the action into the functional integral:

$$W[G, \overline{\psi}, \psi] = \int [dU][d\overline{\psi}, d\psi] e^{-S_{QCD}}, \tag{8.93}$$

where

$$S_{QCD} = S_q + S_G, \tag{8.94}$$

while S_q and S_G are defined by Eqs. (8.81) and (8.86), respectively.

A typical task of the LQCD is to find a vacuum expectation value of some operator, that is, to calculate the integral

$$< 0|\mathcal{O}(U,\overline{\psi},\psi)|0 >= \frac{\int [dU][d\overline{\psi},d\psi]\mathcal{O}(U,\overline{\psi},\psi)e^{-S_{QCD}(U,\overline{\psi},\psi)}}{\int [dU][d\overline{\psi},d\psi]e^{-S_{QCD}(U,\overline{\psi},\psi)}}, \qquad (8.95)$$

where $\mathcal{O}$ is arbitrary operator of field variables U, $\overline{\psi}$, and ψ. Since we integrate fields in every site and in every lattice link, the number of integrations is very big. So, in LQCD, calculating integrals of the kind (8.95) is not a simple task. The first-rate supercomputers are being used and there exists a large community of physicists who develop different algorithms for the fastest computations of similar integrals. Discussion of these algorithms could be found in the review [135]. To gain the insight of calculations volume we consider a specific example. From the beginning we evaluate the lattice size that is required to fulfill calculations in the real LQCD. If we want to describe the baryon structure with an accuracy of 10%, then the linear baryon size must be accounted for of the order of ten lattice spacings. But in QCD, π-mesons exist that are approximately lighter than baryons by a factor of 10, and about 100 lattice spacings are accounted for by these mesons. Therefore, the minimal number of sites is $V = L^4 \approx 100^4$. At the QCD computations on a lattice with a volume L^4, one has to calculate $32L^4$ multiple integrals over gauge fields while integrating over fermion degrees of freedom demands calculation of the determinant of a $12L^4 \times 12L^4$-matrix. Thus, for example, a semi phenomenological expression for LQCD computations taking into consideration the contributions caused by virtual u- and d-quarks has the form [136]:

$$\text{calculations volume} \approx 2.8 \left(\frac{N_{conf}}{1000}\right)\left(\frac{m_\pi/m_\rho}{0.6}\right)^{-6}\left(\frac{R}{3\text{ fm}}\right)^5 \times$$

$$\times \left(\frac{1/a}{2\text{ GeV}}\right)^7 \text{ TeraFLOPS year}, \qquad (8.96)$$

where N_{conf} is the number of gluon field configurations, m_π and m_ρ are the masses of π-meson and ρ-meson, respectively (the ratio m_π/m_ρ defines the quark mass value, $m_u = m_d$) and 1 TeraFLOPS (TFLOPS) is one trillion floating point operations per second (this measure of compute capacity describes how many multiplications can be performed within one second). From (8.96) it follows that for 100 configurations with parameters $m_\pi/m_\rho = 0.6$ (i.e., with a quark mass ~ 50 MeV), $R = 3$ fm, $1/a = 2$ GeV to generate, 100 working days of computer with the capacity 1 TFLOPS are needed. Special attention must be given to high powers in the expression (8.98), which point to fast growth of calculations volume when approaching the continuum or chiral limits. From the relation (8.98) it is inferred that in order to fulfill realistic calculations within LQCD, large computer capacities are required.* The appearance of such large numbers immediately suggests a statistical approach. Really, there are also an enormous number of ways to place molecules of C_2H_5OH into a glass and yet one only needs a few to determine the thermodynamic properties, as well as the killing power, of alcohol. The idea is to provide a small number of configurations that are typical of thermal equilibrium in the statistical analog. Whereas the super-astronomical number of terms appearing in Eq. (8.96) can never be summed exactly, it is straightforward to store the few tens of thousands of numbers characterizing typical configurations that strongly dominate the sum.

---

*By way of approximation the following scheme is often used. Calculations are fulfilled at large quark mass values (50–100 MeV) when the π-meson is not light. The obtained values of observables are extrapolated to the realistic value of the light quark mass (~ 3 MeV) by means of the chiral perturbation theory.

A Monte Carlo method (MCM) is a generally accepted method for calculating the integrals (8.95). Allowing for the wide application of this method in elementary particle physics, we shall expound its fundamental ideas. The MCM or the method of statistic tests is the numerical method of different problems solution with the help of a simulation of random events. When applied to physics, the MCM can be determined as the method of investigation of a physical process by means of the creation and operation of a stochastic model that reflects the dynamics of the process in question. In statistical physics this method began to be applied a considerably long time ago [137]. Let us imagine some process that is described by k variables $p_1, \ldots, p_k$, which may be considered as randomly selected numbers with the distribution density $F(p_1, \ldots, p_k)$. The problem is to evaluate the distribution density of some characteristic f of this process being a function of variables $f = f(p_1, \ldots, p_k)$ or that of the totality of such characteristics $f_1, \ldots, f_m$. To solve this problem on a computer the algorithm is created, the aim of which is to multiply generate the set of the quantities $q_1, \ldots, q_k$ with the probability density F. The procedure of the multiple obtaining of the set $\{q_i\}$ is called a *physical process simulation*; the numbers $\{q_i\}$ are identified with the variables p_i. For every concrete set $\{q_i^j\}$, the quantity $f(q_1^j, \ldots, q_k^j)$ is calculated. Having obtained a large enough number N of sets $\{q_i\}$, the average value of f, its dispersion, and the behavior of the probability distribution density function can be evaluated. This approach is called a *direct simulation*. At the so-called *oblique simulation*, the process is described with one or more equations (differential, integral, and so on) to be solved with the help of the MCM. From the mathematical point of view, both procedures are equivalent to calculating an integral over some multidimensional region.

To realize a randomly selected number, a source of random numbers uniformly distributed between zero and one is traditionally used. Such generators are standard in most high-level computer languages. In calculations on a computer, pseudo-random numbers are employed that are obtained with the help of some algorithm. The purpose of such an algorithm is to generate numbers that are similar to random ones, although, strictly saying, they are determinate. It is necessary to have special investigations and tests to be sure in enough randomness of such numbers (uniformity of distribution, absence of correlations, and so on).

To use the MCM in physics is mainly based on the possibility of its application for computing integrals, solving integral equations, and so on. Thus, for example, we need to evaluate the integral

$$\int_\Omega f(x)dx,$$

where Ω is a finite k-dimensional definition region. In the MCM, the calculation algorithm is based on the mean value theorem:

$$\int_\Omega f(x)dx = V <f>,$$

where V is the volume of the region Ω. Let us choose the k-dimensional parallelepiped with the volume W, which involves the region Ω and assign in a random way a sufficiently large number N of points uniformly distributed in this parallelepiped. Next, the value of the function f is calculated for M points fallen into the region Ω. In that way, the integral evaluation is given by

$$I = \frac{M}{N}W \times \frac{1}{M}\sum_{j=1}^{M} f(x_j) = \frac{W}{N}\sum_{j=1}^{M} f(x_j). \tag{8.97}$$

If in the region Ω points are distributed with the probability density $p(x)$, then, knowing the volume V, one may obtain the following integral evaluation:

$$I = \frac{V}{M} \sum_{j=1}^{M} \frac{f(x_j)}{p(x_j)}. \tag{8.98}$$

So, the MCM allows us to calculate any desired correlation functions. Furthermore, it provides the possibility of performing *experiments* on virtual *crystals* with practically any interaction Hamiltonian. This in principle enables the isolation of various dynamical features and their role in such phenomena as phase transitions. The technique converges well both in high- and in low-temperature regimes and interpolates nicely in between. This latter point is overwhelmingly important to the particle physicist, who desires to relate the Wilson demonstration of confinement in the strong coupling regime to the continuum field theory obtained in the weak coupling limit.

Evidently the insertion of fermions into the procedure of the Monte Carlo calculations is very complicated. The essential difficulty with including quarks in a numerical treatment is that the corresponding functional integral is not an ordinary sum, but rather an intricate linear operation from the space of anticommuting variables into the complex numbers. Really, the exponentiated action is an operator and cannot be directly compared with real random numbers. This problem could be circumvented by first integrating out the anticommuting variables analytically. Since the action is quadratic in the fermion fields, this operation presents no problems. The generating functional for n_f degenerate fermions is given by the expression:

$$W[G, \overline{\psi}, \psi] = \int [dU][d\overline{\psi}, d\psi] \exp[-S_{QCD}] = \int [dU][d\psi, d\psi] \exp\left[-S_G(U) - \sum_{i=1}^{n_f} \overline{\psi}_i M_i(U)\psi_i\right] =$$

$$= \int [dU] \{\det M(U)\}^{n_f/2} \exp[-S_G(U)]. \tag{8.99}$$

To make the positive definiteness of the fermion determinant D_F (Fermi determinant) explicit, we rewrite it in the view:

$$D_F = \det[M(U)^\dagger M(U)]^{n_f/4}.$$

Remembering the determinant to be the product of its eigenvalues, we can express its logarithm as the logarithms sum of the eigenvalues, that is, as the spur:

$$W[G, \overline{\psi}, \psi] = \int [dU] \exp\left\{-S_G(U) - \frac{n_f}{4} \mathrm{Sp} \ln[M(U)^\dagger M(U)]\right\}. \tag{8.100}$$

However, the problem does not disappear completely. The determinant appearing after integration is of an extremely large matrix, the number of rows being the product of the number of sites with the ranges of the spinor index, the internal gauge symmetry index, and the flavor index. For interesting-sized systems, this is a many-thousand-dimensional matrix. As the time required to take a determinant of a matrix grows with the cube of its dimension, such direct calculations would take too much time. Moreover, if the masses of quarks are approaching their real small values, M becomes a poorly defined matrix (eigenvalues of M are between some upper-fixed limit and quark mass so that the region of eigenvalues increases when $m_q \to 0$). Thus, for the Fermi determinant to be determined we need many more computational resources as compared with the purely gauge lattice theory taking into account only gluons. For this reason, by way of an intermediate step, the quark operators are studied on a fluctuating gluon background. In other words, we assume the Fermi determinant to be independent on link variables. This scheme received the name of *quenched quark approximation (quenched approximation)* [138]. To replace M by a constant is equivalent to the disregard of contributions coming from closed-quark loops. In this case, the average value of an operator $\mathcal{O}$ becomes the form:

$$< 0|\mathcal{O}(U, \overline{\psi}, \psi)|0 >= \frac{\int [dU]\mathcal{O}(U, \overline{\psi}, \psi)e^{-S_G(U)}}{\int [dU]e^{-S_G(U)}}. \qquad (8.101)$$

Just in this approximation it was first shown [139] that the confinement of quarks, existing in the region of strong coupling, takes place as well when the lattice spacing decreases. We must emphasize that quenched approximation is used only for saving of machine time. The validity of this approximation is not the same for all dynamical observables, and can be defined only a posteriori. For a long time this approximation was almost unique under calculating the QCD observables. However, it has been recently shown that the quenched approximation leads to systematical errors in the interval of 10-20%. The generation of more powerful computers (at present the speed of the best computers is growing as an exponent of the production year) allowed one to proceed to the inclusion of contributions caused by the interaction between quarks and gluons. If contributions of sea quarks are taken into consideration, this scheme is called a *scheme of unquenched or dynamical quarks*. Notice that only light sea quarks (u, d, and s) give substantial contributions into observables. For sea quarks, heavy c- and b-quarks do not have any effect, while a t-quark must not be considered even as a valence quark because it being unstable does not form bound states. In the real world u, d, and s quarks have a small mass. For this reason, in the past, when the computer capacities were poor, if these quarks were allowed for as sea quarks, then much bigger masses, than they had in reality, were ascribed to them.

8.6 Confinement

To determine whether the quark confinement takes place in QCD, it is necessary to find the energy $E(R)$ of a quark at point $x = (t, \mathbf{0})$ and antiquark at point $x = (t, \mathbf{R})$. In the case of the absence of confinement, we assume

$$\lim_{R \to \infty} E(R) = 2m_q. \qquad (8.102)$$

The presence of the confinement means that the inter-quark potential is infinitely increasing

$$\lim_{R \to \infty} E(R) = \infty. \qquad (8.103)$$

A state $\overline{q}q$ at a time t can be represented in the form [140]

$$|q(t, \mathbf{0})\overline{q}(t, \mathbf{R}) >= \sum_C f(C)\Gamma[(t, \mathbf{R}), (t, \mathbf{0}); C]|0 >, \qquad (8.104)$$

where $\Gamma[(t, \mathbf{R}), (t, \mathbf{0}); C]$ is the gauge-invariant $\overline{q}q$-operator:

$$\Gamma[x', x; C] = \overline{q}(x')U(x', x; C)q(x), \qquad (8.105)$$

$$U(x', x; C) = \exp\left[ig_s \int_x^{x'} \frac{\lambda^j}{2}G_\mu^j(y)dy^\mu\right], \qquad (8.106)$$

C is the trajectory connecting the points x and x', while $f(C)$ is a normalization function the form of which is inessential for subsequent discussions. Next, we consider the overlapping of the $\bar{q}q$ state at the time $t = 0$ and that of $\bar{q}q$ state at the time $t = T$:

$$\Omega(T, R) = < 0|\Gamma^\dagger[(0, \mathbf{0}), (0, \mathbf{R}); C]\Gamma[(T, \mathbf{0}), (T, \mathbf{R}); C]|0 > . \tag{8.107}$$

Inserting the total system of the energy eigenvalue states, we get in the Euclidean space

$$\Omega(T, R) = \sum_n | < 0|\Gamma^\dagger[(0, \mathbf{0}), (0, \mathbf{R}); C]|n > |^2 \exp[-E_n T]. \tag{8.108}$$

At large T, the main contribution comes from the term with the smallest energy E_n. This smallest energy eigenvalue corresponds to the potential energy of the $\bar{q}q$-system in which quark and antiquark are positioned on a distance R from each other

$$\lim_{T \to \infty} \Omega(T, R) \sim \exp[-E(R)T]. \tag{8.109}$$

The overlapping function $\Omega(T, R)$ can be written in terms of quark fields

$$\Omega(T, R) = < 0|\bar{q}(0, \mathbf{R})U[(0, \mathbf{R}), (0, \mathbf{0}); C]q(0, \mathbf{0})\bar{q}(T, \mathbf{0})U[(T, \mathbf{0}), (T, \mathbf{R}); C]q(T, \mathbf{R})|0 > . \tag{8.110}$$

Assuming quarks to be heavy, we consider them as external sources. With the help of the projectors

$$\frac{1}{2}\left(1 + \tilde{\gamma}_0\right), \tag{8.111}$$

we single out the quark ψ_+^α and antiquark ψ_-^β states

$$\psi_+^\alpha = \frac{1}{2}\left(1 + \tilde{\gamma}_0\right)\psi^\alpha, \qquad \psi_-^\beta = \frac{1}{2}\left(1 - \tilde{\gamma}_0\right)\psi^\beta.$$

Then the Lagrangian describing a heavy rest quark takes the form

$$\mathcal{L}_0 = \psi_+^{\alpha\dagger}(iD_0 - m_q)\psi_+^\alpha. \tag{8.112}$$

At that, the free quark propagator is given by the expression:

$$G_0^{\alpha\beta} = < 0|q_0^\alpha(t', \mathbf{r})\bar{q}_0^\beta(t, \mathbf{r})|0 > = \frac{\delta_{\alpha\beta}}{(2\pi)^4}\int d^3k e^{i\mathbf{k}\cdot\mathbf{r}}\int \frac{dk_0}{k_0 - m_q - i\epsilon}e^{-ik_0(t'-t)} =$$

$$= \delta_{\alpha\beta}\delta^{(3)}(0)e^{-m_q|t'-t|}, \tag{8.113}$$

where the residue theorem has been used. Allowing for that interaction of a particle with a gauge field is covered by the phase factor (see Eq. (8.59)), we can find the heavy quark propagator in a background gluon field

$$G^{\alpha\beta} = < 0|q^\alpha(t', \mathbf{r})\bar{q}^\beta(t, \mathbf{r})|0 > = \exp\left[ig_s \int_t^{t'} G_0(\tau, \mathbf{r})d\tau\right]G_0^{\alpha\beta} \sim$$

$$\sim U[(t', \mathbf{r}), (t, \mathbf{r}); C]\delta_{\alpha\beta}e^{-m_q|t'-t|}. \tag{8.114}$$

Combining (8.110) and (8.114), we arrive at:

$$\Omega(T, R) \sim \exp[-2m_q T]\mathcal{W}(C), \tag{8.115}$$

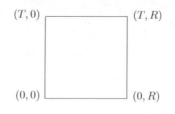

FIGURE 8.7

The Wilson loop.

where

$$\mathcal{W}(C) = < 0|\text{Sp } U[x, x; C]|0 > \tag{8.116}$$

and C is the boundary of the rectangle represented in Fig. 8.7. The correlation function $\mathcal{W}(C)$ is called the *Wilson loop*. Comparing (8.115) with (8.109) leads to the conclusion that $\mathcal{W}(C)$ is a function of $E(R)$

$$\lim_{T\to\infty} \mathcal{W}(C) \sim \exp\left[-T|E(R) - 2m_q|\right]. \tag{8.117}$$

Therefore, the Wilson loop behavior defines the confinement property. By definition $\mathcal{W}(C)$ is a gauge-invariant construction. To move a quark around the contour C generates the external symmetry transformation for its wave function, which is given by the product of the link variables encountered. The Wilson loop essentially measures the response of the gauge fields to an external quark source passing around the loop perimeter. For a time-like loop, this represents the production of a quark-antiquark pair at the initial time, moving them along the world lines dictated by the sides of the loop, and then annihilating at the final time. The Wilson loop is the nonlocal order parameter in the sense that it must exhibit global singularities of a gauge theory (by local we mean the parameter representing a function of gauge variables in a fixed finite domain of the lattice).

To calculate $E(R)$ we address to the quenched approximation. Simplify the gluon field action by discarding the additive normalization constant. Then we obtain

$$S_G(U) = -\frac{\beta}{3}\sum_\Box \text{Re Sp } U_\Box = -\frac{\beta}{6}\sum_\Box [\text{Sp } U_\Box + \text{Sp } U_\Box^*]. \tag{8.118}$$

Consider the rectangular Wilson loop

$$\mathcal{W}(L) = < 0|\text{Sp } U[x, x; L]|0 > = < 0|\text{Sp } \prod_{i,j\in L} U_{ij}|0 > =$$

$$= W^{-1}[G] \int dU e^{-S_G(U)} \text{Sp } \prod_{i,j\in L} U_{ij}, \tag{8.119}$$

where the curve L is the rectangle of dimensions $I \times J$ in lattice units and the group elements U_{ij} are ordered as encountered in circumnavigating the contour. In Fig. 8.8 we display such a contour corresponding to a 3×3-loop. To fulfill the integration over the link variables, we need the following group orthogonality theorems [141]:

$$\int dU U_{ij} = 0, \tag{8.120}$$

$$\int dU U_{ij} U_{kl}^* = \frac{1}{3}\delta_{ik}\delta_{jl}, \tag{8.121}$$

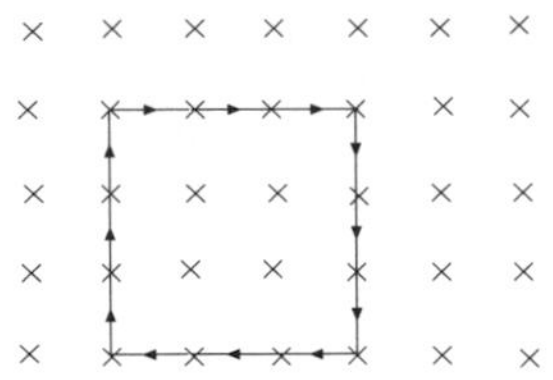

FIGURE 8.8

A 3×3 Wilson loop.

$$\int dU\, U_{ij} U_{kl} U_{mn} = \frac{\epsilon_{ikm}\epsilon_{jln}}{3!}, \qquad (8.122)$$

where we have accepted the normalization

$$\int dU = 1. \qquad (8.123)$$

The theorem (8.121) means that the contributions of links having the opposite directions are mutually cancelled under integration. Therefore, if there are two nearest-neighbor plaquettes of identical orientation, after integration over the variable determined on their common link, they merge into one rectangle (Fig. 8.9). In the strong coupling limit, the

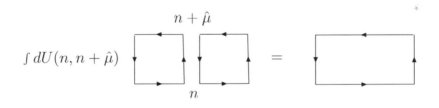

FIGURE 8.9

Merging two adjacent plaquettes into a rectangular path.

quantity β is a small parameter and the exponent in (8.119) may be expanded into the series:

$$\mathcal{W}(L) = W^{-1}[G] \int dU \left\{ 1 + \frac{\beta}{6} \sum_{\square} [\mathrm{Sp}\, U_\square + \mathrm{Sp}\, U_\square^*] \right.$$

$$\left. + \frac{1}{2!}\frac{\beta^2}{36} \left[\sum_{\square}(\mathrm{Sp}\, U_\square + \mathrm{Sp}\, U_\square^*) \right] \left[\sum_{\square'}(\mathrm{Sp}\, U_{\square'} + \mathrm{Sp}\, U_{\square'}^*) \right] + \dots \right\} \mathrm{Sp} \prod_{i,j\in C} U_{ij}. \qquad (8.124)$$

We now observe that owing to the relation (8.120) all Wilson loops will vanish at $\beta \to 0$. Really, for each link in the contour we must bring down at least one corresponding link from an expansion of the exponential of the action if we are to avoid the zeros from Eq. (8.120). Consequently, every link from the action must have a partner, either from the action itself or the inserted loop. The first nonvanishing contribution in the strong coupling series comes from tiling the loop with plaquettes as pictured in Fig. 8.10. It is necessary to pay attention to the orientations of the loops in the figure. To calculate the diagram shown in Fig. 8.10 demands using the integrals (8.123) for links outside the tiled region and those (8.121) within the loop. Unifying all integrals results in a factor 1/3 from each pair of link

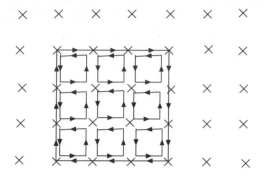

FIGURE 8.10
Tiling the loop with plaquettes.

variables and a factor of 3 from each site on the surface including the boundary. Taking into account the factor $\beta/6$ for each plaquette obtained under expanding the exponential of the action, we find that the lowest-order contribution into $\mathcal{W}(L)$ is given by the term proportional to $(\beta/18)^{N_p}$, where N_p is the minimal number of plaquettes that are required to tile the surface constrained by the contour L:

$$\lim_{\beta \to 0} \mathcal{W}(L) = \left(\frac{\beta}{18}\right)^{N_p}. \tag{8.125}$$

Because the area of the surface constrained by the contour L is equal to

$$S(L) = a^2 N_p, \tag{8.126}$$

then (8.125) could be represented in the form [142]

$$\mathcal{W}(L) \sim (g_s^2)^{-S(L)/a^2} = \exp\left[-KTR\right], \tag{8.127}$$

where

$$K = \frac{\ln g_s^2}{a^2}. \tag{8.128}$$

Comparing (8.127) with (8.117), we get the linearly rising potential

$$E(R) = KR. \tag{8.129}$$

So, $\mathcal{W}(L)$ falls with the exponential of the area of the loop (*area law*) and the coefficient of this area law is the coefficient of the linear potential. From the physical point of view, this area law represents the action of the world sheet of a flux tube connecting the sources. This picture suggests that this area law behavior should hold for arbitrarily shaped loops as long as they are larger than the cross sectional dimensions of a flux tube. Thus, the relations (8.127) and (8.129) will persist for arbitrary shaped contours. In other words, the confinement really exists in LQCD.

The proportionality coefficient K in the area law provides another order parameter for LQCD. It vanishes identically in unconfined phases while remaining nonzero whenever quark sources experience a linear long-range potential.

The qualitative reasoning about the confinement nature is based on a scheme, where quarks are connected with each other by *strings*, narrow tubes of color flux lines, in which all the field energy is concentrated [143]. Let us examine, for example, a $q\bar{q}$-system. The most energy-favorable situation is when the color flux lines connecting quark and anti-quark are strongly located in a bounded region of space. From the translational invariance comes the fact that the density of a gluon field energy is constant along a tube. Consequently, the total field energy is proportional to a distance. It means that, at large distances, quarks are confined by a linearly rising potential

$$E(R) \to KR, \tag{8.130}$$

where K is the constant (*string tension*) whose value is given by the expression (8.128).

A similar phenomenon, the Meissner effect, takes place in superconductors theory. In superconductors, charged condensate ϕ^{--}, Cooper pairs, differs from zero below a phase transition temperature: $< o|\phi^{--}|o > \neq 0$. As a result, a magnetic field cannot penetrate deep into a superconductor. Therefore, if one enters hypothetic magnetic charged particles (monopole g and antimonopole $\bar{g}$) into a superconductor, then in order to minimize a free energy, magnetic field lines are formed into a tube, an Abrikosov tube. The expenditure of energy for the Abrikosov tube production is proportional to its length, that is, to a distance between monopoles. Therefore, in superconductor a linearly increasing potential is acting between monopoles; correspondingly, no matter how large this distance may be the attraction force between them does not decrease with a distance. By analogy, one may assume that in QCD the condensation of color magnetic monopoles* that inhibits the penetration of chromoelectrical field lines into hadronic matter takes place (a dual Meissner effect [144]). In this case, injected color charges are connected by a tube of electric field lines inside which the monopoles condensate proves to be destroyed. As a result, a linear potential appears between color charges.

Let us evaluate the string tension using the string hadrons model. Consider the $q\bar{q}$-pair production in some process. As the distance between q and $\bar{q}$ increases, the processes of breaking the quark string and producing a new $q\bar{q}$-pair and so on happen to be energy-favorable. This is continued until the initial string will be broken to several single strings, which of them has a length typical for hadron size ~ 1 fm (~ 5 GeV^{-1}). Such a picture corresponds to conversion of the initial $q\bar{q}$-pair into hadron jets having typical energies approximately 1 GeV. From the aforesaid it follows that

$$K \sim \frac{1}{5} \text{ GeV}^2. \tag{8.131}$$

Let us examine whether the obtained value of K is in contrast with predictions of the Regge pole theory, which is also based on the string hadron model. In accordance with this theory, hadrons having identical quantum numbers of an external symmetry but different spins lie on the same Regge trajectory [145]

$$J = \alpha_0 + \alpha m_J^2, \tag{8.132}$$

where J is the hadron spin, m_J is the hadron mass, and $\alpha \approx 1$ GeV^{-2}. Consider two massless and spinless quarks that are bound up with each other by a string of a length d and rotated at the light speed [146]. In such a way, every point being located at a distance r from the center moves at the speed $v = 2r/d$. Then, the total mass and total angular momentum are given by the expressions:

$$m = 2 \int_0^{d/2} \frac{K dr}{\sqrt{1 - v^2}} = Kd \int_0^{\pi/2} d\theta = \frac{\pi K d}{2}, \tag{8.133}$$

---

*In QCD gluon field configurations having nature of magnetic monopoles exist.

$$J = 2 \int_0^{d/2} \frac{Krvdr}{\sqrt{1-v^2}} = \frac{\pi K d^2}{8}. \tag{8.134}$$

Therefore, the string tension K entering into (8.130) can immediately be expressed by means of the Regge trajectory slope α:

$$K = \frac{1}{2\pi\alpha} \approx 0.2 \text{ GeV}^2 \tag{8.135}$$

to result in the value that is qualitatively come to an agreement with (8.131). Turning to the ordinary units in (8.135), we get the impressive value $K \approx 14$ tons.

One may also consider the weak coupling expansion for the Wilson loop passing into the continuous limit and replacing the action by the Gauss approximation. Then we obtain the *perimeter law*

$$\mathcal{W}(C) \sim \exp[-E(R)p(C)], \tag{8.136}$$

where $p(C)$ is the perimeter of the contour C. It turns out that this is equivalent to the Coulomb potential $E(R) \sim R^{-1}$. So, the interaction potential of quarks that follows from LQCD is well approximated by the superposition of the Coulomb potential dominating at small distances and the linearly rising potential determining the confinement property at large distances

$$E(R) = \frac{C}{R} + KR. \tag{8.137}$$

As calculations show, the linear growth of the potential takes place at $R > 0.25$ fm. The perturbative dynamics is dominated at smaller distances. In this case, the dependence $g_s^2(a) \sim e^{Ka^2}$, obtained within the strong coupling expansion really goes into the one $g_s^2(a) \sim 1/(\ln a^{-1})$ found in the weak coupling limit when $a \to 0$ (asymptotical freedom).

However, the happy end is still a long away off. The point is that the strong and weak coupling regimes may be separated by one or several discontinuous phase transitions. At present there is no analytical proof that, in QCD, phase transitions are absent at finite coupling constants. Still, considerable study using numerical methods is being given to this problem. Corresponding calculations shows that, in the region of intermediate values of the coupling constant g_s, phase transitions are lacking. These results instill hope in the possibility of proof that the asymptotic freedom and the confinement actually coexist in the unified QCD phase.

8.7   Hadron masses

In this section we shall present the main aspects of calculating hadron masses in LQCD by the example of a meson to be composed of quark and antiquark. Details of similar calculations within the quenched approximation could be found, for example, in Refs. [147, 148], while Refs. [149, 150] illustrate calculations within the dynamical quarks model. Consider an operator

$$\mathcal{O} = h^\dagger(T)h(0) = \left(\overline{\psi}^{a,\alpha,f_2} \Gamma^{\alpha\beta} \psi^{a,\beta,f_1} \right)_T \left(\overline{\psi}^{a,\alpha,f_1} \Gamma^{\alpha\beta} \psi^{a,\beta,f_2} \right)_0$$

that creates a meson at a time $t = 0$ and destroys it at a time $t = T$. Here $\Gamma^{\alpha\beta}$ is some combination of 4×4 γ-matrices that gives the right spin-parity (J^P) quantum numbers to the meson, a is a color index, α and β are spin indices, f_i are flavor indices, and the initial and final points, x and y, are simply denoted by their t coordinates, 0 for the initial point and T for the final point. The meson propagator is defined by the expression:

$$< 0|h^\dagger(T)h(0)|0 >= \frac{\int [dU][d\overline{\psi}][d\psi] \sum_{\mathbf{r}} \left\{\overline{\psi}^{f_1}\Gamma\psi^{f_2}\right\}_T \left\{\overline{\psi}^{f_2}\Gamma\psi^{f_1}\right\}_0 e^{-S_{QCD}}}{\int [dU][d\overline{\psi}][d\psi] e^{-S_{QCD}}}, \qquad (8.138)$$

where for the sake of simplicity we have suppressed color and flavor indices. For the case $f_1 \neq f_2$, to integrate the right side of (8.138) over quark fields results in

$$\frac{\int [dU] \mathrm{Sp}_{spin,color,\mathbf{r}} \left\{M_{f_1}^{-1}(0,T)\Gamma M_{f_2}^{-1}(T,0)\Gamma\right\} (\det M)e^{-S_G}}{\int [dU](\det M)e^{-S_G}}. \qquad (8.139)$$

Evaluating this expression requires us to generate sets of gluon field configurations with probability $e^{-\tilde{S}_{QCD}}$, compute the trace over spin and color of the M^{-1} factors on each configuration, and then overage them over the ensemble. Figure 8.11 illustrates this calculation. It is known as a *two-point function calculation* because there are two operators at

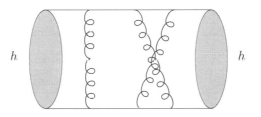

FIGURE 8.11
The two-point function calculation.

times 0 and T, represented by the filled ovals. The straight lines at the top and bottom indicate the valence quark propagators (the M^{-1} factors above that connect the creation and annihilation operators for that particular valence quark). Since the quark propagates, it interacts any number of times with the other quark through the lattice gluon field and this is shown by the curly lines. Some of these gluon lines may involve the effect of a sea quark-antiquark pair (as on the left side of the diagram).

Calculations proceed by evaluating the integrals in Eq. (8.137) by the MCM. The two variables that we have at our disposal are the bare lattice coupling constant g_{sL} and the number of lattice sites. As we shall see, the value of the coupling constant controls the physical size of the lattice spacing. Ideally we would like to have a large number of sites so that the system being modelled fits comfortably on to the lattice. On the other hand, we would like to have a small value of the coupling constant so that the granularity of the lattice is small compared to the object under study. Therefore, for a hadron of radius R we need to require

$$a \ll R \ll L = Na. \qquad (8.140)$$

It should be stressed that the relationship between the value of the coupling constant and the size of lattice spacing cannot be determined a *priori*.

The hadron mass is determined from the results through the usual multi-exponential form

$$< 0|h^\dagger(T)h(0)|0 > = \sum_n C_n e^{-m_n aT}, \qquad (8.141)$$

where n runs over radial excitations of the hadron with the particular set of J^P and flavor quantum numbers, and $m_n a$ are the different hadron masses in lattice units. C_n in Eq. (8.141) is related to the square of the matrix element $< 0|h|n >$ by analogy with Eq. (8.108). The size of C_n then depends on the form of the operator h used to create and destroy the hadron. Any operator with the right quantum numbers can be used and if we make a set of such operators we can calculate the whole matrix of correlators and fit them simultaneously for improved precision. Typically we use the operators made of quark and antiquark fields but separated in space according to some kind of wavefunction (known as *smearing*), for example, $\overline{\psi}_x \phi(x - y)\psi_y$. Of course, the operator of this kind must be gauge-invariant. To this end one may require strings of U to be inserted between x and y. It may be possible to choose ϕ so that a particularly good overlapping with one of the states of the system (for example, the ground state) is obtained, and poor overlapping with the others. Then C_0 would be large in the equation above, and the other C_n small, and this would give improved precision for $m_0 a$. The masses of the radial excitations are of interest as well and smearings that improve overlap with them have also been studied. So, we can calculate the mass $m_0 a$ of some hadron in lattice units. Then, one may choose a specific hadron (customarily as a rule it is a ρ meson) and for it fulfill the estimation of the integrals in Eq. (8.139) at several values of the bare coupling. It gives us the set of points in the plane (a, g_{sL}), that is, it defines the relationship between the masses and the couplings $(a = a(g_{sL}))$. In Fig. 8.12 the lattice spacing as a function of the bare lattice coupling is displayed [148]. The points correspond to the Monte Carlo results obtained by two collaborations in the quenched approximation. Having chosen the mass of the ρ-meson to fix the scale one can then go to make predictions for other physical quantities. From Fig. 8.12 it follows that $g_{sL}^2 = 1$ corresponds to $a^{-1} \approx 2$ GeV, or equivalently $a \approx 0.1$ fm. This lattice spacing should be much less than the mass of the hadron whose mass is calculated. If this condition is satisfied it makes sense to proceed to computing the other hadron masses in terms of m_ρ.

However, the bare coupling must be connected with the lattice spacing by means of the renorm-group equation (RGE). For this reason, we should establish the RGE and verify whether the results of the Monte Carlo calculations are in contrast with the RGE. For simplicity we assume that we are dealing with massless quarks. Then, the only parameter with mass dimension is the inverse lattice spacing, that is, the hadron mass is defined as

$$m_h = \frac{1}{a} f(g_{sL}). \qquad (8.142)$$

To yield a sensible answer the $a \to 0$ limit must clearly be taken with some care. The proper continuum limit is achieved when we take a to zero holding the physical quantities fixed. In practice this means that when $a \to 0$

$$\lim_{a \to 0} \frac{1}{a} f(g_{sL}) \to \text{constant}, \qquad (8.143)$$

that is, g_{sL} has to be changed as a tends to zero. But this corresponds to the actual state of affairs. Thanks to the vacuum effects, g_{sL} is the running coupling constant and its behavior at small distances is predicted by QCD, namely by Eq. (4.43). Using this equation, we shall find the relationship between the coupling and lattice spacing that leads to the sensible continuum limit. So, since physical results cannot depend on the cut-off, in the limit $a \to 0$

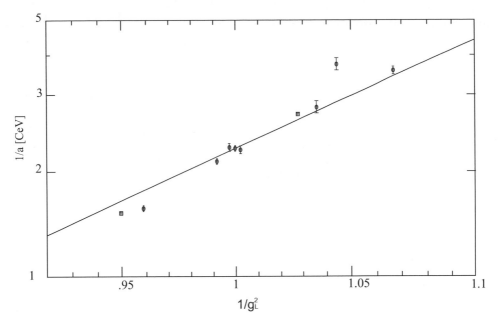

FIGURE 8.12

Lattice spacing determined by m_ρ.

we must have

$$\frac{d}{da}m_h = 0. \tag{8.144}$$

Differentiating Eq. (8.142) and using the chain rule

$$\left[a\frac{\partial}{\partial a}\bigg|_{g_{sL}} + \left(a\frac{\partial g_{sL}}{\partial a} \right) \frac{\partial}{\partial g_{sL}}\bigg|_a \right] m_h = 0, \tag{8.145}$$

we get

$$f(g_{sL}) = \left(a\frac{\partial g_{sL}}{\partial a} \right) \frac{\partial}{\partial g_{sL}} f(g_{sL}). \tag{8.146}$$

According to (4.43), the dependence of the coupling on the scale is determined by the expression

$$a\frac{\partial g_{sL}}{\partial a} \approx \frac{b}{16\pi^2} g_{sL}^3 \left(1 + \beta_1 \frac{g_{sL}^2}{8\pi^2 b} + \dots \right). \tag{8.147}$$

Therefore, we find that all dimensional quantities are proportional to the fixed scale, that is

$$m_h = c\Lambda_L, \tag{8.148}$$

where

$$\Lambda_L = \frac{1}{a}\exp\left(-\frac{8\pi^2}{bg_{sL}^2} \right) \left(\frac{8\pi^2 b + b'g_{sL}^2}{8\pi^2 bg_{sL}^2} \right)^{\beta_1/b^2}. \tag{8.149}$$

So, despite the fact we start with a theory that is scale free we end up with a theory that has one scale, to which all other dimensional quantities are related. This is the phenomenon called *dimensional transmutation.* From Eq. (8.149) we find

$$a \sim \exp\left[-\frac{8\pi^2}{bg_{sL}^2}\right], \qquad (8.150)$$

that is, the lattice spacing goes to zero exponentially in $1/g_{sL}^2$ as g_{sL} tends to zero. By this reason the relationship (8.149) is called *asymptotic scaling.* Now, with its help we can verify the validity of the Monte Carlo simulation presented in Fig. 8.12. At $\Lambda_L = 1.74$ MeV all points fall on the curve shown in this figure. So, the Monte Carlo data are in fine agreement with the slope predicted by asymptotic scaling, which gives the assurance that the results are indeed close to the continuum limit. It should also be noted that the relation (8.149) can be employed to determine Λ_L, which is associated with the Λ_{QCD} parameter measured in the continuum.

In Fig. 8.13 we display the calculation results of light hadron masses in the quenched approximation [151]. From Fig. 8.13 it is seen that the real masses spectrum is described

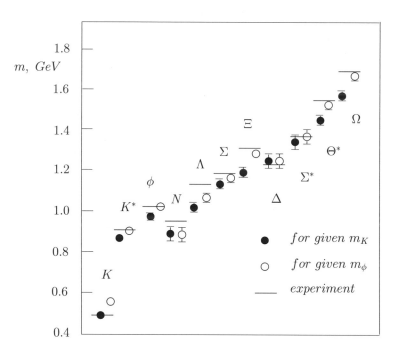

FIGURE 8.13
The hadron mass spectrum obtained by CP-PACS Collaboration.

with much success. One important point to remember is that only information about the QCD Lagrangian was fed into a supercomputer.

It should be stressed that the calculation of the light hadron spectrum, and quark masses as well, is the most fundamental issue in the LQCD simulations. Up to now systematic studies in quenched approximation has been made (see, for example, [151, 152]. It was also shown that within the two flavor full QCD (QCD with two dynamical light quarks) the $O(10\%)$ deviation of the quenched hadron spectrum from the experiment is largely reduced [153]. An analogous circumstance takes place for quark masses too. At present, these problems are being investigated within a 2+1 flavor full QCD in which degenerate u-

and d-quarks and a s-quark are related dynamically (see, for example, [154, 155, 156]).

In a hypothetical unconfined phase, the QCD must contain massless gauge bosons. Using a transfer matrix formalism to determine energies of a system, we define the mass gap as the energy difference between the ground state and the first excited state. This quantity will be exactly equal to zero in an unconfined phase permitting the free gluons existence. Contrary to this, in a phase displaying confinement of massive quarks, we should have a spectrum of massive glueballs and bound states of quarks. Therefore, the mass gap is the order parameter that is expected to vanish in one phase but not be equal to zero in another phase. In statistical mechanics language, the mass gap represents the inverse of the correlation length. The expectation of two separated operators in a statistical system will generally display a correlation between the operators that falls with the distance between them. If, for asymptotic separations, this falloff is exponential, then the coefficient of the decrease is the mass gap m:

$$C(r) \sim \exp(-mr). \tag{8.151}$$

This may be justified employing a transfer matrix along the separation direction r. From the physical point of view, Eq. (8.151) represents the Yukawa exchange of the lightest particles in the theory in question. When the mass gap vanishes, we obtain the power law of decreasing the forces as it was in QED. It must be emphasized that as an order parameter the mass gap is not local in that its definition involves correlations between asymptotically separated operators.

In conclusion in this chapter, we note that the Proceedings of each year's lattice conference provide a useful summary of current results and world averages (see LAT2008, LAT2009, and so on).

9

Quark-gluon plasma

Blessed is a person, who knows nothing:
he does not risk to be not understood.
Confucius

9.1 Aggregate states of matter

Problems of investigating nuclear matter properties at extremely large densities and temperatures were already being raised until QCD was formulated as a theory describing the strong interactions (see, for example, [157, 158]). After the confinement and asymptotic freedom have been discovered it was realized very quickly that in nuclear matter at extreme conditions (strongly compressed and/or heated hadronic matter), a state with the deconfinement of quarks and gluons may exist. The term *quark gluon plasma* itself appears to have been introduced in Ref. [159], where it was shown that this state may come into being under collisions of heavy ions when the energy density of colliding ions reaches the value of the order of 1 GeV/fm^3.

It is well known that any substance (with some exceptions) may be into three aggregate states: solid, liquid and gas. A plasma, a partially or completely ionized gas, is called the *fourth aggregate state of substance*. A plasma can be obtained either from increasing the temperature (thermal ionization) or from irradiating a substance by electromagnetic waves (photoionization) and charged particles. So, a classical or electromagnetic plasma results when molecules are decayed into their constituent atoms; those then turn into ions. At a sufficiently high temperature, nuclear and electron components of atoms are separated because of the thermal ionization. Increasing the temperature and density of such an atomic nuclei gas, one can achieve the kinetic energy of nuclei that exceeds the height of the Coulomb barrier (10-20 MeV) and then the short-range nuclear forces come into effect. Further increasing the temperature and nuclei gas density (electron component and electromagnetic interaction become inessential when mean distance between particles is of the order of 1 fm) results in the dissociation of atomic nuclei into their constituent nucleons, which is accompanied by the production of light hadrons, in the main, of light mesons ($\pi, \rho, \omega, \ldots$) and, to a lesser degree, of baryons. After all nuclei have been dissociated, we get an almost ideal hadronic gas representing the totality of colorless mesons and baryons. It is clear that the dynamics of this gas is completely determined by QCD. However, if we managed to switch off the strong interaction, then the charged hadrons gas would be a classical electromagnetic plasma. The energy density and baryons number in the volume unit are physical variables to characterize a state of an ideal hadronic gas. The corresponding thermodynamical variables are (i) a temperature T as a measure of a mean kinetic energy of particles, which accounts for one degree of freedom, and (ii) a baryon chemical potential μ_B to control a mean baryon density.* As usual in statistical physics, other quantities (pressure, entropy, thermal capacity, and so on) can be represented as functions of T and μ_B. Next, when increasing either temperature or baryon density leads to the fact that the distance between hadrons become commensurable with the confinement radius R_c, quarks from one hadron start to interact with quarks of neighboring hadrons and the border between hadrons disappears. At first, hadronic clusters come into being, then an unified quark-gluon bag of a macroscopic size, inside which color-charged quarks and gluons are moving practically free,* appears. This state of matter is named a *quark-gluon plasma (QGP)*, a *chromoplasma*, a state of a strongly interacting matter that is characterized by the absence of the color confinement [159, 161]. In this state, color quarks and gluons to be captivated by hadrons in a hadron matter become free and may propagate as quasi-free particles over the whole volume of plasma matter. As a result, a colorconductivity appears much as an electroconductivity comes into being in an ordinary electron-ion plasma. In natural conditions, QGP appears to have existed only in the first 10^{-5} s after the Big Bang. It is not inconceivable that QGP exists in the center of the most massive neutron stars too as was first pointed in Ref. [162]. There are also all reasons to believe that atomic nuclei involve small drops of QGP, that is, all nuclei are considered as geterophase systems (two phases are present in a system: in those places of fluctuations of the nuclear density where it vastly exceeds the mean density, the transition of the nucleon phase into the quark-gluon one takes place).

In Fig. 9.1 we display the phase diagram of a hadron matter in the plane $(T, \rho/\rho_0)$, where $\rho_0 \sim 0.15$ baryon/fm^3). There are four regions: (i) nuclear matter, (ii) hadronic gas, (iii) mixed hadronic and quark-gluon phase, and (iv) quark-gluon plasma. In the

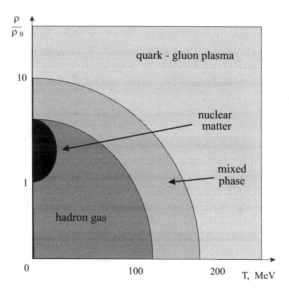

FIGURE 9.1

The phase diagram of the hadronic matter.

---

*The connection of μ_B with the baryon density ρ has a very complicated character and depends on the model in use (see, for example, [160]).

*Interaction intensity decreases when temperature and baryon density increases.

region of the nuclear matter that corresponds to temperatures close to zero and baryon density ρ_0 ($\rho_0 \sim 0.15$ baryon/fm^3), the statistical approach is not applicable because, at such conditions, the hadronic matter is far from ideal, then nucleons in a nucleus cannot be considered as an ideal gas. Here, quantum collective effects and nonperturbative strong interactions have fundamental importance.

A hadronic gas is defined by small values of the baryons density and temperatures up to ~ 170 MeV. As we already mentioned before, a hadronic gas basically consists of mesons. By virtue of the spin-statistics theorem to which a system of identical particles obeys, at $T \sim 0$, the behavior of the mesons system essentially differs from that of the baryons system. Thus, mesons do not make compact formations of the type of atomic nuclei due to the strong interaction.

The region of high temperatures ($T > T_c$) or large baryon densities ($\rho \sim 5 \div 10\rho_0$) is defined as the one of QGP. At large densities of baryons or quarks, the mean distance between quarks is small and the asymptotic freedom predicted by QCD takes place. Because, at high temperatures $T \gg T_c$, typical transferred momenta are $q \sim T$, then $\alpha_s \to 0$. So, we see that the confinement property is not absolute since, at a sufficiently high temperature T_c (deconfinement temperature) or at a sufficiently large baryon density, the phase transition from hadronic matter phase into the QGP phase should occur [163]. Lattice calculations find a quark-antiquark potential as a function of temperature. At the deconfinement temperature, a linearly rising potential becomes constant (Fig. 9.2). It turns out as well that restoration

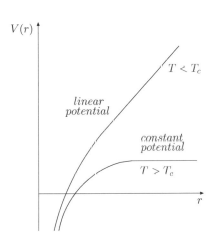

FIGURE 9.2
A quark-antiquark potential as function of temperature and distance.

of the chiral symmetry happens under the confinement-deconfinement phase transition.

9.2 Looking for QGP

Theoretical evidences in favor of the QGP existence stimulate experimental investigations directed at its discovery. The most perspective direction is the advent of QGP by a laboratory way under the collision of heavy nuclei with high energy. During collisions of heavy nuclei involving about 200 protons and neutrons, conditions are more severe than that taking place under collisions of individual protons which are usually used in other experiments of high-energy physics. When colliding heavy ions, instead of a tiny explosion with a great number of fragments flaying apart, a boiling fireball consisting of a thousand particles comes into being. At that, collective properties of particles, like temperature, density, pressure, and viscosity, become essential. Estimations based on the extrapolation of current experimental data show that a strongly interacting system produced in the region of nuclei collision will exist a sufficiently long time and reach both thermodynamic and chemical equilibrium. In this case, energy and the pressing of a system may suffice to get to the QGP phase, making use of functioning accelerators reequipped for accelerating heavy ions.

The search and study of QGP started in early 1980s with the acceleration of the *Au* beam at 1 GeV/A at the Bevalac in Lawrence Berkeley National Laboratory (LBNL). The early success of the experiments in terms of bringing out the collective nature of the matter produced prompted the scientists at Brookhaven National Laboratory (BNL) and CERN to make concrete programs for future accelerator developments for heavy ions. The next milestone came with the acceleration of the *Au* beam at 11.7 GeV/A at Alternating Gradient Synchrotron (AGS) in BNL and the *Pb* beam at 158 GeV/A at Super Proton Synchrotron (SPS) in CERN. In the course of these experiments, a new investigation area at the turn of elementary particle physics (high energies) and traditional nuclear physics (low energies) was discovered. In high-energy physics, particles interaction is described on the basis of first principles (gauge quantum field theories) and hadronic matter is considered to consist of individual particles or field quanta. In contrast, in nuclear physics, the interaction is described phenomenologically and hadronic matter is regarded as a compound medium where collective effects play a vital part.

When colliding relativistic particle beams, the tracks of hundreds of second particles are chaotically distributed in space. This serves as a reason that the statistical approach is applicable to govern such a final state. In other words, we can switch from individual particles to a statistical ensemble to be described by QCD-thermodynamics with the help of macroscopic variables, temperature, pressure, and so on. To ensure that the use of macroscopic variables will be well-founded, a system created under relativistic ions collision must consist of a great number of particles and its size must be much bigger than the strong interaction radius.

A thermodynamical approach may be used for systems being either in a state of thermodynamical equilibrium or in a state close to it, that is, a system lifetime has to be much bigger than a relaxation time. For hadronic matter we have $t \geq 1$ fm. Since the equilibrium may be reached only due to the energy exchange between system particles, then the number of collisions must be sufficiently large. However, only a small number of relativistic ion collisions characterized by a small impact parameter, rather than all collisions, may satisfy the demands mentioned above. The less the impact parameter, the more nucleons from colliding ions have been interacted and the more secondary particles with large transverse momenta relatively to a reaction axis (p_T) have been produced. Consequently, detecting events with a large transverse energy E_T, we select the cases of central collisions of ions.

In high-energy physics, in order to estimate a size of secondary particles source a method of momentum correlations, which was proposed for the determination of stars radius by correlated spectra of radiation as early as 1953, is used. The method is based on a quantum-mechanical connection between (anti)symmetrical wave functions of systems of identical particles in the coordinate and momentum spaces. Roughly speaking, if some correlation function in secondary particles distribution is experimentally measured, this allows one to restore the distribution function of particles in the coordinate space employing the Fourier transformation. So, correlated experiments showed that the region of emitting secondary particles in e^+e^-- or pp-interactions has a size about 1 fm while this region in heavy ion collisions ranges from 3 up to 6 fm, which greatly exceeds the confinement radius of an individual hadron and allows one to talk about the quark deconfinement within the region of emitting secondary particles.

In a thermodynamical equilibrium state, a momentum or energy distribution of fast particles obeys the Boltzmann law

$$N(E) \sim \exp(-E/kT),$$

a mean momentum or energy being proportional to a temperature ($< p >, < E > \sim T$). In relativistic collisions of heavy ions a transverse momentum distribution of secondary particles is not thermolized because of a strong asymmetry in initial conditions ($p_T = 0$, $p_L \gg T$). By this reason, a thermalization degree of system may be judged only by the transverse momentum particles distribution. In relativistic ion collisions, one observes a substantial increase of produced particles with a large p_T as compared with the extrapolation of data for pp-collisions, which might be interpreted as a signal of matter thermalization. However, there exist other explanations of the observed effect as well, for example, multiple rescattering of partons inside a nuclear matter at the initial or final collision stage.

Particles, which do not take part in the strong interaction and are produced at an early stage of forming a hadronic system, carry direct information about the phase transition. Thus, for example, QGP may be revealed by means of its own luminescence. QGP must momentarily be shone like a flash of lightning because it emits high-energy photons that leave QGP practically freely and their spectrum is not distorted by the interaction at later stages of a hadronic system evolution. One may use high-energy photons in order to find out the QGP temperature in the same fashion that astronomers measure the temperature of a distant star investigating its light radiation spectrum. Unfortunately, there are many background processes of producing both photons and leptons that complicate the selection of the efficient signal (the QGP production) and its interpretation.

One of striking signals concerning the QGP production in heavy ion collisions may be suppression of the yield of a J/ψ-meson ($r_{J/\psi} \sim 0.2$ fm) in the processes, when QGP is produced, as compared with those without the QGP production. To create a charmonium is readily detected by means of its lepton decays $J/\psi \to e^+e^-$, $\mu^+\mu^-$. If a $\bar{c}c$-pair comes into being in QGP, then confinement forces between quarks prove to be smaller than in the case when $\bar{c}c$-pairs are created in a vacuum. This is caused by the color charge screening of charmed quarks in QGP (analogue of Debye screening in electromagnetic plasma). Consequently, in processes with the QGP production, a smaller number of J/ψ-mesons must be observed. The experimental evidence on such suppression of the yield of J/ψ-mesons was obtained at CERN in 1989–1990 in investigating collisions of ^{16}O- and ^{32}S-nuclei with ^{238}U-nuclei. In 2000, experiments were carried out in the collisions of lead ions. At that, the total energy of colliding ions had of the order of 33 TeV while for QGP to be produced about 3.5 TeV is needed. The density of produced matter was approximately 20 times larger than that of nuclear matter. Here, the effect became apparent more strikingly and, seemingly, cannot be explained without the strong thermalization hypothesis of hadronic matter and the QGP production. In Fig. 9.3 we display the theoretical curve that extrapolates the results obtained in proton-nucleus interactions to the case of the collisions of two nuclei. It is seen that in the collisions

$$^{16}O + {}^{64}Cu, \qquad {}^{16}O + {}^{238}U, \qquad {}^{32}S + {}^{238}U$$

the extrapolation comes to an agreement with the experimental data. However, in the case of the collisions of $^{207}Pb + {}^{207}Pb$, the deviation of the measurements from theory is sizeable. So, in this case, we deal with a new effect that is not observed in proton-nucleus collisions and even in collisions of light-heavy nuclei.

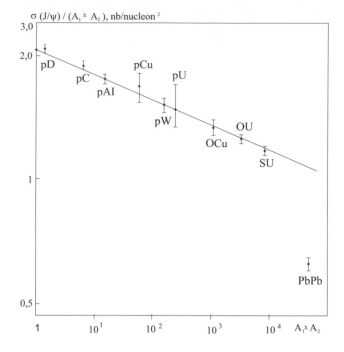

FIGURE 9.3

The cross section of the J/ψ production versus the product of the nucleons number in the colliding nuclei $A_1 \times A_2$. The curve displays the extrapolation of the data obtained in proton-nucleus collisions to the region of nucleus-nucleus collisions.

So then, during 1896–2000 a wide energy interval of ion collisions was investigated: $\sqrt{s_{NN}} = 1$ GeV at Bevalac in LBNL, $\sqrt{s_{NN}} = 5$ GeV at AGS in BNL and $\sqrt{s_{NN}} = 17$ GeV at SPS in CERN. However, at the energies listed above, the unambiguous evidence for the QGP production was never obtained.

More conclusive results were achieved in the course of experiments in 2001 on collisions of gold ions at a relativistic heavy ion collider (RHIC). This collider located at BNL was placed in operation in 2000 and up to now investigates nucleus-nucleus collisions at the highest energies. The RHIC is directed toward studying collisions both of heavy ions (to the point of gold) under maximum collision energy $\sqrt{s_{NN}} = 200$ GeV and of polarized protons up to $\sqrt{s_{pp}} = 500$ GeV. Two lines consisting of 870 superconducting magnets cooled by tons of liquid helium direct ion beams along two overlapping vacuum rings of length 3.8 km. Beams are collided at four points where the rings are crossed. In the intersection points, four detectors—BRAHMS (Broad RAnge Hadron Magnetic Spectrometers), PHENIX (Pioneering High-Energy Nuclear Interactions eXperiment), PHOBOS (put out of operation in 2005), and STAR (the Solenoidal Tracker at RHIC), are located. When two gold nuclei are subjected to a head-on collision at the most high energy accessible at RHIC, they generate altogether more than 2×10^4 GeV into a microscopic plasma small ball, fireball, of a diameter only of 10 fm. Nuclei and their constituent protons and neutrons are literally fused and more multitudes of quarks, antiquarks and gluons appear from the whole energy available. In typical collisions, more than five thousands of elementary particles are momentarily released. The pressure at the collision moment exceeds the atmosphere pressure in 10^{30} times and the temperature inside the fireball reaches a trillion degrees. However, after only 5×10^{-23} s, all quarks, antiquarks, and gluons are again unified into hadrons to fly away in all directions. The first two facilities, BRAHMS and PHOBOS, are relatively small and are built up for the investigation of specific features of fragments. Two other facilities, PHENIX and STAR, are constructed around huge recording general-purpose units that fill up three-storied experimental halls with thousands of tons of magnets, detectors, absorbers, and screens. BRAHMS collaborators study the remainders of initial protons and neutrons flying along a motion direction of colliding nuclei. The PHOBOS group keep a look-out of particles over a wide range of angles and analyze correlations between particles. The STAR facility is built on the basis of a digital chamber, the largest in the world that represents a huge cylinder with a gas where three-dimensional images of trajectories of all charged particles are obtained in a large aperture surrounding a beam axis. PHENIX is looking for specific particles that appear at the most early collision stages and, not changing their states, fly from the arising fireball. As a consequence, a roentgen portrait of internal fireball regions is obtained. Each of the RHIC experiments is directed to the study of observable sets that are characteristic and specific just for it. Nevertheless, in intersecting regions, experimental data received at different facilities are in good agreement. This is an additional argument in favor of the reliability of the results obtained at RHIC.

Early in papers devoted QGP, from model considerations it is assumed that at the critical temperature $T_c \approx 170$ MeV and the energy density of the order of 1 GeV/fm^3, the first-order phase transition takes place (see, for example, [159]). The pointed energy densities are two times greater than the energy density in a nucleon (0.5 GeV/fm^3) and almost one order of magnitude greater than the one in a nucleus (0.14 GeV/fm^3). Lattice calculations fulfilled under the transition from QGP to hadronic matter allow one to concretize these qualitative conceptions on the structure of compressed and heated nuclear matter and specify the phase transition character. Right up to the discovery of a holographic duality to be dealt with at the end of this chapter, lattice QCD (LQCD) was practically the sole method of the theoretical description of QGP. By this reason, before discussing the RHIC results, we consider some questions concerning QCD behavior under high temperatures and large densities. For the detailed inspection of this issue, the reader is referred to the following books [164, 165].

One of the most simple operators to describe the confinement deconfinement transition is the vacuum expectation value of the so-called *Polyakov loop*. This loop is defined as the product of link variables along a time like direction of a lattice closed due to boundary conditions

$$\mathcal{P}(x) = \mathrm{Sp} \prod_{t=1}^{N} U(\mathbf{n}, \mathbf{n} + \hat{t}), \qquad (9.1)$$

where $U(\mathbf{n}, \mathbf{n} + \hat{t})$ is a matrix associated with a link coming from point $\mathbf{r}$ and ending in point $\mathbf{r} + \hat{t}$

$$U(\mathbf{n}, \mathbf{n} + \hat{t}) = \exp\left\{ iag_s \frac{\lambda^\alpha}{2} G_{\mathbf{n},0}^\alpha \right\}.$$

One could show [165] that in the case of a static quark, the vacuum expectation value of the Polyakov loop is given by the expression:

$$< 0|\mathcal{P}(\mathbf{r})|0 > = \exp(-\beta F_0), \qquad (9.2)$$

where F_0 is the free energy of a single static quark. The vacuum expectation value of a pair of Polyakov loops in opposite directions that is associated with a quark-antiquark pair is as follows

$$< 0|\mathcal{P}(\mathbf{r})\mathcal{P}^*(\mathbf{r} + \mathbf{R})|0 > = \exp[-\beta F(\mathbf{R})], \qquad (9.3)$$

where $\mathbf{R}$ is a distance between quark and antiquark while $F(R)$ is the free energy of separation of a pair of static quarks. In the leading strong coupling approximation this vacuum expectation value is

$$F(R) \sim R \log(2g_s^2)$$

at large R. At infinite separation we get a zero expectation value, the proper test of confinement. At high temperature, a phase transition is encountered, leading to a deconfined phase in which the free energy of the Yang–Mills theory on the lattice separation is asymptotically constant. This result can be understood from the strong coupling point of view by considering the entropy of fluctuations in the surface connecting the two Polyakov loops. As the temperature is increased at a fixed strong coupling, the rapidly increasing multiplicity of terms contributing at the same high order offsets the cost of going to high order. The flux tube connecting the static charges fluctuates freely, leading to deconfinement.

To simulate a hadronic ensemble at nonzero baryon number density one introduces a baryon chemical potential. In the continuum limit, it is achieved by entering the term

$$\overline{\psi}(x)\gamma_0[\partial_t - iG_0(x) + \mu_B/3]\psi(x) \tag{9.4}$$

into the fermion Lagrangian. Then, in the case of naive fermions, this leads to the appearance of the term

$$\frac{1}{2a}\sum_n \left[\overline{\psi}_n e^{\mu_B a/3}\gamma_0 U_{n,n+\hat{t}}\psi_{n+\hat{t}} - \overline{\psi}_{n+\hat{t}}e^{-\mu_B a/3}U_{n,n+\hat{t}}^\dagger \gamma_0\psi_n\right] \tag{9.5}$$

in the lattice Lagrangian. For other actions one simply replaces each forward gauge link by $\exp[\mu_B a/3]U_{0,x}$ and each backward link by $\exp[-\mu_B a/3]U_{0,x}^\dagger$.

As a function of the quark masses the phase transition from ordinary matter to QGP has two important limits. When all quark masses are infinite we have, in effect, a pure Yang–Mills theory, which for the $SU(3)_c$ gauge group has a first-order high-temperature phase transition from a confined to a deconfined phase. The order parameter is the asymptotic value of the Polyakov loop correlator introduced by Eq. (9.3). That is, for Eq. (9.3) with $\beta = 1/T$, the order parameter $2F_0 = \lim_{R\to\infty} F(R)$, can be interpreted as (twice) the free energy of a point charge. In the confined phase that free energy is infinite. In the deconfined phase it is finite.

One of the major predictions of QCD in extreme conditions of high temperature or large baryon number density is the existence of a critical point at a particular temperature and density where a sharp transition between the QGP phase and the hadronic phase first appears. It could be shown that at the critical point the Polyakov loop correlator has an inflection point.

It should be noted that the Polyakov loop correlator (9.2) is the order parameter only for the model without dynamical quarks, that is, in the case of infinitely large masses of quarks. In the opposed limit of massless quarks, that is, in the chiral limit of QCD, a chiral condensate $< 0|\overline{\psi}\psi|0 >$ is an order parameter. The phase transition order depends on the number of massless quarks. In QCD, the chiral phase transition with three massless quarks represents the first-order phase transition while in the case of two massless quarks the phase transition is absent and, instead of the transition, a so-called *soft crossover** is observed.

In the real QCD, the phase transition properties are defined by two light quarks (u and d) and one heavier quark (s) while the remaining quarks are much too heavy to have an appreciable impact. At zero and finite values of quark masses neither the Polyakov loop correlator nor the chiral condensate, strictly speaking, are order parameters. However, as calculations show, in the intermediate regime of physical quark masses that are neither zero nor infinite, both quantities make a smooth, but steep transition across the same

---

*It is often called a *continuum phase transition*.

temperature region between $0.9\,T_c$ and $1.1\,T_c$. On the other hand, rather than looking for an inflection point in the Polyakov loop correlator and chiral condensate, it is often easier to look for peaks in corresponding susceptibilities [166]

$$\chi_P = <0|\mathcal{P}^2|0> - (<0|\mathcal{P}|0>)^2, \qquad (9.6)$$

or

$$\chi_c = <0|(\overline{\psi}\psi)^2|0> - (<0|\overline{\psi}\psi|0>)^2. \qquad (9.7)$$

A finite size scaling analysis studies the peak height as a function of increasing lattice volume V. If the peak height is asymptotically constant, there is no phase transition. If it rises linearly with the volume, the phase transition is the first order. If it rises less rapidly, it is a critical point.

It may be possible to access the critical point experimentally by scanning the QCD phase diagram in terms of T and μ_B. This can be accomplished by varying beam energies from about $\sqrt{s_{NN}}$=5 GeV to 100 GeV. The discovery of the critical point would be very important to the QGP study. Such a program has been undertaken at RHIC [167]. Lattice computations reveal that the crossover transition is realized at the baryon chemical potential values corresponding to the RHIC energy (see Fig. 9.4). The first-order phase transition,

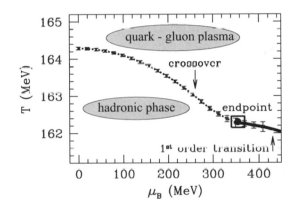

FIGURE 9.4
The critical temperature and the phase transition type versus the chemical potential.

as it was assumed in early works (see, for example, [159]), takes place at higher baryon densities. It should be noted, that experiments at GSI synchrotron [168] are planned to study the critical points as well.

One of the key questions, for which one must answer in investigating the QGP production, is a question of a energy density that could be achieved in central nucleus-nucleus collisions at the RHIC energies. The energy taken from initial particles under their collision may be evaluated with the help of so-called stopping power. The stopping power δy is defined from a rapidity distribution of initial baryons (net-baryons) dN/dy (see [169]):

$$\delta y = y_b - <y>, \qquad <y> = \int_0^{y_b} y\frac{dN}{dy}dy \bigg/ \int_0^{y_b} \frac{dN}{dy}dy, \qquad (9.8)$$

where y_b is a rapidity of the incoming nucleon or nucleus. With allowance made for the net-baryons density distribution and the cross section dependence of produced hadrons on transverse momenta, it was found that each of the colliding nucleons loses, on average, from 50 to 85% of initial energy [170]. The uncertainty of this estimation is primarily connected with that of the extrapolation of the net-baryons distribution in the region of larger rapidities because the rapidity distribution is measured for technical reasons in the region that is narrower than the region assigned by rapidities of colliding nuclei. As this takes place, at different stages of the reaction, one may speak about the following densities that sequentially appear: (i) a maximum energy density in the time when colliding nuclei are completely covered; (ii) a maximum density of an formed energy, that is, an energy that was transferred to the produced particle at the time as soon as a fireball was formed (τ_{form}); and (iii) a maximum energy density in the time of the establishment of local thermalization (τ_{therm}).

The peak value of the energy density at the time of covering the initial nuclei is easily calculated from simple geometrical considerations. At collisions of nuclei with the radius R for the maximum RHIC energy ($\gamma = \sqrt{s_{NN}}/(2m_N)$), it amounts to $\epsilon_i \approx 3000$ GeV/fm^3. It is clear that by no means all of this energy is spent on the formation of a fireball formed in the central rapidities region.

To estimate the formed energy density one may use the results of Ref. [171]:

$$< \epsilon(\tau_{form}) >= \frac{dN(\tau_{form})}{dy} \frac{< m_T >}{A\tau_{form}}, \qquad (9.9)$$

where A is a transverse size of colliding nuclei, m_T is an transverse mass that is determined in terms of parameters of detected particles:

$$< m_T >= \frac{dE_T(\tau_{form})/dy}{dN(\tau_{form})/dy}. \qquad (9.10)$$

It is natural to assume that τ_{form} has the order of 1 fm. Employing this value, we obtain for the formed energy density: $\epsilon = 1.5$ GeV/fm^3—for AGS energy, $\epsilon = 2.9$ GeV/fm^3—for SPS energy and $\epsilon = 5.4$ GeV/fm^3—for maximum RHIC energy. For a more argued choice of the formation time one may use the relation [171]

$$\tau_{form} = t = \frac{1}{m_T}, \qquad (9.11)$$

where t is a time of forming a particle with a transverse mass m_T. The transverse energy and multiplicity distributions are investigated in Ref. [172] for three values of the RHIC energies. It was shown that the transverse energy of charged particles within a wide interval of impact parameters is equal to 0.85 GeV for the total RHIC energy. This value of the transverse mass corresponds to the formation time $\tau_{form} = 0.35$ fm. Then, taking into account a pseudorapidity distribution of charged particles in $Au + Au$-collisions at the total RHIC energy [173], we find that the energy density at the formation time is

$$\epsilon = 15 \text{ GeV/fm}^3. \qquad (9.12)$$

Because this value is get from estimations of final particles density, then we should consider it as an upper estimation. The found value (9.12) substantially exceeds the limit that is needed to form a fireball. However, we should keep in mind that this value was obtained for the state that could not be considered as a state where the thermodynamical (even local) equilibrium has been established. Next, following Ref. [171], one may evaluate the thermalization time and find the energy density

$$\epsilon(\tau_{therm}) \approx 5.4 \text{ GeV/fm}^3.$$

Therefore, the experimental data obtained at RHIC lead to the conclusion that the energy densities reachable at central collisions exceed the necessary limit for the QGP production that is given by the LQCD computations.

Before the RHIC operation, it was expected that a multiplicity, which is defined by the soft particles yield, will be described by models allowing for mini-jets production. The pointed models predict the fast growth of the multiplicity with increasing the initial energy, this growth becoming more drastic with increasing a centrality. Data obtained at RHIC during the first operation year [174] showed that the multiplicity growth is notably weaker than had been expected. To explain such a relatively weak growth of the multiplicity versus the energy, two possible mechanisms were proposed. The former explains the weakening of the multiplicity growth due to "shadowing" of gluons [175]. The latter is based on the assumption of gluon distributions saturation at small values of $x = E_{constituent}/E_{hadron}$ thanks to the nonlinear interaction of gluons (gluon-gluon fission and gluon fusion). A state appearing in the process is called a *color glass condensate*[*] [176]. Current experimental date testify that models allowing for gluon distributions saturation are the most probable candidate for description of the multiplicity of soft particles.[†]

At the time of experiments in 2001, all four detectors of RHIC observed a specific effect caused by a jet quenching. When colliding two ions at ordinary conditions, they give two particle jets to be scattered in opposite directions. However, at a collision of two gold nuclei, the detectors fix from time to time the presence of only one jet. During June–March 2003, test experiments in the course of which gold ions are collided with lighter deuterium ions were fulfilled. Although the energy of gold ions was remained identical to that of basic experiments, the combined energy of a collision was no longer sufficient to obtain QGP. When two raced gold nuclei are subjected to a head-on collision, reachable temperatures are so extremal that, in merging gold nuclei, protons and neutrons are fused to set at liberty their constituent quarks and gluons. In contrast, a small deuterium passes through a large gold nucleus like a bullet not heating and not compressing it. As a result, a gold nucleus remains in its ordinary state, that is, in the state made up of neutrons and protons.

The jet quenching effect is one of the most striking results obtained in nucleus-nucleus collisions at the RHIC energy. First, this effect cannot be observed at lower energies because of a small cross section of jets production. Second, the jet quenching effect belongs to so-called hard probes to provide data concerning early reaction stages. Third, this effect has an obvious interpretation and was predicted long before the RHIC launching [177, 178]. On qualitative grounds the jet quenching effect can be described as follows. At the first reaction stage, partons belonging to colliding nuclei are hard scattered to create partons with large transverse momenta, those next fragment into a hadronic jet with large transverse momenta $p_T \geq 3 - 6$ GeV. Passing through this heated nuclear matter, these partons are losing an energy that results in the jet quenching effect. As this takes place, energy losses of partons depend on properties of environment in which partons are moving. Obviously, energy losses of color parton in QGP are appreciably larger as compared with that in the environment consisting of colorless hadrons. It is natural that the value of energy losses depends on properties of a heated and compressed quark-gluon matter (first of all on gluon density)

---

[*]The color glass condensate (CGC) is a new form of nuclear matter determining properties of the strong interaction at very high energies. It is universal and does not depend on a specific state of a hadron. The CGC must describe the cross sections at high energies, distribution of produced particles, and distributions of hadron constituents at small values of x, as well as initial conditions in relativistic collisions of heavy ions. Just as nuclei and electrons form part of atoms, nucleons involve the CGC fields as fundamental fields at high energies. Therefore, at ultrarelativistic energies, heavy ions collision may be considered as a collision of two sheets of color glasses under which thawing of color glasses takes place and quarks and gluons are produced.

[†]Just these particles give the main contribution to the multiplicity.

through which partons pass. For the RHIC conditions, energy losses of a quark dE/dx reach values of 1 GeV/fm [179]. In Fig. 9.5 the jet productions in the high-energy proton-proton and nucleus-nucleus collisions are displayed. One of the direct signals that allow one to

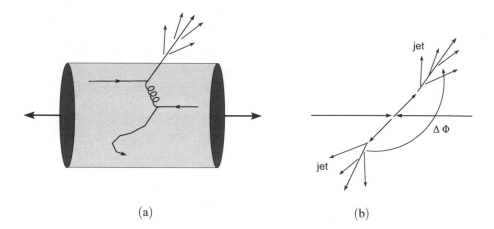

(a)   (b)

FIGURE 9.5

The jet productions: (a) in a nucleus-nucleus collision one of the leading partons escapes the interaction region and yields a jet in the detector, but the other one is absorbed by the surrounding matter; (b) in a proton-proton the leading partons fragment into two back-to-back hadronic jets, which are both observed by the detector.

observe the jets quenching effect is the measurement of the yields of hadrons with large transverse momenta both in nucleus-nucleus collisions and in nucleon-nucleon collisions at identical collision energy. From the scenario described above it follows that the energy losses of partons with large transverse momenta lead to decreasing the yields of high-energy hadrons in nucleus-nucleus collisions. To quantitatively characterize such a suppression one introduces a nuclear modification factor R_{AB}, which is equal to the ratio of the number of detected hadrons to the number of hadrons that must be created with regard to the number of pair collisions between nucleons to be part of the colliding nucleus. As early as the first year of the RHIC operation at the energy $\sqrt{s_{NN}}$=130 GeV, experimental data showed that the yield of neutral pions with large transverse momenta $p_T \geq 2$ GeV in $Au + Au$-collisions was suppressed approximately by a factor of five ($R_{AA} \approx 0.2$) and the suppression value weakly depended on a detected pion momentum. Such a suppression could be explained assuming that a parton produced in a hard scattering loses, on the average, 20% energy when passing through a fireball produced in a central collision.

Now, we touch upon roles of different quarks at the QGP production. Three lightest quarks (u, d, s), and corresponding antiquarks, are created in abundance (and approximately in identical amount) during collisions at RHIC. The strange quark mass $m_s \approx 104 \pm 30$ MeV is comparable to the confinement scale of QCD. For this reason alone it becomes obvious that s-quark must influence the QCD phase diagram. Really, lattice simulations have shown that the location of the QCD critical point is extremely sensitive to the value of the strange quark mass (see [180] and references therein). Moreover, strange quarks may be responsible for the existence of a second critical point on the phase boundary between hadronic and quark matter in the region of large baryon density and low temperature. Since strange quarks are not easily produced below the critical temperature T_c, they thus serve as indicators of quark deconfinement. It should be also noted that the structure of the color superconducting phase of quark matter is critically sensitive to the value of m_s (the color-flavor locked phase, in particular, could not exist without strangeness).

Inasmuch as for heavier quarks to be produced larger energy is needed, in mini-explosions they appear earlier (when energy densities is larger) and considerably rarer. By this reason, heavy quarks are valuable indicators of fluxes form and other properties that are inherent in early stages of the mini-explosion evolution. The PHENIX and STAR facilities fit well for such an investigation because they can detect high-energy electrons and muons that often appear under decays of heavy quarks.

In accordance with the last results one may state that at RHIC the quark-gluon state of matter existing during 10^{-23} s was observed. However, it looks like a liquid rather than a gas. The notion of QGP as a gas of weakly interacting particles that existed before the RHIC operation does not find its confirmation in experimental data. One of the examples is a small value of the equilibrium establishment time. The next example is associated with the distribution form of final hadrons. In heavy ion collisions, an economical way of describing the evolution of the system is provided by relativistic hydrodynamics, which transforms the gradients of the initial parton density into the momentum flow of the produced hadrons. Of particular interest is the azimuthal asymmetry of the hadron momentum distribution parameterized by the *elliptic flow*. In non ideally central collisions (the most frequent case), the distribution of outgoing hadrons that reach a detector has an ellipse form. More energetic hadrons are moving mainly within an interaction plane rather than at a right angle to this plane. The elliptical distribution is evidence of the fact that in quark-gluon matter, substantial pressure gradients act and free quarks and gluons from which outgoing hadrons were produced behave as a particles collective, that is, as a liquid rather than a gas. From a gas hadrons would fly out uniformly in all directions. If quark-gluon matter possesses liquid properties, then liberated particles rather intensively interact with each other. The decrease of this interaction caused by the asymptotical freedom is overridden by a drastic increase of liberated particles. The obtained crowd of strongly coupled particles is almost no different from a liquid. This is contrary to the naive theoretical picture where the environment in question is considered an ideal liquid.

A detailed study of elliptical asymmetry shows that this quantity is sensitive to the value of shear viscosity [181]. It appears that the data favor small values of shear viscosity that are not far from the *perfect liquid* bound (calculated value is approximately one-tenth of shear viscosity of superfluid helium). It is very surprising that the most hot and dense matter of all the available ones is in excess of all known liquids by an ideality extent. Viscous effects contribute not only at the quark-gluon plasma stage of the evolution, but also after the hadronization; these effects can be introduced through a hadron cascade [182].

Theoretical calculations also do not testify in favor of QGP as a system of weakly interacting partons. Let us demonstrate it. Assume that we deal with an ideal quark-gluon gas. From Bose–Einstein and Fermi–Dirac distributions it follows that for massless non-interacting particles at the temperature T the energy density $\epsilon_0 = \pi^2 T^4/30$ is accounted for by one boson degree of freedom while $7\epsilon_0/8$ — by one fermion degree of freedom (these values are an analogue of the Stefan–Boltzmann law for the blackbody). With allowance made for contributions caused by the number of quarks, antiquarks and gluons as well as by flavor, color, and spin degrees of freedom, we get the following expressions for the energy density

$$\epsilon_{SB} = \left[2_f \cdot 2_s \cdot 2_q \cdot 3_c \cdot \frac{7}{8} + 2_s \cdot 8_c \right] \epsilon_0 = 37\epsilon_0 \qquad (9.13)$$

for two active quarks and

$$\epsilon_{SB} = \left[3_f \cdot 2_s \cdot 2_q \cdot 3_c \cdot \frac{7}{8} + 2_s \cdot 8_c \right] \epsilon_0 = 47.5 \epsilon_0 \qquad (9.14)$$

for three active quarks. In Fig. 9.6 the energy density as a function of the temperature is shown [183]. The results of Eqs. (9.13) and (9.14)) are shown by arrows while those of lattice calculations are represented by dots on curves (they are drawn to distinguish calculations with different conditions). Note that for normal hadronic matter (the case $T < T_c$), when

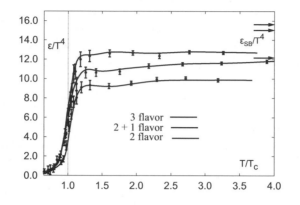

FIGURE 9.6
The energy density versus the temperature.

only pions contribute, the numerical coefficient in front of ϵ_0 is equal to 3. As follows from Fig. 9.6, LQCD gives the energy density less by 30–35% than the corresponding value in the Stefan–Boltzmann limit. By way of other example, confirming the hydrodynamical model of QGP, we mention the LQCD results for the effective value of the coupling constant [184, 185].

9.3 QGP within holographic duality

Much of our physical intuition is based on the quasi-particle description. Indeed, it is difficult to imagine a composite object without well-defined constituents. However, this kind of behavior can be encountered, for example, in studying conformal theories. Conformal theories do not possess dimensional scales, so the spectral density of an excitation should look like a structureless mass distribution—instead of a particle with a well-defined mass we are dealing with a broad amorphous *unparticle* [186]. This situation is somewhat reminiscent of what we expect to happen to QGP at strong coupling—no well-defined quasi-particles exist, just very broad *unquarks* and *ungluons*. We are thus dealing with what may be called an *unplasma* of such unquark and ungluon states.

While we still do not know how to treat QCD analytically at strong coupling, there has been a theoretical breakthrough in understanding the dynamics of maximally supersymmetric conformal $\mathcal{N} = 4$ Yang–Mills theory based on the holographic correspondence between the field theory describing behavior on the $4D$ Minkowski boundary and the superstring theory in the $5D$ bulk Anti-de Sitter (AdS) space, supplemented by a $5D$ sphere, $AdS_5 \times S_5$. Note, that an important feature of the AdS space (an space-time with the negative curvature) implies that it has a boundary where time is well defined. This boundary existed and will eternally be existing while the expanding universe resulting from the Big Bang does not possess such a boundary. The AdS_5 metric is unique if we want to extend the Minkowski boundary space into the fifth dimension preserving the conformal invariance. The AdS/CFT (Ads/conformal field theory) correspondence was originally proposed by J. Maldacena [187] and its important aspects were given in Refs. [188, 189]. This correspondence is sometimes called the *holographic duality (HD)* or the *gauge-string duality*. Let us elucidate what is meant by this correspondence.

We were already faced with different kinds of correspondences. Thus, an ordinary optical hologram is a plain (two-dimensional) picture. However, when looking at it, we see a three-dimensional image. The three-dimensional image proves to be equivalent to a two-dimensional picture, that is, a three-dimensional world appears from the physics of a two-dimensional universe. So, duality is defined as the connection between two, it would seem, absolutely different theories. By analogy, in quantum field theory the connection between the gravitation theory in some world (call it the A theory) and a particular theory of elementary particles living on the border of this world (call it the B theory) is referred to as the HD. At that, the following is important. If $f_A(g_A)$ labels any observable in the A theory, where g_A is a coupling constant in the A theory, then there is a corresponding observable $f_B(g_B)$ in the B theory, such that $f_A(g_A) = f_B(1/g_B)$ and $f_A(1/g_A) = f_B(g_B)$. It means that the HD connects one theory in the regime of the weak (strong) coupling with another theory in the regime of the strong (weak) coupling. Consequently, the HD confronts some physical laws that are acting in a volume with another laws to be being true at a surface bounding this volume. Physics on a border is represented by quantum particles that have color charges and interact in much the same way as quarks and gluons of the standard model. Laws in the interior represent a sort of string theory, including the gravitation interaction that is hard to describe in terms of quantum field theory. However, physics on a surface and in a volume are completely equivalent in spite of totally different ways of description. In other words, the HD states that the elementary particle theory on the surface and the gravitation theory in the volume, despite their disparateness, represent one and the same theory that is simply observed from different directions.

Here are some examples. In the 1970s S. Hawking showed that black holes had temperature and emitted a radiation. However, temperature is a property of totality particles and then one must provide an answer *of which particles do black holes consist?* In the holographic theory, this problem is solved simply: a black hole is equivalent to a swarm of interaction quarks and gluons on a bounding space-time surface. Since the number of particles is very large and all they are in continuous motion, then to determine temperature one may use the ordinary rules of statistical mechanics. The calculated value exactly coincides with the result Hawking obtained using an absolutely different way.

We may employ the holographic correspondence in the opposite direction, and, using known properties of black holes in internal space-time, deduce the behavior of quarks and gluons at very high temperatures on a border. In such a way, investigating the shear viscosity of black holes leads to the conclusion that it is extremely small, namely, it is less than the shear viscosity of any known liquid. Then, because of the HD, the viscosity of strongly interacting quarks and gluons must also be very small at high temperatures.

However, physicists shall have to reply to many questions about holographic theories. In particular, whether there is something like the HD in our universe rather than in the AdS space. The history of physics evolution shows that the number of different types of mathematically self-consistent theories is not so large. Besides, too large coincidences do not happen by chance, a reason must necessarily be found for them. Therefore, if two theories even very different at first glance start to somewhat resemble each other, there appears a chance that they are various sides of an identical mathematical construction. The creation of five string theories and the discovery of the fact that all they are various manifestations of one fundamental theory [190] was the most brilliant confirmation of this principle.

The story of the gauge-string duality turns on the understanding of $D3$-branes.* By definition, *D-branes* are locations in ten-dimensional space-time where strings can end. A $D3$-brane is a D-brane extending in three spatial dimensions as well as in the time direction. The low-energy excitations of a single $D3$-brane are described by $\mathcal{N} = 4$ supersymmetric $U(1)$ gauge theory. If N $D3$-branes are placed on top of one another, one finds instead $\mathcal{N} = 4$ supersymmetric $U(N)$ Yang–Mills gauge theory ($\mathcal{N} = 4$ SYM). This theory splits into a $U(1)$ part, which is free, and relates to the center of mass motions of the stack of $D3$-branes; and an $SU(N)$ part, which is interacting, and relates to the relative motions of the branes. Gluons of these gauge theories can be represented as strings running from one brane in the stack to another. If $N = 3$ and we label the branes as R, G, and B, then a string running from R to B has color quantum numbers $\bar{R}B$, as appropriate for a gluon of the $SU(3)$ gauge theory. The $\mathcal{N} = 4$ supersymmetry of $D3$-branes dictates that this $\bar{R}B$ string can act not only as a gluon, but also as a fermion with spin $1/2$ or as a scalar. The strings describing these superpartners of gluons are distinguished from the ones describing gluons themselves by fermionic excitations on the string. Regardless of spin, all particles in the $\mathcal{N} = 4$ SYM are charged in the adjoint of the gauge group. The precise dynamics of the gauge theory is almost entirely determined, at the level of a renormalizable Lagrangian, by the number of branes N and the supersymmetry. The only freedoms one has are to adjust the gauge coupling constant and a θ parameter that breaks the CP invariance. $D3$-branes have a definite mass per unit volume, and they have a definite charge under a 5-form field strength. Therefore, they deform space-time into a solution of the ten-dimensional Einstein equations coupled to the 5-form. Close to the $D3$-branes, this solution takes the form of a direct product, $AdS_5 \times S^5$. The gauge-string duality is an equivalence between string theory on this geometry and the $\mathcal{N} = 4$ SYM in four dimensions. It is significant that Green functions of gauge-invariant operators, Wilson and Polyakov loops represent the gauge theory constructions that have well-studied translations into the language of string theory in the AdS space.

So, the AdS/CFT conjecture establishes duality between the $\mathcal{N} = 4$ SYM theory with arbitrary values for the parameters g_s, N_c and the type IIB superstring theory [190] living in a $D = 10$ curved space-time, which is $AdS_5 \times S^5$. The duality extends to a finite temperature by adding a black hole to AdS_5. Thus, one obtains the $AdS_5 \times S^5$-Schwarzschild metric, for which a common parametrization reads

$$ds^2 = \frac{r^2}{R^2}[-f(r)dt^2 + d\mathbf{x}^2] + \frac{R^2}{r^2 f(r)}dr^2 + R^2 d\Omega_5^2, \tag{9.15}$$

where t and $\mathbf{x} = (x, y, z)$ are the time and spatial coordinates of the physical Minkowski world, r (with $0 \leq r < \infty$) is the radial coordinate on AdS_5 (or 5th dimension), and $d\Omega_5^2$ is the angular measure on S^5. Furthermore, R is the common radius of AdS_5 and S^5, and

$$f(r) = 1 - \frac{r_0^4}{r^4}, \tag{9.16}$$

where r_0 is the black hole horizon and is related to its temperature T (the same as for the $\mathcal{N} = 4$ SYM plasma) via $r_0 = \pi R^2 T$. Note that this black hole is homogeneous in the

---

*A *brane* is any lengthy object in a string theory. $D1$-brane is called a *string*, $D2$-brane—a *membrane*. In general, the Dp-brane has p spatial and one time dimensions.

four physical dimensions but has a horizon in the fifth dimension, which encloses the real singularity at $r = 0$. When $r \to \infty$, $f(r) \to 1$ and $ds^2 \propto (-dt^2 + d\mathbf{x}^2)$ is conformal to the flat Minkowski metric. In fact, we have $f(r) \approx 1$ whenever $r \gg r_0$, so far away from the horizon the geometry is $AdS_5 \times S^5$. Besides R, the superstring theory involves two more parameters, the dimensionless string coupling constant g_{st} and the string length L_{st}, which is the characteristic scale on which the string structure (as opposed to a point-like particle) can be resolved, and is related to the Planck length in ten dimensions by $L_P = g_{st}^{1/4} L_{st}$. The AdS/CFT correspondence makes the following identification between the free parameters of the two dual descriptions:

$$4\pi g_{st} = g_s^2, \qquad (R/L_{st})^4 = g_s^2 n_c \equiv \lambda \qquad (9.17)$$

The first relation means that when the Yang–Mills coupling g_s^2 is small, so is also the string coupling, hence one can neglect quantum corrections (string loops) on the string theory side. The second relation tells us that when λ is large, the geometry of the string theory is weakly curved, so that the massive string excitations (with mass $m \sim R/L_{st}^2$) can be reliably decoupled from the low-energy ones, and then the superstring theory reduces to type IIB supergravity. Therefore, when we have

$$n_c \to \infty \quad \text{and} \quad \lambda \equiv g_s^2 n_c \to \infty \quad \text{with} \quad g_s^2 \ll 1 \qquad (9.18)$$

(the strong t'Hooft coupling limit), the dual superstring theory reduces to classical supergravity in ten dimensions. After also performing a Kaluza–Klein reduction around S^5 and keeping only the lowest harmonics, one finally gets a classical theory in five dimensions that involves massless fields, among which the 5-dimensional graviton, the dilaton, and a SO(6)∼SU(4) non-Abelian gauge field. The quantum correlation functions in the strongly coupled CFT can now be calculated using solutions of the classical motion equations for these massless fields with appropriate boundary conditions.

So, a particularly useful feature of the holographic correspondence is the duality between the strongly-coupled Yang–Mills theory and the weakly coupled gravity that can be treated by semi-classical methods. On the other hand, the weakly coupled Yang–Mills theory, which is amenable to the perturbation theory analysis is dual to strongly coupled quantum gravity. In particular, there is an intriguing relation between Yang–Mills dynamics at weak coupling and the dynamics of gravitational collapse at strong coupling [191]. Most of the applications, however, deal with the strongly coupled Yang–Mills theory using the duality to translate some hopelessly difficult field-theoretical problems into treatable exercises in classical gravity. Since the finite temperature T is introduced on the gravity side by putting a black hole in the center of the AdS_5, it is thus associated with a Hawking temperature of the black hole. In its turn, the entropy density s of the Yang–Mills plasma is bound up with the black hole event horizon.

One should note a very interesting conclusion obtained with the help of the HD, namely, the transition from hadronic phase to QGP phase must happen in two stages [192, 193]. In accordance with the HD, the behavior of quarks in the strong interaction theory mathematically correspond to that of some $D3$-brane suspended in ten-dimensional space over a black hole. This brane possesses a surface tension, the less the nuclear matter temperature, the stronger being a surface tension of the brane. Computations showed that at low temperatures this surface tension holds back the brane from a fall into the black hole. Giving back to the *ordinary* world, one can prove that this is responsible for the situation with confinement. When increasing the nuclear matter temperature, the brane becomes more and more pliable and at some critical temperature begins to be sucked in the black hole. In Ref. [192] it was proved that a phase transition associated with this suck leads to the deconfinement. However, the properties of this phase transition proved to be very unusual. In the HD picture, hadrons are described by brane oscillations. At low temperatures the surface tension of branes is strong so masses of hadrons are large. According to computations, when increasing temperature, masses of hadrons are decreased because of weakening the surface tension of branes and, at the deconfinement instant, become zero. As this takes place, hadrons themselves do not disappear, only their masses vanish. The final disappearance of hadrons, that is, the absence of brane oscillations, comes only at even higher temperatures. Therefore, by the HD, the deconfinement is under way actually in two stages. When temperature exceeds the first critical value, individual gluons and massless hadrons begins to freely travel along a nuclear matter. Emphasize that hadrons are not broken down into individual quarks in the process. This happens only at exceeding the second temperature threshold after which we may say that the nuclei has finally been fused. It should be noted that all the phenomena described above were obtained within approximation of large color numbers ($n_c \to \infty$) in the Ad-space, rather than within the real QCD.

9.4   Resume

The quark-gluon liquid with corrections for scales and temperatures vastly resembles ordinary liquids, like water. It has a very small shear viscosity and possesses a high degree of homogeneity. The obtained results are evidence of the fact that the interaction force between quarks and gluons is greatly stronger than they considered before. Therefore, the next task is to investigate the state of this liquid, which is one of the most unusual liquids existing in the universe, that is, to find its parameters such as viscosity, heat capacity, sound speed, and so on. However, it turns out that, for the quark-gluon liquid compressed to unimaginable density and sweepingly expanding with a near-light speed in all directions, a fulfillment of any QCD calculations is extremely complicated. At the weak coupling constant g_s, quarks and gluons are the dynamical degrees of freedom in QGP. The decay widths Γ of quark and gluon quasi-particles are small compared to the typical thermal scale given by the temperature T ($\Gamma \ll T$) because they are suppressed by the powers of the coupling constant $g_s \ll 1$. Gluons acquire a dynamical Debye mass $m_D \sim g_s T$, and the color Coulomb potential is screened at distances $\sim 1/m_D$. Powerful resummation techniques have been developed to describe the thermodynamics and transport properties in this regime. Resummation methods extend the applicability of the weak coupling approach down to moderate temperatures, perhaps as low as $T \simeq (2 \div 3)T_c$ (see, for a review, [194]. The parton scattering cross sections in this regime are small, $\sim g_s^4$, and this leads to large shear viscosity η and small bulk viscosity ζ. Things become much more complicated at temperatures $T_c \leq T \leq (2 \div 3)T_c$. In this temperature range (most relevant for RHIC experiments) the coupling constant $g_s^2(T)$ is not small, and the quasi-particle description breaks down. In other words, the gluon clouds of the quark and gluon quasi-particles become so dense and the interactions between the clouds of different partons becomes so intense that the representation of this system as a superposition of individual dressed partons fails, even as a rough approximation to reality.

So, the question arises: *whether the QGP similarity to a liquid, which was revealed at RHIC, will be conserved at higher temperatures and energy densities available at the LHC operation.* There are two points of view. The former is: as soon as a mean energy of quarks exceeds 1 GeV, the force acting between quarks becomes weak and the QGP begins to behave, nevertheless, as a gas. Adherents of the second point of view state that at higher energies the QCD force is decreasing insufficiently fast and, as a result, quarks and gluons keep strongly coupled like water molecules. For this reason, at present two designations are used: wQGP for a gas of weakly interacting partons and sQGP for a plasma of strongly interacting partons. More detailed discussion of these issues may be found in Ref. [185].

In conclusion in this case, we briefly discuss future trends in the QGP investigation. The future on the experimental side appears bright, by virtue of the ongoing upgrades of the STAR and PHENIX detectors at RHIC, combined with the RHIC luminosity upgrade, and the startup of the LHC. Physics of heavy ion collisions is an integral part of the baseline program of the LHC that will accelerate $Pb - Pb$ beams at $\sqrt{s_{NN}} = 5.5$ TeV, a factor 27 higher than the top RHIC energy.* This increase is even larger than the factor 10 in going from the CERN-SPS to RHIC. It will lead to significant extension of the kinematic range in transverse momentum p_T and in x. The ALICE experiment (A Large Ion Collider Experiment) at the LHC [195, 196] will probe a continuous range of x as low as about 10^{-5}. The study of low x regime, especially at forward rapidities, will be most appropriate to study the early stage of nuclear collision.

From the measured spectra and particle ratios, it is possible to estimate the freeze-out temperature and chemical potential by using thermal model fits [197]. The chemical potential at top RHIC energies is between 20–40 MeV and at LHC energies it is expected to be less than 10 MeV. The QCD phase boundary is slowly getting mapped with data points from various experiments. Important points missing from our experimental radar are the QCD critical points. We need to have a good guidance from the lattice calculations with regard to the location of these points. The exact location of the critical points is not known yet. Thus, for example, since various lattice calculations give different values of chemical potentials (from 180 MeV to 500 MeV) for the first critical point, it is obvious that a thorough energy scan is needed from $\sqrt{s_{NN}}$−5 GeV to 100 GeV, in order to probe the region around this critical point.

---

*In November 2010, CMS Collaboration reported about observation of the quark-gluon plasma under collisions of lead ions.

10

QCD vacuum

Ain't never been such sort of thing
and here we go again.
Ex-Prime Minister of Russia V. Tchernomyrdin

10.1 Kinks and breathers

At present, in many regions of physics, different nonlinear equations are intensively investigated. Real physics are actually nonlinear, however, solutions of appropriate problems are extremely complicated. Moreover, since within quantum field theory, matters are considered as point objects, then it leads to unpleasant obstacles in the form of infinities. Thus, even in classical field theory we face the self-energy problem of a point charge. In quantum field theory, these troubles do not disappear and, on the contrary, are redoubled and find their solution only after the renormalization procedure that is not without some affectation share. However, it turns out that in nonlinear classical field theories, there are promising solutions known as *solitons* that represent stable and extended configurations not possessing singularities. Let us proceed to the investigations of these objects.

A soliton is characterized by the following properties: it is localized in a finite region; it propagates without deformation carrying energy, momentum, and angular momentum; it preserves its structure under interactions with other solitons; it can form coupled states and ensembles. In a nonlinear environment the soliton form is defined by two competitive processes: spreading of a wave because of an environment dispersion and breaking of a growing wave front because of nonlinearity.

Until the beginning of 1960s, a soliton denoted a solitary wave—a wave packet of an invariable form that propagates with constant speed either on a surface of liquid having finite depth or in a plasma. At present, a totality of various physical objects comes under the soliton definition. The main classification of solitons is made by means of the number of spacial dimensions along which stationary perturbation localization of nonlinear medium takes place. Among one-dimensional solitons are solitary waves in a cool magnetized plasma and in a plasma without a magnetic field, domain walls in ferro- and antiferromagnetics, quanta of a magnetic flux in Josephson contacts in superconductors, and so on. Among two-dimensional solitons are dislocations in a crystalline lattice, magnetic tubes (Abrikosov vortices) in superconductors of the second kind, anticyclone regions in geophysical hydrodynamics including the big red spot on Jupiter, and so on. Toroidal vortical structures in ferromagnetics and in a thick layer of superfluid helium He^3, black holes in the gravitation theory and soliton models of elementary particles* are examples of three-dimensional soli-

---

*Investigation of the scattering of classical solitons shows their similarity with analogous processes in hadrons physics.

273

tons. In quantum field theory, solitons localized in four-dimensional space-time, instantons, are considered.

From the mathematical point of view, solitons represent localized stationary solutions of nonlinear partial differential equations or their generalizations (differential-difference equations, integro-differential equations, and so on). In many cases different physical situations and phenomena are described by identical equations, for example, the Korteweg–de Vries equation, the sine-Gordon equation, the nonlinear Schrödinger equation, and so on. Recall that linear equations (except one-dimensional wave equation) do not possess localized stationary solutions.

Soliton solutions of gauge field equations are a major focus of theoretical physics interest. The following solution are of special interest: solutions in two spatial dimensions, that is, strings in three-dimensional space—vortices; solutions in three spatial dimensions localized in space rather than in time—magnetic monopoles; solutions in a four-dimensional space-time localized in space and time—instantons. Note that solitons stability arise from the fact that boundary conditions are broken into different classes, a vacuum belonging only to one of them. These boundary conditions are defined by a chosen form of correspondence (mapping) between coordinate and group spaces. In consequence of the fact that these mappings cannot be converted into each other, they are topologically distinguishable.

For the first time a solitary wave on a liquid surface was observed by J. S. Rassell in 1834 and a corresponding equation was deduced by Korteweg and de Vries

$$\frac{\partial u}{\partial t} + 6u\frac{\partial u}{\partial x} + \frac{\partial^3 u}{\partial x^3} = 0. \tag{10.1}$$

It turned out that this equation also describes an oscillatory motion of atoms chain and, in the limit of small amplitude and a large wavelength, has a soliton solution. From the numerical solution of Eq. (10.1) (see, for example, [198]) it follows that solitons possess significant stability and, under collisions, they are elastically scattered holding their shape and amplitude.

As a case in point, we consider one-dimensional a sine-Gordon equation

$$\frac{\partial^2 \psi}{\partial t^2} - \frac{\partial^2 \psi}{\partial x^2} + a\sin(b\psi) = 0, \tag{10.2}$$

where a and b are constant quantities. This equation describing a scalar field has two exact solutions. The former, called a *kink*, represents a solitary wave to move with a speed v ($v^2 < 1$). Let us begin with this very case. In accordance with the causality principle running solutions must have the form:

$$\psi(x,t) = f(x - vt) = f(\xi). \tag{10.3}$$

By the direct substitution one may be convinced in the fact that the function $f(\xi)$ is governed by the expression

$$f(\xi) = \frac{4}{b}\arctan[\exp(\pm\gamma\xi)], \tag{10.4}$$

where $\gamma = 1/\sqrt{(1 - v^2)}$. As we can see, this wave possesses very unusual properties: it moves without changing a form and a size and, as a result, without dissipation (Fig. 10.1). Since kinks are solutions of nonlinear wave equations, then the superposition principle does not work for them. It means that when two kinks are encountered, a wave of a quite complicated form is produced. And at the same time, the surprising thing is that, after collisions, kinks are again divided, that is, they, as it were, pass through each other. The fact that the quantization of kinks represents a nontrivial problem are another corollary of the absence of the superposition principle.

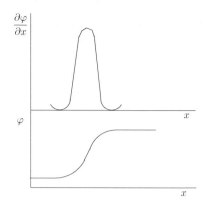

FIGURE 10.1

The solitary wave.

It is obvious that Eq. (10.2) has an infinite large number of constant solutions; those, as will be shown later, possess zero energy:

$$\psi - \frac{2\pi n}{b}, \qquad n - 0, \pm 1, \pm 2, \ldots, \tag{10.5}$$

that is, a degenerated vacuum is associated with the sine-Gordon equation (here by a vacuum is meant a classical field configuration with zero energy). The Lagrangian corresponding to Eq. (10.2) is as follows:

$$\mathcal{L} = \frac{1}{2} \left(\frac{\partial \psi}{\partial t} \right)^2 - \frac{1}{2} \left(\frac{\partial \psi}{\partial x} \right)^2 - V(\psi), \tag{10.6}$$

where

$$V(\psi) = \frac{a}{b} \left[1 - \cos(b\psi) \right],$$

and a constant b is chosen in such a way as to the equality $V = 0$ is fulfilled on the solutions of Eq. (10.5). These solutions are characterized by zero energy because an energy density of a field configuration is

$$H = \frac{1}{2} \left(\frac{\partial \psi}{\partial t} \right)^2 + \frac{1}{2} \left(\frac{\partial \psi}{\partial x} \right)^2 + V(\psi). \tag{10.7}$$

Expanding $V(\psi)$ into the series and introducing the designations $m^2 = ab$ and $\lambda = ab^3$, we get

$$V(\psi) = \frac{m^2}{2} \psi^2 - \frac{\lambda}{4!} \psi^4 + \ldots, \tag{10.8}$$

suggesting that m may be considered as a particle mass and λ as a self-interaction constant. The obtained result is evidence of the fact that the scalar field equation with the self-interaction $\lambda \psi^4$ also possesses the soliton solution.

Construct the following configuration. At $x \to -\infty$ the function $\psi(x)$ approaches to zero of $V(\psi)$ corresponding to $n = 0$ and at $x \to \infty$ it does to zero of $V(\psi)$ associated with $n = 1$. Obviously, between these two zeros there is the region where

$$\psi \neq \frac{2\pi n}{b}, \qquad \frac{\partial \psi}{\partial x} \neq 0, \tag{10.9}$$

and, therefore, an energy density is positive. Assume that the configuration in question is static, that is, $\partial\psi/\partial t = 0$. Owing to the boundary conditions imposed on ψ, a total energy must be finite. Find this energy. We have for the sine-Gordon equation

$$\frac{\partial^2\psi}{\partial x^2} = \frac{\partial V}{\partial\psi}. \tag{10.10}$$

Integrating Eq. (10.10) and reasoning the integration constant to be equal to zero, we obtain

$$\frac{1}{2}\left(\frac{\partial\psi}{\partial x}\right)^2 = V(\psi). \tag{10.11}$$

Then, the stationary state energy is

$$E = \int H dx = \int \left[\frac{1}{2}\left(\frac{\partial\psi}{\partial x}\right)^2 + V(\psi)\right]dx = \int 2V(\psi)dx = \int_0^{2\pi/b}\sqrt{2V(\psi)}d\psi =$$

$$= \sqrt{\frac{2a}{b}}\int_0^{2\pi/b}\sqrt{1-\cos(b\psi)}d\psi = \sqrt{\frac{4a}{b}}\int_0^{2\pi/b}\sin\left(\frac{b\psi}{2}\right)d\psi = 8\sqrt{\frac{a}{b^3}} = \frac{8m^3}{\lambda}. \tag{10.12}$$

So, constructed configuration has a finite energy, that is, it can be realized in nature. Note as well that the energy proves to be inversely proportional to the coupling constant.

The stability of a kink is evidence of the existence of a particular conservation law. It is easy to see that a current of the form

$$J^\mu = \frac{b}{2\pi}\epsilon^{\mu\nu}\frac{\partial\psi}{\partial x^\nu}, \tag{10.13}$$

where $\mu, \nu = 0, 1$, $\epsilon^{\mu\nu}$ is an antisymmetric tensor and $\epsilon^{01} = 1$, has zero divergence. A conserved charge associated with it is given by the expression:

$$Q = \int_{-\infty}^{+\infty}J^0 dx = \frac{b}{2\pi}\int_{-\infty}^{+\infty}\frac{\partial\psi}{\partial x}dx = \frac{b}{2\pi}[\psi(+\infty) - \psi(-\infty)] = n, \tag{10.14}$$

where n is a difference between two integer numbers in Eq. (10.5). The charge Q is named a *topological charge*. Therefore, the topological conservation law (10.14) breaks the entire totality of solutions with a finite energy down separate sectors: $n = 0$ (vacuum), $n = 1$ (kink), $n = -1$ (antikink), and so on. It is obvious that solutions with $n \neq 0$ are stable with respect to the transition into a vacuum. It should be stressed that the existence of the current J^μ does not follow from the invariance of a Lagrangian $\mathcal{L}$ relative to any symmetry transformation, that is, J^μ is not the Noether current.

Two kinks separated by a distance L, which is larger than their typical sizes $\sim \gamma$, may conveniently be represented as two relativistic particles interacting with a potential $U(L) \sim Q_1 Q_2 \exp(-L)$. Therefore, kinks with identical topological charges are repulsed while those with opposite topological charges are attracted.

The pair of kinks with opposite charges (kink and antikink) can generate a coupled oscillating state, the so-called *breather*, which represents the second type of an exact soliton solution of Eq. (10.2):

$$\psi_{br} = 4\arctan\{\tan\beta\cos[t\cos\beta]\cosh^{-1}x\sin\beta\}, \tag{10.15}$$

where, for the simplicity, we set $a = b = 1$. A parameter β is changed within the limits $0 < \beta < \pi/2$ and characterizes the breather coupling energy. A moving breather can be obtained by a corresponding Lorentz transformation from (10.15). Collisions of breathers with each other, as well as with kinks, are also elastic. In real systems a breather is not observed because of dissipation.

In the limit $\psi^2 \ll 1$ the substitution

$$\psi(x,t) = \varphi(x,t)\exp(-it) + \varphi^*(x,t)\exp(it)$$

converts Eq. (10.2) in the nonlinear Schrödinger equation:

$$i\frac{\partial\varphi}{\partial t} + \frac{\partial^2\varphi}{\partial x^2} + 2|\varphi|^2\varphi = 0. \tag{10.16}$$

At that, the breather (10.15) (when $\mu \ll 1$) is transformed into a rest soliton with an amplitude μ:

$$\varphi(x,t) = 2i\mu\exp\left(4i\mu^2 t\right)\cosh^{-1}\left(2\mu x\right).$$

In quantum field theory solitons appear as solutions ensuring local minima of the action. The quantum approach to solitons demands the realization of the quantization procedure of fluctuations around a classical solution (quasi-classical approximation). In quantum field theory, the significance of the sine-Gordon equation is caused by the fact that it is related to equations for which an exact quantum solution for S matrices of solitons could be obtained. As this takes place, as in the classical theory, interactions of solitons do not lead to additional production of particles, that is, it is elastic. As a consequence, the S matrix of multiparticle processes possesses the factorability property, that is, it is represented in the product form of the S matrices of different paired processes.

10.2   Topology and vacuum in gauge theory

In classical physics, a force acting on a particle with a charge q in the electromagnetic field is given by the Lorentz formula.

$$\mathbf{F} = q\{\boldsymbol{\mathcal{E}} + [\mathbf{v} \times \mathbf{B}]\}. \tag{10.17}$$

Then, as might appear at first sight, only the physical influence of the electromagnetic field on a charge is of the Lorentz force, which acting only in those regions where $\boldsymbol{\mathcal{E}}$ and $\mathbf{B}$ are nonzero. However, there exists a quantum-mechanical effect, Aharonov–Bohm effect (ABE), which shows that physical effects occur in the regions where $\boldsymbol{\mathcal{E}}$ and $\mathbf{B}$ vanish but the potential A_μ is nonzero. The presence of such a nonlocal influence of the electromagnetic field on a charged particle, which vanishes in the classical limit, stresses that at the quantum consideration an interaction of a charged particle with the electromagnetic field is not reduced to local action of the Lorentz force on this particle. The first to point to the existence possibility of such an effect were W. Ehrenberg and R. E. Siday in 1949. Independently, the detailed theoretical investigation of the effect was given by Y. Aharonov and D. Bohm in Ref. [199] where a close relation of this effect with quantum theory foundations was noted. The first report concerning an experimental observation of the ABE was represented in Ref. [200].

Proceed to the investigation of this effect. Let us consider diffractions of electrons on two slits (Fig. 10.2) When a path difference Δ of de Broglie waves coming from points A and B to screen point P amounts to an even (odd) number of half-waves ($\lambda/2$), then we have the maximum (minimum) in this point. Because in a real experiment $L \gg D$, then one may consider that the angle $\angle ACB$ is practically right and AC is approximately perpendicular

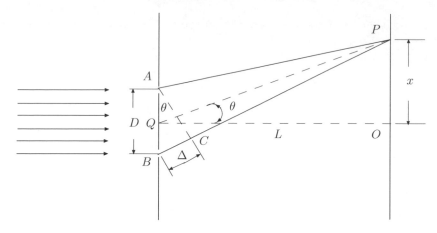

FIGURE 10.2

The electron diffraction on two slits.

to PQ; consequently, $\angle BAC \approx \theta$. This, in turn, gives for the path difference the following expression:

$$\Delta \approx D \sin \theta. \tag{10.18}$$

Next, we take into account that the distance from point P to mean point Q located between slits is approximately equal to L. Therefore,

$$\sin \theta \approx x/L,$$

and then we may write

$$\Delta \approx \frac{Dx}{L}, \tag{10.19}$$

Finally, from (10.19) it follows

$$x \approx \frac{\Delta L}{D} = \frac{\lambda L}{D} \delta, \tag{10.20}$$

where $\delta = n$ corresponds to the maximum in the point P while $\delta = n + 1/2$—to the minimum.

Now, we place an infinitesimally small solenoid behind the two slits somewhere in the vicinity of point Q. Inside the solenoid the magnetic field is available while outside the solenoid it is absent. Therefore, when the solenoid is small enough, then all electrons are moving in the region where the field does not exist. However, despite this fact, the interference picture is changed. Therein lies the idea of the ABE.

In the cylindrical system of coordinates with the z axis directed along the solenoid axis, we have

$$B_r = B_\varphi = 0, \qquad B_z = B \tag{10.21}$$

for the field inside the solenoid and

$$\mathbf{B} = 0 \tag{10.22}$$

for the field outside the solenoid. It is easy to see that if we choose the potentials in the following view

$$A_r = A_z = 0, \qquad A_\varphi = \frac{Br}{2} \tag{10.23}$$

inside and

$$A_r = A_z = 0, \qquad A_\varphi = \frac{BR^2}{2r}, \tag{10.24}$$

(R is the solenoid radius) outside solenoid, then the conditions (10.21) and (10.22) will be fulfilled. Let us examine how this field has an influence upon an electron.

The wave function of the electron in the free case has the form:

$$\psi(\mathbf{r}) = \psi_0 \exp\{i[-Et + (\mathbf{p} \cdot \mathbf{r})]\} == A(t) \exp(i\alpha), \tag{10.25}$$

where $\alpha = (\mathbf{p} \cdot \mathbf{r})$. Under the electromagnetic field presence, the electron momentum $\mathbf{p}$ is changed in accordance with

$$\mathbf{p} \rightarrow \mathbf{p} - e\mathbf{A}.$$

As this takes place, the wave function phase α takes the view:

$$\alpha \rightarrow \alpha - e(\mathbf{r} \cdot \mathbf{A}).$$

Therefore, changing the phase along the whole trajectory equals

$$\Delta\alpha = -e \int_C (\mathbf{A} \cdot d\mathbf{r}), \tag{10.26}$$

where C denotes the electron trajectory. Thus, changing the phases difference in point P owing to the interaction with the solenoid field is

$$(\Delta\alpha)_P = \Delta\alpha_1 \quad \Delta\alpha_2 = e \int_{L_1} (\mathbf{A} \cdot d\mathbf{r}) + e \int_{L_2} (\mathbf{A} \cdot d\mathbf{r}) = e \oint_L (\mathbf{A} \cdot d\mathbf{r}) =$$

$$= e \int_S (\text{rot}\mathbf{A} \cdot d\mathbf{S}) - e \int_S (\mathbf{B} \cdot d\mathbf{S}) - e\Phi, \tag{10.27}$$

where Φ is the magnetic flux through the solenoid. So, under the presence of the solenoid, the interference picture is shifted on a value

$$\Delta x = \frac{\lambda L}{D} (\Delta\alpha)_P = \frac{\lambda L}{D} e\Phi \tag{10.28}$$

although electrons are moving in the region where the magnetic field is equal to zero. The possibility of the ABE is formally caused by the fact that the Schrödinger equation for a charged particle in an external electromagnetic field involves a potential of this field. The potential defines the wave function phase and leads to the observed interference effect even under the absence of direct force action of the field on the particle providing a suitable experiment geometry. This effect does not depend on the gauge of potentials and calls forth the phases difference along different possible ways of particles propagation. It exists both for the scalar and for vector potentials of the electromagnetic field.

It could be shown [199] that after collision on the solenoid, the wave function amplitude will look like:

$$A(\varphi) \sim \frac{1}{\sqrt{2\pi k}} \frac{\sin(\pi\Phi/\Phi_0)}{\sin(\varphi/2)}, \tag{10.29}$$

where φ is the scattering angle counted from the direction of the ingoing plane wave and $\Phi_0 = 2\pi\hbar c/e$ is the magnetic flux quantum. Note that the formula is invalid in the region of small angles where an exact calculation is evidence of the shadow appearance behind the solenoid, a weakening coefficient of an ingoing plane wave being equal to $\cos(\pi\Phi/\Phi_0)$. A characteristic feature of the ABE is the disappearance of a scattered wave, when the magnetic flux in the solenoid is equal to an integer number of the magnetic flux quanta $\Phi = n\Phi_0$. Attention is drawn to the fact that the condition of the ABE absence coincides with that of the Dirac quantization for a magnetic charge (see Eq. (17.5) of *Advanced Particle Physics Volume 1*).

Let us demonstrate that the ABE is caused by a nontrivial vacuum topology in a gauge theory. So, outside the solenoid, we have $\boldsymbol{\mathcal{E}} = \mathbf{B} = 0$, therefore, the energy density of the electromagnetic field is $W_{em} = 0$, that is, we are in a vacuum. But since $\mathbf{A} \neq 0$, then we should assume that a vacuum possesses some structure. Taking into account that rot $\mathbf{A} = 0$, we can write $\mathbf{A} = \nabla\chi$, where χ is a function that could be found using the relation

$$A_\varphi = \frac{1}{r}\frac{\partial\chi}{\partial\varphi} = \frac{BR^2}{2r}. \tag{10.30}$$

Integrating (10.30) and setting the integration constant to be equal to zero, we arrive at

$$\chi = \frac{BR^2}{2}\varphi. \tag{10.31}$$

From (10.31) it immediately follows that the function χ is not uniquely defined because it is increased on the value πBR^2 when $\varphi \to \varphi + 2\pi$. At that, the phases difference $\Delta\alpha$ appears thanks to this unambiguity. Really, from (10.27) and (10.31) it is evident that

$$(\Delta\alpha)_P = e\int(\mathbf{A}\cdot d\mathbf{r}) = e\int(\nabla\chi\cdot d\mathbf{r}) = e\chi(\varphi)\Big|_0^{2\pi} = e\pi BR^2 = e\Phi. \tag{10.32}$$

On the other hand, regular unambiguous functions may be in existence only in not simply connected (multiply connected) spaces. Multiply connected space is taken to mean a space in which not all curves can continuously be contracted into a point. In turn, a simply connected space represents a space in which any closed curve may continuously be contracted into a point. In our case, we deal with a space outside the solenoid ($\mathbf{B} = 0$), that is, with a vacuum space. And, as the analysis shows, this space is multiply connected one. Fig. 10.3 elucidates the aforesaid. Really, the curve L_1 can be contracted into a point while the curve

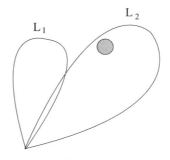

FIGURE 10.3
The multiply connected space in the ABE.

L_2 cannot. Besides, the curve passing around the solenoid n times can be deformed into the one that passes around the solenoid $m \neq n$ times. Therefore, the function is many-valued and this is possible owing to the fact that the one is determined in a multiply connected space. This is equivalent to the inverse conclusion: if the function χ were single-valued, then the equality

$$\mathbf{B} = \text{rot } \mathbf{A} = \text{rot grad } \chi \equiv 0$$

would be fulfilled everywhere and there would be no magnetic flux at all.

So, one may conclude that a multiple connectedness of configuration space of a vacuum is a necessary condition of the ABE existence. Since by definition $A_\mu = \partial_\mu \chi$, then one may say that the function χ is defined in a group space of the $U(1)$ gauge group. This, in turn, leads to the fact that A_μ may be considered by the gauge transformation of the true vacuum $A_\mu = 0$. The corresponding transformation for the electron is

$$\psi' = \exp(-ie\chi)\psi, \qquad \psi^{*'} = \exp(ie\chi)\psi^*, \tag{10.33}$$

that is, it represents an element of the $U(1)$ group. To determine the group space of the $U(1)$ group we report the field ψ in terms of the vector in two-dimensional space with the orthonormalized basic vectors $\mathbf{i}$ and $\mathbf{j}$

$$\psi = \mathbf{i}\psi_1 + \mathbf{j}\psi_2. \tag{10.34}$$

Then, the gauge transformations (10.33) are written in the view:

$$\psi'_1 = \psi_1 \cos \Lambda + \psi_2 \sin \Lambda, \qquad \psi'_2 = -\psi_1 \sin \Lambda + \psi_2 \cos \Lambda, \tag{10.35}$$

where $\Lambda = e\chi$. Obviously, this is nothing more nor less than the rotation of the vector ψ in a two-dimensional plane about the angle Λ. The totality of such rotations generates the $O(2)$ group. Therefore, the $U(1)$ group is locally isomorphous to the $O(2)$ group ($U(1) \approx O(2)$). Indeed, it is easy to see that this is one and the same group. Every element of the $O(2)$ group is unambiguously defined by the rotation angle in the ψ plane. The group space represents a set of values of ψ. At that, one needs to identify values of ψ under $\Lambda = 0, 2\pi, 4\pi, \ldots$ so that they correspond to the same rotation. Consequently, the group space is a circle that will be denoted by S^1. On the other hand, the $U(1)$ group represents a group of all complex numbers of the kind $e^{-i\Lambda} = \cos \Lambda - i \sin \Lambda$, and since $\cos^2 \Lambda + \sin^2 \Lambda = 1$, then the corresponding space is also a circle S^1. The obtained group space of the $U(1)$ group is multiply connected because the path going around a circle twice cannot be continuously deformed (not leaving a circle) into a curve that goes around a circle only once. In reality, there exists an infinite number of equivalent paths in a group space. Attach stricter mathematical meaning to the aforesaid.

To study the topological properties of continuous functions, one can divide them into homotopic classes; each class consists of functions that can be deformed continuously into each other. Consider a space X and define a path a in it as a continuous function $a(s)$ of a real parameter s in such a way that every value of s enclosed within an interval $0 \leq s \leq 1$ is associated with a point $a(s)$ in a space X. If the path a connects points P and Q, then we have $a(0) = P$, $a(1) = Q$. When $a(0) = a(1) = P$, then we deal with a closed path, a loop, in point P. Next, we introduce a conception of homotopic equivalence: two closed curves $a(s)$, $b(s)$ are called *homotopically equivalent* or *homotopic* ($a \sim b$), if and only if there is a continuous function $L(t, s)$ with a parameter $t \in [0, 1]$ that meets the condition

$$L(0, s) = a(s), \qquad L(1, s) = b(s). \tag{10.36}$$

The function $L(t, s)$, which deforms the function $a(s)$ continuously into $b(s)$, is called the *homotopy*. The condition (10.36) divides loops space into homotopic classes: two functions are in the same class if they are homotopic. A path that is inverse regarding a path a is denoted by a^{-1} and defined by the relation:

$$a^{-1}(s) = a(1 - s). \tag{10.37}$$

In that way, it corresponds to an initial path passed in reverse direction. If the final point of path a coincides with the initial point of path b so that $a(1) = b(0)$, then we can determine the paths product $c = ab$ by the relations:

$$c(s) = \begin{cases} a(2s) & \text{when} & 0 \le s \le 1/2, \\ b(2s - 1) & \text{when} & 1/2 \le s \le 1. \end{cases}$$

A zero path represents one point. If $a \sim b$, then a path ab^{-1} is homotopic to zero one. In a space X, we consider a class of paths homotopic to a path a. Denote this class by c_a. It is clear that these paths must have coinciding final points. Homotopy classes may be multiplied using the multiplication law that is determined in the following way:

$$c_a \times c_b = c_{ab}. \tag{10.38}$$

By this law, the so-called fundamental group or the first homotopic group of a space X is defined. In what follows, we shall denote it by $\pi_1(X)$ (as a review on homotopic groups we recommend Ref. [203]). It is easy to verify that $\pi_1(X)$ satisfies to the following group laws:

(i) Completeness: if $c_a \in \pi_1(X)$ and $c_b \in \pi_1(X)$, then, with allowance made for (10.38), $c_a \times c_b \in \pi_1(X)$.
(ii) Associativity: since $(ab)d \sim a(bd)$, then we have $(c_a c_b)c_d = c_a(c_b c_d)$.
(iii) Unit element: this is defined as a class containing zero path c_1 because $c_a c_1 = c_a$.
(iv) Inverse element: $c_a c_{a^{-1}} = c_1$, and, as a result $c_a^{-1} = c_{a^{-1}}$.

A unit circle S^1 is the simplest example of manifold X with a nontrivial fundamental group. By way of a coordinate on a circle it is convenient to take an angle $\theta(s)$, the points $\theta(0) = 0$ and $\theta(1) = 2\pi n$ (n is an arbitrary number) being identified. The continuous functions

$$f(\theta) = \exp[i(N\theta + \lambda)] \tag{10.39}$$

form a homotopic class for different values of λ and a fixed integer N. Really, we can construct a homotopy

$$L(t, \theta) = \exp\{i[N\theta + (1 - t)\theta_0 + t\theta_1]\},$$

such that

$$f_0(\theta) = \exp[i(N\theta + \theta_0)] \tag{10.40}$$

and

$$f_1(\theta) = \exp[i(N\theta + \theta_1)] \tag{10.41}$$

are homotopic. One can visualize $f(\theta)$ of (10.39) as a mapping of a circle onto another circle, $S^1 \to S^1$. In this mapping, N points of the first circle are mapped into one point of the second circle and we can think of this as winding around it N times. So, each homotopic class is characterized by the *winding number* N, which is called the *Pontryargin index* as well. From (10.39) follows that the winding number N for a given mapping $f(\theta)$ can be determined as

$$N = \frac{-i}{2\pi} \int_0^{2\pi} \left[\frac{1}{f(\theta)} \frac{df(\theta)}{d\theta} \right] d\theta. \tag{10.42}$$

It is clear, the fundamental group of the circle $\pi_1(S^1)$ has an infinitely great number of elements that are in one-to-one correspondence with integer numbers N. The product of the classes c_n and c_m consists of loops that initially pass around the circle n times and then do this m times more. By this reason, in $\pi_1(S^1)$, multiplication is associated with the addition of integer numbers:

$$c_n \times c_m = c_{n+m}, \tag{10.43}$$

to lead to the conclusion

$$\pi_1(S^1) = Z,$$

where Z denotes an integer numbers group concerning addition (it is an infinite Abelian group). However, the group space of the $U(1)$ group is exactly S^1. Therefore,

$$\pi_1(U(1)) = Z. \tag{10.44}$$

This, in turn, allows one to conclude: the ABE is caused by the fact that the gauge group of the electrodynamics $U(1)$ is not simply connected (a space X is simply connected if and only if any closed path is homotopic to a zero path so that $\pi_1(X) = 1$).

In the ABE the configuration space represents a plane R^2 with a hole. This plane symbolizes pairs set of real numbers. Topologically this is equivalent to the direct product of the straight line R^1 and the circle S^1: $R^1 \times S^1$. The straight line can be parametrized, for example, by a quantity r and the circle—θ. The gauge function χ is a mapping of a group space G on a configuration space X:

$$\chi: \ G \to X. \tag{10.45}$$

In the case is question G coincides with S^1, while X with $R^1 \times S^1$. Inasmuch as all functions mapping S^1 onto R^1 can be deformed into a constant so that the nontrivial part of the function χ has the form:

$$\chi: \ S^1 \to S^1. \tag{10.46}$$

In that way we obtain a method to define homotopic groups. Let $[X, Y]$ be the set of all homotopy classes for continuous mappings X onto Y. This set forms a group. Therefore, the set

$$[S^1, Y] = \pi_1(Y) \tag{10.47}$$

is the first homotopic group for Y. Analogously, when a d-sphere with $d \geq 2$ (S^d) is a group space of a group, then the set

$$[S^d, Y] = \pi_n(Y) \tag{10.48}$$

is a d-th homotopic group for Y and represents an Abelian group. Note, the first homotopic group is not always Abelian, however, it is an Abelian group in all physically interesting cases. It is evident from (10.46) that the gauge functions χ are divided into individual classes generating a group

$$[S^1, S^1] = \pi_1(S^1) = Z. \tag{10.49}$$

Thus, from the mathematical point of view, the cause of the ABE lies in the fact that the configuration space associated with a zero field (vacuum) represents the plane with a hole, that is, $R^1 \times S^1$. The vector potential A_μ is made up using the gauge function χ that maps the gauge space onto the configuration one. These mappings are divided into individual classes and, in consequence of the fact that

$$[S^1, U(1)] = \pi_1(U(1)) = Z, \tag{10.50}$$

they cannot be deformed into a constant gauge function $\chi = \text{const}$, which obviously leads to the equality $A_\mu = 0$ and to the absence of the ABE.

10.3 Vortices and monopoles

Now we proceed to the consideration of field configurations in two-dimensional space. A circle at infinity, which will be denoted by S^1_∞, is a border of such a space. Consider a scalar field with a value on a border

$$\psi(r \to \infty) = a \exp[in\theta], \tag{10.51}$$

where r and θ are polar coordinates, a is a constant, and n is such an integer number that makes a function ψ single-valued. This field represents the generalization of a constant solution of the sine-Gordon equation (10.5) on a two-dimensional case. Our task is to find an equation that is associated with the asymptotic solution (10.51). Acting on (10.51) by the Nabla operator, we get

$$\boldsymbol{\nabla}\psi = \frac{ina}{r}\exp[in\theta]\mathbf{n}_\theta. \tag{10.52}$$

By analogy with a one-dimensional case, the Lagrangian $\mathcal{L}$ and Hamiltonian H are given by the expressions:

$$\mathcal{L} = \frac{1}{2}\left(\frac{\partial\psi}{\partial t}\right)^2 - \frac{1}{2}\left|\boldsymbol{\nabla}\psi\right|^2 - V(\psi), \tag{10.53}$$

$$H = \frac{1}{2}\left(\frac{\partial\psi}{\partial t}\right)^2 + \frac{1}{2}\left|\boldsymbol{\nabla}\psi\right|^2 + V(\psi). \tag{10.54}$$

Now we choose a static configuration whose potential energy turns into zero at a border

$$V(\psi) = [a^2 - \psi^*\psi]^2. \tag{10.55}$$

When $r \to \infty$, we get

$$H = \frac{1}{2}\left|\boldsymbol{\nabla}\psi\right|^2 = \frac{1}{2}\left|\frac{ina}{r}\exp[in\theta]\mathbf{n}_\theta\right|^2 = \frac{n^2a^2}{2r^2}. \tag{10.56}$$

Then, the energy (mass) of the investigated configuration is

$$E \approx \int_0^\pi d\theta \int_0^\infty Hr dr = \pi n^2 a^2 \int_0^\infty \frac{dr}{r}. \tag{10.57}$$

From (10.57) it immediately follows that a kink, in the form as it was defined, cannot be generalized on a two-dimensional space because its energy equals infinity. One may demonstrate that this conclusion remains valid in a space of higher dimensionality too [201].

Let us see whether the situation is improved in a real theory, that is, in a theory taking into account an interaction with a gauge field. Define a covariant derivative

$$D_\mu\psi = (\partial_\mu + ieA_\mu)\psi, \tag{10.58}$$

where a potential A_μ was chosen in the form

$$\mathbf{A}_{r\to\infty} = \frac{1}{e}\boldsymbol{\nabla}(n\theta). \tag{10.59}$$

The choice (10.59) means

$$A_r \to 0, \qquad A_\theta \to -\frac{n}{er}. \tag{10.60}$$

Allowing for the covariant derivative values

$$D_r\psi = 0, \qquad D_\theta\psi = \frac{1}{r}\frac{\partial\psi}{\partial\theta} + ieA_\theta\psi = 0, \tag{10.61}$$

we arrive at the conclusion: at the infinity $D_\mu\psi \to 0$. The Lagrangian of the system in question is governed by the expression:

$$\mathcal{L} = -\frac{1}{4}F_{\mu\nu}^2 + |D_\mu\psi|^2 - V(\psi). \tag{10.62}$$

Since at $r \to \infty$ the potential is defined by Eg. (10.59) (i.e., the field A_μ is a pure gauge, $A_\mu|_{r\to\infty} \to \partial_\mu\chi$), then $F_{\mu\nu}$ vanishes at the infinity. In the case of the static configuration, we have $H = -\mathcal{L}$ and, when $V(\psi)$ is given by Eq. (10.55), we obtain

$$H_{r\to\infty} = 0. \tag{10.63}$$

So, when switching on interaction, there exist field configurations with a finite energy. Moreover, introducing a gauge field results in the appearance of a quantized soliton magnetic flux Φ. Really,

$$\Phi = \int_S (\mathbf{B} \cdot d\mathbf{S}) = \int_S (\text{rot}\mathbf{A} \cdot d\mathbf{S}) = \oint (\mathbf{A} \cdot d\mathbf{l}) = \oint A_\theta r d\theta = -\frac{2\pi n}{e}. \tag{10.64}$$

Hence, we built up two-dimensional field configuration consisting of the charged scalar and gauge (electromagnetic) fields. It bears the magnetic flux and its energy is finite. Obviously, if the third dimension (z axis) will be added providing the field does not depend on it, then this configuration turns into a vortical line. Let us demonstrate, the vortex stability is ensured by the topology of the $U(1)$ gauge group.

We shall proceed from the Higgs Lagrangian

$$\mathcal{L} = -\frac{1}{4}F_{\mu\nu}^2 + |D_\mu\psi|^2 - m^2\psi^*\psi - \lambda(\psi^*\psi)^2. \tag{10.65}$$

At $m^2 < 0$, we have the spontaneous symmetry breaking. As this takes place, a classical vacuum is defined by the relation

$$|\psi|_{vac} = \left(-\frac{m^2}{2\lambda}\right)^{1/2} = u. \tag{10.66}$$

Varying ψ and A_μ leads to the motion equations

$$D^\mu(D_\mu\psi) = -m^2\psi - 2\lambda\psi|\psi|^2, \tag{10.67}$$

$$ie(\psi\partial_\mu\psi^* - \psi^*\partial_\mu\psi) + 2e^2A_\mu|\psi|^2 = \partial^\mu F_{\mu\nu}. \tag{10.68}$$

At first we make sure that at the infinity the obtained equations admit the solutions (10.51) and (10.60). When $r \to \infty$, then the left side of Eq. (10.67) turns into zero. If ψ takes the vacuum value (10.66), then the same happens with the right side of Eq. (10.67). Owing to the fact that A_μ is a pure gauge, the right side of Eq. (10.68) equals zero. With allowance made for (10.51) and (10.60), the left side of this equation identically turns into zero at $\mu = r$. When $\mu = \theta$, it vanishes under condition that ψ takes the vacuum value (10.66).

When the value of r is finite, it is evident that the values of A_μ and ψ are changed. We shall consider that the system is three-dimensional and possesses a symmetry with respect to the z axis. Further, since the magnet flux exists, then the magnetic field component B_z must be nonzero. We direct it along the z axis and assume the field A to have only the θ-component. Then we get

$$B = B_z = \frac{1}{r}\frac{d}{dr}[rA], \tag{10.69}$$

$A = A_\theta(r)$. From the continuity condition of ψ it follows

$$\psi = \chi(r)\exp(in\theta), \tag{10.70}$$

where

$$\chi(r)\Big|_{r\to 0} \to 0, \qquad \chi(r)\Big|_{r\to\infty} \to a. \tag{10.71}$$

Taking into consideration both the obtained values of A_μ, ψ and the static character of the task, we rewrite Eqs. (10.67) and (10.68) in the following form:

$$\frac{1}{r}\frac{d}{dr}\left(r\frac{d\chi}{dr}\right) - \left[\left(\frac{n}{r} - eA\right)^2 + m^2 + 2\lambda\chi^2\right]\chi = 0, \tag{10.72}$$

$$\frac{d}{dr}\left[\frac{1}{r}\frac{d}{dr}(rA)\right] - 2e\left(\frac{n}{r} + eA\right)\chi^2 = 0. \tag{10.73}$$

Unfortunately, to find an analytical solution of the system of the nonlinear equations (10.72) and (10.73) is not possible in the general case. In order to solve this system, we have to use different approximations. Thus, for example, in approximation when $\chi \approx a$ (i.e., when $r \to \infty$) the following expressions for the potential and magnetic field were obtained [202]

$$A = -\frac{n}{er} - \frac{C}{e}K_1(|e|ar) \stackrel{r\to\infty}{\longrightarrow} -\frac{n}{er} - \frac{C}{e}\left(\frac{\pi}{2|e|ar}\right)^{1/2}e^{|e|ar} + \dots, \tag{10.74}$$

$$B_z = C\chi K_0(|e|ar) \stackrel{r\to\infty}{\longrightarrow} \frac{C}{e}\left(\frac{\pi a}{2|e|r}\right)^{1/2}e^{-|e|ar} + \dots, \tag{10.75}$$

where K_0 and K_1 are modified Bessel functions, and C is an integration constant. Representing the scalar field in the view

$$\chi(r) = a + \rho(r), \tag{10.76}$$

we can get [202]

$$\rho(r) \approx e^{-\sqrt{-m^2}r}. \tag{10.77}$$

The obtained solutions are displayed in Fig. 10.4. Again, as in the one-dimensional case,

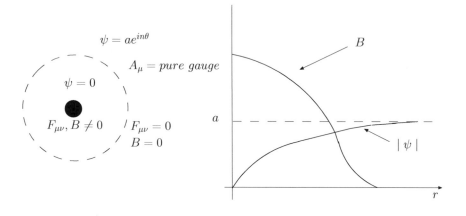

FIGURE 10.4
Nielsen–Olesen solution.

the stability of the solutions obtained is connected with the topology. The Lagrangian (10.65) is invariant with respect to the $U(1)$ gauge group. The field ψ with the boundary condition (10.51) makes a representation of this group. The group space of $U(1)$ represents a circle S^1. The field ψ in Eq. (10.51) is a representation basis of the $U(1)$ group, but it is the boundary value of the field in a two-dimensional space. This boundary, obviously, represents the circle S^1 with $r \to \infty$. Therefore, ψ defines a mapping of the boundary S^1 in the configuration space onto the group space S^1:

$$\psi : S^1 \to S^1,$$

the mapping form being defined by the topological index N. The solution characterized by a particular value of N is stable because it cannot be deformed by the continuous way into the solution associated with another value of N. Using the topology language, we may state: the stability of the obtained soliton solution is caused by the fact that the first homotopic group of the group space S^1 of the symmetry group $U(1)$ is nontrivial:

$$\pi_1(S^1) = Z. \tag{10.78}$$

So, with the help of topology, we can find the conditions of the solitons existence.

By way of exercise, we find out whether string-like solutions of spontaneously broken gauge theories exist in the case of the $SU(2)$ and $O(3)$ gauge groups. Let us begin with the $SU(2)$ group. So, we should consider mappings from a three-sphere to $SU(2)$ space, that is, mappings from the points on S^3, the sphere in four-dimensional Euclidean space labelled by three angles, to the elements of the $SU(2)$ group, which are also determined by three parameters. More explicitly, the manifold of the $SU(2)$ group elements is topologically equivalent to a three-sphere S^3. This is because, any element in the $SU(2)$ group can be written in terms of the Pauli matrices as

$$U - \exp\{i(\boldsymbol{\epsilon} \cdot \boldsymbol{\sigma})\}.$$

Using the relations between the Pauli matrices, we can write the matrix U in the form

$$U = \lambda_0 \times I + i \sum_{j=1}^{3} \lambda_j o_j, \tag{10.79}$$

where I is the identity matrix and real numbers λ_k ($k = 0, 1, 2, 3$) obey the relation

$$\lambda_0^2 + \boldsymbol{\lambda}^2 - 1, \tag{10.80}$$

which follows from $UU^\dagger = U^\dagger U = 1$. This is clearly the equation for the sphere in four-dimensional Euclidean space S^3. Mappings in this case are also characterized by the topological winding number N. Mapping with the lowest nontrivial topological winding number $N = 1$ has the form

$$f_1(\boldsymbol{\theta}) = \exp[i(\boldsymbol{\sigma} \cdot \boldsymbol{\theta})]. \tag{10.81}$$

By taking the powers of this mapping, we can get mappings of higher winding numbers. Thus, for example, the mapping $[f_1]^N$ will have winding number N. In the Cartesian coordinate system, the expression for f_1 could be written as

$$f(x_0, \mathbf{x}) = x_0 + i(\mathbf{x} \cdot \boldsymbol{\sigma}), \qquad \text{where} \qquad x_0^2 + \mathbf{x}^2 = 1. \tag{10.82}$$

We can generalize the domain X of this mapping from the unit circle to the three-dimensional space

$$-\infty \le \mathbf{x} \le \infty,$$

by identifying the endpoints $\mathbf{x} = \infty$ and $\mathbf{x} = -\infty$ to be the same point. In other words, we shall consider the mappings possessing the property

$$f_1(\mathbf{x} = \infty) = f(\mathbf{x} = -\infty).$$

Clearly this has the same topology as the unit circle. Examples of this type of mapping with winding number $N = 1$ are

$$f_1(x) = \exp\left\{ i\pi \frac{(\mathbf{x} \cdot \boldsymbol{\sigma})}{\sqrt{\mathbf{x}^2 + b^2}} \right\}, \tag{10.83}$$

$$f_1'(x) = \frac{(i\mathbf{x} + b\boldsymbol{\sigma})^2}{\mathbf{x}^2 + b^2}, \tag{10.84}$$

where b is an arbitrary number. One can show that the topological winding number N is given by the expression that is analogous to (10.42) [204]

$$N = -\frac{1}{24\pi^2} \int d\theta_1 d\theta_2 d\theta_3 \text{Sp}[\epsilon_{ijk} A_i A_j A_k], \tag{10.85}$$

where

$$A_i = f^{-1}(x_0, \mathbf{x}) \partial_i f(x_0, \mathbf{x})$$

while θ_1, θ_2, and θ_3 are the three angles that parametrize S^3. The expression (10.85) may also be represented in the form of the volume integral

$$N = -\frac{1}{24\pi^2} \int d^3x \text{Sp}[\epsilon_{ijk} A_i A_j A_k] \qquad (A_i = f^{-1}(x) \partial_i f(x)). \tag{10.86}$$

In a gauge theory with the $SU(2)$ group, stable vortices would exist, if the mapping of the group to the border S^1 be broken into different classes. In other words, if the $\pi_1(S^3)$ group were trivial. However, in reality, the $\pi_1(S^3)$ group is trivial, that is, S^3 represents a simply connected space: every closed curve S^1 on S^3 may be contracted into a point. Therefore, any boundary condition can be reduced to the trivial condition $\psi = const$ and vortices are absent in a gauge theory with the $SU(2)$ group.

Now we proceed to the $O(3)$ group. Consider the matrix describing the rotation about the z axis through the angle α

$$R_z = \begin{pmatrix} \cos\alpha & -\sin\alpha & 0 \\ \sin\alpha & \cos\alpha & 0 \\ 0 & 0 & 1 \end{pmatrix}. \tag{10.87}$$

To the matrix R_z corresponds the following matrix of the $SU(2)$ group

$$M_z = \begin{pmatrix} e^{i\alpha/2} & 0 \\ 0 & e^{-i\alpha/2} \end{pmatrix}. \tag{10.88}$$

The case $\alpha = 0$ gives the identity matrix for both groups, however, the value $\alpha = 2\pi$ provides the identity matrix for $O(3)$ ($I_{O(3)}$) again and the identity matrix with the minus sign for $SU(2)$ ($-I_{SU(2)}$). This is a consequence of different behaviors of tensors and spinors under the rotation through the angle 2π. So, to two elements of the $SU(2)$ group ($I_{SU(2)}$ and $-I_{SU(2)}$) correspond only to $I_{O(3)}$, that is, there exists the mapping $\mathbf{2} \rightarrow \mathbf{1}$ of the $SU(2)$ group to $O(3)$ group. The group space of $O(3)$ is obtained from that of $SU(2)$ by means of the identification of opposite points in three-dimensional space S^3 because they conform to one and the same transformation of $O(3)$. Let us show that this space is doubly connected. Consider closed curves S^1 in the group space of the $O(3)$ group. Every curve corresponding to a continuous set of rotations starts, perhaps, from the identical transformation 0 and goes back to it. One of the possible closed curves is a path c_1 displayed in Fig. 10.5. This curve correlates with a set of rotations, none operating on an angle greater than π. If a

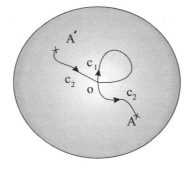

FIGURE 10.5

Two types of paths in the group space of $SO(3)$.

rotation angle exceeds π, then a path in a group space is similar to c_2. Having reached the angle value π in the point A, the path appears anew in the opposite point A' and, in the end, returns in the origin of coordinates O. Obviously, the path c_1 is homotopic to a point while the path c_2 is homotopic to a line. One may go on further and show that a closed path on which a rotation angle exceeds 2π appears twice at opposite points of a surface S^3 and, as a result, is homotopic to a point. By analogy, a path on which an angle exceeds 3π is homotopic to a line. Therefore, in the group space of the $O(3)$ group there exist closed paths S^1 only of two kinds: paths being homotopic to a point and paths being homotopic to a line. This means, in a gauge theory with the $O(3)$ group only one nontrivial vortex exists. Vortices may possess *charges* (fluxes) 1 or 0 with the algebra

$$0 + 0 = 0, \qquad 1 + 0 = 1, \qquad 1 + 1 = 0, \tag{10.89}$$

that is, nontrivial vortices are mutually annihilated. Note, the fact that whether the gauge group is $SU(2)$ or $O(3)$ depends on the particles content of a theory: if particles with the isospins $1/2, 3/2, 5/2$, and so on are available, then the $SU(2)$ group is a gauge group, but if particles possess the integer isospin we deal with the $O(3)$ gauge group.

One of the ways to produce the magnetic flux lines lies in the fact that two opposite magnetic charges are placed close to each other. Consequently, there appears a natural question: if gauge theories permit flux lines, can they permit the existence of magnetic charges? In 1974 t'Hooft [205] and Polyakov [206] revealed that gauge theories of a particular class permit the existence of magnetic charges (t'Hooft–Polyakov monopoles). Like vortices, these monopoles are stable owing to nontrivial topological properties of a gauge group. In this sense, they are entirely distinguished from *ordinary* magnetic monopoles considered in Chapter 17.4 of *Advanced Particle Physics Volume 1* (Dirac monopoles). Another important distinction consists in the fact that the t'Hooft–Polyakov monopole has a definite size defined by its mass while the Dirac monopole presents a point object. However, at large distances, both monopoles behave as point particles.

The simplest example of a non-Abelian gauge theory having monopole solutions is the $SO(3)$ model proposed by Georgi and Glashow in 1972 [207]. However, this model was ruled out experimentally by the discovery of the neutral-current phenomena. In a theory with the $SU(2)_L \times U(1)_Y$ gauge group, t'Hooft–Polyakov monopoles are absent and for them to be obtained theories with larger gauge groups should be considered. We deal with such gauge groups in Grand Unification Theories (GUTs). In GUTs, the electromagnetic field is coupled with a multiplet of charged gauge fields X with the masses $m_X \sim 10^{14}$ GeV. For some gauge groups, stable configurations of fields X localized in a region of size $l < 1/m_X$ exist. Outside this region they produce a spherically symmetric magnetic field. The existence of such configurations depend on topological properties of a group or more precisely, on the embedding way of a symmetry subgroup to be conserved after the spontaneous symmetry breaking. Stability of these monopoles is determined by a particular behavior of fields at large distances from the center. A mass of monopoles can be calculated, it depending on a specific field model. For the wide class of models we have $m_m \sim 10^{16}$ GeV. This means that this type of monopole is out of reach for its production by accelerators. But it could be relevant for physics in the extreme early universe. Such monopoles could be created in a hot universe shortly after the Big Bang during a phase transition concerned with a spontaneous symmetry breaking and appearance of a vacuum condensate of scalar fields. The amount of created monopoles is defined by the evolution of the universe. Therefore, investigating the monopoles distribution in the current universe gives information about both the past and the future of our universe. In fact, attempts to suppress the monopole abundance in the conventional cosmology originally led to the *inflationary universe scenario* that has the promise of solving several fundamental problems in cosmology.

By now, despite intensive searches, monopoles are not discovered. Experiments of their detection are being performed in two directions:

(i) The magnetic monopole could be revealed directly through its magnetic flux. To pass a magnetic charge ng_0 through a superconducting circuit changes a flux on $2n\Phi_0$, where $\Phi_0 \approx 2 \times 10^{-3}$ Gs · m². Then, the electromagnetic induction phenomenon leads to a current jump in a circuit that could be measured with the help of a superconducting quantum interferometer.

(ii) A heavy monopole must have a high penetrating power and produce a powerful ionization on its way. For this reason, to search for monopoles underground detectors built up for the investigations of cosmic neutrino fluxes and looking for a proton decay are used. The search for monopoles captured in magnetic ores (both of terrestrial and of extraterrestrial origins) is also carried. Monopoles are searched for at accelerators, too.

The strongest constraint on a possible number of monopoles in cosmic space is given by the consideration connected with the presence of galactic magnetic fields (GMFs). When moving in GMFs magnetic monopoles would be accelerated, while GMFs would lose energy and be weakened as time goes by. The current estimations give the constraint on a cosmic monopole flux of the order of 10^{-16} cm⁻² · sr⁻¹ · s⁻¹.

10.4 Instantons

We shall study instanton solutions, that is, solutions with topological structure in the Euclidean four-dimensional space-time E^4 ($x^2 = x_0^2 + \mathbf{x}^2$) (detailed introduction to this theme could be found in the book [208]). Like the soliton solutions considered before, they have a finite spatial extension, thus the *-on* in its name, and unlike solitons, they are also structures in time (albeit imaginary time)—thus, the *instant-*. It comes as no surprise that such solutions are available because equations of gauge fields are relativistically invariant and, hence, permit nontrivial topology both in time and in space. Three-dimensional sphere S^3 is a border of E^4. As we already noted, a group space of the $SU(2)$ group, which plays an exceptional role in microworld physics, constitutes a space S^3 too. Therefore, topologically nontrivial solutions of equations for gauge fields with the $SU(2)$ group exist in the case when nontrivial (nonhomotopic) mappings $S^3 \to S^3$ are available, that is, when the $\pi_3(S^3)$ group is nontrivial. Really, this group is nontrivial

$$\pi_3(S^3) = Z. \tag{10.90}$$

From here it follows that instantons may be in existence in a pure gauge theory, that is, the spontaneous symmetry breaking is not needed. This distinguishes instantons from monopoles.

We shall seek the finite-action solution to the classical Yang–Mills theory in E^4. The SU(2) gauge fields:

$$G_\mu = \frac{\sigma^a}{2} G_\mu^a, \qquad G_{\mu\nu} = \frac{\sigma^a}{2} G_{\mu\nu}^a, \qquad a = 1, 2, 3,$$

are described by the Lagrangian

$$\mathcal{L} = \frac{1}{g^2} \mathrm{Sp}\, F_{\mu\nu} F_{\mu\nu}, \tag{10.91}$$

where for notational convenience we have scaled the gauge fields as:

$$G_\mu \to \frac{i}{g} G_\mu, \qquad G_{\mu\nu} \to \frac{i}{g} G_{\mu\nu}. \tag{10.92}$$

Then, under a gauge transformation U, the fields G_μ are transformed by the following way:

$$G'_\mu = U^{-1} G_\mu U + U^{-1} \partial_\mu U. \tag{10.93}$$

We require the solution to satisfy the boundary condition that the Lagrangian vanishes at infinity so that the Euclidean action

$$S_E = \frac{1}{2g^2} \int d^4x \, \mathrm{Sp}(G_{\mu\nu} G_{\mu\nu})$$

is finite. This means that in Euclidean space an asymptotic condition must be fulfilled

$$G_{\mu\nu}(x)\Big|_{|x|\to\infty} \to 0. \tag{10.94}$$

As we already knew, for the condition (10.94) to be performed one sufficiently demands the field G_μ to be pure gauge. In other words, the field G_μ must approach the configuration

$$G_\mu(x)\Big|_{|x|\to\infty} \to U^{-1} \partial_\mu U, \tag{10.95}$$

which is obtained from $G_\mu(x) = 0$ by a gauge transformation (10.93).

One notes that points at infinity ($|x| \to \infty$) in E^4 are three-sphere S^3 and the gauge transformation matrix U in (10.95) represents mappings from S^3 to $SU(2)$ space. Thus, the U functions belong to the $S^3 \to S^3$ type of functions.

First we express the winding number in terms of the gauge fields. For this purpose we introduce an (unobservable) gauge-dependent current:

$$K_\mu = 4\epsilon_{\mu\nu\lambda\rho} \mathrm{Sp} \left[G_\nu \partial_\lambda G_\rho + \frac{2}{3} G_\nu G_\lambda G_\rho \right]. \tag{10.96}$$

Show that the following relation takes place

$$\partial_\mu K_\mu = 2\mathrm{Sp}\,(G_{\mu\nu} \tilde{G}_{\mu\nu}), \tag{10.97}$$

where

$$\tilde{G}_{\mu\nu} = \frac{1}{2} \epsilon_{\mu\nu\lambda\rho} G_{\lambda\rho}. \tag{10.98}$$

Taking into consideration the cyclical spur property, we have

$$\partial_\mu K_\mu = 4\epsilon_{\mu\nu\lambda\rho} \, \text{Sp}\left[(\partial_\mu G_\nu)(\partial_\lambda G_\rho) + 2(\partial_\mu G_\nu)G_\lambda G_\rho\right]. \tag{10.99}$$

On the other hand,

$$\text{Sp}(G_{\mu\nu}\tilde{G}_{\mu\nu}) = \frac{1}{2}\epsilon_{\mu\nu\lambda\rho} \, \text{Sp}\left\{\left(\partial_{[\mu}G_{\nu]} + [G_\mu, G_\nu]\right)\left(\partial_{[\lambda}G_{\rho]} + [G_\lambda, G_\rho]\right)\right\} =$$

$$= 2\epsilon_{\mu\nu\lambda\rho}\left\{\text{Sp}[(\partial_\mu G_\nu)(\partial_\lambda G_\rho)] + \text{Sp}[(G_\mu G_\nu)(\partial_\lambda G_\rho)] + \text{Sp}[(\partial_\mu G_\nu)(G_\lambda G_\rho)] +\right.$$

$$\left. + + \text{Sp}[(G_\mu G_\nu)(G_\lambda G_\rho)]\right\}. \tag{10.100}$$

Owing to the cyclical spur property the second term in Eq. (10.100) is equal to the third while the forth vanishes. So, the relation (10.97) indeed exists.

Next, we consider the volume integral

$$\int d^4x \text{Sp}\left(G_{\mu\nu}\tilde{G}_{\mu\nu}\right) = \frac{1}{2}\int d^4x\partial_\mu K_\mu = \frac{1}{2}\int_S d\sigma_\mu K_\mu, \tag{10.101}$$

where the surface integral is taken over the S^3 at infinity. In this region G_μ is given by Eq. (10.95), therefore, K_μ is

$$K_\mu = \frac{4}{3}\epsilon_{\mu\nu\lambda\rho} \, \text{Sp}[(U^\dagger\partial_\nu U)(U^\dagger\partial_\lambda U)(U^\dagger\partial_\rho U)], \tag{10.102}$$

where we have allowed for the unitarity of U. Substituting (10.102) into (10.101) and comparing it with the expression for the $S^3 \to S^3$ winding number (10.86), we arrive at

$$N = \frac{1}{16\pi^2}\int d^4x \, \text{Sp}(G_{\mu\nu}\tilde{G}_{\mu\nu}). \tag{10.103}$$

Now we define the constraints the demand of the action finiteness imposes on the field tensor. Using the positivity condition in Euclidean space

$$\text{Sp}\left[\int d^4x(G_{\mu\nu} \pm \tilde{G}_{\mu\nu})^2\right] \geq 0 \tag{10.104}$$

and allowing for

$$(G_{\mu\nu} \pm \tilde{G}_{\mu\nu})^2 = 2(G_{\mu\nu}G_{\mu\nu} \pm G_{\mu\nu}\tilde{G}_{\mu\nu}), \tag{10.105}$$

we arrive at the inequality

$$\text{Sp}[\int d^4x(G_{\mu\nu}G_{\mu\nu})] \geq \left|\text{Sp}[\int d^4x(G_{\mu\nu}\tilde{G}_{\mu\nu})]\right| = 16\pi^2 N. \tag{10.106}$$

Consequently, the Euclidean action satisfies the inequality

$$S_E(G) \geq \frac{8\pi^2 N}{g^2}. \tag{10.107}$$

Then, from (10.104) it is evident that the action is minimized (i.e., equality achieved in (10.107)) when the self-duality condition

$$G_{\mu\nu}(x) = \tilde{G}_{\mu\nu}(x) \tag{10.108}$$

or the antiself-duality condition

$$G_{\mu\nu}(x) = -\tilde{G}_{\mu\nu}(x) \tag{10.109}$$

are fulfilled. At that, it turns out that the positive values of N are associated with self-dual solutions while the negative values of N are associated with antiself-dual solutions. It is clear, that the ordinary solution $G_\mu(x) = 0$ with the trivial winding number $N = 0$ satisfies the conditions (10.108) and (10.109). However, nontrivial solutions are of physical interest. Let us solve the self-duality equation for the case $N = 1$ (one-instanton or, simply, an instanton solution). Introduce t'Hooft symbols $\eta^a_{\mu\nu}$:

$$\eta^a_{00} = 0, \qquad \eta^a_{ij} = \epsilon_{aij}, \qquad \eta^a_{0i} = -\delta_{ia}, \qquad \eta^a_{i0} = \delta_{ai}. \tag{10.110}$$

In what follows we need the following relations concerning $\eta^a_{\mu\nu}$:

$$\epsilon^{abc}\eta^b_{\mu\rho}\eta^c_{\nu\sigma} = \delta_{\mu\nu}\eta^a_{\rho\sigma} - \delta_{\mu\sigma}\eta^a_{\rho\nu} + \delta_{\rho\sigma}\eta^a_{\mu\nu} - \delta_{\rho\nu}\eta^a_{\mu\sigma}, \qquad \eta^a_{\mu\sigma}\eta^a_{\mu\sigma} = 12. \tag{10.111}$$

The simplest choice for the field tensor of the self-dual field is

$$G^a_{\mu\nu}(x) = \eta^a_{\mu\nu}B(x), \tag{10.112}$$

where $B(x)$ is a scalar function. However, all are not so simple because not any tensor of the kind (10.112) can play a part of the field tensor, which is expressed in terms of the potentials according to the definition:

$$G^a_{\mu\nu}(x) = \partial_\mu G^a_\nu(x) - \partial_\nu G^a_\mu(x) + f^{abc}G^b_\mu(x)G^c_\nu(x). \tag{10.113}$$

We choose the sampling potential in the view

$$G^a_\mu(x) = \eta^a_{\mu\nu}\partial_\nu \ln \Phi(x) \tag{10.114}$$

and attempt to find conditions at which the corresponding field tensor will be self-dual. Employing the definition (10.113) and the first of the relations (10.111), we obtain

$$G^a_{\mu\nu} = \eta^a_{\nu\rho}[\partial_\mu\partial_\rho \ln \Phi + (\partial_\mu \ln \Phi)(\partial_\rho \ln \Phi)] - (\mu \leftrightarrow \nu) + \eta^a_{\mu\nu}(\partial_\rho \ln \Phi)^2. \tag{10.115}$$

As we can see, only the last term in (10.115) has the desired self-dual form. If one demands the square bracket in the expression (10.116) to be proportional to the Kronecker symbol

$$[\ldots]_{\mu\rho} \sim \delta_{\mu\rho}, \tag{10.116}$$

then the first term turns into zero. It is easy to see, that this will take place under condition [209]

$$\Phi(x) = 1 + \frac{\rho^2}{l^2}, \tag{10.117}$$

where $\rho^2 = x_0^2 + \mathbf{x}^2$ and l is an arbitrary scale parameter, which is usually called the *instanton size.* * In such a way, the instanton solution is governed by the expression:

$$G^a_\mu(x) = \frac{2\eta^a_{\mu\nu}x_\nu}{\rho^2 + l^2}. \tag{10.118}$$

At that, the field tensor is

$$G^a_{\mu\nu}(x) = -\frac{4l^2\eta^a_{\mu\nu}}{(\rho^2 + l^2)^2}. \tag{10.119}$$

---

*We call attention to the fact that a scale parameter may take any values from 0 to ∞. To calculate a pure instanton effect needs integration over an instanton size. But the effective (in the renormalized meaning) gauge coupling constant is changed with the scale of distances, namely, it is increased at large distances. Therefore, the larger the instanton, the stronger the coupling constant and, as a result, the weak coupling condition of the quasi-classical method is violated.

The instanton solution (10.118) may also be written in the form

$$G_\mu(x) = \left[\frac{\rho^2}{\rho^2 + l^2}\right] U^{-1}(x)\partial_\mu U(x), \tag{10.120}$$

where

$$U(x) = \frac{x_0 + i(\mathbf{x} \cdot \boldsymbol{\sigma})}{\rho}.$$

Obviously, at $\rho \gg l$ we have the boundary condition (10.95). In the component-wise notation, the one-instanton solution (10.118) will look like:

$$A_0(x) = \frac{-i(\mathbf{x} \cdot \boldsymbol{\sigma})}{\rho^2 + l^2}, \qquad \mathbf{A}(x) = \frac{-i\{\boldsymbol{\sigma} x_0 + [\boldsymbol{\sigma} \times \mathbf{x}]\}}{\rho^2 + l^2}. \tag{10.121}$$

Next, one may check that the corresponding action integral really has the value $8\pi^2/g^2$.

The solution (10.118) (and (10.119)) is localized in the Euclidean space. It has been written in the particular gauge. The set of other solutions can be obtained with the help of local gauge transformations. The antiself-dual solution with $N = -1$ (anti-instanton solution) follows from (10.118) and (10.119) under the replacement

$$\eta^a_{\mu\nu} \to \overline{\eta}^a_{\mu\nu}, \tag{10.122}$$

where $\overline{\eta}^a_{\mu\nu}$ are distinguished from $\eta^a_{\mu\nu}$ by the sign of components $\overline{\eta}^a_{0i} = -\overline{\eta}^a_{i0} = \delta_{ai}$. The solutions of Eqs. (10.108) and (10.109) can also be found for $N > 1$ (the multiple-instanton solution). The calculation details could be found in Ref. [210].

The physical meaning of the instanton solution becomes apparent under consideration of the classical Yang–Mills Hamiltonian in the gauge $G_0 = 0$:

$$H = \frac{1}{2g^2} \int d^3x [E_i^2 + B_i^2], \tag{10.123}$$

where E_i^2 and B_i^2 accord with the kinetic and potential energies. A classical vacuum includes configurations with zero field strength (for non-Abelian fields this constraint results in the pure gauge (10.95)). Such configurations are defined by a topological number N distinguishing gauge transformations U, which are not coupled by the continuous way with each other. It means, that there is an infinite set of classical vacua with different N. The instantons represent the solutions of the classical motion equations that link various vacua. These solutions have the potential $\mathbf{B}^2 > 0$ and kinetic $\mathbf{E}^2 < 0$ energies, the energies sum being equal to zero at any instant of time. So, the instanton is a particular kind of vacuum oscillation under which a strong gluon field spontaneously flares up to be quenched. This process, being a quantum phenomenon, does not contradict the energy conservation law due to the uncertainty relation. The instanton, like the soliton, owes its existence to strong nonlinear effects. There is, however, one important distinction: solitons are localized in space but are infinitely extended in time. The term *instanton* refers to a vacuum rebuilding process that takes a finite time. Therefore, the instanton field is localized both in space and in time.

In quantum field theory, any process is described by the sum over all possible trajectories that realize the transition from the initial to final states. In classical limit trajectories, being solutions of classical dynamics equations, are separated from this sum. In those cases, when the given transition is classically impossible, it is realized owing to the tunnelling effect.

The physical interpretation of the instanton solutions as quantum-mechanical events corresponding to tunnelling between vacuum states of different topological numbers was first advanced by t'Hooft [211]. Let us show, the instanton solutions really connect vacuum states with different topological numbers.

Employ the analogy between the soliton and instanton solutions. In the case of the kink, when the spacial coordinate is changed from $-\infty$ to $+\infty$, then the field configuration is changed and the boundary conditions at $-\infty$ and $+\infty$ are different (see Fig. 10.1). The obvious way to interpret the instanton solution consists in the fact that we should consider this solution to describe the evolution in time rather than in space. With allowance made for this, we picture the coordinate space border S^3 covering the volume V^4 in Fig. 10.6. The hypersurfaces I and II correspond to $x_0 = +\infty$ and $x_0 = -\infty$, respectively, while the

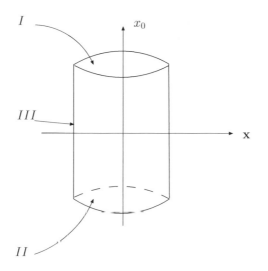

FIGURE 10.6

The border corresponding to the instanton.

hypercylindrical surface III connects them. The topological number of the system under consideration is as follows:

$$N = \frac{1}{32\pi^2} \int_V d^4 x \, \partial_\mu K_\mu = \frac{1}{32\pi^2} \oint_S d\sigma_\mu K_\mu =$$

$$= \frac{1}{24\pi^2} \epsilon_{\mu\nu\lambda\rho} \oint_S d\sigma_\mu \, \mathrm{Sp} \, [(U^{-1}\partial_\nu U)(U^{-1}\partial_\lambda U)(U^{-1}\partial_\rho U)] =$$

$$= \frac{1}{24\pi^2} \left\{ \int_{I-II} \epsilon_{0ijk} d^3\sigma \, \mathrm{Sp} \, [(U^{-1}\partial_i U)(U^{-1}\partial_j U)(U^{-1}\partial_k U)] + \right.$$

$$\left. + \int_{-\infty}^{\infty} dx_0 \int_{III} \epsilon_{i\nu\lambda\rho} d^2\sigma_i \, \mathrm{Sp} \, [(U^{-1}\partial_\nu U)(U^{-1}\partial_\lambda U)(U^{-1}\partial_\rho U)] \right\}. \qquad (10.124)$$

At $x_0 = \pm\infty$ the solution (10.120) turns into the pure gauge, however, inside the four-dimensional volume V^4 it has the required form $G_{\mu\nu} \neq 0$. The expression for G_μ (10.120) must be used in Eq. (10.124). However, since the quantity N is gauge invariant, we may choose any gauge. Physical results becomes more transparent in the light-like gauge

$$G_0'(x) = 0. \qquad (10.125)$$

As this takes place, the integral over the cylindrical surface III in Eq. (10.124) turns into zero because the condition of nonvanishing this integral lies in the fact that one of the indices ν, λ, ρ should be equal to 0.

To find G_i' we shall act in the following way. Address the one-instanton solution (10.121). For this solution to satisfy the light-like gauge we subject it to the gauge transformation:

$$G_\mu(x) \rightarrow G_\mu'(x) = U^{-1}(x)G_\mu(x)U(x) + U^{-1}(x)\partial_\mu U(x). \tag{10.126}$$

From the condition (10.125) follows

$$\frac{\partial}{\partial x_0}U(x) = -G_0(x)U(x) = \frac{i(\mathbf{x} \cdot \boldsymbol{\sigma})}{\rho^2 + l^2}U(x), \tag{10.127}$$

To integrate Eq. (10.127) results in

$$U(x) = \exp\left\{\frac{i(\mathbf{x} \cdot \boldsymbol{\sigma})}{\sqrt{\mathbf{x}^2 + l^2}}\left[\arctan\left(\frac{x_0}{\sqrt{\mathbf{x}^2 + l^2}}\right) + \theta_0\right]\right\}, \tag{10.128}$$

where θ_0 is an integration constant. Choose θ_0 in the form

$$\theta_0 = \left(n + \frac{1}{2}\right)\pi, \qquad n = 1, 2, \ldots \tag{10.129}$$

Then, as can be readily seen, we shall have

$$U(\mathbf{x}, x_0 = -\infty) = U_n(\mathbf{x}) = \exp\left\{i\pi\frac{(\mathbf{x} \cdot \boldsymbol{\sigma})}{\sqrt{\mathbf{x}^2 + l^2}}n\right\}, \tag{10.130}$$

$$U(\mathbf{x}, x_0 = \infty) = U_{n+1}(\mathbf{x}) = \exp\left\{i\pi\frac{(\mathbf{x} \cdot \boldsymbol{\sigma})}{\sqrt{\mathbf{x}^2 + l^2}}(n + 1)\right\}. \tag{10.131}$$

The substitution (10.130) and (10.131) into the expression for the topological number (10.124) gives the result $N = 1$. Now, if one demands the fulfillment of the boundary conditions

$$G_i\Big|_{x_0 \to \infty} = G_i\Big|_{x_0 \to -\infty} = 0,$$

then the required solutions are given by the expressions:

$$\left.\begin{array}{llll} G_i'(\mathbf{x}) = U_n^{-1}(\mathbf{x})\partial_i U_n(\mathbf{x}), & \text{when} & x_0 = -\infty \\ G_i'(\mathbf{x}) = U_{n+1}^{-1}(\mathbf{x})\partial_i U_{n+1}(\mathbf{x}), & \text{when} & x_0 = \infty \end{array}\right\} \tag{10.132}$$

It is evident that U_n presents an element of the $SU(2)$ group, however, U_n and U_m ($m \neq n$) are not homotopic. In particular, U_1 and $U_0 = 1$ are not homotopic, that is, it is impossible to find such a function $L(U_1, t)$ ($t \in [0, 1]$), so that the conditions $L(U_1, 1) = U_1$ and $L(U_1, 0) = 1$ are fulfilled. Consequently, the instanton describes the solution of gauge field equations for which, at changing the x_0 coordinate from $-\infty$ to ∞, a vacuum belonging to the homotopic class n converts to an other vacuum belonging to homotopic class $n + 1$. In the space between these vacua a region exists where the field tensor $G_{\mu\nu}$ is nonzero and, therefore, the field has a positive energy. So, the Yang–Mills vacuum is infinitely degenerate, it includes an infinite number of homotopically nonequivalent vacua. The instanton solution is in line with the transition from the vacuum of one class to the vacuum of another class. Therefore, our next task is to calculate the amplitude of this transition. From the classical point of view, it undoubtedly equals zero as a potential hump between two vacua is available. However, at the quantum-mechanical consideration, we must allow for the tunnelling effect.

The fact that the instanton is the minimum of the Euclidean action, that is, it is the classical path for imaginary time, reminds one of the quasi-classical barrier-penetration amplitudes in nonrelativistic quantum mechanics. In relation to this, it is useful to briefly discuss the elementary quantum-mechanical example of tunnelling between the two ground states of the double-well potential (see Fig. 10.7(a)):

$$V(q) = (q^2 - q_0^2)^2, \tag{10.133}$$

where $q(t)$ is some generalized coordinate. The energy of the system in question is

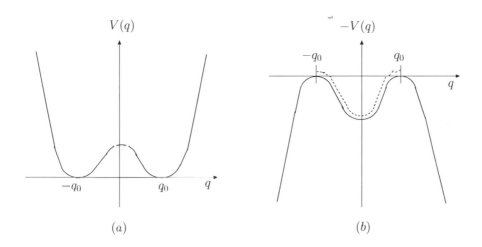

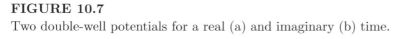

FIGURE 10.7

Two double-well potentials for a real (a) and imaginary (b) time.

$$E = \frac{1}{2}\left(\frac{dq}{dt}\right)^2 + V(q). \tag{10.134}$$

The classical ground state for this system is at $q = q_0$ or $q = -q_0$ with $E = 0$. In classical mechanics undoubtedly there is no $E = 0$ path leading from q_0 to $-q_0$. However, there can be quantum-mechanical tunnelling, so that the true ground state is neither $|q_0 >$ nor $|-q_0 >$, but their superposition

$$|0 >= \frac{1}{2}\left(|q_0 > + |-q_0 >\right). \tag{10.135}$$

The quantum-mechanical tunnelling amplitude can be calculated with the help of the classical particle trajectory in the imaginary time system. This is connected with the fact that in the imaginary time system where $t = -i\tau$ and $(dq/dt)^2 = -(dq/d\tau)^2$, the energy of the system is governed by the expression

$$-E == \frac{1}{2}\left(\frac{dq}{d\tau}\right)^2 - V(q), \tag{10.136}$$

which is equivalent to a particle moving in the potential $-V(q)$ pictured in Fig. 10.7(b). So, in imaginary time there exists a path going between $-q_0$ and q_0 with $E = 0$. Setting $E = 0$ in Eq. (10.136), we can find this trajectory:

$$q(\tau) = q_0 \tanh[q_0\tau\sqrt{2}]. \tag{10.137}$$

The action for this trajectory in the imaginary-time system is finite and can be expressed in the view:

$$S_t = \int_{-\infty}^{+\infty} d\tau \left\{ \frac{1}{2} \left(\frac{dq}{d\tau} \right)^2 - [-V(q)] \right\} = 2 \int_{-\infty}^{+\infty} d\tau V(q) =$$

$$= 2 \int_{-\infty}^{+\infty} d\tau [q^2(\tau) - q_0^2]^2 = \frac{4}{3} q_0^3 \sqrt{2}. \tag{10.138}$$

In the formalism of the functional integrals, the tunnelling amplitude will look like:

$$< q_f | e^{-iHt/\hbar} | q_i > = \int [dq] e^{iS/\hbar}. \tag{10.139}$$

In the Euclidean space this formula takes the form

$$< q_f | e^{-H\tau/\hbar} | q_i > = \int [dq] e^{-S_E/\hbar}, \tag{10.140}$$

where $S_E = -iS$ is the Euclidean action. The right side of Eq. (10.140) can be considered as the summation over all possible paths going from q_i to q_f. In the quasi-classical approximation (expansion in powers of $\hbar$), the main contribution into the integral (10.140) is given by those trajectories for which S_E is stationary. It is easy to verify that the tunnelling amplitude for the case under consideration is of the form

$$\mathcal{A} \sim \exp \left(-\frac{S_t}{\hbar} \right) [1 + O(\hbar)] \approx \exp \left(-\frac{4q_0^3 \sqrt{2}}{3\hbar} \right). \tag{10.141}$$

When we generalize the transition amplitude in (10.140) to quantum field theory (QFT), we will discover that not all field configurations (trajectories) result in the finite action because QFT has infinite degrees of freedom. In the quasi-classical approximation, however, configurations with an infinite action do not contribute to the functional integral as every trajectory on it has the weight factor $\exp(-S_E/\hbar)$. For this reason, the functional integral will be dominated by configurations with an finite action. Therefore, the tunnelling amplitude in the quasi-classical approximation can be calculated in terms of the instanton configuration and has the form

$$\mathcal{A} \sim \exp \left(-S_E \right)) \approx \exp \left(-\frac{8\pi^2}{g^2} \right), \tag{10.142}$$

where the condition (10.107) has been used. The expression (10.142) also indicates clearly that this effect cannot be revealed within the ordinary perturbation theory.

Since vacuum states $|n >$ associated with different topological numbers are separated by finite-energy barriers and there are tunnellings between these states, then one expects the true vacuum state to be a suitable superposition of these states. We note the following: under a gauge transformation T_1 corresponding to the topological number 1, the topological number is increased by 1

$$T_1 | n > = | n+1 >, \tag{10.143}$$

and the gauge invariance means that the operator T_1 commutes with the Hamiltonian H. This situation is very similar to the familiar quantum-mechanical problem with periodical potential where the translation operator plays a part of T_1, the true ground state is the

Bloch wave* and, besides, the conserved quasi-momentum exists. The true vacuum, the θ-vacuum, can be constructed by the following way:

$$|\theta> = \sum_n e^{-in\theta}|n> . \tag{10.144}$$

It is clear that it presents an eigenstate of the gauge transformation operator

$$T_1|\theta> = e^{i\theta}|\theta> . \tag{10.145}$$

Just like the quasi-momentum for the periodic potential case, θ labels the physically inequivalent sectors of the theory and within each sector the propagation of gauge-invariant disturbances may be investigated. As sectors associated with different values of θ are in no way coupled with each other, there is no *a priori* method of determining the value of θ.

In functional integral formalism the vacuum-to-vacuum transition amplitude is generalized in the following way:

$$< n|\exp(-iHt)|m>_J = \int [dG]_{n-m} \exp\left\{-i\int(\mathcal{L} + JG)d^4x\right\}, \tag{10.146}$$

where $|n>$ and $|m>$ are vacuum states with the topological indices n and m, respectively and the sum is taken over all gauge fields belonging to one and the same homotopic class with the topological number $n-m$. Then, in terms of the θ-vacuum, the vacuum-to-vacuum transition amplitude takes the form

$$\mathcal{A} = < 0'|e^{-iHt}|\theta> = \sum_{m,n} e^{im\theta'}e^{-in\theta} < m|e^{-iHt}|n>_J = \sum_{m,n} e^{i(m-n)\theta}e^{im(\theta'-\theta)} \int [dG]_{n \ \ m} \times$$

$$\times \exp\left[-i\int(\mathcal{L}+JG)d^4x\right] = \delta(\theta'-\theta)\sum_\nu e^{-i\nu\theta} \int [dG]_\nu \exp\left[-i\int(\mathcal{L}+JG)d^4x\right] =$$

$$= \delta(\theta'-\theta)\sum_\nu \int [dG]_\nu \exp\left[-i\int(\mathcal{L}_{eff}+JG)\,d^4x\right], \tag{10.147}$$

where

$$\mathcal{L}_{eff} = \mathcal{L} + \frac{\theta}{16\pi^2}\,\mathrm{Sp}(G_{\mu\nu}\tilde{G}^{\mu\nu}) \tag{10.148}$$

and we have take into account the definition of the topological number (10.103).

So, the rich structure of the gauge theory vacuum corresponding to tunnelling between states with different topological numbers gives rise to the effective Lagrangian term $G_{\nu\nu}\tilde{G}^{\mu\nu}$ violating P- and CP- invariance. Because such a term can be expressed as the divergence of current (10.97), then it may be discarded. Nevertheless, as there is a nontrivial instanton gauge field configuration that does not vanish at infinity (10.95), such an abnormal term actually survives in non-Abelian gauge field theories.

We now show that the instanton existence also gives one more possible solution of the $U(1)$-problem. For simplicity we consider the case where there are only two quark flavors

---

*The Schrödinger equation solution in the periodic potential field $V(\mathbf{r} + \mathbf{a}) = V(\mathbf{r})$ (Bloch wave) is given by the expression:

$$\psi_{\mathbf{k}}(\mathbf{r}) = e^{i(\mathbf{k}\cdot\mathbf{r})}u_{\mathbf{k}},$$

where $\mathbf{k}$ is a wave vector of a particle and $u_{\mathbf{k}}$ is a periodic function with a potential period. If $\psi(\mathbf{r})$ is the Schrödinger equation solution corresponding to a stationary particle state, then $\psi(\mathbf{r} + \mathbf{a})$ also presents its solution, the relation $\psi(\mathbf{r} + \mathbf{a}) = \pm\psi(\mathbf{r})$ taking place.

in the theory, u and d quarks. In the limit $m_{u,d} \to 0$, the QCD Lagrangian has the symmetry $SU(2)_L \times SU(2)_R \times U(1)_V \times U(1)_A$. The $U(1)_V$ symmetry connected with the transformation $q_i \to e^{i\alpha} q_i$ leads to the baryon current

$$J_\mu^B = \bar{u}\gamma_\mu u + \bar{d}\gamma_\mu d \tag{10.149}$$

and provides the baryon charge conservation. The $U(1)_A$ symmetry associated with the transformation $q_i \to e^{i\gamma_5 \theta} q_i$ gives rise to the current

$$J_\mu^5 = \bar{u}\gamma_\mu\gamma_5 u + \bar{d}\gamma_\mu\gamma_5 d, \tag{10.150}$$

that, as we already knew, does not correspond to any observed symmetry in the hadron spectra. The divergence of this current was already investigated by us and in the two-flavor approximation it looks like:

$$\partial^\mu J_\mu^5 = 4\frac{g^2}{16\pi^2}\mathrm{Sp}(G_{\mu\nu}\tilde{G}^{\mu\nu}) + 2im_u\bar{u}\gamma_5 u + 2im_u\bar{d}\gamma_5 d, \tag{10.151}$$

where we have refused fields normalization adopted in (10.92). Taking into consideration that the first term in (10.151) can be represented in the form of the four-dimensional divergence, we define a new axial-vector current.

$$\tilde{J}_\mu^5 = J_\mu^5 - K_\mu, \tag{10.152}$$

Its divergence is given by the expression:

$$\partial^\mu \tilde{J}_\mu^5 = 2im_u\bar{u}\gamma_5 u + 2im_u\bar{d}\gamma_5 d. \tag{10.153}$$

Of course the current K_μ and, therefore, $\tilde{J}_\mu^5$ are unobservable, because they are not gauge-invariant. Nevertheless, because the charge

$$\tilde{Q}_5 = \int d^3x \tilde{J}_0^5(x), \tag{10.154}$$

is conserved in the limit $m_{u,d} \to 0$, then this symmetry when realized in the Goldstone mode would demand the existence of a pseudoscalar meson. As this takes place, computations show [55] that its mass m_0 is of the order of the pion mass $m_0 \leq \sqrt{3}m_\pi$. Since experimentally no such isoscalar pseudoscalar meson has been seen, there must exist a theoretical principle to hinder its existence. Let us demonstrate how the structure of the θ-vacuum leads to the spontaneous breaking of the $U(1)_A$ symmetry and why the associated massless particle is not a physical particle.

As the charge $\tilde{Q}_5$ is not a gauge invariant quantity under a gauge transformation having the topological number n, then it is varied in the following way:

$$\tilde{Q}_5 \to \tilde{Q}_5' = T_n^{-1}\tilde{Q}_5 T_n = \tilde{Q}_5 + 2ni_F, \tag{10.155}$$

where i_F is a number of massless quark flavors. To verify this relation we set the initial charge $\tilde{Q}_5$ to equal zero. Then, from the equality

$$\tilde{Q}_5' = \int d^3x K_0'(x), \tag{10.156}$$

where

$$K_0'(x) = \frac{2i_F g^2}{12\pi^2}\epsilon_{0ijk}\,\mathrm{Sp}[G_i(x)G_j(x)G_k(x)], \qquad G_i(x) = U^{-1}(x)\partial_i U(x), \tag{10.157}$$

follows the relation (10.155), if one compares the expression for $\tilde{Q}'_5$ with that for the topological number (10.86). The relation (10.155) means that T_n acts as an operator raising chirality—instantons *eat* massless quark pairs:

$$[\tilde{Q}_5, T_n] = 2ni_F T_n. \tag{10.158}$$

Moreover, the operators T_n changes the topological vacuum number:

$$T_n|0>= |n>. \tag{10.159}$$

However

$$\tilde{Q}_5|n>= 2ni_F|n>, \tag{10.160}$$

owing to (10.158) and, as a result, $\tilde{Q}_5|0>= 0$. So, the vacuum state with a definite topological number also possesses definite chirality, that is, it is an eigenstate of the operator $\tilde{Q}_5$. Since the charge $\tilde{Q}_5$ is conserved

$$[\tilde{Q}_5, H] = 0,$$

then the vacuum-to-vacuum transition amplitude vanishes unless the initial and final states have the same topological numbers:

$$< n|c^{-iHt}|m >\sim \delta_{mn}, \tag{10.161}$$

or, more generally for any operator $P_\nu(x)$ with chirality $2i_F\nu$

$$< n|e^{-iHt}P_\nu(x)|m >\sim \delta_{m-n,\nu}. \tag{10.162}$$

Consequently, in the presence of massless fermions tunnelling between vacua with different topological numbers is suppressed.

However, the θ-vacuum plays an important part as before. Thus, for example, in the vacuum state with $n = 0$, an average value of operators taken in points being positioned at sizeable distances does not vanish because there is an abnormal vacuum with a topological number $n \neq 0$:

$$< 0|P_\nu^\dagger(x)P_\nu(y)|0 >\overset{|x-y|\to\infty}{\longrightarrow} \sum_m < 0|P_\nu^\dagger(x)|m >< m|P_\nu(y)|0 >=$$

$$=< 0|P_\nu^\dagger(x)|\nu >< \nu|P_\nu(y)|0 >\neq 0. \tag{10.163}$$

The relation (10.163) also indicates spontaneous symmetry breaking of the chiral symmetry as the vacuum expectation value of operators carrying chirality is nonzero.

Under a chiral rotation by an angle θ we get

$$e^{-i\alpha\tilde{Q}_5}|\theta >= \sum_n e^{-in\theta-i\alpha 2ni_F}|n >= |\theta + 2\alpha i_F >, \tag{10.164}$$

that is, θ-vacuum converts to another one with $\theta' = \theta + 2\alpha i_F$. So, the conservation of a gauge-variant current means that the theory is invariant under a rotation that changes its vacuum state $|\theta >\to |\theta' >$. To put it differently, in the presence of massless fermions the quantity θ has no physical meaning, that is, all values of θ are equivalent. The given situation is very much like the one we encountered in the nonlinear scalar theory with the spontaneous breaking of the $U(1)$ symmetry (Section 11.1 of *Advanced Particle Physics Volume 1*). In that case, under $U(1)$ rotations, the vacuum states can convert to each other as pictured in Fig. 11.2 of *Advanced Particle Physics Volume 1*. Thus, in QCD the parameter θ could be interpreted as the one that characterizes the direction of symmetry breaking. Nevertheless, there is a crucial difference between these two spontaneously broken $U(1)$ theories. In nonlinear scalar theory with the Lagrangian given by (11.1) (see *Advanced Particle Physics Volume 1*), all the vacuum states connected by rotations are equivalent while in QCD the parameter θ looks like the coupling constant because different values of θ correspond to different Hilbert spaces. In the $U(1)$ theory with (11.1) (see *Advanced Particle Physics Volume 1*), small oscillations around the true vacuum in the angular direction (ξ), which do not require any energy (zero energy excitations) are interpreted as physical massless particles, the Goldstone bosons. Unlike this, in QCD, changes of the θ-parameter are meaningless just as the quasi-momenta cannot be changed in the case of periodic potential. Therefore, small oscillations of the parameter θ in QCD are associated with an unphysical massless particle.

Thus, the existence of instantons apparently resolves the $U(1)$ problem. But, on the other hand, their presence results in the θ-term in the effective Lagrangian which violates P and CP invariance. One may ask how can the exotics of instanton phenomena change the results of the traditional QCD. The stringent experimental upper limit on the neutron dipole moment gives the following bound on the QCD θ parameter $\theta < 10^{-9}$. Thus, the θ term in the QCD Lagrangian is so small that it practically has no influence.

Let us consider a tunnelling amplitude caused by one-instanton field configuration:

$$< 0| \pm 1 > \approx \exp\left[-\frac{2\pi}{\alpha_g}\right]. \tag{10.165}$$

After renormalization procedure the coupling constant α_g must be replaced by the running coupling constant so that, with a precision of logarithmic corrections, the expression for the tunnelling amplitude takes the form

$$< 0| \pm 1 > \approx \left(\frac{\Lambda_{QCD}^2}{Q^2}\right)^{(33-2n_f)/3}. \tag{10.166}$$

This formula shows, at large transferred momenta Q^2, tunnelling effects are negligibly small and the state $|0>$ may be considered as the state of the true vacuum. At that, a mistake caused by the expression (10.166) may surely be neglected. Really, estimations show [212] that instanton corrections to processes of e^+e^--annihilation or deep inelastic scattering are completely negligible when $Q^2 \geq 1$ GeV^2. Therefore, when instanton effects are significant, calculations within the perturbation theory are inapplicable and when the perturbation theory can be used, effects caused by instantons are not observable. It is obvious, there also exists a region of borderline values of α_s where one may use the perturbation theory as before, but with allowance made for instanton effects.

10.5   Instanton physics

Since discovery of instantons, instanton physics develops in three directions. The first direction is associated with the investigation of a fundamental role of instantons in the QCD vacuum structure: in the basis of the instanton mechanism of the spontaneous breaking of the chiral symmetry, in the appearance of gluon condensate, and so on (see, for example, [213]).

The second direction is coupled with application of instanton physics for computations of different hadron properties (masses, formfactors and hadron radii, parton distribution functions, and so on) [214].

The third direction is to investigate instantons within the lattice QCD. In the context of this approach it was shown that instantons are major fluctuations of fields in a vacuum. Estimations of density and mean size of instantons were also obtained [215].

The results of investigations lead to the conclusion that the model in which the vacuum presents an instanton liquid appears to be the most appropriate model for a nonperturbative vacuum of QCD. However, for the final formation of the theory, conclusive proofs that observed properties of hadrons cannot be described without instantons are needed. Spin and flavor hadron physics, where dynamics of quarks and gluons is investigated, represents exactly the most sensible region for testing of different QCD nonperturbative models.

The non-Abelian character of the QCD gauge group, in the final analysis, leads to the existence of the complicated structure of the ground state, the vacuum. Detailed knowledge of this structure is of fundamental importance for the description of hadron properties because hadrons are nothing but particular excitations of the QCD vacuum. The QCD vacuum theory began to develop after the discovery of instantons.

Now we consider in more detail the key moments of the instanton vacuum model. Begin with writing a solution for an instanton with the center in point τ_λ and with size ρ

$$G_\mu^{a} - \frac{2}{g_s} \frac{\eta_{\mu\lambda}^a (x - \tau)_\lambda}{(x - \tau)^2 + \rho^2}, \tag{10.167}$$

$$G_{\mu\nu}^a = -\frac{4}{g_s} \frac{\eta_{\mu\nu}^a \rho^2}{[(x - \tau)^2 + \rho^2]^2}. \tag{10.168}$$

Frequently, it is convenient to use the expression for G_μ^a in the so-called *singular gauge* when the *bad* behavior of G_μ^a is transferred from an infinitely distant space point to the instanton center. Such a transition is fulfilled with the help of a gauge transformation

$$g_s \frac{\sigma^a}{2} G_\mu^{a\prime} = U^\dagger g_s \frac{\sigma^a}{2} G_\mu^a U + i U^\dagger \partial_\mu U, \tag{10.169}$$

where

$$U = \frac{i\sigma_\mu^+ (x - \tau)_\mu}{\sqrt{(x - \tau)^2}}, \qquad \sigma_\mu^\pm = (\mp i, \boldsymbol{\sigma}). \tag{10.170}$$

To substitute (10.169) into (10.170) gives the following expression for the potential in the singular gauge

$$G_\mu^{a} = \frac{2}{g_s} \frac{\overline{\eta}_{\mu\lambda}^a (x - \tau)_\lambda \rho^2}{(x - \tau)^2 [(x - \tau)^2 + \rho^2]}. \tag{10.171}$$

Note, the symbols $\overline{\eta}_{\mu\lambda}^a$ and not $\eta_{\mu\lambda}^a$ enter Eq. (10.171). This circumstance is related with the fact that in the singular gauge the topological number is defined in the vicinity $x = x_0$ rather than at infinity. Instantons lead to a specific multi-quarks vertex, t'Hooft vertex [216], that, in the case of small size instantons ($\rho \to 0$), massless quarks* and $N_f = N_c = 3$, could be presented in the form of the effective Lagrangian [217]

$$\mathcal{L}_{eff} = \int d\rho n(\rho) \left\{ \prod_{i=u,d,s} \left(m_i \rho - \frac{4\pi^2 \rho^3}{3} \overline{q}_{iR} q_{iL} \right) + \frac{3}{32} \left(\frac{4\pi^2 \rho^3}{3} \right)^2 \left[\left(j_u^a j_d^a - \frac{3}{4} j_{u\mu\nu}^a j_{d\mu\nu}^a \right) \times \right.$$

---

*All quarks are light for instantons of sufficiently small sizes.

$$\times \left(m_s\rho - \frac{4}{3}\pi^2\rho^3\overline{q}_{sR}q_{sL} \right) + \frac{9}{40} \left(\frac{4\pi^2\rho^3}{3} \right)^2 d^{abc} j^a_{u\mu\nu} j^b_{d\mu\nu} j^c_s + 2 \text{ perm.} \right] +$$

$$+ \frac{9}{320} \left(\frac{4\pi^2\rho^3}{3} \right)^3 d^{abc} j^a_u j^b_d j^c_s + \frac{ig_s f^{abc}}{256} \left(\frac{4\pi^2\rho^3}{3} \right)^3 j^a_{u\mu\nu} j^b_{d\nu\lambda} j^c_{s\lambda\mu} + (R \leftrightarrow L) \right\}, \quad (10.172)$$

where $j^a_i = \overline{q}_{iR}\lambda^a q_{iL}$, $j^a_{i\mu\nu} = \overline{q}_{iR}\sigma_{\mu\nu}\lambda^a q_{iL}$, $n(\rho)$ is an instanton density and $\sigma_{\mu\nu} = (\gamma_\mu\gamma_\nu - \gamma_\nu\gamma_\mu)/2$. In the limit $m_{q_i} \to 0$ this Lagrangian is symmetric with respect to the $SU(3)_f$ group but violates the axial $U_A(1)$ symmetry. The Lagrangian (10.172) follows from consideration of quarks scattering at the co-called *zero mode* in an instanton field. The quark zero mode was found by t'Hooft [216], who showed that the Dirac equation

$$-\gamma_\mu \left(i\partial_\mu + g_s \frac{\lambda^a}{2} G^a_\mu(x) \right) \psi_n(x) = E_n\psi_n(x) \quad (10.173)$$

in the instanton field (10.171) has the solution with the zero energy ($E_0 = 0$)

$$\psi_0(x - \tau) = \frac{\rho(1 - \gamma_5)}{2\pi[(x - \tau)^2 + \rho^2]^{3/2}} \frac{\hat{x}}{\sqrt{x^2}}\varphi, \quad (10.174)$$

where φ is a two-component spinor, that is, $\varphi^a_m = \epsilon^a_m/\sqrt{2}$, a is an index of the $SU(2)_c$-color subgroup of the $SU(3)_c$-group and m is a spin index.

Note some features of the expression (10.174). First, zero modes in the instanton field have definite helicity (in (10.174) the replacement $(1 - \gamma_5) \to (1 + \gamma_5)$ must be done for an anti-instanton). For this reason, a quark flying into the instanton field must have the right-hand helicity to be scattered in the zero mode. As this takes place, an outgoing quark will have the left-hand helicity. At scattering in an anti-instanton field, the situation is reverse. Second, at the zero mode, the values of the color and spin of the quark are severely correlated by way of the spinor φ^a_m. From these properties and the Pauli principle it follows that only one quark of a definite flavor may be in the zero mode. Moreover, the quark helicity is turned over under scattering by an instanton. These properties play an important part in accounting for different spin and flavor anomalies that are observed in the strong interaction.

There exist two principal distinctions of the effective vertices induced by instantons (10.172) from the quark-gluon perturbative vertex. First, in contrast to the perturbative vertex, they stimulate the spin flip of a quark. And so, the quark helicity for the N_f quark vertex is changed by the value $2N_f$. Second, as a consequence of the Pauli principle, the vertices (10.172) result in the violation of Okubo–Zweig–Iizuka (OZI) rules because they are not equal to zero only for different quark flavors.

Within the instanton model of the QCD vacuum, all instantons have the same size and, therefore, the instanton density is approximated by the delta function:

$$n(\rho, x) = n_0\delta(\rho - \rho_c), \quad (10.175)$$

where $\rho_c \approx 1.6$ GeV$^{-1} \approx 0.3$ fm is a mean size of an instanton in the QCD vacuum while the value of n_0 is defined by the gluon condensate value:

$$n_0 = \frac{\alpha_s}{16\pi} < 0|G^a_{\mu\nu}G^a_{\mu\nu}|0 > . \quad (10.176)$$

From (10.172) it is easy to get the two-particle Lagrangian in the instanton liquid model. To this end it is enough to replace the bilinear combinations of quark fields of a definite flavor in (10.172) by their vacuum expectation values:

$$\overline{q}_i O_{ij} q_j \to < 0|\overline{q}_i O_{ij} q_j|0 > . \quad (10.177)$$

where O_{ij} is an arbitrary matrix, as well as to use the connection between the effective instanton density and quark condensates [214]:

$$\left.\begin{array}{ll} n_{eff} = n_0 m_u^* m_d^* m_s^* \rho_c^3, & m_i^* = m_i + m^*, \\[2mm] m^* = -\dfrac{2\pi^2 \rho_c^2}{3} < 0|\bar{q}q|0 >, & < 0|\bar{q}q|0 >= -\dfrac{\sqrt{3n_{eff}}}{\pi\rho_c}, \end{array}\right\} \qquad (10.178)$$

where m_i, m_i^* are current and constituent masses of quarks, respectively, and $i = u, d, s$. Finally, we get

$$\mathcal{L}_{inst} = \frac{1}{2} \sum_{i \neq j} \eta_{ij} \left[\bar{q}_{iR} q_{iL} \bar{q}_{jR} q_{jL} + \frac{3}{32} \left(\bar{q}_{iR} \lambda^a q_{iL} \bar{q}_{jR} \lambda^a q_{jL} - \right. \right.$$

$$\left. \left. - \frac{3}{4} \bar{q}_{iR} \sigma_{\mu\nu} \lambda^a q_{iL} \bar{q}_{jR} \sigma_{\mu\nu} \lambda^a q_{jL} + (R \leftrightarrow L) \right) \right], \qquad (10.179)$$

where $\eta_{ij} = (4\pi^2 \rho_c^2 m^{*2})/(3m_i^* m_j^*)$. Note some properties of the obtained Lagrangian. First, it has turned out that the effective coupling constant is actually a function only of a single parameter, an instanton size in the QCD vacuum, while the dependence on the instanton density disappears. Second, as compared with the Lagrangian of the perturbative gluon exchange, the expression (10.180) possesses absolutely different structures both in the color and flavor spaces.

With the help of the Lagrangian (10.179) the wide region of phenomena in hadron spectroscopy can be investigated. Let us consider some of them.

We begin with calculation of hadron masses. The QCD vacuum has a complicated structure caused by the existence of quark and gluon nonperturbative fields. When computing nonperturbative contribution into hadron masses, these fields may be divided into two classes [218]. The former consists of vacuum fields that have wave-lengths considerably exceeding the nucleon size. Such fields may be considered to be constant along the nucleon size and their effect are reduced to the interaction of hadron quarks with quark and gluon condensates. The effective Hamiltonian describing this interaction follows from the QCD energy-momentum tensor by means of dividing of the quark and gluon fields into the quantum and classical parts:

$$\psi(x) = q(x) + Q(x) \qquad (10.180$$

for quark fields, and

$$G_\mu^a(x) = b_\mu^a(x) + B_\mu^a(x)$$

for gluon fields. Then the effective Hamiltonian takes the form

$$H_{vac} = \frac{i}{2} \left(\bar{q}\gamma_0 \partial_0 Q - \partial_0 \bar{q}\gamma_0 Q \right) - i\bar{q}\gamma_\mu D_\mu Q + M\bar{q}q + (q \leftrightarrow Q), \qquad (10.181)$$

where M is the quark mass matrix. To calculate the contribution of long-wave fluctuations into the shift of the hadron self-energy it is enough to find the matrix element

$$E_{long} = < \Phi| \int H_{vac} d^3 x |\Phi >, \qquad (10.182)$$

with the hadron wave functions Φ.

The second class describes contributions coming from vacuum fluctuations with the wavelength being smaller than the confinement region size $\rho < R_{conf} \approx 1$ fm. Within the instanton liquid model the contribution into the hadron energy is calculated using the t'Hooft interaction (10.180):

$$E_{inst} = - < \Phi| \int \mathcal{L}_{inst} d^3x |\Phi > .$$

(10.183)

Concrete calculations show [219] that a mass spectrum of ground hadron states could be reproduced. Furthermore, the model also explains peculiarities of spin-spin splitting between hadron multiplets, namely: instantons gives determinative contribution to observed mass splitting between octets of pseudoscalar and vector mesons as well as between octet and decuplet of baryons. The reason leading to large splitting is identical in both cases. Instantons give strong attraction in channels where quark-antiquark or two quarks are in a state with zero spin. In the first case, this leads to a practically massless pion and serves a fundamental reason of observed spontaneous breaking of chiral symmetry, that is, the appearance of a quark condensate [214]. The attraction in a quark-quark channel may bring both to the appearance of a quasi-coupled scalar diquark in a nucleon [220] and to the formation of a color diquark condensate.

Explanation of the violation of the OZI rule represents another example of fruitful activity of the instanton physics. Recall that the OZI rule is an empiric rule forbidding transitions between states with different quark flavors. The analysis of decays of vector and pseudoscalar mesons leads to the conclusion: this rule is well fulfilled in the vector channel where we observe ideally mixed unitary states, that is, $\Phi = \bar{s}s$ and $\omega = (\bar{u}u + \bar{d}d)/\sqrt{2}$. A contrary situation takes place in the pseudoscalar nonet. Here, the octet and nonet are not practically mixed. Such a distinction may exist only in the case when the OZI rule is realized in the vector channel while it is very strongly violated in the pseudoscalar channel. Within the perturbation theory, the QCD suppression of the transition $\bar{s}s \rightarrow \bar{u}u + \bar{d}d$ in the vector case, as compared with the pseudoscalar one, is usually explained by the fact that such a transition happens through a three-gluon exchange for the vector mesons and through a two-gluon exchange for the pseudoscalar mesons. However, this explanation does not fail in line with experimental data. At this mechanism the transition amplitudes $\bar{s}s \rightarrow \bar{u}u + \bar{d}d$ in the pseudoscalar and vector channels must be distinguished on the value of the order of ≈ 0.7, while the experiment gives mixing of the order of few percents in the first case and practically total mixing in the second case. Moreover, in the tensor nonet 2^+ as in the pseudoscalar one, the transition between different kinds of quarks may occur by way of a two-gluon intermediate state. In the experiment the decay $f'(1525) \rightarrow 2\pi$ is suppressed relatively to $f'(1525) \rightarrow 2K$ while the decay $f(1270) \rightarrow 2\pi$ is allowed. Therefore, f' must be a nearly-pure $\bar{s}s$-state and $f = (\bar{u}u + \bar{s}d)/\sqrt{2}$. Because of this, an additional mechanism for explanation of the OZI rules in a tensor channel is needed. It should be pointed out that the QCD perturbation theory results in mixing angles that are less than the experimental ones [221]. The Lagrangien (10.180) leads to the existence of the two-quark diagram determining a mixing between different kinds of quarks at the expense of instanton interaction. This diagram is pictured in Fig. 10.8, where the effective two-quark interaction is obtained from the effective three-quark interaction (10.172) under closing the two-quark line through a quark condensate. This contribution is exactly equal to zero for the vector and tensor mesons because of the selection rule (10.175) forbidding quarks entering a meson to be in a state with the total spin $S = 1$. Then, in these channels the OZI rule is well fulfilled. In the pseudoscalar channel, on the contrary, this diagram gives a large contribution into the $\pi^0 - \eta - \eta'$-mixing [218]. Calculations confirm [219], the instanton liquid model results in small angles of a singlet-octet mixing for the pseudoscalar mesons and, therefore, explains the observed large violation of the OZI rule in this channel. We call attention to the following properties of the model under consideration. On the one hand, a practically massless π-meson appears in the natural way. On the other hand, a massive η'-meson is present

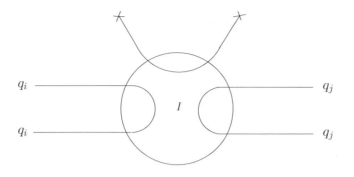

FIGURE 10.8

The contribution of the instanton interaction into transitions between different quark flavors. The symbol I denotes an instanton and the crosses denote quark ends connected through a quark condensate.

in the model. Therefore, the two fundamental problems of hadron spectroscopy are solved simultaneously: the Goldstone π-meson and the $U(1)_A$ problem.

Now we briefly discuss the perspectives of the explanation of a $\Delta I = 1/2$ rule. Experiments have been demonstrating the existence of a large CP violation in the decay $K \to \pi\pi$. This effect is rather difficult explained within the standard model. One of the cornerstones of this problem is a well-known $\Delta I = 1/2$ rule. This phenomenological rule is related to an observed large amplification of weak decays with changing of the isotopic spin on the value $\Delta I = 1/2$ as compared to decays with $\Delta I = 3/2$. Thus, for example, experiments give the following value for the ratio of amplitudes with a different total isospin of final pions

$$|A_0|/|A_2| \approx 22.2,$$

where A_0 is an amplitude with zero isotopic spin and A_2 is an amplitude with isospin 2. There exist several possible mechanisms of this phenomenon. One of them is the well-known perturbative QCD mechanism [222], which appears due to the contribution of small distances, that is, due to large virtualities of quarks and gluons. Maximum amplification this effect gives results in the factor 5 that is approximately five times less than the experimental value. Another possible source of this rule comes from the contributions of quark interactions at large distances and is coupled with effects of interactions in final states. These effects could result in the amplification of the A_0-amplitude to 50% of its experimental value [223]. The nonperturbative multi-quark t'Hooft interaction induced by instantons is very sensitive to flavors and helicities of quarks. As is well known, the weak interaction possesses unique helicity and flavor properties too. This circumstance suggests that there exists amplification mechanisms of weak decays in channels with such quantum numbers that instanton contributions described by the Lagrangian (10.172) become allowed [224]. As we mentioned before, the interaction (10.172) has flavor properties strongly different from the perturbative gluon exchange. Due to these properties, the one-instanton contribution is present in the weak $\Delta I = 1/2$ amplitude and absent in the weak $\Delta I = 3/2$ amplitude. Therefore, the specific flavor dependence of the instanton interaction is the reason for the anomalous amplification of the A_0 amplitude in the decays $K \to \pi\pi$.

In summary we note, at present the large part of calculations in instanton physics have been fulfilled in the chiral limit. However, even these results indicate an extraordinary significance of instanton effects in the standard model.

11

QCD experimental status

And do not stop.
And do not change their feet—
The light illuminates our faces.
And the boots that shine!
V. Vysotsky

In conclusion of our excursus in QCD we summarize the quantitative verification of the theory. It is obvious, as precision tests of any interaction theory, we can use various measurements of its coupling constant. In QED, the best determinations of the coupling constant agree with the theoretical value to eight significant figures. However, in QCD the situation is rather different. Since the QCD perturbation theory works only for hard-scattering processes, with uncertainties specified by soft processes that are difficult to estimate, this theory has not been tested to such an extreme precision. Just the same, it is interesting to compare the best available determinations of α_s to see how well they agree.

Consider a typical QCD cross section that calculated perturbatively. If this process is allowed even in the first order of the perturbation theory, the resulting cross section is given by the expression:

$$\sigma = \sum_{i=0} A_{2i+1} \alpha_s^{2i+1}. \tag{11.1}$$

When the process in question may begin with the second order of the perturbation theory, then, instead of (11.1), we have

$$\sigma = \sum_{i=1} A_{2i} \alpha_s^{2i}. \tag{11.2}$$

The coefficients A_1, A_2, ... come from calculating the appropriate Feynman diagrams. In performing such calculations, various divergences arise, and these must be regulated in a consistent way. This requires a particular renormalization scheme (RS). The most commonly used one is the modified minimal subtraction ($\overline{MS}$) scheme (see Section 3.4). The cross sections and other physical quantities calculated to all orders in the perturbation theory do not depend on the RS. However, the finite coefficients A_n (with $n \geq 2$) and the coupling constant α_s depend implicitly on the renormalization convention employed and explicitly on the scale Q.

Usually, the QCD cross sections are known to the leading order (LO), or to the next-to-leading order (NLO), or in some cases, to the next-to-next-to-leading order (NNLO). Note, only the latter two cases, which have cancelled RS dependence, are useful for precision tests. At NLO the RS dependence is completely given by one condition that can be taken to be the value of the renormalization scale Q. At NNLO this is not sufficient, and Q is no longer equivalent to a choice of scheme; both must now be specified. Therefore, one has to address the question of what is the best choice for Q within a given scheme, usually $\overline{MS}$. There is no definite answer to this question—higher-order corrections do not fix the scale, rather they give the theoretical predictions less sensitive to its variation.

One should assume, choosing a scale Q characteristic of the typical energy scale (E) in the process would be most appropriate. In general, a poor choice of scale generates terms of the order $\ln(E/\mu)$ in the A_n's. Various methods have been proposed including choosing the scale for which the NLO correction turns into zero (Fastest Apparent Convergence [225]); the scale for which the NLO prediction is stationary [226], (that is, the value of Q where $d\sigma/dQ = 0$); or the scale dictated by the effective charge scheme [228] or by the scheme proposed in [227]. By comparing the values of α_s, which different reasonable schemes render, an estimate of theoretical errors can be obtained. One can also attempt to determine the scale from data by allowing it to vary and using a fit to define it. This method can allow a determination of the error due to the scale choice and can give more confidence in the end result [229]. It should be stressed that in many cases this scale uncertainty is the dominant error.

An important practical consequence is that if the higher-order corrections are naturally small, then the additional uncertainties caused by the Q dependence are likely to be small. However, there exist some processes for which the choice of scheme can influence the extracted value of α_s. There is no resolution to this problem other than to try to calculate even more terms in the perturbation series. It is important to note that, since the perturbation series is an asymptotic expansion, there is a limit to the accuracy with which any theoretical quantity can be calculated.

Experiments such as those from deep-inelastic scattering involve a range of energies. After fitting to their measurements to QCD, the resulting fit can be expressed as a value of α_s. Determinations of α_s arise from fits to data employing NLO and NNLO: LO fits are not useful and their values will not be employed in the following. Care must be exerted when comparing results from NLO and NNLO. So as to compare the values of α_s from various experiments, they must be evolved using the renormalization group to a common scale. For convenience, this is taken to be the mass of the Z-boson. The extrapolation is carried out using the same order in the perturbation theory as was used in the analysis. This evolution uses the third-order perturbation theory and can enter additional errors particularly if extrapolation from very small scales is employed. The variation in the charm and bottom quark masses can also introduce errors. For example, using

$$m_b = 4.3 \pm 0.2 \text{ GeV}, \qquad m_c = 1.3 \pm 0.3 \text{ GeV},$$

and working within the perturbation theory one can obtain

$$\Delta\alpha_s(m_Z) = \pm 0.001.$$

Note, there could be additional errors from nonperturbative effects that enter at low energy.

There are many ways in which the perturbative QCD can be tested at high-energy hadron colliders. The quantitative tests are only useful if the process under consideration has been calculated beyond LO in the QCD perturbation theory. The production of hadronic jets with a large transverse momentum in hadron-hadron collisions provides a direct probe of the scattering of quarks and gluons: $qq \to qq$, $qg \to qg$, $gg \to gg$, and so on. Higher-order QCD calculations of the jet rates and shapes are in impressive agreement with the data. This agreement has led to the proposal that these data could be used to provide a determination of α_s [230]. A set of structure functions is assumed and jet data are fitted over a very large range of transverse momenta to the QCD prediction for the underlying scattering process that depends on α_s. The evolution of the coupling over this energy range (40 to 250 GeV) is therefore tested in the analysis. For example, CDF (Collider Detector at Fermilab) obtains [231]

$$\alpha_s(m_Z) = 0.1178 \pm 0.0001 \text{(stat.)} \pm 0.0085 \text{(syst.)}.$$

Estimation of the theoretical errors is not straightforward. The structure functions used depend implicitly on α_s and an iteration procedure must be used to obtain a consistent result; different sets of structure functions yield different correlations between the two values of α_s. CDF includes a scale error of 4% and a structure function error of 5% in the determination of α_s. Ref. [230] estimates the error from unknown higher-order QCD corrections to be ± 0.005. Combining these then gives

$$\alpha_s(m_Z) = 0.118 \pm 0.011.$$

Data are also available on the angular distribution of jets; these are also in agreement with QCD expectations [232].

Measurements of the fragmentation function $D_i(z, E)$ ($i = q, \bar{q}, g$) can be employed to define α_s. As in the case of scaling violations in structure functions, the perturbative QCD predicts only the dependence on the typical energy scale (the E dependence). Therefore, measurements at different energies are needed to extract a value of α_s. Because the QCD evolution mixes the fragmentation functions for each quark flavor with the gluon fragmentation function, it is necessary to define each of these before α_s can be extracted. A global analysis [233] using data from SLC (SLAC Linear Collider), DELPHI (DEtector with Lepton, Photon and Hadron Identification), OPAL (Omni Purpose Apparatus for LEP), ALEPH (Apparatus for LEP PHysics at CERN), and from the LBL TPC (large time projection chamber) collaboration gives the result

$$\alpha_s(m_Z) = 0.1172^{+0.0055+0.0017}_{-0.0069-0.0025}.$$

The second error is a theoretical one arising from the choice of scale.

At lowest order in α_s the ep scattering process produces a final state of $(1 + 1)$ jets, one from the proton fragment and the other from the quark knocked out by the process $e^- q \rightarrow e^- q$. At next order in α_s, a gluon can be radiated, and, as a result, a $(2 + 1)$ jet final state created. By comparing the rates for these $(1 + 1)$ and $(2 + 1)$ or $(2 + 1)$ and $(3 + 1)$ jet processes, a value of α_s can be obtained. A summary of the measurements from HERA (Hadron-Electron Ring Accelerator) can be found in Ref. [234]. Results from H1 and ZEUS detectors at HERA can be combined to give

$$\alpha_s(m_Z) = 0.1178 \pm 0.0033(\text{expt.}) \pm 0.006(\text{theor.}).$$

The theoretical errors arise from the scale choice, structure functions, and hadronization correction.

Photoproduction of two or more jets via processes such as $\gamma g \rightarrow q\bar{q}$ can be observed at HERA as well. The process is similar to jet production in hadron-hadron collisions. Agreement with the perturbative QCD is excellent and the results of ZEUS detector are [235]

$$\alpha_s(m_Z) = 0.1224 \pm 0.0020(\text{expt.}) \pm 0.0050(\text{theor.}).$$

Within the perturbative QCD a value of α_s could also be found in the following cases: at the measurements of the total cross section for e^+e^- annihilation into hadrons, or equivalently, the ratio of the number of observed hadronic and leptonic events; at the measurement of the transverse momentum spectrum of W bosons produced from quark-antiquark annihilation at high-energy $p\bar{p}$ colliders; under the inclusion of gluon radiative corrections to the vertices in deep inelastic neutrino scattering; under studying the rate of the Bjorken scaling violation in deep inelastic scattering processes; under investigation of the decays of the lightest $b\bar{b}$ bound state Υ and $c\bar{c}$ bound state ψ; and under studying the decays of τ lepton and Z boson and so on (see, for review, [236]).

Lattice QCD can be also employed to calculate, using nonperturbative methods, a physical quantity that can be measured experimentally. The value of this quantity can then be used to determine the QCD coupling that enters in the computation. The main quantitative difference between lattice calculations and those discussed above is that the experimental measurements involved, such as the masses of Υ states, are so precise that their uncertainties have almost no impact on the final comparisons. A discussion of the uncertainties that enter into the QCD tests and the determination of α_s is therefore almost exclusively a discussion of the techniques utilized in the computations. To go along with α_s, other physical quantities such as the masses of the light quarks can be calculated. For example, the energy levels of a heavy $q\bar{q}$ resonance can be determined and then used to extract α_s. The masses of the $q\bar{q}$ states depend only on the quark mass and on α_s. For three light quark flavors, calculations of α_s have been done in Ref. [237]. With allowance made for the mass differences of Υ and Υ' and Υ'' and χ_b, the result is

$$\alpha_s(m_Z) = 0.1170 \pm 0.0012.$$

Many other quantities such as the pion decay constant, and the masses of the B_s meson and Ω baryon are used and the overall consistency is excellent. In Fig. 11.1 we represent the values of $\alpha_s(m_Z)$ deduced from the experiments discussed above. The values shown indicate the process and the measured value of α_s extrapolated to $Q = m_Z$. The error presented is the total error including theoretical uncertainties. An average of the values in Fig. 11.1

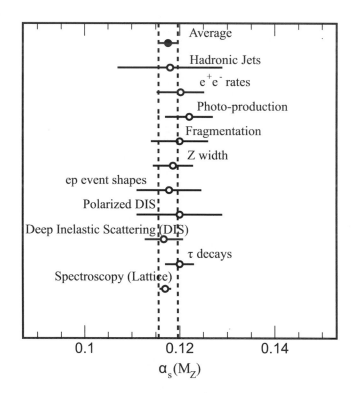

FIGURE 11.1
Summary of the value of $\alpha_s(m_Z)$ from various processes.

results in $\alpha_s(m_Z) = 0.1176$, with a total χ^2 of 9 for eleven fitted points, showing good consistency among the data. The error on this average, assuming that all of the errors in the contributing results are uncorrelated, is ± 0.0009, and may be an underestimate. Almost all of the values used in the average are dominated by systematic, usually theoretical, errors.

Only some of these, notably from the choice of scale, are correlated. The error on the lattice QCD result is the smallest and then there are several results with comparable small errors: these are the ones from τ decay, deep inelastic scattering, Υ decay, and the Z width. Omitting the lattice QCD result from the average changes it to $\alpha_s(m_Z) = 0.1185$ or 1σ. All of the results that dominate the average are from NNLO. The NLO results have little weight; there are no LO results utilized. Almost all of the results have errors that are dominated by theoretical issues, either from unknown higher-order perturbative corrections or estimates of nonperturbative contributions. It is therefore sensible to be conservative and quote an up-to-date average value as $\alpha_s(m_Z) = 0.1176 \pm 0.002$.

In Fig. 11.2 we have plotted the original values of α_s versus the momentum scale Q at which each was obtained. The lines show the central values and the $\pm 1\sigma$ limits of the

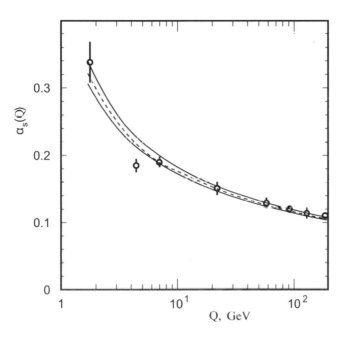

FIGURE 11.2
Summary of the values of $\alpha_s(Q)$ at the values of Q where they are measured.

average. This figure clearly shows the experimental evidence for the variation of $\alpha_s(Q)$ with Q.

Forty years, that have elapsed from the time of the QCD appearance, were filled up by abundant experimental confirmations of this theory. One may state, to date a very important frontier has been traversed in a longstanding history of investigation of the strong interaction. Undoubtedly, subsequent advances and the search for answers to many questions that have yet to be solved, will proceed with allowances made for the previous achievements. The QCD successes allow to claim, that at present it is the only serious candidate regarding the strong interaction theory. However, the following question is lawful: "What is the strong interaction waiting for in the future?" If one believes that the history of physics teaches something, then the forward way will be full of unexpected turns and phenomena about which we are not aware of and which could even lead to changing the ideologies. First of all, one cannot exclude the possibility of the fact that quarks and gluons take part in new strong interactions appearing at an energy scale Λ' that is much bigger than the typical energy Λ_{QCD}. For example, quarks may prove to be bound states of more fundamental fermions interacting with gauge fields: corresponding asymptotically free theories could lead to a strong coupling (creation of quarks) at an anergy of the order of Λ'. In this case, the effective Lagrangian density of quarks for energies $E \ll \Lambda'$ will involve unrenormalizable interactions suppressed only by the powers E/Λ'. Such interactions may become apparent in processes proceeding both with weak violations of symmetries (parity and so on) and with a flavor conservation at ordinary energies. They will also result in qualitative deviations from the QCD predictions for energies approaching to Λ' [238]. In other words, we cannot rule out the possibility that QCD should be considered as an effective field theory, that is, the low-energy approximation of a more fundamental theory that may not prove to be even a field theory.

References

[1] G. t'Hooft, Nucl. Phys. **B33**, 173 (1971).

[2] S. Coleman, D. Gross, Phys. Rev. Lett. **31**, 851 (1973).

[3] E. de Rafael, In *Quantum Chromodynamics* (Alonso and Tarrach, eds.), (Springer, Berlin, 1979).

[4] B. de Witt, *Relativity, Groups and Topology*, (Blakie and Son, London, 1964).

[5] R. P. Feynman, Acta Phys. Polonica **24**, 697 (1967).

[6] L.D. Faddeev, V. N. Popov, Phys. Lett. **B25**, 29 (1967).

[7] C. Becchi, A. Rouet, R. Stora, Phys. Lett. **B52**, 344 (1974).

[8] A. A. Slavnov, Phys. Elemen. Part. and Atom. Nuclei **5**, 193 (1979).

[9] J. C. Taylor, Nucl. Phys. **B33**, 436 (1971).

[10] R. P. Feynman, Rev. Modern Phys. **20**, 367 (1948).

[11] R. P. Feynman, A. R. Hibbs, *Quantum Mechanics and Path Integrals*, (McGraw-Hill, New York, 1965).

[12] N. Christ, T. D. Lee, Phys. Rev. **D22**, 939 (1980).

[13] W. Kummer, Acta Physica Austriaca **41**, 315 (1975).

[14] E. Tomboulis, Phys. Rev. **D8**, 2736 (1973).

[15] A. A. Slavnov, Theor. Math. Phys. **10**, 99 (1972).

[16] J. C. Taylor, Nucl. Phys. **B33**, 436 (1971).

[17] R. P. Feynman, Acta Physica Polonica **24**, 697 (1963).

[18] R. E. Cutkosky, J. Math. Phys. **1**, 429 (1960).

[19] B. W. Lee and J. Zinn-Justin, Phys. Rev. **D5**, 3121 (1972).

[20] G. t'Hooft, M. Veltman, Nucl. Phys. **B44**, 189 (1972).

[21] J. C. Collins, *Renormalization*, (Cambridge University Press, London, 1984).

[22] G. t'Hooft, Nucl. Phys. **B61**, 455 (1973).

[23] S. Weinberg, Phys. Rev. **D8**, 3497 (1973).

[24] D. Gross, F. Wilczek, Phys. Rev. Lett. **30**, 1343 (1973).

[25] D. Gross, F. Wilczek, Phys. Rev. **D8**, 3633 (1973).

[26] H. D. Politzer, Phys. Rev. Lett. **30**, 1346 (1973).

[27] H. D. Politzer, Phys. Rep. **C14**, 129 (1974).

[28] M. Yoshimura, Phys. Rev. Lett. **41**, 281 (1978).

[29] W. Thirring, Phil. Magazine **41**, 113 (1950).

[30] E. C. G. Stueckelberg, A. Peterman, Helv. Phys. Acta **26**, 499 (1953).

[31] M. Gell-Mann, F. E. Low, Phys. Rev. **95**, 1300 (1954).

[32] N. N. Bogoliubov, D. V. Shirkov, *Introduction to the Theory of Quantized Fields*, (Interscience Publishers Inc., New York, 1959).

[33] C. G. Callan, Phys. Rev. **D2**, 1541 (1970).

[34] K. Symanzik, Comm. Math. Phys. **18**, 227 (1970).

[35] W. E. Caswell, Phys. Rev. Lett. **33**, 224 (1974).

[36] O. V. Tarasov et al., Phys. Lett. **B93**, 429 (1980).

[37] R. Tarrach, Nucl. Phys. **B183**, 384 (1981).

[38] S. Narison, Phys. Rep. **C82**, 263 (1982).

[39] J. Schwinger, Phys. Rev. Lett. **3**, 296 (1959).

[40] R. P. Feynman, M. Gell-Mann, Phys. Rev. **109**, 193 (1958).

[41] N. Cabibbo, Phys. Rev. Lett. **10**, 531 (1963).

[42] M. Gell-Mann, Physics **1**, 63 (1964).

[43] V. de Alfaro, S. Fubini, G. Furlan, C. Rosetti, *Currents in Hadron pphysics*, (North-Holland, Amsterdam, 1973).

[44] S. D. Drell, A. C. Hearn, Phys. Rev. Lett. **16**, 908 (1966).

[45] S. Fubini, G. Furlan, Physics **1**, 229 (1965).

[46] S. L. Adler, Phys. Rev. **143**, 1144 (1966).

[47] J. Goldstone, Nuovo Cimento **19**, 154 (1961).

[48] M. Gell-Mann, M. Levy, Nuovo Cimento **16**, 705 (1960).

[49] Y. Nambu, Phys. Rev. Lett. **4**, 380 (1960).

[50] S. L. Adler, Phys. Rev. **B139**, 1638 (1965).

[51] S. L. Adler, R. F. Dashen, *Current Algebra and Applications to Particle Physics*, (Benjamin, New York, 1968).

[52] Y. Tomozawa, Nuovo Cimento **A46**, 707 (1966).

[53] M. L. Goldberger, S. B. Treiman, Phys. Rev. **110**, 1178 (1958).

[54] Y. Nambu and G. Jona-Lasinio, Phys. Rev. **122**, 345 (1961).

[55] S. Weinberg, Phys. Rev. **D11**, 3583 (1975).

[56] J. Schwinger, Phys. Rev. **82**, 664 (1951).

[57] J. S. Bell, R. Jackiw, Nuovo Cimento **A60**, 47 (1969).

[58] S. L. Adler, Phys. Rev. **B177**, 2426 (1969).

[59] J. Ellis, In: *Weak and Electromagnetic Interactions at High Energy*, (North-Holand, Amsterdam, 1976).

[60] S. L. Adler and W. A. Bardeen, Phys. Rev. **B182**, 1517 (1969).

[61] J. Schwinger, Phys. Rev. **128**, 2425 (1962).

[62] K. Fujikawa, Phys. Rev. Lett. **42**, 1195 (1979).

[63] S. Weinberg, *The Quantum Theory of Fields*, Volume 2, (University Press, Cambridge,, 2001).

[64] D. G. Sutherland, Phys. Lett. **23**, 384 (1966).

[65] D. J. Gross, R. Jackiw, Phys. Rev. **96**, 477 (1969).

[66] H. Georgi, S. L. Glashow, Phys. Rev. **D6**, 429 (1972).

[67] M. L. Menta, P. K. Srivastava, J. Math. Phys. **7**, 1833 (1966).

[68] J. Collins, A. Duncan, S. Joglecar, Phys. Rev. **D16**, 438 (1977).

[69] S. Coleman, In: *The Ways of Subnuclear Physics*, (ed. A. Zichichi), (Plenum Press, New York, 1979).

[70] M. E. Peskin, D. V. Schroeder, *An Introduction to Quantum Field Theory*, (Addison-Wesley Publishing Company, Reading MA, 1995).

[71] E. Reya, Phys. Rep. **C69**, 195 (1981).

[72] P. D. B. Collins, A.D. Martin, *Hadron Interactions*, (Adam Hilger Ltd., Bristol, 1984).

[73] T. Kinoshita, J. Math. Phys. **3**, 650 (1962).

[74] T. D. Lee, M. Nauenberg, Phys. Rev. **133**, 1549 (1964).

[75] J. G. H. de Groot, et al. Z. Phys. **C1**, 143 (1979).

[76] G. Sterman et al., Rev. Mod. Phys. **67**, 157 (1995).

[77] C. Amsler et al., Phys. Lett. **B667**, 1 (2008).

[78] A. D. Martin et al., Phys. Lett. **B652**, 292 (2007).

[79] E. Leader et al., Eur. Phys. J. **C23**, 479 (2002).

[80] J. Blümlein and H. Böttcher, Nucl. Phys. **B636**, 225 (2202).

[81] J. Kuti, V. F. Weisskopf, Phys. Rev. **D4**, 3418 (1971).

[82] D. Gross, C. H. Smith Llewellyn, Nucl. Phys. **B14**, 337 (1969).

[83] A. D. Martin et al., Eur. Phys. J. **C4**, 463 (1998).

[84] G. Altarelli, Phys. Rep. **C81**, 1 (1982).

[85] G. Altarelli, G. Parisi, Nucl. Phys. **B126**, 298 (1977).

[86] B. Foster, A. D. Martin, and M. G. Vinter, Phys. Lett. (Review of Particle Physics) **B667**, 188 (2008).

[87] C. Weizsäcker, Zeits. Phys. **88**, 612 (1934); E. Williams, Phys. Rev. **45** 729 (1934).

[88] O. Biebel, D. Milstead, P. Nason, and B. R. Webber, Phys. Lett. (Review of Particle Physics) **B667**, 202 (2008).

[89] G. Curci et al., Nucl. Phys. **B175**, 27 (1980).

[90] J. Kalinowski et al., Nucl. Phys. **B181**, 253 (1981).

[91] P. Nason and B. R. Webber, Nucl. Phys. **B421**, 473 (1994).

[92] O. Biebel, P. Nason, and B. R. Webber, hep-ph/0109282.

[93] S. D. Drell and T. M. Yan, Phys. Rev. Lett. **25**, 316 (1970).

[94] G. Altarelli, R. K. Ellis, G. Martinelli, Nucl. Phys. **B143**, 521 (1978); Erratum, **146** 544 (1978).

[95] B. Humpert, W. L. Neerven, Nucl. Phys. **B184**, 225 (1981).

[96] E. L. Berger et al., Phys. Rev. **D40**, 83 (1989); ibid. **D40** 3789 (erratum).

[97] A. D. Martin, R. G. Roberts, and W. J. Stirling, Phys. Rev. **D50**, 6734 (1994).

[98] H. Abramowicz et al., Zeits. Phys. **C17**, 283 (1983).

[99] F. Halzen and D. M. Scott, Phys. Rev. **D18**, 3378 (1978).

[100] R. W. Brown, and K. O. Mikaelian, Phys. Rev. **D19**, 922 (1979).

[101] R. W. Brown, D. Sahdev, and K. O. Mikaelian, Phys. Rev. **D20**, 1164 (1979).

[102] K. O. Mikaelian, M. A. Samuel, and D. Sahdev, Phys. Rev. Lett. **43**, 746 (1979).

[103] K. O. Mikaelian, Phys. Rev. **D30**, 1115 (1984).

[104] M. A. Samuel, and J. H. Reid, Progr. Theor. Phys. **76**, 184 (1986).

[105] M. Leurer, H. Harari, and R. Barberi, Phys. Lett. **B141**, 455 (1984).

[106] F. Renard, Nucl. Phys. **B196**, 93 (1982).

[107] B. W. Lee, C. Quigg, and H. B. Thacker, Phys. Rev. **D16**, 1519 (1977).

[108] F. Wilczek, Phys. Rev. Lett. **39**, 1304 (1977).

[109] M. Carena, and H. E. Haber, Prog. in Part. Nucl. Phys. **50**, 152 (2003).

[110] H. Georgi et al., Phys. Rev. Lett. **40**, 692 (1978).

[111] L. Resnick, M. K. Sundaresan, and P. J. Watson, Phys. Rev. **D8**, 172 (1973).

[112] D. Graudens, M. Spira, and P. M. Zervas, Phys. Rev. Lett. **70**, 1372 (1993).

[113] S. Dawson, Nucl. Phys. **B359**, 283 (1991).

[114] S. Petcov, and D. R. T. Jones, Phys. Lett. **D84**, 440 (1979).

[115] M. S. Chanovitz, and M. R. Gaillard, Phys. Lett. **B142**, 85 (1984).

[116] R. N. Cahn, and S. Dawson, Phys. Lett. **B136**, 196 (1984).

[117] T. Han, G. Valencia, and S. Willenbrock, Phys. Rev. Lett. **69**, 3274 (1992).

[118] J. Ellis et al., Nucl. Phys. **B106**, 292 (1976).

[119] G. Altarelli, R. K. Ellis, and G. Martinelli, Nucl. Phys. **B143**, 521 (1978); ibid. **B146**, 544 (erratum) (1978); ibid. **B157** 461 (1979).

[120] Z. Kunszt, Nucl. Phys. **B247**, 339 (1984).

[121] S. Dawson, J. F. Gunion, H. E. Haber, and G. L. Kane, *The Higgs Hunter's Guide*, (Addison-Wesley, Reading, MA, 1990).

[122] K. G. Wilson, Phys. Rev. **D10**, 2445 (1974).

[123] M. Creutz, L. Jacobs, C. Rebbi, Phys. Rev. **D20**, 1915 (1979).

[124] D. Weingarten, D. Petcher, Phys. Lett. **B99**, 333 (1981).

[125] H. Hamber, G. Parisi, Phys. Rev. Lett. **47**, 1792 (1981).

[126] K. G. Wilson, Phys. Rev. **D14**, 2455 (1974).

[127] J. Kogut, L. Susskind, Phys. Rev. **D9**, 3501 (1974).

[128] K. Wilson, In *New Phenomena in Subnuclear Physics*, (Plenum, New York, 1977).

[129] J. Kogut and L. Susskind, Phys. Rev. **D11**, 395 (1975).

[130] D. B. Kaplan, Phys. Lett. **B288**, 342 (1992).

[131] P. M. Vranas, Phys. Rev. **D57**, 1415 (1998).

[132] P. Chen et al., Phys. Rev. **D64**, 14503 (2001).

[133] S. Mandelstam, Ann. Phys. **19**, 1 (1962).

[134] C. N. Yang, Phys. Rev. Lett. **33**, 445 (1975).

[135] M. Peardon, Nucl. Phys. B: Proc. Suppl. **106–107**, 3 (2002).

[136] A. Ukawa (CP-PACS and JLQCD Collab.) Nucl. Phys. B: Proc. Suppl. **106–107**, 195 (2002).

[137] N. Metropolis et al. J. Chem. Phys. **21**, 1087 (1953).

[138] D. Weingarten, Phys. Lett. **B109**, 57 (1982).

[139] M. Creutz, Phys. Rev. Lett. **43**, 553 (1979).

[140] M. Bander, Phys. Rep. **C75**, 205 (1981).

[141] M. Creutz, *Quarks, Gluons and Lattices*, (Cambridge University Press, Cambridge, 1983).

[142] M. Creutz, Rev. Mod. Phys. **50**, 561 (1978).

[143] Y. Nambu, Phys. Rev. **D10**, 4262 (1974).

[144] S. Mandelstam, Phys. Rep. **C23**, 245 (1976).

[145] G. Chew, S. C. Frautschi, Phys. Rev. **123**, 1478 (1961).

[146] S. Gasiorowicz, J. L. Rosner, Am. J. Phys. **49**, 954 (1981).

[147] F. Butler et al., Nucl. Phys. **B430**, 179 (1994).

[148] C. R. Allton, Nucl. Phys. **B437**, 641 (1995).

[149] G. Peter Lepage, arXiv:hep-lat/0506036.

[150] C. Davies, arXiv:hep-lat/0509046.

[151] S. Aoki et al. [CP-PACS Collaboration], Phys. Rev. Lett. **84**, 238 (2000).

[152] S. Aoki et al. [CP-PACS Collaboration], Phys. Rev. **D67**, 034503 (2003).

[153] A. Ali Khan et al. [CP-PACS Collaboration], Phys. Rev. Lett. **85**, 4674 (2000); Phys. Rev. **D65**, 054505 (2003); Phys. Rev. **D67**, 059901(E) (2003).

[154] T. Kaneko et al. [CP-PACS and JLQCD Collaborations], Nucl. Phys. (Proc. Suppl.)**B129**, 188 (2004); T. Ishikawa et al. [CP-PACS and JLQCD Collaborations], Nucl. Phys. (Proc. Suppl.) **B140**, 225 (2005); T. Ishikawa et al. [CP-PACS and JLQCD Collaborations], PoS LAT2006:181, 2006.

[155]) U. M. Heller et al. [MILC Collaboration], J. Phys. Conf. Ser. **9**, 248 (2005); A. Bazavov et al. [MILC Collaboration], arXiv:0903.3598 [hep-lat].

[156] D. J. Antonio et al. [RBC and UKQCD] Phys. Rev. **D75**, 114501 (2007); C. Alton et al. [RBC and UKQCD] Phys. Rev. **D78**, 114509 (2008).

[157] G. Chapline et al., Phys. Rev. **D8**, 4302 (1973).

[158] T. D. Lee, G. C. Wick, Phys. Rev. **D9**, 2291 (1974).

[159] E. V. Shuryak, Phys. Rep. **61**, 71 (1980).

[160] N. Xu, M. Kaneta, Nucl. Phys. **306**, 182301 (2002).

[161] H. Satz, Phys. Repts. **88**, 349 (1982).

[162] J. C. Collins, M. J. Perry, Phys. Rev. Lett. **34**, 1353 (1975).

[163] F. Wilczek, arXiv:hep-ph/0003183.

[164] I. Montvay, G. Münster, *Quantum Fields on a Lattice*, (Cambridge University Press, London, 1994).

[165] T. DeGrand, C. DeTar, *Lattice Methods for Quantum Chromodynamics*, (World Scientific, Singapore, 2006).

[166] F. Karsch, E. Laermann, A. Peikert, Nucl. Phys. **B605**, 579 (2001).

[167] T. Ludlum *et al.* BNL-75692-2006, Proceedings of the Workshop on *Can we discover the QCD Critical point at RHIC?*, March 9-10, 2006 (https://www.bnl.gov/riken/QCDRhic/).

[168] C. Hohne *et al.* [CBM Collaboration] Nucl. Phys. **A749**, 141 (2005).

[169] F. Videbaek, O. Hansen, Phys. Rev. **C52**, 26 (1995).

[170] I. G. Bearden et al. [BRAHMS Collab.] Phys. Rev. Lett. **93**, 1020301 (2004).

[171] K. Adcox et al. [PHENIX Collab.], Nucl. Phys. **A757**, 184 (2005).

[172] S. S. Adler et al., Phys. Rev. **C71**, 034908 (2005).

[173] B. B. Back et al., Phys. Rev. Lett. **91**, 052303 (2003).

[174] K. Adcox K. et al. Phys. Rev. Lett. **86**, 3500 (2001); ibid. **87**, 052301 (2001).

[175] S. Li, X. N. Wang, Phys. Lett. **B527**, 85 (2002).

[176] D. Kharzeev, M. Nardi, Phys. Lett. **B507**, 121 (2001); V. T. Pop et al., Phys. Rev. **C68**, 054902 (2003).

[177] D. F. Bjorken, ERMILAB-PUB-82-059-THY. 1982.

[178] M. N. Thoma, M. Gyulassy, Nucl. Phys. **B351**, 491 (1991).

[179] X. N. Wang, Phys. Rev. **C58**, 2321 (1998).

[180] B. Müller, arXiv:0812.4638 [nucl-th].

[181] M. Luzum and P. Romatschke, arXiv:0901.4588 [nucl-th].

[182] T. Hirano et al., J. Phys. **G34**, S879 (2007).

[183] F. Karsch, E. Laermann, and A. Peikert, Phys. Lett. **B478**, 447 (2000)

[184] M. Gyulassy, arXiv:hep-ph/0403032.

[185] E. Shuryak, arXiv:hep-ph/0405066.

[186] H. Georgi, Phys. Rev. Lett. **98**, 221601 (2007).

[187] J. Maldacena, Adv. Theor. Math. Phys. **2**, 231 (1998).

[188] E. Witten, Adv. Theor. Math. Phys. **2**, 253 (1998).

[189] S. S. Gubser, I. R. Klebanov, and A. M. Polyakov, Phys. Lett. **B428**, 105 (1998).

[190] B. Greene, *The Elegant Universe*, (W. W. Norton and Company, Inc. New York, 1999); *The Fabric of the Cosmos*, (Penguin Press Science, 1999).

[191] L. Alvarez-Gaume et al., Nucl. Phys. **B806**, 327 (2009).

[192] D. Mateos, R. C. Myers, and R. M. Thomson, Phys. Rev. Lett. **97**, 091601 (2006); JHEP **0705**, 067 (2007).

[193] D. Mateos, Class. Quant. Grav. **24**, s713 (2007).

[194] J. P. Blaizot, E. Iancu and A. Rebhan, arXiv:hep-ph/0303185.

[195] ALICE Collaboration, J. Phys. **G30**, 1517 (2004).

[196] ALICE Collaboration, J. Phys. **G32**, 1295 (2006).

[197] N. Borghini and U. A. Wiedemann, arXiv:0707.0564 [hep-ph].

[198] G. L. Lamb, *Elements of Soliton Theory*, (Wiley, New York, 1980).

[199] Y. Aharonov, D. Bohm, Phys. Rev. **115**, 485 (1959).

[200] R. G. Chambers, Phys. Rev. Lett. **5**, 3 (1960).

[201] G. H. Derrick, J. Math. Phys. **5**, 1252 (1964).

[202] H. B. Nielsen, P. Olesen, Nucl. Phys. **B61**, 45 (1973).

[203] D. Speiser, In: *Group Theoretical Concepts and Methods in Elementary Particle Physics* (ed. F. Gürsey), (Gordon and Breach, Science Publishers, New York, 1964).

[204] S. Coleman, In: *The Ways of Subnuclear Physics. Proc. 1977 Int. Sch. Subnucl. Phys. "Ettore Majorana"* (ed. A. Zishichi), (Plenum Press, New York, 1977).

[205] G. t'Hooft, Nucl. Phys. **B79**, 276 (1974).

[206] A. M. Polyakov, Pis'ma ZETF **20**, 430 (1974) [JETP Letters **20**, 174 (1974)].

[207] H. Georgi, S. L. Glashow, Phys. Rev. Lett. **28**, 1494 (1972).

[208] R. Rajaraman, *Solitons and Instantons*, (North-Holland Publishing Company, Amsterdam-New York-Oxford, 1982).

[209] A. A. Belavin et al., Phys. Lett. **B59**, 85 (1975).

[210] M. F. Atiyah itet al., Phys. Lett. **A65**, 185 (1978).

[211] G. t'Hooft, Phys. Rev. Lett. **37**, 8 (1976).

[212] L. Baulieu et al., Phys. Lett. **B81**, 41 (1979).

[213] D. L. Dyakonov, V. Y. Petrov, Nucl. Phys. **B272**, 457 (1986.)

[214] T. Schäfer, E. V. Shuryak, Rev. Mod. Phys. **70**, 1323 (1998).

[215] D. . Smith, M. Teper, Phys. Rev **D58**, 014505 (1998).

[216] G. t'Hooft, Phys. Rev. **D14**, 3432 (1976).

[217] M. A. Shifman, A. I. Vainshtein, V. L. Zakharov, Nucl. Phys. **B163**, 43 (1980).

[218] A. E. Dorokhov, Y. A. Zubov, N. I. Kochelev, Sov. J. Part. Nucl **23**, 522 (1992).

[219] N. I. Kochelev, Phys. Part. Nucl. **36**, 608 (2005).

[220] A. E. Dorokhov, N. I. Kochelev, Zeits. Phys. **C46**, 281 (1990).

[221] B. V. Geshkenbein, B. L. Joffe, Nucl. Phys. **B166**, 340 (1980).

[222] M. A. Shifman, V. I. Vainstein, V. I. Zakharov , Nucl. Phys. **B120**, 316 (1977).

[223] E. Pallante, A. Pich, Phys. Rev. Lett. **84**, 319 (2000).

[224] N. I. Kochelev, V. Vento Phys. Rev. Lett. **87**, 11601 (2001)

[225] G. Grunberg, Phys. Lett. **B95**, 70 (1980); Phys. Rev. **D29**, 2315 (1984).

[226] P. M. Stevenson, Phys. Rev. **D23**, 2916 (1981).

[227] S. Brodsky and H. J. Lu, SLAC-PUB-6389 (Nov. 1993).

[228] S. Brodsky, G. P. Lepage, P. B. Mackenzie, Phys. Rev. **D28**, 228 (1983).

[229] P. Abreu et al. Zeits. Phys. **C54**, 55 (1992).

[230] W. T. Giele et al., Phys. Rev. **D53**, 120 (1996).

[231] T. Affolder et al., Phys. Rev. Lett. **88**, 042001 (2002).

[232] UA1 Collaboration: G. Arnison et al., Phys. Lett. **177**, 244 (1986).

[233] B. A. Kniehl et al., Phys. Rev. Lett. **85**, 5288 (2000).

[234] C. Glasman, S. Maxfield, M. Nadolsky AIP Conf. Proc. **792**, 161 (2005).

[235] ZEUS Collaboration: S. Chekanov et al. Phys. Lett. **B558**, 41 (2003).

[236] I. Hinchliffe, Phys. Rev. Lett. (Review of Particle Physics) **B667**, 116 (2008).

[237] C. T. H. Davies et al., Phys. Rev. Lett. **92**, 022001 (2004).

[238] E. Eichten, K. Lane, M. Peskin, Phys. Rev. Lett. **50**, 811 (1983).

Part II

Electroweak interactions

12

Glashow–Weinberg–Salam theory

> *More flowers I noted, yet I none could see*
> *But sweet or colour it had stol'n from thee.*
> William Shakespeare

12.1 Choice of gauge group

The principle ideas of building the Glashow–Weinberg–Salam (GWS) model of the electroweak interaction were stated in Section 11.2 of *Advanced Particle Physics Volume 1* (we recommend a reader to look through this section once more). Here we shall consider this theory in detail. Obviously, the first we must do is to argue the choice of a gauge group in the form of $SU(2)_L \times U(1)_Y$.

By the time of the GWS model making, the weak interaction theory has already passed two stages. At the former, the weak interaction is described by the four-fermion contact Fermi theory. Splendid successes of QED where the interaction is carried by a photon as well as the belief in symmetry of nature suggest that a similar mechanism should lie in the basis of the weak interaction too. These arguments have culminated in an intermediate vector boson (IVB) theory in which two massive charged vector bosons $W^\pm$ serve as carriers of the weak interaction. Since the IVBs possess an electrical charge, they take part in the electromagnetic interaction. Now, we must have not only a good (renormalizable) weak interaction theory for the IVBs, but a renormalizable electrodynamics as well. However, numerous attempts of building the models with the required properties, remaining beyond the scope of the spontaneous symmetry breaking, were unsuccessful (see, for example, [8, 9] and references therein).

So, the correspondence principle demanded that at least three gauge bosons ($W^\pm$ and γ)* must exist in a new theory including the weak and electromagnetic interactions. And these bosons have to interact with the weak and electromagnetic currents (J^W and J^{em}). The simplest group with three generators is the $SU(2)$ group. It turns out, however, that the algebra of the J^W and J^{em} currents is not closed with respect to the commutation operation. For simplicity we consider a theory where there is only an electron and a neutrino. Then, the weak and electric charges are defined in the following way:

$$S_+^W(t) = \frac{1}{2} \int d^3 x J_0^W(x) = \frac{1}{2} \int d^3 x \nu_e^\dagger(x)(1 - \gamma_5)e(x), \qquad S_-^W(t) = S_+^{W\dagger}(t), \qquad (12.1)$$

$$Q(t) = \int d^3 x J_0^{em}(x) = - \int d^3 x e^\dagger(x)e(x). \qquad (12.2)$$

---

*Since QED represents a gauge theory, it is quite reasonable to demand the same from the weak interaction theory too.

Using simultaneous commutation relations for fermions

$$\{\psi_i^\dagger(\mathbf{x}, t), \psi_j(\mathbf{x}', t)\} = \delta_{ij}\delta^{(3)}(\mathbf{x} - \mathbf{x}'),$$

we get

$$[S_+^W(t), S_-^W(t)] = 2S_3^W(t), \tag{12.3}$$

where

$$S_3^W(t) = \frac{1}{4} \int d^3x [\nu_e^\dagger(x)(1 - \gamma_5)\nu_e(x) - e^\dagger(x)(1 - \gamma_5)e(x)]. \tag{12.4}$$

Since $S_3^W \neq Q$, then $S_\pm^W$ and Q do not form a closed algebra. Really, for the operator Q to be a generator of the $SU(2)$ the charges of a complete multiplet must add up to zero, corresponding to the requirement that the generators for $SU(2)$ must be traceless. In the case under consideration, we try to build a doublet from the particles ν_e and e which do not explicitly satisfy this condition. Besides, the operators $S_\pm^W$ have the $V - A$-form while the operator Q is purely vectorial. To correct the state of affairs we have three possibilities:
(i) One may introduce one more gauge boson associated with the operator S_3^W in (12.3). Then, these four generators form the $SU(2) \times U(1)$ group algebra.
(ii) One may enter several gauge bosons. Then, generators connected with these bosons form the algebra of a group that involves the $SU(2)$ group but is wider than in the first case.
(iii) One may add new fermions to the multiplet and thus modify the currents in order that the new set of $S_\pm^W$ and Q will be closed to form $SU(2)$ under commutation.

Experimental data exclude the third possibility. In the GWS model the $SU(2) \times U(1)$ gauge group is used. Now, our next task is to determine quantum numbers of this group. For the sake of simplicity, we are constrained by one fermion generation. Since in gauge interactions the chirality is conserved, we can proceed from independent left- and right-handed fermions. Therefore, the first generation consists of the following 16 two-component spinors:

$$\nu_{eL}, \; e_L, \; \nu_{eR}, \; e_R, \; u_L^\alpha, \; d_L^\alpha, \; u_R^\alpha, \; d_R^\alpha$$

(in what follows the color index α will be omitted). From (12.1), (12.2), and (12.4) it follows that the weak currents

$$S_+^W = \int d^3x(\nu_{eL}^\dagger e_L + u_L^\dagger d_L), \qquad S_+^{W\dagger} = S_-^W, \tag{12.5}$$

$$S_3^W = \frac{1}{2} \int d^3x(\nu_{eL}^\dagger \nu_{eL} - e_L^\dagger e_L + u_L^\dagger u_L - d_L^\dagger d_L) \tag{12.6}$$

are the generators of the $SU(2)$ group. From these expressions for the $SU(2)$ generators, it is evident that

$$l_L \equiv \begin{pmatrix} \nu_{eL} \\ e_L \end{pmatrix}, \qquad q_L \equiv \begin{pmatrix} u_L \\ d_L \end{pmatrix} \tag{12.7}$$

represent the $SU(2)$-doublets and ν_{eR}, e_R, u_R, and d_R—singlets.

Proceed to the $U(1)$ group. This group must be chosen in such a way that the electric charge

$$Q = \int d^3x \left(-e^\dagger e + \frac{2}{3}u^\dagger u - \frac{1}{3}d^\dagger d \right) = \int d^3x \left(-e_L^\dagger e_L - e_R^\dagger e_R + \frac{2}{3}u_L^\dagger u_L + \right.$$

$$\left. + \frac{2}{3}u_R^\dagger u_R - \frac{1}{3}d_L^\dagger d_L - \frac{1}{3}d_L^\dagger d_L \right) \tag{12.8}$$

represents a linear combination of the $U(1)$ group generator and the generator S_3^W belonging to the $SU(2)$ group. Obviously, the combination

$$Q - S_3^W = \int d^3x \left[-\frac{1}{2}\left(\nu_{eL}^\dagger \nu_{eL} + e_L^\dagger e_L\right) + \frac{1}{6}\left(u_L^\dagger u_L + d_L^\dagger d_L\right) - \right.$$

$$\left. -e_R^\dagger e_R + \frac{2}{3}u_R^\dagger u_R - \frac{1}{3}d_R^\dagger d_R \right] \tag{12.9}$$

has the property of giving the same quantum number to all members of an $SU(2)$ doublet in (12.7). The direct calculation shows that it commutes with all the $SU(2)$ generators, that is

$$[Q - S_3^W, S_i^W] = 0, \qquad i = 1, 2, 3. \tag{12.10}$$

This allows us to choose

$$Y^W = 2(Q - S_3^W) \tag{12.11}$$

as the generator of the $U(1)$ group and refer to Y^W as the weak hypercharge. Unlike the S_i^W generators, the Y^W generator does not satisfy any nonlinear commutation relations. Its scale, that is, the proportional constant between it and $Q - S_3^W$, may be chosen at will. To get the correct electric charges for particles we need to use Eqs. (12.9) and (12.11) and set

$$\left.\begin{array}{lll} Y^W(l_L) = -1, & Y^W(e_R) = -2, & Y^W(\nu_R) = 0 \\[2mm] Y^W(q_L) = \dfrac{1}{3}, & Y^W(u_R) = \dfrac{4}{3}, & Y^W(d_R) = \dfrac{2}{3}. \end{array}\right\} \tag{12.12}$$

These hypercharge values coincide with the doubled average charges of each multiplet, as the average value of the S_3^W generator is always zero. So, the aforesaid allows us to denote the electroweak interaction group in the GWS theory in the following way: $SU(2)_L \times U(1)_Y$ (hereafter the superscript W of the hypercharge is omitted).

Using the quantum numbers of the $SU(2)_L \times U(1)_Y$ gauge group, we can rewrite the expressions for the Lagrangians (11.44) and (11.49) (see *Advanced Particle Physics Volume 1*) describing interactions of the charged and neutral currents in the following view:

$$\mathcal{L}_G(x) = \frac{g}{\sqrt{2}}J_\mu^{CC}(x)W^{\mu+}(x) + \text{conj.} + \frac{g}{\cos\theta_W}J_\mu^{NC}(x)Z^\mu(x) + eJ_\mu^{em}(x)A^\mu(x), \tag{12.13}$$

where

$$\left.\begin{array}{l} J_\mu^{CC}(x) = \dfrac{1}{2}\left[\bar{\nu}_l(x)\gamma^\mu(1-\gamma_5)l(x) + \bar{q}_i^u(x)\gamma^\mu(1-\gamma_5)\mathcal{M}_{ik}^{CKM}q_k^d(x)\right], \\[4mm] J_\mu^{NC}(x) = \dfrac{1}{2}\displaystyle\sum_{f=l,q}\left[g_L^f\bar{f}(x)\gamma^\mu(1-\gamma_5)f(x) + g_R^f\bar{f}(x)\gamma^\mu(1+\gamma_5)f(x)\right], \end{array}\right\} \tag{12.14}$$

$$\left.\begin{array}{llll} g_L^\nu = \dfrac{1}{2}, & g_R^\nu = 0, & g_L^e = -\dfrac{1}{2}+s_W^2, & g_R^e = s_W^2, \\[3mm] g_L^u = \dfrac{1}{2}-\dfrac{2}{3}s_W^2, & g_R^u = -\dfrac{2}{3}s_W^2, & g_L^d = -\dfrac{1}{2}+\dfrac{1}{3}s_W^2, & g_R^d = \dfrac{1}{3}s_W^2, \end{array}\right\} \tag{12.15}$$

and $s_W = \sin\theta_W$. The relations (12.15) could be reduced into the single formula

$$g_{L,R}^f = S_3^W(f_{L,R}) - Q(f)s_W^2. \tag{12.16}$$

12.2 Lagrangian in generalized renormalizable gauge

In Section 11.2 of *Advanced Particle Physics Volume 1* the GWS model was formulated in the unitary or U gauge. In this gauge the propagators of the gauge bosons have the form normally expected for a massive vector field

$$D_{\mu\nu}(k) = -\frac{g_{\mu\nu} - k_\mu k_\nu / m_V^2}{k^2 - m_V^2 + i\epsilon}, \tag{12.17}$$

where $V = W, Z$. An enormous disadvantage of this propagator consists in the fact that when $k \to \infty$, we have $D_{\mu\nu} \to$ const, while at large k the propagators of scalar particles and photons behave like k^{-2} and those of fermions—like k^{-1}. Being based on the naive counting of the momentum powers, one may conclude that the renormalizability appears to be lost. However, it should be remembered that the theory might prove to be renormalizable when cancellations among the particular Green functions take place. Fortunately, in the GWS theory such cancellations really occur. From the observation that the original Lagrangian before spontaneous symmetry breaking is renormalizable by power counting, t'Hooft [1, 2] proved that the theory remains renormalizable even after the symmetry breakdown. The key is to choose another set of gauges, the renormalizable gauges, in which the theory has a good high-energy behavior. The point is that in the unitary gauge, although the particle content is simple, renormalizability is not transparent. But the theory should be equivalent to that in the renormalizable gauge, where we obtain propagators with the good high-energy behavior at the expense of introducing fictitious particles (the would-be-Goldstone bosons). Besides the unitary gauge, in the literature R- and t'Hooft–Feynman gauges are often discussed. Each of them represents a special case of a so-called generalized renormalizable R_ξ gauge characterized by continuous real parameters ξ, ξ_Z, ξ_A [1, 2, 3, 4]. As was shown in Ref. [3] the Green functions depend on these parameters but physical results (S matrix elements) do not. At R_ξ gauge the sector of particles, apart from particles being present at the unitary gauge (physical particles), include three nonphysical scalar bosons $\Phi^\pm$ and Φ_3 as well as an isotriplet of scalar-fermion ghosts. The total Lagrangian in the R_ξ gauge will look like:

$$\mathcal{L} = \mathcal{L}_g + \mathcal{L}_f + \mathcal{L}_H + \mathcal{L}_Y + \mathcal{L}_{gf} + \mathcal{L}_{gh}. \tag{12.18}$$

Here $\mathcal{L}_g$ is the free Lagrangian of the gauge fields, $\mathcal{L}_f$ is the fermion Lagrangian involving interaction with the gauge fields, $\mathcal{L}_H$ is the Higgs bosons Lagrangian, $\mathcal{L}_Y$ is the Yukawa Lagrangian, $\mathcal{L}_{gf}$ is the gauge-fixing Lagrangian, and $\mathcal{L}_{gh}$ is the Faddeev–Popov (FP) ghost-field Lagrangian, which compensates contributions coming from nonphysical degrees of freedom.

The Lagrangian of the free gauge fields W_μ^a ($a = 1, 2, 3$) and B_μ is given by the expression

$$\mathcal{L}_g = -\frac{1}{4} W_{\mu\nu}^a(x) W^{a\mu\nu}(x) - \frac{1}{4} B_{\mu\nu}(x) B^{\mu\nu}(x),$$

where $W_{\mu\nu}^a(x)$ is the non-Abelian field tensor

$$W_{\mu\nu}^a(x) = \partial_\mu W_\nu^a(x) - \partial_\nu W_\mu^a(x) + g\varepsilon_{abc} W_\mu^b(x) W_\nu^c(x),$$

$B_{\mu\nu}(x)$ is the Abelian field tensor

$$B_{\mu\nu}(x) = \partial_\mu B_\nu(x) - \partial_\nu B_\mu(x).$$

The fermion Lagrangian $\mathcal{L}_f$ is

$$\mathcal{L}_f = i \sum_f^6 \overline{L}(x) \gamma_\mu D_L^\mu(x) L_f(x) + i \sum_n^{12} \overline{R}_n(x) \gamma_\mu D_R^\mu(x) R_n(x), \tag{12.19}$$

where the sum is taken over all six left-hand doublets L_f

$$\begin{pmatrix} \nu_{eL} \\ e_L^- \end{pmatrix}, \begin{pmatrix} \nu_{\mu L} \\ \mu_L^- \end{pmatrix}, \begin{pmatrix} \nu_{\tau L} \\ \tau_L^- \end{pmatrix}, \qquad \begin{pmatrix} u_L \\ d_L' \end{pmatrix}, \begin{pmatrix} c_L \\ s_L' \end{pmatrix}, \begin{pmatrix} t_L \\ b_L' \end{pmatrix}$$

and twelve right-hand singlets R_n

$$e_R^-, \ \mu_R^-, \ \tau_R^-, \qquad \nu_{eR}, \ \nu_{\mu R}, \ \nu_{\tau R},$$

$$u_R, \ c_R, \ t_R, \qquad d_R', \ s_R', \ b_R',$$

(a neutrino is supposed to be massive) and the covariant derivatives have the form

$$D_{\mu L}(x) = \partial_\mu - ig\frac{\sigma^a}{2}W_\mu^a(x) - ig'\frac{Y_L^W}{2}B_\mu(x), \qquad D_{\mu R}(x) = \partial_\mu - ig'\frac{Y_R^W}{2}B_\mu(x). \quad (12.20)$$

In the Higgs Lagrangian

$$\mathcal{L}_H = |D_\mu\varphi(x)|^2 - V(\varphi), \tag{12.21}$$

where

$$V(\varphi) = \mu^2\varphi^\dagger(x)\varphi(x) + \frac{1}{2}\lambda[\varphi^\dagger(x)\varphi(x)]^2,$$

the scalar doublet φ in an arbitrary gauge is parametrized as:

$$\varphi(x) = \frac{1}{\sqrt{2}}\left(H + v + i\Phi''\sigma''\right)\begin{pmatrix} 0 \\ 1 \end{pmatrix} = \frac{1}{\sqrt{2}}\begin{pmatrix} i\Phi_1(x) + \Phi_2(x) \\ H(x) + 2\dfrac{m_W}{g} - i\Phi_3(x) \end{pmatrix}, \tag{12.22}$$

where Φ_n ($n = 1, 2, 3$) are massless Goldstone bosons (they disappear in the unitary gauge), and we have taken into account that the vacuum expectation value (VEV) v is related with the W boson mass by the relation $v = 2m_W/g$.

The Yukawa Lagrangian $\mathcal{L}_Y$, describing the interaction of fermions with Higgs fields as well as allowing for mixing in the fermion sector, will look like:

$$\mathcal{L}_Y = -\sum_{f=l,q}\left[\sum_{i,j=1}^{3}\left(\kappa_{ij}^{(f)}\overline{L}_i^{(f)}(x)\varphi(x)R_j^{(f)}(x) + \tilde{\kappa}_{ij}^{(f)}\overline{L}_i^{(f)}(x)\varphi^c(x)\tilde{R}_j^{(f)}(x)\right)\right] + \text{conj.}, \quad (12.23)$$

where $\kappa_{ij}^{(f)}$ and $\tilde{\kappa}_{ij}^{(f)}$ are Yukawa coupling constants, φ^c is isospinor charge-conjugate to isospinor φ

$$\varphi^c = i\tau_2\varphi^*, \qquad (\varphi^c)_i = \epsilon_{ik}\varphi^{*k},$$

($i, k = 1, 2$, the summation over the index k is absent and $\epsilon_{12} = -\epsilon_{21} = 1$)* and right-handed singlets are divided into the up (R_j) and down ($\tilde{R}_j$) parts in the following way:

$$R_j^{(l)} = (e_R, \mu_R, \tau_R), \qquad R_j^{(q)} = (d_R', s_R', b_R');$$

$$\tilde{R}_j^{(l)} = (\nu_{eR}, \nu_{\mu R}, \nu_{\tau R}), \qquad \tilde{R}_j^{(q)} = (u_R, c_R, t_R).$$

The fermion masses are defined by the mass term

$$\mathcal{L}_m = \mathcal{L}_m^{(l)} + \mathcal{L}_m^{(q)}, \tag{12.24}$$

---

*The charge-conjugate spinor is needed to give masses to up-fermions.

that follows from (12.23) under the replacement of φ and φ^c by their VEVs

$$< 0|\varphi|0 >= \begin{pmatrix} 0 \\ v \end{pmatrix}, \qquad < 0|\varphi^c|0 >= \begin{pmatrix} v \\ 0 \end{pmatrix},$$

to give the following value of the mass matrices

$$M_{ij}^{(f)} = v\kappa_{ij}^{(f)}, \qquad \tilde{M}_{ij}^{(f)} = v\tilde{\kappa}_{ij}^{(f)}. \tag{12.25}$$

Experimental data demonstrate that e, μ, τ and u, c, t are states with a definite mass, that is, they are physical states. As far as up-fermions are concerned, they represent mixings of physical states. Hence, the mass matrices of down-leptons and up-quarks have to be diagonal while those of neutrinos M_ν and down-quarks M_{q^d} have to be nondiagonal. Diagonalization of M_ν and M_{q^d} is realized by means of the transition to the basis of the mass eigenstates

$$\begin{pmatrix} \nu_e \\ \nu_\mu \\ \nu_\tau \end{pmatrix} = \mathcal{M}^{NM} \begin{pmatrix} \nu_1 \\ \nu_2 \\ \nu_3 \end{pmatrix}, \qquad \begin{pmatrix} d' \\ s' \\ b' \end{pmatrix} = \mathcal{M}^{CKM} \begin{pmatrix} d \\ s \\ b \end{pmatrix}, \tag{12.26}$$

where the mixing matrices for neutrinos and quarks are given by the expressions (12.1) and (11.47) of *Advanced Particle Physics Volume 1*, respectively. As a result, the mass term (12.24) takes the view

$$\mathcal{L}_m = -\sum_f m_f \overline{\psi}_f \psi_f, \tag{12.27}$$

where ψ_f is the Dirac spinor of the fermion $f = u, d, c, s, t, b, e, \nu_1, \mu, \nu_2, \tau, \nu_3$. Note, the mixing existence means the violation of the partial lepton flavor. In view of the importance of the mixing matrices, they will be subjected to the detail discussion in what follows.

Let us now turn from fields W_μ^a, B_μ and Φ_1, Φ_2 to fields $W_\mu^\pm, Z_\mu, A_\mu$ and $\Phi^\pm$

$$\left. \begin{array}{l} W_\mu^\pm(x) = \dfrac{W_\mu^1(x) \mp iW_\mu^2(x)}{\sqrt{2}}, \qquad Z_\mu(x) = -B_\mu(x)s_W + W_\mu^3(x)c_W, \\[3mm] A_\mu(x) = B_\mu(x)c_W + W_\mu^3(x)s_W, \qquad \Phi^\pm(x) = \dfrac{\Phi_1(x) \mp i\Phi_2(x)}{\sqrt{2}}, \end{array} \right\} \tag{12.28}$$

$(c_W = \cos\theta_W)$ that are associated with W and Z bosons, photons, and charged Higgs scalars, respectively. Note that the Goldstone bosons $\Phi^\pm$ correspond to the $W^\pm$ bosons while Φ_3 corresponds to the Z boson (in the unitary gauge $\Phi^\pm, \Phi_3$ are absorbed by the longitudinal components of $W^\pm$ and Z). In terms of the fields (12.28) the fermion and Yukawa Lagrangians are

$$\mathcal{L}_f = \sum \left\{ i\overline{\psi}_f \gamma^\mu \partial_\mu \psi_f + \frac{g}{2\sqrt{2}} \left[\overline{\psi}_i^u \gamma^\mu (1 - \gamma_5) \mathcal{M}_{ij} \psi_j^d W_\mu^+ + \text{conj.} \right] + eQ_f \overline{\psi}_f \gamma^\mu \psi_f A_\mu + \right.$$

$$\left. + \frac{g}{2c_W} \left[\overline{\psi}^f \gamma^\mu \left(S_3^W(f) - 2Q_f s_W^2 - S_3^W(f)\gamma_5 \right) \psi_f \right] Z_\mu \right\}, \tag{12.29}$$

$$\mathcal{L}_Y = \sum \left\{ -m_f \overline{\psi}_f \psi_f + \frac{ig}{\sqrt{2}m_W} \left[\overline{\psi}_i^u \left(m_i^u \frac{1 - \gamma_5}{2} - m_j^d \frac{1 + \gamma_5}{2} \right) \mathcal{M}_{ij} \psi_j^d \Phi^+ - \right. \right.$$

$$\left. \left. -\text{conj.} \right] - \frac{gm_f}{2m_W} \left(\overline{\psi}_f \psi_f H - 2iS_3^W(f)\overline{\psi}_f \gamma_5 \psi_f \Phi_3 \right) \right\}, \tag{12.30}$$

where ψ^u are up-fermions, ψ^d are down-fermions, m_i^u, m_i^d and Q_f are their masses and charges (in units of $|e|$).

Next, we engage in the following three terms in the total Lagrangian (12.18)

$$\mathcal{L}' = \mathcal{L}_g + \mathcal{L}_H + \mathcal{L}_{gf}. \tag{12.31}$$

We present the gauge-fixing Lagrangian as

$$\mathcal{L}_{gf} = -K^+ K^- - \frac{1}{2}\left[(K_Z)^2 + (K_A)^2\right], \tag{12.32}$$

where

$$\left.\begin{array}{cc} K^{\pm} = \dfrac{1}{\xi}\partial^\mu W_\mu^{\pm} + \xi m_W \Phi^{\pm}, & K_A = \dfrac{1}{\xi_A}\partial^\mu A_\mu \\[2mm] K_Z = \dfrac{1}{\xi_Z}\partial^\mu Z_\mu + \xi_Z m_Z \Phi^0. \end{array}\right\} \tag{12.33}$$

In $\mathcal{L}'$ we shall single out the following parts: (i) the part quadratic in fields $\mathcal{L}_{prop}$ that defines particle propagators; (ii) the part $\mathcal{L}_{VV'}$ determining the self-interaction of the gauge bosons; and (iii) the term $\mathcal{L}_{VS}$ describing the interaction of the gauge bosons with scalar fields. So, we have

$$\mathcal{L}' = \mathcal{L}_{prop} + \mathcal{L}_{VV'} - V(\varphi) + \mathcal{L}_{VS}, \tag{12.34}$$

where

$$\mathcal{L}_{prop} = -\partial_\mu W_\nu^+ \partial^\mu W^{-\nu} + \left(1 - \frac{1}{\xi^2}\right)\partial^\mu W_\mu^+ \partial_\nu W^{-\nu} - \frac{1}{2}\left[\partial_\mu Z_\nu \partial^\mu Z^\nu - \left(1 - \frac{1}{\xi_Z^2}\right)(\partial_\mu Z_\mu)^2\right] -$$

$$- \frac{1}{2}\left[\partial_\mu A_\nu \partial^\mu A^\nu - \left(1 - \frac{1}{\xi_A^2}\right)(\partial_\mu A_\mu)^2\right] + \frac{1}{2}\partial_\mu H \partial^\mu H + \partial_\mu \Phi^+ \partial^\mu \Phi^- + \frac{1}{2}\partial_\mu \Phi^0 \partial^\mu \Phi^0 -$$

$$+ m_W^2 W_\mu^+ W^{-\mu} + \frac{1}{2}m_Z^2 Z^\mu Z_\mu - \xi^2 m_W^2 \Phi^+ \Phi^- - \frac{1}{2}\xi_Z^2 m_Z^2 \Phi^0 \Phi^0 - \frac{1}{2}m_H^2 H^2, \tag{12.35}$$

$$\mathcal{L}_{VV'} = -igc_W\left\{\partial^\nu Z^\mu W_\mu^{[+}W_\nu^{-]} - Z^\nu W^{[+\mu}\partial_\nu W_\mu^{-]} + Z^\mu W^{[+\nu}\partial_\nu W_\mu^{-]}\right\} -$$

$$- ie\left\{\partial^\nu A^\mu W_\mu^{[+}W_\nu^{-]} - A^\nu W^{[+\mu}\partial_\nu W_\mu^{-]} + A^\mu W^{[+\nu}\partial_\nu W_\mu^{-]}\right\} + \frac{g^2}{2}\left\{(W_\mu^+ W_\nu^-)^2 -\right.$$

$$\left. - (W_\mu^+ W_\mu^-)^2\right\} + g^2 c_W^2\left\{Z^\mu Z^\nu W_\mu^+ W_\nu^- - Z^\nu Z_\nu W^{+\mu}W_\mu^-\right\} + e^2\left\{A^\mu A^\nu W_\mu^+ W_\nu^- -\right.$$

$$\left. - A^\nu A_\nu W^{+\mu}W_\mu^-\right\} + egc_W\left\{A^\mu Z^\nu W_\mu^{[+}W_\nu^{-]} - 2A^\nu Z_\nu W_\mu^+ W^{-\mu}\right\}, \tag{12.36}$$

$$V(\varphi) = \beta_H\left\{\frac{2m_W}{g}H + \frac{1}{2}\left[H^2 + (\Phi^0)^2 + 2\Phi^+\Phi^-\right]\right\} + g\alpha_H m_W\left[H^3 + H(\Phi^0)^2 + 2H\Phi^+\Phi^-\right] +$$

$$+ \frac{g^2\alpha_H}{8}\left[H^4 + (\Phi^0)^4 + 2H^2(\Phi^0)^2 + 4(\Phi^0)^2\Phi^+\Phi^- + 4H^2\Phi^+\Phi^- + 4(\Phi^+\Phi^-)^2\right], \tag{12.37}$$

$$m_H^2 = \frac{4\lambda m_W^2}{g^2}, \qquad \beta_H = \mu^2 + \frac{2\lambda m_W^2}{g^2}, \qquad \alpha_H = \frac{m_H^2}{4m_W^2},$$

$$-\mathcal{L}_{VS} = -\frac{ig}{2}\left[W_\mu^+(\Phi^0\partial^\mu\Phi^- - \Phi^-\partial^\mu\Phi^0) - \text{h.c.}\right] + \frac{g}{2}\left[W_\mu^+(H\partial^\mu\Phi^- - \Phi^-\partial^\mu H) + \text{h.c.}\right] +$$

$$+\frac{g}{2c_W}Z_\mu(H\partial^\mu\Phi^0 - \Phi^0\partial^\mu H) + \left(ieA^\mu - \frac{igs_W^2}{c_W}Z^\mu\right)m_W[W_\mu^+\Phi^- - \text{h.c.}] +$$

$$+\left(ieA^\mu + \frac{igc_{2W}}{c_W}Z^\mu\right)[\Phi^+\partial_\mu\Phi^- - \text{h.c.}] - \frac{g^2}{4}W_\mu^+W^{-\mu}(HH + \Phi^0\Phi^0 + 2\Phi^+\Phi^-) -$$

$$-\frac{g^2}{8c_W^2}Z^\mu Z_\mu\left[HH + \Phi^0\Phi^0 + 2c_{2W}^2\Phi^+\Phi^-\right] - \frac{e^2}{2c_W}Z^\mu\left[(\Phi^0 + iH)W_\mu^+\Phi^- + \text{h.c.}\right] +$$

$$+\frac{ge}{2}A^\mu\left[(\Phi^0 + iH)W_\mu^+\Phi^- + \text{h.c.}\right] - ge\left(\frac{c_{2W}}{c_W}Z^\mu A_\mu + s_W A^\mu A_\mu\right)\Phi^+\Phi^- +$$

$$+gm_W H\left(W_\mu^+W^{-\mu} + \frac{1}{2c_W^2}Z^\mu Z_\mu\right), \tag{12.38}$$

$$A^{[+}B^{-]} = A^+B^- - A^-B^+,$$

and we have introduced the designations $c_{2W} = \cos 2\theta_W$. From $\mathcal{L}_{prop}$ we easily obtain the boson propagators:

W_μ^+ W_ν^-

$$D_{\mu\nu}^W(p) = -\frac{1}{p^2 - m_W^2}\left\{g_{\mu\nu} + (\xi^2 - 1)\frac{p_\mu p_\nu}{p^2 - \xi^2 m_W^2}\right\}, \tag{12.39}$$

Z_μ Z_ν

$$D_{\mu\nu}^Z(p) = D_{\mu\nu}^W(m_W \to m_Z, \xi \to \xi_Z), \tag{12.40}$$

A_μ A_ν

$$D_{\mu\nu}^A(p) = -\frac{1}{p^2}\left\{g_{\mu\nu} + (\xi_A^2 - 1)\frac{p_\mu p_\nu}{p^2}\right\}, \tag{12.41}$$

Φ^+ Φ^-

$$D^{\Phi^\pm}(p) = \frac{1}{p^2 - \xi^2 m_W^2}, \tag{12.42}$$

Φ^0 Φ^0

$$D^{\Phi^0}(p) = \frac{1}{p^2 - \xi_Z^2 m_Z^2}, \tag{12.43}$$

H H

$$D^H(p) = \frac{1}{p^2 - m_H^2}. \tag{12.44}$$

The procedure of obtaining the ghost Lagrangian completes building the total Lagrangian of the SM. From QCD it has been known that a ghost-antighost pair is needed for every degree of freedom of a gauge field. Since closed cycles of ghost lines are related to contributions involving an additional factor (-1), we may consider ghost scalar fields to be quantized on Fermi–Dirac. Let us introduce the following ghost fields: $Y^\pm$, Y^Z, and Y^A, where $Y^\pm$ (Y^Z) are associated with the $W^\pm$ bosons (Z boson) while Y^A is associated with the photon. For the sake of convenience, corresponding antighost fields will be denoted as

$\overline{Y}^\pm$, $\overline{Y}^Z$, and $\overline{Y}^A$. It should be stressed that $\overline{Y}^+ \neq Y^-$ and $\overline{Y}^- \neq Y^+$, that is, Y^- field has nothing in common with Y^+ field. For example, it is not true that Y^+ field contains a creation operator for a Y^- particle. The way this must be read is as follows: the Y^- field contains the absorption operator for a Y^- and the creation operator for an anti-Y^-. The anti-Y^- field will be denoted by $\overline{Y}^-$ and contains the absorption operator for an anti-Y^- and the creation operator for a Y^-. Thus, for example, on the one loop level, considering Z self-energy diagrams there will be typically two Y-graphs: one with an Y^- and one with Y^+ circulating. By contrast, there is only one graph with a circulating charged gauge boson.

To define the ghost Lagrangian we shall use the method stated in Section 2.4. Thus, we should find the law of changing K^i $(i = \pm, Z, A)$ under the $SU(2) \times U(1)$ gauge transformations. At infinitesimal transformations the initial fields are changed as

$$W_\mu^a \to W_\mu^a + g\epsilon_{abc}\Lambda^b W_\mu^c - \partial_\mu\Lambda^a, \qquad B_\mu \to B_\mu - \partial_\mu\Lambda^0, \qquad (12.45)$$

$$\varphi \to \left(1 - ig\Lambda^a\frac{\sigma^a}{2} + \frac{ie}{2c_W}\Lambda^0\right)\varphi, \qquad (12.46)$$

where Λ^a and Λ^0 are gauge parameters of the $SU(2)$ and $U(1)$ group, respectively. From (12.46) it is easy to get

$$\left.\begin{aligned}
\Phi^0 &\to \Phi^0 - \frac{g}{2}(\Lambda^3 - \tan\theta_W\Lambda^0)\left(H + \frac{2m_W}{g}\right) + \frac{ig}{2}(\Lambda^-\Phi^+ - \Lambda^+\Phi^-), \\
\Phi^- &\to \Phi^- - \frac{g}{2}\Lambda^-\left(H + \frac{2m_W}{g} + i\Phi^0\right) - \frac{ig}{2}(-\Lambda^3 - \tan\theta_W\Lambda^0)\Phi^-,
\end{aligned}\right\} \qquad (12.47)$$

where

$$\Lambda^\pm = \frac{\Lambda^1 \mp i\Lambda^2}{\sqrt{2}}.$$

Employing (12.45), (12.47) and introducing the designations

$$\Lambda^3 - \tan\theta_W\Lambda^0 = \frac{1}{c_W}\Lambda_Z, \qquad \Lambda^3 + \tan\theta_W\Lambda^0 = \frac{c_{2W}}{c_W}\Lambda^Z + 2s_W\Lambda^A, \qquad (12.48)$$

we find the transformation laws of the fields entering into the gauge functions K^i

$$\left.\begin{aligned}
W_\mu^- &\to W_\mu^- - ig\Lambda^-(c_W Z_\mu + s_W A_\mu) + ig(c_W\Lambda^Z + s_W\Lambda^A)W_\mu^- - \partial_\mu\Lambda^-, \\
A_\mu &\to A_\mu + ie(\Lambda^- W_\mu^+ - \Lambda^+ W_\mu^-) - \partial_\mu\Lambda^A, \\
Z_\mu &\to Z_\mu + igc_W(\Lambda^- W_\mu^+ - \Lambda^+ W_\mu^-) - \partial_\mu\Lambda^Z, \\
\Phi^- &\to \Phi^- - \frac{g}{2}\Lambda^-\left(H + \frac{2m_W}{g} + i\Phi^0\right) + \frac{ig}{2}\left(\frac{c_{2W}}{c_W}\Lambda^Z + 2s_W\Lambda^A\right)\Phi^-, \\
\Phi^0 &\to \Phi^0 - \frac{g}{2c_W}\Lambda^Z\left(H + \frac{2m_W}{g}\right) + \frac{ig}{2}(\Lambda^-\Phi^+ - \Lambda^+\Phi^-).
\end{aligned}\right\} \qquad (12.49)$$

The relations obtained allow us to write the transformation laws for the gauge functions K^i. In so doing, we have to single out the parts independent on g (they give the ghost propagators) and those proportional to g (they describe interaction of ghosts with gauge and scalar fields). With allowance made for this, the transformation law of K^i takes the form

$$K^i \to K^i + (M^{ij} + gV^{ij})\Lambda^j, \qquad (12.50)$$

where $i, j = \pm, Z, A$,

$$K^- \to K^- - \frac{1}{\xi}\partial^\mu\partial_\mu\Lambda^- - \xi m_W^2\Lambda^- - \frac{ig}{\xi}\partial^\mu[\Lambda^-(c_W Z_\mu + s_W A_\mu)] + \frac{ig}{\xi}\partial^\mu[(c_W\Lambda^Z +$$

$$+ s_W\Lambda^A)W_\mu^-] - \frac{\xi g m_W}{2}(H + i\Phi^0)\Lambda^- + \frac{i\xi g m_W c_{2W}}{2c_W}\Lambda^Z\Phi^- + i\xi e m_W\Lambda^A\Phi^-, \qquad (12.51)$$

$$K^A \to K^A - \frac{1}{\xi_A}\partial^\mu\partial_\mu\Lambda^A + \frac{ie}{\xi_A}\partial^\mu(\Lambda^- W_\mu^+ - \Lambda^+ W_\mu^-) \qquad (12.52)$$

and

$$K^Z \to K^Z - \frac{1}{\xi_Z}\partial^\mu\partial_\mu\Lambda^Z - \xi_Z m_Z^2\Lambda^Z + \frac{ig c_W}{\xi_Z}\partial^\mu(\Lambda^- W_\mu^+ - \Lambda^+ W_\mu^-) -$$

$$- \frac{\xi_Z g m_Z}{2c_W}\Lambda^Z H + i\xi_Z g m_Z(\Lambda^-\Phi^+ - \Lambda^+\Phi^-). \qquad (12.53)$$

Separating out the terms M^{ii} in (12.51)–(12.53), we get the ghost propagators:

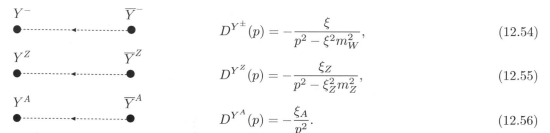

$$D^{Y^\pm}(p) = -\frac{\xi}{p^2 - \xi^2 m_W^2}, \qquad (12.54)$$

$$D^{Y^Z}(p) = -\frac{\xi_Z}{p^2 - \xi_Z^2 m_Z^2}, \qquad (12.55)$$

$$D^{Y^A}(p) = -\frac{\xi_A}{p^2}. \qquad (12.56)$$

Now, is the time to give our agreement about arrow directions on the Feynman diagrams. The arrow convention is as follows:

(i) The arrows occurring in lines are denoting fermion lines, or the flow of the electric charge or the flow of the FP ghost number. An incoming W^+ will, therefore, be denoted by an incoming arrow.

(ii) An arrow pointing inwards implies a positive charge flowing into the vertex. For a negatively charged FP field the flow of the charge is opposite to the direction of the arrow; for a positively charged FP field it is in the direction of the arrow.

The Lagrangian describing the interaction of ghosts with gauge fields $W_\mu^\pm$, Z_μ, A_μ, and Higgses $\Phi^\pm, \Phi_3$ is governed by the relation

$$\mathcal{L}_{gh}^{int} = g\overline{Y}^i V^{ij} Y^j. \qquad (12.57)$$

Making use of (12.50), we obtain without difficulty

$$\mathcal{L}_{gh}^{int} = ig c_W W_\mu^+\left(\frac{1}{\xi}\partial^\mu\overline{Y}^+ Y^Z - \frac{1}{\xi_Z}\partial^\mu\overline{Y}^Z Y^-\right) - ig c_W W_\mu^-\left(\frac{1}{\xi}\partial^\mu\overline{Y}^- Y^Z -\right.$$

$$\left. - \frac{1}{\xi_Z}\partial^\mu\overline{Y}^Z Y^+\right) + ie W_\mu^+\left(\frac{1}{\xi}\partial^\mu\overline{Y}^+ Y^A - \frac{1}{\xi_A}\partial^\mu\overline{Y}^A Y^-\right) - ie W_\mu^-\left(\frac{1}{\xi}\partial^\mu\overline{Y}^- Y^A -\right.$$

$$\left. - \frac{1}{\xi_A}\partial^\mu\overline{Y}^A Y^+\right) - \frac{i}{\xi}(eA_\mu + g c_W Z_\mu)\left(\partial^\mu\overline{Y}^+ Y^+ - \partial^\mu\overline{Y}^- Y^-\right) + i\Phi^+\left(\frac{g\xi_Z m_Z}{2}\overline{Y}^Z Y^- -\right.$$

$$\left. - g\xi m_Z c_{2W}\overline{Y}^+ Y^Z - e\xi m_W\overline{Y}^+ Y^A\right) - i\Phi^-\left(\frac{g\xi_Z m_Z}{2}\overline{Y}^Z Y^+ -\right.$$

$$-g\xi m_Z c_{2W}\overline{Y}^{-}Y^Z - e\xi m_W\overline{Y}^{-}Y^A\Bigg) + \frac{ig\xi m_W}{2}\left(\overline{Y}^{+}Y^{+} - \overline{Y}^{-}Y^{-}\right)\Phi^0. \qquad (12.58)$$

Although the photon propagator (12.41) contains arbitrary parameter ξ_A, this does not influence the S matrix elements because $D^A_{\mu\nu}(p)$ always appears in the plates of conserved currents and the term $(1 - \xi_A)(p_\mu p_\nu)/p^2$ turns into zero in the S matrix elements. One can show that all other propagators behave as p^{-2} for large p^2 when $\xi, \xi_Z \neq \infty$. The naive counting of the momentum powers in the S matrix elements points to renormalizability of the theory. Since the expression for $D^{W^\pm}$ is

$$D^{W^\pm}_{\mu\nu}(p) = -\left[g_{\mu\nu} - \frac{p_\mu p_\nu}{m_W^2}\right]\frac{1}{p^2 - m_W^2 + i\epsilon} - \frac{1}{m_W^2}\frac{p_\mu p_\nu}{p^2 - \xi^2 m_W^2 + i\epsilon},$$

then $D^{W^\pm}$ has an additional pole at

$$p^2 = \xi^2 m_W^2.$$

However, in the S matrix this pole is compensated by the one of the propagator $D^{\Phi^\pm}(p)$. The analogous compensation occurs for the Z boson (photon) propagator thanks to $D^{Y^Z}(p)$ $(D^{Y^A}(p))$.

The limit $\xi, \xi_Z \rightarrow 0$ corresponds to the R gauge. When $\xi, \xi_Z = 1$, we get t'Hooft–Feynman gauge. The unitary gauge follows at $\xi, \xi_Z \rightarrow \infty$. However, the transition to this limit may be fulfilled only after the S matrix elements have been calculated, otherwise ambiguities appear. In the unitary gauge the propagators of the W and Z gauge bosons are given by the expressions (12.17) and the theory renormalizability is not obvious. However, since the S matrix elements do not depend on ξ, ξ_Z, it becomes possible to prove renormalizability for $\xi, \xi_Z \neq \infty$ and then pass on to the limit $\xi, \xi_Z \rightarrow \infty$.

Parity violation in the weak interaction makes additional demands of the theory. A real difficulty for renormalizability appears in the fermion sector and becomes apparent in the existence of triangular diagrams in which all three internal lines are fermionic, external lines are vector gauge lines, and vertices have γ_5 matrices. Divergences of these diagrams cannot be renormalized in such a way as to conserve the gauge invariance. As the reader has already guessed, the case in point is the Adler–Bell–Jackiw (ABJ) anomalies. These anomalies hinder gauge theories to be renormalizable. However, as was shown in Section 6.4, in the model with the $SU(2)_L \times U(1)_Y$ gauge group they are absent providing the number of the quark generations is equal to that of the lepton generations. One can demonstrate, in the total standard $SU(3)_c \times SU(2)_L \times U(1)_Y$ model there are no ABJ anomalies because contributions of additional triangle diagrams with gluons and gauge electroweak bosons are mutually cancelled too. Therefore, the condition of the absence of ABJ anomalies may be considered as one more indication in favor of the quark-lepton symmetry. The total set of the Feynman rules for the SM in the R_ξ gauge is brought to Appendix B.

12.3 Cabibbo–Kobayashi–Maskawa matrix

For the first time a hypothesis of mixing in the quark sector was introduced by Cabibbo in 1963 [5] in order to explain decays of strange particles. As a case in point we consider the decay

$$K^+ \rightarrow \mu^+ + \nu_\mu. \qquad (12.59)$$

Since K^+ meson consists of u and $\overline{s}$ quarks, then a weak current connecting u and $\overline{s}$ quarks must exist (Fig. 12.6). Still, if we assumed that quarks take part in the weak interaction

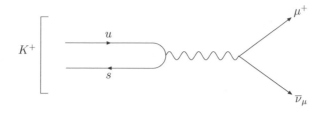

FIGURE 12.1
The decay $K^+ \to \mu^+ + \overline{\nu}_\mu$.

by means of charged $V - A$ currents built up from the following pairs of the left-handed quark states

$$\begin{pmatrix} u \\ d \end{pmatrix}, \qquad \begin{pmatrix} c \\ s \end{pmatrix}, \qquad \ldots, \tag{12.60}$$

then the decay (12.59) would be forbidden because only the transitions $u \leftrightarrow d$ and $c \leftrightarrow s$ are allowed. All fell into place when one makes the assumption that the charged current relates rotated states

$$\begin{pmatrix} u \\ d' \end{pmatrix}, \qquad \begin{pmatrix} c \\ s' \end{pmatrix}, \qquad \ldots, \tag{12.61}$$

where

$$d' = d\cos\theta_c + s\sin\theta_c, \qquad s' = -d\sin\theta_c + s\cos\theta_c. \tag{12.62}$$

The Cabibbo angle θ_c may be defined comparing decays with $\Delta S = 1$ and $\Delta S = 0$. For example,

$$\sin^2\theta_c \sim \frac{\Gamma(K^+ \to \mu^+\nu_\mu)}{\Gamma(\pi^+ \to \mu^+\nu_\mu)}, \qquad \sin^2\theta_c \sim \frac{\Gamma(K^+ \to \pi^0 e^+\nu_e)}{\Gamma(\pi^+ \to \pi^0 e^+\nu_e)}. \tag{12.63}$$

After the inclusion of kinematic factors associated with particle mass differences, experiments show the transitions with $\Delta S = 1$ to be suppressed approximately by a factor 20 relative to those with $\Delta S = 0$. This is consistent with the Cabibbo angle $\theta_c \approx 13°$. So, with the help of mixing in the quark sector, we enter processes *preferred by Cabibbo* (proportional to $\cos\theta_c$) and *suppressed by Cabibbo* (proportional to $\sin\theta_c$).

The weak interaction considered include only u and d' quark states. However, in Eq. (12.61) we related s' with a charmed quark c, therefore, apart from the transitions $d' \leftrightarrow u$, we now have the transitions $s' \leftrightarrow c$ as well. Based upon these considerations, Glashow, Iliopolous, and Maiani [6] (GIM) came out with a suggestion about the existence of a c quark some years before its discovery. The logic of this hypothesis is understandable when one considers the decay

$$K_L^0 \to \mu^+ + \mu^-. \tag{12.64}$$

If there would exist only the transitions $d' \leftrightarrow u$, then the diagram shown in Fig. 12.2a gives the width of the decay (12.64) exceeding many times over the experimental value

$$\frac{\Gamma(K_L^0 \to \mu^+\mu^-)}{\Gamma(K_L^0 \to \text{all modes})} = (6.84 \pm 0.11) \times 10^{-9}.$$

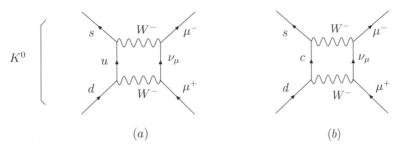

(a) $\qquad\qquad\qquad\qquad\qquad$ (b)

FIGURE 12.2

The decay $K_L^0 \to \mu^+ + \mu^-$ in (a) and (b).

But if one introduces the c-quark, the second diagram (Fig. 12.2(b)) is added. It would be exactly cancelled with the diagram pictured in Fig. 12.2(a), if the masses of the u and c quarks were equal to each other.

The transitions $s' \leftrightarrow c$ are responsible for charmed particle decays. An example is provided by the decay of the D^+ meson consisting of c and $\bar{d}$ quarks. As $\cos^2 \theta_c \gg \sin^2 \theta_c$, then from (12.62) it follows that the decay scheme presented by the quark diagram in Fig. 12.3 is dominant. The corresponding amplitude is

$$\mathcal{A}(c \to su\bar{d}) \sim \cos^2 \theta_c. \tag{12.65}$$

Thus, the K^- meson must be the most preferable fragment of the D^+ meson decay. When

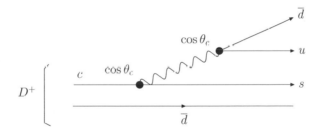

FIGURE 12.3

The D^+ decay.

the K^+ meson is present in the final state, then the decay will be strongly suppressed (doubly Cabibbo-suppressed modes) because

$$\mathcal{A}(c \to \bar{s}ud) \sim \sin^2 \theta_c. \tag{12.66}$$

Therefore, the D^+ meson decay must have very specific peculiarities. Thus, for example,

$$\Gamma(D^+ \to K^- \pi^+ \pi^+) = 0.0922 \pm 0.0021,$$

while

$$\Gamma(D^+ \to K^+ \pi^+ \pi^-) = (6.2 \pm 0.7) \times 10^{-4}.$$

So, the Cabibbo–GIM scheme (in the literature this scheme is simply called the *GIM mechanism*) states that the charged current connects the left-handed quark states $d' \leftrightarrow u$

and $s' \leftrightarrow c$, where d' and s' are orthogonal combinations of the physical quarks d and s. At that, the quark mixing is defined by the single parameter, the *Cabibbo angle*.

Originally the GIM mechanism is intended for the explanation of the lack of the transitions $d \leftrightarrow u$ in which the flavor is changed but the electric charge is conserved. Experimental evidences of the lack of neutral currents changing the strangeness are very impressive. For example, the decays

$$K^0 \to \mu^+ + \mu^-, \qquad K^+ \to \pi^+ + e^- + e^+, \qquad K^+ \to \pi^+ + \nu_l + \bar{\nu}_l,$$

that would occur at the presence of such currents, either are absent at all or are strongly suppressed. In accordance with the GIM mechanism

$$q'_{di} = \mathcal{M}_{ij} q_{dj}, \tag{12.67}$$

where

$$\mathcal{M}_{ij} = \begin{pmatrix} \cos\theta_c & \sin\theta_c \\ -\sin\theta_c & \cos\theta_c \end{pmatrix}$$

and $q_{d1} \equiv d_L$, $q_{d2} \equiv s_L$, we have

$$\bar{q}'_{di} q'_{di} = \bar{q}_{di} \mathcal{M}^\dagger_{ij} \mathcal{M}_{jk} q_{dk} = \bar{q}_{di} q_{di} = \bar{d}d + \bar{s}s. \tag{12.68}$$

When deducing Eq. (12.68), we have allowed for the mixing matrix unitarity. Hence, only the transitions $d \leftrightarrow d$ and $s \leftrightarrow s$ are allowed while the ones $s \leftrightarrow d$ with changing the flavor are forbidden. It should be stressed, the GIM mechanism means much more than the simple cancellation of processes with changing the strangeness at the tree level. It leads to an additional suppression of neutral currents with $\Delta S \neq 0$ induced by higher-order loop diagrams. The need for this additional suppression is that without it these induced amplitudes would be of the order of $G_F \alpha$ while the experimental data on these processes are typically of the order of $G_F^2 m^2$, where m is of the order of few GeV.

Let us generalize the GIM mechanism on N generations. Our task is to define the amount of observed parameters involved in a $N \times N$ mixing matrix. We can independently change a phase of each of $2N$ quark states not changing observables. For this reason, a unitary matrix contains

$$N^2 - (2N - 1)$$

real parameters. One phase has been omitted since the common phase change remains the $\mathcal{M}$ matrix invariant. On the other hand, an orthogonal $N \times N$ matrix describing a rotation in real space has only $N(N-1)/2$ real parameters related with rotation angles. Therefore, generally speaking, one cannot make the $\mathcal{M}$ matrix real by means of redefinition of quark phases because this matrix must hold

$$N^2 = (2N - 1) - \frac{N}{2}(N - 1) = \frac{1}{2}(N - 1)(N - 2)$$

remainder phase parameters. Consequently, in the case of two doublets only one real parameter is available and a phase parameter is absent, whereas in the case of three doublets there are three real parameters and one phase factor. An initial parametrization of the 3×3 mixing matrix was proposed by Koboyashi and Maskawa in Ref. [7]. At present, a somewhat different parametrization is being used. The rotated and physical quarks are related by the following way

$$\begin{pmatrix} d' \\ s' \\ b' \end{pmatrix} = \mathcal{M}^{CKM} \begin{pmatrix} d \\ s \\ b \end{pmatrix} = \begin{pmatrix} \mathcal{M}_{ud} & \mathcal{M}_{us} & \mathcal{M}_{ub} \\ \mathcal{M}_{cd} & \mathcal{M}_{cs} & \mathcal{M}_{cb} \\ \mathcal{M}_{td} & \mathcal{M}_{ts} & \mathcal{M}_{tb} \end{pmatrix} \begin{pmatrix} d \\ s \\ b \end{pmatrix} =$$

$$= \begin{pmatrix} c_{12}c_{13} & s_{12}c_{13} & s_{13}e^{-i\delta_{13}} \\ -s_{12}c_{23} - c_{12}s_{23}s_{13}e^{i\delta_{13}} & c_{12}c_{23} - s_{12}s_{23}s_{13}e^{i\delta_{13}} & s_{23}c_{13} \\ s_{12}s_{23} - c_{12}c_{23}s_{13}e^{i\delta_{13}} & -c_{12}s_{23} - s_{12}c_{23}s_{13}e^{i\delta_{13}} & c_{23}c_{13} \end{pmatrix} \begin{pmatrix} d \\ s \\ b \end{pmatrix}, \quad (12.69)$$

where $c_{ij} = \cos\theta^{CKM}_{ij}$, $s_{ij} = \sin\theta^{CKM}_{ij}$, $i, j = 1, 2$. The matrix $\mathcal{M}^{CKM}$ is called the *Cabibbo–Koboyashi–Maskawa (CKM) matrix*. Clearly one can move δ_{13} to other sectors by redefining the phases of the quark fields. This means that one must involve more than one matrix element in the mixing matrix $\mathcal{M}^{CKM}$ to get the CP violation. From the physical point of view, this corresponds to the fact that violation of the CP invariance is caused by interference between amplitudes with different CP eigenvalues.

The important property of the neutral current proportional to the operator $S^W_3(f_{L,R}) - Q(f)s^2_W$ lies in the fact that it is flavor-diagonal (or flavor-conserving). This follows from the fact that all fermions with the same charge and the same helicity have the same transformation properties under the gauge group $SU(2)_L \times U(1)_Y$, so that the rotation matrices such as the CKM matrix commute with the neutral-current operator $g^f_{L,R} = S^W_3(f_{L,R}) - Q(f)s^2_W$. In other words, a part of the neutral current related with down-quarks is defined by the relation

$$J^{NC}_{\mu d}(x) = \frac{1}{2}(\overline{d}', \overline{s}', \overline{b}') \left[g^q_L \gamma^\mu (1 - \gamma_5) + g^q_R \gamma^\mu (1 + \gamma_5) \right] \begin{pmatrix} d' \\ s' \\ b' \end{pmatrix} =$$

$$= \frac{1}{2}(\overline{d}, \overline{s}, \overline{b}) \left[g^q_L \gamma^\mu (1 - \gamma_5) + g^q_R \gamma^\mu (1 + \gamma_5) \right] \begin{pmatrix} d \\ s \\ b \end{pmatrix}. \quad (12.70)$$

The experimental data available give the following values for the magnitudes of all nine CKM elements:

$$\mathcal{M}^{CKM} = \begin{pmatrix} 0.97419 \pm 0.00022 & 0.2257 \pm 0.0010 & 0.00359 \pm 0.00016 \\ 0.2256 \pm 0.0010 & 0.97334 \pm 0.00023 & 0.0415^{+0.0010}_{-0.0011} \\ 0.00874^{+0.00026}_{-0.00037} & 0.0407 \pm 0.0010 & 0.999133^{+0.000044}_{-0.000043} \end{pmatrix}. \quad (12.71)$$

Most strikingly in the expression (12.71), the diagonal elements $\mathcal{M}_{ud}, \mathcal{M}_{cs}$, and $\mathcal{M}_{tb}$ are dominant. The large value of $|\mathcal{M}_{cs}|$ reflects the experimental fact of preferable decay of charmed particles into strange ones. The relation $|\mathcal{M}_{cb}| > |\mathcal{M}_{ub}|$ corresponds to the fact that B mesons mainly decay into charmed particles. In summary, we list sources of obtaining information about the CKM matrix elements (Table 12.1).

12.4 Four-fermion invariants and Fierz identities

When calculating electroweak processes, a Fierz transformation [10] is frequently used. Usage of this transformation is at times simply a matter of taste and at other times a cruel necessity. By way of example it is useful to consider the following processes:

$$\overline{\nu}_\mu + e^- \rightarrow \overline{\nu}_\mu + e^- \quad (12.72)$$

$$\overline{\nu}_e + e^- \rightarrow \overline{\nu}_e + e^-. \quad (12.73)$$

The corresponding Feynman diagrams are displayed in Figs. 12.4 and 12.5. The process of the elastic scattering $\overline{\nu}_\mu e^-$ proceeds only due to the interaction of neutral currents while that $\overline{\nu}_e e^-$ is caused by the interaction both of neutral currents and of charged currents. In

TABLE 12.1
The processes used for the definition of the CKM elements.

Matrix element	Experimental information
$|\mathcal{M}_{ud}|$	super-allowed $0^+ \to 0^+$ nuclear β-decay
$|\mathcal{M}_{us}|$	decays $K_L^0 \to \pi e \nu$, $K_L^0 \to \pi \mu \nu$, $K^\pm \to \pi^0 e^\pm \nu$, $K^\pm \to \pi^0 \mu^\pm \nu$ and $K_S^0 \to \pi e \nu$
$|\mathcal{M}_{ub}|$	decays $B \to X_u l \overline{\nu}$ $B \to \pi l \overline{\nu}$
$|\mathcal{M}_{cd}|$	decays $D \to K l \nu$ $D \to \pi l \nu$
$|\mathcal{M}_{cs}|$	decays $W \to l \overline{\nu}_l$, lepton and semi-lepton decays D mesons as well as the unitarity condition $\mathcal{M}^{CKM}$ matrix
$|\mathcal{M}_{cb}|$	semi-leptonic decays of B mesons into charmed particles
$|\mathcal{M}_{tb}|$	decays $\Gamma(t \to Wb)$ and $\Gamma(t \to Wq)$
$|\mathcal{M}_{td}|$ and $|\mathcal{M}_{ts}|$	$B - \overline{B}$ oscillations caused by box-diagrams with t quark, as well as rare decays of K and B mesons due to loop diagrams

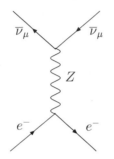

FIGURE 12.4
The Feynman diagram corresponding to the process $\overline{\nu}_\mu + e^- \to \overline{\nu}_\mu + e^-$.

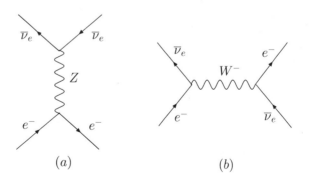

(a) (b)

FIGURE 12.5
The Feynman diagrams (a) and (b) associated with the process $\overline{\nu}_e + e^- \to \overline{\nu}_e + e^-$.

accordance with the Feynman rules the invariant amplitude of the process (12.72) in the unitary gauge is

$$\mathcal{A}(\overline{\nu}_\mu e^-) = C \left\{ \frac{ig}{4c_W} \overline{v}_{\nu_\mu}(k_1) \gamma_\lambda [1 - \gamma_5] v_{\nu_\mu}(k_2) \right\} \frac{g^{\lambda\sigma} - q^\lambda q^\sigma / m_Z^2}{q^2 - m_Z^2} \times$$

$$\times \left\{ \frac{ig}{2c_W} \overline{u}_e(p_2)\gamma_\sigma[g_V - g_A\gamma_5]u_e(p_1) \right\}, \tag{12.74}$$

where $q = k_1 - k_2 = p_2 - p_1$, $g_V = 2s_W^2 - 1/2$, $g_A = -1/2$ and

$$C = \sqrt{\frac{m_{\nu_\mu}^2 m_e^2}{k_{10}p_{10}k_{20}p_{20}}}.$$

At $|q^2| \ll m_Z^2$ this expression takes the form

$$\mathcal{A}(\overline{\nu}_\mu e^-) = C\frac{G_F}{\sqrt{2}}\overline{v}_{\nu_\mu}(k_1)\gamma_\lambda[1 - \gamma_5]v_{\nu_\mu}(k_2)\overline{u}_e(p_2)\gamma^\lambda[g_V - g_A\gamma_5]u_e(p_1), \tag{12.75}$$

where the transverse part of the Z boson propagator has disappeared because we have set the neutrino mass to be equal to zero and taken into consideration $g^2/(8m_W^2) = G_F/\sqrt{2}$.

Now we address the reaction (12.73). Denoting the amplitude caused by the interaction of the neutral currents as $\mathcal{A}_Z(\overline{\nu}_e e^-)$ and that related with the interaction of the charged currents as $\mathcal{A}_W(\overline{\nu}_e e^-)$, we get for the case $|q^2| \ll m_Z^2$

$$\mathcal{A}_Z(\overline{\nu}_e e^-) = C\frac{G_F}{\sqrt{2}}\overline{v}_{\nu_e}(k_1)\gamma_\lambda[1 - \gamma_5]v_{\nu_e}(k_2)\overline{u}_e(p_2)\gamma^\lambda[g_V - g_A\gamma_5]u_e(p_1), \tag{12.76}$$

$$\mathcal{A}_W(\overline{\nu}_e e^-) = -C\frac{G_F}{\sqrt{2}}\overline{v}_{\nu_e}(k_1)\gamma_\lambda[1 - \gamma_5]u_{\nu_e}(p_1)\overline{u}_e(p_2)\gamma^\lambda[1 - \gamma_5]v_{\nu_e}(k_2). \tag{12.77}$$

It is obvious that the calculation of the process (12.73) will be significantly simplified if we manage to rewrite $\mathcal{A}_W(\overline{\nu}_e e^-)$ in the form similar to $\mathcal{A}_Z(\overline{\nu}_e e^-)$. This proves to be possible with the help of the so-called *Fierz transformation*. In the case under consideration, it results in

$$\mathcal{A}_W(\overline{\nu}_e e^-) = C\frac{G_F}{\sqrt{2}}\overline{v}_{\nu_e}(k_1)\gamma_\lambda[1 - \gamma_5]v_{\nu_e}(k_2)\overline{u}_e(p_2)\gamma^\lambda[1 - \gamma_5]u_{\nu_e}(p_1). \tag{12.78}$$

With allowance made for (12.78), the amplitude of the process (12.73) will look like:

$$\mathcal{A}(\overline{\nu}_e e^-) = \mathcal{A}_Z(\overline{\nu}_e e^-) + \mathcal{A}_W(\overline{\nu}_e e^-) = C\frac{G_F}{\sqrt{2}}\overline{v}_{\nu_e}(k_1)\gamma_\lambda[1 - \gamma_5]v_{\nu_e}(k_2)\overline{u}_e(p_2)\gamma^\lambda[g_V' -$$

$$-g_A'\gamma_5]u_{\nu_e}(p_1), \tag{12.79}$$

where $g_V' = g_V + 1$ and $g_A' = g_A + 1$.

Thus, we succeeded in simplifying the calculation of the process (12.73). Moreover, we could relate the cross sections of two different processes, (12.72) and (12.73). Really, having fulfilled the replacement

$$g_V \to g_V', \qquad g_A \to g_A'$$

into the cross section of the process (12.72), we get the cross section of the process (12.73).

Proceed to the derivation of the Fierz transformation. We consider two bispinors $\overline{\Phi}$ and Ψ, which describe two different particles. Using them, one can form 16 bilinear combinations that are grouped into five different Lorentz-invariant quantities:

Covariant quantity	Number of components
$\overline{\Phi}\Psi$ scalar	1
$\overline{\Phi}\gamma_\mu\Psi$ vector	4

$$\overline{\Phi}\sigma_{\mu\nu}\Psi \text{ tensor} \qquad\qquad\qquad 6$$

$$\overline{\Phi}\gamma_\mu\gamma_5\Psi \text{ axial} \qquad\qquad\qquad 4$$

$$\overline{\Phi}\gamma_5\Psi \text{ pseudoscalar} \qquad\qquad\qquad 1$$

Next, introducing two further bispinors $\overline{\varphi}$ and ψ into play, we build all possible Lorentz-invariant scalars:

$$(\overline{\Phi}\Psi)(\overline{\varphi}\psi) \text{ —}S\text{-variant,}$$

$$(\overline{\Phi}\gamma_\mu\Psi)(\overline{\varphi}\gamma^\mu\psi) \text{ — }V\text{-variant,}$$

$$(\overline{\Phi}\sigma_{\mu\nu}\Psi)(\overline{\varphi}\sigma^{\mu\nu}\psi) \text{ — }T\text{-variant,}$$

$$(\overline{\Phi}\gamma_\mu\gamma_5\Psi)(\overline{\varphi}\gamma^\mu\gamma_5\psi) \text{ — }A\text{-variant,}$$

$$(\overline{\Phi}\gamma_5\Psi)(\overline{\varphi}\gamma_5\psi) \text{ — }P\text{-variant.}$$

Obviously, the four-fermion interactions of the general form may be written as follows:

$$I_i = C_i(\overline{\Phi}O_i\Psi)(\overline{\varphi}O^i\psi), \qquad\qquad (12.80)$$

where C_i are constants, $i = S, V, T, A, P$ and 16 matrices $O_S = \mathbf{1}, O_V = \gamma_\mu, O_T = \sigma_{\mu\nu}, O_A = i\gamma_\mu\gamma_5, O_P = \gamma_5$ make a complete system. However, the chosen representation (12.80) violates the symmetry of the four-fermion interaction in the sense that it connects $\overline{\Phi}$ into a pair with Ψ and $\overline{\varphi}$ into a pair with ψ. Of course, we have a right to connect $\overline{\Phi}$ into a pair with ψ and $\overline{\varphi}$ into a pair with Ψ. In this case, we would have arrived at other invariants:

$$I'_j = C_j(\overline{\Phi}O_j\psi)(\overline{\varphi}O^j\Psi). \qquad\qquad (12.81)$$

At that, owing to the completeness of the 16 matrices system, old and new invariants must be connected with each other by a matrix, a Fierz matrix, that is, the coefficients C_i and C'_j must obey the relation

$$C_i = \sum_{k=1}^{5} \Lambda_{ij}C'_j. \qquad\qquad (12.82)$$

So, our task is to find the matrix Λ_{ij}. Write the obvious relation

$$\sum_i C_i(\overline{\Phi}O_i\Psi)(\overline{\varphi}O^i\psi) = \sum_j C'_j(\overline{\Phi}O_j\psi)(\overline{\varphi}O^j\Psi). \qquad\qquad (12.83)$$

Writing down the matrix indices and allowing for the arbitrariness of wave functions, instead of (12.83), we obtain

$$\sum_i^{5} C_i(O_i)_{\alpha\delta}(O^i)_{\gamma\beta} = \sum_j^{5} C'_j(O_j)_{\alpha\beta}(O^j)_{\gamma\delta}. \qquad\qquad (12.84)$$

To multiply two sides of (12.84) by $(O_q)_{\beta\gamma}$ gives

$$\sum_i C_i(O_i)_{\alpha\delta}(O^i)_{\gamma\beta}(O_q)_{\beta\gamma} = \sum_j C'_j(O_j)_{\alpha\beta}(O_q)_{\beta\gamma}(O^j)_{\gamma\delta}. \qquad\qquad (12.85)$$

Summing over the indices β and γ and making use of the general property of the 16 O_i matrices

$$\sum_{\beta,\gamma}(O^i)_{\gamma\beta}(O_q)_{\beta\gamma} = \text{Sp}(O^iO_q) = 4\delta_{iq}, \qquad\qquad (12.86)$$

we get

$$C_i(O_i)_{\alpha\delta} = \frac{1}{4}\sum_j C'_j (O_j O_i O^j)_{\alpha\delta} \tag{12.87}$$

(summation over i is not fulfilled). Multiplying two sides of (12.87) by $(O^i)_{\delta\alpha}$ and summing over δ and α, we arrive at

$$C_i = \frac{1}{16}\sum_j C'_j \mathrm{Sp}(O_j O_i O^j O^i) = \sum_j \Lambda_{ij} C'_j. \tag{12.88}$$

Having done uncomplicated but tedious operations with γ matrices, we finally obtain

$$
\begin{pmatrix} C_S \\ C_V \\ C_T \\ C_A \\ C_P \end{pmatrix}
= \frac{1}{4}
\begin{pmatrix}
1 & 4 & 6 & 4 & 1 \\
1 & -2 & 0 & 2 & -1 \\
1 & 0 & -2 & 0 & 1 \\
1 & 2 & 0 & -2 & -1 \\
1 & -4 & 6 & -4 & 1
\end{pmatrix}
\begin{pmatrix} C'_S \\ C'_V \\ C'_T \\ C'_A \\ C'_P \end{pmatrix}. \tag{12.89}
$$

It is easy to verify that the Fierz matrix satisfies the relation

$$\Lambda^2 = 1 \tag{12.90}$$

which is representative of the fact that the double Fierz transformation is the identity transformation. From (12.90) it also follows that $\Lambda^{-1} = \Lambda$, that is, C_i and C'_i are transformed by the same law. One may easily be checked, if a matrix element has the $V - A$ structure in the form with neutral currents, then it conserves this structure in the form with charged currents too. This invariance with respect to the Fierz transformation is an important property of $V - A$ interaction.

Up to now, we considered the relations among the five Lorentz scalars. It turns out that the same Fierz matrix (12.89) relates five Lorentz pseudoscalars with each other too. We can convince ourselves that is the case replacing, for example, ψ by $\gamma_5\psi$ in the relations found above.

The obtained relation between C_i and C'_j (12.89) is valid for wave functions. If we deal with operators, then in (12.89) a common minus sign that is caused by anticommutation rules for fermions appears.

When computing weak processes, two relations are often encountered:

$$[\overline{\Phi}\gamma_\mu(1-\gamma_5)\Psi][\overline{\varphi}\gamma^\mu(1-\gamma_5)\psi] = -[\overline{\Phi}\gamma_\mu(1-\gamma_5)\psi][\overline{\varphi}\gamma^\mu(1-\gamma_5)\Psi], \tag{12.91}$$

$$[\overline{\Phi}\gamma_\mu(1-\gamma_5)\Psi][\overline{\varphi}\gamma^\mu(1+\gamma_5)\psi] = 2[\overline{\Phi}\gamma_\mu(1+\gamma_5)\psi][\overline{\varphi}\gamma^\mu(1-\gamma_5)\Psi]. \tag{12.92}$$

They are readily obtained from the Fierz expansion for C_V. To get the former, one has to make the replacements

$$\overline{\Phi} \to \overline{\Phi}(1+\gamma_5), \qquad \Psi \to (1-\gamma_5)\Psi, \qquad \overline{\varphi} \to \overline{\varphi}(1+\gamma_5), \qquad \Psi \to (1-\gamma_5)\Psi, \tag{12.93}$$

while obtaining the latter requires the replacements

$$\overline{\Phi} \to \overline{\Phi}(1+\gamma_5), \qquad \Psi \to (1-\gamma_5)\Psi, \qquad \overline{\varphi} \to \overline{\varphi}(1-\gamma_5), \qquad \Psi \to (1+\gamma_5)\Psi. \tag{12.94}$$

12.5 Muon decays

More than seventy years have passed since muons were discovered in cosmic rays. Muon properties are very similar to electron ones. A muon, like an electron, takes part only in electromagnetic and weak interactions. At that, the electromagnetic and weak interactions of the electron and muon are manifested absolutely identically. A muon, as well as an electron, can form an atom-like system consisting of an atomic nucleus and μ^-, a muon-nucleon atom. As this takes place, properties of a muon atom are similar to those of an ordinary hydrogen-like atom. Thanks to these circumstances they used the term *heavy electron* for a muon at the beginning of elementary particle physics. But when the destination of an electron is obvious, then who ever orders a muon? One more question is also appropriate: "Why are muon and electron masses so different?" After all, in accordance with contemporary conceptions, a mass is completely defined by the interactions of particles.

In recent years, the muon has become in the center of attention. This is caused by two factors. The former is related to the construction of muon colliders, the First Muon Collider (FMC) and Next Muon Collider (NMC). The latter is the results of the E821 experiment at Brookhaven National Laboratory (BNL) (1997–2001) [11] in which the anomalous magnetic moment (AMM) of a muon has been measured. The E821 results are not in line with the SM predictions, that is, they are a possible signal of New Physics even at the electroweak scale. Later we shall come back to discussion of the muon AMM. Now we will consider muon decay channels.

Begin with the main channel

$$\mu^- \to e^- + \nu_\mu + \overline{\nu}_e, \tag{12.95}$$

which may serve as a typical example of weak decay reactions. The importance of similar processes for investigation of the weak interaction is connected with the fact that in the lowest order of the perturbation theory these processes do not involve complications caused by the strong interaction and, as a result, their decay probabilities may be calculated precisely.

The value of the Fermi constant G_F can be determined by the comparison of experimental data of the decay (12.95) with theoretical predictions. To study this decay channel also allows us to measure the following characteristics: (i) the energy spectrum of electrons in an unpolarized muon beam; (ii) the polarization of decay electrons; and (iii) the angle correlation between the electron momentum and muon polarization. Note, however, to date an important link is absent: decay neutrinos are not observed and their energy spectrum and polarizations cannot be measured.

Let us calculate the decay probability of (12.95) in the second order of the perturbation theory (Fig. 12.6). The invariant amplitude corresponding to the diagram shown in Fig. 12.6 has the form:

$$\mathcal{A} = \left[\frac{ig}{2\sqrt{2}} \overline{u}_{\nu_\mu}(p_2) \gamma_\lambda (1 - \gamma_5) u_\mu(k) \right] \left(\frac{g^{\lambda\sigma} - q^\lambda q^\sigma / m_W^2}{q^2 - m_W^2} \right) \left[\frac{ig}{2\sqrt{2}} \overline{u}_e(p_e) \gamma_\sigma (1 - \gamma_5) v_{\nu_e}(p_1) \right]. \tag{12.96}$$

It is clear that in the muon decay $|q^2| \ll m_W^2$ and so in very good approximation we have

$$\mathcal{A} = \frac{G_F}{\sqrt{2}} \overline{u}_{\nu_\mu}(p_2) \gamma_\lambda (1 - \gamma_5) u_\mu(k) \overline{u}_e(p_e) \gamma^\lambda (1 - \gamma_5) v_{\nu_e}(p_1). \tag{12.97}$$

The fulfillment of the Fierz transformation does not lead to any simplifications in computations and we keep the amplitude $\mathcal{A}$ in the view (12.97). In accordance with Section 18.12 of

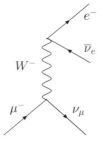

FIGURE 12.6

The decay $\mu^- \rightarrow \nu_\mu + \overline{\nu}_e + e^-$.

Advanced Particle Physics Volume 1 the differential decay probability of the decay (12.95) is given by the expression:

$$d\Gamma = (2\pi)^4 \delta^{(4)}(p_1 + p_2 + p_e - k) \frac{m_e}{E_e} \frac{m_\mu}{E_\mu} \frac{m_{\nu_e}}{E_1} \frac{m_{\nu_\mu}}{E_2} \frac{d^3 p_1}{(2\pi)^3} \frac{d^3 p_2}{(2\pi)^3} |\mathcal{A}|^2, \tag{12.98}$$

where

$$|\mathcal{A}|^2 = \frac{1}{2} G_F^2 \Lambda^{(e)} \Lambda^{(\mu)}, \tag{12.99}$$

$$\Lambda^{(e)} - \mathrm{Sp}[u_e(p_e)\overline{u}_e(p_e)\gamma^\lambda(1 - \gamma_5)v_{\nu_e}(p_1)\overline{v}_{\nu_e}(p_1)\gamma^\sigma(1 - \gamma_5)], \tag{12.100}$$

$$\Lambda^{(\mu)} = \mathrm{Sp}[u_{\nu_\mu}(p_2)\overline{v}_\mu(p_2)\gamma_\lambda(1 - \gamma_5)u_\mu(k)\overline{u}_\mu(k)\gamma_\sigma(1 - \gamma_5)]. \tag{12.101}$$

When the muon and decay electron are polarized, we have

$$u_e(p_e)\overline{u}_e(p_e) = \frac{1}{2}\left(\frac{\hat{p}_e + m_e}{2m_e}\right)(1 + \gamma_5 \hat{s}_e), \qquad u_\mu(k)u_\mu(k) = \frac{1}{2}\left(\frac{\hat{k} + m_\mu}{2m_\mu}\right)(1 + \gamma_5 \hat{s}_\mu), \tag{12.102}$$

where s_e and s_μ are four-dimensional polarization vectors for the electron and muon, respectively. Allowing for the smallness of neutrino masses ($\sim$ few eV), we write

$$u_{\nu_\mu}(p_2)\overline{u}_{\nu_\mu}(p_2) = \frac{\hat{p}_2}{2m_{\nu_\mu}}, \qquad v_{\nu_e}(p_1)\overline{v}_e(p_1) = \frac{\hat{p}_1}{2m_{\nu_e}}. \tag{12.103}$$

The substitution of (12.102) and (12.103) into Eq. (12.100) results in

$$\Lambda^{(e)} = \frac{1}{8m_e m_{\nu_e}} \mathrm{Sp}[(\hat{p}_e + m_e)(1 + \gamma_5 \hat{s}_e)\gamma^\lambda(1 - \gamma_5)\hat{p}_1\gamma^\sigma(1 - \gamma_5)] =$$

$$= \frac{1}{4m_e m_{\nu_e}} \mathrm{Sp}[(\hat{p}_e - m_e \hat{s}_e)\gamma^\lambda \hat{p}_1 \gamma^\sigma(1 - \gamma_5)]. \tag{12.104}$$

By an analogy, we get

$$\Lambda^{(\mu)} = \frac{1}{4m_\mu m_{\nu_\mu}} \mathrm{Sp}[\hat{p}_2 \gamma_\lambda(\hat{k} - m_\mu \hat{s}_\mu)\gamma_\sigma(1 - \gamma_5)]. \tag{12.105}$$

When calculating the product $\Lambda^{(\mu)}\Lambda^{(e)}$, the following formula

$$\mathrm{Sp}[\gamma_\alpha \gamma_\beta \gamma_\rho \gamma_\sigma(1 - \gamma_5)]\mathrm{Sp}[\gamma^\tau \gamma^\beta \gamma^\kappa \gamma^\sigma(1 - \gamma_5)] = 64 g_\alpha^\tau g_\rho^\kappa \tag{12.106}$$

will be useful. With its help, we obtain

$$|\mathcal{A}|^2 = 2G_F^2[(p_e - m_e s_e) \cdot p_2][(k - m_\mu s_\mu) \cdot p_1]. \tag{12.107}$$

Then, the differential decay probability will look like:

$$d\Gamma = \frac{2G_F^2}{(2\pi)^5}(p_e - m_e s_e)^\alpha (k - m_\mu s_\mu)^\beta \frac{p_{1\beta} d^3 p_1}{E_1} \frac{p_{2\alpha} d^3 p_2}{E_2} \delta^{(4)}(p_1 + p_2 + p_e - k). \tag{12.108}$$

Since in all current experiments on the muon decay the neutrino is not detected, we should integrate over neutrino momenta to obtain the expression that may be compared with the experimental data. In so doing, we have to calculate the integral

$$I_{\alpha\beta} = \int d^3 p_2 \int d^3 p_1 \frac{p_{1\beta} p_{2\alpha}}{E_1 E_2} \delta^{(4)}(p_1 + p_2 - q), \tag{12.109}$$

where $q = k - p_e$.

In the general form the tensor of the second rank $I_{\alpha\beta}$ may be written as

$$I_{\alpha\beta} = Aq^2 g_{\alpha\beta} + Bq_\alpha q_\beta, \tag{12.110}$$

where A and B are dimensionless coefficients to be found. Multiplying two sides of (12.110) by $g^{\alpha\beta}$ and $q^\alpha q_\beta$, we arrive at

$$g^{\alpha\beta} I_{\alpha\beta} = 4Aq^2 + Bq^2, \tag{12.111}$$

$$q^\alpha q^\beta I_{\alpha\beta} = Aq^4 + Bq^4. \tag{12.112}$$

From (12.109) it is evident that

$$g^{\alpha\beta} I_{\alpha\beta} = \int d^3 p_2 \int d^3 p_1 \frac{(p_1 p_2)}{E_1 E_2} \delta^{(4)}(p_1 + p_2 - q), \tag{12.113}$$

$$q^\alpha q_\beta I_{\alpha\beta} = \int d^3 p_2 \int d^3 p_1 \frac{(p_1 p_2)^2}{E_1 E_2} \delta^{(4)}(p_1 + p_2 - q). \tag{12.114}$$

Since $g^{\alpha\beta} I_{\alpha\beta}$ and $q^\alpha q^\beta I_{\alpha\beta}$ are invariant, when calculating the integrals (12.113) and (12.114), we may choose any reference frame. Let us take the reference frame where $\mathbf{p}_1 = -\mathbf{p}_2$. Then, with allowance made for

$$\int d^3 p_2 \delta^{(3)}(\mathbf{p}_1 + \mathbf{p}_2 - \mathbf{q}) = 1,$$

we find

$$g^{\alpha\beta} I_{\alpha\beta} = 8\pi \int E_1^2 dE_1 \delta(2E_1 - q_0) = \pi q^2, \tag{12.115}$$

$$q^\alpha q^\beta I_{\alpha\beta} = 16\pi \int E_1^4 dE_1 \delta(2E_1 - q_0) = \frac{\pi q^4}{2}. \tag{12.116}$$

The substitution of the obtained expressions into (12.111) and (12.112) gives

$$A = \frac{\pi}{6}. \qquad B = \frac{\pi}{3}$$

So

$$I_{\alpha\beta} = \frac{\pi}{6} \left(q^2 g_{\alpha\beta} + 2q_\alpha q_\beta \right). \tag{12.117}$$

To substitute (12.117) to the expression for the differential probability of the decay (12.108) results in

$$d\Gamma = \frac{\pi G_F^2}{3(2\pi)^5} \frac{d^3 p_e}{E_e E_\mu} (p_e - m_e s_e)^\alpha (p_\mu - m_\mu s_\mu)^\beta (q^2 g_{\alpha\beta} + 2q_\alpha q_\beta). \tag{12.118}$$

Let us pass on to the muon rest frame where

$$\left. \begin{array}{llll} q_\alpha = (q_0, \mathbf{q}), & q_0 = m_\mu - E_e, & \mathbf{q} = -\mathbf{p}_e, & (s_\mu)_\alpha = (0, \mathbf{s}_\mu), \\[2mm] (s_e)_\alpha = (s_{e0}, \mathbf{s}), & s_{e0} = \dfrac{(\mathbf{p}_e \mathbf{s}_e)}{m_e}, & \mathbf{s} = \mathbf{s}_e + \dfrac{(\mathbf{p}_e \mathbf{s}_e) \mathbf{p}_e}{m_e (E_e + m_e)}, \end{array} \right\} \tag{12.119}$$

($\mathbf{s}_e$ is the polarization vector of the electron in its intrinsic frame of reference). In the spherical coordinate system, where the three-dimensional electron momentum forms the angle θ with the z axis, we have

$$d^3 p_e = \mathbf{p}_e^2 d|\mathbf{p}_e| d\varphi d\cos\theta = \mathbf{p}_e^2 d|\mathbf{p}_e| d\Omega_e \approx E_1^2 dE_1 d\Omega_e. \tag{12.120}$$

The substitution of (12.119) and (12.120) into (12.118) leads to the expression

$$d\Gamma = \frac{\pi G_F^2}{3(2\pi)^5 m_\mu} d\Omega_e |\mathbf{p}_e| dE_e \left\{ (m_\mu^2 + E_e^2 - 2m_\mu E_e - \mathbf{p}_e^2) \left[(E_e - \mathbf{p}_e \cdot \mathbf{s}_e) m_\mu + \right. \right.$$

$$+ m_\mu \left(\mathbf{p}_e - m_e \mathbf{s}_e - \frac{\mathbf{p}_e \cdot \mathbf{s}_e}{E_e + m_e} \mathbf{p}_e \right) \cdot \mathbf{s}_\mu \right] + 2 \left[(E_e - \mathbf{p}_e \cdot \mathbf{s}_e)(m_\mu - E_e) + \right.$$

$$\left. \left. + \left(\mathbf{p}_e - m_e \mathbf{s}_e - \frac{\mathbf{p}_e \cdot \mathbf{s}_e}{E_e + m_e} \mathbf{p}_e \right) \cdot \mathbf{p}_e \right] (m_\mu^2 - m_\mu E_e - m_\mu \mathbf{p}_e \cdot \mathbf{s}_\mu) \right\}. \tag{12.121}$$

The electron has the maximum energy and momentum when two neutrinos are emitted in one direction while the electron is emitted in an opposite direction. In this case

$$(E_e)_{\max} = \frac{(m_\mu^2 + m_e^2)}{2m_\mu}, \qquad |(\mathbf{p}_e)_{\max}| = \frac{(m_\mu^2 - m_e^2)}{2m_\mu}. \tag{12.122}$$

If one neglects the electron mass, then the relations

$$(E_e)_{\max} = \frac{m_\mu}{2}, \qquad \mathbf{p}_e = \mathbf{n} E_e,$$

where $\mathbf{n}$ is the unit vector in the electron motion direction (recall, we are working in the intrinsic reference frame of the muon), will be valid. Next, we direct the muon polarization vector along the z axis of the spherical coordinate system. Then, going on to new variables $\epsilon = E_e/(E_e)_{max} = 2E_e/m_\mu$, we rewrite the expression (12.121) in the following form:

$$d\Gamma = \frac{G_F^2 m_\mu^5}{3(4\pi)^4} \left[2\epsilon^2 (3 - 2\epsilon) \right] \left[1 + \left(\frac{1 - 2\epsilon}{3 - 2\epsilon} \right) \cos\theta \right] \left(\frac{1 - \mathbf{n} \cdot \mathbf{s}_e}{2} \right) d\epsilon d\Omega. \tag{12.123}$$

It is easy to show that for the decay

$$\mu^+ \to e^+ + \nu_e + \overline{\nu}_\mu \tag{12.124}$$

the analogous calculations result in

$$d\Gamma = \frac{G_F^2 m_\mu^5}{3(4\pi)^4} \left[2\epsilon^2 (3 - 2\epsilon) \right] \left[1 - \left(\frac{1 - 2\epsilon}{3 - 2\epsilon} \right) \cos\theta \right] \left(\frac{1 + \mathbf{n} \cdot \mathbf{s}_{\overline{e}}}{2} \right) d\epsilon d\Omega. \tag{12.125}$$

Now we will discuss the structure of the differential muon decay probability (12.123). The first factor in the square brackets represents the normalized energy spectrum of the electrons

$$n(\epsilon) = 2\epsilon^2(3 - 2\epsilon).$$

The second factor in the square brackets describes asymmetry of electrons emitting with respect to the muon polarization direction. The third factor in the parentheses covers the electron helicity ξ, which is equal to -1 independently on the electron energy in the approximation $m_e = 0$. Note, this is in line with the $V - A$ structure of the weak interaction.

The differential decay probability for the unpolarized particles is

$$d\Gamma = \frac{G_F^2 m_\mu^5}{384\pi^4}(3 - 2\epsilon)\epsilon^2 d\epsilon d\Omega_e. \tag{12.126}$$

To integrate the expression (12.126) over exit angles of the electron gives 4π. Then, carrying the integration over the electron spectrum as well, we get the total decay probability

$$\Gamma = \frac{G_F^2 m_\mu^5}{96\pi^3}\int_0^1(3 - 2\epsilon)\epsilon^2 d\epsilon = \frac{G_F^2 m_\mu^5}{192\pi^3}. \tag{12.127}$$

In its turn, the inclusion of the electron mass results in

$$\Gamma = \frac{G_F^2 m_\mu^5}{192\pi^3}F\left(\frac{m_e^2}{m_\mu^2}\right), \tag{12.128}$$

where

$$F\left(\frac{m_e^2}{m_\mu^2}\right) = \left[1 - 8\left(\frac{m_e^2}{m_\mu^2}\right) + 8\left(\frac{m_e^2}{m_\mu^2}\right)^3 - \left(\frac{m_e^2}{m_\mu^2}\right)^4 - 12\left(\frac{m_e^2}{m_\mu^2}\right)^2\ln\left(\frac{m_e^2}{m_\mu^2}\right)\right] =$$

$$= \left(1 - 1.87 \times 10^{-4}\right). \tag{12.129}$$

The main part of the contribution caused by radiative corrections (RCs) to the muon decay is given by diagrams that include the interaction of the charged particles with the electromagnetic field (Fig. 12.7). The bremsstrahlung diagrams have to be included since, owing to the vanishing photon mass, photons with arbitrary small energies may be emitted. On the other hand, because of the limited experimental resolution, it is impossible to distinguish the muon decay accompanied by emission of an extremely soft photon from a decay without radiation. This contribution exactly cancels the divergent terms that appear in the diagrams involving the self-energy loops in the muon and electron lines (Fig. 12.12(b)). The calculation of the RCs leads to a modification of the decay probability Γ by a factor [12, 13]

$$1 - \frac{\alpha}{2\pi}\left(\pi^2 - \frac{25}{4}\right) = 0.9958\ldots. \tag{12.130}$$

Therefore, the electromagnetic RCs are of greater importance than the influence of the finite mass of the electron. The total inclusion of the RCs within the GWS theory that was fulfilled in Refs. [14, 15], gives the result

$$\Gamma = \frac{1}{\tau_\mu} = \frac{G_F^2 m_\mu^5}{192\pi^3}F\left(\frac{m_e^2}{m_\mu^2}\right)\left(1 + \frac{3m_\mu^2}{5m_W^2}\right)\left[1 - \frac{\alpha(m_\mu)}{2\pi}\left(\pi^2 - \frac{25}{4}\right) + C\frac{\alpha^2(m_\mu)}{\pi^2}\right], \tag{12.131}$$

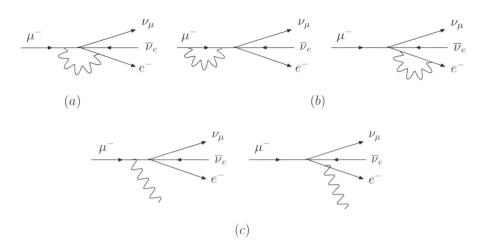

(a)   (b)

(c)

FIGURE 12.7
The high order corrections to the muon decay: (a)—vertex correction, (b)—self-energy corrections, (c)—bremsstrahlung contributions.

where

$$C = \frac{156815}{5184} - \frac{518}{81}\pi^2 - \frac{895}{36}\zeta(3) + \frac{67}{720}\pi^4 + \frac{53}{6}\pi^2\ln(2), \tag{12.132}$$

$$\alpha(m_\mu)^{-1} = \alpha(m_e)^{-1} - \frac{2}{3\pi}\ln\left(\frac{m_\mu}{m_e}\right) + \frac{1}{6\pi} \approx 136 \tag{12.133}$$

and τ_μ is the average lifetime of the muon. Substituting the experimental value of τ_μ

$$\tau_\mu = (2.197019 \pm 0.000021) \times 10^{-6} \text{ s} \tag{12.134}$$

and the muon mass

$$m_\mu = 105.658367 + 0.000004 \text{ MeV} \tag{12.135}$$

into the formula (12.131), we may calculate the value of the Fermi constant G_F

$$G_F = 1.166367(5) \times 10^{-5} \text{ GeV}^{-2}. \tag{12.136}$$

Let us demonstrate that the results obtained can be understood at the purely qualitative level not making any calculations. First of all, since in the natural system of units the decay probability has the mass dimensionality, $[\Gamma] = M$, then from the dimensional reasons follows $\Gamma \sim G_F^2 m_\mu^5$. In doing so, we base only on the fact that the probability is proportional to G_F^2 and that the muon mass is a single dimensional parameter that defines the decay dynamics because the electron and neutrino masses may be neglected.

The angle asymmetry of the electron is both P and C noninvariant: it has different signs in the left- and right-handed coordinate systems; its signs are also opposite for e^- and e^+ in the decays μ^- and μ^+, respectively. It is easy to understand the characteristic features of this asymmetry. Consider the electron with the energy close to the maximum ($\epsilon \sim 1$). At these conditions the neutrino and antineutrino have to fly in the direction opposite to the electron (Fig. 12.8). As the helicities of $\overline{\nu}_e$ and ν_μ are opposite, then they take away the zero angular moment. Hence, the electron must fly out in such a way that its spin is parallel to that of the muon. But since the electron possesses negative helicity, its momentum must be directed mainly against the muon spin. The obtained expression for the angular electron

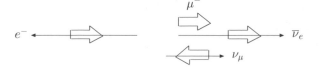

FIGURE 12.8

The scheme of the decay $\mu^- \to e^- \nu_\mu \bar{\nu}_e$ when $\epsilon \sim 1$.

distribution (12.123) is in agreement with this fact because it is proportional to $(1 - \mathbf{n} \cdot \mathbf{s}_e)$ when $\epsilon \sim 1$.

At $\epsilon \ll 1$ the neutrino and antineutrino fly in opposite directions and their total spin equals 1. In this case, due to the angular moment conservation law the electron needs to fly along the muon spin (Fig. 12.9). Without violation of the partial flavor conservation

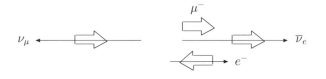

FIGURE 12.9

The scheme of the decay $\mu^- \to e^- \nu_\mu \bar{\nu}_e$ when $\epsilon \ll 1$.

law the muon may also decay through the channel:

$$\mu^- \to e^- + \bar{\nu}_e + \nu_\mu + \gamma. \tag{12.137}$$

Experimental data give the following value for the branching of this channel [16]

$$\mathrm{Br}(\mu^- \to e^- \bar{\nu}_e \nu_\mu \gamma) = \frac{\Gamma(\mu^- \to e^- \bar{\nu}_e \nu_\mu \gamma)}{\Gamma(\mu^- \to e^- \bar{\nu}_e \nu_\mu)} = 0.014 \pm 0.004. \tag{12.138}$$

Inasmuch as the decay (12.137) is hard to distinguish from the decay $\mu^- \to e^- \bar{\nu}_e \nu_\mu$, which is accompanied by a soft photon emission, then (12.137) includes only data with photon energy more than 10 MeV.

The lepton flavor ia also conserved in the decay

$$\mu^- \to e^- + \bar{\nu}_e + \nu_\mu + e^- + e^+. \tag{12.139}$$

One of the diagrams of this decay is displayed in Fig. 12.10. Here the branching value proves to be even smaller [17]

$$\mathrm{Br}(\mu^- \to e^- \bar{\nu}_e \nu_\mu e^- e^+) = (3.4 \pm 0.4) \times 10^{-5}. \tag{12.140}$$

Thanks to the neutrino oscillations, decay channels of the muon that violate the partial lepton flavor are allowed. Among these are

$$\mu \to e^- + \gamma, \tag{12.141}$$

$$\mu \to e^- + e^- + e^+, \tag{12.142}$$

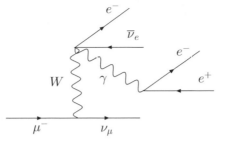

FIGURE 12.10
The diagram associated with the decay $\mu^- \to e^- \overline{\nu}_e \nu_\mu e^- e^+$.

$$\mu \to e^- + \gamma + \gamma, \tag{12.143}$$

$$\mu \to e^- + \nu_e + \overline{\nu}_\mu. \tag{12.144}$$

The first of them, the so-called *radiative muon decay*, will be studied below in greater detail.

In order to seek the radiative muon decay the MEGA Collaboration was organized at the Los Alamos Meson Physics Facility (LAMPF, USA). To date, this collaboration obtained the following upper limit on the branching ratio for the radiative muon decay [18]:

$$\mathrm{Br}(\mu \to e\gamma) = \frac{\Gamma(\mu \to e\gamma)}{\Gamma(\mu \to \nu_\mu e \overline{\nu}_e)} < 1.2 \times 10^{-11}. \tag{12.145}$$

Since 2006, the investigations of the decay (12.141) have being performed by the MEG (Muon to Electron and Gamma) Collaboration at the Pauli Scherer Institute (Switzerland). The final goal of the MEG experiment is to improve the sensitivity in measuring $\mathrm{Br}(\mu \to e\gamma)$ to 10^{-14} [19]. Note, if the decay $\mu \to e\gamma$ is detected in that experiment, it will be possible to measure the angular distribution of decay products and to obtain additional criteria for establishing a true model of the electroweak interaction.

Without an allowance for the electromagnetic properties of the neutrino, this decay width was calculated within the SM (see, for example, [20]), in various supersymmetric Grand Unified Theories [21], and in the model based on the $SU(2)_L \times SU(2)_R \times U(1)_{B-L}$ gauge group with a Majorana neutrino (see, for review, [22]). The mechanism of the decay (12.141) may be presented by the diagram pictured in Fig. 12.11. Note, from the diagram of

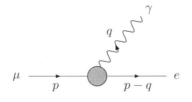

FIGURE 12.11
The radiative muon decay.

Fig. 12.11, it follows that the decay (12.142), which is going even in the fourth order of the perturbation theory, is induced by the process $\mu^- \to e^- \gamma$.

The amplitude of the radiative muon decay can be written

$$\mathcal{A}(\mu^- \rightarrow e^- \gamma) = \epsilon^\rho < e|J_\rho^{em}|\mu >, \tag{12.146}$$

where ϵ^ρ is the photon polarization and the Lorentz vector current operator is represented in the form

$$< e|J_\rho^{em}|\mu >= \overline{u}(p-q)[iq^\nu \sigma_{\rho\nu}(A + B\gamma_5) + \gamma_\rho(C + D\gamma_5) + q_\rho(E + F\gamma_5)]u_\mu(p), \tag{12.147}$$

where A, B, C, D, E, and F are invariant amplitudes. From the electromagnetic current conservation law

$$\partial^\rho J_\rho^{em} = 0$$

it is evident that

$$-m_e(C + D\gamma_5) + m_\mu(C - D\gamma_5) = 0. \tag{12.148}$$

where we have allowed for $q^2 = 0$. Making approximation $m_e = 0$, we get from (12.148)

$$C = D = 0. \tag{12.149}$$

Taking into consideration the transversality of the photon polarization $\epsilon^\rho q_\rho = 0$ as well, we arrive at the conclusion that the $\mu^- \rightarrow e^- \gamma$-decay amplitude corresponds to the magnetic transition

$$\mathcal{A}(\mu^- \rightarrow e^- \gamma) = \epsilon^\rho \overline{u}(p-q)[iq^\nu \sigma_{\rho\nu}(A + B\gamma_5)]u_\mu(p). \tag{12.150}$$

The lowest-order diagrams contributing to the amplitude of the radiative muon decay in the R_ξ gauge are displayed in Fig. 12.12. Since we know that the final amplitude must have the form of the magnetic transition (12.150), our strategy is to ignore all terms that cannot be reduced to the magnetic moment term. This means that there is no need to calculate the diagrams in Fig. 12.12(e) since they are all proportional to $\overline{u}_e \gamma^\rho u_\mu$, and will be cancelled by terms of a similar form coming from the diagrams in Fig. 12.12(a)–(d). Next, for simplicity, we shall neglect the electron mass. Then, because the final-state electron is left-handed (it is e_L that interacts with W^- and Φ^- and the helicity cannot be flipped in the limit $m_e = 0$), then two $\mu^- \rightarrow e^- \gamma$ invariant amplitudes are equal

$$A = B.$$

The attention is drawn to the fact that the virtual neutrinos on the diagrams of Fig. 12.12 represent the ones with definite mass values ν_i ($i = 1, 2, 3$), which are related to the flavor neutrinos by means of the matrix $\mathcal{M}^{NM}$

$$\nu_l(x) = \mathcal{M}_{li}^{NM} \nu_i(x). \tag{12.151}$$

It should be stressed, if the neutrino mixing matrix were diagonal or the neutrino masses were equal to zero, then the decay $\mu^- \rightarrow e^- \gamma$ would be forbidden.

The task may be simplified, when in the resulting matrix element, instead of the term that is proportional to $\epsilon^\rho q^\nu \sigma_{\rho\nu}$, one singles out a simpler combination. This could be realized with the help of the Gordon decomposition

$$\overline{u}(p)\gamma_\mu u(k) = \frac{1}{2m}\overline{u}(p)\left[(p+k)_\mu + i\sigma_{\mu\nu}(p-k)^\nu\right]u(k), \tag{12.152}$$

$$\overline{u}(p)\gamma_\mu\gamma_5 u(k) = \frac{1}{2m}\overline{u}(p)\left[(p-k)_\mu\gamma_5 + i\sigma_{\mu\nu}(p+k)^\nu\gamma_5\right]u(k). \tag{12.153}$$

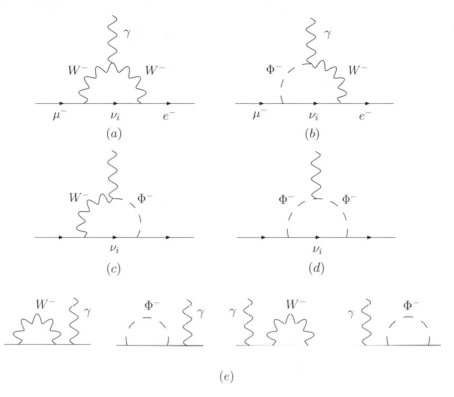

FIGURE 12.12
The diagrams (a)–(e) corresponding to the radiative muon decay in the R_ξ gauge.

These relations can easily be proved using the Dirac equation. Really, acting on

$$\hat{k}u(k) = mu(k) \tag{12.154}$$

by the operator γ_μ, in the left side of (12.154) we get

$$\frac{1}{2}\left([\gamma_\mu, \gamma_\rho] + \{\gamma_\mu, \gamma_\rho\}\right)k^\rho u(k) = \left(-i\sigma_{\mu\rho}k^\rho + k_\mu\right)u(k). \tag{12.155}$$

Combining (12.154) with (12.155), we arrive at

$$\left(-i\sigma_{\mu\rho}k^\rho + k_\mu\right)u(k) = \gamma_\mu mu(k). \tag{12.156}$$

In much the same way, we have for the conjugate spinor $\overline{u}(p)$

$$\overline{u}(p)\left(-i\sigma_{\rho\mu}p^\rho + p_\mu\right) = \overline{u}(p)m\gamma_\mu. \tag{12.157}$$

Multiplying Eq. (12.156) by $\overline{u}(p)$ on the right and Eq. (12.157) by $u(k)$ on the left and adding the results obtained, we lead to the relation (12.152).

Allowing for Eqs. (12.152) and (12.153), we find the expression for the decay amplitude

$$\mathcal{A}(\mu^- \to e^-\gamma) = \overline{u}(p-q)\left(A\epsilon^\rho iq^\nu \sigma_{\rho\nu}(1+\gamma_5)\right)u_\mu(p) =$$

$$= \overline{u}(p - q)\left\{ A(1 + \gamma_5)[2(\epsilon \cdot p) - m_\mu(\epsilon \cdot \gamma)]\right\}u_\mu(p). \tag{12.158}$$

So, in our calculation of the invariant amplitude $\mathcal{A}$ we need only concentrate on the $(\epsilon \cdot p)$ term. The momentum assignments of the diagrams of Fig. 12.12(a)–(d) are pictured in Fig. 12.13. The matrix element describing the diagram shown in Fig. 12.12(a) has the form

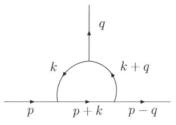

FIGURE 12.13

The momentum assignments of the diagrams associated with the radiative muon decay.

$$\mathcal{A}^{(a)}_{\nu_j} = -i \int \frac{d^4k}{(2\pi)^4}\left[\overline{u}_e(p - q)\left(\frac{ig}{2\sqrt{2}}\right)\mathcal{M}^*_{ej}\gamma^\rho(1 - \gamma_5)\frac{i}{\hat{p} + \hat{k} - m_j}\left(\frac{ig}{2\sqrt{2}}\right)\mathcal{M}_{\mu j}\gamma^\nu \times\right.$$

$$\left. \times (1 - \gamma_5)u_\mu(p)\right]\left[-D^W_{\nu\beta}(k)D^W_{\rho\alpha}(k + q)\right](-ie)\Lambda^{\tau\alpha\beta}\epsilon_\tau, \tag{12.159}$$

where $\Lambda_{\tau\alpha\beta}$ is the $WW\gamma$-vertex

$$\Lambda_{\tau\alpha\beta}(-q, k + q, -k) = g_{\alpha\beta}(2k + q)_\tau + g_{\tau\beta}(q - k)_\alpha - g_{\tau\alpha}(2q + k)_\beta, \tag{12.160}$$

and $D^W_{\nu\beta}$ is the W boson propagator in the R_ξ gauge, which is conveniently taken in the form (see Eq. (12.39))

$$D^W_{\mu\nu}(p) = -\frac{1}{p^2 - m^2_W}\left\{g_{\mu\nu} + (\xi^2 - 1)\frac{p_\mu p_\nu}{p^2 - \xi^2 m^2_W}\right\} = D^W_{\mu\nu}(k) + D^W_{1\mu\nu}(k), \tag{12.161}$$

where

$$D^W_{\mu\nu}(k) = -\frac{1}{p^2 - m^2_W}\left(g_{\mu\nu} - \frac{p_\mu p_\nu}{m^2_W}\right), \qquad D^W_{2\mu\nu}(p) = -\frac{p_\mu p_\nu}{m^2_W(p^2 - \xi^2 m^2_W)}, \tag{12.162}$$

There are three such matrix elements associated with three virtual neutrinos $\nu_{1,2,3}$. Under summation of these diagrams there appears the factor $F_{e\mu}$

$$F_{e\mu} = \sum_j\left[\frac{\mathcal{M}^*_{ej}\mathcal{M}_{\mu j}}{(p + k)^2 - m^2_j}\right].$$

Expanding it as a power series in m^2_j and allowing for

$$\sum_j \mathcal{M}^*_{ej}\mathcal{M}_{\mu j} = 0, \tag{12.163}$$

we get

$$F_{e\mu} = \sum_j \mathcal{M}_{ej}\mathcal{M}_{\mu j}^* \left[\frac{m_j^2}{(p+k)^4} + \dots \right]. \tag{12.164}$$

Note, Eq. (12.163) represents the analog of the GIM mechanism in the neutrino sector. To substitute (12.164) into the amplitudes corresponding to the diagrams of Fig. 12.12(a) gives

$$\mathcal{A}^{(a)} = \sum_j \mathcal{A}_j^{(a)} = \frac{ig^2 e}{4} \sum_j \mathcal{M}_{ej}^* \mathcal{M}_{\mu j} m_j^2 \int \frac{d^4 k}{(2\pi)^4} \frac{1}{(p+k)^4} \left[D_{\nu\beta}^W(k) D_{\rho\alpha}^W(k+q) N^{\rho\nu} \Lambda^{\alpha\beta} \right], \tag{12.165}$$

where

$$\Lambda^{\alpha\beta} = \epsilon_\tau \Lambda^{\tau\alpha\beta}(-q, k+q, -k)$$

and

$$N^{\rho\nu} = \bar{u}_e(p-q)\gamma^\rho(\hat{p}+\hat{k})\gamma^\nu(1-\gamma_5)u_\mu(p).$$

Taking into consideration

$$D_{2\nu\beta}^W(k) D_{2\rho\alpha}^W(k+q)\Lambda^{\alpha\beta} = 0, \tag{12.166}$$

we obtain for the matrix element (12.165)

$$\mathcal{A}^{(a)} = \frac{ig^2 c}{4} \sum_j \mathcal{M}_{ej}^* \mathcal{M}_{\mu j} m_j^2 \int \frac{d^4 k}{(2\pi)^4} \frac{1}{(p+k)^4} \left\{ \frac{F_1 - F_2 - F_3}{(k^2 - m_W^2)[(k+q)^2 - m_W^2]} + \right.$$

$$\left. + \frac{F_2}{(k^2 - \xi^2 m_W^2)[(k+q)^2 - \xi^2 m_W^2]} + \frac{F_3}{(k^2 - m_W^2)[(k+q)^2 - \xi^2 m_W^2]} \right\}, \tag{12.167}$$

where

$$F_1 = \Lambda^{\rho\nu} N_{\rho\nu}, \qquad F_2 = \frac{k^\lambda \Lambda_\lambda^\rho k^\nu N_{\rho\nu}}{m_W^2}, \qquad F_3 = \frac{(k+q)^\lambda \Lambda_\lambda^\rho (k+q)^\nu N_{\rho\nu}}{m_W^2}.$$

In order to calculate the integral in (12.167) we use the Feynman parametrization. To differentiate the basic formula

$$\frac{1}{\alpha_1 \dots \alpha_n} = (n-1)! \int_0^1 d\eta_1 \int_0^1 d\eta_2 \dots \int_0^1 d\eta_n \frac{\delta(\eta_1 + \eta_2 + \dots + \eta_n - 1)}{[\alpha_1 \eta_1 + \alpha_2 \eta_2 + \dots + \alpha_n \eta_n]^n} \tag{12.168}$$

with respect to α_1 results in

$$\frac{1}{\alpha_1^2 \dots \alpha_n} =$$

$$= n! \int_0^1 \eta_1 d\eta_1 \int_0^1 d\eta_2 \dots \int_0^1 \frac{d\eta_{n-1}}{[\alpha_1 \eta_1 + \alpha_2 \eta_2 + \dots + \alpha_n(1 - \eta_1 - \eta_2 - \dots - \eta_{n-1})]^{n+1}}. \tag{12.169}$$

Making use of (12.169), shifting the integration variable k and keeping only terms proportional to $(p \cdot \epsilon)$, we arrive at

$$\mathcal{A}^{(a)} = \frac{3! ig^2 e}{4} \sum_j \mathcal{M}_{ej}^* \mathcal{M}_{\mu j} m_j^2 \int_0^1 \eta_1 d\eta_1 \int_0^1 d\eta_2 \left\{ \int \frac{d^4 k}{(2\pi)^4} \left[\frac{F_1' - F_2' - F_3'}{(k^2 - a^2)^4} + \right. \right.$$

$$\left. \left. + \frac{F_2'}{(k^2 - b^2)^4} + \frac{F_3'}{(k^2 - c^2)^4} \right] \right\}, \tag{12.170}$$

where

$$\left.\begin{array}{l} a^2 = (1 - \eta_1)m_W^2 + \ldots, \\ b^2 = [(1 - \eta_1 - \eta_2)\xi^2 + \eta_2]m_W^2 + \ldots, \\ c^2 = [(1 - \eta_1 - \eta_2) + \xi^2\eta_2]m_W^2 + \ldots \end{array}\right\} \tag{12.171}$$

(since the final result does not depend on the quantities a, b, and c, there is no need to give concrete expression to them),

$$\left.\begin{array}{l} F_1' = 2m_\mu(p \cdot \epsilon)[\bar{u}_e(1 + \gamma_5)u_\mu][2(1 - \eta_1)^2 + \eta_2(2\eta_1 - 1)], \\[2mm] F_2' = -\dfrac{k^2 m_\mu}{m_W^2}(p \cdot \epsilon)[\bar{u}_e(1 + \gamma_5)u_\mu]\{(3\eta_2 - 1) + [2\eta_1^2 - \eta_1 + \eta_2(2\eta_1 - \tfrac{1}{2})]\}, \\[2mm] F_3' = -\dfrac{k^2 m_\mu}{m_W^2}(p \cdot \epsilon)[\bar{u}_e(1 + \gamma_5)u_\mu][2\eta_1^2 + \eta_1 + \eta_2(2\eta_1 - \tfrac{1}{2})]. \end{array}\right\} \tag{12.172}$$

Carrying the integration over the four-dimensional momenta

$$\int \frac{d^4k}{(2\pi)^4}\frac{1}{(k^2 - a^2)^4} = \frac{i}{96\pi^2 a^4}, \qquad \int \frac{d^4k}{(2\pi)^4}\frac{k^2}{(k^2 - a^2)^4} = \frac{-i}{48\pi^2 a^2}, \tag{12.173}$$

and then over the Feynman parameters η_1 and η_2, we find that the contribution caused by the diagrams shown in Fig. 12.12a into the invariant amplitude $\mathcal{A}$ is governed by

$$\mathcal{A}^{(a)} = \frac{m_\mu g^2 e}{256\pi^2 m_W^4}\sum_j \mathcal{M}_{ej}^* \mathcal{M}_{\mu j}m_j^2\left[1 - \frac{2\ln\xi}{3(\xi^2 - 1)} + \frac{1}{\xi^2 - 1}\left(\frac{2\xi^2\ln\xi}{\xi^2 - 1} - 1\right)\right]. \tag{12.174}$$

Now we find the contribution coming from the diagrams of Fig. 12.12(b). The matrix element associated with these diagrams has the form

$$\mathcal{A}^{(b)} = -i\sum_j \int \frac{d^4k}{(2\pi)^4}\left[\bar{u}_e(p - q)\left(\frac{ig}{2\sqrt{2}}\right)\mathcal{M}_{ej}^*\gamma^\rho(1 - \gamma_5)\frac{i}{\hat{p} + \hat{k} - m_j}\left(\frac{-ig}{2\sqrt{2}m_W}\right)\mathcal{M}_{\mu j}\times\right.$$

$$\left.\times[m_j(1 - \gamma_5) - m_\mu(1 + \gamma_5)]u_\mu(p)\right]\left(-iD_{\rho\tau}^W(k + q)\frac{1}{k^2 - \xi^2 m_W^2}\right)(iem_W\epsilon^\tau). \tag{12.175}$$

With allowance made for Eqs. (12.164) and (12.166), as well as the approximate relation

$$\sum_j \frac{\mathcal{M}_{ej}^* \mathcal{M}_{\mu j}m_j^2}{(p + k)^2 - m_j^2} \approx \sum_j \frac{\mathcal{M}_{ej}^* \mathcal{M}_{\mu j}m_j^2}{(p + k)^2}, \tag{12.176}$$

we rewrite Eq. (12.175) in the following manner:

$$\mathcal{A}^{(b)} = \frac{ig^2 e}{4}\sum_j \mathcal{M}_{ej}^* \mathcal{M}_{\mu j}m_j^2 \int \frac{d^4k}{(2\pi)^4}\frac{N_\lambda}{(p + k)^4}\left(\frac{1}{k^2 - \xi^2 m_W^2}\right)\left\{\frac{[\epsilon^\lambda - (\epsilon \cdot k)(k + q)^\lambda/m_W^2]}{(k + q)^2 - m_W^2} + \right.$$

$$\left. + \frac{(\epsilon \cdot k)(k + q)^\lambda/m_W^2}{(k + q)^2 - \xi^2 m_W^2}\right\}, \tag{12.177}$$

where

$$N_\lambda = \bar{u}_e(p - q)(1 + \gamma_5)\gamma_\lambda[(p + k)^2 - m_\mu(\hat{k} + m_\mu)]u_\mu(p). \tag{12.178}$$

Combining the denominators with the help of the Feynman parametrization, shifting the integration variable k and being constrained by terms proportional to $(p \cdot \epsilon)$, we get

$$\mathcal{A}^{(b)} = \frac{3ig^2e}{2} \sum_j \mathcal{M}_{ej}^* \mathcal{M}_{\mu j} m_j^2 \int_0^1 \eta_1 d\eta_1 \int_0^1 d\eta_2 \int \frac{d^4k}{(2\pi)^4} \left[\frac{N_1' - N_2'}{(k^2 - b^2)^4} + \frac{N_2'}{(k^2 - a^2)^4} \right], \tag{12.179}$$

$$\left. \begin{aligned} N_1' &= -2m_\mu(p \cdot \epsilon)[\overline{u}_e(1 + \gamma_5)u_\mu]\eta_2, \\ N_2' &= \frac{k^2 m_\mu(p \cdot \epsilon)}{2}[\overline{u}_e(1 + \gamma_5)u_\mu][(1 - 4\eta_1)(1 - \eta_1 - \eta_2) - (1 - 3\eta_1)]. \end{aligned} \right\} \tag{12.180}$$

Now we remain only to fulfill the integration over the four-dimensional momenta and Feynman parameters. As a result of these operations, we have

$$\mathcal{A}^{(b)} = \frac{m_\mu g^2 e}{256\pi^2 m_W^4} \sum_j \mathcal{M}_{ej}^* \mathcal{M}_{\mu j} m_j^2 \left[\frac{5}{6\xi^2} + \frac{8\ln\xi}{3(\xi^2 - 1)} - \frac{7}{3(\xi^2 - 1)}\left(\frac{2\xi^2 \ln\xi}{\xi^2 - 1} - 1 \right) \right]. \tag{12.181}$$

The calculation of the matrix elements corresponding to the diagrams pictured in Fig. 12.12(c) are carried out by the total analogy with the calculations of the previous diagrams set. The final result is

$$\mathcal{A}^{(J)} = \frac{m_\mu g^2 e}{256\pi^2 m_W^4} \sum_j \mathcal{M}_{ej}^* \mathcal{M}_{\mu j} m_j^2 \left[\frac{5}{6\xi^2} - \frac{2\ln\xi}{\xi^2 - 1} + \frac{1}{3(\xi^2 - 1)}\left(\frac{2\xi^2 \ln\xi}{\xi^2 - 1} - 1 \right) \right]. \tag{12.182}$$

Proceed to the calculation of the last diagram (Fig. 12.12(d)). The corresponding matrix element is given by the expression:

$$\mathcal{A}^{(c)} = -i \sum_j \int \frac{d^4k}{(2\pi)^4} \left[\overline{u}_e(p - q)\left(\frac{ig}{2\sqrt{2}m_W} \right) \mathcal{M}_{ej}^*[m_j(1 + \gamma_5) - m_e(1 - \gamma_5)] \frac{i}{\hat{p} + \hat{k} - m_j} \times \right.$$

$$\left. \times \left(\frac{-ig}{2\sqrt{2}m_W} \right) \mathcal{M}_{\mu j}[m_j(1 - \gamma_5) - m_\mu(1 + \gamma_5)]u_\mu(p)\right]\left(\frac{i}{k^2 - \xi^2 m_W^2} \right) \times$$

$$\times \frac{i}{(k + q)^2 - \xi^2 m_W^2} ie[\epsilon \cdot (2k + q)]. \tag{12.183}$$

Neglecting the electron mass and employing the relation (12.176), we obtain

$$\mathcal{A}^{(c)} = \frac{-ig^2e}{4m_W^2} \sum_j \mathcal{M}_{ej}^* \mathcal{M}_{\mu j} m_j^2 \int \frac{d^4k}{(2\pi)^4} [\overline{u}_e(p - q)(1 + \gamma_5)\hat{k}u_\mu(p)] \frac{2(k \cdot \epsilon)}{(p + k)^2} \times$$

$$\times \left(\frac{1}{k^2 - \xi^2 m_W^2} \right) \frac{1}{(k + q)^2 - \xi^2 m_W^2}. \tag{12.184}$$

Again, combining the denominators, shifting the integration variable k, keeping only a term proportional to $(p \cdot \epsilon)$, and integrating over the four-dimensional momenta, we lead to the result

$$\mathcal{A}^{(c)} = \frac{-ig^2 e m_\mu(p \cdot \epsilon)}{m_W^2} \sum_j \mathcal{M}_{ej}^* \mathcal{M}_{\mu j} m_j^2 [\overline{u}_e(p - q)(1 + \gamma_5)u_\mu(p)] \int_0^1 \eta_1 d\eta_1 \times$$

$$\times \int_0^1 d\eta_2(\eta_1 + \eta_2)\left(\frac{-i}{32\pi^2 d^2} \right), \tag{12.185}$$

where $d^2 = (1 - \eta_1)\xi^2 m_W^2$. Having done the integration over the Feynman parameters, we find

$$A^{(d)} = -\frac{m_\mu g^2 e}{128\pi^2 m_W^4} \sum_j \mathcal{M}_{ej}^* \mathcal{M}_{\mu j} m_j^2 \left(\frac{5}{6\xi^2}\right). \tag{12.186}$$

To sum up Eqs. (12.174), (12.181), (12.182), and (12.186) gives the final expression for the invariant amplitude $\mathcal{A}$

$$\mathcal{A} = \frac{eg^2 m_\mu}{256\pi^2 m_W^2}\delta_\nu, \tag{12.187}$$

where δ_ν is the GIM-suppressed factor:

$$\delta_\nu = \sum_j \mathcal{M}_{ej}^* \mathcal{M}_{\mu j} \left(\frac{m_j^2}{m_W^2}\right). \tag{12.188}$$

It is interesting to note that all the diagrams give different contributions in various gauges. However, the total contribution of all the considered lowest-order diagrams is gauge-invariant as it must. Making use of (12.150), we find that the total width of the radiative muon decay is given by the expression:

$$\Gamma(\mu^- \to e^-\gamma) = \frac{m_\mu^3}{8\pi}(|A|^2 + |B|^2). \tag{12.189}$$

With allowance made for (12.187), the expression for the radiative muon decay branching takes the form

$$\mathrm{Br}(\mu^- \to e^-\gamma) = \frac{\Gamma(\mu^- \to e^-\gamma)}{\Gamma(\mu^- \to e^-\nu_\mu\bar{\nu}_e)} = \frac{3\alpha}{32\pi}\delta_\nu^2. \tag{12.190}$$

We recall the existing limits on the light-neutrino masses. Measurement of the endpoint of the tritium-beta-decay spectrum yields an absolute upper limit of 2 eV on the electron-antineutrino mass (at a 95% C.L.(confidence level)) [23]. This value, together with the results of investigations of neutrino oscillations in the scheme involving three CPT-invariant light neutrinos, also provides an upper limit on the mass of all active neutrinos. Using this upper limit gives a hopelessly small value of $\mathrm{Br}(\mu^- \to e^-\gamma) < 10^{-48}$.

One could expect that the existence of the neutrino dipole magnetic moment (DMM), which it acquires owing to the neutrino interaction with the vacuum, whose structure is determined by the choice of the electroweak interaction model, changes the situation. The current experimental date allow sufficiently large values for the neutrino DMM

$$\mu_\nu \sim 10^{-10}\mu_B, \tag{12.191}$$

where μ_B is the Bohr magneton. However, as the calculations show [24], the inclusion of the neutrino DMM does not change the state of affairs and we again have the infinitesimally small branching

$$(\mathrm{Br}^n)^{SM}(\mu^- \to e^-\gamma) \approx 4 \times 10^{-47}.$$

One may consider the radiative muon decay in the model relying on the same gauge group as the SM and involving two doublets of Higgs fields (two Higgs doublet model or 2HDM) [25]. However, in this case, the inclusion of the neutrino DMM leads to the similar result [24]. It follows that, from the point of view of the SM, the radiative muon decay is an unobservable effect. The detection of this process will be a signal of physics beyond the SM.

Until recently, the muon was a basic probe of the weak interaction theory. Obviously, searches of weak interaction deviations from the $V - A$ structure are basically related with studying muon properties. Let us briefly discuss one of directions of these investigations. We shall take into account the experiments both on the main decay channel

$$\mu^- \to e^- + \nu_\mu + \overline{\nu}_e \tag{12.192}$$

(we call it a direct muon decay) and on the so-called inverse muon decay

$$\nu_\mu + e^- \to \mu^- + \nu_e \tag{12.193}$$

which will be studied in Section 12.6. Write the most general form of the amplitudes of the reactions (12.192), (12.193) and ask: *To what extent do the experimental data constrain the $V - A$ structure of the charged currents?* When building the amplitudes of these processes, we shall assume that they are Lorentz-invariant, linear on four Dirac spinors (or conjugate ones), and does not contain derivatives of these spinors. Then, it turns out that all the results in the direct muon decay (energy spectra of electrons and neutrinos, polarizations, and angular distributions) and in the inverse muon decay (the reaction cross section) at energies well below m_W^2, may be parametrized in terms of amplitudes $g_{\epsilon\tau}^i$ and the Fermi coupling constant G_F, using the matrix element

$$\mathcal{A} = \frac{4G_F}{\sqrt{2}} \sum_i g_{\epsilon\tau}^i < \overline{e}_\epsilon |\Gamma^i|(\nu_e)_n >< (\overline{\nu}_\mu)_m |\Gamma_i|\mu_\tau >, \tag{12.194}$$

where $i = S, V, T$ indicates a scalar, vector, or tensor interaction, $\epsilon, \tau = L, R$ indicate the right- or left-handed chirality of the electron or muon. The chiralities n and m of the ν_e and $\overline{\nu}_\mu$ are then determined by the values of i, ϵ, and τ. In Ref. [26] it was shown that the inclusion of the lepton flavor violation also leads to the matrix element equivalent to (12.194). If one does not a priori assume the invariance with respect to the time inversion, then all the amplitudes $g_{\epsilon\tau}^i$ may be complex. The ten amplitudes g_{RR}^T and g_{LL}^T are identically zero. Thus, we are left with 19 independent real parameters (including G_F) to be determined by experiment. Note, the SM interaction corresponds to one single amplitude g_{LL}^V being unity and all the others being zero. Below we give the current experimental limits on the amplitudes $g_{\epsilon\tau}^i$ taken from Refs. [27] and [28].

$$
\left.
\begin{array}{lll}
|g_{RR}^S| < 0.067 & |g_{RR}^V| < 0.034 & |g_{RR}^T| \equiv 0 \\[4pt]
|g_{LR}^S| < 0.088 & |g_{LR}^V| < 0.036 & |g_{LR}^T| < 0.025 \\[4pt]
|g_{RL}^S| < 0.417 & |g_{RL}^V| < 0.104 & |g_{RL}^T| < 0.104 \\[4pt]
|g_{LL}^S| < 0.550 & |g_{LL}^V| > 0.960 & |g_{LL}^T| \equiv 0 \\[8pt]
\quad |g_{LR}^S + 6g_{LR}^T| < 0.143 & |g_{RL}^S + 6g_{RL}^T| < 0.418 & \\[4pt]
\quad |g_{LR}^S + 2g_{LR}^T| < 0.108 & |g_{RL}^S + 2g_{RL}^T| < 0.417 & \\[4pt]
\quad |g_{LR}^S - 2g_{LR}^T| < 0.070 & |g_{RL}^S - 2g_{RL}^T| < 0.418.
\end{array}
\right\} \tag{12.195}
$$

12.6 $(g-2)_\mu$ anomaly

The anomalous magnetic moment (AMM) of particles, $a_p = (g-2)_p/2$, arise from quantum effects. Their precise measurement has historically played an important role in the development of particle physics. For example, precise measurements of the electron AMM, together with measurements of the hyperfine structure of hydrogen and the Lamb shift, played an important role in the development of QED and the theory of renormalizations.

The detection of nucleon AMMs proved to be a compelling argument in favor of the Yukawa π-meson theory of nuclear forces.

The electron AMM a_e has been measured to within about four parts per billion (ppb), and is thus among the most accurately known quantities in physics. All three charged leptons, e, μ, and τ, of the SM can be treated on the same footing, except that the very different values of their masses will induce different sensitivities with respect to the mass scales involved in the higher-order radiative corrections. The electron AMM is almost only sensitive to the electromagnetic interactions of the leptons, and its value is barely affected by the strong or weak interactions. On the other hand, the strengths of the latter two types of corrections are enhanced by a considerable factor $\sim (m_\tau/m_e)^2 = 1.2 \times 10^7$ and $\sim (m_\mu/m_e)^2 = 4 \times 10^4$ as compared to the electron in the case of the τ and the muon, respectively. Because the same huge enhancement factor would also affect the contributions coming from degrees of freedom beyond the SM, then the measurement of the τ AMM would represent the best opportunity for detecting a new physics signal. Unfortunately, the very short lifetime of the τ makes such a measurement impossible at present. The muon lies somewhat in the intermediate range of mass scales and its lifetime still makes a measurement of its AMM possible.

It was expected that measurements of the muon AMM would also be a crucial point in particle physics. A compilation of the major experimental efforts in measuring a_μ over the last five decades is given in Table 12.2. Starting from the experiment at the Columbia-Nevis cyclotron, where the spin rotation of a muon in a magnetic field was observed for the first time, the experimental precision of a_μ has seen constant improvement first through three experiments at CERN in the 1960s and 1970s and more recently with E821 at Brookhaven National Laboratory (BNL) on Alternating Gradient Synchrotron. The current world average value reaches a relative precision of 0.54 ppm [29]. The muon AMM has been measured

TABLE 12.2
The basic experiments in measuring a_μ.

Experiment	Beam	Measurement	$\delta a_\mu/a_\mu$
Columbia-Nevis (1957)	μ^+	$g = 2.00 \pm 0.10$	
Columbia-Nevis (1959)	μ^+	$0.001\,13^{+(16)}_{-(12)}$	12.4%
CERN 1 (1961)	μ^+	$0.001\,145(22)$	1.9%
CERN 1 (1962)	μ^+	$0.001\,162(5)$	0.43%
CERN 2 (1968)	$\mu^\pm$	$0.001\,166\,16(31)$	265 ppm
CERN 3 (1975)	$\mu^\pm$	$0.001\,165\,895(27)$	23 ppm
CERN 3 (1979)	$\mu^\pm$	$0.001\,165\,911(11)$	7.3 ppm
BNL E821 (2000)	μ^+	$0.001\,165\,919\,1(59)$	5 ppm
BNL E821 (2001)	μ^+	$0.001\,165\,920\,2(16)$	1.3 ppm
BNL E821 (2002)	μ^+	$0.001\,165\,920\,3(8)$	0.7 ppm
BNL E821 (2004)	μ^-	$0.001\,165\,921\,4(8)(3)$	0.7 ppm
World Average (2004)	$\mu^\pm$	$0.001\,165\,920\,80(63)$	0.54 ppm

at CERN with the help of a muon storage ring. The experiments E821 are based on the same concept. Pions are produced by sending a proton beam on a target. The pions subsequently decay into longitudinally polarized muons, which are captured inside a storage ring, where they follow a circular orbit in the presence of both a uniform magnetic field and a quadrupole electric field, the latter serving the purpose of stabilizing the orbits. The cyclotron (orbit) ω_c and spin precession ω_s frequencies for a muon moving in the horizontal plane of a magnetic storage ring are given by:

$$\boldsymbol{\omega}_c = -\frac{q\mathbf{B}}{m_\mu \gamma}, \qquad \boldsymbol{\omega}_s = -\frac{qg\mathbf{B}}{2m_\mu} - (1-\gamma)\frac{e\mathbf{B}}{m_\mu \gamma},$$

where q is the muon charge, g is the gyromagnetic ratio, $\gamma = 1/\sqrt{1-\mathbf{v}^2}$ and $\mathbf{v}$ is the muon velocity. The anomalous precession frequency ω_a is determined from the difference

$$\boldsymbol{\omega}_a = \boldsymbol{\omega}_s - \boldsymbol{\omega}_c = -\left(\frac{g-2}{2}\right)\frac{q\mathbf{B}}{m_\mu} = -a_\mu \frac{q\mathbf{B}}{m_\mu}. \tag{12.196}$$

Because electric quadropoles are used to provide vertical focusing in the storage ring, their electric field is seen in the muon rest frame as a motional magnetic field that can affect the spin precession frequency. In the presence of both $\mathbf{E}$ and $\mathbf{B}$ fields, and in the case that $\mathbf{v}$ is perpendicular to both $\mathbf{E}$ and $\mathbf{B}$, the expression for the anomalous precession frequency becomes

$$\boldsymbol{\omega}_a = \boldsymbol{\omega}_s - \boldsymbol{\omega}_c = -\frac{q}{m_\mu}\left\{ a_\mu \mathbf{B} - [\mathbf{v}\times\mathbf{E}](a_\mu + \frac{1}{1-\gamma^2}) \right\}. \tag{12.197}$$

The term in parentheses multiplying $[\mathbf{v}\times\mathbf{E}]$ vanishes at $\gamma_{mag} = 29.3$ (which corresponds to a muon momentum $p = 3.094$ GeV) and the electrostatic focusing does not affect the spin (except for a correction necessary to account for the finite momentum range $\Delta p/p \approx \pm 0.14\%$ around the magic momentum). The magic energy $E_{mag} = \gamma_{mag} m_\mu$ is the energy for which

$$\frac{1}{\gamma_{mag}^2 - 1} = a_\mu.$$

The existence of a solution is due to the fact that a_μ is a positive constant in competition with an energy dependent factor of the opposite sign (as $\gamma \geq 1$). The second miracle, which is crucial for the feasibility of the experiment, is the fact that $\gamma_{mag} = \sqrt{(1+a_\mu)/a_\mu} \approx 29.378$ is large enough to provide the time dilatation factor for the unstable muon boosting the lifetime $\tau_\mu \approx 2.197 \times 10^6$ sec to $\tau_{in\ flight} = \gamma\tau_\mu \approx 6.454 \times 10^{-5}$ sec, which allows the muons, travelling at $v = 0.99942\ldots$, to be stored in a ring of reasonable size (diameter ~ 14 m). Eq. (12.197) can be rearranged to isolate a_μ, giving $\omega_a/ <B> $ ($<B>$—average magnetic field) multiplied by physical quantities that are known for high precision.

The spin direction of the muon is determined by detecting the electrons or positrons produced in the decay of the muons with an energy greater than threshold energy E_T. The number of detected electrons or positrons $N_e(t)$ decreases exponentially with time and is modulated by the frequency ω_a,

$$N_e(t) = N_0(E_T)\left\{1 + A(E_T)\cos[\omega_a t + \phi(E_T)]\right\}\exp[-\frac{t}{\gamma\tau_\mu}], \tag{12.198}$$

where the normalization N_0, asymmetry A, and phase ϕ_a vary with the chosen energy threshold. The observation of this time dependence thus provides the required measurement of ω_a. The key to the experiment is to determine this frequency to a high precision and to measure the average magnetic field to an equal or better precision.

At BNL measurements of nearly equal samples of positive and negative muons were used. It enabled one to obtain the world-average value of the muon AMM with an accuracy of 0.54 ppm, which represents a 14-fold improvement compared to previous measurements at CERN. In order for a theory to match such an accurate measurement, calculations in the

SM have to be pushed to their very limits. It means that the contributions of all sectors of the SM (or its extension) have to be known very precisely. Therefore, the SM expression for the muon AMM, a_μ^{SM}, should include the following terms

$$a_\mu^{SM} = a_\mu^{QED} + a_\mu^{EW} + a_\mu^{had}, \tag{12.199}$$

where a_μ^{QED}, a_μ^{EW}, and a_μ^{had} are contributions from electromagnetic, electroweak, and hadron sectors, respectively.

The electromagnetic corrections produce the main contributions, their accuracy exceeding the experimental accuracy. The Feynman diagrams determining a_μ^{QED} at a given order of the perturbation theory expansion (powers of α/π) can be divided into four classes: (i) diagrams with virtual photon and muon loops; (ii) vacuum polarization diagrams from electron loops; (iii) vacuum polarization diagrams from tau loops; and (iv) light-by-light scattering diagrams from lepton loops. All together, the purely QED contributions to the muon AMM have been computed through four-loops, that is, with the precision of $(\alpha/\pi)^4$ and estimated at the five-loop level. It should be noted that the calculations with such a precision are very complicated. For example, under calculation of (i) at the four-loop level, there are 891 Feynman diagrams. Some of them are already known analytically, but in general one has to resort to numerous methods for a complete evaluation. This impressive calculation, which requires many technical skills, is under constant updating due to advances in computing technology. Since the QED contribution depends directly on a power series expansion in the fine-structure constant α, a meaningful comparison requires that α is known from an independent experiment with the same relative precision as a_μ^{QED}. The most recent determination of α from the comparison between QED and a new measurement of the electron g-factor yields

$$\alpha^{-1} = 137.035999070(98).$$

The numerical result for the whole QED contribution using this precise value of α as an input is given by

$$a_\mu^{QED} = 116\,584\,718.10(0.16) \times 10^{-11}, \tag{12.200}$$

where the error results from uncertainties in theoretical calculation of a_μ^{QED} and in the determination of α.

The next contribution a_μ^{EW} is related with loops involving the gauge bosons and physical Higgs boson. As we already knew (see Section 21.7) of *Advanced Particle Physics Volume 1*, for AMM of a particle to be calculated, it is enough to consider vertex functions and isolate a term that is proportional to $i\sigma^{\nu\mu}q_\nu/(2m)$. In Fig. 12.14 the corresponding Feynman lowest-order diagrams in the R_ξ gauge are displayed. The vertex function $\Lambda_\beta(p, p')$ associated with the diagram shown in Fig. 12.14(a) is defined by the expression

$$\Lambda_\beta^{(Z)}(p, p') = \frac{-ig^2}{8\cos^2\theta_W} \int \frac{d^4k}{(4\pi)^4} \gamma_\sigma(1 - 4\sin^2\theta_W - \gamma_5) \frac{[(\hat{p}' - \hat{k}) + m]\gamma_\beta[(\hat{p} - \hat{k}) + m]}{[(p' - k)^2 - m^2][(p - k)^2 - m^2]} \times$$

$$\times \gamma_\nu(1 - 4\sin^2\theta_W - \gamma_5)\left[g^{\sigma\nu} - (1 - \xi_Z^2)\frac{k^\sigma k^\nu}{k^2 - \xi_Z^2 m_Z^2}\right]\frac{1}{k^2 - m_Z^2}, \tag{12.201}$$

where $m \equiv m_\mu$. It is clear that in the R_ξ the vertex function is finite. However, the calculations are significantly simplified in the unitary gauge where only contributions caused by physical particles are taken into consideration. In this gauge the Z boson propagator involves the quantity $k_\sigma k_\nu/m_Z^2$, which leads to a linear divergence of the vertex function. Obviously, infinities in $\Lambda_\beta(p, p')$ cancel upon allowing for all divergent diagrams. It should

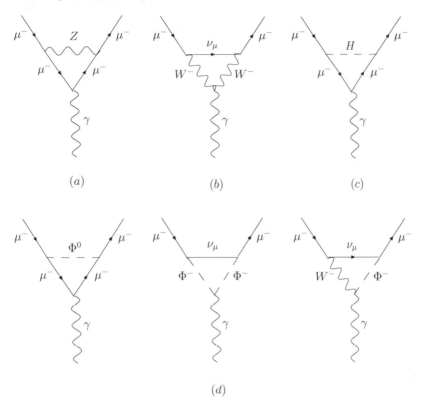

FIGURE 12.14

The diagrams (a)–(d) contributing to the $(g-2)_\mu$ anomaly.

be recalled here that the finite part of a divergent diagram also changes in the renormalization process, that is, it would be illegitimate merely to discard the divergent part in the expression under consideration. In other words, a specific procedure for removing infinities must be employed to determine the finite part of $\Lambda_\beta(p, p')$ in the unitary gauge. For example, one can apply the scheme proposed in Ref. [30] based on the assumption that the Z boson lines do not transfer any external momenta and on the use of the ξ-limiting formalism in which the replacement

$$\frac{k_\mu k_\nu}{m_Z^2} \frac{1}{k^2 - m_Z^2} \to \frac{k_\mu k_\nu}{m_Z^2} \frac{m_Z^2 - \Lambda^2}{(k^2 - m_Z^2)(k^2 - \Lambda^2)}$$

is fulfilled and the limit $\Lambda \to \infty$ is taken at the end of calculations. We choose a simpler method, namely, we use Dyson's procedure [31], in which the expansion of the integrand in a power series in external momenta is followed by the subtraction of divergent quantities.

Combining the propagator denominators, carrying out the integration over the four-dimensional momentum, fulfilling the regularization (subtracting the quantity $\Lambda_\beta^{(Z)}(p_0, p_0)$, where p_0 is a momentum of a free muon, $p_0^2 = m^2$) and separating out terms linear relative to $q = p - p'$, we get the contribution coming from the diagram of Fig. 12.14(a) to the muon AMM:

$$a_\mu^{(Z)} = \frac{m_\mu^2 g^2}{64\pi^2 c_W^2} \int_0^1 dx \left\{ (1 - 4s_W^2)^2 \left[x(x - x^2) \right] + \left[(x - x^2)(x - 4) - \frac{2m_\mu^2 x^3}{m_Z^2} \right] \right\} \times$$

$$\times \frac{1}{m_\mu^2 x^2 + m_Z^2 (1 - x)}. \tag{12.202}$$

The analogous procedure gives the contributions caused by the diagrams displayed in Figs. 12.14(b),(c)

$$a_\mu^{(W)} = \frac{m_\mu^2 g^2}{16\pi^2} \int_0^1 dx \left[(x + 1) + \frac{m_\mu^2 (x - 1)}{2m_W^2} \right] \frac{x^2}{m_\mu^2 x^2 + (m_W^2 - m_\mu^2)x} \tag{12.203}$$

and

$$a_\mu^{(H)} = \frac{m_\mu^4 G_F}{4\sqrt{2}\pi^2} \int_0^1 dx \frac{2x^2 - x^3}{m_\mu^2 x^2 + m_H^2 (1 - x)}. \tag{12.204}$$

When calculating the diagrams connected with the exchange of the virtual W-boson, we have neglected the neutrino mass as compared to the muon mass.

It is evident that the main contribution comes from the diagrams of Fig. 12.14(a),(b). Expanding the integrand of (12.202) and (12.203) in a power series in small parameters (m_μ/m_Z and m_μ/m_W), integrating and having been constrained by the first-order terms, we arrive at the following expressions:

$$a_\mu^{(Z)} = \frac{g^2}{64\pi^2 c_W^2} \left(\frac{m_\mu}{m_Z} \right)^2 \left[\frac{1}{3}(1 - 4s_W^2)^2 - \frac{5}{3} \right], \tag{12.205}$$

$$a_\mu^{(W)} = \frac{g^2}{32\pi^2} \frac{5}{3} \left(\frac{m_\mu}{m_W} \right)^2. \tag{12.206}$$

From the obtained formula it follows that in the one-loop approximation the electroweak sector contribution is governed as [30, 32]

$$a_\mu^{EW}[\text{1 -loop}] = \frac{G_F m_\mu^2}{8\sqrt{2}\pi^2} \left[\frac{5}{3} + \frac{1}{3}(1 - 4s_W^2)^2 + O\left(\frac{m_\mu^2}{m_Z^2} \right) + O\left(\frac{m_\mu^2}{m_W^2} \right) + \right.$$

$$\left. + O\left(\frac{m_\mu^2}{m_H^2} \right) \right] = 194.8 \times 10^{-11}, \tag{12.207}$$

for $s_W^2 = 0.223$. Two-loop corrections are relatively large and negative [33]

$$a_\mu^{EW}[\text{2 -loop}] = -40.7(1.0)(1.8) \times 10^{-11}, \tag{12.208}$$

where the errors stem from quark triangle loops and the assumed Higgs boson mass between 100 and 500 GeV. As calculations demonstrate [33, 34] three-loop corrections are negligible $\sim O(10^{-12})$. Thus, we finally have

$$a_\mu^{EW} = 154(1)(2) \times 10^{-11}. \tag{12.209}$$

Now we proceed to the discussion of the hadron corrections. The term a_μ^{had} is mainly defined by virtual hadronic contributions to the photon propagator in the 4th ($a_\mu^{had}(VP1)$) and 6th order, where the latter includes hadronic vacuum polarization ($a_\mu^{had}(VP2)$) and light-by-light scattering ($a_\mu^{had}(LbyL)$). The hadronic contributions on the level of Feynman diagrams arise through loops of virtual quarks and gluons. These loops also involve the soft scales, and therefore cannot be computed reliably in the perturbative QCD. It is important to keep in mind that all the estimations of the $a_\mu^{had}(LbyL)$ made so far are model dependent. The calculations are based on the chiral perturbation or extended Nambu–Jona-Lasinio model. Also, vector meson dominance is assumed and the phenomenological parametrization of the pion form factor $\pi\gamma^*\gamma^*$ is introduced to regularize the divergence. Note that the main contribution, as well as the largest uncertainty, in a_μ^{had} comes from the term $a_\mu^{had}(VP1)$. All calculations of the lowest-order hadronic vacuum polarization contribution to the muon anomaly are based on the spectral representation

$$a_\mu^{had}(VP1) = \frac{\alpha}{\pi} \int_0^\infty dt \frac{\pi^{-1}\text{Im } \Pi^{had}(t)}{t} \int_0^1 dx \frac{x^2(1-x)}{x^2 + tm_\mu^{-2}(1-x)},$$

where the hadronic spectral function $\text{Im } \Pi^{had}(t)$ is connected with the one-photon annihilation e^-e^+ cross section into hadrons by the relation (e^-e^+-data):

$$\sigma_{e^-e^+ \to \gamma^* \to \text{hadrons}}(t) = \frac{4\pi\alpha\text{Im } \Pi^{had}(t)}{t}.$$

It is also possible to measure $\text{Im } \Pi^{had}(t)$ in the region $0 \leq t \leq m_\tau^2$ using hadronic tau-decays (τ-data). However, the analysis made in Ref. [35] showed a discrepancy between the $a_\mu^{had}(VP1)$ value obtained exclusively from e^+e^--data and that which arises if only τ-data are used. We recall that the inclusion of τ-decay data introduces systematic uncertainties originating from isospin symmetry breaking effects that are difficult to estimate. As additional e^-e^+- and τ-data have become available this discrepancy has increased. This prevents one from using a straightforward average of these different determinations. The recent estimations of the hadron corrections based on the e^+e^--data are given by the expressions:

$$a_\mu^{had}[LO] = 6\ 894(42)(18) \times 10^{-11} \qquad [36], \tag{12.210}$$

where the first error is experimental and the second due to the QED radiative corrections to the data, and

$$a_\mu^{had}[NLO] = 22(35) \times 10^{-11} \qquad [37], \tag{12.211}$$

where the error is dominated by hadronic light-by-light uncertainties. As we can see, the uncertainties of hadronic corrections are much bigger than the experimental ones. Advances in calculating the hadronic corrections are expected due primarily to the application of the lattice QCD. More detailed information concerning the state of affairs in this region could be found in the book [38].

To add the expressions (12.200), (12.209), (12.210), and (12.211) give the following SM prediction:

$$a_\mu^{CM} = 116\ 591\ 788(2)(46)(35) \times 10^{-11}, \tag{12.212}$$

where the errors are due to the electroweak, lowest-order hadronic and higher-order hadronic contributions, respectively. Then, introducing the quantity

$$\delta a_\mu \equiv a_\mu^{exp} - a_\mu^{SM},$$

we get

$$\delta a_\mu = 292(63)(58) \times 10^{-11}. \tag{12.213}$$

Thus, the SM predictions prove to be 3.4σ below the experimental value. Since the experimental data on all the QED effects are in excellent agreement, without any exception, with the theoretical predictions (in cases, when other interactions either are inessential or can be accounted for in these effects), then the deviation between the theory and experiment can be caused by the contribution either from the electroweak or from hadron sectors. We note, in some papers (see, for example, [39, 40]) calculations based on the SM yield the somewhat different results, but all of them lead to deviations from the BNL result in excess of 3σ. It should be stressed that the scatter of the a_μ^{SM} values is due to the uncertainties in calculating hadron corrections.

Since the BNL data have been perfectly collected and investigated over many years, it is highly improbable that this discrepancy could be also accounted for as a mere statistical fluctuation, as several earlier deviations from the SM turned out to be. The presence of the extremely small variation of the muon AMM central value in all the BNL results is also the serious argument in favor of a trustworthiness of the E821 experiment. So, the $(g-2)_\mu$ anomaly may be examined as the New Physics (NP) signal already at the weak scale. Usually, the NP contributions are expected to produce contributions proportional to m_μ^2/M_{NP}^2, where M_{NP} is the mass of a particle that is additional to the sector of the SM particles.

Suggestions made in the literature for explaining $\delta a_\mu/\mu_0$ include supersymmetry [41], additional gauge bosons [42], anomalous gauge boson couplings [43], leptoquarks [44], extra dimensions [45], muon substructure [46], exotic flavor-changing interactions [47], exotic vector-like fermions [48], possible nonperturbative effects at the 1 TeV order [49], the violation of CPT and Lorentz invariance [50], the Higgs sector extension [51], and so on.

Some of explanations of E821-experiment prove to be excluded by the current experimental data. To cite some examples: The possibility of the muon substructure can be immediately ruled out since the necessary compositeness scale of muon should already have been seen in processes involving high-energetic muons at LEP, HERA, and TEVATRON.

For anomalous W-boson dipole magnetic moment

$$\mu_W = \frac{e}{2m_W}(1 + \kappa_\gamma) \tag{12.214}$$

the additional one-loop contribution to a_μ is given by the expression

$$a_\mu(\kappa_\gamma) \approx \frac{G_F m_\mu^2}{4\sqrt{2}\pi^2} \ln\left(\frac{\Lambda^2}{m_W^2}\right)(\kappa_\gamma - 1), \tag{12.215}$$

where Λ is the high momentum cutoff required to give a finite result. For $\Lambda \approx 1$ TeV, in order to obtain the accord between theory and observation one should demand

$$\delta\kappa_\gamma \equiv \kappa_\gamma - 1 \approx 0.4.$$

However, such a big value of $\delta\kappa_\gamma$ is already eliminated by the data at LEP II and TEVATRON, which gives [52]

$$\kappa_\gamma = 0.973^{+0.044}_{-0.045}. \tag{12.216}$$

In this manner, at the moment the $(g-2)_\mu$ anomaly plays the role of an Occam's razor* for the existing SM extensions.

We are already at a compelling moment. The SM gives ~ 3 standard deviations from the BNL results, providing a strong hint of the NP. Only experiment will show what SM extension happens to be true. At present the new BNL muon experiments are being discussed. It is scheduled to improve the data by at least a factor of 2. Of course, the theoretical results concerning calculations of the hadronic corrections should be improved as well. To compare the theoretical and experimental results will give a more precise $(g-2)_\mu$ value that, for its turn, allow us to obtain a more trustworthy constraints on the parameters of the NP.

---

*The essences should not be multiplied without necessity.

12.7 Effects of the interference of weak and electromagnetic interactions

Shortly after the hypothesis of the weak neutral current existence had first been suggested, Y. Zel'dovich proposed the following simple formula to calculate asymmetry appearing as a result of the interference of the electromagnetic amplitude $\mathcal{A}^{em} \sim e^2/s$ (s is an energy in the center-of-mass frame) with weak neutral current amplitude $\mathcal{A}^{NC} \sim G_F$

$$\frac{|\mathcal{A}^{em}\mathcal{A}^{NC}|}{|\mathcal{A}^{em}|^2} \approx \frac{G_F}{e^2/s} \approx \frac{10^{-4}s}{m_p^2}. \tag{12.217}$$

High-energy e^-e^+ colliders are ideally appropriate for an experimental test of these effects. Thus, even at the energies $s = 40$ GeV Eq. (12.217) predicts the effect $\sim 16\%$, which can easily be measured.

Let us study the electroweak interference effect in e^-e^+ annihilation in detail. We consider the reaction

$$e^- + e^+ \to \mu^- + \mu^+. \tag{12.218}$$

In the second order of the perturbation theory this process is presented by the diagrams shown in Fig. 12.15. The corresponding amplitudes are given by the expressions:

FIGURE 12.15
The Feynman diagrams describing the process $e^-e^+ \to \mu^-\mu^+$.

$$\mathcal{A}^{(\gamma)} = ie^2\sqrt{\frac{m_e^2 m_\mu^2}{p_0 k_0 p_0' k_0'}}\overline{u}(p')\gamma^\lambda v(k')\frac{g_{\lambda\beta}}{(p+k)^2}\overline{v}(k)\gamma^\beta u(p), \tag{12.219}$$

$$\mathcal{A}^{(Z)} = \frac{ig^2}{c_W^2}\sqrt{\frac{m_e^2 m_\mu^2}{p_0 k_0 p_0' k_0'}}\overline{u}(p')\gamma^\lambda(g_V - g_A\gamma_5)v(k')\times$$

$$\times\frac{g_{\lambda\beta} - (p+k)_\lambda(p+k)_\beta/m_Z^2}{(p+k)^2 - m_Z^2}\overline{v}(k)\gamma^\beta(g_V - g_A\gamma_5)u(p), \tag{12.220}$$

$$\mathcal{A}^{(H)} = -\frac{ig^2 m_\mu m_e}{4m_W^2}\sqrt{\frac{m_e^2 m_\mu^2}{p_0 k_0 p_0' k_0'}}\overline{u}(p')v(k')\frac{1}{(p+k)^2 - m_H^2}\overline{v}(k)u(p), \tag{12.221}$$

where $g_V = 2s_W^2 - 1/2$, $g_A = -1/2$.

We first compare the relative magnitudes of the coupling constants in the three matrix elements:

$$\mathcal{A}^{(\gamma)} \sim e^2 = g^2 s_W^2, \qquad \mathcal{A}^{(Z)}) \sim \frac{g^2}{c_W^2}, \qquad \mathcal{A}^{(H)} \sim \frac{g^2 m_\mu m_e}{4m_W^2} \approx 10^{-8}g^2. \tag{12.222}$$

From (12.222) it immediately follows that the contribution from the Higgs particle is totally negligible, whereas the matrix elements $\mathcal{A}^{(\gamma)}$ and $\mathcal{A}^{(Z)}$ are of the same order of magnitude—at least at the energies in the range of the Z boson mass. We shall not be interested in polarizations of particles in the final and initial states. Then, carrying out the average over the initial particle spins and summing over the final particle spins, we find

$$\frac{1}{4} \sum |\mathcal{A}^{(\gamma)} + \mathcal{A}^{(Z)}|^2 = C\Bigg\{ \frac{e^4}{s^2} \mathrm{Sp}[(\hat{p}' + m_\mu)\gamma^\alpha(\hat{k}' - m_\mu)\gamma^\beta] \mathrm{Sp}[(\hat{k} - m_e)\gamma_\alpha(\hat{p} + m_e)\gamma_\beta] +$$

$$+ \frac{e^2 g^2}{s c_W^2 (s - m_Z^2)} \left[g_{\alpha\beta} - \frac{(p+k)_\alpha(p+k)_\beta}{m_Z^2} \right] \mathrm{Sp}[(\hat{p}' + m_\mu)\gamma_\tau(\hat{k}' - m_\mu)\times$$

$$\times \gamma^\alpha(g_V + g_A\gamma_5)] \mathrm{Sp}[(\hat{k} - m_e)\gamma^\tau(\hat{p} + m_e)\gamma^\beta(g_V - g_A\gamma_5)] +$$

$$+ \frac{e^2 g^2}{s c_W^2 (s - m_Z^2)} \left[g_{\alpha\beta} - \frac{(p+k)_\alpha(p+k)_\beta}{m_Z^2} \right] \mathrm{Sp}[(\hat{p}' + m_\mu)\gamma_\tau(g_V - g_A\gamma_5)\times$$

$$\times (\hat{k}' - m_\mu)\gamma^\alpha] \mathrm{Sp}[(\hat{k} - m_e)\gamma^\tau(g_V - g_A\gamma_5)(\hat{p} + m_e)\gamma^\beta] +$$

$$+ \frac{g^4}{c_W^4 (s - m_Z^2)^2} \left[g_{\alpha\beta} - \frac{(p+k)_\alpha(p+k)_\beta}{m_Z^2} \right] \mathrm{Sp}[(\hat{p}' + m_\mu)\gamma_\tau(g_V - g_A\gamma_5)\times$$

$$\times (\hat{k}' - m_\mu)\gamma^\alpha(g_V - g_A\gamma_5)] \mathrm{Sp}[(\hat{k} - m_e)\gamma^\tau(g_V - g_A\gamma_5)(\hat{p} + m_e)\gamma^\beta(g_V - g_A\gamma_5)] \Bigg\}, \quad (12.223)$$

where

$$C = \frac{1}{64 p_0 k_0 p_0' k_0'}.$$

In order to simplify our calculation we take into consideration that the most interesting scattering energies are far above 10 GeV. Hence, we can safely neglect all terms including lepton masses. The simplification goes even further: we may omit all contributions in the numerators of the Z boson propagators containing the four-dimensional vectors $(p+k)_\alpha(p+k)_\beta$. Really, we have, for instance,

$$(p+k)_\beta \, \mathrm{Sp}[(\hat{k} - m_e)\gamma^\tau(\hat{p} + m_e)\gamma^\beta] \approx (p+k)_\beta \, \mathrm{Sp}[\hat{k}\gamma^\tau\hat{p}\gamma^\beta] =$$

$$= 4[(\hat{p} \cdot \hat{k})k^\tau - (\hat{p} \cdot \hat{k})(k+p)^\tau + (\hat{p} \cdot \hat{k})p^\tau] = 0. \quad (12.224)$$

With allowance made for the assumptions above, the expression (12.223) takes the form

$$\frac{1}{4} \sum |\mathcal{A}^{(\gamma)} + \mathcal{A}^{(Z)}|^2 \approx C\Bigg\{ \frac{e^4}{s^2} \, \mathrm{Sp}[\hat{p}'\gamma^\alpha\hat{k}'\gamma^\beta]\mathrm{Sp}[\hat{k}\gamma_\alpha\hat{p}\gamma_\beta] + \frac{e^2 g^2}{2 s c_W^2 (s - m_Z^2)} \times$$

$$\times \mathrm{Sp}[\hat{p}'\gamma_\tau\hat{k}'\gamma^\alpha(g_V' + \gamma_5)] \, \mathrm{Sp}[\hat{k}\gamma^\tau\hat{p}\gamma_\alpha(g_V' + \gamma_5)] + \frac{g^4}{256 c_W^4 (s - m_Z^2)^2} \times$$

$$\times \mathrm{Sp}[\hat{p}'\gamma_\tau(g_V' + \gamma_5)\hat{k}'\gamma^\alpha(g_V' + \gamma_5)] \, \mathrm{Sp}[\hat{k}\gamma^\tau(g_V' + \gamma_5)\hat{p}\gamma_\alpha(g_V' + \gamma_5)] \Bigg\}, \quad (12.225)$$

where we have accounted for the equality

$$\sum \mathcal{A}^{(\gamma)}\mathcal{A}^{(Z)*} = \sum \mathcal{A}^{(\gamma)*}\mathcal{A}^{(Z)}$$

being true when $m_e = m_\mu = 0$ and, for the sake of convenience, introduced

$$g'_V = 2g_V.$$

The scheme of further calculations is simple enough and does not harbor any news. Here, we are constrained only by the detailed computation of the interference term in (12.225) (the term proportional to $e^2 g^2$). Taking the spurs, we get

$$\text{Sp}[\hat{p}'\gamma^\tau \hat{k}'\gamma^\alpha(g'_V + \gamma_5)] = 4g'_V[p'^\tau k'^\alpha + p'^\alpha k'^\tau - (p' \cdot k')g^{\tau\alpha}] + 4i\epsilon^{\mu\tau\nu\alpha}p'_\mu k'_\nu \qquad (12.226)$$

and

$$\text{Sp}[\hat{k}\gamma_\tau \hat{p}\gamma_\alpha(g'_V + \gamma_5)] = 4g'_V[k_\tau p_\alpha + p_\alpha k_\tau - (p \cdot k)g_{\tau\alpha}] + 4i\epsilon_{\beta\tau\kappa\alpha}p^\beta k^\kappa. \qquad (12.227)$$

Multiplying (12.226) and (12.227) and making use of the relation

$$\epsilon^{\mu\tau\nu\alpha}\epsilon_{\beta\tau\kappa\alpha} = 2(g^\mu_\kappa g^\nu_\beta - g^\mu_\beta g^\nu_\kappa) \qquad (12.228)$$

we arrive at

$$32\{g'^2_V[(p' \cdot p)(k' \cdot k) + (p' \cdot k)(k' \cdot p)] + [(p' \cdot k)(k' \cdot p) - (p' \cdot p)(k' \cdot k)]\} -$$
$$= 32(g'^2_V - 1)(p' \cdot p)(k' \cdot k) + 32(g'^2_V + 1)(p' \cdot k)(k' \cdot p). \qquad (12.229)$$

Fulfilling the analogous calculations for

$$\sum(\mathcal{A}^{(\gamma)}\mathcal{A}^{(\gamma)*} + \mathcal{A}^{(Z)}\mathcal{A}^{(Z)*}, \qquad (12.230)$$

and summing the results obtained, we get

$$\frac{1}{4}\sum|\mathcal{A}^{(\gamma)} + \mathcal{A}^{(Z)}|^2 = C\left\{\left[\frac{32e^4}{s^2} + \frac{4e^2g^2(g'^2_V + 1)}{sc^2_W(s - m^2_Z)} + \frac{g^4(g'^4_V + 6g'^2_V + 1)}{8c^4_W(s - m^2_Z)^2}\right](p' \cdot k)(k' \cdot p)+\right.$$

$$\left.+\left[\frac{32e^4}{s^2} + \frac{4e^2g^2(g'^2_V - 1)}{sc^2_W(s - m^2_Z)} + \frac{g^4(g'^2_V - 1)}{8c^4_W(s - m^2_Z)^2}\right](p' \cdot p)(k' \cdot k)\right\}. \qquad (12.231)$$

In the center-of-mass frame we have

$$p_0 = k_0 = p'_0 = k'_0 = \frac{\sqrt{s}}{2}$$

and

$$(p' \cdot p) = (k' \cdot k) \approx \frac{s}{4}(1 - \cos\vartheta), \qquad (12.232)$$

$$(p' \cdot k) = (k' \cdot p) \approx \frac{s}{4}(1 + \cos\vartheta), \qquad (12.233)$$

where ϑ is the angle between the three-dimensional momenta of e^- and μ^-.

In accordance with the formula (18.178) of *Advanced Particle Physics Volume 1* in order to obtain the differential cross section we must multiply the quantity (12.231) by

$$(2\pi)^4 \frac{p_0 k_0}{\sqrt{(p \cdot k)^2 - m^4_e}}\delta^{(4)}(p + k - p' - k')\frac{d^3p'}{(2\pi)^3}\frac{d^3k'}{(2\pi)^3}$$

and integrate over d^3p' and d^3k'. Because

$$\frac{p_0 k_0}{\sqrt{(p \cdot k)^2 - m^4_e}} = \frac{p_0}{2|\mathbf{p}|},$$

then the differential cross section is

$$\frac{d\sigma}{d\Omega}(e^-e^+ \to \mu^-\mu^+) = \frac{(2\pi)^4}{128k_0|\mathbf{p}|} \int_0^\infty \frac{|\mathbf{p}'|^2 dp'}{(2\pi)^3 p'_0} \int_0^\infty \frac{d^3k'}{(2\pi)^3 k'_0} \times$$

$$\times \frac{1}{4}\sum |\mathcal{A}^{(\gamma)} + \mathcal{A}^{(Z)}|^2 \delta^{(4)}(p + k - p' - k'). \tag{12.234}$$

The integral in (12.234) is taken with no trouble

$$\int_0^\infty \frac{|\mathbf{p}'|^2 d|\mathbf{p}'|}{(2\pi)^3 p'_0} \int_0^\infty \frac{d^3k'}{(2\pi)^3 k'_0} \delta^{(3)}(\mathbf{p}' + \mathbf{k}')\delta(p'_0 + k'_0 - \sqrt{s}) \approx$$

$$\approx \int_0^\infty \frac{p'_0 dp'_0}{(2\pi)^6 p'_0} \delta(2p'_0 - \sqrt{s}) = \frac{1}{2(2\pi)^6}. \tag{12.235}$$

Next, we should account for the Z boson instability. This is achieved by the following replacement of its propagator:

$$\frac{g_{\alpha\beta} - p_\alpha p_\beta/m_Z^2}{p^2 - m_Z^2} \to \frac{g_{\alpha\beta} - p_\alpha p_\beta/m_Z^2}{p^2 - m_Z^2 + i\Gamma_Z m_z}, \tag{12.236}$$

where Γ_Z is the total width of the Z boson decay

$$\Gamma_Z = 2.4952 \pm 0.0023 \text{ GeV}. \tag{12.237}$$

With allowance made for these circumstances, the differential cross section of the process (12.218) will look like:

$$\frac{d\sigma}{d\Omega}(e^-e^+ \to \mu^-\mu^+) = \frac{\alpha^2}{4s}\left\{(1 + \cos^2\vartheta)\left[1 + \frac{2g_V^2 s(s - m_Z^2)}{\sin^2 2\theta_W[(s - m_Z^2)^2 + \Gamma_Z^2 m_Z^2]} + \right.\right.$$

$$\left.+ \frac{s^2(4g_V^2 + 1)^2}{16\sin^4 2\theta_W[(s - m_Z^2)^2 + \Gamma_Z^2 m_Z^2]}\right] + \cos\vartheta\left[\frac{s(s - m_Z^2)\sin^2 2\theta_W + 2s^2 g_V^2}{\sin^4 2\theta_W[(s - m_Z^2)^2 + \Gamma_Z^2 m_Z^2]}\right]\right\}. \tag{12.238}$$

Before 1989 the maximum available energy was $s \ll m_Z^2$. In this approximation we may set

$$\frac{s(s - m_Z^2)}{[(s - m_Z^2)^2 + \Gamma_Z^2 m_Z^2]} \approx -\frac{s}{m_Z^2}, \qquad \frac{s^2}{[(s - m_Z^2)^2 + \Gamma_Z^2 m_Z^2]} \approx 0, \tag{12.239}$$

to give

$$\frac{d\sigma}{d\Omega}(e^-e^+ \to \mu^-\mu^+) = \frac{\alpha^2}{4s}\left\{(1 + \cos^2\vartheta)\left[\left(1 - \frac{sG_F g_V^2}{\sqrt{2}\pi\alpha}\right) - \frac{sG_F \cos\vartheta}{2\sqrt{2}\pi\alpha(1 + \cos^2\vartheta)}\right]\right\}, . \tag{12.240}$$

where we have taken into account

$$m_Z^2 = \frac{g^2\sqrt{2}}{8c_W^2 G_F} = \frac{2\pi\alpha\sqrt{2}}{G_F s_{2W}^2},$$

$s_{2W} = \sin 2\theta_W$. It is interesting to compare this expression with that which is obtained in the limit $G_F = 0$ (pure QED). Because

$$g_V = 2s_W^2 - \frac{1}{2} \ll 1,$$

then the main correction from the weak current is proportional to $\cos \vartheta$. In the range of forward scattering, that is, for $0 < \vartheta < \pi/2$ this correction suppresses the cross section, whereas in the case of backward scattering, for $\pi/2 < \vartheta < \pi$ it enhances the cross section. So, the neutral weak current causes the back-forward asymmetry (the asymmetry of the angular distribution around $\vartheta = 90^0$). Let us calculate the scale of the integral asymmetry A_{FB}:

$$A_{FB} \equiv \frac{\sigma_F - \sigma_B}{\sigma_F + \sigma_B}, \qquad \sigma_F = \int_0^{\pi/2} \frac{d\sigma}{d\Omega} d\Omega, \qquad \sigma_F = \int_{\pi/2}^{\pi} \frac{d\sigma}{d\Omega} d\Omega. \qquad (12.241)$$

Having integrated (12.241), we get

$$A_{FB} = \frac{3}{8} \left[\left(-\frac{sG_F}{2\sqrt{2}\pi\alpha} \right) : \left(1 - \frac{sG_F g_V^2}{\sqrt{2}\pi\alpha} \right) \right] \approx -\frac{3}{16\sqrt{2}\pi} \left(\frac{sG_F}{\alpha} \right). \qquad (12.242)$$

This result is in line with the naive estimation giving by Eq. (12.217). From (12.242) it is also follows that the asymmetry is quadratically increased versus the $e^- e^+$ beam energy provided $s \ll m_Z^2$. The A_{FB} asymmetry was first clearly observed in an experiment at the accelerator PETRA (Positron-Electron Tandem Ring Accelerator) at DESY (see Fig. 12.16). The full line represents the prediction of the GWS model, while the dashed line

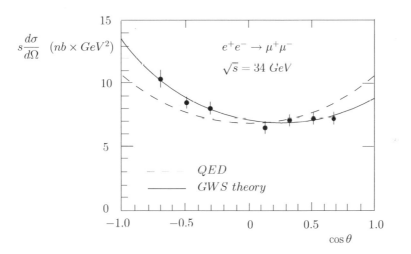

FIGURE 12.16
The angular distribution of $e^+ e^- \to \mu^+ \mu^-$ scattering for $\sqrt{s} = 34$ GeV.

demonstrates the prediction of the pure QED. Although there are also small asymmetric contributions in the framework of QED when higher-order Feynman diagrams are taken into account, these effects are of the order of 1% and hence much smaller than the effect of the neutral weak current.

To integrate the expression (12.238) over $d\Omega$ results in

$$\sigma(e^- e^+ \to \mu^- \mu^+) = \sigma_0 \left[1 + \frac{2g_V^2 s(s - m_Z^2)}{\sin^2 2\theta_W [(s - m_Z^2)^2 + \Gamma_Z^2 m_Z^2]} + \frac{s^2 (4g_V^2 + 1)^2}{16 s_{2W}^4 [(s - m_Z^2)^2 + \Gamma_Z^2 m_Z^2]} \right], \qquad (12.243)$$

where σ_0 is the cross section of the process (12.218) in QED

$$\sigma_0 = \frac{4\pi\alpha^2}{3s}.$$

It is obvious that studying (12.218) defines the neutral weak current parameters.

In the resonance point $s = m_Z^2$ the cross section (12.243) is maximum and given by the expression:

$$\sigma_{res}(e^- e^+ \to \mu^- \mu^+) \approx \frac{\pi\alpha^2}{12 s_{2W}^4 \Gamma_Z^2}, \qquad (12.244)$$

where we have set $g_V \approx 0$. From a precise measurement of the resonance curve the mass and width of the Z boson can be determined with great accuracy. The experiments performed in 1989–1990 at LEP (Large Electron Positron collider) and SLC (Stanford Linear Collider) have confirmed the predictions of the GWS theory with unprecedented precision.

Among this class of experiments are also measurements of an asymmetry in the inelastic scattering of longitudinally polarized electrons (or muons) by nuclear targets. In this case an asymmetry is defined by the quantity

$$A_{LR} = \frac{\sigma_R - \sigma_L}{\sigma_R + \sigma_L}, \qquad (12.245)$$

where $\sigma_{R,L}$ is the cross section of the reaction $e_{R,L} N \to e_{R,L} X$, e_R (e_L) is a left(right)-handed polarized electron. Nonvanishing A_{LR} is evidence of the presence of the parity violation effects.

There is one more class of experiments on the discovery of the effects caused by the interference between the neutral weak and electromagnetic currents. Among this is the precise measurement of the parity violation effect in atomic transitions. The basic idea of this experiment consists in the fact that the Z boson exchange between electron and nucleus must lead to modification of the Coulomb potential. In accordance with the naive estimation (12.217), the discovery of such an effect is an insoluble task. Really, Eq. (12.217) predicts the effect of the order of

$$\frac{G_F s}{e^2} \approx \frac{10^{-4} s}{m_p^2} \approx \frac{10^{-4}}{m_p^2 R^2} \approx 10^{-15}, \qquad (12.246)$$

where R is a typical atom radius ($\sim 10^{-8}$ cm). However, when one investigates strongly forbidden electromagnetic transitions in atoms with large z (an effect is proportional to z^3), then a significantly greater effect could be obtained. The general method lies in the fact that the polarization plane of a laser radiation when passing through a medium is rotated. This rotation is due to interference terms that violate the parity conservation law and are proportional to $g_V^e g_A^q$ and $g_A^e g_V^q$. The first to observe the parity violation in atomic transitions were Barkov and Zolotorev at the beginning of 1978 (Novosibirsk) [53]. Passing linearly polarized laser light through atomic bismuth vapors, they measured the angle of the polarization plane rotation

$$\Delta\varphi = (-3 \pm 0.5) \times 10^{-8} \text{ rad.}$$

The magnitude and sign of the observed optical activity of bismuth vapors were in excellent agreement with the predictions the GWS model. At the end of 1978 the parity violation effect was detected in Berkeley when passing a circularly-polarized laser light through thallium vapors: the absorbtion cross section for right-handed polarized photons proves to be greater than for left-handed ones, which is again in harmony with the GWS theory. Note, if in bismuth the parity violation is measured in the real part of a refraction coefficient, in thallium the one is measured in the imaginary part of the refraction coefficient.

12.8 Neutrino-electron scattering

We start with the processes of elastic electron-neutrino scattering. Consider the following reactions:

$$\overline{\nu}_\mu + e^- \to \overline{\nu}_\mu + e^-. \tag{12.247}$$

and

$$\nu_\mu + e^- \to \nu_\mu + e^- \tag{12.248}$$

Both processes can be going only through the interaction of neutral weak currents (Fig. 12.17). Because the reactions (12.247) and (12.248) are realized at a fixed target, then the

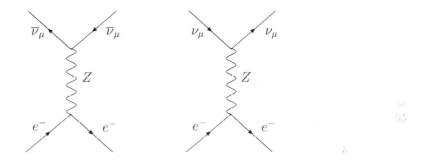

FIGURE 12.17
The Feynman diagrams describing the processes $\overline{\nu}_\mu e^- \to \overline{\nu}_\mu e^-$ and $\nu_\mu e^- \to \nu_\mu e^-$.

following relations take place even for neutrino energies of the order of a few hundreds GeV

$$s = (p_e + p_\nu)^2 \ll m_{W,Z}^2, \tag{12.249}$$

$$(p_e + p_\nu)^2 = (m_e + E_\nu)^2 - \mathbf{p}_\nu^2 \approx 2m_e E_\nu.$$

The other Mandelstam's variables t, u satisfy the analogous conditions too. This allows us to neglect these quantities in the Z-boson propagators. Then, the amplitude of the reaction (12.247) will be governed by the expression (see Eq. (12.75))

$$\mathcal{A}(\overline{\nu}_\mu e^-) = \frac{G_F}{\sqrt{2}} \sqrt{\frac{m_{\nu_\mu}^2 m_e^2}{k_1^0 p_1^0 k_2^0 p_2^0}} \left\{ \overline{v}_{\nu_\mu}(k_1)\gamma_\lambda[1 - \gamma_5]v_{\nu_\mu}(k_2)\overline{u}_e(p_2)\gamma^\lambda[g_V - g_A\gamma_5]u_e(p_1) \right\}. \tag{12.250}$$

Averaging $|\mathcal{A}|^2$ over the initial electron spin and summing over the final one (an antineutrino could be considered as a practically right-handed fermion as its admixture of a left-handed polarization is proportional to m_ν/E_ν), we arrive at

$$\frac{1}{2}\sum|\mathcal{A}(\overline{\nu}_\mu e^-)|^2 = \frac{G_F^2}{64 E_{\nu_1} E_{e_1} E_{\nu_2} E_{e_2}} N_\nu N_e, \tag{12.251}$$

where

$$N_\nu = \text{Sp}[\hat{k}_1\gamma^\lambda(1-\gamma_5)\hat{k}_2\gamma^\sigma(1-\gamma_5)] = 8[k_1^\lambda k_2^\sigma + k_2^\lambda k_1^\sigma - (k_1 \cdot k_2)g^{\lambda\sigma} - i\epsilon^{\alpha\lambda\beta\sigma}k_{1\alpha}k_{2\beta}], \tag{12.252}$$

$$N_e = \mathrm{Sp}[(\hat{p}_2 + m_e)\gamma_\lambda(g_V - g_A\gamma_5)(\hat{p}_1 + m_e)\gamma_\sigma(g_V - g_A\gamma_5)] = \mathrm{Sp}[\hat{p}_2\gamma_\lambda\hat{p}_1\gamma_\sigma(g_V^2 + g_A^2 - 2g_Vg_A\gamma_5)] +$$

$$+ m_e^2\mathrm{Sp}[\gamma_\lambda\gamma_\sigma(g_V^2 - g_A^2)] = 4[p_{2\lambda}p_{1\sigma} + p_{1\lambda}p_{2\sigma} - (p_1 \cdot p_2)g_{\lambda\sigma}](g_V^2 + g_A^2) -$$

$$- 8i\epsilon_{\alpha\lambda\beta\sigma}p_2^\alpha p_1^\beta(g_Vg_A) + 4m_e^2g_{\lambda\sigma}(g_V^2 - g_A^2). \tag{12.253}$$

Multiplying N_ν and N_e, we find

$$\frac{1}{2}\sum|\mathcal{A}(\overline{\nu}_\mu e^-)|^2 = \frac{G_F^2}{E_{\nu_1}E_{e_1}E_{\nu_2}E_{e_2}}[(k_1 \cdot p_2)(k_2 \cdot p_1)(g_V + g_A)^2 + (k_1 \cdot p_1)(k_2 \cdot p_2)(g_V - g_A)^2 +$$

$$+ m_e^2(k_1 \cdot k_2)(g_A^2 - g_V^2)]. \tag{12.254}$$

It should be noted that the first two terms in the right side of (12.254) are invariant with respect to $g_V \leftrightarrow g_A$. The third term does not possess this property, but, without a doubt, it could be neglected as compared with the first two terms when the neutrino energy is high.

Obviously, the corresponding expression for the process (12.248) follows from (12.254) under the replacement $k_1 \leftrightarrow k_2$. This replacement corresponds to the fact that these reactions are related by the crossing operation: emitting $\overline{\nu}_\mu$ is replaced by absorbing ν_μ, absorbing $\overline{\nu}_\mu$—by emitting ν_μ. Therefore,

$$\frac{1}{2}\sum|\mathcal{A}(\nu_\mu e^-)|^2 = \frac{G_F^2}{E_{\nu_1}E_{e_1}E_{\nu_2}E_{e_2}}[(k_2 \cdot p_2)(k_1 \cdot p_1)(g_V + g_A)^2 + (k_2 \cdot p_1)(k_1 \cdot p_2)(g_V - g_A)^2 +$$

$$+ m_e^2(k_1 \cdot k_2)(g_A^2 - g_V^2)]. \tag{12.255}$$

Next, we examine the processes

$$\overline{\nu}_e + e^- \rightarrow \overline{\nu}_e + e^-, \tag{12.256}$$

$$\nu_e + e^- \rightarrow \nu_e + e^-, \tag{12.257}$$

those are already going both through the charged and through neutral weak currents. The amplitude for (12.256) was found before (see Eq. (12.79)). It is related with the amplitude of the reaction (12.247) by means of the substitution

$$g_V \rightarrow g_V' = g_V + 1, \qquad g_A \rightarrow g_A' = g_A + 1.$$

Consequently, the amplitude of the reaction (12.256) is given by the expression

$$\frac{1}{2}\sum|\mathcal{A}(\overline{\nu}_e e^-)|^2 = \frac{G_F^2}{E_{\nu_1}E_{e_1}E_{\nu_2}E_{e_2}}[(k_1 \cdot p_2)(k_2 \cdot p_1)(g_V' + g_A')^2 + (k_1 \cdot p_1)(k_2 \cdot p_2)(g_V' - g_A')^2 +$$

$$+ m_e^2(k_1 \cdot k_2)(g_A'^2 - g_V'^2)]. \tag{12.258}$$

It should be stressed that in reactions with reactor antineutrinos the third term is not small in comparison with first two terms. It is easy to verify that the expression for $\frac{1}{2}\sum|\mathcal{A}(\nu_e e^-)|^2$ also follows from (12.258) under the replacement $k_1 \leftrightarrow k_2$.

Fulfill the calculations of the cross sections for the reaction in question in the laboratory reference frame. In all cases, the following kinematic relations are true

$$\left.\begin{array}{ll}(k_1 \cdot p_1) = (k_2 \cdot p_2) = m_eE_{\nu_1}, & (k_1 \cdot p_2) = (k_2 \cdot p_1) = \\ = m_eE_{\nu_2} = m_e(m_e + E_{\nu_1} - E_{\nu_2}) = m_e(E_{\nu_1} - T_e) = m_eE_{\nu_1}(1 - y),\end{array}\right\} \tag{12.259}$$

TABLE 12.3
The constants C_i entering into (12.260).

Process	C_1	C_2	C_3
$\bar{\nu}_e e^- \to \bar{\nu}_e e^-$	$(g'_V - g'_A)^2$	$(g'_V + g'_A)^2$	$g'^2_A - g'^2_V$
$\nu_e e^- \to \nu_e e^-$	$(g'_V + g'_A)^2$	$(g'_V - g'_A)^2$	$g'^2_A - g'^2_V$
$\bar{\nu}_\mu e^- \to \bar{\nu}_\mu e^-$	$(g_V - g_A)^2$	$(g_V + g_A)^2$	$g^2_A - g^2_V$
$\nu_\mu e^- \to \nu_\mu e^-$	$(g_V + g_A)^2$	$(g_V - g_A)^2$	$g^2_A - g^2_V$

where $y = T_e/E_{\nu_1}$ and $(k_1 \cdot k_2) = m_e E_{\nu_1} y$. Then, the differential cross sections of all four reactions (12.247), (12.248), (12.256), and (12.257) could be unified into the single expression

$$d\sigma = \frac{G_F^2}{4\pi^2} \frac{m_e^2 E_{\nu_1}^2 [C_1 + C_2(1-y)^2] + C_3 m_e^3 E_{\nu_1} y}{m_e E_{\nu_1} E_{e_2} E_{\nu_2}} d^3 p_{e_2} d^3 p_{\nu_2} \delta^{(4)}(p_2 + k_2 - p_1 - k_1), \quad (12.260)$$

where the constants C_i ($i = 1, 2, 3$) are listed in Table 12.3. To get rid of the delta-function we carry out the integrations over $d^3 p_2$ and the solid angle $d\Omega_{\nu_2}$ in (12.260). The calculations produce

$$\int d^3 p_{e_2} \int d\Omega_{\nu_2} \delta^{(3)}(\mathbf{k}_1 - \mathbf{k}_2 - \mathbf{p}_2)\delta(E_{\nu_2} + E_{e_2} \quad m_e \quad E_{\nu_1}) =$$

$$- \int d\Omega_{\nu_2} \delta(E_{\nu_2} - m_e - E_{\nu_1} + \sqrt{m_e^2 + E'^2_{\nu_1} + E'^2_{\nu_2} - 2E_{\nu_1} E_{\nu_2} \cos \vartheta}) =$$

$$= \frac{2\pi}{E_{\nu_1} E_{\nu_2}} \int \delta(E_{\nu_2} - m_e - E_{\nu_1} + x)x dx = \frac{2\pi E_e}{E_{\nu_1} E_{\nu_2}}. \quad (12.261)$$

Including (12.261), we get for the differential cross section of (12.260)

$$\frac{d\sigma}{dy} = \frac{G_F^2 m_e E_{\nu_1}}{2\pi} \left[C_1 + C_2(1-y)^2 + C_3 \frac{m_e y}{E_{\nu_1}} \right], \quad (12.262)$$

where $dE_{\nu_2} = dT_e$ has been taken into consideration. To integrate (12.262) over dy from 0 to 1 provides the total cross section

$$\sigma = \frac{G_F^2 s}{4\pi} \left[C_1 + C_2 \frac{1}{3} + C_3 \frac{m_e^2}{s} \right], \quad (12.263)$$

where $s = (k_1 + p_1)^2$

To give an insight of the values of the electron-neutrino cross sections we cite the following formulae:

$$\sigma(\nu_e e^- \to \nu_e e^-) = 0.952 \times 10^{-43} \left(\frac{E_\nu}{10 \text{ MeV}} \right) \text{cm}^2, \quad (12.264)$$

$$\sigma(\bar{\nu}_e e^- \to \bar{\nu}_e e^-) = 0.399 \times 10^{-43} \left(\frac{E_\nu}{10 \text{ MeV}} \right) \text{cm}^2, \quad (12.265)$$

$$\sigma(\nu_\mu e^- \to \nu_\mu e^-) = 0.155 \times 10^{-43} \left(\frac{E_\nu}{10 \text{ MeV}} \right) \text{cm}^2, \quad (12.266)$$

$$\sigma(\overline{\nu}_\mu e^- \to \overline{\nu}_\mu e^-) = 0.134 \times 10^{-43} \left(\frac{E_\nu}{10 \text{ MeV}}\right) \text{cm}^2, \qquad (12.267)$$

were we have used $s_W^2 = 0.2312$ and disregarded the third term in the right side of (12.263) ($E_\nu \gg m_e$).

The ratio of the cross sections $\sigma(\nu_\mu e^- \to \nu_\mu e^-)$ and $\sigma(\overline{\nu}_\mu e^- \to \overline{\nu}_\mu e^-)$

$$R_N \equiv \frac{\sigma(\nu_\mu e^- \to \nu_\mu e^-)}{\sigma(\overline{\nu}_\mu e^- \to \overline{\nu}_\mu e^-)} = 3\frac{1 - 4s_W^2 + 16s_W^4/3}{1 - 4s_W^2 + 16s_W^4} \qquad (12.268)$$

gives us the opportunity to measure the Weinberg angle. By this method, the electron detection efficiencies cancel in the ratio, and an absolute neutrino flux is not needed. Systematic errors are significantly reduced, resulting in an improvement at the determination of s_W^2.

From (12.263) it might be assumed that the cross sections of the elastic neutrino-electron scattering reveal an unrestrained growth versus the energy of the initial particles. This circumstance astonishes because, working with a renormalizable theory, we have to obtain finite values of physical quantities. To be specific, let us consider the processes

$$\nu_l + e^- \to \nu_l + e^-. \qquad (12.269)$$

The calculations showed that the corresponding cross sections are completely isotropic. Therefore, only partial waves with the total angular moment being equal to zero contribute to scattering. This is the completely prospective fact since, when calculating the cross sections, we actually worked within the contact four-fermion Fermi theory by which the scattering process occurs in a single point. As a result, the collision is central and the total angular moment equals zero. That, this process goes through the channel with the fixed value of the angular moment, leads to the fact that the linear growth of the cross section versus the energy comes into conflict with the unitarity condition. The energy, at which the unitarity violation for the processes (12.269) in the Fermi theory happens, has the order

$$\sqrt{s} \approx 10^3 \text{ GeV}. \qquad (12.270)$$

It is called the *unitary limit energy*. The unitary limit energy for the $\overline{\nu}_l e^-$-scattering is around the same order. The unitary limit is reached when every particle in the center-of-mass frame has the energy of the order of 500 GeV. Of course, to date we are far from reaching these energies, since $s \approx 2m_2 E_\nu^{lab}$, and the unitary limit energy in the laboratory reference frame will be obtained provided $E_\nu^{lab} \approx 10^9$ GeV. However, long before reaching the unitary limit, nonlocality effects of weak interaction will have become significant. It means that the terms thrown away in the gauge boson propagators begin to work. Let us take into account these terms. We will address to the elastic $\nu_\mu e^-$-scattering process. With allowance made for the expression (12.255), the cross section takes the form

$$d\sigma = \frac{g^4}{128\pi^2 c_W^4} \frac{m_e^2 E_{\nu_1}^2 [(g_V + g_A)^2 + (g_V - g_A)^2(1-y)^2] + (g_A^2 - g_V^2)m_e^3 E_{\nu_1}y}{m_e E_{\nu_1} E_{e_2} E_{\nu_2}} \times$$

$$\times \frac{d^3 p_2 d^3 k_2}{[(k_1 - k_2)^2 - m_Z^2]^2 + \Gamma_Z^2 m_Z^2}\delta^{(4)}(p_2 + k_2 - p_1 - k_1). \qquad (12.271)$$

Carry out the integrations over the three-dimensional momentum of the scattered electron and the solid angle Ω_{ν_2}

$$\int d^3 p_2 \delta^{(3)}(\mathbf{k}_1 - \mathbf{k}_2 - \mathbf{p}_2) \int d\Omega_{\nu_2} \frac{\delta(E_{\nu_2} + E_{e_2} - m_e - E_{\nu_1})}{[(k_1 - k_2)^2 - m_Z^2]^2 + \Gamma_Z^2 m_Z^2} =$$

$$= \int d\Omega_{\nu_2} \frac{\delta(E_{\nu_2} - m_e - E_{\nu_1} + \sqrt{m_e^2 + E_{\nu_1}^2 + E_{\nu_2}^2 - 2E_{\nu_1}E_{\nu_2}\cos\vartheta})}{[2E_{\nu_1}E_{\nu_2}(\cos\vartheta - 1) - m_Z^2]^2 + \Gamma_Z^2 m_Z^2} =$$

$$= \frac{2\pi}{E_{\nu_1}E_{\nu_2}} \int \delta(x - E_{e_2}) \frac{x\,dx}{[m_e^2 + E_{\nu_1}^2 + E_{\nu_2}^2 - x^2 - 2E_{\nu_1}E_{\nu_2} - m_Z^2]^2 + \Gamma_Z^2 m_Z^2} =$$

$$= \frac{2\pi E_{e2}}{E_{\nu_1}E_{\nu_2}\{[2m_e T_e - m_Z^2]^2 + \Gamma_Z^2 m_Z^2\}}. \tag{12.272}$$

To substitute (12.272) into (12.271) results in

$$\frac{d\sigma}{dy} = \left(\frac{g^4}{128\pi c_W^4}\right) \frac{s[(g_V + g_A)^2 + (g_V - g_A)^2(1 - y)^2 + 2(g_A^2 - g_V^2)m_e^2 y/s]}{[sy - m_Z^2]^2 + \Gamma_Z^2 m_Z^2}. \tag{12.273}$$

From (12.273) it follows that, when increasing the energy, the cross section is decreased by the law s^{-1} as it must be in a renormalizable theory.

A cross section cannot increase forever as a linear function of energy without violating the unitarity of the S-matrix. Based on the most general properties of the S-matrix, Froissart [54] and Martin [55] showed that a total cross section cannot grow asymptotically faster than $(\ln s)^2$. At low energies, the linear dependence of (12.263) on s is only approximate; actually, at high energies the cross sections of the electron-neutrino scattering tends to a constant in accordance with the asymptotic theorem (Froissart bound).

The integrated cross sections of the reactions (12.247), (12.248), (12.256), and (12.257) can be represented in the (g_V^e, g_A^e) plane by four ellipses. Their intersections give two solutions for g_V^e, g_A^e since the equations are symmetric with respect to $(g_V^e, g_A^e) \leftrightarrow -(g_V^e, g_A^e)$. Precise measurements of the purely leptonic cross sections, investigations of Z and $W^\pm$ pole observables, and so on (see, for review, [56]) result in

$$\sin^2\theta_W = 0.231\ 19(14). \tag{12.274}$$

As we already noted in Section 12.5, the inverse muon decay

$$\nu_\mu + e^- \to \nu_e + \mu^-, \tag{12.275}$$

which is caused only by the charged current (Fig. 12.18), could be used for searches of deviations from the $V - A$-structure of the weak interaction. Obviously, within the SM the

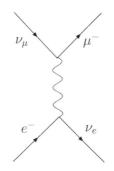

FIGURE 12.18
The Feynman diagram associated with the inverse muon decay.

cross section of (12.275) could be obtained from that of the elastic $\nu_\mu e^-$-scattering provided

$g_A = g_V = 1$ and at the inclusion of the muon mass. Consequently, at the energies $s \ll m_W^2$ the cross section of the inverse muon decay is

$$\sigma(\nu_\mu + e^- \to \nu_e + \mu^-) = \frac{2G_F^2 m_e E_\nu}{\pi} \left[1 - \frac{m_\mu^2}{2m_e E_\nu}\right]. \qquad (12.276)$$

The last factor $[1 - m_\mu^2/(2m_e E_\nu)]$ is purely kinematic.

In the region of ultrarelativistic energies the elastic $\bar{\nu}_e e^-$-scattering process is of great interest. Let us find the corresponding cross section. Uncomplicated but cumbersome calculations result in

$$\frac{1}{2}\sum |\mathcal{A}^Z|^2 = \frac{g^4}{32c_W^4 E_{\nu_1} E_{e_1} E_{\nu_2} E_{e_2}} \left[(k_1 \cdot p_2)(k_2 \cdot p_1)(g_V + g_A)^2 + (k_1 \cdot p_1)(k_2 \cdot p_2)(g_V - g_A)^2 + \right.$$

$$\left. + m_e^2(k_1 \cdot k_2)(g_A^2 - g_V^2)\right] \frac{1}{(u - m_Z^2)^2 + \Gamma_Z^2 m_Z^2}, \qquad (12.277)$$

$$\frac{1}{2}\sum |\mathcal{A}^W|^2 = \frac{g^4}{8E_{\nu_1} E_{e_1} E_{\nu_2} E_{e_2}} \left[\frac{(k_1 \cdot p_2)(k_2 \cdot p_1)}{(s - m_W^2)^2 + \Gamma_W^2 m_W^2}\right], \qquad (12.278)$$

$$\sum (\mathcal{A}^Z)^* \mathcal{A}^W = \frac{g^4}{16c_W^2 E_{\nu_1} E_{e_1} E_{\nu_2} E_{e_2}} \left[2(k_1 \cdot p_2)(k_2 \cdot p_1)(g_V + g_A) + m_e^2(k_1 \cdot k_2)(g_A - g_V)\right] \times$$

$$\times \frac{(u - m_Z^2)(s - m_W^2) + \Gamma_Z \Gamma_W m_Z m_W}{[(u - m_Z^2)^2 + \Gamma_Z^2 m_Z^2][(s - m_W^2)^2 + \Gamma_W^2 m_W^2]}, \qquad (12.279)$$

where $u = (k_1 - k_2)^2$. Making use of (12.259) and carrying out the integration over $\mathbf{p}_2$ and Ω_{ν_2}, we obtain the following expression for the differential cross section:

$$\frac{d\sigma}{dy}(\bar{\nu}_e e^-) = \left(\frac{G_F^2 m_W^4 s}{2\pi}\right) \left\{\frac{4(1-y)^2}{(s - m_W^2)^2 + \Gamma_W^2 m_W^2} + \right.$$

$$+ \frac{[4(g_A + g_V)(1-y)^2 + 4m_e^2 y(g_A - g_V)/s][(sy - m_Z^2)(s - m_W^2) + \Gamma_Z \Gamma_W m_Z m_W]}{c_W^2[(sy - m_Z^2)^2 + \Gamma_Z^2 m_Z^2][(s - m_W^2)^2 + \Gamma_W^2 m_W^2]} +$$

$$\left. + \frac{(g_V + g_A)^2(1-y)^2 + (g_V - g_A)^2 + 2m_e^2 y(g_A^2 - g_V^2)/s}{c_W^4[(sy - m_Z^2)^2 + \Gamma_Z^2 m_Z^2]}\right\}. \qquad (12.280)$$

From (12.280) it is evident that the cross section is linearly increased versus the energy. At $s = m_W^2$ the resonance peak (Glashow resonance [57]) takes place, then the cross section falls down and goes into a constant. In Fig. 12.19 the theoretical curves that are associated with the total cross sections of the electron-antineutrino and electron-neutrino scatterings are presented. For the Glashow resonance to be reached, the neutrino energy must be of the order of 5×10^6 GeV. At present, cosmic sources are the most powerful accelerators of neutrinos. Several sources of ultrahigh-energy neutrinos are known. Active galaxy cores (AGCs) represent one of them. Since a typical luminosity of the AGCs is between 10^{44} and 10^{47} erg/s, we can assume that the evolution of the AGCs is determined by gravity, that is, by the accretion of matter to a supermassive ($M \geq 10^6 M_\odot$) black hole. In the vicinity of the AGC, protons accelerated to ultrahigh energies interact either with matter or with radiation, generating pions, whose decay products include photons and neutrinos. The maximum energy of neutrinos from the AGCs is on the order of 10^{10} GeV. The products of

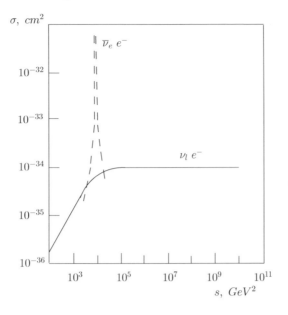

FIGURE 12.19

The cross sections of the neutrino and antineutrino scatterings by electrons at $s \geq m_W^2$.

the decay of pions generated in inelastic collisions of protons with photons that constitute the cosmic microwave radiation background (CRB) appear to be another source of ultrahigh-energy neutrinos. The energy of the neutrinos generated by the CRB may be as high as 10^{12} GeV.

The Glashow resonance is also present in the reaction

$$\overline{\nu}_e + e^- \to \text{hadrons}. \tag{12.281}$$

The total cross section in the resonance vicinity is described by the Breit–Wigner formula

$$\sigma(\overline{\nu}_e e^- \to \text{hadrons}) = \frac{2\pi}{m_W^2} \frac{3\Gamma_l \Gamma_h}{(\sqrt{s} - m_W)^2 + \Gamma^2/4}, \tag{12.282}$$

where Γ_l is the width of the lepton decay channel $W^- \to \overline{\nu}_e e^-$, Γ_h is the width of the hadronic decay, and Γ is the total decay width. Searches of the Glashow resonance are under way at existing neutrino telescopes, as for instance, BAIKAL NT-200, NESTOR (Neutrino Extended Submarine Telescope with Oceanographic Research), AMANDA (Antarctic Muon And Neutrino Detector Array), and so on (see, for review, [58]). It is clear that observation of this resonance allows us to obtain unique information concerning the value of the cosmic neutrino flux.

12.9 Neutrino-nucleon scattering

The pure lepton processes considered above have the advantage of being free of strong interaction complications. However, their usefulness is diminished somewhat because the experimental data in this area generally have poor statistics (at high energies the cross sections of lepton processes are less than those of hadronic processes by a factor of $10^3 - 10^4$). This is in contrast to the neutrino-nucleon scatterings where we have abundant and precise data, especially those of inclusive ν-scatterings from isoscalar targets.* Let us briefly discuss problems of measuring the weak interaction constants for the quarks of the first generation. It is convenient to switch from g_V^q and g_A^q to $g_{L,R}^q$

$$g_L^q = S_3^W(q_L) - Q(q)s_W^2, \qquad g_R^q = S_3^W(q_R) - Q(q)s_W^2. \tag{12.283}$$

Then, using the $\nu_\mu e^-$ cross section, we can get the expression for the cross section of the deep-inelastic scattering:

$$\nu_l + N \to \nu_l + X \tag{12.284}$$

from an isoscalar target. The corresponding subprocess is due to neutral currents. Accounting for that $g_L^\nu = 1/2$, we get

$$\frac{d\sigma}{dy}(\nu_l N \to \nu_l X) = \frac{G_F^2 m_N E_\nu F}{\pi} \left\{ [(g_L^u)^2 + (g_L^d)^2] + [(g_R^u)^2 + (g_R^d)^2](1-y)^2 \right\}, \tag{12.285}$$

where

$$F = \int_0^1 x[f_u(x) + f_d(x)]dx.$$

On the other hand, the cross section of the reaction

$$\nu_l + N \to l^- + X, \tag{12.286}$$

which is caused by the charged weak currents, is

$$\frac{d\sigma}{dy}(\nu_l N \to l^- X) = \frac{G_F^2 m_N E_\nu F}{\pi}. \tag{12.287}$$

As a consequence, the relation

$$\frac{\sigma(\nu_l N \to \nu_l X)}{\sigma(\nu_l N \to l^- X)} = [(g_L^u)^2 + (g_L^d)^2] + \frac{1}{3}[(g_R^u)^2 + (g_R^d)^2]$$

depends only on the sought constants of weak interactions. By an analogy, employing the obtained values for the electron-neutrino scatterings, we find

$$\sigma(\overline{\nu}_l N \to \overline{\nu}_l X) = \frac{G_F^2 m_N E_\nu F}{\pi} \left\{ \frac{1}{3}[(g_L^u)^2 + (g_L^d)^2] + [(g_R^u)^2 + (g_R^d)^2] \right\}, \tag{12.288}$$

$$\sigma(\overline{\nu}_l N \to l^+ X) = \frac{G_F^2 m_N E_\nu F}{3\pi}. \tag{12.289}$$

As a result, we have

$$\frac{\sigma(\nu_l N \to \nu_l X) + \sigma(\overline{\nu}_l N \to \overline{\nu}_l X)}{\sigma(\nu_l N \to l^- X) + \sigma(\overline{\nu}_l N \to l^+ X)} = (g_L^u)^2 + (g_L^d)^2 + (g_R^u)^2 + (g_R^d)^2, \tag{12.290}$$

---

*An isoscalar target consists of protons and neutrons in half $N = (n + p)/2$. As a result, the following relation takes place for such a target

$$f_u(x) = f_d(x), \qquad f_{\overline{u}}(x) = f_{\overline{d}}(x), \qquad f_s(x) = f_{\overline{s}}(x).$$

$$\frac{\sigma(\nu_l N \to \nu_l X) - \sigma(\overline{\nu}_l N \to \overline{\nu}_l X)}{\sigma(\nu_l N \to l^- X) - \sigma(\overline{\nu}_l N \to l^+ X)} = (g_L^u)^2 + (g_L^d)^2 - (g_R^u)^2 - (g_R^d)^2. \qquad (12.291)$$

Just as in the cases of the electron-neutrino scatterings, these data are insufficient in order to obtain complete and single-valued information about the weak interaction constants of the quarks. We need to supplement these measurements with data from proton and neutron targets, or from semi-inclusive pion production on isoscalar targets, to resolve the sign and isospin ambiguities. We shall not give the details of this here. We only note that all the results are again in accord with the SM. The attention is drawn to the relation (12.291) which in the SM takes the form

$$\frac{\sigma(\nu_l N \to \nu_l X) - \sigma(\overline{\nu}_l N \to \overline{\nu}_l X)}{\sigma(\nu_l N \to l^- X) - \sigma(\overline{\nu}_l N \to l^+ X)} = \frac{1}{2}(1 - 2s_W^2). \qquad (12.292)$$

Therefore, it could be used for the Weinberg angle definition as well. It should be mentioned that in the actual phenomenological analysis much more detailed calculations of sea-quark contributions, QCD corrections, etc., are taken into account. The reader is referred to the monograph [86] for a more detailed discussion of these problems.

13

Physics beyond the standard model

If only you knew what trash gives rise
To verse, without a tinge of shame,
Like bright dandelions by a fence,
Like burdock and like cocklebur.
M. Zvetaeva

13.1 Models, models. . .

At present, in elementary particle physics a very interesting situation has arisen. It is related to the situation that we had at the beginning of the twentieth century before the discovery of quantum mechanics and the special theory of relativity. During the last decade many impressive experimental achievements have been done: top-quark discovery, measurement of the direct CP violation in the K-meson system as well as the CP violation in the B meson system, evidence of accelerated universe expansion, determination of portions of dark matter and dark energy in the universe, the establishment of relict radiation anisotropy, and so on. These results strengthen the status of the standard model (SM) as the model which successfully describes nature. However, we again see some small clouds in the clear sky of the SM—the experimental facts that have not found satisfactory explanations within the SM. This is first of all a neutrino mass smallness, an observation of solar and atmospheric neutrino oscillations, the value of the muon AMM, and the prediction of an equal amount of matter and antimatter in the universe. Moreover, experiments showed that the density of matter entering the SM constitutes approximately 5% of the matter density in the universe. From a theoretical point of view, the SM is also far from perfection because it leaves without answering many fundamental questions associated with its structure. The SM may be divided into three sectors: a gauge sector, a flavor sector, and a sector in which a gauge symmetry is breaking. Whereas the two first sectors have been studied in accelerating experiments (LEP at CERN, SLD* and BABAR† at SLAC, BELLE at KEK (Japan), and so on), now the sector of the spontaneous symmetry breaking attracts rapt attention, and not only because physicists hope to discover the Higgs boson at LHC, but because this sector can give the first hints on the existence of New Physics beyond the SM. The Higgs mechanism in the SM represents only the description of the electroweak symmetry breaking. It does not give any explanations of the symmetry violation. In particular, dynamics that could explain the reason of the Higgs potential instability appearance at zero is absent. Consequently, the standard Higgs mechanism calls in question the contemporary understanding of the SM at

---

*Subject Loading Device

†The name of the experiment is derived from the nomenclature for the B meson (symbol **B**) and its antiparticle (symbol $\overline{B}$, pronounced **B bar**).

the quantum level. There is a need to introduce additional structures (new particles, new symmetries, additional dimensions, and so on) in order to stabilize the electroweak scale. All this stimulates searches and investigations of models that lead to the same results like the SM in the low-energy region, while in the high-energy region their predictions are different from the SM predictions. There are at least three principle ways for the SM extension. The first consists in the building of models with a composite Higgs boson and a dynamical symmetry breaking. Nonobservation of the Higgs boson the SM predicts generates a class of Higgsless models. The third direction involves the Higgs mechanism and is based on the idea that the SM is considered as a low-energy approximation of some grand unification theory (GUT). For a symmetry group of the GUT, $SO(10)-$, E_6-groups, and higher dimensionality groups that lead either to the extension of the electroweak group by factors $SU(2)$ [88] and $U(1)$ (see, for review, [59]) or to the replacement $SU(2)_L$ by $SU(3)_L$ (see, for example, [60]) may be employed. In this section we briefly discuss the first two classes of models.

An abundance of elementary particles gives impetus to the assumption of a new level of matter structure—quarks, leptons, and gauge bosons may be built from even smaller particles named *preons*.* Preons must be coupled by a superstrong interaction of a new type that should lead to the formation of quarks, leptons, and so on from preons. Such a hypothetical interaction has different names: hypercolor, technicolor, and the like. The interest peak to preon (or composite) models had fallen in the 1980s. Interest distinctly dropped because many of these models conflicted with the experimental data obtained at accelerators. Besides, after the first superstring revolution many physicist-theorists were inclined to the fact that the superstring theory is more logical and very promising. Accordingly, they focused their efforts in this direction. In the last few years, however, optimism concerning the superstring theory has waned, resulting in a regenerating of interest in composite models.

Models of an extended technicolor are typical representatives of models with a composite Higgs boson. The basic idea consists in the fact that preons (they are called *technifermions* here) are coupled by gauge interactions built by an analogy with QCD [61, 62]. At that, the existence of a set of new gauge charges, the technicolors, is postulated. The goal is to have a spontaneous symmetry breaking (SSB) theory with gauge interactions alone: there is no elementary scalar with its self-couplings and Yukawa couplings. The notion of a composite Higgs scalar is really not a new one. The idea of the Higgs phenomena was first suggested by the theory of superconductivity. There the electromagnetic gauge symmetry is spontaneously broken by the condensate (i.e., nonvanishing ground-state expectation value) of the electron pairs, which acts as an effective composite Higgs scalar. Thus, the SSB is brought on dynamically through the interactions of the electrons with the lattice phonons. One naturally wonders whether the QCD strong force that binds the colored quarks can be the interaction responsible for the SSB in the electroweak interaction? It turns out that such is not the case. However, the analysis of the way it fails suggests possible candidate theories.

Let us consider the standard $SU(3)_c \times SU(2)_L \times U(1)_Y$ model and, for simplicity, restrict ourselves to one generation of fermions. The Lagrangian is given by the expression:

$$\mathcal{L} = -\frac{1}{4}G^{a\mu\nu}G^a_{\mu\nu} - \frac{1}{4}W^{i\mu\nu}W^i_{\mu\nu} - \frac{1}{4}B^{\mu\nu}B_{\mu\nu} + i(\overline{q}\gamma_\mu D^\mu q + \overline{l}\gamma_\mu D^\mu l). \tag{13.1}$$

As there is no Higgs vacuum expectation value (VEV) to break the $SU(2)_L \times U(1)_Y$ gauge symmetry, it would seem that all fermions and all gauge bosons, including $W^\pm$ and Z, will remain massless. We shall see, this is actually not the case.

---

*The term *preon* comes from prequarks. For the first time it was used by Pati and Salam in 1974.

The QCD with two massless u- and d-quarks has a global $SU(2)_L \times SU(2)_R$ symmetry represented by the doublets

$$q_L = \begin{pmatrix} u_L \\ d_L \end{pmatrix}, \qquad q_R = \begin{pmatrix} u_R \\ d_R \end{pmatrix}$$

which can be independently rotated in their respective $SU(2)$ spaces. These two $SU(2)_L$ and $SU(2)_R$ groups are linked by the pairing of q and $\bar{q}$ in the vacuum so that the operator $q\bar{q}$ acquires a nonvanishing VEV. The overall symmetry $SU(2)_V$ corresponding to $L \oplus R$ is unbroken and gives the isospin symmetry of QCD. The other $SU(2)_A$ (from $L \ominus R$) associated with the axial current $j_\mu^{Ai} = \bar{q}\gamma_\mu\gamma_5\sigma^i q$ is spontaneously broken, resulting in three Goldstone bosons, or pions. The matrix element of the current j_μ^{Ai} between the pion and the vacuum is as follows:

$$< 0|j_\mu^{Ai}|\pi^j(k) >= if_\pi k_\mu \delta^{ij}, \qquad f_\pi \approx 95 \text{ MeV},$$

where $i = 1, 2, 3$ and the constant f_π in turn must be related to the QCD scale parameter $\Lambda_{QCD} \approx 200$ MeV as it is the only scale in our theory. Even without the Higgs mechanism, the massless boson W^i, when coupled to the current $J_\mu^i = (j_\mu^{Vi} - j_\mu^{Ai})/2$ built up by the u- and d-quark fields, allows the creation of a pion with amplitude

$$\mathcal{A} = ig\left(-\frac{1}{2}\right)\left(if_\pi k_\mu\right), \tag{13.2}$$

in which the factor $(-1/2)$ is caused by the coefficient of j_μ^{Ai} in J_μ^i.

The contribution of the pion to the vacuum polarization $\Pi_{\mu\nu}(k)$ of the W^i boson as displayed in Fig. 13.1 has a singularity $1/k^2$ near $k^2 = 0$. The residue at $k^2 = 0$ pole is

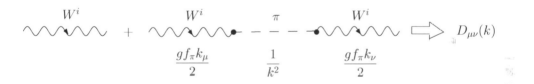

FIGURE 13.1
The Goldstone pion of QCD gives mass to the gauge boson W^i.

$\rho = g^2 f_\pi^2/4$. Together with the conservation law of the current J_μ^i, the vacuum polarization must satisfy

$$k^\mu \Pi_{\mu\nu}(k) = 0,$$

so that near the $k^2 = 0$ pole, the vacuum polarization $\Pi_{\mu\nu}(k)$ is

$$\Pi_{\mu\nu}^{ij}(k) = \left(g_{\mu\nu} - \frac{k_\mu k_\nu}{k^2}\right)\left(\frac{gf_\pi}{2}\right)^2 \delta^{ij} = \left(g_{\mu\nu} - \frac{k_\mu k_\nu}{k^2}\right)\rho\delta^{ij}. \tag{13.3}$$

For the W^i-boson the dressed boson propagator $D_{\mu\nu}(k)$ is generated by summing the geometric series of $\Pi_{\mu\nu}(k)$ (this summation is visualized in Fig. 13.2). The propagator of the W^i-boson is dressed by the pion, and we have

$$\frac{-ig_{\mu\nu}}{k^2} \to D_{\mu\nu}(k) = \frac{-ig_{\mu\nu}}{k^2[1 - \Pi(k^2)]} = \frac{-ig_{\mu\nu}}{k^2[1 - \rho/k^2]}. \tag{13.4}$$

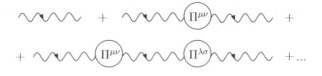

FIGURE 13.2
Dressed boson propagator $D_{\mu\nu}(k)$.

It has a pole at $k^2 = \rho$, that is, the massless W^i-boson gets a mass $gf_\pi/2$ by absorbing the Goldstone pion coming from the spontaneous $SU(2)_A$ symmetry breaking of QCD. Thus, QCD already can give a mass $gf_\pi/2 \approx 31$ MeV to the W^i-boson, which may eventually emerge as the W boson of the electroweak interaction. So, the VEV $< 0|\bar{q}q|0 >$ breaks the $SU(2)_L \times U(1)_Y$ symmetry down to the electromagnetic $U(1)_Q$ with $\boldsymbol{\pi}$ being eaten by the three gauge bosons to become $W^\pm$ and Z. In other words, the QCD interaction violates the electroweak gauge group in just the right pattern. However, it falls short of being a realistic possibility because (i) the scale is all wrong, we obtain $m_W \approx 31$ MeV instead of 80 GeV as required; and (ii) fermions remain massless.

It is relatively straightforward to overcome the first problem. We may postulate the existence of another QCD-like interaction, called the *technicolor interaction (TC)*, which has a scale parameter Λ_{TC} such that it produces the phenomenologically correct mass for the W boson. As the true W boson mass is $gv/2$, Susskind and Weinberg proposed technicolor as a copy of QCD with a technipion having a decay constant $F_\pi = v = 246$ GeV. This technipion is the Goldstone boson built up from U and D techniquarks and would be responsible for the weak bosons masses

$$m_W = \frac{1}{2}g_{TC}F_\pi \approx 80 \text{ GeV}.$$

Then, the scale parameter of the TC theory Λ_{TC} has the order of 1 TeV. To put it differently, the TC interaction, with a gauge group $SU(3)$ for example, is in every way similar to QCD except that it produces the VEV at energy three orders of magnitude larger than the QCD. In this theory fermions carry technicolors (the techniquarks Q) and have the following quantum number of the $SU(3)_{TC} \times SU(3)_c \times SU(2)_L \times U(1)$ group:

$$Q_L(3,1,2,1/3) = \begin{pmatrix} U_L \\ D_L \end{pmatrix},$$

$$U_R(3,1,1,4/3), \qquad D_R(3,1,1,-2/3).$$

The familiar quarks and leptons represent TC singlets. These technicolor quarks form bound states just like ordinary quarks under ordinary color interactions, the TC chiral symmetry being also spontaneously broken.

To make fermions massive, we should enlarge the technicolor group G_{TC} to an extended technicolour (ETC) gauge group G_{ETC} by placing technifermions F (having, say, three technicolors) and ordinary fermions f (technicolor singlet) in a single irreducible representation of G_{ETC}. Next, the ETC symmetry breaks down to technicolor symmetry at some energy scale μ. The vector gauge boson E being in the ETC but not in the TC acquires mass

$$m_E \approx g_{ETC}\mu, \tag{13.5}$$

and interacts with currents of the form $\overline{F}\gamma_\mu f$. The effective four-fermion interactions mediated by the E boson is

$$H_E = \frac{1}{2}\left(\frac{g_{ETC}}{m_E}\right)^2 (\overline{F}_L \gamma_\mu f_L)(\overline{f}_R \gamma_\mu F_R). \tag{13.6}$$

Using the Fierz transformation and (13.5), we get

$$H_E = -\frac{1}{2\mu^2}[(\overline{F}F)(ff) - (\overline{F}\gamma_5 F)(\overline{f}\gamma_5 f) + \ldots]. \tag{13.7}$$

Then, the technifermion condensate $< 0|\overline{F}F|0 > \neq 0$ gives a mass for the ordinary fermion

$$m_f = \frac{1}{2\mu^2} < 0|\overline{F}F|0 > . \tag{13.8}$$

As $< 0|\overline{F}F|0 > \sim (1 \text{ TeV})^3$, it is necessary to have $\mu \approx 30$ TeV to produce $m_f \sim 1$ GeV.

Inasmuch as the vector bosons E are $SU(3)_c \times SU(2)_L \times U(1)$ singlets, we need a set of technifermions for each ordinary fermion in order to give the latter a mass. For one generation of fermions, we must have eight sets of technifermions. As a result, the flavor symmetry of the TC interaction is then at least as large as $SU(8)_L \times SU(8)_R$. When this chiral symmetry is spontaneously violated, three of the Goldstone bosons combine with $W^\pm$- and Z-bosons, and we are left with a large number of relatively light (on the TeV mass scale) pseudo-Goldstone bosons.

The TC hypothesis has appeal because it represents the natural mechanism of the electroweak symmetry breaking at a characteristic scale being significantly lower than the scale that is usually accepted as fundamental.* One needs only to assume that directly below the fundamental scale the unbroken gauge group $SU(3)_c \times SU(2)_L \times U(1)_Y$ associated with the strong and electroweak interactions as well as the TC gauge group (all groups have sufficiently small coupling constants) exist. If the TC interactions possess the property of the asymptotic freedom, at decreasing an energy the TC coupling constant will slowly grow, as does the QCD coupling constant, and becomes large under energies being considerably lower than the fundamental scale. The energy at which the TC interactions become strong determines the characteristic scale F_π. Because, when decreasing an energy, the coupling constant is logarithmically increased, then moderate distinction in the beta-functions for the color and technicolor may easily lead to distinction on three orders of magnitude between the scales at which the corresponding interactions become strong. Among phenomenological consequences of the TC theories, we also note the predictions of the nonstandard triple and quartic gauge boson vertices as well as the resonant contributions to the processes

$$e^+ + e^- \to l^+ + l^-, \qquad e^+ + e^- \to W^+ + W^-$$

$$e^+ + e^- \to \gamma + \pi_T^0, \qquad e^+ + e^- \to \gamma + Z,$$

where π_T^0 is a neutral technipion (discussion of a low-scale TC phenomenology could be found in Ref. [63]). For a detailed study of the TC theories, we recommend the following reviews [64, 65, 66].

The next way of the SM extension is building Higgsless models. Models with additional dimensions can be a good example. Some types of models include the Higgs mechanism.[†]

---

*In superstring theory this scale is given by the Planck mass 10^{18} GeV.

[†]The reader finds a comprehensive description of these models in review [67].

Let us concentrate our attention on the Higgsless version of models in five-dimensional space-time (5HLMs). The principal idea on which the 5HLMs are based lies in the fact that a particle momentum component along an additional dimension is equivalent to a mass in four dimensions [68, 69]. For this reason, one may introduce a mass of a particle imparting a momentum along some additional dimension to it. Similar to that, as happens in quantum mechanics, nonzero momentum along a compact direction may appear as a result of imposing nontrivial boundary conditions. Therefore, the task is to find an additional dimension geometry and select an appropriate boundary condition, so that the spectrum of SM particles can be reproduced. The 5HLMs are nonrenormalizable and become strongly coupled at a cutoff scale Λ_{cut}. However, Higgsless models was proposed just in order to make the scale Λ_{cut} sufficiently large and in order to avoid difficulties that might appear when model predictions are confronted with the results of precision measurements of electroweak observables (this means that a sector with a strong coupling cannot be observed at LHC). And still increasing Λ_{cut} needs the introduction of weakly coupled states in addition to the spectrum of the SM particles. Just these states represent one of the objects of searches at LHC.

There exists many realizations of Higgsless models that are distinguished by the way of introducing fermions or additional dimensions into the theory. However, the fundamental mechanism used for increasing the scale Λ_{cut} is common for all these models. It is as follows. At energies of the order of few TeV, the existence of new massive particles with the spin 1 and having the same quantum number as the SM gauge bosons (Kaluza–Klein bosons)* is needed. Constants determining interactions of the Kaluza–Klein (KK) bosons with $W^{\pm}$, Z, and γ have to obey the unitary sum rules to call forth cancellation of contributions, which come from longitudinal $W^{\pm}$ and Z bosons and increase versus an energy. Therefore, scattering processes of the vector gauge bosons represent the test for the Higgsless scenario, which is independent on a model.

A scale at which the unitarity is violated is defined by a cutoff scale. The analysis shows that the cutoff scale is given by the formula [70]:

$$\Lambda_{cut} \sim \frac{12\pi^4 m_W^2}{g^2 m_{W'}}. \tag{13.9}$$

From (13.9) it follows that the heavier a KK boson, the lower is the scale at which the perturbative unitarity is broken. This also gives a rough estimation, with an accuracy of a numeral coefficient, of a nonperturbative physics scale. Explicit calculations of scattering amplitude that take into consideration inelastic channels show that this estimation is really valid and the numeral coefficient is approximately equal to $1/2$ [70].

In general, the sum rules are saturated even at the inclusion of the first or of several the lowest resonances (the KK bosons). Moreover, the demand of smallness of oblique corrections results in the fact that interaction constants of the gauge KK bosons with light fermions of the SM are small too. Consequently, not depending on a model, one may predict the existence of narrow and light resonances appearing in the scattering of the $W^{\pm}$ and Z bosons. As this takes place, at least one of them must appear at lower energies, approximately at 1 TeV, otherwise it will be ineffective to restore the unitarity. In Ref. [71] the elastic scattering reaction

$$W_L^{\pm} + Z_L \to W_L^{\pm} + Z_L \tag{13.10}$$

---

*The KK bosons appear as a result of compactification and their masses usually equal the integer or half-integer number of inverse compactification radii.

was investigated within the 5HLM. It was shown that with 10 fb^{-1} of data, corresponding to one year of running at low luminosity, LHC will probe a Higgsless resonance W' up to 550 GeV, while covering the whole preferred range up to 1 TeV will require 60 fb^{-1}. Though the Higgsless resonance W' could hardly escape discovery even at LHC, we will have to wait experiments at linear colliders ILC (International Linear Collider) and CLIC (Compact Linear Collider) to precisely measure its interaction constants and thus to experimentally verifies the saturation of the unitary sum rules.

Another method of looking-for W' and Z' is based on the Drell–Yan processes [72]. However, the results of such an analysis are model-dependent to a greater extent because they need the knowledge of the way of introducing the SM fermions into a theory. It is also worthy of note that the KK gluons could be easily revealed as resonances in distributions of events with two hadron jets [72].

In conclusion, one more interesting prediction of Higgsless models lies in the presence of anomalous three- and four-boson interaction constants. Really, in the SM the sum rules those provide for cancellation of terms growing as the fourth power of energy are automatically satisfied because of the gauge invariance. To correctly include contributions of new states, constants of interaction between the SM gauge bosons must be changed. If one assumes the sum rules to be fulfilled already for the first resonance, then the value of these deviations for the WWZ vertices is given by the expression:

$$\Delta = \frac{\delta(g_{WWZ})}{g_{WWZ}} \sim \frac{1}{3} \frac{m_Z^2}{m_{Z'}^2}, \tag{13.11}$$

where Δ is the shift in the interaction constant, and the deviation is estimated with the help of the sum rule for elastic scattering of the W bosons. The deviations in the three-boson interaction constants (trilinear gauge coupling constants) are expected in the range from 1% to 3%. These values are very close to experimental bounds obtained by LEP and they could be tested at LHC. Without question, the linear colliders ILC and CLIC will be able to measure such shifts. It should be stressed that these deviations are a solid prediction of the Higgsless mechanism and are independent on the details of the specific Higgsless model. Nonobservation of a physical Higgs boson would be the first indication on the existence of the Higgsless scenario. However, *the absence of proof is not the proof of absence* and, of course, there are other models, except for the aforementioned ones, in which a Higgs boson cannot be observed at LHC.

It is useful to formulate common features of the Higgsless models and models with the composite Higgs boson. Their common properties are: (i) the nonstandard trilinear and quartic gauge boson couplings; (ii) the nonstandard couplings of the ordinary fermions with the $W^{\pm}$- and Z-boson; and (iii) the resonant contributions to the cross sections caused by particles additional to the SM.

13.2   Multipole moments of gauge bosons

So, the existence of the fundamental scale Λ is of a common property for models belonging to the first two classes of the SM extensions. When the energy $\sqrt{s}$ exceeds Λ, the deviations from the SM are easily observed. Manifestations of deviations from the SM predictions are less obvious when $\sqrt{s} < \Lambda$. Let us attract attention to looking for the New Physics signals at $\sqrt{s} < \Lambda$. Since the models in question predict anomalous interactions between W, Z, and

γ, trilinear and quartic gauge boson couplings, then it is worthwhile to consider a process in which these interactions play a determinative part.

When investigating processes in the range of energies smaller than Λ, it is enough to use the effective Lagrangian technique. In so doing, we have to determine the content of the particles sector and demand fulfillment of particular symmetries. For the sake of simplicity, we assume the following: (i) there are only the SM particles in the theory; (ii) the theory possesses the $SU(2)_L \times U(1)$ global gauge symmetry; and (iii) any exotic-fermion contact interactions are absent. This leads to the fact that the charged and neutral currents have the same form as in the SM. As regards the Lagrangian describing trilinear gauge boson couplings (TBCs), in the most general case it will look like [73]:

$$\mathcal{L}_{WWV} = g_{WWV}\left[ig_1^V(W_{\mu\nu}^*W^\mu V^\nu - W^{*\mu}W_{\mu\nu}V^\nu) + ik_V W_\mu^* W_\nu V^{\mu\nu} + \frac{i\lambda_V}{m_W^2}W_{\lambda\mu}^* W^{\mu\nu}V_\nu^\lambda - \right.$$

$$-g_2^V W_\mu^* W_\nu(\partial^\mu V^\nu + \partial^\nu V^\mu) + g_3^V \varepsilon^{\mu\nu\rho\sigma}(W_\mu^* \partial_\rho W_\nu - \partial_\rho W_\mu^* W_\nu)V_\sigma + i\tilde{k}_V W_\mu^* W_\nu \tilde{V}^{\mu\nu} +$$

$$\left. +\frac{i\tilde{\lambda}_V}{m_W^2}W_{\lambda\mu}^* W^{\mu\nu}\tilde{V}_\nu)\lambda\right], \tag{13.12}$$

where $V = Z, \gamma$, $W_{\mu\nu} = \partial_\mu W_\nu - \partial_\nu W_\mu$, $V_{\mu\nu} = \partial_\mu V_\nu - \partial_\nu V_\mu$, $\tilde{V}_{\mu\nu} = \frac{1}{2}\varepsilon_{\mu\nu\lambda\sigma}V_{\lambda\sigma}$, $g_{WW\gamma} = -e$, $g_{WWZ} = -e\cot\theta_W$. The coefficients k_γ and λ_γ determine the magnetic dipole and electric quadrupole moments of the W boson (for the sake of simplicity we shall assume $g_1^V = 1$) in accordance with the relations

$$\mu_\gamma = \frac{e}{2m_W}(1 + k_\gamma + \lambda_\gamma), \tag{13.13}$$

$$Q_\gamma = -\frac{e}{m_W^2}(k_\gamma - \lambda_\gamma). \tag{13.14}$$

The expressions for their Z analogues follows from (13.13) and (13.14) by the replacements

$$e \to e\cot\theta_W, \qquad \gamma \to Z. \tag{13.15}$$

The first three terms in (13.12) are invariant under the P, C, and T transformations. The terms proportional to g_2^V and g_3^V are invariant with respect to the P and CP operations, respectively. However, they both violate the symmetry with respect to the charge conjugation operation. The parameters $\tilde{k}_\gamma$ and $\tilde{\lambda}_\gamma$ fix the values of electric dipole and magnetic quadrupole moments to be determined by the relations:

$$d_\gamma = \frac{e}{2m_W}(\tilde{k}_\gamma + \tilde{\lambda}_\gamma), \tag{13.16}$$

$$\tilde{Q}_\gamma = -\frac{e}{m_W^2}(\tilde{k}_\gamma - \tilde{\lambda}_\gamma). \tag{13.17}$$

Again, to obtain the expressions for the Z analogues we should carry out the replacement (13.15) in (13.16) and (13.17). In the SM and its extensions belonging to the third class, we have at the tree level:

$$k_V = 1, \qquad \tilde{\lambda}_V = \lambda_V = \tilde{k}_V = g_2^V = g_3^V = 0. \tag{13.18}$$

The radiative corrections (RCs) increase the value of the electromagnetic and weak multipole moments (MMs). Thus, in the one-loop approximation, these corrections to k_γ give the value less than 10^{-2} [74]. Note that the electric dipole moment remains equal to zero in the one-loop approximation too. The introduction either of extra fermion generations or of extra Higgs doublets (HDs) increase the values of $k_{\gamma,Z}$ and $\lambda_{\gamma,Z}$ even at the tree level. For instance, in the case of two HDs with masses of charged and neutral Higgs bosons in the range 1 TeV, we have [75]

$$\Delta k_\gamma = 0.1\%, \qquad \Delta\lambda_\gamma = 0.03\%.$$

Each extra fermion $SU(2)$-doublet gives contribution to the Δk_γ less than 0.4%. Thus, we see that the substantial deviation of the MMs from their SM values could be obtained by a ridiculously large number of Higgs bosons or fermion families.

A good test for definition of the MMs is provided by the process

$$e^+ + e^- \to W^- + W^+. \tag{13.19}$$

In the second order of the perturbation theory the diagrams of this reaction are shown in Fig. 13.3. Next, we shall assume that only k_V and λ_V take anomalous values. In the case

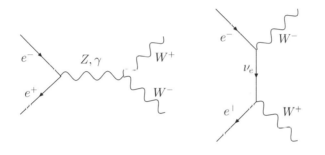

FIGURE 13.3
The Feynman diagrams corresponding to the process $e^+e^- \to W^-W^+$.

when the initial and final particles are unpolarized, the differential cross section is governed by the expression [76, 77]:

$$\frac{d\sigma}{d\Omega} = \frac{a^2\beta}{32s_W^4 s}\left[M_{\gamma\gamma} + M_{ZZ} + M_{\nu\nu} + +M_{\gamma Z} + M_{\nu Z} + M_{\gamma\nu}\right], \tag{13.20}$$

where

$$M_{\gamma\gamma} = 4s_W^4 N_1(k_\gamma, k_\gamma; \lambda_\gamma, \lambda_\gamma), \tag{13.21}$$

$$M_{ZZ} = \frac{s^2[1 + 4s_W^2(2s_W^2 - 1)]}{2[(s - m_Z^2)^2 + \Gamma_Z^2 m_Z^2]}N_1(k_Z, k_Z; \lambda_Z, \lambda_Z), \tag{13.22}$$

$$M_{\gamma Z} = \frac{2s(s - m_Z^2)[s_W^2(1 - 4s_W^2)]}{[(s - m_Z^2)^2 + \Gamma_Z^2 m_Z^2]}N_1(k_\gamma, k_Z; \lambda_\gamma, \lambda_Z), \tag{13.23}$$

$$M_{Z\nu} = \frac{s(s - m_Z^2)(2s_W^2 - 1)}{2[(s - m_Z^2)^2 + \Gamma_Z^2 m_Z^2]}N_2(k_Z, \lambda_Z), \tag{13.24}$$

$$M_{\gamma\nu} = -s_W^2 N_2(k_\gamma, \lambda_\gamma), \qquad M_{\nu\nu} = N_3, \tag{13.25}$$

$$N_1(a, b; c, d) = F\left[(ab + 2cd)y^2 - [1 - d - c + (a + c)(b + d)]y + 3\right] +$$

$$+2(1+a+c)(1+b+d)\beta^2 y,$$

$$N_2(a,b) = \frac{F}{2}\left(4ay^2 - 4y - \frac{8ym_W^2}{t}\right) + 8(1+a+b)\left(1+\beta^2 y + \frac{m_W^2}{t}\right),$$

$$N_3 = 4y + \frac{F}{2}\left(y^2 + \frac{4y^2 m_W^4}{t^2}\right).$$

p_+^W (p_-^W) is the momentum of W^+ (W^-), p_+^e (p_-^e) is the momentum of e^+ (e^-), and

$$s = (p_+^W + p_-^W)^2, \qquad t = (p_+^W - p_+^e)^2, \qquad u = (p_+^W - p_-^e)^2, \qquad \beta = (1-2/y)^{1/2},$$

$$y = s/2m_W^2, \qquad F = \frac{1}{y^2}\left(\frac{ut}{m_W^4} - 1\right).$$

If one neglects the contributions coming from the diagrams with the Z boson and neutrino exchanges, then the cross section obtained repeats the result of the classical work by Cabibbo and Gatto [78]. Setting

$$\Delta k_V = 0, \qquad \lambda_V = 0, \tag{13.26}$$

where Δk_V and $\Delta\lambda_V$ are the MMs deviations from their values in the SM, into (13.20), we lead to the differential cross section of the SM, which was first obtained in Ref. [79].

The total cross section follows from (13.20) by means of the multiplication by 2π and the replacement

$$N_{1,2,3} \to D_{1,2,3}, \tag{13.27}$$

where

$$D_1(a,b;c,d) = \beta^2\left\{\frac{4y^2}{3}(ab+2cd) + \frac{8y}{3}\left[1+2d+2c+(a+c)(b+d)+\frac{3}{2}(a+b)\right]+4\right\},$$

$$D_2(a,b) = 2\beta^2 y\left[4\left(2a+2b+\frac{5}{3}\right)+\frac{4ay}{3}\right]+16(a+b+1)\left(1-\frac{1}{\beta y}\ln\frac{1+\beta}{1-\beta}\right)-$$

$$-\frac{8}{\beta y^2}\ln\frac{1+\beta}{1-\beta}+4(1+\beta^2),$$

$$D_3 = 8+\frac{4(1+\beta^2)}{\beta}\ln\frac{1+\beta}{1-\beta}+2\beta^2 y\left(4+\frac{y}{3}\right).$$

Under fulfillment of (13.26), the obtained expression turns into the SM total cross section

$$\sigma^{CM} = \frac{\pi\alpha^2\beta}{2s_W^4 s}\left\{\frac{1}{\beta}\left(1+\frac{2}{y}+\frac{2}{y^2}\right)\ln\frac{1+\beta}{1-\beta}+\frac{m_Z^2(1-2s_W^2)}{s-m_Z^2}\left[\frac{2}{\beta y^2}(1+2y)\ln\frac{1+\beta}{1-\beta}-\right.\right.$$

$$\left.\left.-\frac{y}{12}-\frac{5}{3}-\frac{1}{y}\right]+\frac{m_Z^4(8s_W^4-4s_W^2+1)\beta^2}{48(s-m_Z^2)^2}(y^2+20y+12)\right\}, \tag{13.28}$$

where for the sake of simplicity we have set Γ_Z equal to zero. As follows from (13.28), each diagram in Fig. 13.3 taken separately results in the cross section that increases versus an energy and violates the unitarity. However, the interference terms lead to cancellations, both of linear increasing items and of the items that tend to constant values. As a result, the total cross section with the growth of s reaches the maximum (somewhere in the range of 200 GeV) and then begins to decrease by the law:

$$\sigma^{CM} \simeq s^{-1} \ln s. \tag{13.29}$$

In models with the anomalous MMs the situation is different. In the range of energies at least less than the fundamental scale Λ, the cross section linearly increases versus an energy. However, it should be noted that these models could also give $\sigma(e^+e^- \to W^-W^+)$ with the asymptotic behavior that is just the same as in the SM. It is possible when the MMs have a structure very similar to that of the hadron formfactors. Thus, for example, we could make the simplest possible *ansatz** "that all the MMs are defined as [76]:

$$s^{-n}c_i\theta(\sqrt{s} - \Lambda) + q_i, \tag{13.30}$$

where q_i are the values in the SM, c_i are arbitrary constants, $n \geq 1$ for λ_V and $n \geq 2$ for k_V."

Next, we shall assume all the MMs are constant quantities and deal with energies $\sqrt{s} < \Lambda$. Then, the asymptotic behavior of the reaction (13.12) will be defined by the formula:

$$\sigma^{MM}(e^+e^- \to W^-W^+) \simeq s. \tag{13.31}$$

One would think that at these circumstances a model with anomalous values of the MMs could be easily distinguished from some other models in any range of energies under consideration. Really, in the SM, the total cross section having reached its maximum at $\sqrt{s} \simeq 200$ GeV starts to decrease according to (13.29). The total cross sections of models with extended gauge groups of the electroweak symmetry display the analogous behavior with the only difference that they have one more maximum in the vicinity of the Z' resonance. In a model with the anomalous MMs, we are expecting that σ^{MM} will grow as a linear function of s over the whole range of energies. However, under certain values of the MMs, $\sigma^{MM}(e^+e^- \to W^-W^+)$ exposes an interesting property easily visible from its analytical expression (13.27). The total cross section increases until energy 200 GeV or so, then it falls down up to its minimum and only after that it starts to grow linearly on s [80].

The conditions

$$\Delta\lambda_\gamma = \Delta\lambda_Z = \Delta k_\gamma = 0, \qquad -0.7 < \Delta k_Z < 0.83 \tag{13.32}$$

is one of the examples of the minimum existence. When $|\Delta k_Z|$ increases, then the minimum is shifted towards smaller values of $\sqrt{s}$, the minimal value of σ^{MM} increasing. Giving the SM values to the three MM parameters and varying only a single one, we find four simple conditions for the minimum existence:

$$\left.\begin{array}{l} -0.78 < \Delta k_\gamma < 1.10, \\ -0.61 < \Delta\lambda_\gamma < 0.77, \\ -0.57 < \Delta\lambda_Z < 0.59. \end{array}\right\} \tag{13.33}$$

Of course, there exist the minimum conditions under deviations of all four MMs from the SM values, but, in what follows, we shall constrain ourselves by the variation of one of them.

It is evident, if we work in the range of energies less than $(\sqrt{s})_{min}$, then the deviations of the σ^{MM} values from the SM predictions are small. To obtain more precise constraints on

---

*An ansatz is a German noun with several meanings in the English language. In physics and mathematics, an ansatz is an educated guess that is verified later by its results.

the MM values, an excess of a collider energy over $(\sqrt{s})_{min}$ is needed because, as this takes place, σ^{MM} turns to the range of a linear rise on s. When using polarized electron-positron beams, the reaction (13.19) gives the information concerning the ratio of $W^+W^-\gamma$- and W^+W^-Z-couplings. To this end one needs to compare the differential cross sections for unpolarized and transverse-polarized electron-positron beams [81].

Let us see whether the property of the minima existence in the cross sections are common for processes with the vector bosons participation for models with the anomalous MMs. We address reactions that are sensitive only to the electromagnetic MMs. Consider the W boson scattering from the Coulomb nucleus field. At an arbitrary value of the anomalous magnetic moment, the differential cross section has the form [82]

$$\frac{d\sigma}{d\Omega} = \left(\frac{Ze^2}{2pv\sin^2\frac{\theta}{2}}\right)^2 \left\{1 + \frac{2v^2\sin^2\frac{\theta}{2}}{3m_W^2}\left(p^2 - 2k_\gamma m_W^2 + k_\gamma^2 p^2\right) - \right.$$

$$\left. - \frac{2p^2v^2\sin^4\frac{\theta}{2}}{3m_W^2}\left[1 + 2k_\gamma - k_\gamma^2\left(1 + \frac{2p^2}{m_W^2}\right)\right]\right\}, \tag{13.34}$$

where p and v are the momentum and velocity of W boson, θ is the angle between the boson momenta in initial and final states. As follows from (13.34), in the whole range of energies of the ingoing W boson ($\sqrt{s}$), the cross section increases $\sim s$ and has no minima at any values of k_γ.

Analogously, the analysis of the process

$$\gamma + \gamma \to W^+ + W^-. \tag{13.35}$$

demonstrates [83] that its total cross section is rising as a linear function of s and does not possess minima whatever the values k_γ has.

We introduce the quantity δ, which characterizes the experimental sensibility to the deviations from the SM

$$\delta = \frac{\sigma^{MM} - \sigma^{SM}}{\sqrt{\sigma^{SM}}}\sqrt{LT}, \tag{13.36}$$

where σ^{MM} is the cross section of the process (13.19) in a model with the anomalous MMs and LT is the integrated luminosity of the collider in units of pb^{-1}. From (13.36) it is evident that δ is an observable of the effect caused by the New Physics and it gives the deviations from the SM expressed in the standard error units. The notation $\delta(\Delta k_Z)$ means that δ is a function of Δk_Z at $\Delta k_\gamma = \Delta\lambda_Z = \Delta\lambda_\gamma = 0$ and so on. In Fig. 13.4 we present the graphs of the functions $\delta(\Delta k_\gamma)$, $\delta(\Delta k_Z)$, $\delta(\Delta\lambda_Z)$, and $\delta(\Delta\lambda_\gamma)$ under $LT = 500 \text{ pb}^{-1}$ and $\sqrt{s} = 196$ GeV.

It is known that linear e^-e^+-colliders (ILC, CLIC) can operate as $\gamma\gamma$-colliders. This is caused both by the classical photon bremsstrahlung of e^-e^+ beam and by the Compton scattering of laser photons from e^- and e^+. At such colliders we can investigate the reaction

$$e^-e^+ \to e^-e^+W^-W^+. \tag{13.37}$$

Its differential cross section could be obtained in the following way:

$$\frac{d\sigma}{d\cos\theta} \simeq \int_{4m_W^2/s}^1 d\tau \frac{dL_{\gamma\gamma}(\tau)}{d\tau}\frac{d\sigma_{\gamma\gamma \to W^-W^+}}{d\cos\theta}, \tag{13.38}$$

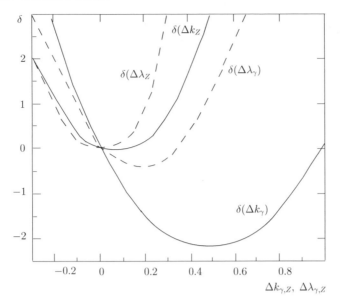

FIGURE 13.4
The graphs of the functions $\delta(\Delta k_{\gamma,Z})$ and $\delta(\Delta\lambda_{\gamma,Z})$.

where $dL_{\gamma\gamma}/d\tau$ is the differential photon luminosity

$$\frac{dL_{\gamma\gamma}}{d\tau} = \int_{\tau/x_m}^{x_m} \frac{dx}{x} f_\gamma(x) f_\gamma(\tau/x),$$

$x_m \leq 1$, $\tau = \hat{s}/s$, s and $\hat{s}$ is the square of the total energy in the center-of-mass frame for e^-e^+ and $\gamma\gamma$, respectively. The expressions for f_γ both in the case of the photon bremsstrahlung and of laser radiation are well known and may be found in Ref. [84]. In the case of laser photons the spectrum is hard and the total luminosity has the same order as does the e^-e^+-beam luminosity. This circumstance makes use of laser radiation under construction of a $\gamma\gamma$ collider more perspective.

Thanks to the W exchange in the t-channel, the cross section of the process (13.35) in the SM asymptotically tends to the value ~ 80 pb. Therefore, $\gamma\gamma$ colliders at the luminosity $10 - 20$ fb^{-1} will produce about one million W^+W^- pairs. This means that, using the reaction (13.35), one could obtain more precise constraints on the electromagnetic MMs than would give the reaction (13.19). In fact, at the following parameters of ILC

$$\sqrt{s} = 500 \text{ GeV}, \qquad \int L_{e^-e^+} dt = 10 \text{ fb}^{-1},$$

the bounds on $WW\gamma$ couplings (at 90% CL) obtained with the help of laser photons will look like [84]:

$$-0,02 \leq \Delta k_\gamma \leq 0,04, \qquad -0,06 \leq \lambda_\gamma \leq 0,1. \tag{13.39}$$

There is a substantial distinction in investigating the MMs at e^+e^- and hadron colliders. Because at LEP, ILC, and CLIC a detailed study of individual helicity amplitudes is possible, then contributions of formfactors could be divided at any energy in the center-of-mass frame. In addition, specificity of these machines is such that the cross section of the W pair production may be measured with the precision of a few percents, that is, deviations of the MMs from their SM values may be detected at the level of $O(10^{-2})$. The hadron colliders (Tevatron, LHC) investigate all processes of boson-pair production, namely, W^+W^--, $W^\pm\gamma$- and $W^\pm Z$-processes. In the last two cases, one may measure the $WW\gamma$- and WWZ-couplings independently. Due to more complicated situations associated with the background isolation, incomplete knowledge of the QCD radiative corrections, effects of structure functions and so on, comparing the measurement results with theory predictions at the level of $O(10^{-2})$ is impossible unfortunately. However, on the other hand, at hadron colliders, in comparison with e^+e^- machines, larger values of $\sqrt{s}$ are available and, as a result, the investigation of the MMs behavior under higher energy scales is possible.

In conclusion in this case, we list the current estimations for possible deviations of the MMs from the SM predictions [85]

$$\left.\begin{array}{ll} k_\gamma = 0.973^{+0.044}_{-0.045}, & k_Z = 0.924^{+0.059}_{-0.056} \pm 0.024 \\[2mm] \lambda_\gamma = -0.028^{+0.020}_{-0.021}, & \lambda_Z = -0.088^{+0.060}_{-0.057} \pm 0.023. \end{array}\right\} \quad (13.40)$$

13.3 Left-right model

> *If it is impossible to go straight, one should*
> *go either to the left or to the right.*
> Inviolable truth

From all the data available it follows, that neutrino masses are by many orders less than masses of other fundamental fermions—charged leptons and quarks. An explanation for such a giant suppression of the neutrino mass is provided by a *seesaw* mechanism that comprises a necessary component of many grand unified theories (GUTs). Because the GUTs predict connection between the quark and lepton Yukawa coupling constants at the unification scale, then within these theories it has been possible to obtain the following relation between the mixing angles in the quark and lepton sectors [87]

$$\theta_{12}^{NMM} + \theta_{12}^{CKM} = \frac{\pi}{4}, \qquad \theta_{23}^{CKM} + \theta_{23}^{NMM} = \frac{\pi}{4}, \qquad (13.41)$$

$$\theta_{13}^{NMM} \sim \theta_{13}^{CKM} \sim O(\theta_C^3), \qquad (13.42)$$

where θ_{ij}^{NMM} are the elements of the neutrino mixing matrix, θ_{ij}^{CKM} are the elements of the Cabibbo–Kobayashi–Maskawa matrix and $\theta_{12}^{CKM} = \theta_C$ is the Cabibbo angle, which are in good agreement with the experimental data. Thus, for example, Eqs. (13.41) and (13.42) could be obtained in the GUT based on the $SO(10)$ group. There is no doubt that these predictions can be considered as new experimental evidence for the idea of grand unification.

An interesting aspect of the $SO(10)$ model lies in the fact that it contains the $SU(2)_L \times SU(2)_R \times U(1)_{B-L}$ gauge group, that is, a group of electroweak symmetry is expanded by the factor $SU(2)$ compared to the SM. This SM extension received the name the left-right model (LRM). It is known that many troubles appearing in the SM are taken away in the LRM. It is useful to remember the basic advantages of this SM extension. They are:

(i) in the LRM all the fundamental fermions enter the theory in a symmetric way, (they form left- and right-handed doublets according to weak isospin), that is, the initial Lagrangian is P-invariant;

(ii) an almost maximum parity violation observed at a low-energy weak interaction can be interpreted within the LRM as appearing spontaneously and is connected with a nonzero value of neutrino masses;

(iii) the LRM naturally involves the seesaw mechanism;

(iv) quantum numbers of the $U(1)$ group is identified with $B-L$ that enables one to connect the parity violation with the local $B-L$-symmetry breaking;

(v) both in the quark and lepton sectors, the CP-violation effect is also caused by the spontaneous symmetry breaking and its parameters (the amplitudes ratio of decays of K_L^0- and K_S^0-mesons into pions, a charged asymmetry parameter of lepton decays of K_L^0-mesons, electric dipole moments of neutron and charged leptons, and so on) are of the same order as in experiments;

(vi) the Higgs sector in the LRM contains common elements with such popular SM extensions as the modified SM with two Higgs doublets, the minimal supersymmetric SM and the model, based on $SU(3)_L \times U(1)_N$ gauge group (N is a number of generations);

(vii) since the LRM belongs to a number of models, in which constants of coupling between Higgs bosons and charged leptons define the neutrino sector structure, then parameters of neutrino oscillations could be measured in processes without direct participation of neutrinos; and

(viii) at definite limiting transitions, the cross sections of many non-supersymmetric SM extensions could be obtained from the corresponding cross sections of the LRM.

For the first time the LRM was proposed at the beginning of the 1970s [88]. Then several versions of this model, which are distinguished by the choice of the transformation to the mass eigenstate basis in the space of neutral gauge bosons [89, 90, 91, 92, 93], appeared. This choice is determined both by the Higgs sector structure and by the gauge coupling constant values of the $SU(2)_L$, $SU(2)_R$, and $U(1)_{B-L}$ gauge groups. In Refs. [94, 95] it was shown that all versions of the LRMs can be unified into the so-called *continuous* LRM, which is characterized by the orientation angle of the $SU(2)_R$ generator in the group space.

The LRM is based on the gauge group $SU(2)_L \times SU(2)_R \times U(1)_{B-L}$ with quarks and leptons entering into the left- and right-handed doublets

$$\left.\begin{array}{l} Q_L^a(\frac{1}{2}, 0, \frac{1}{3}) - \begin{pmatrix} u_L^a \\ d_L^a \end{pmatrix} : \begin{pmatrix} u_L \\ d_L \end{pmatrix}, \begin{pmatrix} c_L \\ s_L \end{pmatrix}, \begin{pmatrix} t_L \\ b_L \end{pmatrix}, \\[12pt] Q_R^a(0, \frac{1}{2}, \frac{1}{3}) = \begin{pmatrix} u_R^a \\ d_R^a \end{pmatrix} : \begin{pmatrix} u_R \\ d_R \end{pmatrix}, \begin{pmatrix} c_R \\ s_R \end{pmatrix}, \begin{pmatrix} t_R \\ b_R \end{pmatrix}, \\[12pt] \Psi_L^a(\frac{1}{2}, 0, -1) = \begin{pmatrix} \nu_{aL} \\ l_{\alpha L} \end{pmatrix} : \begin{pmatrix} \nu_{eL} \\ e_L^- \end{pmatrix}, \begin{pmatrix} \nu_{\mu L} \\ \mu_L^- \end{pmatrix}, \begin{pmatrix} \nu_{\tau L} \\ \tau_L^- \end{pmatrix}, \\[12pt] \Psi_R^a(0, \frac{1}{2}, -1) = \begin{pmatrix} N_{aR} \\ l_{aR} \end{pmatrix} : \begin{pmatrix} N_{eR} \\ e_R^- \end{pmatrix}, \begin{pmatrix} N_{\mu R} \\ \mu_R^- \end{pmatrix}, \begin{pmatrix} N_{\tau R} \\ \tau_R^- \end{pmatrix}, \end{array}\right\} \quad (13.43)$$

where $a = 1, 2, 3$, in brackets the values of S_L^W, S_R^W and $B-L$ are given, S_L^W (S_R^W) is the weak left (right) isospin while B and L are the baryon and lepton numbers. The LRM has three gauge coupling constants: g_L, g_R, and g' for the $SU(2)_L$, $SU(2)_R$, and $U(1)_{B-L}$ groups, respectively. In the two-dimensional space of the functions (13.43), two representations of the $SU(2)_L$ and $SU(2)_R$ groups with the weight $j = \frac{1}{2}$ and one representation of the $U(1)_{B-L}$ group with the weight $j = 0$ are realized, that is,

$$F'_{L,R} = U(\omega(x))F_{L,R} \quad (13.44)$$

where

$$U(x) = U(\omega(x)_L)_L \times U(\omega(x)_R)_R \times U(\beta(x))_{B-L} = \exp{-i[g_L\omega_L^a(x)T_L^a + g_R\omega_R^a(x)T_R^a +}$$

$$+g'\beta(x)Y_{B-L}],$$

$F_{L,R}$ denotes the fermion doublet, $\omega_L^a(x), \omega_R^a(x), \beta(x)$ are transformation parameters, T_L^a, T_R^a, Y_{B-L} are the generators of the subgroups $SU(2)_L, SU(2)_R, U(1)_{B-L}$, which satisfy the commutation relations:

$$[T_L^a, T_L^b] = i\epsilon_{abc}T_L^c, \qquad [T_R^a, T_R^b] = i\epsilon_{abc}T_R^c, \qquad (13.45)$$

$$[T_L^a, T_R^b] = 0, \qquad [T_L^a, Y_{B-L}] = 0, \qquad [T_R^a, Y_{B-L}] = 0. \qquad (13.46)$$

and ϵ_{abc} are the structural parameters of the $SU(2)$ group. In the LRM scheme, there exist several possibilities for choosing a Higgs sector; however, any construction contains the bidoublet $\Phi(1/2, 1/2, 0)$. Nonzero VEVs of the electrically neutral components of the field Φ generate quark and lepton masses. Further, we can introduce either two triplets Δ_L $(1, 0, 2)$ and Δ_R $(0, 1, 2)$, or two doublets $X_L(1/2, 0, 1)$ and $X_R(0, 1/2, 1)$. The neutrino is a Majorana particle in the first case and a Dirac particle in the second one. Let us discuss in detail the model with the bidoublet and two triplets.

The transformation properties of the scalar Higgs fields are given by the relations:

$$\left. \begin{array}{c} \Phi' = U(\omega_L(x))_L \Phi U^*(\omega_R(x))_R, \qquad \Delta_L' = U(\omega_L(x))_L \Delta_L U^*(\omega_L(x))_L, \\ \Delta_R' = U(\omega_R(x))_R \Delta_R U^*(\omega_R(x))_R. \end{array} \right\} \qquad (13.47)$$

The total Lagrangian of the model is

$$L = -\frac{1}{4}(W_{L\mu\nu}^a W_{L\mu\nu}^a + W_{R\mu\nu}^a W_{R\mu\nu}^a + B_{\mu\nu}B_{\mu\nu}) + i\sum_a (\overline{\Psi}_L^a \gamma_\mu D_\mu \Psi_L^a +$$

$$+\overline{\Psi}_R^a \gamma_\mu D_\mu \Psi_R^a + \overline{Q}_L^a \gamma_\mu D_\mu Q_L^a + \overline{Q}_R^a \gamma_\mu D_\mu Q_R^a) + \sum_i \mid D_\mu \varphi_i \mid^2 + L_Y - V, \qquad (13.48)$$

where

$$W_{L,R\mu\nu}^a = \partial_\mu W_{L,R\nu}^a - \partial_\nu W_{L,R\mu}^a - g_{L,R}\epsilon_{abc}W_{L,R\mu}^b W_{L,R\nu}^c, \qquad B_{\mu\nu} = \partial_\mu B_\nu - \partial_\nu B_\mu,$$

$W_{L\mu}^a$ $(W_{L\mu}^a)$ form the gauge fields triplet associated with the $SU(2)_L$ $(SU(2)_R)$ group, B_μ is the gauge field singlet of the $U(1)$ group, L_Y is the Yukawa Lagrangian describing the gauge-invariant interaction between the Higgs particles φ_i and fermions, V is the Higgs potential. The gauge-covariant derivative entering to (13.48) is defined by the expression:

$$D_\mu = \partial_\mu - ig_L T_L^a W_{L\mu}^\alpha - ig_R T_R^\alpha A_{R\mu}^\alpha - ig' Y_{B-L}B_\mu.$$

Thus, for example, its action on the Higgs fields gives

$$D_\mu\Phi = \partial_\mu\Phi - i\left(g_L \frac{\sigma^a}{2}W_{\mu L}^a\Phi - g_R\Phi\frac{\sigma^a}{2}W_{\mu R}^a\right),$$

$$D_\mu\Delta_{L,R} = \partial_\mu\Delta_{L,R} - ig_{L,R}\frac{\sigma^a}{2}W_{\mu L,R}^a\Delta_{L,R} - ig'B_\mu\Delta_{L,R}.$$

where $\sigma_{1,2,3}$ are the Pauli matrices. It is known that the invariance with respect to the local gauge transformations U will be ensured providing

$$D_\mu(x) = U^{-1}(x)D_\mu'(x)U(x). \qquad (13.49)$$

Then, the demand (13.49) leads to the following transformation law of the gauge vector fields:

$$W_{L\mu}'^a(x) = W_{L\mu}^a(x) - g_L\epsilon_{cba}W_{L\mu}^c\omega_L^b(x) - \partial_\mu\omega_L^a(x), \qquad (13.50)$$

$$W_{R\mu}^{'a}(x) = W_{R\mu}^a(x) - g_R\epsilon_{cba}W_{R\mu}^c\omega_R^b(x) - \partial_\mu\omega_R^a(x), \tag{13.51}$$

$$B_\mu'(x) = B_\mu(x) - \partial_\mu\beta(x). \tag{13.52}$$

The symmetry of the theory with respect to the $SU(2)_L \times SU(2)_R \times U(1)_{B-L}$ gauge group demands the invariance of the Lagrangian (13.48) under the transformations (13.44), (13.47), (13.50)—(13.52). From the viewpoint of the Noether theorem, this means that the weak left and right isospins as well as the quantum number $B - L$ are conserved before the SSB.

The structure of the potential V is the essential element of the theory because it defines the physical states basis of Higgs bosons, Higgs masses, and interactions between Higgses. In Ref. [96] it was demonstrated that the most general form of the Higgs potential for the symmetric version of the LRM ($g_L = g_R$) is

$$V = -\mu_1^2[\mathrm{Sp}(\Phi^\dagger\Phi)] - \mu_2^2[\mathrm{Sp}(\tilde\Phi\Phi^\dagger) + \mathrm{Sp}(\tilde\Phi^\dagger\Phi)] - \mu_3^2[\mathrm{Sp}(\Delta_L\Delta_L^\dagger) + \mathrm{Sp}(\Delta_R\Delta_R^\dagger)] +$$
$$+\lambda_1\{[\mathrm{Sp}(\Phi\Phi^\dagger)]^2\} + \lambda_2\{[\mathrm{Sp}(\tilde\Phi\Phi^\dagger)]^2 + [\mathrm{Sp}(\tilde\Phi^\dagger\Phi)]^2\} + \lambda_3[\mathrm{Sp}(\tilde\Phi\Phi^\dagger)\mathrm{Sp}(\tilde\Phi^\dagger\Phi)] +$$
$$+\lambda_4\{\mathrm{Sp}(\Phi\Phi^\dagger)[\mathrm{Sp}(\tilde\Phi\Phi^\dagger) + \mathrm{Sp}(\tilde\Phi^\dagger\Phi)]\} + \rho_1\{[\mathrm{Sp}(\Delta_L\Delta_L^\dagger)]^2 + [\mathrm{Sp}(\Delta_R\Delta_R^\dagger)]^2\} +$$
$$+\rho_2\{[\mathrm{Sp}(\Delta_L\Delta_L)\mathrm{Sp}(\Delta_L^\dagger\Delta_L^\dagger) + \mathrm{Sp}(\Delta_R\Delta_R)\mathrm{Sp}(\Delta_R^\dagger\Delta_R^\dagger)]\} + \rho_3[\mathrm{Sp}(\Delta_L\Delta_L^\dagger)\mathrm{Sp}(\Delta_R\Delta_R^\dagger)] +$$
$$+\rho_4[\mathrm{Sp}(\Delta_L\Delta_L)\mathrm{Sp}(\Delta_R^\dagger\Delta_R^\dagger) + \mathrm{Sp}(\Delta_L^\dagger\Delta_L^\dagger)\mathrm{Sp}(\Delta_R\Delta_R)] + \alpha_1\{\mathrm{Sp}(\Phi\Phi^\dagger)[\mathrm{Sp}(\Delta_L\Delta_L^\dagger) +$$
$$+\mathrm{Sp}(\Delta_R\Delta_R^\dagger)]\} + \alpha_2[\mathrm{Sp}(\Phi\tilde\Phi^\dagger)\mathrm{Sp}(\Delta_R\Delta_R^\dagger) + \mathrm{Sp}(\Phi^\dagger\tilde\Phi)\mathrm{Sp}(\Delta_L\Delta_L^\dagger)] +$$
$$+\alpha_2^*[\mathrm{Sp}(\Phi^\dagger\tilde\Phi)\mathrm{Sp}(\Lambda_R\Lambda_R^\dagger) + \mathrm{Sp}(\tilde\Phi^\dagger\Phi)\mathrm{Sp}(\Lambda_L\Lambda_L^\dagger)] + \alpha_3[\mathrm{Sp}(\Phi\Phi^\dagger\Lambda_L\Lambda_L^\dagger) |$$
$$+\mathrm{Sp}(\Phi^\dagger\Phi\Lambda_R\Lambda_R^\dagger)] + \beta_1[\mathrm{Sp}(\Phi\Lambda_R\Phi^\dagger\Lambda_L^\dagger) + \mathrm{Sp}(\Phi^\dagger\Lambda_L\Phi\Lambda_R^\dagger)] + \beta_2[\mathrm{Sp}(\tilde\Phi\Lambda_R\Phi^\dagger\Lambda_L^\dagger) +$$
$$+\mathrm{Sp}(\tilde\Phi^\dagger\Delta_L\Phi\Delta_R^\dagger)] + \beta_3[\mathrm{Sp}(\Phi\Delta_R\tilde\Phi^\dagger\Delta_L^\dagger) + \mathrm{Sp}(\Phi^\dagger\Delta_L\tilde\Phi\Delta_R^\dagger)]. \tag{13.53}$$

Next, we shall use the Higgs potential in this form. Masses of fermions and their interactions with physical Higgs bosons are governed by the Yukawa Lagrangian. In the lepton sector this Lagrangian will look like:

$$\mathcal{L}_Y = -\sum_{a,b}\{h_{ab}\overline\Psi_{aL}\Phi\Psi_{bR} + h_{ab}'\overline\Psi_{aL}\tilde\Phi\Psi_{b,R} + if_{ab}[\Psi_{aL}^T C\sigma_2(\boldsymbol\sigma\cdot\boldsymbol\Delta_L)\Psi_{bL} + (L\to R)] + \mathrm{h.c.}\},$$

$$\tag{13.54}$$

where C is the charge conjugation matrix, $\tilde\Phi = \sigma_2\Phi^*\sigma_2$, $a, b = e, \mu, \tau$, h_{ab}, h_{ab}', h.c. denotes the Hermitian conjugate terms, and $f_{ab} = f_{ba}$ are the bidoublet and triplet Yukawa coupling constants (YCCs), respectively. Later, we shall discuss the possible variants of the Yukawa Lagrangian for quarks too.

In the LRM the Gell-Mann–Nishijima formula takes the form:

$$Q = S_{3L}^W + S_{3R}^W + \frac{B-L}{2}, \tag{13.55}$$

where S_{3L}^W (S_{3R}^W) denotes the third components of the weak left (right) isospin. In terms of the components, the Higgs multiplets can be represented as

$$\Phi = \begin{pmatrix} \Phi_1^0 & \Phi_2^+ \\ \Phi_1^- & \Phi_2^0 \end{pmatrix}, \tag{13.56}$$

and

$$(\boldsymbol\sigma\cdot\boldsymbol\Delta_L) = \begin{pmatrix} \delta_L^+/\sqrt2 & \delta_L^{++} \\ \delta_L^0 & -\delta_L^+/\sqrt2 \end{pmatrix}, \qquad (\boldsymbol\sigma\cdot\boldsymbol\Delta_R) = \begin{pmatrix} \delta_R^+/\sqrt2 & \delta_R^{++} \\ \delta_R^0 & -\delta_R^+/\sqrt2 \end{pmatrix}. \tag{13.57}$$

The spontaneous breakdown of symmetry according to the chain

$$SU(2)_L \times SU(2)_R \times U(1)_{B-L} \rightarrow SU(2)_L \times U(1)_Y \rightarrow U(1)_Q \qquad (13.58)$$

is realized for the following choice of the VEVs:

$$< \delta^0_{L,R} >= \frac{v_{L,R}}{\sqrt{2}}, \qquad < \Phi^0_1 >= k_1, \qquad < \Phi^0_2 >= k_2. \qquad (13.59)$$

To achieve agreement with experimental data, it is necessary to ensure fulfillment of the conditions

$$v_L << \max(k_1, k_2) << v_R. \qquad (13.60)$$

Constraints on the possible choice of left-handed Higgs multiplets could be obtained investigating the parameter ρ_0

$$\rho_0 = \frac{m^2_W}{m^2_Z c^2_W \rho_t}, \qquad (13.61)$$

where ρ_t includes the RCs connected with the t quark. In the lowest order of the perturbation theory ρ_t is given by the expression*

$$\rho_t = \frac{3 G_F m^2_t}{8 \sqrt{2} \pi} = 0.00915 \left(\frac{m_t}{170.9 \text{ GeV}} \right)^2. \qquad (13.62)$$

If we denote the VEV of an electrically neutral member of the i-th Higgs multiplet by λ_i, then the formula [97]

$$\rho_0 = \frac{2 \sum_i (S^W_{3L})^2_i \mid \lambda_i \mid^2}{\sum_i \{(S^W_L)_i [(S^W_L)_i + 1] - (S^W_{3L})^2_i\} \mid \lambda_i \mid^2} \qquad (13.63)$$

takes place. The global fitting the experimental data gives

$$\rho_0 = 1.004^{+0,0008}_{-0.0004}, \qquad (13.64)$$

where we have used

$$114.4 \text{ GeV} \leq m_H \leq 215 \text{ GeV}, \qquad m_t = 171.2 \pm 1.9 \text{ GeV}.$$

In our case, Eq. (13.63) results in

$$\rho_0 = \frac{1 + 4x}{1 + 2x},$$

where $x = \mid v_L / k_+ \mid^2$ and $k^2_+ = k^2_1 + k^2_2$, that is, x has to be much less than 1. However, one should treat such a type of constraints carefully. To compare the theoretical and experimental expressions demands the concrete definition of a theoretical model. The estimations (13.64) have been fulfilled using the SM. Within the LRM, for instance, we shall have the RCs caused by particles additional relative to the SM (heavy neutrinos, additional gauge and Higgs bosons), which influence the values of parameters being in the right side of Eq. (13.61).

---

*Two-loop corrections modify ρ_t into $\hat{\rho}_t$

$$\rho_t \rightarrow \hat{\rho}_t = \rho_t [1 + \frac{1}{3} R(m_H, m_t) \rho_t],$$

where the analytical form of $R(m_H, m_t)$ could be found in Ref. [98].

In accordance with the general rules, we must turn to new Higgs fields

$$\Phi' = \begin{pmatrix} \Phi_1^0 - k_1 & \Phi_2^+ \\ \Phi_1^- & \Phi_2^0 - k_2 \end{pmatrix}, \tag{13.65}$$

$$\Delta_L' = \begin{pmatrix} & \delta_L^{++} \\ & \delta_L^+ \\ \delta_L^0 - v_L/\sqrt{2} & \end{pmatrix}, \qquad \Delta_R = \begin{pmatrix} & \delta_R^{++} \\ & \delta_R^+ \\ \delta_R^0 - v_R/\sqrt{2} & \end{pmatrix} \tag{13.66}$$

providing the correct definition of a vacuum state, namely

$$< \Delta_{L,R}' >= 0, \qquad < \Phi' >= 0.$$

Investigating the mass matrices of Higgs bosons gives us the answer as to which of these bosons are physical and which are nonphysical. To obtain the Lagrangians describing interactions of physical particles we should pass on from the gauge basis (GB) to that of states with definite mass (physical basis). With that end in view in the GB we find a mass matrix which, in turn, consists of fermion M_f and boson M_b matrices. The M_f matrix follows from the Yukawa Lagrangian and it is linear on fermion masses. The form of M_b is defined both by the Yukawa potential and by the item $\sum_i |\mathcal{D}_\mu \varphi_i|^2$ in the total Lagrangian. This matrix determines the squared boson masses. Both matrices have a box-diagonal form, that is, they consist of square matrices (blocks) that are placed along the main diagonal and have zeros for the rest of the elements. Next, we find eigenvalues and eigenvectors (eigenstates) for every square matrix. At the last stage, the matrix of the transformation from the GB to the physical basis (PB) is built from the eigenvectors.

Let us demonstrate this procedure in detail by the example of the block corresponding to the sector of singly charged Higgs bosons [99]. With the help of the existence conditions of the potential minima

$$\frac{\partial V}{\partial \Phi_1^0} = \frac{\partial V}{\partial \Phi_2^0} = \frac{\partial V}{\partial \delta_L^0} = \frac{\partial V}{\partial \delta_R^0} = 0, \tag{13.67}$$

we can get the following equations system:

$$\mu_1^2 = \frac{2v_L v_R(\beta_2 k_1^2 - \beta_3 k_2^2) + (v_L^2 + v_R^2)(\alpha_1 k_-^2 - \alpha_3 k_2^2)}{2k_-^2} + k_+^2 \lambda_1 + 2k_1 k_2 \lambda_4, \tag{13.68}$$

$$\mu_2^2 = \frac{v_L v_R(\beta_1 k_-^2 - 2k_1 k_2(\beta_2 - \beta_3)] + (v_L^2 + v_R^2)(2\alpha_2 k_-^2 + \alpha_3 k_1 k_2)}{4k_-^2} +$$

$$+ \frac{k_+^2 \lambda_4}{2} + k_1 k_2(2\lambda_2 + \lambda_3), \tag{13.69}$$

$$\mu_3^2 = \frac{\alpha_1 k_+^2 + 4\alpha_2 k_1 k_2 + \alpha_3 k_2^2 + 2\rho_1(v_L^2 + v_R^2)}{2}, \tag{13.70}$$

$$\beta_2 = \frac{v_L v_R(2\rho_1 - \rho_3) - \beta_1 k_1 k_2 - \beta_3 k_2^2}{k_1^2}. \tag{13.71}$$

where $k_-^2 = k_1^2 - k_2^2$. Further, going over from basis $\{\Phi_1^\dagger, \Phi_2^\dagger, \delta_R^\dagger, \delta_L^\dagger\}$ to basis $\{\Phi_+^+, \Phi_-^+, \delta_R^+, \delta_L^+\}$ where

$$\Phi_+^+ = \frac{k_1 \Phi_1^+ + k_2 \Phi_2^+}{k_+}, \qquad \Phi_-^+ = \frac{k_1 \Phi_2^+ - k_2 \Phi_1^+}{k_+}, \tag{13.72}$$

and ruling out the parameters $\mu_{1,2,3}^2$ and β_3 with the help of Eqs. (13.68)–(13.71), we obtain for the mass matrix elements

$$M_{11} = \frac{\alpha_3 k_+^2 v_R^2}{2k_-^2}, \qquad M_{12} = M_{21} = 0, \qquad M_{13} = M_{31} = \frac{\alpha_3 k_+ v_R}{2\sqrt{2}},$$

$$M_{14} = M_{41} = \frac{v_R k_+ (\beta_1 k_1 + 2\beta_3 k_2)}{2\sqrt{2}k_1}, \qquad M_{22} = M_{23} = M_{32} = M_{24} = M_{42},$$

$$M_{33} = \frac{\alpha_3 k_-^2}{4}, \qquad M_{34} = M_{43} = \frac{k_-^2(\beta_1 k_1 + 2\beta_3 k_2)}{4k_1}, \qquad M_{44} = \frac{\alpha_3 k_-^2 - 2v_R^2(2\rho_1 - \rho_3)}{4}.$$

Diagonalizing this matrix and determining its eigenstates, we arrive at the following conclusions. Four states

$$G_{\tilde\delta}^{(\pm)} = -\frac{bk_0}{v_R}\Phi_+^{(\pm)} + a\Delta_R^{(\pm)} + \frac{d\beta k_0}{(\alpha + \rho_1 - \rho_3/2)v_R}\Delta_L^{(\pm)}, \qquad G_h^{(\pm)} = \Phi_-^{(\pm)} \tag{13.73}$$

describe nonphysical (massless) Goldstone fields that can be excluded from the theory due to the local gauge invariance with respect to the $SU(2)_L \times SU(2)_R \times U(1)_{B-L}$ group. These fields cause an appearance of longitudinal components in $W_1^\pm$ and $W_2^\pm$ bosons (we recall that in the LRM W_1 plays a part of W in the SM). Four Higgs bosons

$$h^{(\pm)} = b\Phi_+^{(\pm)} + \frac{ak_0}{v_R}\delta_R^\pm + \frac{d\beta k_0^2}{(\alpha + \rho_1 - \rho_3/2)v_R^2}\delta_L^\pm, \tag{13.74}$$

$$\tilde\delta^{(\pm)} = \frac{a\beta k_0}{(\alpha + \rho_1 - \rho_3/2)v_R}\delta_R^\pm - d\delta_L^\pm, \tag{13.75}$$

where

$$k_0 = \frac{k_-^2}{\sqrt{2}k_+}, \qquad \alpha = \frac{\alpha_3 k_+^2}{2k_-^2}, \qquad \beta = \frac{k_+^2(\beta_1 k_1 + 2\beta_3 k_2)}{2k_-^2 k_0}, \qquad b = \left(1 + \frac{k_0^2}{v_R^2}\right)^{-1/2},$$

$$a = \left\{1 + \left[1 + \frac{\beta^2}{(\alpha + \rho_1 - \frac{\rho_3}{2})^2}\right]\frac{k_0^2}{v_R^2}\right\}^{-1/2}, \qquad d = \left[1 + \frac{\beta^2 k_0^2}{(\alpha + \rho_1 - \frac{\rho_3}{2})^2 v_R^2}\right]^{-1/2},$$

acquire a mass in the result of the SSB and represent physical Higgs bosons, that is, they must be observed in experiments. The squares of their masses are given by

$$m_h^2 = \alpha(v_R^2 + k_0^2) + \frac{\beta^2 k_0^2}{\alpha + \rho_1 - \rho_3/2}, \tag{13.76}$$

$$m_{\tilde\delta}^2 = (\rho_3/2 - \rho_1)v_R^2 - \frac{\beta^2 k_0^2}{\alpha + \rho_1 - \rho_3/2}. \tag{13.77}$$

Having done the analogous operations in the sector of doubly charged Higgs bosons, we come to recognize that there are four physical scalars $\Delta_{1,2}^{(\pm\pm)}$ which are orthogonal mixtures of $\delta_L^{(\pm\pm)}$ and $\delta_R^{(\pm\pm)}$ states

$$\Delta_1^{(\pm\pm)} = c_{\theta_d}\delta_L^{(\pm\pm)} + s_{\theta_d}\delta_R^{(\pm\pm)}, \tag{13.78}$$

$$\Delta_2^{(\pm\pm)} = -s_{\theta_d}\delta_L^{(\pm\pm)} + c_{\theta_d}\delta_R^{(\pm\pm)}, \tag{13.79}$$

where $c_{\theta_d} = \cos\theta_d$, $s_{\theta_d} = \sin\theta_d$ and

$$\tan 2\theta_d = \frac{2k_-^2[\beta_3 k_+^2 + \beta_1 k_1 k_2)}{k_1^2(2\rho_1 - \rho_3 - 4\rho_2)v_R^2}.$$

The squared masses of the doubly charged Higgs bosons are given by

$$m_{\Delta_1}^2 = \frac{\alpha_3 k_-^2 + 4\rho_2 v_R^2}{2} + \frac{k_-^4(\beta_3 k_+^2 + \beta_1 k_1 k_2)^2}{2k_1^4(4\rho_2 + \rho_3 - 2\rho_1)v_R^2}, \tag{13.80}$$

$$m^2_{\Delta_2} = \frac{\alpha_3 k^2_- - (2\rho_1 - \rho_3)v^2_R}{2} - \frac{k^4_- (\beta_3 k^2_+ + \beta_1 k_1 k_2)^2}{2k^4_1(4\rho_2 + \rho_3 - 2\rho_1)v^2_R}. \tag{13.81}$$

Next, we assume that the mixing in the charged leptons sector is absent, that is, the gauge basis coincides with the physical one. Because of the historic reasons the charged and neutral currents involve a neutrino with the definite flavor value (when introducing into the theory the neutrino has been considered to be massless), then we also use the field neutrino functions only in the gauge basis under obtaining the corresponding Lagrangians. Hence, in (13.54) we shall realize the transition to the physical basis for the Higgs bosons only. Making use of (13.74) and (13.75), we obtain the Lagrangian to describe the interaction between leptons and singly charged Higgs bosons

$$\mathcal{L}^{sc}_l = \sum_{a,b} \left\{ b \left[\frac{h'_{ab}k_2 - h_{ab}k_1}{2k_+} \overline{\nu}_a (1 - \gamma_5) l_b - \frac{h_{ab}k_2 - h'_{ab}k_1}{2k_+} \overline{N}_a (1 + \gamma_5) l_b \right] h^{(+)} + \right.$$

$$+ \frac{f_{ab}}{\sqrt{2}} \overline{l^c_a} (1 + \gamma_5) \nu_b \left[\left(\frac{d\beta k^2_0}{(\alpha + \rho_1 - \rho_3/2)v^2_R} h^{(+)} - d\tilde{\delta}^{(+)} \right) + \overline{l^c_a} (1 - \gamma_5) N_b \left(\frac{a k_0}{v_R} h^{(+)} + \right. \right.$$

$$\left. \left. + \frac{a\beta k_0}{(\alpha + \rho_1 - \rho_3/2)v_R} \tilde{\delta}^{(+)} \right) \right] + \text{h.c.} \right\}, \tag{13.82}$$

where the superscript c denotes the charge conjugation operation. With the help of the identity

$$\overline{\nu}_a(1 - \gamma_5)l_b = \overline{(\nu^c_u)^c}(1 - \gamma_5)(l^c_b)^c = -(\nu^c_u)^T C^{-1}(1 - \gamma_5)C(\overline{l^c_b})^T = \overline{l^c_b}(1 - \gamma_5)\nu^c_u,$$

and self-conjugation condition for a Majorana spinor

$$\nu^c_a - \lambda^*_{\nu a}\nu_a,$$

where $\lambda_{\nu a}$ is the creation phase factor of the field ν_a, the first two terms in (13.82) can be rewritten in the view

$$\sum_{a,b} b \left[\lambda^*_{\nu a} \frac{h'_{ab}k_2 - h_{ab}k_1}{2k_+} \overline{l^c_b}(1 - \gamma_5)\nu_a - \lambda^*_{N a} \frac{h_{ab}k_2 - h'_{ab}k_1}{2k_+} \overline{l^c_b}(1 + \gamma_5)N_a \right] h^{(+)}. \tag{13.83}$$

To determine the Lagrangian that governs the interaction of the physical Higgs bosons with gauge bosons we also need to have formulae expressing the connection between the gauge and physical bases in the gauge boson sector. The mass matrix for charged gauge bosons follows from the part of the total Lagrangian $\sum_i | D_\mu \varphi_i |^2$. Calculations result in

$$\begin{pmatrix} \frac{g^2_L}{2}(k^2_+ + 2v^2_L) & g_L g_R k_1 k_2 \\ g_R g_L k_1 k_2 & \frac{g^2_R}{2}(k^2_+ + 2v^2_R) \end{pmatrix} = \begin{pmatrix} M^2_L & M^2_{LR} \\ M^2_{LR} & M^2_R \end{pmatrix}. \tag{13.84}$$

The eigenvalues and eigenstates of this matrix are as follows:

$$W_1 = W_L \cos\xi + W_R \sin\xi, \qquad W_2 = -W_L \sin\xi + W_R \cos\xi, \tag{13.85}$$

$$m^2_{W_{1,2}} = \frac{1}{2} \left(M^2_L + M^2_R \mp \sqrt{(M^2_L - M^2_R)^2 + 4M^2_{L,R}} \right), \tag{13.86}$$

where

$$\tan 2\xi = \frac{4g_L g_R k_1 k_2}{g_R^2(k_+^2 + 2v_R^2) - g_L^2(k_+^2 + 2v_L^2)} \tag{13.87}$$

and ξ is the mixing angle in the charged gauge boson sector.

The source of the appearance of the mass matrix for neutral gauge bosons is also the quantity $\sum_i |D_\mu \varphi_i|^2$. However, here the situation is not quite unambiguous. The mass matrix in the basis $(W_{3\mu}^L, W_{3\mu}^R, B_\mu)$ has the form

$$\begin{pmatrix} A_L & D & C_L \\ D & A_R & C_R \\ C_L & C_R & G \end{pmatrix}, \tag{13.88}$$

where

$$\left. \begin{aligned} A_{L,R} &= \frac{g_{L,R}^2}{2}\left(k_+^2 + 4v_{L,R}^2\right), & G &= 2g'^2\left(v_L^2 + v_R^2\right), \\ D &= -\frac{g_L g_R}{2}k_+^2, & C_{L,R} &= -2g' g_{L,R}v_{L,R}^2. \end{aligned} \right\} \tag{13.89}$$

Diagonalizing the matrix (13.88) is carried by two stages. In doing so, to change the initial basis to the final one is realized by the chain:

$$\begin{pmatrix} W_{3L\mu} \\ W_{3R\mu} \\ B_\mu \end{pmatrix} \xrightarrow{\Lambda} \begin{pmatrix} Z_{L\mu} \\ Z_{R\mu} \\ A_\mu \end{pmatrix} \xrightarrow{U_N} \begin{pmatrix} Z_{1\mu} \\ Z_{2\mu} \\ A_\mu \end{pmatrix} \tag{13.90}$$

(Z_1 is the analog of the Z boson in the SM). Consequently, the matrix of transformation from the gauge basis to the physical one represents the product of two matrices Λ and U_N, where U_N has the same form in all versions of the LRMs

$$U_N = \begin{pmatrix} \cos\phi & \sin\phi & 0 \\ -\sin\phi & \cos\phi & 0 \\ 0 & 0 & 1 \end{pmatrix}, \tag{13.91}$$

(ϕ is the mixing angle in the neutral gauge boson sector), while the form of the matrix Λ defines one of the possible versions of the LRMs. In the most general case the matrix Λ is given by the expression:

$$\Lambda = \begin{pmatrix} eb(g'^{-2}c_\varphi + a_- g_R^{-1}) & eb(g'^{-2}s_\varphi - a_- g_L^{-1}) & -ebg'^{-1}a_+ \\ -bg'^{-1}s_\varphi & bg'^{-1}c_\varphi & -ba_- \\ eg_L^{-1} & eg_R^{-1} & eg'^{-1} \end{pmatrix}, \tag{13.92}$$

where

$$\left. \begin{aligned} a_+ &= g_R^{-1}s_\varphi + g_L^{-1}c_\varphi, & a_- &= g_R^{-1}c_\varphi - g_L^{-1}s_\varphi, & b^{-1} &= \sqrt{g'^{-2} + a_-^2}, \\ & & e^{-2} &= g_L^{-2} + g_R^{-2} + g'^{-2}, \end{aligned} \right\} \tag{13.93}$$

$c_\varphi = \cos\varphi, s_\varphi = \sin\varphi$. Thus, photon A and Z_1, Z_2 bosons are related with neutral gauge bosons W_{3L}, W_{3R}, and B in the gauge basis in the following way:

$$\begin{pmatrix} Z_{1\mu} \\ Z_{2\mu} \\ A_\mu \end{pmatrix} = U_N \Lambda \begin{pmatrix} W_{3L\mu} \\ W_{3R\mu} \\ B_\mu \end{pmatrix}. \tag{13.94}$$

When varying the angle φ, this model [94, 95] (it received the name the *continuous LRM*) reproduces all possible versions of the LRMs. Below we choose φ equal to zero. We recall that in this case the continuous LRM coincides with the model proposed by Mohapatra [93]. Then, from (13.94) it follows that

$$W_{3L\mu} = c_W c_\phi Z_{1\mu} - c_W s_\phi Z_{2\mu} + s_W A_\mu, \tag{13.95}$$

$$W_{3R\mu} = e c_W^{-1}(-g_R^{-1}s_W c_\phi + g'^{-1}s_\phi)Z_{1\mu} + e c_W^{-1}(g_R^{-1}s_W s_\phi + g'^{-1}c_\phi)Z_{2\mu} + e g_R^{-1}A_\mu. \tag{13.96}$$

$$B_\mu = -e c_W^{-1}(g'^{-1}s_W c_\phi + g_R^{-1}s_\phi)Z_{1\mu} + e c_W^{-1}(g'^{-1}s_W s_\phi - g_R^{-1}c_\phi)Z_{2\mu} + e g'^{-1}A_\mu. \tag{13.97}$$

At $g_L = g_R$ (the symmetric LRM) the expressions for the neutral gauge boson masses are quite simple

$$m_{Z_{1,2}} = \frac{1}{2}\left[Y \mp \sqrt{Y^2 - 4g_L^2(g_L^2 + g'^2)X}\right], \tag{13.98}$$

where

$$Y = k_+^2 g_L^2 + 2(g_L^2 + g'^2)(v_L^2 + v_R^2), \qquad X = k_+^2(v_L^2 + v_R^2) + 4v_L^2 v_R^2.$$

At that the mixing angle in the neutral gauge boson sector is defined as

$$\tan 2\phi \simeq \frac{k_+^2 \sqrt{c_{2W}^3}}{2c_W^4 v_R^2}, \tag{13.99}$$

where for the sake of simplicity we have set $v_L = 0$.

Using the formula of the transition to the mass basis, one can get all the Lagrangians we are interested in. Thus, for example, employing the relations (13.74), (13.75), (13.85), (13.95)–(13.97), we find the Lagrangian to describe the interaction between the singly charged Higgs bosons and W_1 and Z_1 bosons

$$\mathcal{L}_V^{sc} = \left\{ -\left[\frac{g_L g_R c_W c_\phi b k_+^2}{\sqrt{2}k_+}h^{(+)} + \frac{g_L g_R g' s_W(g'^{-1}s_W c_\phi + g_R^{-1}s_\phi)}{c_W}\left(a k_0 h^{(+)} + \right.\right.\right.$$

$$\left.\left.\left. + \frac{a\beta k_0}{(\alpha + \rho_1 - \rho_3/2)}\delta^{(+)}\right)\right]s_\xi - \right.$$

$$\left. - \left[\frac{e g_L s_\phi v_L(g'g_R^{-1} + g_R g'^{-1})}{c_W}\left(\frac{d\beta k_0^2}{(\alpha + \rho_1 - \rho_3/2)v_R^2}h^{(+)} - d\tilde{\delta}^{(+)}\right)\right]c_\xi\right\}W_{1\mu}Z_{1\mu} + \text{h.c.}. \tag{13.100}$$

Next, taking into consideration Eqs. (13.78) and (13.79), we obtain the Lagrangians governing the interaction of leptons and gauge bosons with $\Delta_{1,2}^{(\pm\pm)}$ scalars

$$\mathcal{L}_l^{dc} = -\sum_{a,b}\frac{f_{ab}}{2}\left[\overline{l_a^c}(1 + \gamma_5)l_b c_{\theta_d} - \overline{l_a^c}(1 - \gamma_5)l_b s_{\theta_d}\right]\Delta_1^{(++)} +$$

$$+ (\Delta_1 \to \Delta_2, \theta_d \to \theta_d - \frac{\pi}{2}) + \text{h.c.}, \tag{13.101}$$

$$\mathcal{L}_W^{dc} = -g_L^2 v_L[c_{\theta_d}\Delta_1^{(++)} - s_{\theta_d}\Delta_2^{(++)}]W_{L\mu}W_L^\mu - g_R^2 v_R[s_{\theta_d}\Delta_1^{(++)} + c_{\theta_d}\Delta_2^{(++)}]W_{R\mu}W_R^\mu, \tag{13.102}$$

$$\mathcal{L}_{Z_1}^{dc} = \left[(\alpha_L c_{\theta_d}^2 + \alpha_R s_{\theta_d}^2)\Delta_1^{(--)}\partial_\mu\Delta_1^{(++)} + (\alpha_L^2 c_{\theta_d}^2 + \alpha_R^2 s_{\theta_d}^2)\Delta_1^{(--)}\Delta_1^{(++)}Z_{1\mu} + \right.$$

$$+ (\Delta_1 \to \Delta_2, \theta_d \to \theta_d + \pi/2) - s_{\theta_d}c_{\theta_d}(\alpha_L - \alpha_R)\Delta_1^{(--)}\partial_\mu\Delta_2^{(++)} -$$

$$-s_{\theta_d}c_{\theta_d}(\alpha_L^2 - \alpha_R^2)\Delta_1^{(--)}\Delta_2^{(++)}Z_{1\mu} + \text{h.c.}\Bigg]Z_{1\mu}, \qquad (13.103)$$

where

$$\alpha_L = e[2\cot 2\theta_W c_\phi - g's_\phi c_W^{-1}g_R^{-1}], \qquad \alpha_R = e[-2c_W^{-1}s_W c_\phi + s_\phi c_W^{-1}g'^{-1}g_R^{-1}(g_R^2 - g'^2)].$$

For the Higgs potential (13.53) the matrix of the transition to the mass eigenstate basis for the neutral scalar Higgs boson is too cumbersome. This causes us to make some simplifying assumptions concerning the form of V. Let us suppose, that the following is true:

$$\alpha_1 = -\frac{2\alpha_2 k_2}{k_1}, \qquad \alpha_3 = -\frac{2\alpha_2 k_-^2}{k_1 k_2}, \qquad \beta_1 = -\frac{2\beta_3 k_2}{k_1}. \qquad (13.104)$$

The analysis of the mass matrix shows, at $v_L = 0$ there are two nonphysical Goldstone bosons ϕ_-^{0i} and δ_R^{0i} whose degrees of freedom are spent on the longitudinal components production of the Z_1 and Z_2 gauge bosons. In the physical Higgs boson sector we have 4 scalar

$$S_1 = \phi_-^{0r}c_{\theta_0} + \phi_+^{0r}s_{\theta_0}, \qquad S_2 = -\phi_-^{0r}s_{\theta_0} + \phi_+^{0r}c_{\theta_0}, \qquad S_3 = \delta_R^{0r}, \qquad S_4 = \delta_L^{0r}, \quad (13.105)$$

and 2 pseudoscalar

$$P_1 = \phi_+^{0i}, \qquad P_2 = \delta_L^{0i} \qquad (13.106)$$

neutral bosons. In Eqs. (13.105) and (13.106) the superscript r (i) denotes the real (imaginary) part of the corresponding quantity, $c_{\theta_0} = \cos\theta_0$, $s_{\theta_0} = \sin\theta_0$, and

$$\tan 2\theta_0 = \frac{4k_1 k_2 k_+^2[2(2\lambda_2 + \lambda_3)k_1 k_2 + \lambda_4 k_+^2]}{k_1 k_2[(4\lambda_2 + 2\lambda_3)(k_-^4 - 4k_1^2 k_2^2) - k_+^2(2\lambda_1 k_+^2 + 8\lambda_4 k_1 k_2)] - (\alpha_2 - \alpha_4)v_R^2 k_+^4}.$$

The squared masses of these bosons are defined by

$$m_{S_1}^2 = 2\lambda_1 k_+^2 + 8k_1^2 k_2^2(2\lambda_2 + \lambda_3)/k_+^2 + 8\lambda_4 k_1 k_2 +$$

$$+\frac{4k_1 k_2 k_+^4[2(2\lambda_2 + \lambda_3)k_1 k_2/k_+^2 + \lambda_4]^2}{(\alpha_2 - \alpha_4)v_R^2 k_+^2}, \qquad (13.107)$$

$$m_{S_2}^2 = -\frac{\alpha_2 v_R^2 k_+^2}{k_1 k_2} - \frac{4k_1 k_2 k_+^4[2(2\lambda_2 + \lambda_3)k_1 k_2/k_+^2 + \lambda_4]^2}{(\alpha_2 - \alpha_4)v_R^2 k_+^2}, \qquad (13.108)$$

$$m_{S_3}^2 = 2\rho_1 v_R^2, \qquad m_{S_4} = (\rho_3/2 - \rho_1)v_R^2, \qquad (13.109)$$

$$m_{P_1}^2 = 2k_+^2(\lambda_3 - 2\lambda_2) - \frac{\alpha_2 v_R^2 k_+^2}{k_1 k_2}, \qquad m_{P_2}^2 = (\rho_3/2 - \rho_1)v_R^2. \qquad (13.110)$$

The Lagrangians that determine the interaction of the physical neutral Higgs bosons with leptons and gauge bosons are given by the expressions

$$\mathcal{L}_l^n = -\frac{1}{\sqrt{2}k_+}\sum_a m_a \bar{l}_{aR}l_{aL}(S_1 c_{\theta_0} - S_2 s_{\theta_0}) - \frac{1}{\sqrt{2}k_+}\sum_{a,b}\bar{l}_{aR}l_{bL}[(h_{ab}k_1 - h'_{ab}k_2)(S_1 s_{\theta_0} + S_2 c_{\theta_0}) +$$

$$i(h_{ab}k_1 + h'_{ab}k_2)P_1] - \frac{1}{\sqrt{2}k_+}\sum_{a,b}\{\overline{N}_{aR}\nu_{bL}[h_{ab}(k_1 c_{\theta_0} - k_2 s_{\theta_0}) + h'_{ab}(k_2 c_{\theta_0} +$$

$$k_1 s_{\theta_0})]S_1 - [h_{ab}(k_1 s_{\theta_0} + k_2 c_{\theta_0}) + h'_{ab}(k_2 s_{\theta_0} - k_1 c_{\theta_0})]S_2 - i(h_{ab}k_2 +$$

$$h'_{ab}k_1)P_1\} - \frac{1}{\sqrt{2}}\sum_{a,b} f_{ab}[\overline{\nu}^c_{aL}\nu_{bL}(S_4 + iP_2) + \overline{N}^c_{aR}N_{bR}S_3] + \text{h.c.}, \tag{13.111}$$

$$2\mathcal{L}^n_W = k_+[W^*_{1\mu}W_{1\mu}(g^2_L c^2_\xi + g^2_R s^2_\xi) + W^*_{2\mu}W_{2\mu}(g^2_L s^2_\xi +$$

$$g^2_R c^2_\xi) + \frac{1}{2}s_{2\xi}(g^2_L - g^2_R)(W^*_{1\mu}W_{2\mu} + W^*_{2\mu}W_{1\mu})](S_1 c_{\theta_0} - S_2 s_{\theta_0}) -$$

$$\frac{g_L g_R}{k_+}\{[c_{2\xi}(W^*_{2\mu}W_{1\mu} + W^*_{1\mu}W_{2\mu}) + s_{2\xi}(W^*_{2\mu}W_{2\mu} - W^*_{1\mu}W_{1\mu})][(2k_1 k_2 c_{\theta_0} + k^2_- s_{\theta_0})S_1 -$$

$$(2k_1 k_2 s_{\theta_0} - k^2_- c_{\theta_0})S_2] - ig_L g_R k_+(W^*_{2\mu}W_{1\mu} - W^*_{1\mu}W_{2\mu})P_1 + g^2_L v_L[W^*_{1\mu}W_{1\mu}c^2_\xi +$$

$$W^*_{2\mu}W_{2\mu}s^2_\xi + \frac{1}{2}s_{2\xi}(W^*_{1\mu}W_{2\mu} + W^*_{2\mu}W_{1\mu})]S_4 + g^2_R v_R[W^*_{1\mu}W_{1\mu}s^2_\xi +$$

$$W^*_{2\mu}W_{2\mu}c^2_\xi - \frac{1}{2}s_{2\xi}(W^*_{1\mu}W_{2\mu} + W^*_{2\mu}W_{1\mu})]S_3 + \text{h.c.}. \tag{13.112}$$

When building the interaction Lagrangians of quarks, the following difficulties emerge. Let us assume, in the quark sector we use the traditional expression for the Yukawa Lagrangian

$$\mathcal{L}_Y = -\sum_{a,b}(h^{(q)}_{ab}\overline{Q}_{aL}\phi Q_{bR} + h^{(q)'}_{ab}\overline{Q}_{aL}\tilde{\phi}Q_{b,R} + \text{h.c.}). \tag{13.113}$$

Then, if in the quark sector the bidoublet VEVs are degenerated (DBV)

$$k_1 = k_2$$

or quasi-degenerated (QGBV)

$$k_1 - k - \frac{\Delta k}{2}, \qquad k_2 = k + \frac{\Delta k}{2}, \qquad \frac{\Delta k}{k} \ll 1,$$

in the case of the symmetric LRM we have

$$\mathcal{M}_u = \mathcal{M}_d, \qquad \text{or} \qquad \mathcal{M}_u \approx \mathcal{M}_d, \tag{13.114}$$

where $\mathcal{M}_u$ ($\mathcal{M}_d$) is the diagonal matrix for the up (down) quarks. To avoid (13.114) one can introduce the additional Higgs triplets $\Delta'_L(1,0,2/3)$ and $\Delta'_R(0,1,2/3)$ and supplement the Lagrangian (13.113) with the term [100]

$$-\sum_{a,b}\{f^{(q)}_{ab}Q^T_{aL}C\tau_2(\boldsymbol{\sigma}\cdot\boldsymbol{\Delta}'_L)Q_{bL} + (L \to R) + \text{h.c.}\}.$$

There is also another way—introduce the additional bidoublet $\phi_u(1/2, 1/2, 0)$ that interacts both with up-, and down-quarks, but which contributes only to the masses of up-quarks [92]. However, in both approaches the undesirable increase in the number of the physical Higgs bosons takes place. In the asymmetric version of the LRM it is not at all necessary to complicate the Higgs sector to obtain $\mathcal{M}_u \neq \mathcal{M}_d$ at the DBV. Instead of (13.112) one could take the Lagrangian similar in its structure to the corresponding Lagrangian of the SM

$$\mathcal{L}^{(q)}_Y = -\sum_{a,b}\{h^{(q)}_{ab}\overline{Q}_{aL}\sigma_-\phi\sigma_+ Q_{bR} + h^{(q)'}_{ab}\overline{Q}_{aL}\sigma_+\phi^*\sigma_- Q_{b,R} + \text{h.c.}\}, \tag{13.115}$$

where $\sigma_\pm = \frac{1}{2}(\sigma_1 \pm i\sigma_2)$ and one obtains from it all the requisite Lagrangians. Thus, for example, we find the interaction Lagrangian of quarks with neutral Higgs bosons

$$\mathcal{L}_q^n = -\frac{1}{\sqrt{2}k_+} \sum_a \overline{u}_a \{ m_{u_a}[(\cos\theta_0 + \frac{k_1}{k_2}\sin\theta_0)S_1 - (\sin\theta_0 - \frac{k_1}{k_2}\cos\theta_0)S_2] +$$

$$\frac{im_{u_a}k_1}{k_2}\gamma_5 P_1 \} u_a + (u_a \to d_a, \theta_0 \to -\theta_0). \tag{13.116}$$

The wave neutrino functions in the flavor and physical bases are related by the mixing matrix $\mathcal{U}$

$$\Omega^l = \begin{pmatrix} \nu_{eL} \\ \nu_{\mu L} \\ \nu_{\tau L} \\ N_{eR} \\ N_{\mu R} \\ N_{\tau R} \end{pmatrix} = \mathcal{U}\Omega^m = \mathcal{U} \begin{pmatrix} \nu_1 \\ \nu_2 \\ \nu_3 \\ N_1 \\ N_2 \\ N_3 \end{pmatrix}. \tag{13.117}$$

In the most general case $\mathcal{U}$ contains 15 oscillation angles. For simplicity, we shall assume that the mixing between light and heavy neutrinos belonging to different generations is absent. This reduces the oscillation angles number to 10 and the matrix $\mathcal{U}$ is defined by the expression:

$$\mathcal{U} = \mathcal{M}^{\nu N} \begin{pmatrix} \mathcal{D}^{\nu\nu} & 0 \\ 0 & \mathcal{D}^{NN} \end{pmatrix}, \tag{13.118}$$

where

$$\mathcal{M}^{\nu N} = \begin{pmatrix} c_e & 0 & 0 & s_e & 0 & 0 \\ 0 & c_\mu & 0 & 0 & s_\mu & 0 \\ 0 & 0 & c_\tau & 0 & 0 & s_\tau \\ -s_e & 0 & 0 & c_e & 0 & 0 \\ 0 & -s_\mu & 0 & 0 & c_\mu & 0 \\ 0 & 0 & -s_\tau & 0 & 0 & c_\tau \end{pmatrix},$$

$$\mathcal{D}^{\alpha\alpha} = \begin{pmatrix} c_{e\mu}^\alpha c_{e\tau}^\alpha & s_{e\mu}^\alpha c_{e\tau}^\alpha & s_{e\tau}^\alpha \\ -s_{e\mu}^\alpha c_{\mu\tau}^\alpha - c_{e\mu}^\alpha s_{\mu\tau}^\alpha s_{e\tau}^\alpha & c_{e\mu}^\alpha c_{\mu\tau}^\alpha - s_{e\mu}^\alpha s_{\mu\tau}^\alpha s_{e\tau}^\alpha & s_{\mu\tau}^\alpha c_{e\tau}^\alpha \\ s_{e\mu}^\alpha s_{\mu\tau}^\alpha - c_{e\mu}^\alpha c_{\mu\tau}^\alpha s_{e\tau}^\alpha & -c_{e\mu}^\alpha s_{\mu\tau}^\alpha - s_{e\mu}^\alpha c_{\mu\tau}^\alpha s_{e\tau}^\alpha & c_{\mu\tau}^\alpha c_{e\tau}^\alpha \end{pmatrix}, \tag{13.119}$$

$c_{ab}^\alpha = \cos\theta_{ab}^\alpha$, $s_{ab}^\alpha = \sin\theta_{ab}^\alpha$, $a, b = e, \mu, \tau$, $\alpha = \nu, N$. In such a way, the Lagrangian that governs the interaction of the charged gauge bosons with leptons has the form:

$$\mathcal{L}_l^{CC} = \frac{g_L}{2\sqrt{2}} \sum_l \overline{l}(x)\gamma^\mu(1 - \gamma_5)\nu_l(x)W_{L\mu}(x) + \frac{g_R}{2\sqrt{2}} \sum_l \overline{l}(x)\gamma^\mu(1 + \gamma_5)N_l(x)W_{R\mu}(x),$$

$$\tag{13.120}$$

where

$$\nu_{eL}(x) = \sum_k \mathcal{U}_{1k}\Omega_k^m(x), \qquad \nu_{\mu L}(x) = \sum_k \mathcal{U}_{2k}\Omega_k^m(x), \qquad \nu_{\tau L}(x) = \sum_k \mathcal{U}_{3k}\Omega_k^m(x),$$

$$N_{eR}(x) = \sum_k \mathcal{U}_{4k}\Omega_k^m(x), \qquad N_{\mu R}(x) = \sum_k \mathcal{U}_{5k}\Omega_k^m(x), \qquad N_{\tau R}(x) = \sum_k \mathcal{U}_{6k}\Omega_k^m(x).$$

The interaction Lagrangian of the charged gauge bosons with quarks has the analogous form:

$$\mathcal{L}_q^{CC} = \frac{g_L}{2\sqrt{2}} \sum \overline{q}_i^u(x)\gamma^\mu(1 - \gamma_5)\mathcal{M}_{ik}^{CKM}q_k^d(x)W_{L\mu}(x) +$$

$$+\frac{g_R}{2\sqrt{2}}\sum \overline{q}_i^u(x)\gamma^\mu(1+\gamma_5)\mathcal{M}_{ik}^{CKM}q_k^d(x)W_{R\mu}(x). \tag{13.121}$$

Our list is completed by the interaction Lagrangian of the neutral gauge bosons with the fundamental fermions

$$\mathcal{L}_f^{NC} = \sum_f \overline{\psi}_f(x)\gamma^\mu\left[Q(f)A_\mu(x)+\frac{1}{2}\sum_{n=1}^2(g_{Vn}^f - g_{An}^f\gamma_5)Z_{n\mu}\right]\psi_f(x), \tag{13.122}$$

where

$$g_{V1}^f = ec_\phi c_W^{-1}s_W^{-1}\left[S_3^W(f_L)-2Q(f)s_W^2\right]+$$

$$+\frac{es_\phi c_W^{-1}}{\sqrt{e^{-2}g_R^2 c_W^2 - 1}}\left[e^{-2}g_R^2 c_W^2 S_3^W(f_R)+S_3^W(f_L)-2Q(f)\right], \tag{13.123}$$

$$g_{A1}^f = ec_\phi c_W^{-1}s_W^{-1}S_3^W(f_L) - \frac{es_\phi c_W^{-1}}{\sqrt{e^{-2}g_R^2 c_W^2 - 1}}\left[e^{-2}g_R^2 c_W^2 S_3^W(f_R)-S_3^W(f_L)\right], \tag{13.124}$$

$$g_{V2}^f = g_{V1}^f\left(\phi\to\phi+\frac{\pi}{2}\right),\qquad g_{A2}^f = g_{A1}^f\left(\phi\to\phi+\frac{\pi}{2}\right) \tag{13.125}$$

and by the Lagrangian describing the trilinear gauge boson couplings

$$\mathcal{L}^{TBC} = i\rho_{kl}^{(V)}[W_k^{\mu\nu*}V_\nu + W_{k\nu}^*V^{\nu\mu}]W_{l\mu}, \tag{13.126}$$

where

$$\rho_{ll}^{(Z_1)} = \cos^2\left(\xi+\frac{\pi}{2}\delta_{l2}\right)g_L M_{11} + \sin^2\left(\xi+\frac{\pi}{2}\delta_{l2}\right)g_R M_{12}, \tag{13.127}$$

$$\rho_{kl}^{(Z_1)} = \rho_{lk}^{(Z_1)} = \frac{1}{2}\sin 2\xi(g_L M_{11}-g_R M_{12}),\qquad k\neq l,. \tag{13.128}$$

$$\rho_{kl}^{(\gamma)} = e\delta_{kl}, \tag{13.129}$$

$$V_{\mu\nu} = \partial_\mu V_\nu - \partial_\nu V_\mu,\qquad V=\gamma, Z_1, Z_2,$$

M_{ij} are the elements of the matrix $U_N\Lambda$ (see Eqs. (13.91) and (13.92))* and the expressions for $\rho_{ll}^{(Z_2)}, \rho_{kl}^{(Z_2)}$ follow from (13.127) and (13.128) under the substitutions

$$g_L M_{11} \to g_R M_{22},\qquad g_L M_{12}\to g_R M_{21},\qquad \xi\to\xi+\frac{\pi}{2}.$$

It may be shown that the correspondence principle takes place, that is, there is the limiting transition under which the LRM formulae turn into the corresponding expressions of the SM. The LRM reproduces the SM, providing

$$\varphi = \phi = \xi = 0,\qquad g_L = g_R = es_W^{-1},\qquad g' = \frac{e}{\sqrt{c_W^2 - s_W^2}} \tag{13.130}$$

and all masses of additional particles standing in the denominators (numerators) tend to infinity (zero).

The LRM is sufficiently popular and looking-for new particles predicted by this model enters the working program of every accelerator. The limits on the masses of these particles

---

*Recall, we have set the angle of the $SU(2)_R$-generator orientation φ equal to zero.

become higher and higher with increasing energy. Thus, the lower limit on the W_R boson mass reaches 715 GeV while that on the Z_R boson mass is equal to 630 GeV. The mixing angles in the charged and neutral gauge boson sectors have the order of 10^{-2}. As far as the charged Higgs bosons are concerned, their lower limits on the masses lie at the electroweak scale

$$m_{h^\pm, \tilde{\delta}^\pm} \geq 75 \text{ GeV}, \qquad m_{\Delta^{\pm\pm}} \geq 119 \text{ GeV}.$$

13.4   Left-right model in the making

Now we discuss the phenomenological status of the LRM. One of more drastic predictions is the presence of the additional gauge bosons, $W_2^\pm$ and Z_2. In connection with this, studying the reaction

$$e^- + e^+ \rightarrow W_k^- + W_n^+, \qquad (k, n = 1, 2), \tag{13.131}$$

is of great importance. The corresponding Feynman diagrams are pictured in Fig. 13.5, where in brackets the momenta of particles are indicated. Note that the cross section

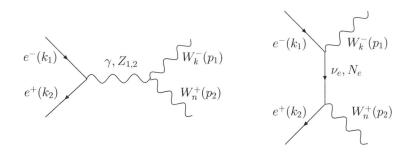

FIGURE 13.5
The Feynman diagrams corresponding to the process $e^- e^+ \rightarrow W_k^- W_n^+$.

of the process (13.131) does not depend on the assumed neutrino nature. For the sake of simplicity, we shall suppose the mixing in the neutrino sector to be absent. We shall consider that electron-positron beams are polarized while the final $W_k^\pm$ bosons are unpolarized. The radiative corrections (RCs) will be accounted for at the level of the improved Born approximation [101]. Then, the differential cross section of the reaction (13.131) is defined by the expression [95, 102]:

$$\frac{d\sigma^{(kn)}}{dt} = \frac{\rho}{(4\pi s)^2} \left\{ \left[\frac{1 - \lambda\bar{\lambda}}{4} \left(\left| e^2[1 + (-1)^{k+n}] - 2s \sum_{l=1}^{2} \rho_{kn}^{(Z_l)} g_{Vl}^e d_{Z_l} \right|^2 + \right. \right. \right.$$

$$\left. + \left| 2s \sum_{l=1}^{2} \rho_{kn}^{(Z_l)} g_{Al}^e d_{Z_l} \right|^2 \right) - \frac{\lambda - \bar{\lambda}}{4} \left(2se^2[1 + (-1)^{k+n}] \sum_{l=1}^{2} \rho_{kn}^{(Z_l)} g_{Al}^e \text{Re}(d_{Z_l}) - \right.$$

$$- 2s^2 \sum_{l=1}^{2} \rho_{kn}^{(Z_l)} g_{Al}^e d_{Z_l} \sum_{j=1}^{2} \rho_{kn}^{(Z_j)} g_{Vl}^e d_{Z_j}^\star \Bigg) \Bigg] B_1^{(kn)}(s,u,t) -$$

$$- \Bigg[\frac{(1-\overline{\lambda})(1+\lambda)}{4} g_L^2 a_+^{kn} \left(e\rho_{kn}^{(\gamma)} - \frac{s}{2} \sum_{l=1}^{2} \rho_{kn}^{(Z_l)}(g_{Vl}^e + g_{Al}^e)\mathrm{Re}(d_{Z_l}) \right) -$$

$$- \frac{(1+\overline{\lambda})(1-\lambda)}{4} g_R^2 a_-^{kn} \left(e\rho_{kn}^{(A)} - \frac{s}{2} \sum_{l=1}^{2} \rho_{kn}^{(Z_l)}(g_{Vl}^e - g_{Al}^e)\mathrm{Re}(d_{Z_l}) \right) t d_{N_e} \Bigg] B_2^{(kn)}(s,t,u) +$$

$$+ \Bigg[\frac{(1+\overline{\lambda})(1-\lambda)}{8} g_R^4 b_-^{kn} t^2 d_{N_e}^2 + \frac{(1-\overline{\lambda})(1+\lambda)}{8} g_L^4 b_+^{kn} \Bigg] B_3^{kn}(s,t,u) \Bigg\} , \qquad (13.132)$$

where

$$a_\pm^{kn} = \tfrac{1}{4} \left\{ 1 + (-1)^{k+n} \mp \left[(-1)^k + (-1)^n \right] \cos 2\xi \pm \left[1 - (-1)^{k+n} \right] \sin 2\xi \right\} , \Bigg\}$$
$$b_\pm = \tfrac{1}{4} \left\{ 1 + (-1)^{k+n} \cos^2 2\xi \mp \left[(-1)^k + (-1)^n \right] \cos 2\xi \right\} , \qquad\qquad (13.133)$$

$$m_{N_1} = m_{N_e}, \qquad d_{N_e} = (t - m_{N_e}^2)^{-1}, \qquad d_{Z_j} = (s - m_{Z_j}^2 + i\Gamma_{Z_j} m_{Z_j})^{-1},$$

$$\beta = \left[\left(1 - \frac{m_{W_n}^2 + m_{W_k}^2}{s} \right)^2 - \frac{4 m_{W_n}^2 m_{W_k}^2}{s^2} \right]^{1/2} , \qquad F = \frac{4(ut - m_{W_n}^2 m_{W_k}^2)}{s^2}, \qquad (13.134)$$

$\lambda(\overline{\lambda})$ is the helicity of the electron (positron), $s, t,$ and u are the Mandelstam variables, and the functions $B_i^{(kn)}(s,t,u)$ have the form:

$$B_1^{(kn)}(s,t,u) = \frac{\beta^2 s(m_{W_n}^2 + m_{W_k}^2)}{2 m_{W_n}^2 m_{W_k}^2} + F \left[\left(\frac{s - m_{W_n}^2 - m_{W_k}^2}{4 m_{W_n} m_{W_k}} \right)^2 + \frac{1}{2} \right], \qquad (13.135)$$

$$B_2^{(kn)}(s,u,t) = \frac{m_{W_n}^2 + m_{W_k}^2}{m_{W_n}^2 m_{W_k}^2} \left[s - m_{W_n}^2 - m_{W_k}^2 + \frac{2 m_{W_n}^2 m_{W_k}^2}{t} \right] -$$

$$+ \frac{Fs}{8 m_{W_n}^2 m_{W_k}^2} \left[u + t + \frac{4 m_{W_n}^2 m_{W_k}^2}{t} \right], \qquad (13.136)$$

$$B_3^{(kn)}(s,t,u) = F \left[\left(\frac{s}{4 m_{W_n}^2 m_{W_k}^2} \right)^2 + \frac{s^2}{4 t^2} \right] + \frac{s(m_{W_n}^2 + m_{W_k}^2)}{2 m_{W_n}^2 m_{W_k}^2}. \qquad (13.137)$$

The quantity ρ entering into (13.132) is caused by the RCs coming from all heavy particles. Its value is determined by

$$\rho = 1 + \Delta\rho_t + \Delta\rho_M + \ldots, \qquad (13.138)$$

where $\Delta\rho_t \simeq 3 G_F m_t^2/(8\sqrt{2}\pi)$. $\Delta\rho_M$ arises due to the mixing in the gauge boson sector and has the general asymptotic form [103]:

$$\Delta\rho_M = c_0^2 (m_{Z_1}/m_{Z_2})^2 - c_1^2 (m_{W_1}/m_{W_2})^2 \qquad (13.139)$$

with c_0 and c_1 constants depending on the Higgs particle VEVs and the gauge coupling constants. Additional contributions to ρ (dots in the ρ definition) come from Higgs particles, heavy right-handed neutrinos, etc. They could be both positive and negative. For example, the contribution from the standard Higgs boson is logarithmic on the Higgs boson mass and for $m_H >> m_{W_1}$ it has a negative sign. In its turn, the contribution connected with the presence of weak isotriplet Higgs bosons is [104]

$$(\Delta\rho_{WI})_{L,R} = -\frac{2v_{L,R}^2}{\nu^2},$$

where $\nu = 246$ GeV is the standard Higgs-doublet VEV.

Rather than use the relation (13.139) involving the bulk of the LRM parameters, one may address to experiments. Let us employ, as the upper and lower limits on $\Delta\rho_M$, the following relation

$$(\Delta\rho_M)_{upper,lower} = (\Delta\rho)_{upper,lower} - \Delta\rho_t - 1,$$

where the values of $(\Delta\rho)_{upper,lower}$ are taken from the experimental data.* The importance of the quantity $\Delta\rho_M$ is called for by the fact that it redefines the effective s_W^2 in the LRM in accordance with the relation

$$s_W^2 = \bar{s}_W^2 - \frac{\bar{s}_W^2 \bar{c}_W^2}{\bar{c}_W^2 - \bar{s}_W^2}\Delta\rho_M. \tag{13.140}$$

where $\bar{s}_W^2 \equiv (s_W^2)^{SM}$.

It is easy to verify that the cross section $d\sigma^{(11)}$ turns into the corresponding cross section of the SM under the fulfillment of (13.130). From (13.132) it is also follows that all the effects associated with N_e in the differential cross section are negligibly small. However, for the total cross section σ^{kn} the situation with N_e is drastically altered. The expression for σ^{kn} follows from (13.132) after multiplication by $\beta s/2$ and the replacement

$$\left.\begin{array}{ll} B_i^{(kn)}(s,u,t) \to D_i^{(kn)}, & i = 1,2,3, \\ B_j^{(kn)}(s,u,t)(td_{N_e})^{j-1} \to D_j^{(kn)}, & j = 2,3, \end{array}\right\} \tag{13.141}$$

where

$$D_1^{(kn)} = \frac{\beta^2}{12m_{W_n}^2 m_{W_k}^2}\left[(s - m_{W_n}^2 - m_{W_k}^2)^2 + 12s(m_{W_n}^2 + m_{W_k}^2) + 8m_{W_n}^2 m_{W_k}^2\right], \quad (13.142)$$

$$D_{2\nu}^{(kn)} = \frac{(s - m_{W_n}^2 - m_{W_k}^2)}{m_{W_n}^2 m_{W_k}^2}\left\{2(m_{W_n}^2 + m_{W_k}^2) + \frac{s\beta^2}{6} + \frac{2m_{W_n}^2 m_{W_k}^2}{s} - \frac{m_{N_e}^2}{2s}\left[2m_{N_e}^2 - m_{W_n}^2 - \right.\right.$$

$$\left.\left. -m_{W_k}^2 + s + \frac{8m_{W_n}^2 m_{W_k}^2}{m_{W_n}^2 + m_{W_n}^2 - s}\right]\right\} + \frac{4L}{\beta s}\left[-m_{W_n}^2 - m_{W_k}^2 - \frac{m_{W_n}^2 m_{W_k}^2}{s} + \right.$$

$$+m_{N_e}^2(m_{W_n}^2 + m_{W_k}^2 - s)\left[\frac{m_{W_n}^2 + m_{W_k}^2}{2m_{W_n}^2 m_{W_k}^2} + \frac{3}{4s} - \frac{m_{N_e}^4}{4sm_{W_n}^2 m_{W_k}^2}\right] + \frac{\beta^2 sm_{N_e}^4}{4m_{W_n}^2 m_{W_k}^2}\right], \quad (13.143)$$

$$D_{3\nu}^{(kn)} = \frac{1}{12m_{W_n}^2 m_{W_k}^2}\left[s^2\beta^2 - 48m_{W_n}^2 m_{W_k}^2 + 12s(m_{W_n}^2 + m_{W_k}^2) - 6m_{N_e}^2(s - m_{W_k}^2 - m_{W_n}^2 + \right.$$

$$\left. +4m_{N_e}^2)\right] + C_\nu \frac{sm_{N_e}^4(m_{W_n}^2 + m_{W_k}^2)}{m_{W_n}^2 m_{W_k}^2} + \frac{L}{\beta s}\left\{2(s - m_{W_n}^2 - m_{W_k}^2) + 5m_{N_e}^2 + \right.$$

---

*We have assumed the additional contributions in (13.137) to be mutually cancelled.

$$+ \frac{m^2_{N_e}}{2m^2_{W_n} m^2_{W_k}} \left[m^2_{N_e}(4m^2_{N_e} - 3m^2_{W_n} - 3m^2_{W_k} + 3s) - 4s(m^2_{W_n} + m^2_{W_k})\right] \bigg\} , \qquad (13.144)$$

$$D_2^{(kn)} = D_{2\nu}^{(kn)}\bigg|_{m_{N_e}=0}, \qquad D_3^{(kn)} = D_{3\nu}^{(kn)}\bigg|_{m_{N_e}=0}, \qquad (13.145)$$

$$\left.\begin{array}{c} C_\nu = \left[m^2_{W_n} m^2_{W_k} - m^2_{N_e}(m^2_{W_n} + m^2_{W_k} - m^2_{N_e} - s)\right]^{-1}, \\[2mm] L = \ln\left| \dfrac{m^2_{W_n} + m^2_{W_k} - 2m^2_{N_e} - \beta s - s}{m^2_{W_n} + m^2_{W_k} - 2m^2_{N_e} + \beta s - s}\right|. \end{array}\right\} \qquad (13.146)$$

As we see from (13.143) and (13.144) in σ^{kn} there are the terms that are proportional to $m^2_{N_e}$ and $m^4_{N_e}$. They arise when we integrate in (13.132) the expressions containing (td_{N_e}) and $(td_{N_e})^2$. In the case of unpolarized e^+e^- beams, their contribution to $\sigma^{(11)}$ is very small because the dominant terms caused by the ν_{eL} exchange enter into $\sigma^{(11)}$ as $D_2^{(11)}\cos^2\xi$ and $D_3^{(11)}\cos^4\xi$, while the ones determined by the N_e exchange are $D_{2\nu}^{(11)}\sin^2\xi$ and $D_{3\nu}^{(11)}\sin^4\xi$. The contributions from the heavy neutrino will be considerable under the $W_1^- W_2^+$- and $W_2^- W_2^+$-pair production for $\overline{\lambda} = 1$ and $\lambda = -1$. As the analysis demonstrates [105], when the $e^- e^+$ beams energy increases, the total cross sections $\sigma^{(12)}$ and $\sigma^{(22)}$ as functions of m_{N_e} behave in the following way. Until reaching the maximum,* the total cross sections corresponding to larger values of m_{N_e} are increasing slower than those with smaller values of m_{N_e}. After reaching the maximum, the situation is changed. Now, the total cross sections with a smaller value of m_{N_e} are decreasing faster than those with larger value of m_{N_e}. Therefore, investigating this process, we can extract information not only on the $W_2^\pm$ bosons, but on the heavy neutrino as well. Note, in order to obtain the SM total cross section from $\sigma^{(11)}$ we have to carry out the replacement (13.130) and set m_{N_e} equal to zero.

The choice of a gauge group of the electroweak interaction unambiguously fixes the multipole moments (MMs) of the $W^\pm$ bosons. At that, if the electromagnetic MMs determined by k_γ, λ_γ have the same values as in the SM, then this is not the case for the weak MMs determined by k_Z, λ_Z. Thus, for the $W_1^\pm$ boson the value of k_{Z_1} is defined as

$$(k_{Z_1})^{LRM} = -(k_{Z_1})^{SM} \tan\theta_W \rho_{11}^{(Z_1)}.$$

One important point to remember is that the contributions from the electromagnetic and weak MMs can be distinguished by means of polarized electron-positron beams. So, just as in the models of the first two classes (models with the composite Higgs and Higgsless models), in models with an extended gauge group of the electroweak interaction, the $W^\pm$ bosons possess anomalous weak MMs. However, despite this discouraging circumstance, we can easily distinguish these models from each other even at identical values of the MMs. In the models belonging to the first two classes, the total cross section $\sigma^{(11)}$ increases as a linear function of s until reaching the maximum, further, at a small energy interval somewhere in the range of 200 GeV, it may go to the minimum and then it again increases $\sim s$ to the extent of the fundamental scale Λ. In the LRM case, before reaching the maximum, the cross section $\sigma^{(11)}$ increases by the law $\sim s$ and then begins to fall down by the law $s^{-1}\ln s$.

Anew, to observe the New Physics effects one may introduce the quantity

$$\delta^{\lambda\overline{\lambda}}(x) = \frac{(\sigma^{(11)})^{LRM} - \sigma^{SM}}{\sqrt{\sigma^{SM}}}\sqrt{LT}$$

---

*Here, we are dealing with the cross section maximum that is not connected with the Z_2 resonance.

and carry out the investigations in the region of the LEP II energies. The analysis shows that, in the region of the energies until the Z_2 resonance, the deviations from the SM predictions are small and have the order of 10^{-3} [105].

Next, we apply to the neutrino sector. Let us for the sake of simplicity work in the two flavor approximation. Then, in the basis

$$\Psi = \begin{pmatrix} \nu_{aL} \\ \nu_{bL} \\ N_{aR} \\ N_{bR} \end{pmatrix}$$

the neutrino mass matrix takes the form

$$\mathcal{M} = \begin{pmatrix} \mathcal{M}_\nu & \mathcal{M}_D \\ \mathcal{M}_D^T & \mathcal{M}_N \end{pmatrix}, \tag{13.147}$$

where

$$\mathcal{M}_\nu = \begin{pmatrix} f_{aa}v_L & f_{ab}v_L \\ f_{ab}v_L & f_{bb}v_L \end{pmatrix}, \qquad \mathcal{M}_N = \mathcal{M}_\nu(v_L \to v_R), \qquad \mathcal{M}_D = \begin{pmatrix} m_D^a & M_D \\ M_D' & m_D^b \end{pmatrix}.$$

$$\left. \begin{array}{ll} m_D^a = h_{aa}k_1 + h'_{aa}k_2, & m_D^b = m_D^a(a \to b), \\ M_D = h_{ab}k_1 + h'_{ab}k_2, & M_D' = M_D(a \leftrightarrow b) \end{array} \right\} \tag{13.148}$$

and $a, b = e, \mu, \tau$. Recall that the bidoublet Yukawa coupling constants h_{aa} and h'_{aa} also define the charged lepton masses according to the relation

$$m_a = h_{aa}k_2 + h'_{aa}k_1.$$

The existence of three additional particles in the neutrino sector enables one to explain the neutrino mass smallness as compared with masses of charged leptons and quarks. Assume that $v_L = 0$ and $\mathcal{M}_N \gg \mathcal{M}_D$. Then, one can block diagonalize the matrix (13.147) by a similarity transformation and obtain the light neutrino mass matrix M_ν to be

$$M_\nu = \mathcal{M}_D^T \mathcal{M}_N^{-1} \mathcal{M}_D. \tag{13.149}$$

Of course, this matrix needs further diagonalization to obtain the light neutrino eigenvalues and eigenstates. The relation (13.149) is called the *seesaw relation* . It is clear that the roughly estimation of the heavy neutrino sector parameters could be done using the seesaw relation. To this end, it is necessary to determine the obvious form of the matrix M_ν. The form of M_ν can be restored studying weak decays. However, since decay neutrinos do not represent mass eigenstates, then we are obliged to do particular assumptions concerning the mixing value between the flavor and physical neutrinos. Of course, one may assume that ν_1, ν_2, ν_3 are predominantly electron, muon, and tau-lepton light neutrinos, respectively (this is true only when all mixing angles are very small). Then, the results of the laboratory experiments lead to the following limits:

$$m_{\nu_e} < 2 \text{ eV}, \qquad m_{\nu_\mu} < 0.19 \text{ MeV} \qquad m_{\nu_\tau} < 18.2 \text{ MeV}. \tag{13.150}$$

The inequalities (13.150) do not depend on the fact which of the existing models of the electroweak interactions were used under the analysis of experiments, since they follow from kinematics for the most part. The upper limit on the electron neutrino mass (to be specific, the electron antineutrino mass) was determined studying the tritium β^- decay

$$^3_1H \to {}^3_2He + e^- + \overline{\nu}_e.$$

The limit on the muon neutrino mass was found using the decay

$$\pi^+ \to \mu^+ + \nu_\mu.$$

Information on the ν_τ mass was obtained when investigating τ lepton decays

$$\tau^- \to 2\pi^- + \pi^+ + \nu_\tau, \qquad \tau^- \to 3\pi^- + 2\pi^+ + \nu_\tau, \qquad \tau^- \to 2\pi^- + \pi^+ + \pi^0 + \nu_\tau,$$

and so on. Subsequently, we shall show that the heavy neutrino masses are mainly determined by f_{ab} and v_R. Then, working in the one-flavor approximation, assuming

$$\mathcal{M}_D \approx m_\mu$$

(this takes place when $k_1 \approx k_2$ or $h_{\mu\mu} \approx h'_{\mu\mu}$) and setting

$$m_{\nu_\mu} = 0.19 \text{ MeV},$$

we obtain

$$f_{\mu\mu} v_R \approx 5.8 \text{ TeV}.$$

Thus, for the neutrino mass smallness to be successively explained, the presence of two widely separated scales of the SSB is needed.

The mass matrix $\mathcal{M}$ is diagonalized with the help of the orthogonal matrix U having the form:

$$U = \begin{pmatrix} c_{\varphi_a} & 0 & s_{\varphi_a} & 0 \\ 0 & c_{\varphi_b} & 0 & s_{\varphi_b} \\ -s_{\varphi_a} & 0 & c_{\varphi_a} & 0 \\ 0 & -s_{\varphi_b} & 0 & c_{\varphi_b} \end{pmatrix} \begin{pmatrix} c_{\theta_\nu} & s_{\theta_\nu} & 0 & 0 \\ -s_{\theta_\nu} & c_{\theta_\nu} & 0 & 0 \\ 0 & 0 & s_{\theta_N} & s_{\theta_N} \\ 0 & 0 & -s_{\theta_N} & c_{\theta_N} \end{pmatrix},$$

where we have taken the following designations: φ_a (φ_b) is the mixing angle in the a (b) generation between the light and heavy neutrinos, θ_ν (θ_N) is the mixing angle between ν_{aL} neutrino and ν_{bL} neutrino (N_{aR} and N_{bRL}), $c_\varphi = \cos\varphi$, $s_\varphi = \sin\varphi$, and so on. Further on, we assume, that the mixing takes place in the $e - \mu$ sector. Using the eigenvalues equation for the mass matrix we obtain the relations between the bidoublet and triplet Yukawa coupling constants (YCCs) and the neutrino sector parameters. Calculations result in [106]

$$m_D^e = \sum_i (\Lambda_{ee}^{LR})_i m_i = c_{\varphi_e} s_{\varphi_e} (-m_1 c_{\theta_\nu}^2 + m_2 s_{\theta_\nu}^2 + m_3 c_{\theta_N}^2 + m_4 s_{\theta_N}^2), \tag{13.151}$$

$$M_D = \sum_i (\Lambda_{e\mu}^{LR})_i m_i = c_{\varphi_e} s_{\varphi_\mu} c_{\theta_\nu} s_{\theta_\nu} (m_1 - m_2) + s_{\varphi_e} c_{\varphi_\mu} c_{\theta_N} s_{\theta_N} (m_4 - m_3), \tag{13.152}$$

$$f_{e\mu} v_R = \sum_i (\Lambda_{e\mu}^{RR})_i m_i = s_{\varphi_e} s_{\varphi_\mu} c_{\theta_\nu} s_{\theta_\nu} (m_2 - m_1) + c_{\varphi_e} c_{\varphi_\mu} c_{\theta_N} s_{\theta_N} (m_4 - m_3), \tag{13.153}$$

$$f_{ee} v_R = \sum_i (\Lambda_{ee}^{RR})_i m_i = (s_{\varphi_e} c_{\theta_\nu})^2 m_1 + (c_{\varphi_e} c_{\theta_N})^2 m_3 + (s_{\varphi_e} s_{\theta_\nu})^2 m_2 + (c_{\varphi_e} s_{\theta_N})^2 m_4, \tag{13.154}$$

$$f_{\mu\mu} v_R = \sum_i (\Lambda_{\mu\mu}^{RR})_i m_i = (s_{\varphi_\mu} s_{\theta_\nu})^2 m_1 + (c_{\varphi_\mu} s_{\theta_N})^2 m_3 + (s_{\varphi_\mu} c_{\theta_\nu})^2 m_2 + (c_{\varphi_\mu} c_{\theta_N})^2 m_4, \tag{13.155}$$

$$m_D^\mu = \sum_i (\Lambda_{\mu\mu}^{LR})_i m_i = m_D^e(\varphi_e \to \varphi_\mu, \theta_{\nu,N} \to \theta_{\nu,N} + \frac{\pi}{2}), \qquad M_D' = M_D(\varphi_e \leftrightarrow \varphi_\mu), \tag{13.156}$$

$$f_{ll'} v_L = f_{ll'} v_R(\varphi_{l,l'} \to \varphi_{l,l'} + \frac{\pi}{2}), \qquad l, l' = e, \mu, \tag{13.157}$$

where

$$\mathcal{M}_{ab} = \left(UMU^{-1}\right)_{ab} = \sum_{i=1}^{4} \left(\Lambda_{ab}\right)_i m_i, \qquad a, b = \nu_{eL}, \nu_{\mu L}, N_{eR}, N_{\mu R},$$

$$M = \text{diag}\left(m_1, m_2, m_3, m_4\right).$$

The importance of the formulae obtained lies in the fact that their left sides can be measured in accelerator experiments without the neutrino in the initial and final states. Since the lepton YCCs define the interaction of the physical Higgs bosons with leptons, then reactions with the virtual Higgs bosons in the s channels prove to be the most perspective ones. Therefore, the relations (13.151)–(13.157) lead us to the conclusion that the neutrino oscillation parameters in the LRM could be measured by the indirect way as well.

Our next task is to find the heavy neutrino masses as functions of the YCCs and VEVs. To do it we first determine the connection between the mixing angles θ_N, $\varphi_{e,\mu}$, and the triplet YCCs. The use of Eqs. (13.154), (13.155), and (13.157) produces the formulae we were searching for

$$s_{\theta_N}^2 = \frac{f_{\mu\mu}(v_R + v_L) - (m_{\nu_1} s_{\theta_\nu}^2 + m_{\nu_2} c_{\theta_\nu}^2) - m_{N_2}}{m_{N_1} - m_{N_2}}, \tag{13.158}$$

$$c_{\theta_N}^2 = \frac{f_{ee}(v_R + v_L) - (m_{\nu_1} c_{\theta_\nu}^2 + m_{\nu_2} s_{\theta_\nu}^2) - m_{N_2}}{m_{N_1} - m_{N_2}}, \tag{13.159}$$

$$\sin 2\varphi_e = \frac{2\sqrt{f_{ee}^2 v_R v_L - [f_{ee}(v_R + v_L) - m_{\nu_1} c_{\theta_\nu}^2 - m_{\nu_2} s_{\theta_\nu}^2](m_{\nu_1} c_{\theta_\nu}^2 + m_{\nu_2} s_{\theta_\nu}^2)}}{f_{ee}(v_R + v_L) - 2(m_{\nu_1} c_{\theta_\nu}^2 + m_{\nu_2} s_{\theta_\nu}^2)}. \tag{13.160}$$

$$\sin 2\varphi_\mu = \sin 2\varphi_e \left(f_{ee} \to f_{\mu\mu}, \theta_\nu \to \theta_\nu + \frac{\pi}{2}\right). \tag{13.161}$$

Note, that when one neglects the light neutrinos masses, then for the mixing angles φ_e and φ_μ the simple relation

$$\sin 2\varphi_e = \sin 2\varphi_\mu = \frac{2\sqrt{v_R v_L}}{v_R + v_L} \tag{13.162}$$

takes place. Using Eqs. (13.153) and (13.157), it is possible to obtain

$$f_{e\mu}^2 v_R v_L = \frac{1}{4}(\sin 2\varphi_e)(\sin 2\varphi_\mu)c_{\theta_N}^2 s_{\theta_N}^2 (m_{N_2} - m_{N_1})^2 = \frac{1}{4}(\sin 2\varphi_e)(\sin 2\varphi_\mu)[f_{\mu\mu}(v_R + v_L) -$$

$$-m_{\nu_1} s_{\theta_\nu}^2 - m_{\nu_2} c_{\theta_\nu}^2 - m_{N_2}][f_{ee}(v_R + v_L) - m_{\nu_1} c_{\theta_\nu}^2 - m_{\nu_2} s_{\theta_\nu}^2 - m_{N_2}]. \tag{13.163}$$

With the help of Eq. (13.163) one can show, when $m_{N_2} \gg m_{N_1}$, then $f_{e\mu}$ reaches its maximal value

$$(f_{e\mu})_{max} \approx \sqrt{f_{ee} f_{\mu\mu}}, \tag{13.164}$$

while in the case of (quasi)degeneration of the heavy neutrino masses $f_{e\mu}$ turns to zero, that is,

$$f_{e\mu} \in [0, \sqrt{f_{ee} f_{\mu\mu}}]. \tag{13.165}$$

Two roots of Eq. (13.163) are defined by the expression

$$(m_{N_2})_{1,2} = \frac{1}{2}\left[(f_{ee} + f_{\mu\mu})(v_R + v_L) - (m_{\nu_1} + m_{\nu_2})\right] \pm \Omega, \tag{13.166}$$

where

$$\Omega = \sqrt{\frac{1}{4}[(f_{ee} + f_{\mu\mu})(v_R + v_L) - m_{\nu_1} - m_{\nu_2}]^2 - f_{ee}f_{\mu\mu}(v_R + v_L)^2 + \frac{4f_{e\mu}^2 v_R v_L}{\sin 2\varphi_e \sin 2\varphi_\mu}}. \tag{13.167}$$

When $m_{\nu_1} = m_{\nu_2} = 0$, the expression (13.167) is significantly simplified

$$\Omega = \frac{v_R + v_L}{2}\sqrt{(f_{\mu\mu} - f_{ee})^2 + 4f_{e\mu}^2}. \tag{13.168}$$

Using Eqs. (13.154), (13.155), and (13.157), it is easy to obtain

$$f_{ee} + f_{\mu\mu} = \frac{m_{\nu_1} + m_{\nu_2} + m_{N_1} + m_{N_2}}{v_R + v_L}. \tag{13.169}$$

Combining this relation with Eq. (13.166), we arrive at

$$(m_{N_1})_{1,2} = \frac{1}{2}[(f_{ee} + f_{\mu\mu})(v_R + v_L) - (m_{\nu_1} + m_{\nu_2})] \mp \Omega. \tag{13.170}$$

So, there are the two sets of the symmetric values for m_{N_1} and m_{N_2}. It means that both the direct $(m_{N_1} > m_{N_2})$ and inverse $(m_{N_1} < m_{N_2})$ hierarchies are possible. Owing to Eq. (13.168) the sum of neutrino masses is constant for every set of f_{ab}. The bounds on the values of f_{ab} and $v_{L,R}$ could be found from the experimental data. To obtain the value of v_R we can use either relation

$$v_R = \sqrt{\frac{(m_{W_2}^2 - m_{W_1}^2)\cos 2\xi}{g_L^2}}, \tag{13.171}$$

or relation

$$v_R = \sqrt{\frac{\cos 2\theta_W[m_{Z_1}^2(\sin^2\phi - \cos^2\phi\cos 2\theta_W) + m_{Z_2}^2(\cos^2\phi - \sin^2\phi\cos 2\theta_W)]}{2g_L^2\cos^2\theta_W}}, \tag{13.172}$$

which follow from formulae defining the masses and mixing angles of the gauge bosons. To estimate v_L could be done with the help of the quantity ρ_0. With this aim in view, we should compare its theoretical value

$$\rho_0^{theor} = \frac{1 + 4v_L^2/k_+^2}{1 + 2v_L^2/k_+^2}, \tag{13.173}$$

with the experimental one (see Eq. (13.64)). For the YCCs to be found, the contributions from Higgs bosons into the muon AMM, the cross sections of processes with lepton flavor violation

$$\mu^- \to e^+ + e^- + e^-, \qquad \mu^- \to e^- + \nu_e + \bar{\nu}_\mu, \qquad \mu^- e^- + \gamma, \tag{13.174}$$

and the cross sections of low-energy elastic scattering of a light neutrino were investigated in Ref. [107]. Using the obtained data leads to the conclusion that the lower bound on the heavy neutrino mass lies at the electroweak scale ≈ 100 GeV. Note, that the heavy neutrino masses are mainly defined by f_{ab} and v_R. The corrections to $m_{N_{1,2}}$ due to v_L could reach of few$\times$ GeV while those caused by the light neutrino masses have the order of eV.

It should also be stressed, that the scheme of the heavy neutrino masses estimations, based on the investigation of the Higgs sector parameters, can be used for any gauge electroweak theory with an extended Higgs sector and the seesaw mechanism. In the LRM the suggested scheme will have produced the exact $m_{N_{1,2}}$ values under the direct measurement of the $\Delta_{1,2}^{(--)}$, $\tilde{\delta}^{(-)}$ bosons parameters and the W_2 (or Z_2) gauge boson mass. This, for example, can be done under the registration of the $\Delta_{1,2}^{(--)}$ bosons in the reactions

$$e^- e^- \to e^- e^-, \mu^- \mu^-, \qquad e^- \mu^- \to e^- \mu^- \tag{13.175}$$

and after the discovery of the W_2 boson.

Thus, the information about the heavy neutrinos can be obtained under investigation of the processes without their direct participation. The most important thing is, that now the subject of investigation is the charged, not neutral particles and this makes the experimental part of the task much more easy. Definition of the heavy neutrinos masses will allow us to start their searching in collider experiments. This, for example, can be done at the Tevatron, LHC, or at the $e^- \gamma$ colliders under investigations of the reactions

$$p\bar{p} \to Z_{1,2}^* \to N_l \overline{N}_l, \qquad e^- \gamma \to W_1^{-*} \to W_1^- N_e, \tag{13.176}$$

where the heavy neutrinos will be identified through the decay channels

$$N_l \to l W_1^+, \ l W_2^+, \ l\tilde{\delta}^{(+)}, \ l h^{(+)}. \tag{13.177}$$

In the neutrino sector the problem associated with the neutrino nature exists. Until now the question remains open as to whether the neutrino is a Majorana or Dirac particle. One of the key experiments that may confirm the Majorana nature of the neutrino will be the observation of the *neutrinoless double-β decay* ($0\nu2\beta$). However, one can establish the neutrino nature in collider experiments too. In the case of the Majorana neutrino the processes of a inverse neutrinoless double-β decay

$$e^- + e^- \to W^- + W^- \tag{13.178}$$

which is connected (by time reversal symmetry) to ($0\nu2\beta$) and an inverse neutrinoless double-μ decay

$$\mu^- + \mu^- \to W^- + W^-.$$

prove to be allowed. These processes will be investigated at the lepton colliders of the next generation (ILC, CLIC, FMC, NMC). Within the LRM, being restricted by the second order of the perturbation theory, we calculate the cross section of the reaction [108]

$$e^- + e^- \to W_n^- + W_k^- \tag{13.179}$$

In doing so, we shall assume that only the mixing angles between the neutrinos of one and the same generation are nonzero. The corresponding Feynman diagrams are presented in Fig. 13.6. The main difference of the LRM from the SM is that if in the latter the reaction (13.178) occurs for left-polarized electrons ($e_L^- e_L^-$) only, whereas in the LRM the contribution to the cross section arise both from $e_{L,R}^- e_{L,R}^-$- and from $e_{L,R}^- e_{R,L}^-$-beams. We shall consider the initial particles to be polarized and the final particles to be unpolarized. Then, the differential cross section of the process

$$e_L^- + e_R^- \to W_k^- + W_n^- \tag{13.180}$$

takes the form

$$\frac{d\sigma_{LR}^{(k,n)}}{dt} = \frac{(g_L g_R m_{W_n}^2 \sin 2\alpha)^2}{512\pi s^2} \left[\left(\frac{b_R^{(n)} b_L^{(k)}}{t - m_{N_e}^2} \right)^2 B_3^{(kn)}(s,t,u) + \left(\frac{b_R^{(k)} b_L^{(n)}}{u - m_{N_e}^2} \right)^2 B_3^{(kn)}(s,u,t) + $$

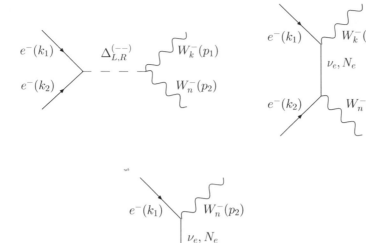

FIGURE 13.6

The Feynman diagrams describing the inverse neutrinoless double-β decay.

$$+ \frac{b_L^{(k)} b_L^{(n)} b_R^{(k)} b_R^{(n)}}{(t - m_{N_e}^2)(u - m_{N_e}^2)} B_4^{(kn)}(s,t,u) \Bigg], \tag{13.181}$$

where

$$B_4^{(kn)}(s,t,u) = s(m_{W_k}^2 + m_{W_n}^2)\left(\frac{2}{ut} - \frac{1}{m_{W_k}^2 m_{W_n}^2}\right) - \frac{F}{8} \frac{s^2}{m_{W_k}^2 m_{W_n}^2}, \tag{13.182}$$

the function $B_3^{(kn)}(s,u,t)$ is given by the expression (13.137), $b_{L,R}^{(k)}$ has been defined by (13.133) while F and β are governed by Eqs. (13.134). When writing (13.181), we have set the mass of ν_{eL} equal to zero and taken into consideration the approximate relation

$$m_{N_e} \approx f_{ee} v_R$$

which follows from (13.154). It should be particularly emphasized that in the modified SM without the scalar triplet, ν_{eL} has the seesaw mass $\sim m_l^2/m_N$ (m_l is a charged lepton mass) and the mixing angle $\alpha \sim m_l/m_N$ in the one-flavor approximation. In such a way, assuming the neutrino mass to be equal to zero, we obtain the zero mixing angle too. However, in the LRM, the light neutrino obtains the mass from $\Delta_L^{(--)}$ as well. Therefore, it is now possible to get zero mass for ν_{eL} and, at the same time, to keep nonzero value of the mixing angle inside the neutrino generation.

Corresponding calculations for the process

$$e_L^- + e_L^- \to W_n^- + W_k^- \tag{13.183}$$

gives the differential cross section in the view

$$\frac{d\sigma_{LL}^{(kn)}}{dt} = \frac{(g_L^2 m_{N_e} b_L^{(k)} b_L^{(n)} \sin^2 \alpha)^2}{512\pi s^2}\left\{\left[3s - \frac{Fs^2(s - 2m_{W_n}^2 - 2m_{W_k}^2)}{4m_{W_n}^2 m_{W_k}^2} - \frac{st(s - m_{W_n}^2 - m_{W_k}^2)}{m_{W_n}^2 m_{W_k}^2}\right] \times\right.$$

$$\times \frac{1}{(t - m_{N_e}^2)^2} + \frac{4s}{(s - m_{\Delta_L}^2)^2 + \Gamma_{\Delta_L}^2 m_{\Delta_L}^2} \left[8 + \frac{(s - m_{W_n}^2 - m_{W_k}^2)^2}{m_{W_n}^2 m_{W_k}^2} + \frac{s - m_{\Delta_L}^2}{t - m_{N_e}^2} \left(8 + \right. \right.$$

$$\left. \left. + \frac{2t(m_{W_n}^2 + m_{W_k}^2 - s)}{m_{W_n}^2 m_{W_k}^2} \right) \right] + (u \leftrightarrow t) +$$

$$+ \left[2s + \frac{F s^2 (s - 2m_{W_n}^2 - 2m_{W_k}^2)}{2 m_{W_n}^2 m_{W_k}^2} \right] \frac{1}{(t - m_{N_e}^2)(u - m_{N_e}^2)} \Bigg\}, \tag{13.184}$$

where Γ_{Δ_L} is the decay width of the $\Delta_L^{(--)}$ boson. Considering only the prominent decay mode of $\Delta_{L,R}^{(--)}$ into $\tau^- \tau^-$, we obtain

$$\Gamma_{\Delta_{L,R}} = \frac{m_{N_e}^2}{16 \pi v_{L,R}^2} m_{\Delta_{L,R}} \left(\frac{m_\tau}{m_{\Delta_{L,R}}} \right)^2 \sqrt{1 - \frac{4 m_\tau^2}{m_{\Delta_{L,R}}^2}},$$

where m_τ is the τ-lepton mass. The differential cross section of the reaction

$$e_R^- + e_R^- \to W_k^- + W_n^- \tag{13.185}$$

follows from (13.184) by the substitution

$$g_L \to g_R, \qquad b_L^{(k)} \to b_R^{(k)}, \qquad \alpha \to \alpha + \frac{\pi}{2}, \qquad m_{\Delta_L} \to m_{\Delta_R}. \tag{13.186}$$

The total cross section of the reaction (13.180) is obtained from (13.181) after multiplication by $\beta s / 2$ and substitution

$$\left. \begin{array}{c} \dfrac{B_3^{(kn)}(s, t, u)}{(t - m_{N_e}^2)^2} \\[3mm] \dfrac{B_3^{(kn)}(s, u, t)}{(u - m_{N_e}^2)^2} \end{array} \right\} \to D_3^{(kn)}(s), \tag{13.187}$$

$$\frac{B_4^{(kn)}(s, t, u)}{(t - m_{N_e}^2)(u - m_{N_e}^2)} \to D_4^{(kn)}(s), \tag{13.188}$$

where

$$D_3^{(kn)}(s) = \frac{1}{m_{W_n}^2 m_{W_k}^2} \left[1 + \frac{(m_{W_n}^2 + m_{W_k}^2 - s - 2 m_{N_e}^2) \ln L}{2 \beta s} \right] +$$

$$+ \frac{1}{m_{N_e}^4} \left\{ \frac{[4 m_{W_n}^2 m_{W_k}^2 - 2 m_{N_e}^2 (m_{W_n}^2 + m_{W_k}^2 - s)](\ln L - \ln L_\nu)}{\beta s m_{N_e}^2} - 4 \right\} + \frac{s(m_{W_n}^2 + m_{W_k}^2) C_\nu}{m_{W_n}^2 m_{W_k}^2}, \tag{13.189}$$

$$D_4^{(kn)}(s) = - \frac{2[2 s(m_{W_n}^2 + m_{W_k}^2) - C_\nu^{-1}] \ln L}{\beta s m_{W_n}^2 m_{W_k}^2 (m_{W_n}^2 + m_{W_k}^2 - s - 2 m_{N_e}^2)} - \frac{1}{m_{W_n}^2 m_{W_k}^2} -$$

$$- \frac{8(m_{W_n}^2 + m_{W_k}^2)}{\beta m_{N_e}^2 (m_{N_e}^2 - m_{W_n}^2 - m_{W_k}^2 + s)} \left[\frac{\ln L}{m_{W_n}^2 + m_{W_k}^2 - s - 2 m_{N_e}^2} - \frac{\ln L_\nu}{m_{W_n}^2 + m_{W_k}^2 - s} \right], \tag{13.190}$$

C_ν and L are defined by the formulae (13.146) and $L_\nu = L \big|_{m_{N_e} = 0}$. For the total cross section of (13.183) we find the expression

$$\sigma_{LL}^{(kn)} = \beta \frac{(g_L^2 m_{N_e} b_L^{(k)} b_L^{(n)} \sin^2 \alpha)^2}{256 \pi s} \left\{ \frac{s - 4(m_{W_n}^2 + m_{W_k}^2)}{m_{W_n}^2 m_{W_k}^2} + s C_\nu \left(3 + \frac{m_{N_e}^4 + m_{W_n}^2 m_{W_k}^2}{m_{W_n}^2 m_{W_k}^2} \right) + \right.$$

$$+\frac{\ln L}{\beta s}\left[\frac{2sm_{N_e}^2+2(m_{W_n}^2+m_{W_k}^2-s-2m_{N_e}^2)(m_{W_n}^2+m_{W_k}^2)}{m_{W_n}^2 m_{W_k}^2}+\right.$$

$$\left.+\frac{2}{m_{W_n}^2+m_{W_k}^2-s-2m_{N_e}^2}\left(s-\frac{s-2m_{W_n}^2-2m_{W_k}^2}{C_\nu m_{W_n}^2 m_{W_k}^2}\right)\right]+$$

$$+\frac{1}{(s-m_{\Delta_L}^2)^2+\Gamma_{\Delta_L}^2 m_{\Delta_L}^2}\left[16s+\frac{2s(s-m_{W_n}^2-m_{W_k}^2)^2}{m_{W_n}^2 m_{W_k}^2}+16(s-m_{\Delta_L}^2)\left(\frac{\ln L}{\beta}+\right.\right.$$

$$\left.\left.\left.+\frac{s(m_{W_n}^2+m_{W_k}^2-s)}{4m_{W_n}^2 m_{W_k}^2}(1+\frac{m_{N_e}^2 \ln L}{\beta s})\right)\right]\right\}. \tag{13.191}$$

Having done the replacement (13.186) in the expression (13.191), we get the total cross section of the reaction (13.185).

Now we discuss the asymptotic behavior of the cross sections obtained above. From the expressions (13.187), (13.188), and (13.191) we can see that their partial contributions violate the unitary limit. To be definite, in the case of LR electron beam they increase as a linear function of s, while for LL or RR electron beams those tend to constant values when $s \to \infty$. However, in all cases the total cross section resulting from the sum of those contributions have a right asymptotic behavior

$$\sigma \sim s^{-1}\ln s. \tag{13.192}$$

It is caused by the cancellation among the partial contributions. We should stress that the reasons leading to (13.192) are quite different for LR and LL (or RR) polarized electron beams. In the former case the $\Delta^{(--)}$ contribution to the cross section is absent and the cancellation is connected with the spin behavior of the virtual neutrino. If the neutrino flips the helicity between the acts of emission and absorbtion then the amplitudes coming from ν_e and N_e exchanges are proportional to the neutrino masses. Since we neglected m_{ν_e} only the N_e exchange term will contribute. On the contrary, when the neutrino does not flip its helicity (this case is realized for LR electron beams) we have two nonvanishing terms. They are caused by ν_e and N_e exchanges and have opposite signs. As the calculations show, disregarding the contribution of the light neutrino would then lead to a cross section proportional to s.

As it follows from the expression for the cross section of the reaction $e^- + e^- \to W_k^- + W_n^-$ this reaction could be a good tool for the definition of such LRM parameters as m_{N_e}, g_R, α, and ξ. Thus, for example, the values of $\sigma^{(kn)}$ strongly depend on m_{N_e}. Let us set the following values for the LRM parameters:

$$m_{W_2} = 715 \text{ GeV}, \qquad m_{Z_2} = 800 \text{ GeV}, \qquad \phi = 9.6 \times 10^{-3}, \qquad \xi = 3.1 \times 10^{-2},$$

$$m_{\Delta_L} = 100 \text{ GeV}, \qquad m_{\Delta_R} = 110 \text{ GeV}, \qquad \alpha = 10^{-2}.$$

Then, at $m_{N_e} = 100$ GeV and $g_L = g_R$ in the energy region up to 200 GeV we have $\sigma_L^{(11)} \approx 8\sigma_R^{(11)} \approx 8 \times 10^{-2}$ fb whereas $\sigma_{LR}^{(11)}$ is about few $\times 10^{-3}$ fb. When increasing m_{N_e} the contribution from $\sigma_{LR}^{(11)}$ becomes dominant. At $m_{N_e} = 1$ TeV and $\sqrt{s} = 200$ GeV we have

$$\sigma_R^{(11)} \approx 3.3 \cdot 10^{-3} \text{ pb}, \qquad \sigma_L^{(11)} \approx 2.6 \cdot 10^{-2} \text{ pb}, \qquad \sigma_{LR}^{(11)} \approx 5 \cdot 10^{-2} \text{ pb}.$$

It should be noted here that at the sufficiently great m_{N_e} ($m_{N_e} > 5$ TeV) the cross section of the reaction $e^- + e^- \to W_1^- + W_1^-$ could reach the values that are compatible and even greater than the cross section of the reaction

$$e^- + e^+ \to W_1^- + W_1^+.$$

It might be well to point out that in the case

$$m_{\Delta_{L,R}} < (m_{W_n} + m_{W_k})$$

the cross section of the inverse neutrinoless β-decay weakly depends on a supposed value of $m_{\Delta_{L,R}}$. When violating this inequality, we shall have the s-channel resonance ($\Delta_{L,R}$-resonance) giving the increase of the total cross section for $e_L^- e_L^-$ and $e_R^- e_R^-$ beams. Note, in particular, that the signature of doubly charged Higgs bosons would strongly support the idea of a left-right approach.

Now we should pay some attention to the question of detectability of the processes:

$$e^- e^- \to W_k^- W_n^- \to l_i^- l_j^- + \text{neutrinos}, \qquad (13.193)$$

$$e^- e^- \to W_k^- W_n^- \to l_i^- + \text{jet} + \text{neutrinos}, \qquad (13.194)$$

where i and j are the fermion flavors. The most clean signature is the case $i, j = e, \mu$ plus missing momentum carried away by neutrinos. In order to eliminate the major background coming from the RCs to elastic $e^- e^-$ scattering we should demand $i \neq j$ or the sizable $p_\perp$ of outgoing leptons. The cut on $p_\perp$ helps to reduce the QED background caused by radiative pair production

$$e^- e^- \to e^- e^- \mu^- \mu^+ \tau^- \tau^+. \qquad (13.195)$$

Other relevant background processes are

$$e^- e^- \to e^- e^- Z_k \longrightarrow e^- e^- l_i^- l_i^+, \qquad (13.196)$$

$$e^- e^- \to e^- e^- Z_k \to e^- e^- \nu_l \overline{\nu_l}, \qquad (13.197)$$

$$e^- e^- \to e^- e^- W_k^- W_n^+ \to e^- e^- l_i^- l_j^+ \nu_i \overline{\nu_j}. \qquad (13.198)$$

The main difference between the final states of the reactions (13.193),(13.194), and (13.196)–(13.198) is that in the former case they consist of like-sign leptons pair only. Therefore, it is easy to distinguish the $W^- W^-$ signal from these backgrounds by means of the charged final-state leptons identification. We also have the background caused by multihadronic events

$$e^- e^- \longrightarrow e^- e^- W_k^- W_n^+ q_i \overline{q_i}. \qquad (13.199)$$

For this reaction there is no possibility to observe the high $p_\perp$ of the final like-sign leptons. Hence, we conclude that the cut on $p_\perp$ and the charged lepton identification of the outgoing leptons will be very useful in order to reduce backgrounds for the $W_k^- W_n^-$ signal.

One more interesting prediction of the LRM is connected with investigating ultrahigh-energy (UHE) cosmic neutrinos. Let us show that these neutrinos could be used for the definition of properties of the singly charged Higgs bosons, $h^{(\pm)}$ and $\tilde{\delta}^{(\pm)}$. At that, the reactions that have the resonance splash related with the $h^{(\pm)}$- and $\tilde{\delta}^{(\pm)}$-bosons, predicted by G. Boyarkina [99], are of the most perspective. The process

$$e^- + \nu_e \to \mu^- + \nu_\mu, \qquad (13.200)$$

is an example of similar reactions. It could be studied with the help of such neutrino telescopes as BAIKAL NT-200, NESTOR, AMANDA, and so on. The corresponding Feynman diagrams are shown in Fig. 13.7, where we have neglected the contribution caused by the W_2^- boson. We are reminded that because the neutrino is considered to be the Majorana particle, then the neutrino does not carry any lepton flavor. Therefore, the process (13.200) goes with the partial lepton flavor violation and the arrows on the neutrino lines point only

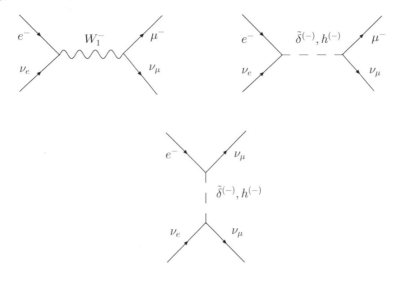

FIGURE 13.7
The Feynman diagrams associated with the process $e^-\nu_e \to \mu^-\nu_\mu$.

to the fact that we are dealing either with the neutrino incoming to an electron target or with the neutrino in the final state. The total cross sections are given by

$$\sigma_L = \frac{1}{32\pi s^2}\left\{(1-\lambda_\nu)\left[\frac{g_L^4 c_\xi^4 s^3}{3[(s-m_{W_1}^2)^2+\Gamma_{W_1}^2 m_{W_1}^2]} + \frac{2g_L^2 c_\xi^2 f_{e\mu}^2(s-m_{W_1}^2)}{[(s-m_{W_1}^2)^2+\Gamma_{W_1}^2 m_{W_1}^2]}\left(\frac{s^2}{2}\right.\right.\right.$$
$$\left.-m_{\tilde\delta}^2 s + \frac{m_{\tilde\delta}^2\Gamma_{\tilde\delta}^2 - m_{\tilde\delta}^4}{2}N_{\tilde\delta} + 2m_{\tilde\delta}^3\Gamma_{\tilde\delta}Q_{\tilde\delta}\right) + f_{e\mu}^2\left(4f_{e\mu}^2 - \frac{2g_L^2 c_\xi^2\Gamma_{W_1}\Gamma_{\tilde\delta}m_{W_1}m_{\tilde\delta}}{(s-m_{W_1}^2)^2+\Gamma_{W_1}^2 m_{W_1}^2}\right)\times$$
$$\left.\left.\times\left(s+m_{\tilde\delta}^2 N_{\tilde\delta} + \frac{m_{\tilde\delta}^3 - m_{\tilde\delta}\Gamma_{\tilde\delta}^2}{\Gamma_{\tilde\delta}}Q_{\tilde\delta}\right)\right] + (1+\lambda_\nu)\frac{4f_{ee}^2 f_{\mu\mu}^2 s^3}{(s-m_{\tilde\delta}^2)^2+\Gamma_{\tilde\delta}^2 m_{\tilde\delta}^2}\right\}, \tag{13.201}$$

for the case of the left-polarized electrons, and

$$\sigma_R = \frac{1}{128\pi s^2}\left[(1-\lambda_\nu)\frac{(\alpha_{ee}\alpha_{\mu\mu})^2 s^3}{(s-m_h^2)^2+\Gamma_h^2 m_h^2} + (1+\lambda_\nu)\alpha_{e\mu}^4\left(s+m_h^2 N_h + \frac{m_h^3 - m_h\Gamma_h^{'2}}{\Gamma_h}Q_h\right)\right], \tag{13.202}$$

for the case of the right-polarized electrons, where

$$N_k = \ln\left|\frac{m_k^4+\Gamma_k^2 m_k^2}{(s+m_k^2)^2+\Gamma_k^2 m_k^2}\right|, \qquad Q_k = \arctan\left(\frac{s+m_k^2}{\Gamma_k m_k}\right) - \arctan\left(\frac{m_k}{\Gamma_k}\right),$$

$k = \tilde\delta, h$ and the quantity λ_ν denotes the neutrino helicity.

For the left-polarized electrons and the right-polarized neutrinos the deviations from the SM predictions are very small. For example, at $f_{e\mu} = 3 \times 10^{-2}$ they are of the order of 0.1%. Recall that in this case there is the W-resonance peak (Glashow resonance) and its height (σ_W) reaches the values of the order of 10^4 pb.

From the definitions of m_{W_1} and M_D we could obtain [99]

$$\alpha_{e\mu} = \frac{\sqrt{(1+\rho_t\Delta\rho_0 - 4k_1^2/v_R^2)[2g_L^{-2}m_{W_1}^2 - k_1^2(1+\rho_t\Delta\rho_0)]}M_D}{2g_L^{-2}m_{W_1}^2 - 2k_1^2(1+\rho_t\Delta\rho_0) + 4k_1^4/v_R^2}\frac{k_+}{k_2}, \tag{13.203}$$

When M_D is non-zero, then even for large values of φ_e and φ_μ the quantity $\alpha_{e\mu}$ is very small. For example, at $\varphi_e = 2.5 \times 10^{-2}$, $\varphi_\mu = 3 \times 10^{-2}$ and $k_1 = 70$ GeV, we have $\alpha_{e\mu} = 1.8 \times 10^{-5}$. Since the coupling constant α_{ll} is defined by the equation

$$\alpha_{ll} = \frac{2k_1 k_2 m_l - k_+^2 m_D^l}{k_-^2 k_+} \tag{13.204}$$

which follows from the determination of m_D^l and m_l (see, Eqs. (13.148), (13.149)), then the same is true for the case $l = e$. On the other hand, the quantity $\alpha_{\mu\mu}$ could be as large as 10^{-2} even at the small angles φ_μ. In the case of the degeneracy of the bidoublet vacuum expectation values ($k_1 = k_2 = k_g$) we have

$$\alpha_{ab} = 2h'_{ab}.$$

However, even at abnormally large values of $\alpha_{e\mu}$, say 2×10^{-2}, the cross section for the right-polarized electrons and the left-polarized electron neutrinos can reach only the values of the order of 3×10^{-3} fb. Therefore, of primary interest are the cases with the initial $e_L^- \nu_L$ and $e_R^- \nu_R$.

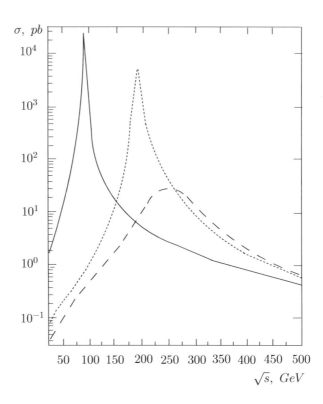

FIGURE 13.8

The σ_{RR} versus $\sqrt{s}$ at $\lambda_\nu = -1$ for the cases: $m_h = 90$ GeV—solid line; $m_h = 175$ GeV—dotted line; $m_h = 200$ GeV—dashed line. In all three cases we adopt $\varphi_e = 8 \times 10^{-3}$, $\varphi_\mu = 9 \times 10^{-3}$, $f_{\tau\tau} = 0.5$, and $k_1 = 115$ GeV.

In Fig. 13.8 we represent the total cross section for the latter case. It is well to bear in mind that the following possibility could take place. The cross section on the singly charged

Higgs boson resonances $(\sigma_{e^-\nu_e \to \mu^-\nu_\mu})_{\tilde{\delta},h}$ proves to be bigger than σ_W. For example, when $f_{\tau\tau} = 0.6$, $f_{\mu\mu} = 0.75$, $f_{ee} = 0.09$, and $m_{\tilde{\delta}} = 74$ GeV, then we have $(\sigma_{e^-\nu_e \to \mu^-\nu_\mu})_{\tilde{\delta}} \sim 10^5$ pb. Note that the resonances associated with the singly charged Higgs bosons could be also observed in the reaction [99]

$$e^- + \nu_e \to W_1^- + Z_1. \tag{13.205}$$

To the successful LRM predictions must be attributed the explanation of the observed value of the muon anomalous magnetic moment (AMM). The AMM of a particle may be considered either as the notion on its composite structure or as a result of the RCs inclusion. However, the possibility of muon substructure has been ruled out. Therefore, the explanation of the $(g-2)_\mu$ anomaly consists in accounting for the muon interaction with the physical vacuum the theory determines. Since the SM is not able to predict the observed value of the muon AMM, then one should look for the answer when studying the contributions caused by particles that are additional relatively to the SM. The influence of the W_2- and Z_2-gauge bosons was investigated in Ref. [109], where it was shown that the Z_2 contribution is negative, while in order to explain the $(g-2)_\mu$ anomaly the W_2-gauge boson mass must lie around 100 GeV that is at variance with the experimental data. The Higgs bosons are the following candidates for the $(g-2)_\mu$ anomaly explanation. In Figs. 13.9–13.11 we display one-loop diagrams contributing to the muon AMM due to the physical Higgs bosons in the unitary gauge.

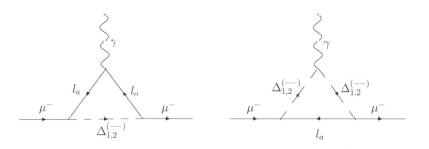

FIGURE 13.9
The diagrams contributing to the muon AMM due to the doubly charged Higgs bosons.

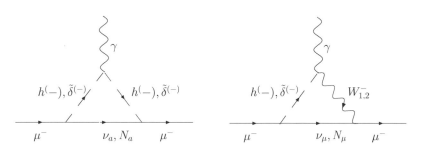

FIGURE 13.10
The diagrams contributing to the muon AMM due to the singly charged Higgs bosons.

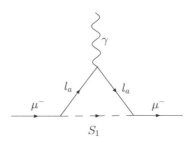

FIGURE 13.11

The diagrams contributing to the muon AMM due to the neutral Higgs bosons S_1.

Let us assume the following scenario: the dominant contribution comes from the $\Delta_1^{(--)}$ boson. A negligible value of the corrections from the $S_1, \Delta_2^{(--)}, \tilde{\delta}^{(-)}$, and $h^{(-)}$ bosons could be caused either by the large values of their masses or by the small values of their coupling constants.

The diagrams caused by $\Delta_{1,2}^{(--)}$ induce the following corrections [51]:

$$\delta a_\mu^\Delta = \frac{1}{8\pi^2} \left(4f_{\mu e}^2 \sum_{i=1}^2 I_e^{\Delta_i} + f_{\mu\mu}^2 \sum_{i=1}^2 I_\mu^{\Delta_i} + 4f_{\mu\tau}^2 \sum_{i=1}^2 I_\tau^{\Delta_i} \right), \qquad (13.206)$$

where

$$I_{l_a}^{\Delta_i} = \int_0^1 \left(\frac{2m_\mu^2(z^2 - z^3)}{m_\mu^2(z^2 - z) + m_{\Delta_i}^2 z + m_{l_a}^2(1 - z)} + \frac{m_\mu^2(z^2 - z^3)}{m_\mu^2(z^2 - z) + m_{\Delta_i}^2(1 - z) + m_{l_a}^2 z} \right) dz.$$

If the difference between the experiment and SM prediction is 3.4 σ, then at 90% CL, δa_μ must lie in the range

$$182 \times 10^{-11} \le \delta a_\mu \le 407 \times 10^{-11}. \qquad (13.207)$$

In Fig. 13.12, in the m_{Δ_1} vs $f_{\mu\mu}$ parameter space we display the solid and dotted curves, which correspond to the values of δa_μ^Δ equal to 407×10^{-11} (dotted line) and 182×10^{-11} (solid line), respectively. The range of the Higgs boson sector parameters allowed by the E821-experiment lies between the solid and dotted curves.

One could consider the scenarios with the contributions from the singly charged and neutral Higgs bosons as well. In all cases it has been possible to find the region of the LRM parameters under which the theory is in line with the E821 results [110].

But besides the Higgs and gauge bosons the LRM contains another three additional particles, three heavy neutrinos, at that the one of $N_{1,2,3}$ might have the mass lying on the electroweak scale [111]. So, we should investigate the influence of heavy neutrino properties on the muon AMM. Let us direct our attention to neutrino multipole moments (MMs). The neutrino is a neutral particle, and its total bare Lagrangian does not involve any MMs. The appearance of the neutrino MMs is due to neutrino interaction with the vacuum, whose structure is determined by the choice of model.* The conventional choice of the LRM

---

*Detailed discussion of the neutrino MMs will be fulfilled in Section 14.2.

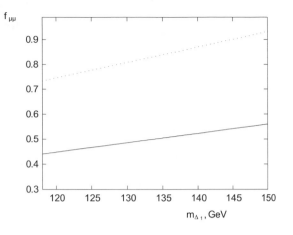

FIGURE 13.12
The contours of the one-loop contribution from the $\Delta_1^{(--)}$ boson to the muon AMM. In numerical calculations we have used $m_{W_2} = 2500$ GeV and $\xi = 10^{-2}$.

Higgs sector is given by Eqs. (13.65) and (13.66). Such a choice ensures the Majorana nature of the neutrinos. Then, if we are dealing with diagonal elements of the MMs, the only nonvanishing MM is an anapole moment. However, this is not true for nondiagonal elements. For example, nondiagonal elements of the dipole magnetic moment (DMM) could be nonzero. Then the neutrino DMM induced by the RCs leads to the appearance of the following terms in the effective interaction Lagrangian

$$H_{add} = \sum_{i \neq j} \left[\mu_{\nu_i \nu_j} \overline{\nu}_i(p) \sigma_{\lambda\tau} q_\tau \nu_j(p') + \mu_{N_i N_j} \overline{N}_i(p) \sigma_{\lambda\tau} q_\tau N_j(p') + \mu_{\nu_i N_j} \overline{\nu}_i(p) \sigma_{\lambda\tau} q_\tau N_j(p') + \right.$$

$$\left. + \mu_{N_i \nu_j} \overline{N}_i(p) \sigma_{\lambda\tau} q_\tau \nu_j(p') \right] A_\lambda(q), \tag{13.208}$$

where $q_\tau = (p' - p)_\tau$. For the sake of simplicity, let us suppose that the mixing takes place between the muon and tau neutrino only. In Fig. 13.13 we represent the Feynman diagrams with the light and heavy neutrinos making a contribution into the muon AMM.

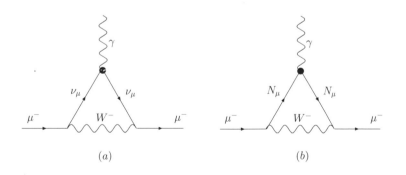

FIGURE 13.13
The Feynman diagrams (a) and (b) giving a contribution to the muon AMM.

The analysis show [112], that in the Majorana case, in order to explain the BNL result, the neutrino DMM must be as large as $\times 10^{-7}\mu_B$. The obtained value proves to be three orders of magnitude higher than that calculated in Ref. [113]. However, it should be remembered that there are presently no experimental limits on the neutrino dipole magnetic moment of heavy neutrinos.

In the LRM, the neutrino can possess a Dirac nature too. In this case, the Higgs sector includes two doublets $\chi_L(2,1,1)$, $\chi_R(1,2,1)$ and one pseudoscalar real singlet $\sigma(1,1,0)$. Upon spontaneous symmetry breaking, the theory has two singly charged and six neutral physical Higgs bosons. In this version of the theory, the diagonal elements of the neutrino DMM are no longer zero and can be considerably larger than off-diagonal elements. The neutrino DMM associated with the heavy neutrino is given by [113]

$$\mu_{N_a N_a} = \frac{3eg_L^2 m_{N_a}}{64\pi^2}\left(\frac{s_\xi^2}{m_{W_1}^2} + \frac{c_\xi^2}{m_{W_2}^2}\right) + \frac{eg_L^2}{8\pi^2 m_{W_1}^2}s_\xi c_\xi m_\tau \text{Re}(e^{-i\psi}V_{a\tau}^\dagger V_{\tau a}), \qquad (13.209)$$

where $N_a = \sum_l V_{al}N_l$ and ψ is an angle associated with the CP-violation. Substituting the following numerical values

$$m_{N_a} = m_{N_b} = 1 \text{ TeV}, \qquad \xi = 3\times 10^{-3}, \qquad m_{W_2} = 736 \text{ GeV}$$

into the above formula, we obtain

$$\mu_{N_a N_a} \approx 3.7\times 10^{-9}\mu_B. \qquad (13.210)$$

So, we see that in this case the influence of the neutrino DMM on the muon AMM is negligibly small as well. Then, we come to the conclusion: among the additional particles of the LRM only the Higgs bosons could give explanation for the $(g-2)_\mu$ anomaly.

In conclusion in this case, we briefly discuss the radiative muon decay

$$\mu^\pm \to e^\pm + \gamma \qquad (13.211)$$

within the LRM with a Dirac neutrino. Further, we shall assume that only the electron and muon neutrinos are mixed. Contributions to the muon radiative decay will be caused by the diagrams taking into account the neutrino DMM (they follow from Fig. 13.13 when the final muon line is replaced by the electron line and the external electromagnetic line is replaced by the photon line) and the diagrams pictured in Fig. 13.14.

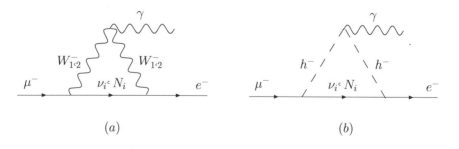

FIGURE 13.14
The Feynman diagrams (a) and (b) associated with the radiative muon decay.

Following Ref. [114], we estimate these contributions. Let us start with the diagrams that are due to the neutrino DMM. The results of the calculations show that a dominant contribution comes from the diagrams in which the virtual lines are associated with heavy neutrinos. Moreover, only the diagrams involving W_2-boson exchange contribute substantially among all the diagrams of the type shown in Fig. 13.13. Summing the contributions of the diagrams pictured in Figs. 13.13a and 13.13b, we obtain the following expression for the decay width:

$$\Gamma^n(\mu \to e\gamma) = \frac{32m_\mu^5}{(4\pi)^5}\left(I_N^{W_2}c_\xi^2 + I_N^h\right)^2 (c_{\theta_N}s_{\theta_N})^2. \tag{13.212}$$

Here, $I_N^{W_2}$ and I_N^h are the contributions of the W_2-boson and the Higgs h-boson, respectively:

$$I_N^{W_2} = \frac{g_L^2}{8}\left(D_{N_1}^{W_2}m_{N_1}\mu_{N_1N_1}c_{\varphi_\mu}^2 - D_{N_2}^{W_2}m_{N_2}\mu_{N_2N_2}c_{\varphi_e}^2\right),$$

$$I_N^h = \alpha_{ehN_e}\alpha_{\mu hN_\mu}\left(D_{N_1}^h m_{N_1}\mu_{N_1N_1}c_{\varphi_\mu}^2 - D_{N_2}^h m_{N_2}\mu_{N_2N_2}c_{\varphi_e}^2\right),$$

$$D_j^{W_k} = \int_0^1 \left\{\frac{2x(1-x)}{b+m_\mu^2x^2}\ln\left|\frac{L_j^{W_k}+b+m_\mu^2x^2}{L_j^{W_k}}\right| + \frac{x(3x-1)}{m_{W_k}^2}\left[\frac{L_j^{W_k}}{b}\ln\left|\frac{L_j^{W_k}+b}{L_j^{W_k}}\right| - \right.\right.$$
$$\left.- \frac{L_j^{W_k}}{b+m_\mu^2x^2}\ln\left|\frac{L_j^{W_k}+b+m_\mu^2x^2}{L_j^{W_k}}\right| + \ln\left|\frac{L_j^{W_k}+b}{L_j^{W_k}+b+m_\mu^2x^2}\right|\right] +$$
$$\left.+ \frac{m_\mu^2x^3(1-x)}{m_{W_k}^2(b+m_\mu^2x^2)^2}\left(b+m_\mu^2x^2 - L_j^{W_k}\ln\left|\frac{L_j^{W_k}+b+m_\mu^2x^2}{L_j^{W_k}}\right|\right)\right\}dx,$$

$$b = x(m_e^2 - m_\mu^2), \qquad L_j^{W_k} = (m_j^2 - m_{W_k}^2)x + m_{W_k}^2, \qquad k = 1,2, \qquad j = N_1, N_2,$$

$$D_j^h = \int_0^1 \frac{x(1-x)}{b+m_\mu^2x^2}\ln\left|\frac{L_j^h+b+m_\mu^2x^2}{L_j^h}\right|dx, \qquad L_j^h = L_j^{W_k}(m_{W_k} \to m_h)$$

where α_{lhN_l} is the $lh^{(-)}N_l$ coupling constant and the electron mass is taken into account in the denominators in order to avoid the emergence of divergences at the upper limit of integration. We note that the quantities $I_N^{W_2}$ and I_N^h are of the same sign.

Now we evaluate contributions from the diagrams in Fig. 13.14. Among these, dominant contributions come from the diagrams involving heavy-neutrino exchanges. The corresponding decay width is given by:

$$\Gamma^b(\mu \to e\gamma) = \frac{e^2m_\mu^5}{(4\pi)^5}\left(A_N^{W_2}c_\xi^2 + A_N^h\right)^2 (c_{\theta_N}s_{\theta_N})^2, \tag{13.213}$$

where $A_N^{W_2}$ and A_N^h are determined by the diagrams involving the W_2^--boson and Higgs boson $h^{(-)}$, respectively

$$A_N^{W_2} = \frac{g_L^2}{2}\left(B_{N_1}^{W_2}c_{\varphi_\mu}^2 - B_{N_2}^{W_2}c_{\varphi_e}^2\right), \qquad A_N^h = 4\alpha_{ehN_e}\alpha_{\mu hN_\mu}\left(B_{N_1}^h c_{\varphi_\mu}^2 - B_{N_2}^h c_{\varphi_e}^2\right),$$

$$B_j^{W_k} = -\int_0^1\left[\frac{x^2}{b+m_\mu^2x^2}\ln\left|\frac{L_{W_k}^j+b+m_\mu^2x^2}{L_{W_k}^j}\right| + \frac{x^2(1-2x)}{(b+m_\mu^2x^2)^2}\left(b+m_\mu^2x^2 -\right.\right.$$
$$\left.\left.- L_{W_k}^j\ln\left|\frac{L_{W_k}^j+b+m_\mu^2x^2}{L_{W_k}^j}\right|\right)\right]dx, \qquad L_{W_k}^j = (m_{W_k}^2 - m_j^2)x + m_j^2,$$

$$B_j^h = \int_0^1 \frac{x^2(1-x)}{(b+m_\mu^2 x^2)^2} \left(b + m_\mu^2 x^2 - L_h^j \ln \left| \frac{L_h^j + b + m_\mu^2 x^2}{L_h^j} \right| \right) dx,$$

$$L_h^j = L_{W_k}^j (m_{W_k} \to m_h).$$

Because the sign of $\mathcal{A}_N^{W_2}$ is opposite to that of $\mathcal{A}_N^h$, the interference between the diagrams associated with charged gauge bosons and the diagrams generated by singly charged Higgs bosons is destructive.

From expressions (13.212) and (13.213), it follows that the widths $\Gamma^n(\mu \to e\gamma)$ and $\Gamma^b(\mu \to e\gamma)$ are functions of $\Delta m_{ij} = m_i - m_j$ that decrease with decreasing Δm_{ij} and are zero in the limit of degenerate heavy-neutrino masses.

Let us assume temporarily that the coupling constants α_{ehN_e} and/or $\alpha_{\mu h N_\mu}$ are very small. Therefore, the contributions of the charged Higgs boson could be neglected. It turns out that, at identical values of the model parameters, the contributions of the gauge bosons are on average four orders of magnitude greater than the contributions coming from the neutrino DMM. In order to obtain values in the range $\mathrm{Br}(\mu \to e\gamma) < 1.2 \times 10^{-11}$, it is necessary in this case to have quasi-degeneracy in the heavy-neutrino sector. If, for example,

$$m_{W_2} = 800 \text{ GeV}, \qquad \varphi_\mu = 10^{-2}, \qquad \varphi_e = 2 \times 10^{-2}, \qquad m_{N_1} + m_{N_2} = 1200 \text{ GeV},$$

the upper experimental limit on the branching ratio for radiative muons decay is obtained at $m_{N_1} = 613$ GeV and $m_{N_2} = 587$ GeV.

However, there is one important case where the contributions of the neutrino DMM play a leading role. Since $\mathcal{A}_N^{W_2}$ and $\mathcal{A}_N^h$ have opposite signs, they can cancel each other. One can readily show that, under the condition

$$4\alpha_{ehN_e}\alpha_{\mu h N_\mu} \int_0^1 (x^2 - x^3) \left[\frac{c_{\varphi_\mu}^2}{L_h^{N_1}} - \frac{c_{\varphi_e}^2}{L_h^{N_2}} \right] dx \approx g_L^2 \int_0^1 (x^2 - x^3) \left[\frac{c_{\varphi_\mu}^2}{L_{W_2}^{N_1}} - \frac{c_{\varphi_e}^2}{L_{W_2}^{N_2}} \right] dx \quad (13.214)$$

the width $\Gamma^b(\mu \to e\gamma)$ vanishes. In this case, the analysis reveals that, if only the term $I_N^{W_2}$ is taken into account in (13.212), the value of 10^{-12} or less can be obtained for the branching ratio for the radiative muon decay in the experimentally allowed region of the parameters of the LRM [111]. But if I_N^h is of the same order of magnitude as $I_N^{W_2}$, $\mathrm{Br}^n(\mu \to e\gamma)$ may reach values of the order of 10^{-11}.

Appendix: Feynman rules for the GWS model

Gauge boson three-vertices:

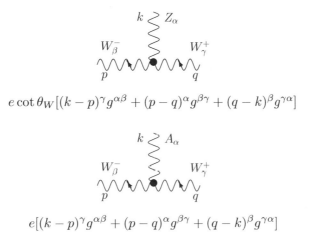

$$e \cot \theta_W [(k-p)^\gamma g^{\alpha\beta} + (p-q)^\alpha g^{\beta\gamma} + (q-k)^\beta g^{\gamma\alpha}]$$

$$e[(k-p)^\gamma g^{\alpha\beta} + (p-q)^\alpha g^{\beta\gamma} + (q-k)^\beta g^{\gamma\alpha}]$$

Gauge boson four-vertices:

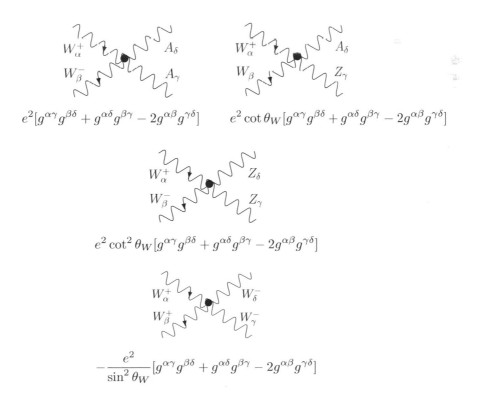

$$e^2[g^{\alpha\gamma}g^{\beta\delta} + g^{\alpha\delta}g^{\beta\gamma} - 2g^{\alpha\beta}g^{\gamma\delta}] \qquad e^2 \cot \theta_W [g^{\alpha\gamma}g^{\beta\delta} + g^{\alpha\delta}g^{\beta\gamma} - 2g^{\alpha\beta}g^{\gamma\delta}]$$

$$e^2 \cot^2 \theta_W [g^{\alpha\gamma}g^{\beta\delta} + g^{\alpha\delta}g^{\beta\gamma} - 2g^{\alpha\beta}g^{\gamma\delta}]$$

$$-\frac{e^2}{\sin^2 \theta_W}[g^{\alpha\gamma}g^{\beta\delta} + g^{\alpha\delta}g^{\beta\gamma} - 2g^{\alpha\beta}g^{\gamma\delta}]$$

Gauge-boson-fermion vertices:

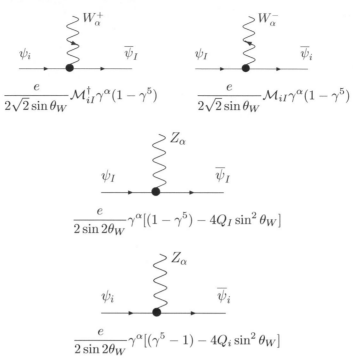

$$\frac{e}{2\sqrt{2}\sin\theta_W}\mathcal{M}_{iI}^{\dagger}\gamma^{\alpha}(1-\gamma^5)$$

$$\frac{e}{2\sqrt{2}\sin\theta_W}\mathcal{M}_{iI}\gamma^{\alpha}(1-\gamma^5)$$

$$\frac{e}{2\sin 2\theta_W}\gamma^{\alpha}[(1-\gamma^5)-4Q_I\sin^2\theta_W]$$

$$\frac{e}{2\sin 2\theta_W}\gamma^{\alpha}[(\gamma^5-1)-4Q_i\sin^2\theta_W]$$

Gauge-boson-Higgs three vertices:

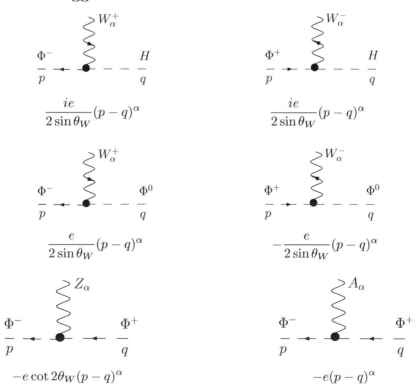

$$\frac{ie}{2\sin\theta_W}(p-q)^{\alpha}$$

$$\frac{ie}{2\sin\theta_W}(p-q)^{\alpha}$$

$$\frac{e}{2\sin\theta_W}(p-q)^{\alpha}$$

$$-\frac{e}{2\sin\theta_W}(p-q)^{\alpha}$$

$$-e\cot 2\theta_W(p-q)^{\alpha}$$

$$-e(p-q)^{\alpha}$$

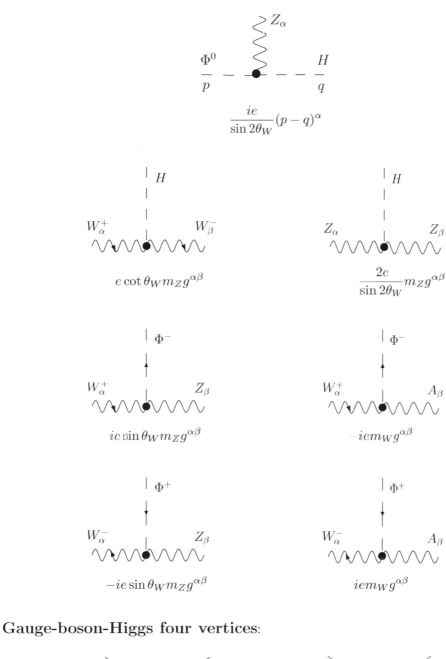

$$\frac{ie}{\sin 2\theta_W}(p-q)^\alpha$$

$$e\cot\theta_W m_Z g^{\alpha\beta}$$

$$\frac{2c}{\sin 2\theta_W}m_Z g^{\alpha\beta}$$

$$ic\sin\theta_W m_Z g^{\alpha\beta}$$

$$-iem_W g^{\alpha\beta}$$

$$-ie\sin\theta_W m_Z g^{\alpha\beta}$$

$$iem_W g^{\alpha\beta}$$

Gauge-boson-Higgs four vertices:

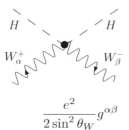

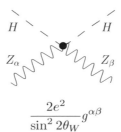

$$\frac{e^2}{2\sin^2\theta_W}g^{\alpha\beta}$$

$$\frac{2e^2}{\sin^2 2\theta_W}g^{\alpha\beta}$$

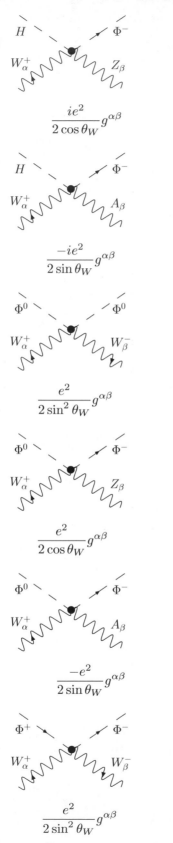

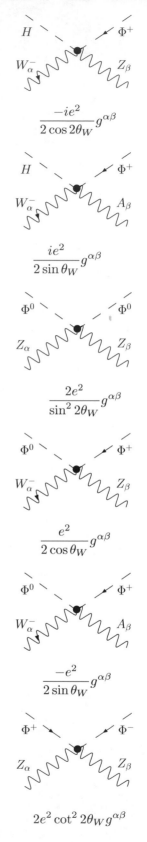

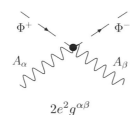

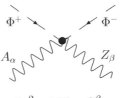

$$2e^2 g^{\alpha\beta}$$

$$2e^2 \cot 2\theta_W g^{\alpha\beta}$$

Higgs three vertices:

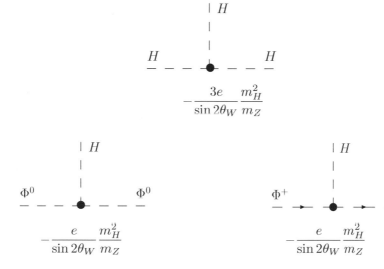

$$-\frac{3e}{\sin 2\theta_W} \frac{m_H^2}{m_Z}$$

$$-\frac{e}{\sin 2\theta_W} \frac{m_H^2}{m_Z}$$

$$-\frac{e}{\sin 2\theta_W} \frac{m_H^2}{m_Z}$$

Higgs four vertices:

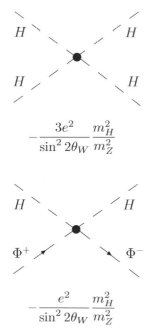

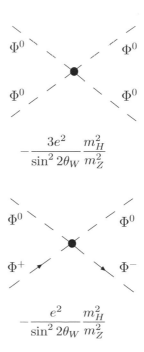

$$-\frac{3e^2}{\sin^2 2\theta_W} \frac{m_H^2}{m_Z^2}$$

$$-\frac{3e^2}{\sin^2 2\theta_W} \frac{m_H^2}{m_Z^2}$$

$$-\frac{e^2}{\sin^2 2\theta_W} \frac{m_H^2}{m_Z^2}$$

$$-\frac{e^2}{\sin^2 2\theta_W} \frac{m_H^2}{m_Z^2}$$

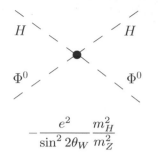

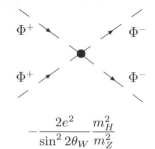

$$-\frac{e^2}{\sin^2 2\theta_W}\frac{m_H^2}{m_Z^2}$$

$$-\frac{2e^2}{\sin^2 2\theta_W}\frac{m_H^2}{m_Z^2}$$

Higgs-boson-fermion vertices:

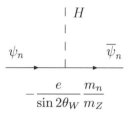

$$-\frac{e}{\sin 2\theta_W}\frac{m_n}{m_Z}$$

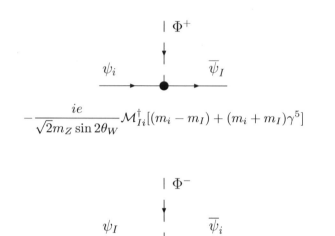

$$-\frac{ie}{\sqrt{2}m_Z\sin 2\theta_W}\mathcal{M}_{Ii}^{\dagger}[(m_i - m_I) + (m_i + m_I)\gamma^5]$$

$$-\frac{ie}{\sqrt{2}m_Z\sin 2\theta_W}\mathcal{M}_{iI}[(m_I - m_i) + (m_i + m_I)\gamma^5]$$

$$-\frac{iem_I}{m_Z\sin 2\theta_W}\gamma^5$$

$$\frac{iem_i}{m_Z\sin 2\theta_W}\gamma^5$$

Gauge-boson-ghost vertices:

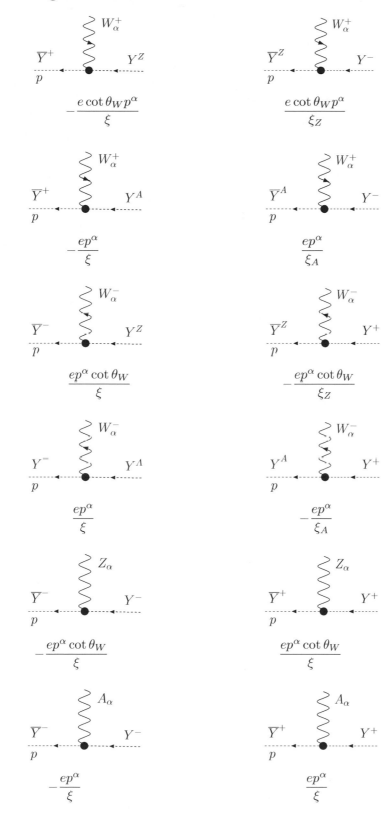

Higgs-ghost vertices:

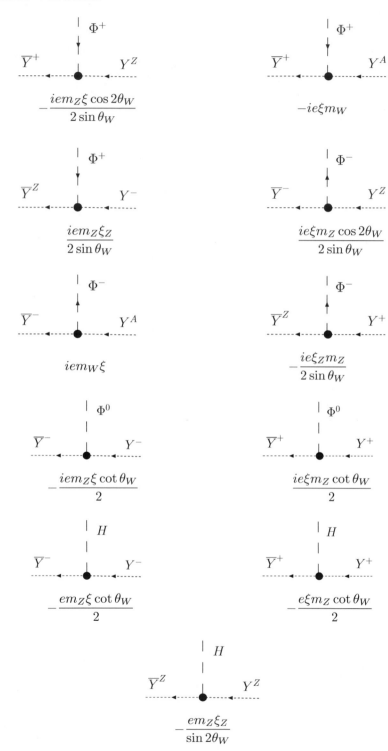

References

[1] G. t'Hooft, Nucl. Phys. **B33**, 173 (1971).

[2] G. t'Hooft, Nucl. Phys. **B35**, 167 (1971).

[3] K. Fujikawa, B. W. Lee, and A. I. Sanda, Phys. Rev. **D6**, 2923 (1972).

[4] Y. P. Yao, Phys. Rev. **D7**, 1647 (1973).

[5] N. Cabibbo, Phys. Rev. Lett. **10**, 531 (1963).

[6] S. L. Glashow, J. Iliopoulos, L. Maiani, Phys. Rev. **D2**, 1285 (1970).

[7] M. Kobayashi, M. Maskawa, Prog. Theor. Phys. **49**, 652 (1973).

[8] O. M. Boyarkin, Zh. Eksp. Teor. Fiz. **75**, 26 (1978).

[9] O. M. Boyarkin, J. Phys. G: Nucl. Phys. **8**, 161 (1982); Sov. Phys. J. **32**, 828 (1989).

[10] M. Fierz, Zeits. Phys. **104**, 553 (1937).

[11] G. W. Bennct et al., Phys. Rev. **D73**, 072003 (2006).

[12] M. Roos and A. Sirling, Nucl. Phys. **B29**, 296 (1971).

[13] A. Sirling, Rev. Mod. Phys. **50**, 573 (1978).

[14] W. J. Marciano and A. Sirlin, Phys. Rev. Lett. **61**, 1815 (1988).

[15] T. van Ritbergen and R. G. Stuart, Phys. Rev. Lett. **82**, 488 (1999).

[16] R. R. Grittenden, W. D. Walker, J. Ballam, Phys. Rev. **121**, 1823 (1961).

[17] W. Bertl et al., Nucl. Phys. **B260**, 1 (1985).

[18] M. L. Brooks et al. (MEGA Collaboration), Phys. Rev. Lett. **83**, 1521 (1999).

[19] Y. Hisamatsu, Eur. Phys. J. **C52**, 477 (2007).

[20] T. P. Cheng and L. F. Li, Phys. Rev. **D16**, 1425 (1977); S. T. Petcov, Yad. Fiz. **25**, 641 (1977).

[21] F. Borzumati and A. Masiero, Phys. Rev. Lett. **57**, 961 (1986); K. Huitu, J. Maalampi, M. Raidal and A. Santamaria, Phys. Lett. **B430**, 355 (1998).

[22] A. G. Akeroyd, M. Aoki, and Y. Okada, Phys. Rev. **D76**, 013004 (2007).

[23] W. M. Yao et al., J. Phys. (The Particle Data Group) **G33**, 1 (2006).

[24] O. M. Boyarkin and G. G. Boyarkina, Phys. Atom. Nucl. **72**, 607 (2009).

[25] M. Sher, Phys. Rep. **179**, 273 (1989).

[26] P. Langacker and D. London, Phys. Rev. **D39**, 266 (1989).

[27] S. R. Mishra et al. Phys. Lett. **B252**, 170 (1990).

[28] C. A. Gagliardi, R. E. Tribble, and N. J. Williams, Phys. Rev. **D72**, 073002 (2005).

[29] G. N. Bennett et al., Phys. Rev. **D73**, 072003 (2006).

[30] R. Jackiw and S. Weinberg, Phys. Rev. **D5**, 2396 (1972).

[31] F. J. Dyson, Phys. Rev. **75**, 1736 (1949).

[32] G. Altarelli et al., Phys. Lett. **B40**, 415 (1972).

[33] A. Czarnecki et al., Phys. Rev. **D67**, 073006 (2003).

[34] G. Degrassi and G. F.Giudice, Phys. Rev. **D58**, 053007 (1998).

[35] M. Davier, S. Eidelman, A. Hocker, and Z. Zhang, Eur. Phys. J. **C27**, 497 (2003).

[36] K. Hagiwara *at al.*, Phys. Lett. **B649**, 173 (2007).

[37] J. Bijnens and J. Prades, Mod. Phys. Lett. **A22**, 767 (2007).

[38] F. Jegerlehner, *The Anomalous Magnetic Moment of the Muon*, (Springer, 2007).

[39] J. P. Miller, E. de Rafael, and B. L. Roberts, Rep. Prog. Phys. **70**, 795 (2007).

[40] M. Davier, Nucl. Phys. (Proc. Suppl.) **B169**, 288 (2007).

[41] A. Czarnecki, W. J. Marciano, Phys. Rev. **D64**, 013014 (2001).

[42] P. Das et al., hep-ph/0102242; S. N. Gninenko and N. V. Krasnikov, Phys. Lett. **B513**, 119 (2001).

[43] M. Beccaria et al., Phys. Lett. **B448**, 129 (1999).

[44] D. Chakraverty et al., Phys. Lett. **B506**, 103 (2001); U. Mahanta, ibid **B535**, 111 (2001).

[45] R. Casadio et al., Phys. Lett. **B495**, 378, (2000); M. L. Graesser, Phys. Rev. **D61**, 074019 (2000).

[46] M. C. Gonzales-Garcia and S. F. Novaes, Phys. Lett. **B389**, 707 (1996); K. Lane, hep-ph/0102131.

[47] R. T. Huang et al., Phys. Rev. **D64**, 071301(R) (2001).

[48] D. Choudhury et al., Phys. Lett. **B507**, 219 (2001).

[49] U. Mahanta, Phys. Lett. **B511**, 235 (2001).

[50] D. Colladay and V. A. Kostelecky, Phys. Rev. **D58**, 116002 (1998).

[51] G. G Boyarkina and O. M. Boyarkin, Phys. Rev. **D67**, 073023 (2003).

[52] C. Caso and A. Gurtu, Phys. Lett. (Review of Particle Physics) **B667**, 389 (2008).

[53] L. M. Barkov, M. S. Zolotorev, Phys. Lett. **B85**, 308 (1979).

[54] M. Froissart, Phys. Rev. **123**, 1053 (1961).

[55] A. Martin, Nuovo Cimento **42**, 930 (1966).

[56] J. Erler and P. Langacker, Phys. Lett. **B667**, 125 (2008).

[57] S. Glashow, Phys. Rev. **118**, 316 (1960).

[58] D. Fargion, In: *Proc. Third Int. Workshop on Neutrino Oscillations, 7-10 Feb., Venice, 2006* (Ed. M. Baldo Ceolin), 515 (2006) (arXiv:astro-ph/0604430v2).

[59] L. Hewett, T. Y. Rizzo, Phys. Rep. **183**, 193 (1989).

[60] F. Pisano and V. Pleitez, Phys. Rev. **D46**, 410 (1992); P. H. Frampton, Phys. Rev. Lett. **69**, 2887 (1992).

[61] S. Weinberg, Phys. Rev. **D19**, 1277 (1979).

[62] L. Susskind, Phys. Rev. **D20**, 2619 (1979).

[63] K. Lane, A. Martin, arXiv:0907.3737 [hep-ph].

[64] K. Lane, hep-ph/0202255.

[65] C. T. Hill and E. H. Simmons, Phys. Rept. **381**, 235 (2003).

[66] A. Belyaev et al., Phys. Rev. **D79**, 035006 (2009).

[67] G. Cacciapaglia, In: *Les Houches "Physics at TeV Colliders 2005" Beyond the Standard Model Working Group: Summary Report* (Eds. B.C. Allanach, C. Grojean, P. Skands); hep-ph/0602198.

[68] C. Csaki et al., Phys. Rev. **D69**, 055006 (2004); C. Csaki et al., Phys. Rev. Lett. **92**, 101802 (2004).

[69] C. Csaki, J. Hubisz, P. Meade, hep-ph/0510275.

[70] G. Cacciapaglia et al., Phys. Rev. **D71**, 035015 (2005).

[71] A. Birkedal, R. Matchev, M. Perelstein, Phys. Rev. Lett. **94**, 191803 (2005).

[72] H. Davoudiasl et al., J. High Energy Phys. (JHEP05), 015 (2004); J. L. Hewett, B. Lillie, T. G. Rizzo, J. High Energy Phys. (JHEP10), 014 (2004).

[73] K. J. Gaemers, G. J. Gounaris, Z. Phys. **C1**, 259 (1979).

[74] E. N. Argyres et al., Nucl. Phys. **B391**, 23 (1993).

[75] G. Couture et al., Phys. Rev. **D36**, 859 (1987).

[76] O. M. Boyarkin, In: *"Physics of Elementary Interactions", Proc. the XIII Warsaw Symposium on Elementary Interactions, 1990, 28 May-1 June, Kazimeirz, Poland,* (Eds. Z. Ajduk, S. Pokorski and A. K. Wroblewski), (World Scientific, Singapore, 1991); Sov. Phys. J. **9**, 773 (1990).

[77] O. M. Boyarkin, Acta Physica Polonica **B25**, 287 (1994).

[78] N. Cabibbo, R. Gatto, Phys. Rev. **124**, 1577 (1961).

[79] W. Alles, C. Boyer, and A. J. Buras, Nucl. Phys. **B119**, 125 (1977).

[80] O. M. Boyarkin, Sov. J. Nucl. Phys. **54**, 640 (1991).

[81] D. Zeppenfeld, Phys. Lett. **B183**, 380 (1987).

[82] O. M. Boyarkin, Sov. Phys. J. **10**, 828 (1989).

[83] G. Tupper and M. A. Samuel, Phys. Rev. **D23**, 1933 (1981).

[84] S. Y. Choi, F. Schrempp, Phys. Lett. **B272**, 149 (1991).

[85] C. Caso and A. Gurtu, Phys. Lett. (Review of Particle Physics) **B667**, 389 (2008).

[86] D. Bardin and G. Passarino, *The Standard Model in the Making*, (Clarendon Press, Oxford, 1999).

[87] M. Raidal, Phys. Rev. Lett. **93**, 161801 (2004).

[88] J. C. Pati and A. Salam, Phys. Rev. **D10**, 275 (1974); R. N. Mohapatra and J. C. Pati, ibid. **D11**, 566 (1975).

[89] R. N. Mohapatra, D. Sidhu, Phys. Rev. Lett. **38**, 667 (1977).

[90] R. N. Mohapatra, G. Senjanovic, Phys. Rev. **D23**, 165 (1981).

[91] S. Rajpoot, Phys. Rev. **D40**, 3795 (1989).

[92] R. M. Frankis et al., Phys. Rev. **D41**, 2369 (1991).

[93] R. N. Mohapatra, Progr. Part. Nucl. Phys. **26**, 1 (1991).

[94] O. M. Boyarkin, Acta Physica Polonica **B23**, 1031 (1992); Phys. Atom. Nucl. **56**, 77 (1993).

[95] O. M. Boyarkin, Phys. Rev. **D50**, 2247 (1994).

[96] N. G. Deshpande, J. F. Gunion, B. Kayser, F. Olness, Phys. Rev. **D44**, 837 (1991).

[97] T. G. Rizzo, Phys. Rev. **D25**, 1355 (1982).

[98] J. Fleischer, O. V. Tarasov, and F. Jegerlehner, Phys. Lett. **B319**, 249 (1993).

[99] G. G. Boyarkina, O. M. Boyarkin, Eur. Phys. J. **C13**, 99 (2000).

[100] R. N. Mohapatra, Fortschr. Phys. **31**, 185 (1983).

[101] W. Beenakker, W. Hollik, Zeits. Phys. **C40**, 141 (1988); G. Altarelli *et al.*, Phys. Lett. **B263**, 459 (1991).

[102] O. M. Boyarkin, Sov. J. Nucl. Phys. **54**, 506 (1991).

[103] G. Altarelli *et al.*, Nucl. Phys. **B342**, 15 (1990).

[104] D. A. Ross, M. Veltman, Nucl. Phys. **B75**, 135 (1975).

[105] O. M. Boyarkin, In: *"Puzzles on the Electroweak Scale", Proc. the XIV Inter. Warsaw Meeting on Elementary Particle Physics, 1991, 27-31-May, Warsaw, Poland,* (Eds. Z. Ajduk, S. Pokorski and A.K. Wroblewski), (World Scientific, Singapore, 1992); O. M. Boyarkin, Phys. Atom. Nucl. **57**, 2151 (1994).

[106] G. G. Boyarkina, O. M. Boyarkin, Phys. Atom. Nucl. **60**, 601 (1997).

[107] O. M. Boyarkin, G. G. Boyarkina, and T. I. Bakanova, Phys. Rev. **D70**, 113010 (2004).

[108] O. M. Boyarkin, D. Rein, Phys. Rev. **D53**, 361 (1996).

[109] J. P. Leveille, Nucl. Phys. **B137**, 63 (1978).

[110] G. G. Boyarkina, O. M. Boyarkin, V. V. Mahknach, Phys. Atom. Nucl. **66**, 281 (2003).

[111] O. M. Boyarkin, G. G. Boyarkina, Phys. Atom. Nucl. **68**, 2047 (2005).

[112] O. M. Boyarkin, G. G. Boyarkina and V. V. Makhnach, Phys. Rev. **D77**, 033004 (2008).

[113] R. E. Shrok, Nucl. Phys. **B206**, 359 (1982); M. A. Beg, W. J. Marciano, M. Ruderman, Phys. Rev. **D17**, 1395 (1978).

[114] O. M. Boyarkin, G. G. Boyarkina, Phys. Atom. Nucl. **72**, 607-618 (2009).

Part III

Neutrino physics

14

Neutrino properties

> *One going forward loses peace,*
> *one standing still loses himself.*
> Confucius

14.1 Neutrino mass in models with the $SU(2)_L \times U(1)_Y$ gauge group

We consider introducing the neutrino mass within the SM in more detail. The SM structure is such that after the spontaneous symmetry breaking (SSB) the neutrino remains a massless particle (a Dirac mass $\sim \bar{\nu}_L \nu_R$ will not appear because of the absence of a left-handed neutrino singlet in the theory, and the appearance a Majorana mass $\sim \nu_L^T C \sigma_2 \nu_L$ is forbidden by the $SU(2)_L$ invariance). The SM is based on the $SU(2)_L \times U(1)_Y$ gauge group. However, this fixes only gauge bosons of the theory while Higgs and fermion sectors remain arbitrary. According to this, the SM modifications where neutrinos prove to be massive are divided into three classes: (i) models with additional leptons; (ii) models with extended Higgs sector; and (iii) models including both new leptons and new Higgs bosons.

a. Models with additional leptons. The SM contains left- and right-handed chiral projection of all fermions except the neutrino. To remedy this discrimination we introduce right-handed neutral fields N_{lR} ($l = e^-, \mu^-, \tau^-$). Similar to other right handed fields, they are assumed to be $SU(2)_L$ singlets. Then, from the equation

$$Q = I_{3L} + \frac{Y}{2}$$

follows that they have $Y = 0$. Therefore, we deal with N_{lR} (1,0), where in parentheses the quantum number of the $SU(2)_L \times U(1)_Y$ gauge group are given. Since N_{lR} are singlets with respect to $SU(2)_L \times U(1)_Y$, they do not interact with the gauge bosons. However, N_{lR} enter into the Yukawa Lagrangian $\mathcal{L}$ and then the lepton part of $\mathcal{L}$ takes the form

$$-\mathcal{L}_Y = \sum_l h_{ll} \bar{\psi}_{lL} \varphi \psi_{lR} + \sum_{l,l'} h'_{ll'} \bar{\psi}_{lL} \varphi^c N_{l'R} + \text{h.c..}, \tag{14.1}$$

where φ denotes the SM Higgs doublet. This, in its turn, causes not only the appearance of the neutrino mass, but leads to the interaction of the neutrino with the physical Higgs boson as well.

From (14.1) the main disadvantage of such an approach becomes evident—there are no explanations of tremendous difference (10^6 and more) in the values of Yukawa constants that define masses of charged leptons ($h_{ll'}$) and neutrinos ($h'_{ll'}$).

Next, two variants are possible depending on whether we want the neutrino to be Majorana or Dirac particles. In the case of the Dirac neutrino we should identify N_{lR} with the right-handed component of the ordinary neutrino, that is, $N_{lR} = \nu_{lR}$. After the SSB we may already disregard those Higgs degrees of freedom that have been spent on giving the masses to gauge bosons and use the following expressions for the Higgs doublets

$$\varphi(x) = \frac{1}{\sqrt{2}} \begin{pmatrix} 0 \\ H(x) + v \end{pmatrix}, \qquad \varphi^c(x) = \frac{1}{\sqrt{2}} \begin{pmatrix} H(x) + v \\ 0 \end{pmatrix}. \tag{14.2}$$

To substitute (14.2) into $\mathcal{L}_Y$ brings into existence the mass term in the free neutrino Lagrangian $\mathcal{L}_0(x)$, which have the form now

$$\mathcal{L}_0(x) = \frac{i}{2} \sum_l [\bar{\nu}_{lL}(x)\gamma^\lambda \partial_\lambda \nu_{lL}(x) + \bar{\nu}_{lR}(x)\gamma^\lambda \partial_\lambda \nu_{lR}(x)] - \frac{1}{2} \sum_{ll'} [\bar{\nu}_{lR}(x)M_{ll'}\nu_{l'L}(x) +$$

$$+ \bar{\nu}_{l'L}(x)M_{l'l}\nu_{lR}(x)], \tag{14.3}$$

where the neutrino mass matrix M in the flavor basis is defined by the expression:

$$M = \frac{1}{\sqrt{2}} \begin{pmatrix} h'_{ee}v & h'_{e\mu}v & h'_{e\tau}v \\ h'_{\mu e}v & h'_{\mu\mu}v & h'_{\mu\tau}v \\ h'_{\tau e}v & h'_{\tau\mu}v & h'_{\tau\tau}v \end{pmatrix}. \tag{14.4}$$

In what follows we shall assume $h'_{ll'}$ to be equal to $h'_{l'l}$ to provide $M = M^T$. As we can see, the mass matrix in the flavor basis is not diagonal. The responsibility for this rests with those Yukawa constants that lead to the lepton flavor (LF) violation at the tree level because of the interaction

$$\mathcal{L}^\eta_{int}(x) = -\frac{1}{\sqrt{2}} \sum_{ll'} h_{ll'} \bar{\nu}_{lL}(x) N_{l'R}(x) H(x) + \text{h.c.}. \tag{14.5}$$

The matrix M could be diagonalized using the unitary transformation

$$UMU^{-1} = \text{diag}(m_1, m_2, m_3).$$

This, in turn, means that the physical states, that is, the states with definite values of masses, are $\nu_j(x)$ $(j = 1, 2, 3)$

$$\nu_j(x) = \sum_l U_{jl}\nu_l(x) \tag{14.6}$$

rather than $\nu_l(x)$. Eq. (14.6) gives some concern and we should check whether we come into conflict with the physical reality. Following the apparitions chronology of neutrinos to experimentalists, we remember our definitions of these particles. A particle that appears in the $\beta^{(+)}$ decay

$$p \to n + e^+ + \nu_e, \tag{14.7}$$

was called an *electron neutrino*, a *muon neutrino* was identified with a satellite of a muon in the decay channel of a π^+ meson

$$\pi^+ \to \mu^+ + \nu_\mu, \tag{14.8}$$

and we defined a *tau-lepton neutrino* as a particle that presents in three-lepton channels of τ^- decays, for instance

$$\tau^- \to \nu_\tau + \mu^- + \bar{\nu}_\mu. \tag{14.9}$$

One may unify these definitions having bound them with some nonstable particle—in its decay modes three known neutrinos are present. As a such a particle let us choose the W

boson. Then, we agree to call particles accompanying an electron, muon and tau-lepton in the decays

$$W^+ \rightarrow \bar{l} + \nu_l \tag{14.10}$$

by electron, muon, and tau-lepton neutrinos, respectively. Under such definitions there is no need to demand ν_e, ν_μ and ν_τ to be physical states. It means that they may lack properties of physical particles. However, the reaction (14.10) obeys conservation laws and this leads to the fact that the sole poetic licence we could allow implies that electron, muon, and tau-lepton neutrinos are not states with definite mass values. To put it differently, these neutrinos represent mixing of physical states ν_1, ν_2, and ν_3. Recall, similar mixing takes place in the quark sector as well, where the role of the U matrix plays the Cabibbo–Kobayashi–Maskawa matrix. However, in the quark case, down members of weak isospin doublets are mixed while, in the lepton case, upper members of weak isospin doublets take part in this mixture. So, the flavor basis for neutrinos loses the physical status because it is not a basis of mass eigenstates.

For a neutrino to be a Majorana particle we must assume that N_{lR} is not already a right-handed partner of ν_{lL} but it is some other particle. Then, the mass term takes the form

$$-\mathcal{L}'_m = \frac{1}{2} \sum_{l,l'} M_{ll'} \overline{\nu}_{lL} N_{l'R} + \text{h.c..} \tag{14.11}$$

Since N_{lR} is invariant with respect to the $SU(2)_L \times U(1)_Y$ transformations, then the neutrino Lagrangian may also involve the gauge invariant quantity

$$-\mathcal{L}''_m = \frac{1}{2} \sum_{l,l'} F_{ll'} \overline{N^c_{lR}} N_{l'R} + \text{h.c..} \tag{14.12}$$

Of course, introducing this term results in the violation of the $B - L$ symmetry. However, in the SM, it is only the global symmetry and the refusal from it does not result in any changes in the SM structure apart from the fact that $B - L$ ceases to be a conserved quantity. It is evident that nothing prevents us to enter the term

$$\frac{1}{2} \sum_{l,l'} D_{ll'} \overline{\nu^c_{lL}} \nu_{l'L} + \text{h.c.} \tag{14.13}$$

into the neutrino Lagrangian in addition to (14.12). The terms (14.12) and (14.13) are named the mass terms in the Majorana basis. It is easy to see, the mass matrix ($F_{ll'}$ or $D_{ll'}$) in the Majorana basis has to be symmetric. Really, using the definition of the charge conjugate spinor, we have

$$\frac{1}{2} \sum_{l,l'} F_{ll'} \overline{N}_{lL} N^c_{l'R} = \frac{1}{2} \sum_{l,l'} F_{ll'} [N^c_l]^T C P_R C \gamma_0 N^*_{l'} = -\frac{1}{2} \sum_{l,l'} F_{ll'} N^\dagger_{l'} \gamma_0 C^T P_R C^T N^c_l =$$

$$= \frac{1}{2} \sum_{l,l'} F_{ll'} \overline{N}_{l'L} N^c_{lR} = \frac{1}{2} \sum_{l,l'} F_{l'l} \overline{N}_{lL} N^c_{l'R}, \tag{14.14}$$

where we have accounted for the anti-commutativity of neutrino field operators and the relation

$$C^T \gamma_\mu C^T = -\gamma^T_\mu.$$

Setting $D_{ll'} = 0$ for simplicity, summing (14.12) and (14.13) as well as making use of the identity

$$\overline{\nu}_{aL} N_{bR} = \overline{N^c_{bL}} \nu^c_{aR}, \tag{14.15}$$

we get the following mass term for a neutrino

$$-\mathcal{L}_m = \frac{1}{2} \left(\bar{\nu}_L, \overline{N^c}_L \right) \begin{pmatrix} 0 & M \\ M & F \end{pmatrix} \begin{pmatrix} \nu_R^c \\ N_R \end{pmatrix} + \text{h.c.}, \tag{14.16}$$

where M and F are 3×3 matrices while ν_L and N_R are three-component column-matrices. In the simple case of one generation, the mass matrix $\mathcal{M}$ has the form

$$\mathcal{M} = \begin{pmatrix} 0 & M \\ M & F \end{pmatrix}, \tag{14.17}$$

where now M and F are simply numbers. For the sake of simplicity, we assume M and F to be real and $F > 0$. Introducing an orthogonal matrix O

$$O = \begin{pmatrix} \cos\theta & -\sin\theta \\ \sin\theta & \cos\theta \end{pmatrix},$$

$(\tan 2\theta = 2M/F)$, we get

$$O\mathcal{M}O^T = \mathcal{M}' = \begin{pmatrix} -m_1 & 0 \\ 0 & m_2 \end{pmatrix}, \tag{14.18}$$

where

$$m_{1,2} = \frac{1}{2} \left[\sqrt{F^2 + 4M^2} \mp F \right]. \tag{14.19}$$

Because $m_{1,2} \geq 0$, then we cannot interpret the states in the basis

$$\begin{pmatrix} \nu_R^c \\ N_R \end{pmatrix}' = O \begin{pmatrix} \nu_R^c \\ N_R \end{pmatrix}$$

as physical ones. For a way out we pass on to the basis where the phase $\nu_R'^c$ must be changed by π. Rewrite the matrix $\mathcal{M}'$ in the view

$$\mathcal{M}' = \mathcal{M}_+ D, \tag{14.20}$$

where

$$\mathcal{M}_+ = \begin{pmatrix} m_1 & 0 \\ 0 & m_2 \end{pmatrix}, \qquad D = \begin{pmatrix} -1 & 0 \\ 0 & 1 \end{pmatrix}. \tag{14.21}$$

Next, we introduce two auxiliary states

$$\begin{pmatrix} \xi_{1L} \\ \xi_{2L} \end{pmatrix} = O \begin{pmatrix} \nu_L \\ N_L^c \end{pmatrix}, \qquad \begin{pmatrix} \xi_{1R} \\ \xi_{2R} \end{pmatrix} = DO \begin{pmatrix} \nu_R^c \\ N_R \end{pmatrix}. \tag{14.22}$$

Then the mass term will look like:

$$-\mathcal{L}_{m_\nu} = m_1 \bar{\xi}_{1L} \xi_{1R} + m_2 \bar{\xi}_{2L} \xi_{2R} + \text{h.c.}. \tag{14.23}$$

From the definitions (14.22) it follows that

$$\left. \begin{array}{l} \xi_1 = \xi_{1L} + \xi_{1R} = (\nu_L - \nu_R^c)\cos\theta - (N_L^c - N_R)\sin\theta, \\ \xi_2 = \xi_{2L} + \xi_{2R} = (\nu_L + \nu_R^c)\sin\theta + (N_L^c + N_R)\cos\theta, \end{array} \right\} \tag{14.24}$$

and we find without trouble

$$\xi_1 = -\xi_1^c, \qquad \xi_2 = \xi_2^c, \tag{14.25}$$

that is, the states ξ_1 and ξ_2 describe the Majorana particles that have the opposite phases of the charge conjugation. Evidently, in the three-generation case, we shall already have six Majorana neutrinos.

Let us try to obtain the seesaw relation using (14.19). In the mass matrix, the quantity M arises from the ordinary coupling with the Higgs boson and therefore it is natural to assume that M has the same order as the mass of the charged fermion in the same generation. Then, in order to achieve our goal we have to assume that $F \gg m_l$. In this case Eq. (14.19) results in

$$m_1 \approx \frac{m_l^2}{F}, \qquad m_2 \approx F \tag{14.26}$$

and the seesaw relation

$$m_{\nu_l} m_{N_l} = m_l^2 \tag{14.27}$$

takes place. However, the obtained relation does not remove the neutrino mass problem once and for all in the SM. Really, if one sets $m_{\nu_\mu} \approx 0.11$ MeV, then the obtained value for m_{N_μ} will exceed the electroweak scale over the three orders of magnitude. So, in the SM the problem is not totally solved because the way of generating a large mass for a heavy neutrino is absent. Successive solutions may be obtained only in models with an extended gauge group where there are two scales of breaking an underlying symmetry. The model based on the $SU(3)_L \times SU(3)_R \times U(1)_{B-L}$ gauge group is an example.

b. Models with an extended Higgs sector. In these models additional leptons are absent and the neutrino sector involves two degrees of freedom, namely, ν_{lL} and ν_{lL}^c. With their help, one may construct the mass term of the Majorana type only. In the SM the source of particle masses is the Yukawa Lagrangian. The initial $SU(2)_L \times U(1)_Y$ invariance provides the conservation of the $B - L$ quantum number. If after the SSB the mass term of the Majorana type appears in the free neutrino Lagrangian, then one or several $B - L$-noninvariant terms must be present in the interaction Lagrangian. The latter describe either the interaction of a neutrino with (physical) Higgs bosons or that between Higgs bosons. Therefore, Higgs bosons inducing the neutrino mass have to possess the nonzero $B - L$ value. Now, to build the $B - L$-invariant Lagrangian, we should find fermion bilinear combinations with $B - L = C \neq 0$ and multiply them by wave functions of Higgs bosons having $B - L = -C$. Below we give the quantum numbers of doublets and singlets in the SM

$$\begin{aligned}
\psi_L = \begin{pmatrix} \nu_{lL} \\ l_L \end{pmatrix} &\quad : \quad (1/2, -1) \\
l_R &\quad : \quad (0, -2) \\
\psi_R^c = iC\gamma_0\tau_2\psi_L^* &\quad : \quad (1/2, 1) \\
l_L^c = C\gamma_0 l_R^* &\quad : \quad (0, 2).
\end{aligned} \tag{14.28}$$

Note that in the third line of Eq. (14.28) we have taken the conjugation with respect to the spinor index ($\psi^c = C\gamma_0\psi^*$) and the $SU(2)_L$ index ($\psi^c = i\tau_2\psi^*$). From (14.28) follows that we can construct two lepton bilinear combinations with a definite value of $B - L$

$$\overline{\psi}_L \psi_R^c \quad : \quad (1/2, 1) \bigotimes (1/2, 1) = (0, 2) \bigoplus (1, 2), \tag{14.29}$$

$$\overline{l^c}_L l_R \quad : \quad (0, -2) \bigotimes (0, -2) = (0, -4). \tag{14.30}$$

The Higgs multiplets that can directly couple to these bilinears to form gauge invariant Yukawa couplings are as follows [1]: (i) a triplet Δ : $(1, -2)$; (ii) a singly charged singlet $h^{(-)}$: $(0, -2)$; and (iii) a doubly charged singlet $H^{(--)}$: $(0, 4)$.

Let us consider in detail all three possibilities. With the value of Y equal to -2, the electric charge of the components of the triplet Δ are

$$\Delta = \begin{pmatrix} \Delta_0 \\ \Delta^{(-)} \\ \Delta^{(--)} \end{pmatrix}, \tag{14.31}$$

and we have an additional term in the Yukawa Lagrangian of the SM:

$$-\mathcal{L}'_Y = \frac{1}{\sqrt{2}} \sum_{l,l'} f_{ll'} \overline{\psi}_{lL} (\boldsymbol{\sigma} \cdot \boldsymbol{\Delta}) \psi^c_{l'R} + \text{h.c.}. \tag{14.32}$$

The triplet Yukawa coupling constants (YCCs) $f_{ll'}$ are symmetric quantities. To make sure of that, we write the Hermitian-conjugate quantity from Eq. (14.32) in the explicit form

$$\frac{i}{\sqrt{2}} \sum_{ll'} f^*_{ll'} \psi^T_{l'L} C \tau_2 (\boldsymbol{\sigma} \cdot \boldsymbol{\Delta}^*) \psi_{lL}. \tag{14.33}$$

It is evident, since the fermion bilinear combination consists of two doublets (rather than from a doublet and its conjugate doublet), it must obey Fermi statistics. The $SU(2)_L$ indices are contracted by the $i\sigma_2\boldsymbol{\sigma}$ matrix that is symmetric. The spinor indices are contracted by the matrix C that is antisymmetric. The residuary indices are the flavor indices, and they have to be symmetric in order that the entire combination is antisymmetric, that is, $f_{ll'} = f_{l'l}$.

We assume the Higgs potential parameters to be such that the minimum takes place providing:

$$< \varphi_0 > = \frac{v}{\sqrt{2}}, \qquad < \Delta_0 > = \frac{v_L}{\sqrt{2}}, \tag{14.34}$$

where v and v_L are VEVs of neutral components of the SM doublet Φ and the triplet Δ, respectively. In the result of the SSB the neutrino acquires the mass, that is

$$-\mathcal{L}_m = \sum_{l,l'} \overline{\nu}_{lL} M_{ll'} \nu^c_{l'R} + \text{conj.}, \tag{14.35}$$

where

$$M_{ll'} = \frac{f_{ll'} v_L}{\sqrt{2}}. \tag{14.36}$$

Note, since the YCCs $f_{ll'}$ are unknown, then we may not say anything definite concerning masses and mixing angles of neutrinos. However, the advantage of this model is the fact that it provides the explanation of why neutrino masses are at least three orders of magnitude smaller than charged fermion masses in the same generation. To make this evident we account for the following. Since the Δ_0 boson is related with the gauge bosons W and Z, there is the contribution from the VEV v_L into m_W and m_Z

$$m_W^2 = \frac{g^2}{4}(v^2 + 2v_L^2), \qquad m_Z^2 = \frac{g^2 + g'^2}{4}(v^2 + 4v_L^2). \tag{14.37}$$

Now, taking into account the definition of the parameter ρ_0 and its experimental value, we come to the conclusion

$$v_L \ll v.$$

Among exotic phenomena of the given SM modification we note the prediction of processes with the $B - L$ violation on two units and the existence of doubly charged physical Higgs bosons.

A model with a singly charged scalar singlet $h^{(-)}$ (Zee's model) was proposed in Ref. [2]. Here the neutrino mass is completely caused by the radiative corrections (RCs). The lepton part of the Yukawa Lagrangian

$$-\mathcal{L}_Y = \sum_l h_{ll} \overline{\psi}_{lL} \varphi \psi_{lR} + \sum_{l,l'} f_{ll'} \overline{\psi}_{lL} \psi_{l'R}^c h^{(-)} + \text{h.c.}, \tag{14.38}$$

cannot lead to the appearance of the neutrino mass. The reason is evident, $h^{(-)}$ has the electric charge and its nonzero VEV results in the spontaneous breakdown of the $U(1)_Q$ symmetry. Since the Lagrangian parts containing kinetic energy and gauge interaction cannot violate any $U(1)$ quantum numbers, only the Higgs potential V_X may pretend to the role of the source that generates the $B - L$ violating interaction. However, if apart from $h^{(-)}$ we have just one doublet φ as in the SM, it is impossible to introduce any $B - L$ violating terms into V_X. To improve the situation we should add a doublet φ' with the same quantum numbers as the Higgs doublet of the SM, that is, with $(1/2, 1)$. Therefore, φ' possesses a nonzero value of $B - L$. Then, we may enter the $B - L$ noninvariant term into the Higgs potential

$$i\mu \varphi^T \sigma_2 \varphi' h^{(-)} + \text{h.c.} \tag{14.39}$$

It should be stressed, it is impossible to produce the similar term from one scalar doublet and a scalar singlet because $i\varphi^T \sigma_2 \varphi$ proves to be equal to zero.

After the SSB we are left with four physical singly charged Higgs bosons $S_1^{(\pm)}$ and $S_2^{(\pm)}$. They represent linear combinations of scalars $h^{(\pm)}$, $\varphi^{(\pm)}$ $\varphi^{(\pm)'}$ and, therefore, are not the $B - L$ eigenstates. $S_1^{(\pm)}$ and $S_2^{(\pm)}$ bosons will induce the Majorana neutrino mass already at the one-loop level (Fig. 14.1). Stress that the diagram of Fig. 14.1 does not involve

FIGURE 14.1
The Feynman diagram giving neutrino masses in Zee's model.

divergencies whereas in a theory with a bare mass they would contain logarithmic and linear divergencies.

If one assumes that only one of the scalar doublets is related to leptons, then calculations lead to the following neutrino mass matrix [3]:

$$M_{ll'} = A f_{ll'} (m_l^2 - m_{l'}^2), \tag{14.40}$$

where A is a constant. Since $f_{ll'} = -f_{l'l}$ (it follows from Eq. (14.38)), then the matrix $M_{ll'}$ is symmetric. The main advantage of this model consists in the fact that it not only gives a pattern of neutrino masses and mixing but it relates the neutrino oscillation parameters to the Higgs sector parameters as well. To show that such is the case, we introduce the designations

$$\tan \alpha = \frac{f_{\mu\tau}}{f_{e\tau}} \left(1 - \frac{m_\mu^2}{m_\tau^2}\right), \qquad \beta = \frac{f_{e\mu} m_\mu^2}{f_{e\tau} m_\tau^2} \cos \alpha \tag{14.41}$$

and set $m_e = 0$. Then the mass matrix of neutrinos in the basis $\psi^T = (\nu_e^T, \nu_\mu^T, \nu_\tau^T)$ takes the form

$$M = m_0 \begin{pmatrix} 0 & \beta & \cos\alpha \\ \beta & 0 & \sin\alpha \\ \cos\alpha & \sin\alpha & 0 \end{pmatrix}, \qquad (14.42)$$

where $m_0 = Am_\tau^2 f_{\tau e}/\cos\alpha$. It is quite reasonable to assume that $\beta \ll 1$. Fulfilling the diagonalization of the mass matrix (14.42) and having been constrained by the first order in β, we get

$$M' = \text{diag}[m_1, m_2, m_3] = m_0 \text{diag}[-\beta\sin 2\alpha, (\beta\sin 2\alpha - 2)/2, (\beta\sin 2\alpha + 2)/2]. \quad (14.43)$$

Again, in order to eliminate the negative eigenvalue m_2, we should change the phase of the wave function ν_2 by π. We already carried out a similar operation using the matrix D. If m_1 proves to be negative as well, then the analogous phase change of the wave function ν_1 corrects the state of affairs. From Eq. (14.43) it is evident that

$$m_2 \approx m_3 \gg m_1. \qquad (14.44)$$

So, in this model, the information about neutrino oscillation parameters could be obtained when studying the processes without their direct participation. The most important thing is that now the subject of investigation is not only the neutral particles but the charged particles as well. It is evident, that this circumstance makes the experimental part of the task much more easy.

c. Models with two Higgs doublets. By way of a typical representative of the SM modifications including both new Higgs bosons and new leptons, we consider models with two Higgs doublets (2HDMs). For the first time the model based on the $SU(2)_L \times U(1)_Y$ gauge group involving the two Higgs doublets Φ_j : $(1/2, 1)$ $(j = 1, 2)$ was proposed by T. D. Lee [4]. In general, such a model will generate the fermion flavor (FF) violation at the tree level. To protect the model from such terms, the *ad hoc* discrete symmetry [5] on the 2HDM scalar potential and the Yukawa interaction is proposed and there appears two different versions of the 2HDM depending on whether up-type and down-type fermions couple to the same or different scalar doublets. In model I, the up-type and down-type fermions get mass via VEV of only one Higgs field. In model II, which coincides with the minimal supersymmetric SM (MSSM) in the Higgs sector, the up-type and down-type fermions get mass via the VEV of the Higgs fields Φ_1 and Φ_2, respectively. The model wherein the mentioned discrete symmetry is absent and, as a result, both the neutral and charged flavor-changing currents appear at the tree level is called as the *model III*, or the *general 2HDM (G2HDM)*. During the last twenty years the investigation of low-energy phenomena from the point of view of the G2HDM was done and the strong constraints on the flavor-changing (FC) currents existence at the tree level was placed. However, these investigations mainly concerned the quark sector of the G2HDM. Recall that just the nondiagonal YCCs leading to the flavor violation in the lepton sector are responsible for the neutrino mixing between generations. In the 2HDM right-handed neutrinos ν_{lR} figure as new fermions. They are introduced only into the Yukawa Lagrangian. It is clear that neutrinos in these models are Dirac particles. The 2HDM also represent the example of models in which there is the linkage between the neutrino oscillation parameters and the Higgs sector parameters. We shall demonstrate this for the G2HDM in the explicit form.

The Yukawa Lagrangian of the G2HDM is defined by the expression:

$$-\mathcal{L}_Y = \sum_{l,l'} (h_{ll'}^u \overline{\Psi}_{lL} \tilde{\Phi}_1 \nu_{l'R} + h_{ll'}^d \overline{\Psi}_{lL} \Phi_1 l'_R + h_{ll'}^{'u} \overline{\Psi}_{lL} \tilde{\Phi}_2 \nu_{l'R} + h_{ll'}^{'d} \overline{\Psi}_{lL} \Phi_2 l'_R) + \text{h.c.}, \quad (14.45)$$

where $\tilde{\Phi}_{1,2} = i\sigma_2\Phi_{1,2}^*$. We shall employ the Higgs potential that does not lead to the spontaneous breaking of the CP-invariance [6]:

$$V_X = -\mu_1^2 x_1 - \mu_2^2 x_2 + \lambda_1^2 x_1^2 + \lambda_2^2 x_2^2 + \lambda_3^2 x_3^2 + \lambda_4^2 x_4^2 + \lambda_5^2 x_1 x_2, \qquad (14.46)$$

where the following designations

$$x_1 = \Phi_1^\dagger\Phi_1, \qquad x_2 = \Phi_2^\dagger\Phi_2, \qquad x_3 = \mathrm{Re}\{\Phi_1^\dagger\Phi_2\}, \qquad x_4 = \mathrm{Im}\{\Phi_1^\dagger\Phi_2\}$$

have been introduced. The potential minimum is realized providing

$$\Phi_j = \frac{1}{\sqrt{2}}\begin{pmatrix} 0 \\ v_j \end{pmatrix},$$

where

$$v_1^2 + v_2^2 = v^2 = (246 \text{ GeV})^2.$$

Next, for the sake of simplicity, we are limited by the two-flavor approximation. Then, after the SSB the neutrino mass matrix in the flavor basis takes the form

$$\mathcal{M} = \begin{pmatrix} (h_{ee}^u v_1 + h_{ee}^{\prime u} v_2)/\sqrt{2} & (h_{eX}^u v_1 + h_{eX}^{\prime u} v_2)/\sqrt{2} \\ (h_{eX}^u v_1 + h_{eX}^{\prime u} v_2)/\sqrt{2} & (h_{XX}^u v_1 + h_{XX}^{\prime u} v_2)/\sqrt{2} \end{pmatrix}, \qquad (14.47)$$

where $X = \mu, \tau$. Using the relation

$$O\mathcal{M}O^T = \begin{pmatrix} m_1 & 0 \\ 0 & m_2 \end{pmatrix}, \qquad (14.48)$$

we obtain the desired formulae

$$\left.\begin{array}{c} (h_{ee}^u v_1 + h_{ee}^{\prime u} v_2)/\sqrt{2} = m_1 c_\theta^2 + m_2 s_\theta^2, \qquad (h_{eX}^u v_1 + h_{eX}^{\prime u} v_2)/\sqrt{2} = c_\theta s_\theta (m_1 - m_2), \\ (h_{XX}^u v_1 + h_{XX}^{\prime u} v_2)/\sqrt{2} = m_1 s_\theta^2 + m_2 c_\theta^2, \end{array}\right\} \tag{14.49}$$

where θ is the neutrino oscillation angle.

After the SSB the Higgs mass matrices appear too. Their diagonalization establishes the spectrum of physical Higgs bosons. So, we have:
two neutral CP-even scalar

$$\begin{pmatrix} H \\ h \end{pmatrix} = \begin{pmatrix} c_\alpha & s_\alpha \\ -s_\alpha & c_\alpha \end{pmatrix}\begin{pmatrix} \Phi_1^{0r} \\ \Phi_2^{0r} \end{pmatrix}, \qquad (14.50)$$

one neutral CP-odd scalar

$$A = -s_\beta\Phi_1^{0i} + c_\beta\Phi_2^{0i}, \qquad (14.51)$$

and two singly charged scalars

$$h^{(\pm)} = -s_\beta\Phi_1^{\pm} + c_\beta\Phi_2^{\pm}, \qquad (14.52)$$

where

$$\tan 2\alpha = \frac{v_1 v_2(\lambda_3 + \lambda_5)}{\lambda_2 v_2^2 - \lambda_1 v_1^2}, \qquad \tan\beta = \frac{v_2}{v_1},$$

the superscript r (i) means the real (imaginary) part of the corresponding quantity. Note, the spectra of physical Higgs bosons in the G2HDM and MSSM completely coincide.

Carrying out the transition from gauge basis to mass basis in the Yukawa Lagrangian, we can get the Lagrangians describing the interaction between the Higgs bosons and leptons

$$\mathcal{L}_{h^{(\pm)}} = -\sqrt{2}v^{-1}\tan\beta \sum_l m_l \bar{\nu}_{lL} l_R h^{(+)} + \sum_{l,l'} \kappa_{ll'} \bar{\nu}_{lL} l'_R h^{(+)} + \text{h.c.}, \tag{14.53}$$

$$\mathcal{L}_{H,h,A} = \frac{1}{\sqrt{2}} \sum_{l,l'} \{\bar{l}_L l'_R[(h^d_{ll'}c_\alpha + h'^d_{ll'}s_\alpha)H + (h'^d_{ll'}c_\alpha - h^d_{ll'}s_\alpha)h+$$

$$+i(h'^d_{ll'}c_\beta - h^d_{ll'}s_\beta)A] + (l \to \nu, d \to u) + \text{h.c..}\}, \tag{14.54}$$

where $\kappa_{ll'} = h'^d_{ll'}c_\beta^{-1} + h^u_{ll'}s_\beta - h'^u_{ll'}c_\beta$. From Eqs. (14.49), (14.53), and (14.54) follows that in the G2HDM the oscillation parameters could be estimated by the indirect route at the accelerators. However, in this case the whole complex of the experiments should also include the reactions with the neutrino beam in the initial or in the final states.

14.2   Multipole moments of neutrinos

Anomalous values of multipole moments (MMs) of neutrinos are caused by the RCs. In the one-loop approximation for the SM the neutrino interaction with the electromagnetic field is presented by the diagrams shown in Fig. 14.2. The most general form of the matrix

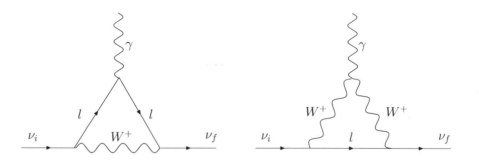

FIGURE 14.2
The Feynman diagrams giving the contributions to the neutrino MMs.

element for the neutrino electromagnetic current $< \nu(p_f, s_f)|J_\mu|\nu(p_i, s_i) > \equiv (J_\mu)_{if}$ where a four-dimensional momentum p_i (p_f) and a helicity s_i (s_f) of a neutrino in the initial (final) state, could be obtained from the demands of the relativistic and gauge invariances. A similar operation was already fulfilled in *Advanced Particle Physics Volume 1* (see Eq. (10.43)). It is easy to show that $(J_\mu)_{if}$ must have the form

$$(J_\mu)_{if} = \bar{u}_f \left\{ f_Q(q^2)\gamma_\mu + f_A(q^2)[q^2\gamma_\mu - \gamma_\lambda q^\lambda q_\mu]\gamma_5 + if_M(q^2)\sigma_{\mu\nu}q^\nu + f_E(q^2)\sigma_{\mu\nu}q^\nu\gamma_5 \right\} u_i, \tag{14.55}$$

where u_i is the bispinor with the four-dimensional momentum p_i and helicity s_i, $q = p_f - p_i$, $f_n(q^2)$ are the formfactors describing the distribution of the electric charge ($n = Q$), axial charge ($n = A$), magnetic dipole moment ($n = M$) and electric dipole moment ($n = E$), respectively. The view of the formfactors depends on the neutrino nature, that is, on whether the neutrino is a Dirac (ν^D) or Majorana (ν^M) particle. Because the electric charge is equal to zero for both neutrino types, then $f_Q(0) = 0$. From the CPT invariance of the vector $| \nu^M >$ it is evident that all the formfactors, except the axial one f_A, are identically equal to zero for a Majorana neutrino. Therefore, ν^M does not have both the magnetic and electric dipole moments. It should be stressed, we are dealing with diagonal elements only (in the initial and final states neutrinos are identical particles). Matrix elements describing the transitions between neutrinos either with different flavors or with different helicities are nonzero for Majorana neutrinos too. The single nonvanishing diagonal element for ν^M, which is related to f_A is called an *anapole moment*. In the case of a Dirac neutrino $f_A(0)$ and $f_M(0)$ are not equal to zero while the static electric dipole moment turns into zero because of the the CPT invariance.

Let us focus our attention on the anomalous dipole magnetic moment (AMM) of a neutrino. First, we consider the SM with $m_\nu \neq 0$ and a Dirac neutrino. The presence of the AMM, which is due to the RCs could be reduced to the appearance of the following term in the interaction Lagrangian

$$\mathcal{L}_{AMM} = \frac{i}{2}\mu_{jj'}\overline{\nu}_j(1+\gamma_5)\sigma^{\mu\nu}F_{\mu\nu}(1+\gamma_5)\nu_{j'}+\text{h.c.} = \frac{i}{2}\mu_{ll'}\overline{\nu}_l(1+\gamma_5)\sigma^{\mu\nu}F_{\mu\nu}(1+\gamma_5)\nu_{l'}+\text{h.c.}, \tag{14.56}$$

where real (imaginary) $\mu_{ll'}$ and $\mu_{jj'}$ represent the magnetic (electric) dipole moments of a neutrino in the flavor and mass bases, respectively. In the third order of the perturbation theory $\mu_{\nu_a\nu_b}$ is defined by the expression [7]

$$\mu_{\nu_j\nu_{j'}} = \frac{3G_F}{16\sqrt{2}\pi^2}(m_{\nu_j} + m_{\nu_{j'}})\sum_l U_{jl}^\dagger U_{lj'}\mu_B. \tag{14.57}$$

The reason of the proportionality of $\mu_{\nu_j\nu_j}$ to the neutrino mass m_ν (that is, the smallness of $\mu_{\nu_j\nu_{j'}}$) in the SM consists in the fact that the W boson in the diagram of Fig. 14.2 interacts with left-handed currents only. Therefore, the chirality flip on the internal neutrino line that is needed for the existence of nonvanishing values of $\mu_{\nu_j\nu_{j'}}$ has to be done at the external neutrino line, $\hat{p}\nu_L = m_{\nu_e}\nu_R$.

Now we proceed to the discussion of the constraints on the matrix $\mu_{\nu\nu}$, which follow from laboratory experiments and astrophysical estimations. Consider the process of elastic scattering of neutrino from electron

$$\overline{\nu}_e + e^- \rightarrow \overline{\nu}_e + e^-. \tag{14.58}$$

In the second order of the perturbation theory this process is going only due to the weak interaction and, in accordance with the SM, the differential cross section is given by the expression:

$$\frac{d\sigma_{CM}}{dT} = \frac{G_F^2 m_e}{2\pi}\left\{4s_W^4 + (1 + 2s_W^2)\left[\left(1 - \frac{T}{E_\nu}\right)^2(1 + 2s_W^2) - \frac{Tm_e}{E_\nu}s_W^2\right]\right\}, \tag{14.59}$$

where E_ν is the energy of the incoming antineutrino and T is the kinetic energy of the recoil electron that is detected in the experiment in fact. The electromagnetic interaction (14.56) stipulates for the additional contribution to the cross section (14.59), which is as follows:

$$\frac{d\sigma_{em}}{dT} = \frac{\pi\alpha^2}{m_e^2\mu_B^2}\left(\sum_j |\mu_{\nu_e\nu_j}|^2\right)\left[\frac{1}{T} - \frac{1}{E_\nu}\right]. \tag{14.60}$$

In the case in question, the electromagnetic and weak amplitudes do not interfere, therefore, the resultant cross section is simply obtained by means of summing Eqs. (14.59) and (14.60). At $T \ll E_\nu$ the cross section (14.60) goes on a constant whereas the cross section (14.59) behaves as T^{-1} and becomes comparable with the weak cross section providing

$$T \approx 0.3 \text{ MeV}, \qquad \mu_{eff} = \left(\sum_j |\mu_{\nu_e \nu_j}|^2 \right)^{1/2} = 10^{-10} \; \mu_B. \qquad (14.61)$$

Consequently, the problem of lowering the experimental boundary on μ_{eff} lies in decreasing the detection threshold of recoil electrons. However, this is complicated by the growth of background in the range of small electron energies. From experiments with reactor antineutrinos were found:

$$|\mu_\nu| < 1.0 \times 10^{-10} \; \mu_B \qquad (14.62)$$

at 90% CL in MUNU experiment at Bugey (France), and

$$|\mu_\nu| < 3.2 \times 10^{-11} \; \mu_B \qquad (14.63)$$

at 90% CL in GEMMA experiment (Germanium Experiment on the measurement of Magnetic Moment of Antineutrino) at Kalinin Nuclear Power Plant. To consider cooling a young white dwarf under the decay of a plasmon γ^* owing to the electromagnetic interaction

$$\gamma^* \to \nu + \bar{\nu}$$

could give constraints on the matrix μ_{ij} as well. The decay width is

$$\Gamma_{\gamma^* \to \nu \bar{\nu}} = \frac{\alpha \omega_p^3}{24 m_e^2 \mu_B^2} \left(\sum_{i,j} |\mu_{ij}|^2 \right), \qquad (14.64)$$

where ω_p is the plasmic frequency in a star. The analysis of astrophysical data results in

$$\left(\sum_{i,j} |\mu_{ij}|^2 \right)^{1/2} < (0.8 - 0.1) \times 10^{-11} \; \mu_B. \qquad (14.65)$$

The solar neutrino data obtained by BOREXINO* lead to the inequality

$$|\mu_\nu| < 0.84 \times 10^{-10} \; \mu_B. \qquad (14.66)$$

The analysis of observations of SN 1987A results in the following boundary

$$\mu_\nu < (0.7 - 1.5) \times 10^{-12}) \; \mu_B. \qquad (14.67)$$

It should be stressed that the above-mentioned astrophysical estimations belong to Dirac neutrinos. An upper limit on the dipole magnetic moment of the Majorana neutrino was obtained on the basis of analyzing the emissivity of a red giant before and after a helium burst and is

$$\mu_{exp} \geq 3 \times 10^{-12} \; \mu_B. \qquad (14.68)$$

---

*BOREXINO is short for BORon EXperiment

However, in the SM even for the τ-lepton neutrino whose mass may be in the range of few MeV, we get a negligibly small value

$$\mu_{\nu_{e(\mu)}\nu_\tau} \sim 10^{-16}\mu_B.$$

Therefore, the detection of the values of the neutrino AMM in the interval

$$\mu_{exp} \sim (10^{-10} - 10^{-12})\mu_B, \tag{14.69}$$

will demand a way out beyond the SM or, at least, a cosmetic reconstruction of the SM.

Let us consider one of the SM modifications in which, along with the right-handed neutrino ν_{aR}, a charged scalar $h^{(-)}$ is introduced into a theory [8]. This scalar represents a singlet with respect to the $SU(2)$ transformations. It has the $B - L$ quantum number equal to 2 and possesses zero VEV. Since the $SU(2)$ singlet is not related with quarks, then the $h^{(-)}$ boson does not induce currents that change the flavor at the tree level in the hadron sector. This, in turn, does not lead to hard limits on its mass. In this SM modification, the Yukawa Lagrangian determining the interaction of the $h^{(-)}$ boson with leptons is governed by the expression:

$$\mathcal{L}_Y = -\sum_{a,b} \left(ig_{ab}\overline{\psi^c_{aL}}\sigma_2\psi_{bL} + f_{ab}\overline{\nu^c_{aR}}l_{bR}\right)h^\dagger + \text{h.c.}. \tag{14.70}$$

The invariance of (14.70) with respect to the $SU(2)_L$ transformations and the Fermi statistics ensure the fulfillment of the condition

$$g_{ab} = -g_{ba}. \tag{14.71}$$

Often, for the sake of simplicity, one imposes the condition $f_{ab} = -f_{ba}$ too. Note that the neutrino in this model is a Dirac particle. Now, apart from the diagram of Fig. 14.2, there exist additional diagrams with the exchanges of the $h^{(\pm)}$ bosons that contribute to the neutrino magnetic moment (Fig. 14.3). Taking into account only the dominating

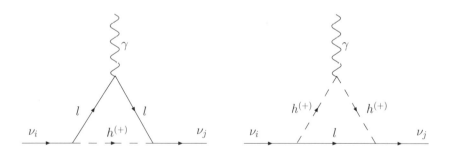

FIGURE 14.3
The Feynman diagrams associated with the neutrino magnetic moment.

contribution from the diagram with the τ lepton in the intermediate state, we arrive at the expression:

$$\mu_\nu = \frac{f_{13}g_{13}m_e m_\tau}{16\pi^2 m_h^2}\left[\ln\frac{m_h^2}{m_\tau^2} - 1\right]\mu_B. \tag{14.72}$$

Setting $f_{13}g_{13} \approx 10^{-1}$ and $m_h \approx 100$ GeV, what does not contradict the experimental data, we get $\mu_\nu \approx 10^{-10}\mu_B$. A disadvantage of the model lies in the fact that the neutrino mass remains a free parameter, that is, the neutrino mass smallness is a mystery as before. Many attempts have been made to improve the model. At that, the main difficulty consisted in the following. Even if one assumes the bare neutrino mass to be equal zero, the neutrino mass will arise due to the RCs already at the one-loop level. Because the neutrino magnetic moment is induced by attaching a photon line to the internal lines of self-energy neutrino diagrams, then there is the linkage between the mass and magnetic moment. Thus, for example, the model of Ref. [9] predicts the relation:

$$\mu_{\nu_e \nu_e} \approx \frac{2 m_e m_{\nu_e}}{M^2} \mu_B,$$

where M denotes the typical masses in the loop. Then for $M \approx 100$ GeV, in order to obtain $\mu_{\nu_e \nu_e} \approx 10^{-10} \mu_B$, we must have m_{ν_e} to be of the order of tens KeV to contradict the experiment.

Let us show how to avoid this difficulty. The most general form for the mass neutrino term is

$$\mathcal{L}_{m_\nu} = -\frac{1}{2} m_{ij} \overline{\psi_{iL}^c} \psi_{jL} + \text{h.c.,} \tag{14.73}$$

where ψ are the Weyl spinors, $i, j = 1, 2, \ldots 2N$, N is the number of the neutrino generation. For simplicity we shall discuss the case with $N = 2$. Then, assuming $m_{11} = m_{22} = 0$, we ensure the lepton number conservation $L_1 - L_2$. If one determines two fields ψ_L and ψ_R as

$$\psi_L = \psi_{1L}, \qquad \psi_R = [\psi_{2L}]^c, \tag{14.74}$$

then (14.73) results in the Dirac neutrino mass

$$\mathcal{L}_{m_\nu} = -m(\overline{\psi}_L \psi_R + \overline{\psi}_R \psi_L) = -m \overline{\psi} \psi, \tag{14.75}$$

where

$$\psi = \begin{pmatrix} \psi_R \\ \psi_L \end{pmatrix}.$$

Note, to get the Majorana mass we sufficiently demand that at least one diagonal element of the mass matrix would not be equal to zero. We can conveniently rewrite the neutrino electromagnetic interaction in the form of the Weyl's spinors

$$\mathcal{L}_{em} = \frac{1}{2} \mu_{ij} \overline{\psi_{iL}^c} \sigma_{\mu\nu} \psi_{jL} F^{\mu\nu},$$

where $\mu_{ij} = -\mu_{ji}$. The next step consists in the assumption that the spinors ψ_{1L} and ψ_{2L} form the doublet with respect to some group $SU(2)$ which will be called a $SU(2)_\nu$ group. Then, the quantity $\mathcal{L}_{m_\nu}$ is a triplet under $SU(2)_\nu$ whereas the magnetic moment term $\mathcal{L}_{em}$ is a singlet. Therefore, the exact invariance of the theory with respect to the $SU(2)_\nu$ transformations forbids the appearance of the mass term and does not prevent the existence of the neutrino MMs. The proposal of Voloshin [10] was to place ν_L and $[\nu_L]^c$ (that is, an antiparticle of ν_R) in a $SU(2)_\nu$ doublet and demand the invariance of the theory under the $SU(2)_\nu$ transformations. However, we cannot impose the $SU(2)_\nu$ symmetry for no particular reason because ψ_{1L} and ψ_{2L} belong to different multiplets of the $SU(2)_L$ group (ν_L belongs to a doublet and ν_R belongs to a singlet). The successive solution of the problem was realized in the following directions:

(i) **Extension of the electroweak group.** The electroweak $SU(2)_L \times U(1)_Y$ symmetry is enlarged to $SU(3)_L \times U(1)_Y$, so that it may include Voloshin's $SU(2)_\nu$ group [11]. In so doing, the lepton multiplet has the form

$$\begin{pmatrix} \nu_{aL}^c \\ \nu_{aL} \\ l_a \end{pmatrix}.$$

(ii) **Introduction of horizontal symmetry (symmetry between generations).** One may use some symmetry $SU(2)_F$ that commutes with the electroweak $SU(2)_L \times U(1)_Y$ group. Usually the horizontal symmetry is chosen between the electron and muon generations, that is, ν_e and ν_μ form a doublet of $SU(2)_F = SU(2)_\nu$.

The main problem inherent in the models with the $SU(2)_\nu$ symmetry lies in the hard settling of the Higgs potential. Under the spontaneous breaking of the $SU(2)_\nu$ symmetry, a neutrino mass is induced. Then, it is not easy to keep the $SU(2)_\nu$ breaking scale sufficiently low without a fine-tuning of the parameters in the Higgs potential to give a large magnetic moment of $\mu_e = 10^{-10} - 10^{-11}\mu_B$ with the neutrino mass kept small enough.

15

Neutrino oscillation in a vacuum

> *And that which rushes in the misty outlines,*
> *it is attached in the durable thoughts.*
> V. von Goethe

15.1   Description in the formalism of de Broglie's waves

Let us consider a neutrino within the SM. To make a neutrino massive we choose the most simple way, that is, add a right-handed neutrino singlet ν_{lR} to the lepton sector. Then, the Yukawa Lagrangian takes the form

$$\mathcal{L}_Y = -\sum_a h_{aa}\bar{l}_{aL}\varphi l_{aR} - \sum_{a,b} h'_{ab}\bar{\nu}_{aL}\varphi^c \nu_{bR} + \text{h.c.}. \tag{15.1}$$

The nondiagonal YCCs h'_{ab} $(a \neq b)$ in (15.1) call forth not only mixing of neutrino generations, but lead to the existence of neutrino processes with the lepton flavor violation at the tree level as well. To date the processes that violate both the total and partial lepton flavors are nor observed. However, we do not have any fundamental principle that demands the existence of similar phenomena. This circumstance prompts experimentalists to look for such processes. Because the accuracy of any experiment is limited, then the absence of positive results in searches of processes with the lepton flavor violation should be interpreted no more than the establishment of upper boundaries on the cross sections of these processes, which, in turn, results in constraints on values of nondiagonal YCCs.

We consider the one-dimensional case and assume that at time $t = 0$ we get a ν_l neutrino with momentum p perfectly defined. Then, in the absence of external fields, the set of the de Broglie's waves with the constant amplitude $\nu_j(0)$ is associated with this neutrino

$$\nu_l(z, t = 0) = \sum_{j=1}^{3} U_{lj}\nu_j(0)\exp{(ipz)}. \tag{15.2}$$

After a time t we shall have instead of (15.2)

$$\nu_l(z, t) = \sum_{j=1}^{3} U_{lj}\nu_j(0)\exp{[i(pz - E_j t)]}. \tag{15.3}$$

In most cases we shall deal with, the neutrino energy belongs to the ultrarelativistic region, that is, $E_j \gg m_j$. At that, it is difficult to resist the temptation and not expand all available expressions as a power series in the infinitesimally small parameter m_j^2/p^2 $(p =| \mathbf{p} |)$. Then, the neutrino energy and velocity are given by the expressions:

$$E_j = p\sqrt{1 + \frac{m_j^2}{p^2}} \simeq p + \frac{m_j^2}{2p}, \tag{15.4}$$

$$v_j \simeq 1 - \frac{m_j^2}{2p^2} \simeq 1. \tag{15.5}$$

All this allows us to rewrite the neutrino wave function in the following form: ($t \simeq z$)

$$\nu_l(z) = \sum_{j=1}^{3} U_{lj} \exp\left(-im_j^2 z/2p\right) U_{l'j}^* \nu_{l'}(0), \tag{15.6}$$

where $\nu_l(z, z) \equiv \nu_l(z)$ and we have returned to the flavor basis (recall that $UU^\dagger = 1$). So, after traveling a distance z, the wave function of the neutrino, born with the flavor l, represents a superposition of states with all the neutrino flavors. It means that there is a nonvanishing probability of finding the $\nu_{l' \neq l}$ components in the neutrino beams, at a distance z from its source, if originally this beam consisted of ν_l particles. To put it differently, neutrinos can exhibit oscillation transitions. The neutrino oscillation phenomenon resembles microparticles diffraction on several slits (in our case the number of slits is equal to that of the neutrino generations), which, in accordance with the Huygens–Fresnel principle, is explained by the appearance of secondary de Broglie waves with amplitudes $U_{lj}U_{l'j}^*\nu_{l'}(0)$ and their subsequent interference.

Let us examine whether the expression (15.6) is changed when the neutrino was born with the definite energy, that is, when all the states in the physical basis have the same energy. Then, in the ν_j state the momentum is defined by the expression

$$p_j = \sqrt{E^2 - m_j^2} \approx E - \frac{m_j^2}{2E} \tag{15.7}$$

and, again, we have the same phase shift for different mass eigenstates as in Eq. (15.6). So, in the ultrarelativistic case, there is no distinction between the situations when the neutrino is born in the state with the definite momentum value or when the one is born in the state with the definite energy value.

Using the expression for the wave function (15.6), we can find the probabilities of transitions between states with different flavor values, that is, the quantities measured directly in neutrino oscillation experiments. There are two kinds of such experiments, namely, *appearance experiments* and *disappearance experiments*. In both cases, the total number and spectrum of charged leptons produced by a neutrino beam at different distances from a source are measured. In appearance experiments, they are charged leptons that do not correspond to an initial neutrino type. In the latter, they are charged leptons belonging to the same generation l as an initial neutrino ν_l, that is, the weakening of a beam is looked for.

The probability of finding the neutrino to have flavor l' at a distance z from its source, if originally it had flavor l will look like:

$$\mathcal{P}_{l \to l'}(z) = \left| \sum_j U_{lj} \exp\left(-im_j^2 z/2p\right) U_{l'j}^* \right|^2 = \sum_j |U_{lj}|^2 |U_{l'j}|^2 +$$

$$+ \sum_{j \neq j'} \left[\text{Re}(U_{lj}U_{lj'}^* U_{l'j'} U_{l'j}^*) \cos\left(\frac{\Delta m_{jj'}^2}{2E} z\right) + \text{Im}(U_{lj}U_{lj'}^* U_{l'j'} U_{l'j}^*) \sin\left(\frac{\Delta m_{jj'}^2}{2E} z\right) \right], \tag{15.8}$$

where we have set $E \approx p$ and introduced the designation $\Delta m_{jj'}^2 = m_j^2 - m_{j'}^2$ (hereafter Greek letters denote physical states). If one assumes that the CP parity is the conserved quantum number, then the matrix U could be chosen as real. Then, the expression (15.8) takes the form:

$$\mathcal{P}_{l \to l'}(z) = \sum_j U_{lj}^2 U_{l'j}^2 + \sum_{j \neq j'} U_{lj}U_{lj'}U_{l'j'}U_{l'j} \cos\left(\frac{2\pi}{l_{jj'}} z\right), \tag{15.9}$$

where the quantities

$$|l_{jj'}| = \frac{4\pi E}{\Delta m_{jj'}^2} \tag{15.10}$$

are called the *oscillation lengths* between the ν_j and $\nu_{j'}$ states; those give a distance scale over which the oscillation effects can be appreciable.

In disappearance experiments the probability of surviving a neutrino emitted by a source is measured. Assume that we are dealing with an electron neutrino. If we fulfill the observation of a neutrino flux at a distance L from its source, then the survival probability of ν_e will be

$$\mathcal{P}_{\nu_e \nu_e} = 1 - \sum_{j,j'} 4U_{ej}^2 U_{ej'}^2 \sin^2\left(\frac{\pi L}{l_{jj'}}\right). \tag{15.11}$$

When $L \ll l_{jj'}$, then the second term in (15.11) may be neglected and $\mathcal{P}_{\nu_e \nu_e} = 1$, that is, oscillations will be unobservable. This puts some idea into us how to define the sensitivity to measuring $\Delta m_{jj'}^2$ for one or another neutrino oscillation experiments. In the sine argument of the expression (15.11) we pass on from a natural system of units to ordinary units

$$\sin^2\left(\frac{\Delta m_{jj'}^2 L}{\hbar c E}\right) = \sin^2\left(\frac{1.27 \Delta m_{jj'}^2 (\text{eV}^2) L(\text{km})}{E(\text{GeV})}\right). \tag{15.12}$$

Consequently, to improve the experiment accuracy (measuring small values of $\Delta m_{jj'}^2$) the ratio L/E should be increased. The most optimal solution is to use low-energy neutrinos and place the detector at large distances from a source. Thus, for example, in the experiment with a neutrino beam having the energy $E \approx 1$ GeV at the distance $L = 1$ km, one can measure $\Delta m_{jj'}^2$ with an accuracy of 1 eV^2.

Under the fulfillment

$$L \gg l_{jj'}, \tag{15.13}$$

or

$$\Delta m_{jj'}^2 \gg \frac{4\pi E}{L}, \qquad \text{for all } (j,j') \text{ pairs} \tag{15.14}$$

the sines in (15.12) quickly oscillate allowing us to set

$$\sin^2\left(\frac{\pi L}{l_{jj'}}\right) \approx \frac{1}{2} \tag{15.15}$$

and time-averaged (distance-averaged) oscillations are obtained. Making use of the explicit form of the matrix U, we find

$$\mathcal{P}_{\nu_e \nu_e} \geq \frac{1}{3}. \tag{15.16}$$

The analysis shows, in the case of N neutrino generations, the expression for $\mathcal{P}_{\nu_e \nu_e}$ takes the form

$$\mathcal{P}_{\nu_e \nu_e} \geq \frac{1}{N}. \tag{15.17}$$

From the obtained expressions it follows that the appearance of transitions between different kinds of neutrinos, as they travel in a vacuum, is possible only under the realization of the following conditions: (i) neutrinos must have masses and a mass-degeneration multiplicity does not equal to the number of neutrino generations; (ii) some or all mixing angles must be nonvanishing; and (iii) for oscillations to appear a neutrino beam must pass a distance comparable with an oscillation length.

Frequently, when analyzing neutrino experiments, one assumes the mixing only between two neutrino generations (two-flavor approximation). Let us believe that it takes place for ν_μ and ν_τ. Then, the matrix U is

$$U = \begin{pmatrix} \cos\theta_{\mu\tau} & \sin\theta_{\mu\tau} \\ -\sin\theta_{\mu\tau} & \cos\theta_{\mu\tau} \end{pmatrix}. \tag{15.18}$$

The probability $\nu_\mu \to \nu_\tau$ becomes the form

$$\mathcal{P}_{\nu_\mu\nu_\tau} = \sin^2 2\theta_{\mu\tau} \sin^2\left(\frac{\Delta m_{23}^2 L}{4E}\right). \tag{15.19}$$

The survival probability of the muon neutrino may be found from the probability conservation law

$$\mathcal{P}_{\nu_\mu\nu_\mu} = 1 - \mathcal{P}_{\nu_\mu\nu_\tau}. \tag{15.20}$$

If the mixing occurs for three neutrino generations, but a double mass-degeneration exists, then the two-flavor approximation could be used again.

Consider the following situation. The neutrino masses satisfy the relations

$$\mid \Delta m_{21}^2 \mid \ll \mid \Delta m_{31}^2 \mid \cong \mid \Delta m_{32}^2 \mid, \tag{15.21}$$

and the experimental conditions are

$$\frac{\mid \Delta m_{31}^2 \mid L}{E} \approx 1. \tag{15.22}$$

From (15.22) the inequality

$$\frac{\mid \Delta m_{21}^2 \mid L}{E} \ll 1 \tag{15.23}$$

follows. Then, the transition probability $\nu_l \to \nu_{l' \neq l}$ will look like:

$$\mathcal{P}_{\nu_l \nu_{l'}} \approx 4 U_{l3}^2 U_{l'1}^2 \sin^2\left(\frac{\Delta m_{31}^2 L}{4E}\right). \tag{15.24}$$

Thanks to (15.24), the experiment does not feel the mass difference of the states ν_1 and ν_2, that is, it interprets the both states as a single state. This explains the resemblance of the obtained expression (15.24) with the analogous formula of the two-flavor approximation.

In a real experiment the quantities E and L are changed in some intervals and, by this reason, we should carry out the average both over neutrino energies and over distances they travel. Consider these operations in detail. Let the quantity $a = L/(4E)$ be described by a Gaussian distribution with the standard deviation σ_a relative to the central value a_0. Then, the theoretical value of the averaged survival probability of the muon neutrino is as follows:

$$< \mathcal{P}_{\nu_\mu\nu_\mu} > = \frac{1}{\sigma_a\sqrt{2\pi}} \int_0^\infty \left\{1 - \frac{1}{2}\sin^2 2\theta_{\mu\tau}[1 - \cos(2a\Delta m_{23}^2)]\right\} \times$$

$$\times \exp\left[-\frac{(a-a_0)^2}{2\sigma_a^2}\right] da = 1 - \frac{1}{2}\sin^2 2\theta_{\mu\tau}\{1 - \cos(2a_0\Delta m_{23}^2)\exp[-2\sigma_a(\Delta m_{23}^2)^2]\}. \tag{15.25}$$

Obviously, the form of the expressions that are compared with experimental data defines an experiment design. What can we say looking at Eq. (15.25)? To measure Δm_{23}^2 the sine argument in (15.25) needs to have a finite value. At small (large) values of Δm_{23}^2, it is necessary to increase (decrease) the ratio $L/(4E)$. However, in doing so, one should keep a particular balance between $L/(4E)$ and Δm_{23}^2, namely, one cannot permit a drastic excess of one quantity over another because, as this would take place, the loss of an experiment sensitivity to Δm_{23}^2 happens. Thus, when $\Delta m_{23}^2 L/(4E) \gg 1$,

$$\cos\left(\frac{\Delta m_{23}^2 L}{4E}\right) \to \frac{1}{2}, \tag{15.26}$$

while at $\Delta m_{23}^2 L/(4E) \ll 1$

$$\cos\left(\frac{\Delta m_{23}^2 L}{4E}\right) \to 1. \tag{15.27}$$

It is evident too, under small $\theta_{\mu\tau}$ and/or small $\Delta m_{23}^2 L/(4E)$, to detect oscillations needs very high statistics.

In the SM with $\mathcal{L}_Y$ defined by the expression (15.1), the Lagrangian which describes the interaction between the Higgs boson H and leptons has the form

$$\mathcal{L}_H = -[\sum_a h_{aa}\bar{l}_a(x)l_a(x) + \sum_{ab} h'_{ab}\bar{\nu}_a(x)\nu_b(x)]H(x), \tag{15.28}$$

where $h_{aa} = m_a/v$. Remembering the mass matrix to be diagonalized with the help of the matrix U, we may establish the linkage between the YCCs h'_{ab} and the neutrino oscillation parameters. In the case of $\nu_\mu \to \nu_\tau$ mixture, this linkage is

$$h'_{\mu\mu}v = m_2 \cos^2\theta_{\mu\tau} + m_3 \sin^2\theta_{\mu\tau}, \qquad h'_{\tau\tau}v = m_2 \sin^2\theta_{\mu\tau} + m_3 \cos^2\theta_{\mu\tau}, \tag{15.29}$$

$$h'_{\mu\tau}v = \sin\theta_{\mu\tau}\cos\theta_{\mu\tau}(m_3 - m_2). \tag{15.30}$$

From (15.30) it is evident, if the masses m_3 and m_2 are very close to each other, then the nondiagonal Yukawa neutrino coupling constants are extremely small. If such is the case, in the SM we have no chance to determine the neutrino oscillation parameters in collider experiments.

15.2 Oscillations and the uncertainty relation

Consider a typical appearance experiment. There is a source that generates a neutrino beam due to the decays

$$\pi^+ \to \mu^+ + \nu_\mu. \tag{15.31}$$

A detector whose construction allows one to keep a record of neutrinos appearing in the initial beam is located at a distance L. The experiment gives the result $\mathcal{P}_{\nu_\mu\nu_e} \neq 0$. Suppose, further, that we can detect the muons from the pion decays and measure their momenta accurately enough, that is, we know which mass eigenstate ν_i $(i = 1, 2)$ was actually emitted in each decay. This is, of course, impossible in practice (for the moment at least), and we attach the status of a mental experiment to the aforesaid. The neutrino beam with one and the same value of a momentum is no longer a coherent superposition of mass eigenstates ν_1 and ν_2. Therefore, the probability $\mathcal{P}_{\nu_\mu\nu_e}$ will be equal to zero, since oscillations occur under the interference between ν_1 and ν_2, whereas only one of them is present in the beam. From the point of view of quantum mechanics, the result is quite prospective. Every measurement is an interaction between device and object. In a microworld this interaction cannot be reduced to the level at which it could be neglected. As a result of measurement, a microsystem either turns into an other state or is destroyed. This is the severe price of knowledge in a microworld. In our case the process of exact measuring the momentum results in the disappearance of the interference picture. The Heizenberg uncertainty relations elucidate the essence of this phenomenon.

The error in the measurement of the neutrino mass $m = \sqrt{E^2 - p^2}$ is given by the expression

$$\delta(m^2) = \sqrt{(2E)^2(\delta E)^2 + (2p)^2(\delta p)^2}, \tag{15.32}$$

where δE and δp are the errors in measuring the energy and momentum of the neutrino (we assume them to be uncorrelated). Now, it will be possible to define which mass eigenstate is emitted in each decay only if $\delta(m^2)$ is less than the smallest mass squared difference $| m_1^2 - m_2^2 |$. The uncertainty relation reads

$$\delta z \delta p \geq \frac{1}{2}, \tag{15.33}$$

where δz is the error in the measurement of the position of the neutrino source. From (15.32) and (15.33) it is evident that

$$\delta z > \frac{2p}{| \Delta_{12} |} = \frac{l_v}{2\pi}. \tag{15.34}$$

In such a way, in the exact measurement of the neutrino mass, uncertainty in measuring the coordinate of the neutrino birth (or detection) point must be larger than the oscillation length. On the other hand, the states ν_1 and ν_2 are born in a source and recorded in a detector coherently, oscillations existing only at their interference. When ν_1 and ν_2 are separated by a distance greater than $| l_v |$, oscillations disappear, that is, the necessary condition of the oscillations observation is the condition of localizing both a neutrino source and a detector in regions whose sizes are vastly less than the oscillation length

$$\delta z \ll | l_v |. \tag{15.35}$$

Since the inequalities (15.34) and (15.35) are at variance with each other, then the absence of oscillations becomes evident providing

$$\delta(m^2) \leq | m_1^2 - m_2^2 |. \tag{15.36}$$

When measuring the mass with the precision of

$$\delta(m^2) > | m_1^2 - m_2^2 |, \tag{15.37}$$

the influence on the neutrino system will be less than in the situation associated with Eq. (15.36). In this case, oscillations exist as before though their character had been changed under the influence of the measurement. From the inequalities (15.37) we also have

$$\delta p > \frac{| m_1^2 - m_2^2 |}{2E}. \tag{15.38}$$

Hence, the presence of the momentum or energy spread in the neutrino beam is one of the oscillation existence conditions. This, in turn, means that the successive treatment of neutrino oscillations must be based on using the wave packet formalism rather than the plane de Broglie waves formalism.

15.3   Wave packet treatment of neutrino oscillations

We assume, at time $t = 0$ a neutrino whose momentum spread is described by a Gaussian distribution is produced. For the sake of simplicity, we are constrained by a one-dimensional case. Then, the produced neutrino is associated with the wave packet of the form

$$\nu_l(p_a, 0) = \frac{1}{\sqrt{2\pi}\sigma_{pS})^{3/2}} \sum_{a=1}^{3} U_{la} \int dp \exp\left[-\frac{(p - p_a)^2}{4\sigma_{pS}^2}\right]\nu_a(p, 0), \qquad (15.39)$$

where σ_{pS} is the Gaussian width of the wave packet in a source, p_a is the mean momentum of the ν_a neutrino. In coordinate space at time t, our wave packet is determined by the expression:

$$\nu_l(t) = \sum_{a=1}^{3} U_{la} \int dz \mathcal{F}(\sigma_{zS}, z, t)\nu_a(z, 0), \qquad (15.40)$$

where

$$\mathcal{F}(\sigma_{zS}, z, t) = \frac{1}{(2\pi\sqrt{2\pi}\sigma_{pS})^{3/2}} \int dp \exp\left[-\frac{(p - p_a)^2}{4\sigma_{pS}^2} + i(pz - E_a(p)t)\right].$$

Since the average energy value is a function of a momentum, we have a right to expand $E_a(p)$ as a power series in $p - p_0$. Recall that wave packets are stable for particles moving with the light speed as a result of the coincidence of the phase v_p and group v_g speeds. For massive particles $v_p \neq v_g$ and wave packets spread. In this case, the fact of instability becomes evident when taking into account the third and subsequent terms in the expansion of $E_a(p)$. Since we assume the neutrino to be ultrarelativistic ($v \sim 1$), in the series for $E_a(p)$, we may keep only the two first terms

$$E_a(p) \simeq E_a(p)|_{p=p_a} + \frac{\partial E_a}{\partial p}|_{p=p_a}(p - p_a) = E_a + v_a(p - p_a), \qquad (15.41)$$

where v_a denotes the group packet speed, which is equal to the speed of the ν_a neutrino. To substitute (15.41) into (15.40) results in

$$\nu_l(t) = \frac{1}{(\sqrt{2\pi}\sigma_{zS})^{3/2}} \exp\left[i(p_a z - E_a t) - \frac{(z - v_a t)^2}{4\sigma_{zS}^2}\right], \qquad (15.42)$$

where the Gaussian packet width in coordinate space σ_{zS} is equal to $(2\sigma_{pS})^{-1}$.

A neutrino detector is also characterized by uncertainties under measurements of energy and momentum. Therefore, a flavor state detected at a distance L from a source will be described by a wave packet as well

$$\nu_{l'}(L) = \sum_{b=1}^{3} U_{l'b} \frac{1}{(\sqrt{2\pi}\sigma_{zD})^{3/2}} \int dz_D \exp\left[ip_b z_D - \frac{z^2}{4\sigma_{zD}^2}\right]\nu_b(z_D, 0), \qquad (15.43)$$

where the subscript D ascribes all quantities to a detector. Then, allowing for the normalization condition

$$\nu_b(z_D, 0)^* \nu_a(z, 0) = \delta_{ab}\delta(z_D + L - z),$$

the amplitude of the $l \to l'$ transition takes the form

$$A_{ll'}(L, t) = \sqrt{\frac{2\sigma_{zS}\sigma_{zD}}{\sigma_z^2}} \sum_{a=1}^{3} U_{l'a}^* U_{la} \exp\left[i(p_a L - E_a t) - \frac{(L - v_a t)^2}{4\sigma_z^2}\right], \qquad (15.44)$$

where $\sigma_z^2 = \sigma_{zS}^2 + \sigma_{zD}^2$.

In neutrino production processes the neutrino energy E is determined by kinematics for a massless neutrino. Let us pass on from the average values of energy E_a and momentum p_a to E. This will be achieved with the help of the relations

$$E_a \simeq E + \xi \frac{m_a^2}{2E}, \qquad p_a \simeq E - (1-\xi)\frac{m_a^2}{2E}, \tag{15.45}$$

where the dimensionless parameter ξ, being close to 1, accounts for corrections to the nonzero neutrino mass. The neutrino velocity could be found using the relation

$$v_a \simeq 1 - \frac{m_a^2}{2E^2}. \tag{15.46}$$

When time t is not measured in an experiment, then in order to get the flavor transition probability the amplitude $A_{ll'}(L,t)$ has to be integrated over time

$$\mathcal{P}_{ll'}(L) = \int_0^\infty \mid A_{ll'}(L,t) \mid^2 dt =$$

$$= \sum_j \mid U_{lj} \mid^2 \mid U_{l'j} \mid^2 + \sum_{j \neq j'} \{\mathrm{Re}(U_{lj}U_{lj'}^* U_{l'j'} U_{l'j}^*) \cos\left(\frac{2\pi L}{l_{jj'}}\right) +$$

$$+\mathrm{Im}(U_{lj}U_{lj'}^* U_{l'j'} U_{l'j}^*) \sin\left(\frac{2\pi L}{l_{jj'}}\right)\} \exp\left[-2\pi^2\xi^2 \left(\frac{\sigma_z}{l_{jj'}}\right)^2\right] \exp\left[-\left(\frac{L}{l_{jj'}^c}\right)^2\right], \tag{15.47}$$

where the quantities

$$l_{jj'}^c = \frac{4\sqrt{2}E^2\sigma_z}{\Delta m_{jj'}^2}$$

are known as *coherence lengths* between the ν_j and $\nu_{j'}$ states. So, the wave packet formalism leads to the expression for $\mathcal{P}_{ll'}(L)$, which differs from the analogous expression obtained within the plane wave formalism by the presence of two additional terms. The former

$$\exp\left[-2\pi^2\xi^2 \left(\frac{\sigma_z}{l_{jj'}}\right)^2\right] \tag{15.48}$$

guarantees the condition of oscillations observation. Really, when

$$\sigma_z \ll \mid l_{jj'} \mid,$$

the expression (15.48) tends to 1 while at

$$\sigma_z > \mid l_{jj'} \mid$$

tends to zero. This is nothing more nor less than the well-known statement, the localization of the source and the detector must be much better determined than the oscillation length. The appearance of the latter term is connected with the fact that two wave packets, associated with ν_j and $\nu_{j'}$, each with different momentum and energy, have slightly different group velocities. It means that after some time the mass eigenstate wave packets no longer overlap and cannot interfere to produce oscillations. It is easy to estimate the value of coherent length $l_{jj'}^c$ from pure classical arguments. If both wave packets have the Gaussian width σ_z and move with different group velocities v_j and $v_{j'}$ ($v_j > v_{j'}$), then they cease to overlap each other when the packet corresponding to $\nu_{j'}$ traveled a distance $L \approx l_{jj'}^c$

$$L = \frac{\sigma_z v_j}{v_j - v_{j'}}. \tag{15.49}$$

For ultrarelativistic neutrinos, the relation (15.49) coincides with the definition $l_{jj'}^c$ within a factor of $2\sqrt{2}$. One can show that this increase of $l_{jj'}^c$ is caused by the spreading of the wave packets.

The presence of the factor $\exp[-\left(L/l_{jj'}^c\right)^2]$ in the expression for the $\nu_l \to \nu_{l'}$ transition probability illustrates once again the measurements role in a microworld. Precise measurements of momenta of all particles appearing in the detection process implies the decrease of σ_{pD} and, therefore, the increase both of σ_{zD} and of $l_{jj'}^c$. Let us assume that without measuring the momenta, the wave packets did not overlap in hitting the detector, that is, their interference was absent. If the measurement act increases the coherent length to such an extent that the packets overlap, then, as a consequence, oscillations appear. At

$$z \ll \frac{\sigma_z}{2\Delta m_{jj'}^2 E^2} \simeq \frac{\sigma_z}{\mid v_j - v_{j'} \mid} \tag{15.50}$$

we get

$$\mathcal{P}_{ll'}(L) = \sum_j \mid U_{lj} \mid^2 \mid U_{l'j} \mid^2 + \sum_{j \neq j'} \{\mathrm{Re}(U_{lj}U_{lj'}^* U_{l'j'}U_{l'j}^*) \cos\left(\frac{2\pi L}{l_{jj'}}\right) +$$

$$+\mathrm{Im}(U_{lj}U_{lj'}^* U_{l'j'}U_{l'j}^*) \sin\left(\frac{2\pi L}{l_{jj'}}\right)\}, \tag{15.51}$$

that is, the same expression as in the plane wave formalism. In the case

$$z > \frac{\sigma_z}{\mid v_j - v_{j'} \mid} \tag{15.52}$$

the damping factor

$$\exp\left[-\left(\frac{L}{l_{jj'}^c}\right)^2\right]$$

begins to work in the expression (15.47) and we come to the time-averaged oscillations

$$\mathcal{P}_{ll'}(L) = \sum_j \mid U_{lj} \mid^2 \mid U_{l'j} \mid^2. \tag{15.53}$$

So, both formalisms lead to the same physical results though, without question, using the wave packets is more defensible from the physical point of view. However, both descriptions possess one grave disadvantage, namely, they are suited for the neutrino motion in a vacuum only. To switch on the interaction we need to address more successive treatments, namely, to the Lagrangian or Hamiltonian approaches.

15.4 Evolution equation for the neutrino

We shall work in two-flavor approximation and, for the sake of definiteness, assume that the electron and muon neutrinos are dealt with. Evidently, for the stationary states, the evolution equation in the physical basis is

$$i\frac{\partial}{\partial t}\nu_i(\mathbf{r}, t) = H\nu_i(\mathbf{r}, t). \tag{15.54}$$

where the Hamiltonian operator H is diagonal

$$H = \begin{pmatrix} E_1 & 0 \\ 0 & E_2 \end{pmatrix},$$

In ultrarelativistic case the Hamiltonian ($|\mathbf{p}| \approx E$) takes the following view:

$$H = |\mathbf{p}| + \frac{1}{2|\mathbf{p}|} \begin{pmatrix} m_1^2 & 0 \\ 0 & m_2^2 \end{pmatrix} = H_d - \frac{\Delta m_{12}^2}{4|\mathbf{p}|} \sigma_3, \qquad (15.55)$$

where

$$\Delta m_{12}^2 = m_1^2 - m_2^2, \qquad H_d = \left(|\mathbf{p}| + \frac{m_1^2 + m_2^2}{4|\mathbf{p}|} \right) I,$$

and I is the unit matrix. For simplicity, we shall consider a one-dimensional case. Then, taking into consideration

$$z_i = v_i t \approx (1 - \frac{m_i^2}{E^2}) t \approx t = z,$$

Eq. (15.54) could be rewritten as follows:

$$i\frac{d}{dz}\nu_i(z) = H\nu_i(z), \qquad (15.56)$$

where we have denoted $\nu_i(z, z) \equiv \nu_i(z)$. In all experiments we deal with the electron and muon neutrinos (recall that all our Lagrangians involving fermions have been written in the flavor basis). For this reason, we should pass on to the flavor basis in the evolution equation. Multiplying Eq. (15.54) by U on the right, we get

$$i\frac{d}{dz}\nu_l(z) = H'\nu_l(z), \qquad (15.57)$$

where the new Hamiltonian $H' = UHU^{-1}$ is

$$H' = H_d + \frac{\Delta m_{12}^2}{4E} \begin{pmatrix} -\cos 2\theta_0 & \sin 2\theta_0 \\ \sin 2\theta_0 & \cos 2\theta_0 \end{pmatrix} =$$

$$= H_d + \frac{\Delta m_{12}^2}{4E} (\sigma_1 \sin 2\theta_0 - \sigma_3 \cos 2\theta_0). \qquad (15.58)$$

Accounting for the fact that H' does not depend on z, we find the solution of Eq. (15.57) without any trouble

$$\nu_l(z) = \exp(-iH'z)\nu_l(0), \qquad (15.59)$$

where

$$\nu_e(0) = \begin{pmatrix} 1 \\ 0 \end{pmatrix}, \qquad \nu_\mu(0) = \begin{pmatrix} 0 \\ 1 \end{pmatrix}.$$

Since the presence of the quantity that is proportional to I in the Hamiltonian leads to a common phase for all flavors and only the squared module of the wave function has a physical sense, then the quantity H_d in (15.58) may be omitted. Note that now the consequence of Eq. (15.58) is

$$\tan 2\theta_0 = \frac{2H_{12}'}{H_{22}' - H_{11}'}. \qquad (15.60)$$

Let us get rid of the matrices in the exponent of the equation obtained. It could be achieved under expanding (15.59) into a power series

$$\nu_l(z) = \exp\left[-\frac{i\Delta m_{12}^2 z}{4E} (\sigma_1 \sin 2\theta_0 - \sigma_3 \cos 2\theta_0) \right] \nu_l(0) =$$

$$= \left[\cos\left(\frac{\Delta m_{12}^2 z}{4E} \right) - i(\sigma_1 \sin 2\theta_0 - \sigma_3 \cos 2\theta_0) \sin\left(\frac{\Delta m_{12}^2 z}{4E} \right) \right] \nu_l(0). \qquad (15.61)$$

Using the found solution, we can determine the transition probabilities between different neutrino states and make sure that the results coincide with the previous ones. Thus, for example, when in an initial state only electron neutrinos are present, then the probabilities of finding ν_μ and ν_e in a point z are defined by the expressions

$$\mathcal{P}_{\nu_e \nu_\mu} = \left| (0,1) \left[\cos\left(\frac{\Delta m_{12}^2 z}{4E} \right) - i(\sigma_1 \sin 2\theta_0 - \sigma_3 \cos 2\theta_0) \sin\left(\frac{\Delta m_{12}^2 z}{4E} \right) \right] \begin{pmatrix} 1 \\ 0 \end{pmatrix} \right|^2 =$$

$$= \sin^2 2\theta_0 \sin^2\left(\frac{\Delta m_{12}^2 z}{4E} \right), \qquad (15.62)$$

$$\mathcal{P}_{\nu_e \nu_e} = \left| (1,0) \left[\cos\left(\frac{\Delta m_{12}^2 z}{4E} \right) - i(\sigma_1 \sin 2\theta_0 - \sigma_3 \cos 2\theta_0) \sin\left(\frac{\Delta m_{12}^2 z}{4E} \right) \right] \begin{pmatrix} 1 \\ 0 \end{pmatrix} \right|^2 =$$

$$- 1 - \sin^2 2\theta_0 \sin^2\left(\frac{\Delta m_{12}^2 z}{4E} \right), \qquad (15.63)$$

respectively.

16

Neutrino oscillation in matter

Science! True daughter of Old Time thou art!
Who alterest all things with thy peering eyes.
Why preyest thou thus upon the poet's heart,
Vulture, whose wings are dull realities?
Edgar Elan Poe

16.1 Neutrino motion in condensate matter

The inclusion of the neutrino interaction with matter particles will be the next stage of our analysis. The basic idea of our approach consists in the reduction of the totality of the neutrino interactions in matter to the motion in a field with a potential energy. Doubts because of the appearance of a potential energy coupled with forces acting on the neutrino in the problem under consideration may arise. These forces are caused by the presence of matter particles (electrons, neutrons, protons) and a consistent theory demands a quantum-field description of the whole system. In such a fundamental description the neutrino interaction with matter does not represent a direct process: it is realized by a quantized electroweak field, that is, it happens owing to exchanges of the electroweak interaction carriers. And we want to work within a phenomenological theory. We are interested in only the motion of the single particle, the neutrino, the action of matter particles being reasonably described with the help of an effective potential U_{eff}. We are already faced with the effective potential idea before. Recall the introduction of a refractive index in classical optics. It is well known, that in microscopic scale, the glass that consists of atoms is not an homogeneous medium. Describing the propagation of a light wave, a photon, within the fundamental theory, we would consider the light wave interaction with all atoms of the glass. However, if one is constrained by a phenomenological description of propagating a light through the glass, then the summary effect of elementary interactions could be changed by some effective refractive index. In nonrelativistic quantum mechanics, the particle evolution description is also based on the effective potential idea. However, the description of electromagnetic properties of a solid using the refractive index has its limitations. Analogously, interactions between elementary particles may not always be described by a potential function. It is possible only in the description of interactions with nonrelativistic particles providing the validity of the particle number conservation law.

The procedure of introducing the potential energy describing the neutrino interaction is as follows. First, the neutrino interaction with single electron, neutron and proton is considered. Then, the average over all matter particles is fulfilled.

In Fig. 16.1 the Feynman diagram for the process of elastic scattering of the electron neutrino from the electron due to the charged (weak) current

$$\nu_e + e^- \rightarrow W^{-*} \rightarrow \nu_e + e^- \tag{16.1}$$

is shown. The matrix element corresponding to this diagram will look like:

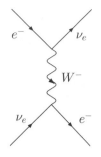

FIGURE 16.1
The Feynman diagram describing the process $\nu_e e^- \to W^{-*} \to \nu_e e^-$.

$$\mathcal{A}^{(a)} = \frac{g_L^2}{8} [\bar{e}(p_1)\gamma^\lambda(1 + \gamma_5)\nu_e(p_2)] \frac{g_{\lambda\sigma} - (p_2 - p_1)_\lambda (p_2 - p_1)_\sigma / m_W^2}{(p_2 - p_1)^2 - m_W^2} [\bar{\nu}_e(p_3)\gamma^\sigma(1 + \gamma_5)e(p_4)].$$
(16.2)

The contribution from the second term in the W boson propagator is proportional to the electron mass and may be neglected. We shall also assume that

$$(p_2 - p_1)^2 \ll m_W^2$$

takes place. Then, taking into account the relation

$$\frac{G_F}{\sqrt{2}} = \frac{g^2}{8m_W^2},$$

we can reduce the interaction with the W boson to the effective Lagrangian

$$\mathcal{L}_{eff} = -\frac{G_F}{\sqrt{2}} [\bar{e}(p_1)\gamma^\lambda(1 + \gamma_5)\nu_e(p_2)][\bar{\nu}_e(p_3)\gamma_\lambda(1 + \gamma_5)e(p_4)] =$$

$$= -\frac{G_F}{\sqrt{2}} [\bar{e}(p_1)\gamma^\lambda(1 + \gamma_5)e(p_4)][\bar{\nu}_e(p_3)\gamma_\lambda(1 + \gamma_5)\nu_e(p_2)],$$
(16.3)

where in the last line we have used the Fierz transformation. We shall be interested in the elastic neutrino forward-scattering. Then, neglecting recoil momenta of matter electrons, we get

$$p_2 = p_3 = p.$$

Consequently, the neutrino behaves in such a way as if no scattering existed. Now, according to our strategy, we should choose a particular volume V and carry out the average over all the electron states. Making use of the nonrelativistic expansion of the Dirac spinors

$$e(x) = \begin{pmatrix} \varphi(x) \\ -i(\nabla \cdot \boldsymbol{\sigma})\varphi(x)/m_e \end{pmatrix},$$
(16.4)

it is easy to make certain that the following formulae

$$\int \bar{e}(x)\gamma^\lambda\gamma_5 e(x)dV = \begin{cases} < s_k >, & \lambda = k \\ 0, & \lambda = 4 \end{cases}$$

$$\int \bar{e}(x)\gamma^k e(x)dV = <v_k>,$$

$$\int \bar{e}(x)\gamma^0 e(x)dV = N_e,$$

where the symbol $<>$ has been used to denote the average values of the spin projections and the velocity while N_e defines the electron density in the volume V, are true. We shall be constrained by the case of unpolarized matter to provide $<s>= 0$ and assume that $<\mathbf{v}>= 0$. Then, the ν_e neutrino interaction with matter electrons owing to charged currents will be described by the effective interaction Lagrangian

$$\mathcal{L}^c_{eff} = -\nu^\dagger_{eL}(x)U^c\nu_{eL}(x), \tag{16.5}$$

where $U^c = \sqrt{2}G_F N_e$.

Now we investigate the contributions caused by processes with the Z boson exchange. Neutral (weak) currents induce the neutrino interactions both with matter electrons and with matter nucleons. The diagram associated with neutrino scattering by the electron

$$\nu_e \mid e^- \quad \rangle \ Z^* \quad \rangle \ \nu_e \mid e^- \tag{16.6}$$

is displayed in Fig. 16.2. As for processes of neutrino scattering from nucleons, then in the

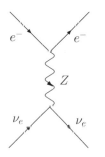

FIGURE 16.2
The Feynman diagram corresponding to the process $\nu_e e^- \to Z^* \to \nu_e e^-$.

fundamental theory we should take into account the composite structure of the nucleons, that is, we should consider the subprocesses

$$\nu_e + q \to Z^* \to \nu_e + q, \qquad (q = u, d). \tag{16.7}$$

and only then, using the quark distribution function in the proton and neutron, proceed to the analysis of the reactions

$$\nu_e + N \to \nu_e + N, \qquad (N = p, n). \tag{16.8}$$

In what follows, we are going to use our results for solar neutrinos whose maximum energy does not exceed 18.8 MeV. de Broglie wavelengths of these neutrinos are merely 10^{-11} cm, which is two orders of magnitude greater than the nucleon size. The inclusion of nucleon motions does not change the state of affairs, since nucleons in the solar matter represent nonrelativistic objects. Therefore, at such conditions, neutrinos cannot see quarks into nucleons and we have the right to pass on from quark diagrams to diagrams of the processes (16.8).

Attention is drawn to the fact that neutral currents cause identical interactions of all kinds of neutrinos with matter particles. Carrying out the Fierz transformation and averaging over matter particles, we obtain the effective interaction Lagrangian of the neutral currents

$$\mathcal{L}^n_{eff} = -\sqrt{2}G_F[\sum_f N_f(S^W_3(f_L) - 2\sin^2\theta_W Q^f][\sum_l \nu^\dagger_{lL}(x)\nu_{lL}(x)], \qquad (16.9)$$

where $f = e^-, p, n$. Because

$$S^W_3(e_L) = S^W_3(n_L) = -S^W_3(p_L) = -\frac{1}{2},$$

then in a neutral matter ($N_e = N_p$), the quantity $\mathcal{L}^n_{eff}$ takes the view

$$\mathcal{L}^n_{eff} = \sum_l \nu^\dagger_{lL}(x)U^n\nu_{lL}(x), \qquad (16.10)$$

where

$$U^n = \frac{1}{\sqrt{2}}G_F N_n.$$

Therefore, the interaction of the neutrino beam with matter is described by the interaction Lagrangian

$$\mathcal{L}_{int} = \mathcal{L}^c_{eff} + \mathcal{L}^n_{eff} = -\sum_l \nu^\dagger_{lL}(x)V_{\nu_l}\nu_{lL}(x), \qquad (16.11)$$

where

$$V_{\nu_e} = \sqrt{2}G_F\left(N_e - \frac{1}{2}N_n\right), \qquad V_{\nu_\mu} = V_{\nu_\tau} = -\frac{1}{\sqrt{2}}G_F N_n.$$

Next, we pass on to the flavor basis and assume $e\mu$-mixing. Employing the expression for the free and interaction Lagrangians and repeating the procedure of deducing the neutrino evolution equation in the free and one-dimensional case, we arrive at the equation

$$i\frac{d}{dz}\nu_l(z) = H(z)\nu_l(z), \qquad (16.12)$$

with the Hamiltonian

$$H(z) = E + \frac{m_1^2 + m_2^2}{4E} - \frac{1}{\sqrt{2}}G_F N_n(z) + \frac{\tilde{M}^2}{2E}, \qquad (16.13)$$

where

$$\tilde{M}^2 = \frac{1}{2}\begin{pmatrix} -\Delta m^2\cos 2\theta_0 + 2A & \Delta m^2\sin 2\theta_0 \\ \Delta m^2\sin 2\theta_0 & \Delta m^2\cos 2\theta_0 \end{pmatrix}$$

and we have entered the quantity $A = 2\sqrt{2}G_F EN_e$. When obtaining Eq. (16.13), one has taken into account that in the most general case the matter density may be changed under transition from one volume to another when the average over the neutrino track is fulfilled.

The effective mixing angle of the neutrino in matter is

$$\tan 2\theta_m = \frac{2H_{12}}{H_{22} - H_{11}} = \frac{\Delta m^2\sin 2\theta_0}{\Delta m^2\cos 2\theta_0 - A}, \qquad (16.14)$$

The behavior character of θ_m becomes more evident when we rewrite the relation (16.14) in the form

$$\sin^2 2\theta_m = \frac{(\Delta m^2)^2 \sin^2 2\theta_0}{(\Delta m^2 \cos 2\theta_0 - A)^2 + (\Delta m^2)^2 \sin^2 2\theta_0}. \tag{16.15}$$

From (16.15) it immediately follows, in matter with variable electron density the dependence of θ_m on N_e has a resonance character. At

$$A = \Delta m^2 \cos 2\theta_0 \tag{16.16}$$

the mixing angle in matter reaches its maximum value $\pi/4$. To deeper realize consequences of such a behavior, we assume the neutrino flux motion to occur in a matter with a constant electron density. We continue to simplify the situation supposing that we are dealing with the time-averaged oscillations. Then, the $\nu_e \to \nu_\mu$ transition probability is given by the expression

$$\mathcal{P}_{\nu_e \to \nu_\mu} = \frac{1}{2} \sin^2 2\theta_m. \tag{16.17}$$

Because $\mathcal{P}_{\nu_e \to \nu_\mu}$ is essentially the quantity that is proportional to the differential cross section of the reaction $\nu_e e^- \to \nu_\mu e^-$ ($d\sigma_{\nu_e e^- \to \nu_\mu e^-}|_{\theta=0}$, where θ is the neutrino mixing angle), then at the resonance, that is, when the condition (16.16) is satisfied, its behavior is described by the Breit–Wigner formula

$$\mathcal{P}_{\nu_e \to \nu_\mu} = \frac{\text{const}}{(N_e(z) - N_R)^2 + \Gamma^2}, \tag{16.18}$$

where N_R is the density of matter electrons at the resonance point and $\Gamma = \delta N_e$ is the resonance width that is equal to $N_R \tan 2\theta_0$ in the case in question. So, when moving the neutrino flux in matter with a variable density, the sharp increase of the transition probability between the neutrino states with different flavor states takes place. The effect of the resonance increase of oscillations in matter was predicted by Wolfenstein [12], Mikheyev and Smirnov [13] and received the name Mikheyev–Smirnov–Wolfenstein (MSW) effect.

In matter the linkage between the physical and flavor states is defined by the same expressions as in a vacuum but with the replacement $\theta_0 \to \theta_m$

$$\left. \begin{aligned} \tilde{\nu}_1 &= \nu_e \cos \theta_m + \nu_\mu \sin \theta_m, \\ \tilde{\nu}_2 &= -\nu_e \sin \theta_m + \nu_\mu \cos \theta_m \end{aligned} \right\}. \tag{16.19}$$

It is clear that the case $\theta_0 \sim \pi/2$ should be excluded. Really, if it would be true, then in a vacuum $\nu_1 \approx \nu_\mu$ and $\nu_2 \approx \nu_e$, but at entering into matter, because of nearness of θ_m to zero, $\tilde{\nu}_1$ become the state consisting mainly of ν_e while $\tilde{\nu}_2 \simeq \nu_\mu$. However, this result remains valid for matter with the vanishingly small density where no reasons for the sharp change of the neutrino flux composition exist. So, it is reasonable to believe the mixing angles to be confined in the interval from 0 to $\pi/4$.

Let us look at the formula (16.15) through the eyes of an experimentalist. Rewrite the quantity N_R

$$N_R = \frac{\Delta m^2 \cos 2\theta_0}{2\sqrt{2} E G_F}, \tag{16.20}$$

in ordinary units

$$\left(\rho(\text{g/cm}^3) n_e \right)_R = 7.07 \times 10^6 \cos 2\theta_0 \frac{\Delta m^2(\text{eV}^2)}{E(\text{MeV})}, \tag{16.21}$$

where ρ is the matter density and n_e is the number of matter electrons per atomic mass unit (a.m.u.). In the central part of the Sun $\rho \sim 150$ g/cm^3 and monotonically goes down to the periphery reaching, for example, the value ~ 15 g/cm^3 at the distance $R_\odot/3$ ($R_\odot$ is the solar radius). Under transition from the solar center to the solar surface, the quantity n_e varies from 0.7 to 0.9. Then, if Δm^2 has the order of 10^{-4} eV^2, at the solar conditions where the neutrino energies reach the values 18.8 MeV, the MSW effect will be possible (for $E = 10$ MeV it will occur providing $\rho n_e \sim 100$). For the Earth, n_e has the same order $\sim 1/2$ while the matter densities are much less, from ~ 13 g/cm^3 at the center to ~ 3g/cm^3 at the surface. Now, for the MSW effect to appear at the same value of Δm^2, the neutrino must have the energy on an order of magnitude higher than in the previous case. However, for this effect to be reliably detected we must know exactly not only the solar neutrino spectrum but the localization of the neutrino birth region as well. Most likely, in the latter case we could never get rid of sizable uncertainties. It is clear, that so-called *long-baseline* oscillation experiments are an ideal tool for the MSW effect observation. Here, a neutrino flux produced at an accelerator propagates through the Earth to be detected by an underground detector at a distance L. Varying the neutrino energy within a wide interval, we have 100% chance of the MSW effect observation.

Return to the analysis of Eq. (16.12). The transition from ν_e to ν_μ is nothing more nor less than the transition between two energy levels. Solving the eigenvalues equation for the operator $H(z)$, we obtain the effective energy values for the states in the physical basis

$$\tilde{E}_i = E - \frac{1}{\sqrt{2}}G_F N_n(z) + \frac{\tilde{m}_i^2}{2E}, \tag{16.22}$$

$$\tilde{m}_{1,2}^2 = \frac{1}{2}[m_1^2 + m_2^2 + A \mp \sqrt{(\Delta m^2 \cos 2\theta_0 - A)^2 + (\Delta m^2)^2 \sin^2 2\theta_0}]. \tag{16.23}$$

The effective neutrino energies in matter can be represented as a function of the parameter $\varepsilon = A/(\Delta m^2 \cos 2\theta_0)$. For the sake of simplicity, we are limited by the case when the vacuum mixing angle θ_0 is small. Now, in a vacuum, $\tilde{\nu}_1$ may be considered as an electron neutrino while $\tilde{\nu}_2$ is considered as a muon neutrino. In Fig. 16.3 we present $E_{1,2}(\varepsilon)$ (solid lines) for $\theta_0 = 0.1$. This figure also shows the average energy values of the ν_e and ν_μ states

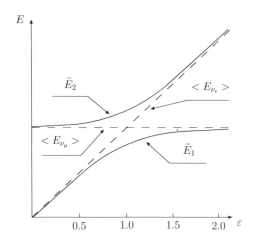

FIGURE 16.3
The effective energies of neutrinos in matter. The scale of the vertical axis is arbitrary.

($< E_{\nu_e}(\varepsilon) >$ and $< E_{\nu_\mu}(\varepsilon) >$) (dashed lines); those are nothing more nor less than the diagonal elements of the matrix $H(z)$.[*] As we can see, at the resonance point $< E_{\nu_e}(\varepsilon) >$ and $< E_{\nu_\mu}(\varepsilon) >$ are intersected, that is, a well-known level crossing phenomenon of a two-level problem of quantum mechanics takes place. From Fig. 16.3 it is easily seen that $\tilde{E}_1(\varepsilon)$ ($\tilde{E}_2(\varepsilon)$) is close to $< E_{\nu_e}(\varepsilon) >$ ($< E_{\nu_\mu}(\varepsilon) >$) for lower densities, whereas it is close to $< E_{\nu_\mu}(\varepsilon) >$ ($< E_{\nu_e}(\varepsilon) >$) for high densities. As should be expected, the analysis of the relations (16.19) defining the states mixing in a medium gives the same result. Indeed, if $N_e \to 0$, then $\theta_m \to \theta_0$ and $\tilde{\nu}_1 \simeq \nu_e$ while $\tilde{\nu}_2 \simeq \nu_\mu$. On the other hand, when $N_e \to \infty$, then $\theta_m \to \pi/2$ and $\tilde{\nu}_1 \simeq \nu_\mu$ while $\tilde{\nu}_2 \simeq \nu_e$. Note that the oscillation length in a medium

$$l_m = \frac{4\pi E}{\tilde{m}_2^2 - \tilde{m}_1^2} = \frac{4\pi E}{\Delta m^2} \frac{1}{\sqrt{(\cos 2\theta_0 - A/\Delta m^2)^2 + \sin^2 2\theta_0}} \tag{16.24}$$

is decreased as compared with that in a vacuum.

To obtain the transition probability we should address the evolution equation for the flavor states. Passing on from the matrix equation to the two equations system, excluding $\nu_e(z)$ and allowing for that only $\tilde{H}_{11}$ depends on z, we get

$$\frac{d^2\nu_\mu}{dz^2} + i(\tilde{H}_{11} + \tilde{H}_{22})\frac{d\nu_\mu}{dz} + (\tilde{H}_{12}^2 - \tilde{H}_{11}\tilde{H}_{22})\nu_\mu(z) = 0. \tag{16.25}$$

Eq. (16.25) can be considerably simplified using the new function

$$\psi_\mu(z) = [\exp\left(-i\int_0^z \tilde{H}_{22}(z')dz'\right)]\nu_\mu(z)$$

which is distinguished only by the phase from $\nu_\mu(z)$ to exert no influence on the form of the expressions for the probabilities. After this procedure, Eq. (16.25) will look like:

$$[\frac{d^2}{dz^2} + i(\tilde{H}_{11} - \tilde{H}_{22})\frac{d}{dz} + \tilde{H}_{12}^2]\psi_\mu(z) = 0. \tag{16.26}$$

Now, we should substitute the expression for $N_e(z)$ into $\tilde{H}_{11}$ and, using the boundary conditions

$$|\nu_e(-\infty)| = 1, \qquad \nu_\mu(-\infty) = 0,$$

solve the equation obtained. The required transition probability for the electron neutrino is defined as $|\nu_\mu(z)|^2$. However, to analyze the neutrino flux behavior in a medium we should have an analytical solution without concrete definition of the $N_e(z)$ form. It is evident that only approximate methods of solving the equations in quantum theory could help us to achieve this goal.

16.2   Adiabatic approximation

Omitting the quantity proportional to the unit matrix in the right side of Eq. (16.12), we arrive at

$$i\frac{d}{dz}\nu_l(z) = \frac{\tilde{M}^2}{2E}\nu_l(z). \tag{16.27}$$

---

[*]To make sure of this, we should define the average value of the Hamiltonian operator over macroscopic volumes (which are so small that N_e and N_n may be considered as constants) along the motion of the neutrino flux.

Carry out the transition to the physical basis in this equation and take into account the fact that θ_m depends on the coordinate. It results in

$$i\frac{d}{dz}\tilde{\nu}_i(z) = \left(\frac{1}{2E}\tilde{U}^\dagger \tilde{M}^2 \tilde{U} - i\tilde{U}^\dagger \frac{d}{dz}\tilde{U}\right)\nu_i(z), \qquad (16.28)$$

where

$$\tilde{U} = \begin{pmatrix} \cos\theta_m & \sin\theta_m \\ -\sin\theta_m & \cos\theta_m \end{pmatrix}.$$

The first term in the parentheses of the right side of Eq. (16.28) gives the quantity $\mathrm{diag}(\tilde{m}_1^2, \tilde{m}_2^2)/2E$. Then, making use of the explicit form of the matrix $\tilde{U}$, we get

$$i\frac{d}{dz}\begin{pmatrix} \tilde{\nu}_1(z) \\ \tilde{\nu}_1(z) \end{pmatrix} = \begin{pmatrix} \dfrac{\tilde{m}_1^2 - \tilde{m}_2^2}{4E} & i\dfrac{d\theta_m}{dz} \\ -i\dfrac{d\theta_m}{dz} & \dfrac{\tilde{m}_2^2 - \tilde{m}_1^2}{4E} \end{pmatrix}\begin{pmatrix} \tilde{\nu}_1(z) \\ \tilde{\nu}_1(z) \end{pmatrix}, \qquad (16.29)$$

where we, following our tradition, have omitted the quantity

$$\frac{\tilde{m}_1^2 + \tilde{m}_2^2}{4E}I.$$

It is evident that we can easily solve Eq. (16.29) when the conditions of the problem will be such that the nondiagonal elements of the Hamiltonian may be neglected as compared with the diagonal ones. This will be possible providing

$$\left|\frac{d\theta_m}{dz}\right| \ll \frac{|\tilde{m}_2^2 - \tilde{m}_1^2|}{4E}. \qquad (16.30)$$

The obtained inequality demands the mixing angle in a medium to be slowly varying parameter. Because this variation is caused by the neutrino interaction with a medium, then the interaction has the character of the adiabatic perturbation. So, all indicate that the condition (16.30) proves to be fulfilled in the adiabatic case. Let us check whether such is the case.

The perturbation may be considered as adiabatic when its action on a system results in the transition between states with energies E_1 and E_2 during a time δt providing

$$\delta E \delta t \gg 1, \qquad (16.31)$$

where $\delta E = (\tilde{E}_2 - \tilde{E}_1)_{min}$. The difference $\tilde{E}_2 - \tilde{E}_1$ reaches its minimum at the resonance. Therefore, if the condition (16.31) is fulfilled in the resonance vicinity, then it will be true in all other regions. From the expression (16.22) we find

$$\delta E = \frac{\Delta m^2}{2E}\sin 2\theta_0. \qquad (16.32)$$

The transition time could be presented in the following way

$$\delta t = \delta z = \left(\frac{1}{N_e}\frac{dN_e}{dz}\right)^{-1}\frac{\delta N_e}{N_e} = \left(\frac{1}{N_e}\frac{dN_e}{dz}\right)^{-1}\frac{\Gamma}{N_e}. \qquad (16.33)$$

Assuming $\Gamma \approx (N_e)_R \tan 2\theta_0$, where

$$(N_e)_R = \frac{\Delta m^2 \cos 2\theta_0}{2\sqrt{2}G_F E},$$

we obtain the adiabaticity condition in the resonance region

$$1 \ll \frac{\Delta m^2 \sin^2 2\theta_0}{2E \cos 2\theta_0} \left(\left| \frac{d}{dz} \ln N_e \right| \right)_R^{-1}. \tag{16.34}$$

Let us check whether the smallness condition of the nondiagonal Hamiltonian elements in Eq. (16.29) gives the same result. Using the expression (16.14), we find

$$\left| \frac{d\theta_m}{dz} \right| = \frac{\sqrt{2} G_F E \Delta m^2 \sin 2\theta_0}{(\Delta m^2 \cos 2\theta_0 - A)^2 + (\Delta m^2)^2 \sin^2 2\theta_0} \frac{dN_e}{dz}. \tag{16.35}$$

With allowance made for the relation (16.35) and the definitions of $\tilde{m}_{1,2}^2$, the inequality (16.30) could be presented in the view

$$\left| \frac{dN_e}{dz} \right| \ll \frac{[(\Delta m^2 \cos 2\theta_0 - A)^2 + (\Delta m^2)^2 \sin^2 2\theta_0]^{3/2}}{4\sqrt{2} G_F E^2 \Delta m^2 \sin 2\theta_0}, \tag{16.36}$$

or, introducing the parameter $\gamma(z)$ as

$$1 \ll \gamma(z), \tag{16.37}$$

where

$$\gamma(z) = \frac{[(\Delta m^2 \cos 2\theta_0 - A)^2 + (\Delta m^2)^2 \sin^2 2\theta_0]^{3/2}}{4\sqrt{2} G_F E^2 \Delta m^2 \sin 2\theta_0} \left(\left| \frac{dN_e}{dz} \right| \right)^{-1}$$

$$= \frac{(\Delta m^2)^2 \sin^2 2\theta_0}{4\sqrt{2} G_F E^2 \sin^3 2\theta_m} \left(\left| \frac{dN_e}{dz} \right| \right)^{-1}. \tag{16.38}$$

At the resonance the quantity $\gamma(z)$ is defined by the relation

$$\gamma_R = \frac{\Delta m^2 \sin^2 2\theta_0}{2E \cos 2\theta_0} \left(\left| \frac{d}{dz} \ln N_e \right| \right)^1. \tag{16.39}$$

Therefore, at the resonance, the adiabaticity condition is nothing more nor less than the condition at which the nondiagonal Hamiltonian elements are much less than the diagonal ones. Obviously, the quantity $\gamma(z)$ makes sense of the *adiabaticity parameter*.

If the electron density is very high at some point, then $\theta_m \to \pi/2$ and $\gamma \gg 1$. If matter density is vanishing somewhere, then $\theta_m \to \theta_0$, so that once again the adiabaticity parameter is large. However, in the resonance vicinity where $\theta_m \to \pi/4$ the quantity γ_R may take any value.

Solving the evolution equation (16.29) in the adiabatic approximation and turning to the flavor basis, we obtain

$$\nu_e(z)^{ad} = \nu_1(0)\{\exp[i \int_0^z \tilde{E}_1(z')dz']\}\cos\theta_s + \nu_2(0)\{\exp[i \int_0^z \tilde{E}_2(z')dz']\}\sin\theta_s, \tag{16.40}$$

$$\nu_\mu(z)^{ad} = -\nu_1(0)\{\exp[i \int_0^z \tilde{E}_1(z')dz']\}\sin\theta_s + \nu_2(0)\{\exp[i \int_0^z \tilde{E}_2(z')dz']\}\cos\theta_s, \tag{16.41}$$

where we have assumed the neutrino to be born in a matter and the subscript s denotes the mixing angle in the birth point. In the relations (16.40) and (16.41) it has been also allowed for, when one takes into account the quantities proportional to I, which were omitted before, then the diagonal Hamiltonian elements present the energy eigenvalues. Using the expressions for the neutrino wave functions, we can determine the required probabilities with no difficulty. Thus, if the distance from the source generating the neutrino flux to the detector is equal to L, then the survival probability for the electron neutrino will look like:

$$\mathcal{P}_{\nu_e\nu_e}^{ad}(L) = \left| \exp\left[i \int_0^L \tilde{E}_1(z')dz'\right] \cos\theta_s \cos\theta_D + \exp\left[i \int_0^L \tilde{E}_2(z')dz'\right] \sin\theta_s \sin\theta_D \right|^2 =$$

$$= \frac{1}{2}\{1 + \cos 2\theta_s \cos 2\theta_D + \sin 2\theta_s \sin 2\theta_D \cos\left[\int_0^L (\tilde{E}_2(z') - \tilde{E}_1(z'))dz'\right]\}, \tag{16.42}$$

where θ_D is the mixing angle in the region of the detector localization. From the perturbation adiabaticity condition it follows that the cosine argument in the third term, which contains the energy integral, is not small. Therefore, under the average, this oscillating term may be neglected to give us the well-known expression for the time-averaged survival probability of ν_e.

16.3 Nonadiabatic effects

Because in the nonadiabatic case the nondiagonal Hamiltonian elements in the physical basis are nonzero, then the transitions between the $\tilde{\nu}_1$ and $\tilde{\nu}_2$ states with the probability W_{12} (level-crossing probability) will be possible. First, we find changes in the expression for the survival probability of ν_e caused by the inclusion of nonadiabatic effects and then define the quantity W_{12}.

Suppose a ν_e produced at the point z_s goes through the resonance at the point z_r, to be detected in a vacuum at the point z_d. At the birth moment, the wave function of the system is

$$\psi(z_s) = \tilde{\nu}_1(z_s)\cos\theta_s + \tilde{\nu}_2(z_s)\sin\theta_s, \tag{16.43}$$

where θ_s is the mixing angle in the point z_s. At $z_r - \delta z$, this function is now determined by the expression

$$\psi(z_r - \delta z) = \tilde{\nu}_1(z_r)\cos\theta_s \exp\left[i \int_{z_s}^{z_r} \tilde{E}_1(z')dz'\right] + \tilde{\nu}_2(z_r)\sin\theta_s \exp\left[i \int_{z_s}^{z_r} \tilde{E}_2(z')dz'\right]. \tag{16.44}$$

At the point $z \simeq z_r$, the $\tilde{\nu}_1 \leftrightarrow \tilde{\nu}_2$ transition is possible, that is,

$$\left.\begin{array}{l} \tilde{\nu}_1(z_r) \to \alpha\tilde{\nu}_1(z_r) + \beta\tilde{\nu}_2(z_r) \\ \tilde{\nu}_2(z_r) \to -\beta^*\tilde{\nu}_1(z_r) + \alpha^*\tilde{\nu}_2(z_r), \end{array}\right\} \tag{16.45}$$

where the normalization condition gives

$$\mid \alpha \mid^2 + \mid \beta \mid^2 = 1.$$

Therefore, after passing through the resonance point, the wave function will look like:

$$\psi(z_r + \delta z) = [\alpha\tilde{\nu}_1(z_r) + \beta\tilde{\nu}_2(z_r)]\cos\theta_s \exp\left[i \int_{z_s}^{z_r} \tilde{E}_1(z')dz'\right] + [-\beta^*\tilde{\nu}_1(z_r)+$$

$$+\alpha^*\tilde{\nu}_2(z_r)]\sin\theta_s \exp\left[i \int_{z_s}^{z_r} \tilde{E}_2(z')dz'\right] \equiv A(z_s, z_r)\tilde{\nu}_1(z_r) + B(z_s, z_r)\tilde{\nu}_2(z_r). \tag{16.46}$$

When $z > z_r$, then the neutrino beam propagation is described in the following way

$$\psi(z) = A(z_s, z_r)\tilde{\nu}_1(z) \exp\left[i\int_{z_r}^z \tilde{F}_1 dz'\right] + B(z_s, z_r)\tilde{\nu}_2(z) \exp\left[i\int_{z_r}^z \tilde{E}_2 dz'\right]. \tag{16.47}$$

Having gone a distance L to a detector, the wave function takes the form

$$\nu_e(L) = A(z_s, z_r)[\nu_e \cos\theta_0 - \nu_\mu \sin\theta_0] \exp\left[i\int_{z_r}^L \tilde{E}_1 dz'\right]+$$

$$+B(z_s, z_r)[\nu_e \sin\theta_0 + \nu_\mu \cos\theta_0] \exp\left[i\int_{z_r}^L \tilde{E}_2 dz'\right]. \tag{16.48}$$

Then, the survival amplitude for the electron neutrino will be defined as

$$\mathcal{A}_{\nu_e\nu_e} = A(z_s, z_r) \cos\theta_0 \exp\left[i\int_{z_r}^L \tilde{E}_1(z') dz'\right] + B(z_s, z_r) \sin\theta_0 \exp\left[i\int_{z_r}^L \tilde{E}_2(z') dz'\right]. \tag{16.49}$$

Using the expression obtained, we find the corresponding probability

$$\mathcal{P}_{\nu_e\nu_e} =\mid A(z_s, z_r) \mid^2 \cos^2\theta_0+ \mid B(z_s, z_r) \mid^2 \sin^2\theta_0 + 2\mid A(z_s, z_r)B(z_s, z_r) \mid \times$$

$$\times \sin\theta_0 \cos\theta_0 \cos\left\{\int_{z_r}^L [\tilde{E}_1(z') - \tilde{E}_2(z')] dz' + \arg[A^*(z_s, z_r)B(z_s, z_r)]\right\}. \tag{16.50}$$

When averaging over the detector location, the last term in (16.50) vanishes and we have

$$\mathcal{P}_{\nu_e\nu_e} =\mid A(z_s, z_r) \mid^2 \cos^2\theta_0+ \mid B(z_s, z_r) \mid^2 \sin^2\theta_0 = \frac{1}{2} + \frac{1}{2}\left(\mid\alpha\mid^2 - \mid\beta\mid^2\right) \cos 2\theta_s \cos 2\theta_0-$$

$$- \mid\alpha\beta\mid \sin 2\theta_s \cos 2\theta_0 \cos\left[\int_{z_s}^{z_r} [\tilde{E}_1(z') - \tilde{E}_2(z')] dz' + \arg(\alpha^*\beta)\right]. \tag{16.51}$$

Now, we should fulfill the average over the neutrino birth region as well. In so doing, the last term in (16.51) turns into zero and we obtain the final expression for the survival probability of the electron neutrino in the general case

$$\mathcal{P}_{\nu_e\nu_e} = \frac{1}{2} + \left(\frac{1}{2} - W_{12}\right) \cos 2\theta_s \cos 2\theta_0, \tag{16.52}$$

where we have accounted for $\mid\beta\mid^2 \equiv W_{12}$. In the adiabatic case $W_{12} = 0$ and we come to the well-known expression for $\mathcal{P}_{\nu_e\nu_e}^{ad}$. When $N_e \to \infty$, then $\cos 2\theta_s \to -1$ and we get

$$(\mathcal{P}_{\nu_e\nu_e})_\infty = \sin^2\theta_0 + W_{12}\cos 2\theta_0. \tag{16.53}$$

From (16.53) it is evident that the survival probability of ν_e is equal to the level-crossing probability under small values of θ_0.

Our next task is to find the quantity W_{12}. Since the quasi-classical condition

$$\left|\frac{d\lambda_{\nu_l}}{dz}\right| \ll 1, \tag{16.54}$$

where λ_{ν_l} is the de Broglie wavelength of ν_l, proves to be fulfilled, one may use the quasi-classical approximation. In the general case, the quasi-classical formula for crossing of energy levels was obtained by Landau [14]. Sometimes, in the literature, the method used by Landau is referred to as the *complex trajectory method*. For the solar neutrinos it was applied in Refs. [15]. The neutrino wave functions in the mass eigenstates basis with an accuracy of a pre-exponential factor are

$$\tilde{\nu}_1 \approx \exp[iS_1(z_i, z_0)], \qquad \tilde{\nu}_2 \approx \exp[iS_2(z_0, z_f)], \tag{16.55}$$

where S_1 and S_2 denote the actions of the neutrinos, z_i and z_f are the points of the entrance and exit from the nonadiabatic region and z_0 is the transition point where crossing the energy levels of the states $\tilde{\nu}_1$ and $\tilde{\nu}_2$ takes place. S_1 and S_2 are complex quantities that reflect the fact of the process impracticability from the point of view of classical physics. In particular, the transition point z_0 happens to be complex. Using (16.55) we understand that the level-crossing probability is governed by the expression

$$W_{12} \approx \exp\left\{-2\mathrm{Im}[S_1(z_i, z_0) + S_2(z_0, z_f)]\right\}. \tag{16.56}$$

Because the points z_i and z_f lie on the real axis, then the imaginary part of the action remains unaffected if we take $z_i = z_f = z_r$. Then (16.56) is rewritten in the form

$$W_{12} \approx \exp\left\{-2\mathrm{Im}\int_{z_r}^{z_0}[\tilde{E}_2(z') - \tilde{E}_1(z']dz'\right\}. \tag{16.57}$$

Passing on to the variable A, we obtain the expression

$$W_{12} \approx \exp\left\{-\frac{1}{E}\mathrm{Im}\left[\int_{A_R}^{A_0} dA\left(\frac{dA}{dz'}\right)^{-1}\sqrt{(\Delta m^2\cos 2\theta_0 - A)^2 + (\Delta m^2)^2\sin^2 2\theta_0}\right]\right\}, \tag{16.58}$$

where $A_R = \Delta m^2\cos 2\theta_0$ and the upper integration limit A_0 represents the value of A in the transition point z_0, that is, this is the value of A_0 for which the two eigenvalues coincide

$$\tilde{E}_1 = \tilde{E}_2. \tag{16.59}$$

However, this equation has no solutions for the real values of A, since, as follows from Fig. 16.4, the energy levels $\tilde{E}_1$ and $\tilde{E}_2$ are not intersected. Substituting the values $\tilde{E}_{1,2}$ into (16.59), we arrive at the equation

$$\sqrt{(\Delta m^2\cos 2\theta_0 - A_0)^2 + (\Delta m^2)^2\sin^2 2\theta_0} = 0, \tag{16.60}$$

which has the following solution

$$A_0 = \Delta m^2\exp\left(\pm 2i\theta_0\right). \tag{16.61}$$

To estimate the integral (16.58) we need to know how $N_e(z)$ behaves as a function of z in the resonance region. Let us consider the case when the variation is linear

$$N_e(z) = N_e(z)|_{z_r} + (z - z_r)\left(\frac{dN_e(z)}{dz}\right)_{z_r}. \tag{16.62}$$

This makes the quantity dA/dz constant and we can take it outside the integral. Inasmuch as W_{12} must be less or equal to 1, then we should take care of the negative sign conservation in the exponent. Then, if the quantity dA/dz is positive (negative), one should take the negative (positive) value of the exponent for A_0. Taking into account the two possibilities and passing on to the new variable

$$y = \frac{A - \Delta m^2\cos 2\theta_0}{\Delta m^2\sin 2\theta_0}$$

we get

$$W_{12} \approx \exp\left\{ -\frac{(\Delta m^2)^2 \sin^2 2\theta_0}{E \mid dA/dz \mid_r} \text{Im}\left[\int_0^i dy \sqrt{1+y^2}\right] \right\} = \exp\left[-\frac{(\Delta m^2)^2 \sin^2 2\theta_0}{E \mid dA/dz \mid_r} \cdot \frac{\pi}{4} \right] =$$

$$= \exp\left[-\frac{\Delta m^2 \sin^2 2\theta_0}{E \cos 2\theta_0} \left(\mid \frac{d}{dz} \ln A \mid_r \right)^{-1} \cdot \frac{\pi}{4} \right] = \exp\left(-\frac{\gamma_R \pi}{4} \right). \tag{16.63}$$

The formula for level-crossing probability in the case of the linear electron density could be obtained directly from the Schrödinger equation solution as it was first done in Ref. [16]. For the sake of simplicity, we assume the resonance to occur in the point $z = 0$. Introducing the new function

$$\psi_\mu(z) = f(z) \exp[-\frac{1}{2} \int_0^z (\tilde{H}_{22} - \tilde{H}_{11}) dz']$$

and new variable

$$y = z \exp[-i\frac{\pi}{4}] \sqrt{\sqrt{2} G_F \dot{N}_e(0)},$$

we come from (16.26) to the equation for the parabolic cylinder function

$$[\frac{d^2}{dy^2} + (n + \frac{1}{2} - \frac{y^2}{4})] f(y) = 0, \tag{16.64}$$

where

$$n = \frac{i\tilde{H}_{12}^2}{\sqrt{\sqrt{2} G_F \dot{N}_e(0)}}, \qquad \dot{N}_e(0) \equiv \left(\frac{dN_e}{dz} \right)_0.$$

From the four solutions of Eq. (16.64)

$$f(y) = D_n(y), \ D_n(-y), \ D_{-n-1}(iy), \ D_{-n-1}(-iy)$$

only $D_{-n-1}(-iz)$ satisfies the initial condition $\nu_\mu(-\infty) = 0$ providing $z = Re^{3\pi i/4}$ (recall, the asymptotic behavior of the function D does not depend on the phase value). Therefore, the sought solution has the view

$$f(y) = CD_{-n-1}(iRe^{3\pi i/4}), \tag{16.65}$$

where C is a normalization constant. To define C we should return to Eq. (16.26) and rewrite it for $\nu_e(z)$. The parabolic cylinder function $D_{-n-1}(iRe^{-\pi i/4})$ will be the solution of this equation. Keeping only the leading term in the asymptotical expansion of the solution, we find

$$D_{-n-1}(iRe^{-\pi i/4}) \sim e^{-i\pi(n+1)/4} e^{-iR^2/4} R^{-n-1}. \tag{16.66}$$

Demanding the fulfillment of (16.66), we arrive at the result

$$C = \sqrt{\kappa} \exp[\frac{-\pi\kappa}{4}], \tag{16.67}$$

where we have set

$$\kappa = -in.$$

Next, in the expansion of the function $D_{-n-1}(iRe^{3\pi i/4})$ at $R \to \infty$ we allow for only the two first dominating terms

$$D_{-n-1}(iRe^{3\pi i/4}) \sim \frac{\sqrt{2\pi}}{\Gamma(n+1)} R^n e^{i\pi n/4} e^{iR^2/4} + R^{-n-1} e^{3i\pi(n+1)/4} e^{-iR^2/4}. \tag{16.68}$$

Then, with allowance made for (16.67) and (16.68), the sought probability is determined by the relation

$$W_{12} \equiv |\nu_\mu(\infty)|^2 = 4\kappa e^{-\pi\kappa/2} \frac{\sqrt{2\pi}}{[\Gamma(i\kappa+1)\Gamma(-i\kappa+1)]^{1/2}} e^{-\pi\kappa/2} = 2\mathrm{sh}(\pi\kappa)e^{-\pi\kappa} = 1 - e^{-2\pi\kappa},$$

(16.69)

where

$$\kappa = \frac{(\Delta m^2)^2 \sin^2 2\theta}{8E\cos 2\theta}\left(\frac{d}{dr}\ln N_e\right)_0^{-1} = \frac{1}{4}\gamma_R.$$

(16.70)

So, we have obtained the same expression as in the semiclassical case (16.63). For this reason, Eq. (16.63) is called the *Landau–Zener formula*. This formula is used when the electron density in the resonance region varies linearly and it is exact providing the mixing angle is small while the adiabaticity parameter is not infinitesimal. The Landau–Zener formula is considered to give the needed accuracy in the description of the solar neutrino behavior.

The usage of the quasi-classical approximation when calculating the level-crossing probability under the exponential dependence of the density of the matter electrons leads to the expression [17]

$$W_{12} \approx \exp\left[-\frac{\gamma_R\pi}{4}(1-\tan^2\theta_0)\right].$$

(16.71)

We see, at small θ_0, the obtained expression only slightly differs from the case when $N_e(z)$ varies linearly.

The method described above is a semiclassical one, and gives the leading term in the limit of large values of the exponent. When $\gamma \ll 1$ the formula (16.63) is unsatisfactory. To understand this we address the simple example [18]. Consider a neutrino beam travelling through the boundary of a uniform medium into the vacuum. The propagation is obviously adiabatic in the medium as well as in the vacuum, since both have uniform density. However, at the boundary, there is an abrupt density change so that

$$\frac{dN_e}{dz} \to \infty,$$

(16.72)

that is, the situation is highly nonadiabatic. In order to distinguish the wave neutrino functions in the medium and vacuum we shall denote the latter by the tilde. Introduce into consideration the following points as well: two points in the medium (vacuum), where the former x (y) is a point deep inside the medium (vacuum) and the latter x' (y') is a point just inside the medium (vacuum) on the boundary plane. Then the level-crossing probability W_{12} may be represented in the form

$$W_{12} \equiv |<\nu_2(y)|\tilde{\nu}_1(x)>|^2 = |\sum_l <\nu_2(y)|\nu_2(y')><\nu_2(y')|\nu_l(y')> \times$$

$$\times <\nu_l(y')|\nu_l(x')><\nu_l(x')|\tilde{\nu}_1(x')><\tilde{\nu}_1(x')|\tilde{\nu}_1(x)>|^2.$$

(16.73)

Since the flavor states are continuous across the boundary, then

$$<\nu_l(y')|\nu_l(x')>=1.$$

(16.74)

Using the mixing matrix in the medium and in vacuum and neglecting all interference terms, we get

$$W_{12} = \sin^2(\tilde{\theta} - \theta_0).$$

(16.75)

If the medium is very dense $\tilde{\theta} \to \pi/2$, then we have

$$W_{12} = \cos^2 \theta_0. \tag{16.76}$$

However, from Eq. (16.63) it follows that

$$W_{12} \to 1 \qquad \gamma_R \to 0. \tag{16.77}$$

The difference between (16.76) and (16.77) becomes significant if the mixing angle θ is not small. As a matter of principle, this result is quite forthcoming because the formula (16.63) is approximate.

The exact solutions of Eq. (16.26) for $N_e \sim z$, $N_e \sim \exp[cz]$ and $N_e \sim z^{-1}$ were found in Refs. [18, 19, 20]. All these solutions have the general form

$$W_{12} = \frac{\exp[-\gamma_R F] - \exp[-\gamma_R F/\sin^2 \theta_0]}{1 - \exp[-\gamma_R F/\sin^2 \theta_0]}, \tag{16.78}$$

where F is calculated with the help of the quasi-classical approximation. When $\gamma_R \to 0$, this formula gives the result coinciding with the expression (16.76). One may assume [18] that (16.78) works irrespective of the nature of the density variation.

Before summing the results obtained in the two flavor-approximation, we take into consideration one more possible situation. Assume the neutrino to be produced in the far half of the Sun. There is a certain probability that it will travel towards the Earth. If it does, the neutrino will pass through the core region. When it is created in the region where the density is lower than the resonance density, it crosses the resonance region once while coming towards the core. After that, the neutrino crosses the resonance region again on its way out. Let us find the survival probability for the electron neutrino in the case in question. Let us assume that the propagation is adiabatic, that is, the resonance transition $\nu_e \to \nu_\mu$ is absent. The corresponding probability is denoted by $\mathcal{P}^{ad}_{\nu_e \nu_e}$. In the nonadiabatic case, the probability that the electron neutrino remains the electron neutrino as before is

$$W' = \mathcal{P}^{ad}_{\nu_e \nu_e} w_{\nu_e \nu_e}, \tag{16.79}$$

where $w_{\nu_e \nu_e}$ is the transition absence probability in passing the resonances. In the adiabatic case, there exists the probability that ν_e can convert into ν_μ $(\mathcal{P}^{ad}_{\nu_e \nu_\mu})$.* Then, owing to the resonance transitions, the muon neutrino can again convert into the electron one. The probability of this event will look like

$$W'' = \mathcal{P}^{ad}_{\nu_e \nu_\mu} w_{\nu_e \nu_\mu}, \tag{3.80}$$

where $w_{\nu_e \nu_\mu}$ is the probability of the transition $\nu_e \to \nu_\mu$ under travelling through the resonance regions. Therefore, the sought probability is defined by the expression

$$\mathcal{P}_{\nu_e \nu_e} = \mathcal{P}^{ad}_{\nu_e \nu_e} w_{\nu_e \nu_e} + \mathcal{P}^{ad}_{\nu_e \nu_\mu} w_{\nu_e \nu_\mu}. \tag{16.81}$$

When the two resonances occur, we have

$$w_{\nu_e \nu_e} = (1 - W_{12})^2 + W_{12}^2, \qquad w_{\nu_e \nu_\mu} = 2W_{12}(1 - W_{12}). \tag{16.82}$$

Substituting (16.82) into (16.81) and allowing for

$$\mathcal{P}^{ad}_{\nu_e \nu_e} = \frac{1}{2}[1 + \cos 2\theta_m \cos 2\theta_0],$$

---

*Sometimes, the transition probability is called the *jumping* or *conversion probability*.

we arrive at the final formula for the survival probability of the electron neutrino

$$\mathcal{P}_{\nu_e\nu_e} = \frac{1}{2}[1 + (1 - 2W_{12})^n \cos 2\theta_m \cos 2\theta_0], \tag{16.83}$$

where n defines the number of the resonances ($n = 1, 2$).

So, for the survival probability of the electron neutrino to be calculated we need the following information. First, the electron density in a medium must be known. Second, one has to define concretely a model of the electroweak interaction because it fixes the value of an interaction potential of a neutrino with a matter (V). The knowledge of V allows us to determine the neutrino mixing angle at the neutrino birth point, the localization region of crossing the energy levels and, at the same time, the value $| d\ln N_e(z)/dz |_r$. Third, one must know the neutrino oscillation parameters, namely, the mixing angles in a vacuum θ_{ij} and the squared mass differences Δm_{ij}^2.

16.4 Neutrino oscillations in the magnetic field

Consider ultrarelativistic neutrinos when they pass through the electromagnetic field. For the sake of simplicity, we shall neglect the neutrino interaction with a medium. If one assumes the field to be weakly changed on a distance $|\mathcal{E}|^{-1}$, where $\mathcal{E}$ is the electric field strength, then the evolution equation in the mass eigenstate basis takes the form

$$i\frac{d}{dz}\nu_j = H\nu_j, \tag{16.84}$$

where the Hamiltonian H is governed by the expression

$$H = \{\frac{m_j^2}{2E} - \mu_{jj'}[P_+\boldsymbol{\sigma}(\mathcal{H} - i\mathcal{E})P_- + P_-\boldsymbol{\sigma}(\mathcal{H} + i\mathcal{E})P_+]\}I \tag{16.85}$$

$\mathcal{H}$ is the magnetic field strength and $P_\pm = (1 \pm \sigma_3)/2$ are the projectors on the states ν_L and ν_R, respectively. Since, each of the elements of the column ν_j is a spinor:

$$\begin{pmatrix} \nu_R \\ \nu_L \end{pmatrix}, \tag{16.86}$$

(the quantization axis is chosen along the neutrino motion direction), then the matrix I has the $2n_f \times 2n_f$-dimension, where n_f is the number of the neutrino generations. Attention is drawn to the fact that thanks to the condition

$$P_\pm\sigma_3 P_\mp = 0,$$

the longitudinal components $\mathcal{H}$ and $\mathcal{E}$ are absent in the Hamiltonian H. The reason is as follows. When passing on to the neutrino rest frame, the longitudinal components of the electromagnetic field remains unchanged while the transversal ones involve the factor $|\mathcal{E}|/m \gg 1$.

Let us go over to the flavor basis and constrain ourselves by the case $n_f = 2$. Then, in the magnetic field the neutrino Hamiltonian takes the form

$$H = \{\frac{\Delta m_{12}^2}{4E}(\sigma_1 \sin 2\theta_0 - \sigma_3 \cos 2\theta_0) - \mu_{ll'}[P_+(\boldsymbol{\sigma}\mathcal{H})P_- + P_-(\boldsymbol{\sigma}\mathcal{H})P_+]\}I. \tag{16.87}$$

From Eq. (16.87) it follows, in the transversal magnetic field $\mathbf{H}_\perp$ the state ν_l undergoes oscillations between the left- and right-handed components with the frequency $\omega(z) = |\mu_{ll}||\mathcal{H}_\perp|$ (spin precession). If we produce a flux of ν_{lL} and let it travel through the magnetic field that does not depend on z, then the number of ν_{lL} and ν_{lR} will be defined as

$$N_{lL}(z) = N_0 \cos^2\left[\int_0^z \omega(z')dz'\right], \qquad N_{lR}(z) = N_0 \sin^2\left[\int_0^z \omega(z')dz'\right]. \qquad (16.88)$$

In the case under consideration, the spin precession occurs for the arbitrarily weak transversal field, since the left- and right-handed neutrino of the same flavor are energy degenerate. The spin flips accompanied by the transitions between states with different flavor are also allowed. The nondiagonal elements of the magnetic moment matrix are responsible for the $\nu_{lL} \leftrightarrow \nu_{l'R}$ transitions. In the two-flavor approximation, these oscillations will have a sizable amplitude providing

$$|\mu_{12}||\mathcal{H}_\perp| \geq \left|\frac{(\Delta m_{12}^2)^2}{2E}\cos 2\theta_0\right|. \qquad (16.89)$$

To estimate the order of magnitude of the quantities in (16.89), we set $E = 10$ MeV, $|\mathcal{H}_\perp| = 10^3$ Gs, $|\mu_{12}| = 10^{-10}$ μ_B and $\cos 2\theta_0 \sim 1$. Then, (16.89) will be true when $|(\Delta m_{12}^2)^2| \leq 10^{-7}$ eV^2. The spin precession phenomenon was predicted in Ref. [21] and derived the name the *Voloshin–Vysotskii–Okun effect*. Note that the magnetic moment for the Dirac neutrino predicted by the SM is proportional to the neutrino mass

$$\mu_\nu = \frac{3eG_F m_\nu}{8\sqrt{2}\pi^2} = 10^{-19}\mu_B\left(\frac{m_\nu}{\text{eV}}\right)$$

and, as a consequence, cannot lead to any observable effects in real fields. In models that contain right-handed charged currents and/or charged Higgs bosons, μ_ν is proportional to the charged lepton mass and may prove to be $10^4 - 10^5$ more. For the Majorana neutrino, owing to the CPT invariance $\mu_\nu = 0$, but the nondiagonal magnetic moments may be nonzero.

16.5   Neutrino oscillations in the left-right model

Now we discuss the neutrino flux evolution within the left-right model (LRM) [22]. The analysis of the neutrino oscillations in the solar matter will be fulfilled for the case when the neutrino transition magnetic dipole moment (NMM) is very small. We shall also neglect the influence of heavy neutrinos (N_l) on the oscillation pattern of light neutrinos. The interaction between the left- and right-handed light neutrinos can arise either from the NMM or from lepton-flavor-violating currents (LFVCs). Since in the LRM the LFVCs do not mix the neutrino states with the different helicity, the evolution equation for the neutrino flux consisting of the left- and right-handed light neutrinos is decoupled on the two independent systems, which of them describes the neutrinos with the same helicity. The subject of our investigation will be the neutrino flux including the ν_{eL}, $\nu_{\mu L}$, and $\nu_{\tau L}$. We shall neglect the CP violation phase and choose the mixing scheme between the generations in the form:

$$\begin{pmatrix} \nu_e \\ \nu_\mu \\ \nu_\tau \end{pmatrix} = \mathcal{U}\begin{pmatrix} \nu_1 \\ \nu_2 \\ \nu_3 \end{pmatrix} =$$

$$= \begin{pmatrix} 1 & 0 & 0 \\ 0 & c_{23} & s_{23} \\ 0 & -s_{23} & c_{23} \end{pmatrix} \begin{pmatrix} c_{13} & 0 & s_{13} \\ 0 & 1 & 0 \\ -s_{13} & 0 & c_{13} \end{pmatrix} \begin{pmatrix} c_{12} & s_{12} & 0 \\ -s_{12} & c_{12} & 0 \\ 0 & 0 & 1 \end{pmatrix} \begin{pmatrix} \nu_1 \\ \nu_2 \\ \nu_3 \end{pmatrix}, \tag{16.90}$$

where φ_{ik} is the mixing angle between ν_i and ν_k in a vacuum and $c_{ik} = \cos \varphi_{ik}$, $s_{ik} = \sin \varphi_{ik}$.

In a vacuum the neutrino flux evolution in ultrarelativistic limit $E \gg max(m_1, m_2, m_3)$ is governed by the Schrödinger-like equation

$$i\frac{d}{dz} \begin{pmatrix} \nu_{eL} \\ \nu_{\mu L} \\ \nu_{\tau L} \end{pmatrix} = H^v \begin{pmatrix} \nu_{eL} \\ \nu_{\mu L} \\ \nu_{\tau L} \end{pmatrix}, \tag{16.91}$$

where

$$H_{11}^v = \frac{1}{6E}[(3s_{13}^2 - 1)\Delta m_{32}^2 + (1 - 3c_{12}^2 c_{13}^2)\Delta m_{21}^2],$$

$$H_{22}^v = \frac{1}{6E}[(3c_{13}^2 s_{23}^2 - 1)\Delta m_{32}^2 + (1 - 3b^2)\Delta m_{21}^2],$$

$$H_{12}^v = \frac{c_{13}}{2E}(s_{13}s_{23}\Delta m_{32}^2 + c_{12}b\Delta m_{21}^2), \qquad H_{23}^v = \frac{1}{2E}(c_{13}^2 c_{23}s_{23}\Delta m_{32}^2 + bb'\Delta m_{21}^2),$$

$$b = c_{23}s_{12} + c_{12}s_{23}s_{13}, \qquad b' = b(\theta_{23} \to \theta_{23} - \frac{\pi}{2}),$$

$$H_{33}^v = H_{22}^v(\theta_{23} \to \theta_{23} - \frac{\pi}{2}), \qquad H_{13}^v = H_{12}^v(\theta_{23} \to \theta_{23} + \frac{\pi}{2}), \qquad H_{ik}^v = H_{ki}^v.$$

Owing to the charged current contributions to the $\nu_{aL}e^-$ forward-scattering, the solar matter modifies both the diagonal and nondiagonal elements of the Hamiltonian H^v. The corresponding Feynman diagrams are shown in Figs. 16.4 and 16.5. The usual direction of

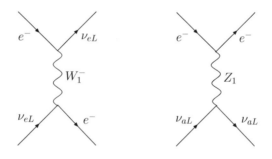

FIGURE 16.4
The diagrams caused by the gauge boson exchanges.

the fermion line corresponds to the description of the charged fermion in terms of the spinor u, while the inverse one means the use of the charge conjugate spinor v. We have not taken into consideration the diagrams with the exchanges of the additional, relative to the SM, gauge bosons because their corrections to the solar matter potential (SMP) are of the order of one percent [23]. Rewriting the amplitudes with the help of the Fierz transformation and averaging the electron field bilinear over the background, we find that the diagrams displayed in Fig. 16.5 give the following contributions to the SMP:

$$V_{ab}^H = \left(\frac{\alpha_{ea}\alpha_{eb}}{2m_h^2} - \frac{f_{ea}f_{eb}}{m_\delta^2} \right) N_e \tag{16.92}$$

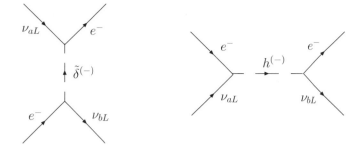

FIGURE 16.5

The diagrams caused by the singly charged Higgs boson exchanges.

where

$$\alpha_{ab} = \frac{h'_{ab}k_2 - h_{ab}k_1}{k_+}.$$

Adding this contribution to that coming from the diagrams shown in Fig. 16.4 we obtain

$$V_{ee} = \frac{g_L^2}{4m_W^2}\left(\frac{1}{2}N_n - N_e\right) + V_{ee}^H, \qquad V_{aa} = \frac{g_L^2}{8m_W^2}N_n + V_{aa}^H, \qquad V_{ab} = V_{ab}^H. \qquad (16.93)$$

The evolution equation in the flavor basis will look like:

$$i\frac{d}{dz}\Psi = H^m\Psi, \qquad (16.94)$$

where

$$H_{12}^m = H_{12}^v + V_{e\mu}^H, \qquad H_{13}^m = H_{13}^v + V_{e\tau}^H,$$

$$H_{23}^m = H_{23}^v + V_{\mu\tau}^H, \qquad V^{SM} = -\sqrt{2}G_F N_a$$

$$H_{11}^m = H_{11}^v + V^{SM} + V_{ee}^H, \qquad H_{22}^m = H_{22}^v + V_{\mu\mu}^H, \qquad H_{33}^m = H_{33}^v + V_{\tau\tau}^H,$$

and we have taken the liberty of omitting a term proportional to the unit matrix.

We are coming to the investigation of all the resonant properties of the neutrino system under study. Three resonance transitions could occur when the neutrino flux moves through the solar matter. They take place between the states i and k under the densities $N_e(z)$ corresponding to the minimum of the difference

$$\Delta M_{ik}^2 = M_i^2 - M_k^2, \qquad (16.95)$$

where

$$M_1^2 = m_1^2 + \frac{C_1}{3} - \frac{1}{3}C_2\left(\cos q - \sqrt{3}\sin q\right),$$

$$M_2^2 = m_1^2 + \frac{C_1}{3} - \frac{1}{3}C_2\left(\cos q + \sqrt{3}\sin q\right), \qquad M_3^2 = m_1^2 + \frac{C_1}{3} + \frac{2}{3}C_2\cos q,$$

$$D_{11} = 2E[c_{12}^2 c_{13}^2 V_{ee} + b_1^2 V_{\mu\mu} + b_1'^2 V_{\tau\tau} - 2c_{12}c_{13}(b_1 V_{e\mu} - b_1' V_{e\tau}) - 2b_1 b_1' V_{\mu\tau}],$$

$$D_{23} = 2E[c_{13}s_{13}s_{12}V_{ee} + s_{23}c_{13}b_2 V_{\mu\mu} - c_{23}c_{13}b_2' V_{\tau\tau} + (s_{13}b_2 + c_{13}^2 s_{23}s_{12})V_{e\mu} -$$

$$-(s_{13}b_2' + c_{13}^2 c_{23}s_{12})V_{e\tau} - (c_{13}s_{23}b_2' - c_{13}c_{23}b_2)V_{\mu\tau}],$$

$$D_{12} = 2E\{c_{13}[c_{13}c_{12}s_{12}V_{ee} + (c_{12}b_2 - s_{12}b_1)V_{e\mu} + (s_{12}b_1' - c_{12}b_2')V_{e\tau}] - b_1 b_2 V_{\mu\mu} -$$

$$-b_1'b_2'V_{\tau\tau} + (b_1b_2' + b_2b_1')V_{\mu\tau}\},$$

$$D_{33} = 2E[s_{13}^2 V_{ee} + c_{13}^2(s_{23}^2 V_{\mu\mu} + c_{23}^2 V_{\tau\tau}) + 2c_{13}(s_{13}s_{23}V_{e\mu} + s_{13}c_{23}V_{e\tau} + c_{13}c_{23}s_{23}V_{\mu\tau})],$$

$$D_{22} = D_{11}(\varphi_{12} \to \varphi_{12} + \frac{\pi}{2}), \qquad D_{13} = D_{23}(\varphi_{12} \to \varphi_{12} - \frac{\pi}{2}),$$

$$b_2 = b_1(\varphi_{12} \to \varphi_{12} + \frac{\pi}{2}), \qquad b_s = b_s'(\varphi_{23} \to \varphi_{23} - \frac{\pi}{2}) \ (s = 1, 2),$$

$$S_{ik} = D_{ii} + D_{kk}, \qquad \Lambda_{ik} = D_{ii}D_{kk} - D_{ik}^2, \qquad i, k = 1, 2, 3,$$

$$C_0 = \Delta m_{21}^2 \Lambda_{13} + \Delta m_{31}^2 \Lambda_{12} + \Delta m_{21}^2 \Delta m_{31}^2 D_{11} + (2E)^3 \sum_{a,b,c} \varepsilon_{abc} V_{ea} V_{\mu b} V_{\tau c},$$

$$a, b, c, = e, \mu, \tau, \qquad C_1 = \sum_i m_i^2 + 2E \sum_a V_{aa}, \qquad C_3 = \Delta m_{21}^2 \Delta m_{31}^2 + \Delta m_{21}^2 S_{31} +$$

$$+ \Delta m_{31}^2 S_{21} + \sum_{i<k} \Lambda_{ik}, \qquad C_2 = \pm\sqrt{|C_1^2 - 3C_3|},$$

$$q = \frac{1}{3}\arccos\left(\frac{9C_1C_3 - 2C_1^3 - 27C_0}{C_2^3}\right),$$

and the signs of C_3 and $9C_1C_3 - 2C_1^3 - 27C_0$ have to coincide.

Recall that the mass eigenvalues M_i^2 do not intersect with each other for any real values of $n_e(z)$. So, depending on the choice of the model parameters, only two of three possible resonances could exist. The resonance densities $N_e(z_r)$ can be approximately found by equating pairs of diagonal elements of the Hamiltonian H^m:

$$\frac{1}{2E}[(s_{13}^2 - c_{13}^2 s_{23}^2)\Delta m_{32}^2 + (b^2 - c_{12}^2 c_{13}^2)\Delta m_{21}^2] = V_{\mu\mu}^H - \sqrt{2}G_F N_e - V_{ee}^H, \tag{16.96}$$

$$\frac{1}{2E}[(s_{13}^2 - c_{13}^2 c_{23}^2)\Delta m_{32}^2 + (b'^2 - c_{12}^2 c_{13}^2)\Delta m_{21}^2] = V_{\tau\tau}^H - \sqrt{2}G_F N_e - V_{ee}^H, \tag{16.97}$$

$$\frac{1}{2E}[c_{13}^2(s_{23}^2 - c_{23}^2)\Delta m_{32}^2 + (b'^2 - b^2)\Delta m_{21}^2] = V_{\tau\tau}^H - V_{\mu\mu}^H. \tag{16.98}$$

Eqs. (16.96)–(16.98) represent the resonance conditions for the channels $\nu_{eL} \leftrightarrow \nu_{\mu L}$, $\nu_{eL} \leftrightarrow \nu_{\tau L}$, and $\nu_{\mu L} \leftrightarrow \nu_{\tau L}$, respectively. The dependence of the nondiagonal Hamiltonian elements H_{ab}^m on $N_e(z)$ could lead to interesting physical consequences. Assume that the conditions (16.96) and (16.97) are fulfilled but H_{23}^m turns into zero. In this case the conversion $\nu_\mu \to \nu_\tau$ does not take place. Then, even at nonvanishing mixing angle θ_{23}, the ν_τ will be absent in the solar neutrino flux as before. In the case under consideration, the expression for the survival probability of the left-handed electron neutrino $\mathcal{P}(\nu_{eL} \to \nu_{eL})$ can be found in the analytical form. To this end we primarily need the values of the mixing angles in the ν_{eL} birth point. Calculations result in

$$\sin^2 2\varphi_{12}^m = \frac{\mathcal{A}_2 \Delta M_{32}^2(\mathcal{A}_1 \Delta M_{12}^2 - \mathcal{A}_2 \Delta M_{32}^2)}{4\mathcal{A}_1^2 \Delta M_{12}^4}, \tag{16.99}$$

$$\sin^2 2\varphi_{13}^m = \frac{\mathcal{A}_1(\Delta M_{32}^2 \Delta M_{13}^2 - \mathcal{A}_1)}{4\Delta M_{32}^4 \Delta M_{13}^4}, \tag{16.100}$$

$$\sin^2 2\varphi_{23}^m = \frac{4\Delta M_{23}^4[\mathcal{B}_1 + \mathcal{B}_2(\mathcal{B}_3 - M_1^2 - M_2^2)]^2}{\mathcal{A}_1^2(\Delta M_{32}^2 - \Delta M_{21}^2)^2}, \tag{16.101}$$

where

$$a_1 = c_{13}^2 c_{12} s_{12} \Delta m_{12}^2 - b_1' b_2' \Delta m_{32}^2 + 2E V_{e\mu}, \qquad a_2 = c_{12}c_{13}s_{13}\Delta m_{12}^2 +$$

$$+c_{23}c_{13}b_1'\Delta m_{32}^2 + 2EV_{e\tau}, \qquad \mathcal{B}_1 = a_1 a_2, \qquad \mathcal{B}_2 = c_{13}s_{13}s_{12}\Delta m_{12}^2-$$

$$-c_{23}c_{13}b_2\Delta m_{32}^2 + 2EV_{\mu\tau}, \qquad \mathcal{B}_3 = (s_{12}^2 c_{13}^2 + s_{13}^2)\Delta m_{12}^2 + (b_2^2 + c_{13}^2 c_{23}^2)\Delta m_{32}^2-$$

$$-2m_2^2 + 2E(V_{\mu\mu} + V_{\tau\tau}), \qquad \mathcal{B}_4 = a_1^2 + a_2^2,$$

$$\mathcal{A}_1 = \mathcal{B}_4 + (\mathcal{B}_3 - M_1^2 - M_2^2)(\mathcal{B}_3 - 2M_3^2), \qquad \mathcal{A}_2 = \mathcal{A}_1 + \Delta M_{31}^2(\mathcal{B}_3 - M_1^2 - M_2^2).$$

It should be stressed, in the LRM both the Hamiltonian eigenvalues and the mixing angles in matter depend on φ_{23}. This directly follows from the fact that ν_e, ν_μ, and ν_τ interact differently with physical Higgs bosons. Recall, in the SM the quantities M_i^2 and φ_{ik}^m are not functions of φ_{23}. Further on, taking into account that at the production and detection positions the propagation is adiabatic; while around the level-crossing points, the propagation is nonadiabatic, we can write down the expression for $\mathcal{P}(\nu_{eL} \to \nu_{eL})$

$$\mathcal{P}(\nu_{eL} \to \nu_{eL}) = \sum_{a,b} |\,\mathcal{U}_{ae}(\varphi_{nk})\,|^2 |\,\mathcal{U}_{be}(\varphi_{nk}^m)\,|^2 \,(p_{ab}^S + p_{ab}^D), \tag{16.102}$$

where p_{ab}^S ($p_{ab}^D = \sum_r p_{ar}^S p_{rb}^S$) is the probability of the resonance transition between ν_{aL} and ν_{bL} for the single (double) passage of the resonance region. The correction to the adiabatic approximation is wholly put into these level-crossing probabilities. Using the Landau method, one can obtain the expression for p_{ab}^S in the following view:

$$p_{ab}^S = \exp[-\frac{1}{E}\mathrm{Im}\int_{N_{ab}}^{\delta N_{ab}} \frac{\Delta M_{ab}^2(N_e^{ab})}{|\,\dot{N}_{ab}\,|}dN_e], \tag{16.103}$$

where the integration has to be done in the upper half of the complex $N_e(z)$-plane, the values of $N_e^{ab} = N_{ab} + i\delta N_{ab}$ are found as complex roots of the equations

$$\Delta M_{ab}^2(N_e^{ab}) = 0,$$

while the quantities N_{ab} and δN_{ab} are identified as the resonance electron density and the width of the corresponding resonance transition, respectively.

When the regions of the resonance transitions $\nu_a \to \nu_b$ and $\nu_c \to \nu_d$ are well separated, that is, the condition

$$N_{ab} + \delta N_{ab} < N_{cd} - \delta N_{cd} \tag{16.104}$$

takes place, then we may use the approximate formula

$$p_{ab}^S = \exp\left(-\frac{\pi\delta N_{ab}^2}{16E\,|\,\dot{N}_{ab}\,|}\right) \tag{16.105},$$

where $\dot{N}_{ab} = dN_{ab}/dt$. In a rough approximation the resonance width and the electron resonance density can be defined with the help of the relations

$$\delta N_{ab} \sim \frac{2H_{ab}^m}{V_{aa}^0 - V_{bb}^0}, \qquad N_{ab} \sim \frac{H_{aa}^v - H_{bb}^v}{V_{aa}^0 - V_{bb}^0},$$

where

$$V_{aa}^0 = (\sqrt{2}G_F)\delta_{ae} - \frac{\alpha_{ea}^2}{2m_h^2} + \frac{f_{ea}^2}{m_{\bar{\delta}}^2}.$$

Note that all the formulae obtained will reproduce the corresponding SM expression when the following conditions are fulfilled:

$$m_{\bar{\delta}} \to \infty, \qquad m_h \to \infty.$$

To have an idea about the corrections to the SMP connected with the Higgs bosons in the LRM we should estimate the upper limits on the YCCs. Necessary information may be obtained from the results of the low-energy experiments investigating the elastic scattering of neutrinos by electrons. To finally establish the $V - A$ structure of the amplitudes in these processes it is necessary to detect the charged lepton and the neutrino in the final state. Since, nowadays it is impossible experimentally, we cannot insist that the corresponding scattering amplitudes really have the $V - A$ structure. The situation is the same as in the case of the direct lepton decays

$$\mu^- \to e^- + \overline{\nu}_e + \nu_\mu, \qquad \tau^- \to e^- + \overline{\nu}_e + \nu_\tau \tag{16.106}$$

and the inverse lepton decays

$$\nu_\mu + e^- \to \mu^- + \nu_e, \qquad \nu_\tau + e^- \to \tau^- + \nu_e. \tag{16.107}$$

In Section 12.5 we wrote the most general form of the amplitude of the direct and inverse muon decays (see Eq. (12.194)). The analogous formula is valid for the τ lepton decays. Let us use these relations. In the LRM the inverse muon (tau-lepton) decays is described by the Feynman diagrams pictured in Fig. 16.6. The appearance of the diagrams with

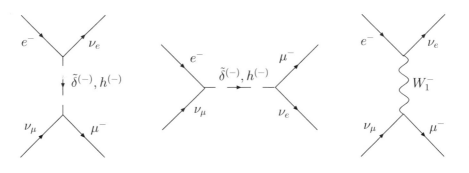

FIGURE 16.6
The diagrams corresponding to the inverse muon decay.

the $h^{(-)}$-exchange in the s-channel is caused by the Majorana nature of the neutrino. The analysis of the corresponding matrix element reads that the nonvanishing amplitudes are: g_{LL}^V, g_{RR}^V, g_{LL}^S and g_{RR}^S. Using the current experimental bounds on these quantities (see Eq. (12.195)) and the expressions for the inverse muon decay, we get the following limits on the values of the LRM parameters:

$$\frac{\alpha_{ee}\alpha_{\mu\mu}}{m_h^2} < 0.218 \times 10^{-5} \text{ GeV}^{-2} \tag{16.108}$$

$$\frac{\alpha_{e\mu}^2}{m_h^2} < 0.264 \times 10^{-5} \text{ GeV}^{-2}, \tag{16.109}$$

$$\frac{f_{ee}f_{\mu\mu}}{m_{\tilde{\delta}}^2} < 0.907 \times 10^{-5} \text{ GeV}^{-2}, \tag{16.110}$$

$$\frac{f_{e\mu}^2}{m_{\tilde{\delta}}^2} < 0.109 \times 10^{-5} \text{ GeV}^{-2}. \tag{16.111}$$

The experimental bounds for the τ-lepton which were obtained by CLEO* Collaboration under measuring the Michel parameters are less severe than for the muon case. At 90% CL the bounds on the amplitudes $g_{\lambda\lambda'}^\gamma$ have the form

$$| g_{LL}^V |< 1.0, \qquad | g_{RR}^V |< 0.2, \qquad | g_{LL}^S |< 1.0, \qquad | g_{RR}^S |< 0.2. \tag{16.112}$$

From (16.112) we find

$$\frac{\alpha_{ee}\alpha_{\tau\tau}}{m_h^2} < 0.66 \times 10^{-5} \text{ GeV}^{-2} \tag{16.113}$$

$$\alpha_{e\tau} \approx 0, \tag{16.114}$$

$$\frac{f_{ee}f_{\tau\tau}}{m_{\tilde\delta}^2} < 3.3 \times 10^{-5} \text{ GeV}^{-2}, \tag{16.115}$$

$$\frac{f_{e\tau}^2}{m_{\tilde\delta}^2} < 0.66 \times 10^{-5} \text{ GeV}^{-2}. \tag{16.116}$$

Usually, when analyzing the neutrino oscillations, the SM modification, in which the neutrino massiveness is provided due to the presence of a right-handed neutrino singlet, is employed. In this case the Yukawa Lagrangian induces the interaction between neutrino and physical Higgs boson that is absent in the ordinary version of the SM. As this interaction is proportional to m_{ν_l}/v ($v \sim 246$ GeV), that is, it proves to be very small, then the whole distinction from the SM is reduced to the kinematic neutrino properties. The SMP value in this version of the SM is defined by the expression

$$V = \sqrt{2}G_F N_e.$$

Let us compare the SMP value predicted by the LRM with this very modification of the SM. If one assumes that $\alpha_{ee} \approx \alpha_{\mu\mu}$ and $f_{ee} \approx f_{\mu\mu}$, then V_{ee}^H may reach the values of the order of 15.5% V^{SM} when the amplitudes g_{RR}^S and g_{LL}^S have the same signs while V_{ee}^H could be as large as 22% V^{SM} when these amplitudes have the opposite signs. Since to date no information concerning the quantities α_{eX}^2/m_h^2 exists, then there is no other way out except the assumption

$$\alpha_{e\mu} = \alpha_{e\tau} = 0.$$

Then, we get

$$V_{\mu\mu}^H < 6.6\% \; V^{SM}, \qquad V_{\tau\tau}^H < 40\% \; V^{SM}.$$

For the nondiagonal Hamiltonian elements the values of the corrections are constrained by the inequalities

$$V_{e\mu}^H < 10\% \; V^{SM}, \qquad V_{e\tau}^H < 25\% \; V^{SM}, \qquad V_{\mu\tau}^H < 16\% \; V^{SM}.$$

So, in the LRM the interaction of the Higgs bosons with leptons could lead to the radical revision of the neutrino parameter values, as compared with the SM predictions.

---

*The name CLEO is not an acronym; it is short for Cleopatra.

17

Solar neutrinos

Oh, my life!
Have I been sleeping fast?
Well, it feels like early in the morning
On a rosy horse I've galloped past.
S. Esenin

17.1 Some information about Sun structure

As the Sun is an ordinary star of our galaxy, then such problems, for example, as energy sources, structure, and spectrum formation are common for the physics of the Sun and stars. Therefore, the whole variety of already established solar phenomena: a granular structure of the surface, complex changes of brightness and motions in its separate active centers, processes in outer rare atmospheric layers, in particular, solar flares, solar prominences, solar wind—are inherent not only in the Sun but in other stars as well. For these reasons, solar physics is of great importance for the development of astrophysics as a whole.

The Sun represents a plasmic sphere with the equatorial radius $R_\odot = 6.961 \times 10^8$ m and with the mass $M_\odot = 1.9889 \times 10^{30}$ kg. The solar matter contains over 70% of hydrogen, over 20% of helium, and about 2% of other elements. The free fall acceleration at the level of visible solar surface is

$$g_\odot = \frac{G_N M_\odot}{R_\odot^2} = 2.74 \times 10^4 \text{ cm/s}^2.$$

The rotation of the Sun has the differential character: the equatorial zone is rotated quicker ($14.4°$ during the day) than high-latitudinal zones ($\sim 10°$ during the day at the poles). The mean period of the solar rotation is 25.38 days. The radiation power of the Sun, the luminosity, is $L_\odot = 3.86 \times 10^{33}$ Erg/s and the effective temperature of the surface is $T_e = 5780$ K.

The analysis of the star position on the spectrum-luminosity diagram, the *Herzsprung–Rassel diagram*, represents a traditional method of investigating stars. Stars form on it some quite clear sequences and the main task of cosmology lies in explaining the existence of these sequences. About 90% of stars on this diagram are located within the relatively narrow band ($\delta \lg L \leq 0.4$), the so-called *main sequence* (MS), which is stretched from stars with $L \sim 10^6 L_\odot$, $M \sim 10^2 M_\odot$ and $R \sim 30 R_\odot$ to stars with $L \sim 10^{-3} L_\odot$, $M \sim 10^{-1} M_\odot$ and $R \sim 10^{-1} R_\odot$. It has been determined, for MS stars their luminosities, radii, and lifetimes t_f are of single-valued mass functions

$$\frac{L}{L_\odot} \approx \left(\frac{M}{M_\odot}\right)^4, \qquad \frac{R}{R_\odot} \approx \left(\frac{M}{M_\odot}\right)^{0.7}$$

and at $M = (1\text{--}10)\, M_{\odot}$

$$t_f \approx 10^{10} \left(\frac{M_{\odot}}{M} \right)^3.$$

On the Herzsprung–Rassel diagram, the Sun is located in the mean part of the MS on which stationary stars those do not practically change its luminosity during many milliard years lie. Under the influence of gravitation, the Sun, as any star, seeks to be compressed, however, the pressure drop, which appears because of high temperature and the density of interior solar layers, hinders it from this. In the center of the Sun $T \approx 1.6 \times 10^7$ K and $\rho \approx 160$ g/cm^3. So high temperature could be maintained only by nuclear reactions of hydrogen-helium fusion, which are primary source of the solar energy. Then, from the Planck radiation law it follows that the primary electromagnetic energy accounts for the Roentgen range. This radiation, because of multiple absorbtion and reradiation, travels the distance from solar center to solar surface during time $\sim 10^6$ years, its spectrum changing drastically. Unlike photons, solar neutrinos, which appear as a result of nuclear reactions in the center of the Sun, not being practically absorbed, reach a terrestrial observer and, therefore, keep the whole information concerning the Sun's interior domains.

In the solar core, atoms (for the most part, they are hydrogen atoms) are in an ionized state. If the hydrogen is completely ionized, then the radiation absorbtion is mainly caused with the ejection of electrons from ions of heavier elements. However, the amount of such elements in the Sun's core is small. Photons travelling from the center of the Sun are partially scattered and absorbed by free electrons. In ionized gas of the central solar part, the total absorbtion is nevertheless relatively small. When a distance from the Sun's center increases, temperature and solar matter density decrease and, at distances greater than $0.7\, R_{\odot}$, neutral atoms (in deeper layers they are helium atoms while closer to surface they are hydrogen atoms) may already exist. With the appearance of neutral atoms (especially large in number of hydrogen atoms), the absorbtion connected with their photoionization sharply increase. The energy transfer by means of radiation is strongly hampered. Another mechanism of the energy transfer is switched on, namely, large-scale convective motions develop and radiative transfer is changed by convection. All these allow us to conditionally introduce three regions defining the Sun's structure: (i) the core, where thermonuclear reactions being the source of the Sun's energy occur; (ii) the radiative zone or the radiative energy transfer zone $(0.1 < r/R_{\odot} < 0.7)$; and (iii) the convective zone $(0.7 < r/R_{\odot} < 1)$, where the matter density on the bottom is as small as $\approx 10^{-2}$g/cm^3. Above the convective zone, in the Sun's atmosphere, the radiative energy transfer again becomes dominant. The radiation coming from the Sun to an external observer appears in an extremely thin surface layer, photosphere, having the thickness nothing but $R_{\odot}/2000 \sim 350$ km and the density $\approx 2 \times 10^{-7}$ g/cm^3. Higher layers of the solar atmosphere, chromosphere $(\rho \approx 3 \times 10^{-12}g/cm^3)$, and corona $(\rho \approx 10^{-15}$g/cm$^3)$ practically freely transmit continuous optical photosphere radiation. Note that the analogous three-layer structure exists in atmospheres of other stars.

In the Sun, a very complicated system of magnetic fields exists, which is changed both in time and in space. In the central part of the Sun's core, the magnetic field must not exceed the value $B_c = 5 \times 10^7$ Gs. Otherwise, as calculations show, at $B > B_c$ this magnetic field would be lost by the Sun due to the effect of *floating to the surface* during its existence. In the radiative zone, the magnetic field value could be as strong as $10^4 - -10^5$ Gs. Both in the center core and in the radiative zone the field does not display the time dependence. In the convective zone the magnetic field module has a 11.2-yr cycle. During the years of the active Sun, in the bottom of the convective zone in the region of 10^3 km the field has the value of 10^5 Gs. With the increase of r the field decreases and its value on the surface totally depends on the existence on the surface of so-called active regions (ARs). The AR first appears as a developing magnetic field, preferably within or close to an old expanding magnetic region whose field has fallen to about 1 Gs or less. Its characteristic sizes on the surface is $R_\odot$ in diameter, and its height reaches the corona level. The number of these regions and their location on the disk changes within the solar cycle. Thus, for example, in the period of the highest maximum (1957–1958), the activity involved practically the whole solar disk. In those places of the AR where the field value reaches 500 Gs the process of sunspot formation begins. The field strength of a developed sunspot is maximal in the center (B_1). It is directed along the solar radius, near periphery it falls down (B_2) and force lines are more strongly inclined to the surface. Fields $B_1 \sim 5 \times 10^3$ Gs and $B_2 \sim 2 \times 10^3$ Gs are typical for big spots ($d \sim 2 \times 10^5$ km). According to modern viewpoints, the geometrical depth of a spot is approximately 300 km. The magnetic field above a spot slowly decreases with the growing of height. Thus, for example, it may reach the values of 10^3 Gs in upper levels of the chromosphere ($\sim 10^3$ km), while beyond a spot region the field value is only 1 Gs. Usually sunspots form a group. Such a group exists for about a month, and the most big spots live up to one hundred days. At the beginning of every 11.2 cycle, sunspots basically emerge in high latitudes ($\sim 35°$). As time goes by, sunspots start to approach the equator and, at the cycle end, they can reach latitudes $\sim 5°$. In addition to the strong local magnetic fields, there is a weaker large-scale magnetic field ($B \sim 1$ Gs) in the high latitude range ($< 50°$). This field changes the sign with the period ~ 22.4 years and turns into zero in the vicinity of the poles at the maximum of the Sun's activity.

The field in the convective zone is characterized by a geometrical phase $\Phi(z)$, defined by

$$B_x \pm iB_y = B_\perp e^{i\Phi(z)},$$

and its first derivative $\dot\Phi(z)$ (we have chosen a coordinate system with the z axis along the solar radius). Nonzero values of $\Phi(z)$ and $\dot\Phi(z)$ also exist both in the photosphere and in the chromosphere in regions above sunspots. The magnetic field above and under a spot has the nonpotential character [24]

$$(\text{rot } \mathbf{B})_z = 4\pi j_z \neq 0.$$

The data concerning centimeter radiation above a spot testify of a gas heating up to the temperature of the coronal order. Thus, for example, at the height $\sim 2 \times 10^2$ km the temperature reaches the values of the order of 10^6 K, which results in a great value of solar plasma conductivity ($\sigma \sim T^{3/2}$). That allows us to suppose that the density of longitudinal electric current might be large enough in a region above a spot.

The Sun's evolution is defined by the change of its chemical mixture as the result of thermonuclear reactions. In accordance with calculations, at present the hydrogen portion is about 35%, whereas at the beginning of the evolution the hydrogen formed nearly 73%. Conversion of hydrogen into helium gradually increases the mean molecular matter weight. Therefore, the equilibrium in the Sun's core is being maintained at increasingly higher temperature and larger density. Because the rates of thermonuclear reactions quickly increase versus the temperature, then, in spite of decreasing the hydrogen content, the energy release in the Sun's interior increases. Consequently, in the course of time, the Sun's luminosity slightly grows. During the evolution the solar core is compressed and the solar shell is expanded, that is, $R_\odot$ increases. The star evolution theory predicts that when the Sun will have reached the age of 9×10^9 years, the hydrogen in the core will be exhausted and thermonuclear reaction will have gone in a layer surrounding the core. At this evolutionary stage with the duration $\approx 5 \times 10^8$ years, the solar radius has been appreciably increased and the effective temperature of the solar surface has been decreased. As a result, the Sun will have been the *red giant*. Next, a quick stage of the burning of helium and heavier elements

($\approx 5 \times 10^7$ years) will have followed, which will have been entailed by throwing off the Sun's shell. As a result, the Sun will have turned into a slowly cooling *white dwarf*.

Different models are suggested to investigate the Sun's structure. The so-called standard solar model (SSM) is the most popular model at present. The following statements lie in its foundation: (i)) the Sun is spherically-symmetrical and is found in hydrodynamical equilibrium; (ii) the Sun constantly was in a state of thermal equilibrium except for some periods during which insignificant entropy change happened; (iii) a chemical mixture change is caused by nuclear reactions in hydrogen and carbon-nitrogen cycles; (iv) a matter is being mixed only in the convective zone; and (v) chemical mixture of the Sun was initially uniform and the Sun evolved without a mass change during 4.7×10^9 years to the present-day values of radius and luminosity.

The statements (i)–(v) are translated in the language of differential equations and supplemented with boundary conditions. Solutions of these equations must define the Sun's parameters (luminosity, radius, convective zone depth, age, chemical mixture, and so on). One of the principal information sources about the solar structure is the investigation of oscillations of the solar surface (*helioseismology*). All helioseismological observations fulfilled up to date are in fine agreement with the SSM predictions. Neutrino astronomy can shed light on the structure and evolution of the Sun. However, here the situation is not so simple. To date we do not exactly know the neutrino parameters (masses, mixing angles, neutrino nature, multipole moments, interaction character between neutrinos and Higgs bosons, and so on). For this reason, the analysis of solar neutrino fluxes is realized in the direction of determination of neutrino properties, rather than in the direction of checking up of the SSM. Of course, when in the near future we shall know about neutrinos, for example, as much as about electrons or photons, then neutrino astronomy turns into a reliable information source concerning the Sun in particular and stars in general.

17.2   Sources of solar neutrinos

The main totality of thermonuclear fusion reactions in the Sun's core converting hydrogen into helium without catalysts is a hydrogen cycle (proton-proton chain). It could be represented in the form of a multistage process

$$4p \rightarrow {}^4He + 2\nu_e + 2e^+ + 26.73 \text{ MeV} - E_\nu, \tag{17.1}$$

where E_ν is the energy carried by the electron neutrino with the mean value ~ 0.6 MeV. In the first stage protons synthesize to the deuterium nucleus (^2D). This occurs in two parallel reactions, called the *pp* and *pep* reactions by the particles that go into them

$$p + p \rightarrow {}^2D + e^+ + \nu_e, \qquad (pp), \tag{17.2}$$

$$p + e^- + p \rightarrow {}^2D + \nu_e, \qquad (pep). \tag{17.3}$$

Once a deuteron is produced, it quickly synthesizes to ^3He

$$^2D + p \rightarrow {}^3He + \gamma. \tag{17.4}$$

And then two ^3He can form ^4He by the strong interaction

$$^3He + {}^3He \rightarrow {}^4He + p + p. \tag{17.5}$$

However, in very few cases, ^3He interacts weakly with protons and produce ^4He, emitting neutrinos in the process

$$^3He + p \rightarrow {}^4He + e^+ + \nu_e, \qquad (Hep). \qquad (17.6)$$

At the next stage, heavier nuclei like ^7Be can be synthesized

$$^3He + {}^4He \rightarrow {}^7Be + \gamma. \qquad (17.7)$$

The appearance of ^7Be induces both the neutrino reactions

$$^7Be + e^- \rightarrow {}^7Li + \nu_e, \qquad (^7Be), \qquad (17.8)$$

with the subsequent lithium decay through the channel

$$^7Li + p \rightarrow {}^4He + He, \qquad (17.9)$$

and neutrinoless reactions

$$^7Be + p \rightarrow {}^8B + \gamma. \qquad (17.10)$$

^8B generates a neutrino

$$^8B \rightarrow {}^8B^* + e^+ + \nu_e, \qquad (^8B), \qquad (17.11)$$

and the appearing $^8B^*$ decays

$$^8B^* \rightarrow {}^4He + {}^4He. \qquad (17.12)$$

Therefore, the *pp* chain is completed by the production of helium nucleus ^3He, two positrons and two electron neutrinos. Note that the hydrogen cycle is the main energy source for stars with the mass $M < 1.2\ M_\odot$ at the initial stage of their existence.

There is also the CNO (carbon-nitrogen) cycle that is responsible for the production of neutrinos. This is the sequence of reactions, resulting in helium production from hydrogen with the participation of carbon, nitrogen, and fluorine as catalysts. The reactions of this cycle are as follows:

$$^{15}C + p \rightarrow {}^{17}N + \gamma$$
$$^{13}N \rightarrow {}^{13}C + e^+ + \nu_e, \qquad (^{13}N)$$
$$^{13}C + p \rightarrow {}^{14}N + \gamma$$
$$^{14}N + p \rightarrow {}^{15}O + \gamma$$
$$^{15}O \rightarrow {}^{15}N + e^+ + \nu_e, \qquad (^{15}O)$$
$$^{15}N + p \rightarrow {}^{12}C + {}^4He$$
$$^{15}N + p \rightarrow {}^{16}O + \gamma$$
$$^{16}O + p \rightarrow {}^{17}F + \gamma$$
$$^{17}F \rightarrow {}^{17}O + e^+ + \nu_e, \qquad (^{17}F)$$
$$^{17}O + p \rightarrow {}^{14}N + {}^4He. \qquad (17.13)$$

The CNO represents the main energy source of massive stars ($M \geq 1.2 M_\odot$) at the initial evolution stages. This cycle dominates over the *pp* chain only if the temperature exceeds $T \geq 18 \times 10^6$ K. For the Sun, this condition is not met and therefore, this contributes only 1.5% to neutrino production.

From Eq. (17.1) one may make rough estimations of the total neutrino flux Φ_ν which falls on a terrestrial surface. Since in every reaction two neutrinos are produced and the energy $E_\gamma \approx 27$ MeV is radiated, then Φ_ν is defined by the expression

$$\Phi_\nu \approx \frac{L_\odot}{4\pi D^2 \frac{1}{2} E_\gamma},\tag{17.14}$$

where D is the distance between the Sun's core and the Earth. To substitute numerical values results in $\Phi_\nu \approx 6 \times 10^{10}$ cm^{-2}s^{-1}. Real calculations must be based on using a concrete solar model and take into account the cross section values of reactions to produce neutrinos. In Table 17.1 the calculation results of the total neutrino fluxes from the pp and CNO cycles that are based on the SSM are given.

TABLE 17.1
The total neutrino fluxes coming from the Sun.

source	flux in cm^{-2}s^{-1}
pp	$5.94 \left(1.00^{+0.01}_{-0.01}\right) \times 10^{10}$
pep	$1.39 \left(1.00^{+0.01}_{-0.01}\right) \times 10^{8}$
hep	2.1×10^{3}
^7Be	$4.8 \left(1.00^{+0.09}_{-0.09}\right) \times 10^{9}$
^8B	$5.15 \left(1.00^{+0.19}_{-0.14}\right) \times 10^{6}$
^{13}N	$6.05 \left(1.00^{+0.19}_{-0.13}\right) \times 10^{8}$
^{15}O	$5.32 \left(1.00^{+0.22}_{-0.15}\right) \times 10^{8}$
^{17}F	$6.48 \left(1.00^{+0.12}_{-0.11}\right) \times 10^{6}$

In Fig. 17.1 the energy spectrum of solar neutrinos predicted by the SSM is displayed. The abscissa is the value of the flux of solar neutrinos on the distance in one astronomic unit (1 astronomic unit (AU)=1.496×10^{11} m). The fluxes from continuum sources are given in the units cm^{-2}s^{-1}MeV^{-1}. For reactions in which the neutrino radiation occurs monochromatically, the fluxes are defined in units cm^{-2}s^{-1}.

17.3   Detection of solar neutrinos

At the end of 2001 seven neutrino telescopes (NT) were in operation for investigations of the solar neutrino fluxes. Three of them were based on radiochemical methods (Homestake, Gallium Experiment (GALLEX), Soviet-American Gallium Experiment (SAGE)). The Homestake experiment, the first experiment with solar neutrinos (1967–2001), used a chlorine-argon method, that is, the process

$$^{37}Cl + \nu_e \rightarrow {}^{37}Ar + e^-\tag{17.15}$$

and had the very low threshold 814 keV, taking into account the contributions from ^8B, ^7Be, pep, ^{13} and ^{15}O neutrinos. Since the cross section of the reaction grows quadratically versus the

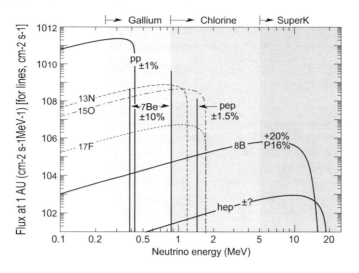

FIGURE 17.1

The energy spectrum of solar neutrinos predicted by the SSM.

neutrino energy, then the most energetic ^8B neutrinos give the main contribution to the capture rate. Already the first measurements showed the disagreement with the SSM predictions. Thus, for example, by 1981 the measured value of the total capture rate of ν_e was defined as

$$\Phi_{exp} = (1.8 \pm 0.3) \text{ SNU}, \tag{17.16}$$

while the calculations within the SSM gave

$$\Phi_{theor} = (7.6 \pm 3.3) \text{ SNU}. \tag{17.17}$$

Here, the SNU (solar neutrino unit) is defined to be 1 capture per 10^{36} target atoms per second, that is, in the situation under consideration, the SNU corresponds to the neutrino flux that produces a single ^{37}Ar nucleus in a target with 10^{36} ^{37}Cl nuclei per second. It should be remembered that the SNU is the unit that measures the product of the flux by the cross section having integrated over neutrino energies, rather than the flux simply. The distinction between the theoretical and measured values of the solar neutrino flux received the name *solar neutrino puzzle*.

The NTs GALLEX (Italy, 1991–1997) and SAGE (Russia, 1991–2006) based their operations on the reaction

$$^{71}Ga + \nu_e \rightarrow {}^{71}Ge + e^-. \tag{17.18}$$

Since the threshold is even lower 233 keV here, then the contributions caused by the most low energetic (but abundant) pp neutrino flux are measurable. The GALLEX Collaboration formally finished their measurements in 1997 and was transformed into the Gallium Neutrino Observatory (GNO) in April 1998. The GNO progressively worked until 2002 constantly lowering the total uncertainty of the measured neutrino flux to 5% and less. As a consequence, the GNO could measure different time variations of the solar neutrino flux predicted by some models. The results of the first set of measurements was published in Ref. [25]. The measured yield of the detected process amounted to

$$\Phi_{exp} = 65.8^{+10.2}_{-9.6}(\text{stat})^{+3.4}_{-3.6}(\text{syst}) \text{ SNU}.$$

For comparison we note that the SSM predictions allows the interval from 115 to 135 SNU. Similar results were obtained at the NT SAGE.

Super-Kamiokande (Japan, 1996) belongs to the second type of the NT (the NT that operates in the discrete counting mode). Thanks to high threshold (5.5 MeV), the ^8B neutrino fluxes alone are detected. Super-Kamiokande confirmed the results obtained by previous Japan Kamiokande Collaboration with the high statistic and systematic precision. The measured neutrino flux proved to be smaller than the SSM prediction by a factor of 2.

The facilities SNO (Sudbury Neutrino Observatory) (Canada, 1999), KamLAND (Japan, 2001) and BOREXINO (Italy, 2001) are members of the second type of the NTs as well. Homestake, GALLEX, SAGE, and Kamiokande represent the NT of the first generation while Super-Kamiokande, SNO, BOREXINO, and KamLAND belong already to the NT of the second generation. One of the advantages of the latter is a high statistics of detected events (SNO and BOREXINO can record up to 6×10^3 events per year and 150 events per day, respectively).

Ideally, to obtain the exhaustive answer on the solar neutrino puzzle one should pose an experiment that allows us to divide contributions from different cycles: pp, ^7Be, pep, ^8B, and CNO. The experiment must provide the sensitivity to the neutrino flavor too. One of the perspective proposals for achievement of this goal is the project LENS (Low Energy Neutrino Spectroscopy). The reaction

$$\nu_e + {}^{176}Yb \rightarrow e^- + {}^{176}Lu^* \tag{17.19}$$

lies in its base. In the process the electron and the returned γ-quantum used as a marker of the decay of the exited Lu-state are detected. The ground state of ^{176}Lu is higher than that of ^{176}Yb to make this nuclide stable relative to the β^- decay. The stability of the target nucleus is mandatory. If it were unstable, then the β^- decay electrons being indistinguishable from captured electrons, which are the signal, would produce an unacceptable background level. This circumstance with the extremely low threshold of the neutrino capture reaction (301 keV) make the nucleus of ^{176}Yb practically unique. Captured electrons and marked quanta are detected by flashes in the liquid scintillator involving nuclei of ^{176}Yb. When the SSM is true, then the detector on the base of 20 t of natural ^{176}Yb provides the following yield of the events per year: 200 for pp, 280 for ^7Be, and 16 for pep cycles. The presence of the γ marker alleviates the hard conditions on the radio frequency level as compared with the BOREXINO experiment. The results of the LENS operation promise to be of great importance. The prospective yield strongly depends on the mixing neutrino parameters that makes it possible to determine them exactly. Note, in the range of the ^7Be neutrinos the LENS is sensitive only to electron neutrinos, whereas the BOREXINO is sensitive to all three neutrino flavors. Therefore, their combination provides a high general sensitivity to all the neutrino flavors.

The solar neutrino puzzle remained unsolved from 1967 until 2001. During this time both the measured neutrino flux and its theoretical estimations are changing very slightly. In Table 17.2 we present the experimental results concerning the solar neutrino observations by 2005. In the same place the theoretical values of the corresponding neutrino fluxes obtained within the generally accepted version of the SSM [26] are listed. Radiochemical results are given in the SNU while fluxes from water Cherenkov detectors are given in units 10^6 cm^{-2}s^{-1}.

Note that only SNO is sensitive to the all neutrino flavors. So, the results of all experiments except for SNO show significant weakness of the neutrino flux in comparison with the SSM predictions. There are solar models that try to solve the solar neutrino puzzle not resorting to anomalous neutrino behavior from the viewpoint of the standard model (vacuum neutrino oscillations, amplification effects of neutrino oscillations in matter and

TABLE 17.2

The experimental and theoretical results for the solar neutrinos.

Experiment	Target	Result	The SSM prediction
Homestake	^{37}Cl	2.56 ± 0.23	7.6 ± 1.2
GALLEX/GNO	^{71}Ga	69.3 ± 5.5	127 ± 10
SAGE	^{71}Ga	$66.9^{+5.3}_{-5.0}$	127 ± 10
Super-Kamiokande	H_2O	2.35 ± 0.10	5.1 ± 0.2
SNO	D_2O	5.21 ± 0.47	5.1 ± 0.2

magnetic field, and neutrino oscillations caused by lepton flavor violating currents). Such direction is usually connected with an astrophysical solution. Cite some of these hypotheses. In Ref. [27] the assumption that to date there is an abundance of ^{3}He atoms in the Sun was made. Authors of Ref. [28] considered that the neutrino deficit may be associated with a fast rotation of the Sun's core that decreases the central density and temperature. It was assumed as well that the Sun is at the later stage of its evolution when hydrogen had been almost burnt and the core consisted of helium [29]. Such an exotic hypothesis as a solar energy generation caused by a black hole accretion in the Sun's center are suggested [30]. Let us demonstrate that the current data exclude an astrophysical solution.

The neutrino flux measured at GALLEX and SAGE represents mainly the sum of contributions from pp, ^7Be, and ^8B. If the pp flux calculated on the base of the Sun's luminosity and the ^8B flux measured at Super-Kamiokande are subtracted from the total flux measured at GALLEX and SAGE, then no place remains for ^7Be neutrinos. We do not know any astrophysical process that could explain the absence of these neutrinos. Therefore, the neutrino disappearance on the way from the Sun's core to the Earth has remained a single explanation. It is evident that the neutrino oscillations in a vacuum and matter are the most attractive and plausible hypotheses. This hypothesis received the first support in 1998, when Super-Kamiokande reported the discovery of oscillations of atmosphere neutrinos [31], which are produced in wide atmosphere showers. In 2002 during the SNO experiment with solar neutrinos [32] they obtained a direct evidence of transitions of the ν_e into ν_μ and ν_τ. This result proved the hypothesis of neutrino oscillations, which was used to explain a deficit of solar neutrinos measured in Homestake, SAGE, GALLEX, GNO, and Super-Kamiokande. Thus, the solar neutrino problem that was the driving force of developments in neutrino physics during the last 35 years has been essentially resolved. The same year the SNO results were confirmed by the experiments with terrestrial neutrinos, in which a well-controlled initial beam was used, namely, in the KamLAND experiment with reactor antineutrinos [33]. All of this enabled us to draw a final conclusion, that the neutrino has mass and there exist mixing in the lepton sector, and call the year 2002 the *annus mirabilis* of solar neutrino physics. However, it was not the end of the story about the neutrino. Until now neutrino physics contains many unsolved problems, the most important of which are a smallness of neutrino mass and a connection between oscillation parameters in the neutrino and quark sectors.

17.4   Analysis of neutrino observation data

Let us study the procedure of obtaining the constraints on the neutrino oscillation parameters with the help of solar neutrino experiments. At first we are limited by the two-flavor approximation. Make two simplifying assumptions: (i) the electron density distribution in the solar matter is described by the function

$$N_e(z) = 200 N_A \exp\left(-\frac{10z}{R_\odot}\right), \qquad (17.20)$$

where $N_A = 6.03 \times 10^{23}$ cm^{-3} is the Avogadro number and we have directed the axis z along the solar radius; and (ii) all neutrinos are created at a point where the electron density is $N_e = 99.8 N_A$.

Note that both of these assumptions are quite believable. The expression (17.20) only slightly differs from the density predicted by the SSM for distances $0 < z < 0.15 R_\odot$ and most neutrinos are really produced near the Sun's core. Since in the expression for the survival probability of the electron neutrino $\mathcal{P}_{\nu_e\nu_e}$ the quantities Δm^2 and E always appear in the combination $y = \Delta m^2/E$, then it is better to consider $y = \Delta m^2/E$ as an independent parameter. In Fig. 17.2, we present the nature of variation of $\mathcal{P}_{\nu_e\nu_e}(y)$ with $y = \Delta m^2/E$ for different values of the vacuum mixing angle θ_0. As we can see, the behavior of the survival

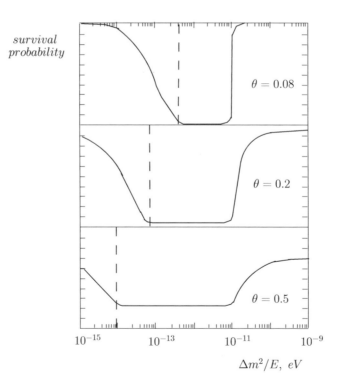

FIGURE 17.2
The survival probability versus $\Delta m^2/E$ for different θ_0.

probability with the decrease of y is sufficiently interesting. For big values of y, $\mathcal{P}_{\nu_e \nu_e}(y)$ is high. As y decreases, there comes a point where the probability begins to fall until it reaches a flat basin. It stays there for a while until, for even small values of y, it rises again and reaches a plateau. We try to explain this not resorting to mathematical details.

When moving away from the Sun's core the denominator of the expression

$$\tan 2\theta_m = \frac{y \sin 2\theta_0}{y \cos 2\theta_0 - 2\sqrt{2} G_F N_e(z)}, \tag{17.21}$$

decreases and turns into zero at the resonance point to give $(\theta_m)_r = \pi/4$. Next, the denominator again increases, resulting in the decrease of the angle θ_m up to the value $\theta_m = \theta_0$ when the neutrino comes into a vacuum. Therefore, for the resonance transition to exist it is necessary that θ_m exceeds $\pi/4$ in the neutrino birth region. As it follows from (17.21) this is possible providing

$$y \cos 2\theta_0 < 2\sqrt{2} G_F N_e(z_s) = 1.5 \times 10^{-11} \text{ eV}. \tag{17.22}$$

The quantity W_{12} allowing for nonadiabatic effects also enters into Eq. (16.83) for $\mathcal{P}_{\nu_e \nu_e}(y)$. Obviously, its influence will be inessential when $W_{12} \ll 1/2$. But how to define the value of W_{12} when it should be taken into consideration? In physics, it is silently meant that 10^{-2} and less is an infinitesimal quantity as compared with 1. Then, from the expression for $\mathcal{P}_{\nu_e \nu_e}(y)$ we could conclude that the threshold value for W_{12}, above which we should take into consideration the nonadiabaticity, is 5×10^{-2}. At our choice of the density profile $\gamma_R \approx 3.8$ corresponds to this value of W_{12}. Then, from the definition of γ_R and (17.20) we get

$$y \frac{\sin^2 2\theta_0}{\cos 2\theta_0} \geq 10^{-14} \text{ eV}. \tag{17.23}$$

In all the plots of Fig. 17.2, we have marked the value of y corresponding to the equality sign in Eq. (17.23) by the dashed line. The nonadiabatic effects are important to the left of the dashed line. At the right end of the plots shown in Fig. 17.2, the value of y is so large that neither the resonance condition, Eq. (16.16), nor the nonadiabaticity condition, Eq. (16.37), is satisfied. Thus,

$$W_{12} \approx 0, \qquad \theta_m \approx \theta_0$$

and we have the time-average oscillations

$$\mathcal{P}_{\nu_e \nu_e} \approx \frac{1}{2}(1 + \cos^2 2\theta_0).$$

As y decreases, we approach the resonance point. For values $\theta \ll 1$, this point is reached when

$$y \approx 1.5 \times 10^{-11} \text{ eV}.$$

Around this point, the survival probability drops down because of the resonance conversion of ν_e into ν_μ. The range of values of y over which this fall occurs, is defined by the resonance width which increases with increasing θ_0

$$\Gamma = N_R \tan 2\theta_0.$$

For y substantially smaller than the resonance value the survival probability is described by the adiabatic approximation, that is $\mathcal{P}_{\nu_e \nu_e} \approx \sin^2 \theta_0$. If there were no nonadiabatic effects, the survival probability would have stayed in that basin for all lower values of y. However, at some point depending on the value of θ_0, nonadiabatic effects become important, resulting in the increase of $\mathcal{P}_{\nu_e \nu_e}$. As it follows from (16.83), the onset of this effect takes place for lower and lower values of y and for larger and larger values of θ_0.

Thus, if $\mathcal{P}_{\nu_e \nu_e}$ is known from an experiment, then our task is to determine the region of the oscillation parameters Δm^2 and θ_0 in which the theory coincides with the experiment. Let us consider the electron neutrino flux with the fixed energy. The results of the analysis are conveniently presented in the plane of variables Δm^2 and $\eta = \sin^2 2\theta_0 / \cos 2\theta_0$ because they enter as parameters to the expression for the survival probability. As we found out earlier, for the conversion $\nu_e \to \nu_\mu$ to take place the fulfillment of three conditions are needed. Every condition is associated with the line in the variables plane $(\Delta m^2, \eta)$. The first condition $(\theta_m)_s > \pi/4$ leads to the horizontal line in Fig. 17.3. Here the neutrino

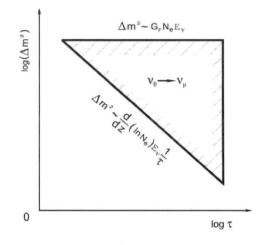

FIGURE 17.3
The MSW triangle. 100% $\nu_e \to \nu_\mu$-conversion takes place inside this triangle.

motion is adiabatic. Since the inequality $\theta_0 \ll 1$ is fulfilled along a greater part of the horizontal branch, then its position in the plane $(\Delta m^2, \eta)$ is governed by the relation

$$\Delta m^2 \sim 2\sqrt{2} G_F N_e(z_s) E. \tag{17.24}$$

The second condition is connected with the minimum existence of $\mathcal{P}_{\nu_e \nu_e}$ in Fig. 17.2. In this case the neutrino motion is adiabatic and the survival probability is proportional to $\sin^2 \theta_0$. On the plane $(\Delta m^2, \eta)$ this condition is associated with the vertical branch. The nonadiabaticity effects cause the third constraint on the variables Δm^2 and η. The condition of the adiabaticity parameter finiteness results in the diagonal branch on which we have

$$\Delta m^2 \sim \mid \frac{d}{dz} \ln N_e(z) \mid_r \frac{E}{\eta}.$$

When the values of the variables Δm^2 and η lie inside the obtained triangle, the theory coincides with the experiment. Similar triangles always appear, when we analyze the neutrino behavior in a matter being based on the MSW effect. For this reason, they are called the *MSW triangles*.

Attention is drawn to the connection between the electron neutrino energy and the character of their conversion along each of three branches. To this end, we again address Fig. 17.2. The horizontal branch of the MSW triangle corresponds to the resonance conversion in the right side of Fig. 17.2 where the survival probability fast decreases with decreasing $\Delta m^2/E$ or with increasing E. Therefore, for the horizontal branch, low-energy neutrinos predominantly survive while high-energy neutrinos change the flavor. The vertical branch is associated with a flat basin of Fig. 17.2 where the survival probability does not depend on the energy ($\mathcal{P}_{\nu_e\nu_e} \sim \sin^2\theta_0$). The diagonal branch maps the nonadiabatic region in the left side of Fig. 17.2. In this case, the resonance conversion probability is higher for low-energy neutrinos and the inclination to survival dominates in the behavior of high-energy neutrinos.

Make one more step toward approaching the conditions of a real experiment. Assume that the measured survival probability for the electron neutrino flux with the energy 5 MeV proved to be in the interval from $(\mathcal{P}_{\nu_e\nu_e})_{min} = 0.19$ to $(\mathcal{P}_{\nu_e\nu_e})_{max} = 0.35$. In Fig. 17.4, two contours corresponding to the values $(\mathcal{P}_{\nu_e\nu_e})_{min}$ and $(\mathcal{P}_{\nu_e\nu_e})_{max}$ are pictured using the dashed lines. The allowed region for the parameters Δm^2 and η is located between these

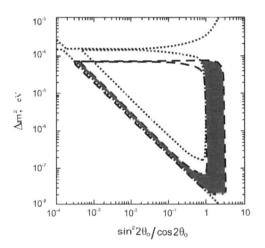

FIGURE 17.4
Δm^2 vs $\eta = \sin^2 2\theta_0 / \cos 2\theta_0$.

contours. Next, we suppose that another experiment with $E = 10$ MeV was fulfilled. The obtained result is

$$0.18 \leq \mathcal{P}'_{\nu_e\nu_e} \leq 0.55. \tag{17.25}$$

The inequality (17.25) gives the allowed region S' displayed in Fig. 17.4 with the help of the dotted lines. Then, from the point of view of the both experiments, the allowed regions represent overlap zones S and S'. They are pictured by the shaded regions. Note that no overlaps are observed for the horizontal part of the MSW triangle. It is obvious that the analogous procedure of obtaining the constraints on the oscillation parameters can be generalized in the case of three and more experiments.

However, in an experiment the survival probability of electron neutrinos is not directly measured. Consider, for example, the process of solar neutrino observation at Super-

Kamiokande (SK) [34]. Neutrinos were detected in the reaction of elastic scattering (ES) by electrons

$$\nu_a + e^- \to \nu_a + e^-. \tag{17.26}$$

The total rate of the ES events is defined by

$$R_{\nu_e}^{ES} = <\sigma_{\nu e}> \Phi_{\nu_e}^{ES}, \tag{17.27}$$

where $<\sigma_{\nu e}>$ is the cross section of the process $\nu_e e^- \to \nu_e e^-$ averaged over the initial spectrum of ^8B neutrinos, and $\Phi_{\nu_e}^{ES}$ is the ν_e flux on the Earth (since, in the SK experiment, the sensitivity to ν_μ and ν_τ is less than that to ν_e by a factor 6, then the SK mainly detects ν_e). The flux $\Phi_{\nu_e}^{ES}$ is specified by the relation

$$\Phi_{\nu_e}^{ES} = <\mathcal{P}_{\nu_e \to \nu_e}^{2\nu}> \Phi_{\nu_e}^0, \tag{17.28}$$

where $<\mathcal{P}_{\nu_e \to \nu_e}^{2\nu}>$ is the averaged survival probability of ν_e, and $\Phi_{\nu_e}^0$ is the total initial flux of solar neutrinos. During 1496 working days, at the total recoil energy threshold 5 MeV, 22 400±800 events were detected. In the most simple case, when the electron and muon neutrinos are mixed and adiabatic effects are negligibly small, the survival probability of the left-handed electron neutrino during daylight hours is

$$\mathcal{P}_{\nu_e \to \nu_e}^{2\nu}(\Delta m_{12}^2, \theta_{12}; N_e) = \frac{1}{2}\left[1 + \cos 2\tilde{\theta}_{12} \cos 2\theta_{12}\right], \tag{17.29}$$

where, for simplicity, the CP phase has been omitted and $\tilde{\theta}_{12}$ represents the effective mixing angle in the neutrino birth point. Then, using (17.28) and (17.29), we can find the theoretical values for $R_{\nu_e}^{ES}$. To compare the obtained expression with the experimental value gives the allowed regions of the oscillation parameters.

The analysis is very complicated when we proceed to the three-flavor approximation. In this approximation the expression for the survival probability of the electron neutrino $\mathcal{P}_{\nu_e \to \nu_e}^{3\nu}$ has a more involved form. (see, for example, [35]). However, for the SM $\mathcal{P}_{\nu_e \to \nu_e}^{3\nu}$ could be related with $\mathcal{P}_{\nu_e \to \nu_e}^{2\nu}$ by rather a simple way [36]

$$\mathcal{P}_{\nu_e \to \nu_e}^{3\nu} = \cos^4\theta_{13}\mathcal{P}_{\nu_e \to \nu_e}^{2\nu}(\Delta m_{12}^2, \theta_{12}; N_e \cos^2\theta_{13}) + \sin^4\theta_{13}. \tag{17.30}$$

It proves to be possible because the Hamiltonian in the flavor basis is reduced to a 2 × 2-matrix with $n_e(r) \to n_e(r)\cos^2\theta_{13}$ at the $\nu_e \to \nu_\mu$ resonance. It should be stressed that the similar simplification does not occur for models where the solar matter potential is modified due to the neutrino interaction with particles that are additional as regards to the SM. Thus, for example, in the left-right model, the interaction Hamiltonian is not diagonal in the flavor basis. Even if the relation

$$V_{e\tau}^H = V_{\mu\tau}^H = 0$$

takes place, we would not manage to connect $\mathcal{P}_{\nu_e \to \nu_e}^{2\nu}$

$$\mathcal{P}_{\nu_e \to \nu_e}^{2\nu}(\Delta m_{12}^2, \theta_{12}; N_e) = \frac{1}{2}\left[1 + \cos 2\tilde{\theta}_{12}\cos 2\theta_{12}\right],$$

where

$$\tan 2\tilde{\theta}_{12} = \frac{\Delta m_{12}^2 \sin 2\theta_{12} + 4EV_{e\mu}^H}{[\Delta m_{12}^2 \cos 2\theta_{12} - 2EV^{SM}] + 2E[V_{\mu\mu}^H - V_{ee}^H]}. \tag{17.31}$$

with $\mathcal{P}_{\nu_e \to \nu_e}^{3\nu}$ by the simple way as it was in the SM.

17.5   Neutrino propagation through the Earth

Solar neutrinos travelling during the night propagate through the Earth before they hit into the NT. This fact must be taken into account in a real solar neutrino experiment too. For definiteness sake, we consider muon and tau-lepton neutrino beams travelling from the Sun that appear under the $\nu_e \to \nu_\mu$ and $\nu_e \to \nu_\tau$ transitions. When propagating the Earth, these neutrinos could be again converted into electron neutrinos due to the MSW effect. Let us determine the conditions when this phenomenon could take place. On the one hand, at the resonance point the relation

$$\Delta m^2(\text{eV}^2) = 1.41 \times 10^{-7} \frac{\rho(\text{g/cm}^3)N_e E(\text{MeV})}{\cos 2\theta_0} \qquad (17.32)$$

must be fulfilled. On the other hand, the necessary condition of the oscillations development is

$$(l_m)_r > 2\pi R_\oplus, \qquad (17.33)$$

where $R_\oplus$ is the Earth's radius. To substitute the expression $(l_m)_r$ into the inequality (17.33) gives

$$\Delta m^2(\text{eV}^2) > 6.2 \times 10^{-8} \frac{E(\text{MeV})}{\sin 2\theta_0}. \qquad (17.34)$$

The region Δm^2 limited by the relations (17.32) and (17.34) represents the action region of the MSW effect for neutrinos that come to the detector during the night. The regeneration effect of the solar neutrino flux propagating through the Earth was predicted in Ref. [12]. It is called a *day-night effect*. In the experiment the quantity

$$A_{dn} = \frac{\Phi_{night} - \Phi_{day}}{\Phi_{night} + \Phi_{day}}, \qquad (17.35)$$

where Φ_{night} (Φ_{day}) is the night (day) flux of the ν_e neutrino, is measured.

Let us find the analytical expression for the regeneration probability of the electron neutrinos. Remember some information concerning the Earth's structure. In accordance with seismic data, the Earth's interior is divided into three main regions: crust, mantle, and core. The crust is separated from the mantle by a sharp boundary on which speeds of seismic longitudinal v_l and transverse v_t waves, as well as a density ρ, spasdomically increase. The effective crust thickness is about 35 km. In the interval of depths 35–2885 km a silicate envelope, mantle, is located. The seismic boundary between mantle and core is the most sharp boundary in the Earth's interior. On it v_l jump-like falls from 13.6 km/s in the mantle to 8.1 km/s in the core while v_t is decreased from 7.3 km/s to zero. The Earth's core may be conditionally divided into three zones as well, namely, exterior core (2885–4980 km), transition zone (4980–5120 km), and internal core (5120–6371 km). Not to pass transverse waves through the core means that the rigidity core modulus is zero, that is, the exterior core ($30\% M_\oplus$) is in a liquid state and, according to current concepts, consists of sulfur (12%) and iron (88%). The interior core ($1.7\% M_\oplus$) involves iron-nickel alloy (20% Ni, 80% Fe). The density distribution of the Earth is displayed in Fig. 17.5. The curve form $\rho(z)$ suggests that in a good approximation the neutrino motion in the Earth's interior could be considered as the motion in two uniform mediums, the mantle ($< \rho_m >$) and the core ($< \rho_n >$). Then, our task is reduced to solving the evolution equation for a uniform medium.

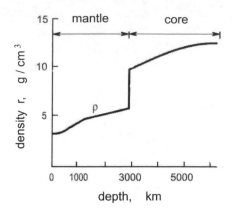

FIGURE 17.5
The density distribution of the Earth.

If one is constrained by the two-flavor approximation, on the Earth's surface the neutrino flux represents the mixture of ν_e and ν_X $(X = \mu, \tau)$

$$\psi(0) = a\nu_e(0) + b\nu_X(0) = a \begin{pmatrix} 1 \\ 0 \end{pmatrix} + b \begin{pmatrix} 0 \\ 1 \end{pmatrix} \qquad (a^2 + b^2 = 1).$$

When travelling in the medium, it is convenient to pass on to a new wave function

$$\begin{pmatrix} \nu'_e(z) \\ \nu'_X(z) \end{pmatrix} = \exp\left[i \int_0^z \sqrt{2} G_F N_e dz'\right] \begin{pmatrix} \nu_e(z) \\ \nu_X(z) \end{pmatrix}. \tag{17.36}$$

Then, the evolution equation will look like:

$$i\frac{d}{dz} \begin{pmatrix} \nu'_e(z) \\ \nu'_X(z) \end{pmatrix} = (A\sigma_3 + B\sigma_1) \begin{pmatrix} \nu'_e(z) \\ \nu'_X(z) \end{pmatrix}, \tag{17.37}$$

where

$$A = \frac{1}{2}\left(\sqrt{2}G_F N_e - \frac{\Delta m^2 \cos 2\theta_0}{2E}\right), \qquad B = \frac{\Delta m^2 \sin 2\theta_0}{4E}.$$

The solution of Eq. (17.37) in the point with the coordinate L satisfying the boundary conditions

$$\nu'_e(0) = 1, \qquad \nu'_X(0) = 0$$

has the form

$$\nu'_e(L) = \cos\left(\frac{\pi L}{l_m}\right) - i\left(\frac{l_m}{l_0}\right)\sin\left(\frac{\pi L}{l_m}\right)\left(\frac{l_0}{l_N} - \cos 2\theta_0\right), \tag{17.38}$$

$$\nu'_X(L) = -i\left(\frac{l_m}{l_0}\right)\sin\left(\frac{\pi L}{l_m}\right)\sin 2\theta_0, \tag{17.39}$$

where

$$l_N = \frac{\sqrt{2}\pi}{G_F N_e}.$$

However, the solutions (17.38) and (17.39) do not form the complete system of orthonormalized functions because there is one more solution

$$\begin{pmatrix} -\nu_X'^*(z) \\ \nu_e'^*(z) \end{pmatrix},$$

being orthogonal to obtained solutions. Really, fulfilling the complex conjugate of Eq. (17.37) and then multiplying it by σ_2 on the left, we get

$$i\frac{d}{dz}\begin{pmatrix} -\nu_X'^*(z) \\ \nu_e'^*(z) \end{pmatrix} = (A\sigma_3 + B\sigma_1)\begin{pmatrix} -\nu_X'^*(z) \\ \nu_e'^*(z) \end{pmatrix}. \tag{17.40}$$

With allowance made for this circumstance, the probability of finding ν_e after propagating through the Earth is given by the expression

$$\mathcal{P}_{\nu_e \nu_e} = \mid [a\begin{pmatrix} \nu_e'(L) \\ \nu_X'(L) \end{pmatrix} + b\begin{pmatrix} -\nu_X'^*(L) \\ \nu_e'^*(L) \end{pmatrix}]\begin{pmatrix} 1 \\ 0 \end{pmatrix}\mid^2 = \mid a\nu_e'(L)\mid^2 \mid b\nu_\mu'(L)\mid^2 -$$
$$- ab^*\nu_\mu'(L)\nu_e'(L) - a^*b\nu_\mu'^*(L)\nu_e'^*(L). \tag{17.41}$$

In the parameters region that concerns us, however, solar neutrinos that leave the Sun are almost in the mass eigenstate ν_2. For this case, the coefficients a and b are defined as

$$a = \sin\theta_0 \exp[iE_2 z], \qquad b = \cos\theta_0 \exp[iE_2 z]. \tag{17.42}$$

To substitute (17.42) into the expression for $\mathcal{P}_{\nu_e \nu_e}$ results in

$$\mathcal{P}_{\nu_e \nu_e} = \sin^2\theta_0 \mid \nu_e'(L)\mid^2 + \cos^2\theta_0 \mid b\nu_\mu'(L)\mid^2 - \frac{1}{2}\sin 2\theta_0[\nu_\mu'(L)\nu_e'(L) - \text{h.c.}]. \tag{17.43}$$

If the resonance condition is approximately realized, then $\nu_e'(L)$ and $\nu_\mu'(L)$ takes the form

$$\nu_e'(L) \approx \cos\left(\frac{\pi L \sin 2\theta_0}{l_0}\right), \qquad \nu_\mu'(L) \approx \sin\left(\frac{\pi L \sin 2\theta_0}{l_0}\right). \tag{17.44}$$

Then the regeneration probability of the electron neutrino will be determined by the quite simple expression

$$\mathcal{P}_{\nu_e \nu_e} = \frac{1}{2} + \frac{1}{2}\cos 2\theta_0 \cos\left(\frac{2\pi L \sin 2\theta_0}{l_0}\right). \tag{17.45}$$

Below we cite the results of looking for the day-night asymmetry for the solar neutrinos with allowing for the data of the second phase of SNO. The second phase was started in June 2001 and finished in October 2003. The experiment used 1000 tons of ultra-pure heavy water D_2O with 2000 kg of dissolved NaCl. The presence of salt in the target strengthens the capture and the neutron detection effectiveness in the reaction

$$\nu_l + d \rightarrow p + n + \nu_l.$$

In Ref. [37] a global χ^2 analysis of the ^8B neutrino data received both by the Super-Kamiokande and by the SNO was made. In so doing, the standard methods of data processing for neutrino experiments proposed in Ref. [38] was used. The obtained results for day-night asymmetry has the form

$$A_{dn} = \frac{\Phi_{night} - \Phi_{day}}{\Phi_{night} + \Phi_{day}} = 0.035 \pm 0.027. \tag{17.46}$$

From (17.46) it is evident that no significant day-night flux asymmetries for solar boron neutrinos are observed within uncertainties. To put it differently, the resonance conversions for ^8B neutrinos travelling in the Earth's interior are absent. The resonance points are not reachable because of a small neutrino energy. It is clear, using collider neutrino beams and varying their energy, we may observe the resonance transitions of neutrinos travelling through the Earth's interior.

18

Atmospheric neutrinos

> *Either the well was very deep, or she fell*
> *very slowly, for she had plenty of time*
> *as she went down to look about her and*
> *to wonder what was going to happen next.*
> L. Carroll, Alice's Adventures in Wonderland

18.1 Cosmic rays

Cosmic radiation was discovered by Hess in 1912. The composition of a cosmic-ray flux is as follows: (i) nuclear component consist of $\sim 90\%$ protons, $\sim 10\%$ helium nuclei, and $\sim 1\%$ heavier nuclei; (ii) electrons make up $\sim 1\%$ of the nuclei number; (iii) positrons form $\sim 10\%$ of the electrons number; and (iv) the portion of antiprotons is $\geq 10^{-4}$ relative to protons. The composition of cosmic rays was experimentally defined in the energy region from few MeV to several TeV [39]. A good agreement between relative occurrences of elements on the Sun and elements in cosmic rays was revealed to point on an unified star mechanism of their production. However, in cosmic radiation, the content of Li, Be, and B nuclei as well as nuclei located immediately before iron are heightened. This excess could be explained assuming the abundance of these elements to have appeared in cosmic rays under motion in cosmic space thanks to spallation reactions on C, N, and Fe nuclei. Then, the ratio of the primary (C, N, and Fe) to secondary (Li, Be, and B) nuclei makes it possible to estimate the time during which cosmic rays are within our galaxy. The result is $\tau \sim 10^6$ years. The relative occurrence distribution of elements proves to be almost independent on the energy (except for Fe).

The first calculations of the spectrum of protons and α-particles were fulfilled in Ref. [40] in the 1–10 GeV range. Now, the calculated spectrum is reached from 1 to 1000 GeV and has the form

$$\frac{dN}{dE_k} \sim E_k^{\alpha}, \tag{18.1}$$

where $\alpha \approx -2.75$ for the wide kinetic energy (E_k) range except for a very low-energy region. However, majority of experiments on the observation of primary cosmic particles that have been fulfilled to date (MASS [41], CAPRICE (Cosmic AntiParticle Ring Imaging Cherenkov Experiment) [42], BESS (Balloon-borne Experiment with a Superconducting Spectrometer) [43], and AMS (Alpha Magnetic Spectrometer) [44]) are not in line both with theoretical predictions and with each other.* This forces us to use formulae different

---

*The existing experimental technology cannot determine the radiation composition at energies above few TeV.

from (18.1). Thus, for a crude estimation of proton flux, one may employ

$$\frac{dN_p}{dE_k} \sim 1.2 \times 10^4 E_k^{-2.7} \text{ m}^{-2}\text{s}^{-1}\text{sr}^{-1}\text{GeV}^{-1}$$

for the 5–100 GeV region (E_k in GeV units).

The helium nuclei flux is measured together with the proton one. The ratio $^2He/^1H$ is $\simeq 0.06$ for ≥ 10 GeV and it increases to 0.1 at 100 GeV.

The next abundant nuclei belong to the CNO group. For example, at 10 GeV/nucleon the ratio $^1H : {}^2He : CNO$ is roughly 0.94:0.06:0.003. The CNO abundance displays a slow increase with energy.

When calculating the cosmic ray flux one should take into account the following two factors to influence its value: (i) modulations caused by the solar wind (this effect is correlated with the Sun's activity and displays strong time-dependence); and (ii) cut-off of the low-energy spectrum part at the expense of geomagnetic fields (this effect depends on the location of the NT and weakly depends on time).

The solar wind decelerates cosmic rays. The effect is usually described by the diffusion-convection model, with which the observed spectrum is given by the expression [45]

$$i(p, r, t) = i_0(p) \exp\left[-\int_{r_{min}}^{r_{max}} \frac{v(t)}{D(p, r', t)} dr'\right], \tag{18.2}$$

where r_{min} is the distance from the Earth (1AU), r_{max} is the boundary of the solar wind from the Sun (33–43 AU), $v(t)$ is the solar wind speed (≈ 430 km/h), D is the diffusion coefficient, $i_0(p)$ is the interstellar spectrum. This effect becomes maximum at the Sun's activity maximum. Thus, for example, the proton flux at 1 GeV in the Sun's activity minimum is twice as large as that in the Sun's activity maximum. The effect is reduced to $< 10\%$ at $E_k = 10$ GeV.

Low-energy cosmic rays do not reach the Earth owing to the dipolar magnetic field that acts as a shield. For particles, an approximate cut-off momentum for the dipolar geomagnetic field is governed by [46]

$$p_{cut} = p_0 \frac{\cos^4 \lambda}{[1 + (1 - \cos^3 \lambda \sin \theta \sin \phi)^{1/2}]^2}, \tag{18.3}$$

where $p_0 = eR_\oplus B/c \approx 59$ GeV, B is the magnetic field at the magnetic equator, $R_\oplus$ is the radius of the Earth, λ is the geomagnetic latitude, θ is the zenith angle, and ϕ is the azimuth measured clockwise from the magnetic north. For nuclei, the momentum is replaced with the rigidity.[*]

At energies above few TeV, the cosmic radiation fluxes becomes too small to be detected by direct methods. Indirect methods are based on the interaction of the primary radiation with the Earth's atmosphere. The necessary information is extracted by the subsequent analysis of the produced secondary radiation. At the interaction of the cosmic rays with nuclei of the Earth's atmosphere, pions and kaons are produced. Decays of these particles lead to the appearance of photons, muons, and neutrinos that are main detected products of the cosmic rays. Because the secondary particle flux is formed in the interval of heights above sea level 10–20 km, then practically all muons with energies above 10 GeV reach the Earth's surface before they decay.

---

[*]In cosmic-ray physics the energy of nuclei is expressed in *rigidity* $p/(ze)$, where p is the momentum and z is the charge of nuclei given in units of GV.

18.2 Atmospheric neutrino production

The main sources of $\nu_e, \overline{\nu}_e$ and $\nu_\mu, \overline{\nu}_\mu$ are

$$\pi^\pm \to \mu^\pm \nu_\mu(\overline{\nu}_\mu), \qquad (100\%), \tag{18.4}$$

$$K^\pm \to \mu^\pm \nu_\mu(\overline{\nu}_\mu), \qquad (63.5\%), \tag{18.5}$$

$$K^0_L \to \pi^\pm e^\mp \overline{\nu}_e(\nu_e), \qquad (38.8\%), \tag{18.6}$$

$$K^0_L \to \pi^\pm \mu^\mp \overline{\nu}_\mu(\nu_\mu), \qquad (27.2\%), \tag{18.7}$$

$$\mu^\pm \to e^\pm \overline{\nu}_\mu \nu_e(\overline{\nu}_e \nu_\mu), \qquad (100\%), \tag{18.8}$$

where in parentheses the branching of the corresponding decay is given.

The calculation of the neutrino flux to fall down to the Earth consists of three stages: (i) estimation of the primary cosmic ray flux at the top of the atmosphere for all directions and the location of the detector; (ii) spectrum calculation of particles produced in the result of the interaction of cosmic rays with the air nuclei; and (iii) computation of neutrino fluxes to appear in decays K^0_L, $K^\pm$, $\pi^\pm$, and $\mu^\pm$. Calculation complications of kaon, muon, and pion fluxes are caused by three circumstances. First, processes of their productions compete with interaction processes. Second, it is necessary to take into consideration the energy loss of the primary cosmic particles when they travel in the atmosphere. Third, before decaying, they also lose energy thanks to the interaction. The master equation that describes the propagation of cosmic ray particles has the form [47]

$$\frac{\partial I_i(E, x, \theta)}{\partial x} = -\mu_i(E) I_i(E, x, \theta) - \frac{m_i}{E \tau_i \rho(x)} I_i(E, x, \theta) +$$

$$+ \sum_j \int dE' d\theta' \mu_i(E') S_{ji}(E', \theta', E, \theta) I_j(E', x, \theta'), \tag{18.9}$$

where $I_i(E, x, \theta)$ is the flux of cosmic rays of i-particles (i=nuclei, π, $K^\pm$, $\mu^\pm$, and so on) at the position x measured in units of the amount of materials traversed ($g\ cm^{-2}$) from the top of the atmosphere, with the zenith angle θ and energy E, m_i and τ_i are mass and decay lifetime, respectively; μ_i is the absorbtion coefficient of particle i, ρ is the atmosphere density, and S_{ji} is the probability of particle production from $i \to j$. The first term represents a loss of flux by absorbtion, the second term is the loss by decay, and the third term is the production of the particle considered. In Eq. (18.9) the energy loss term that is important only for muon propagation was omitted. So, for muons, we have instead of (18.9)

$$\frac{\partial I_\mu(E, x, \theta)}{\partial x} = -\frac{m_\mu}{E \tau_\mu \rho(x)} I_\mu(E, x, \theta) + S_\mu(E, x, \theta) + \frac{\partial}{\partial x}[B(E) I_\mu(E, x, \theta)], \tag{18.10}$$

where $S_\mu(E, x, \theta)$ describes the muon production,

$$B(E) = \frac{dE}{dx} \simeq (a + bE) \tag{18.11}$$

represents the energy loss rate, $a \simeq 2.2\ MeV\ cm^2\ g^{-1}$ is the ionization energy loss, and bE is the sum of the energy losses by bremsstrahlung, $e^- e^+$ pair production, and nuclear interaction, that is,

$$b = b_{brems} + b_{pair} + b_{nucl}.$$

Calculations show that $b_{pair} > b_{brems} > b_{nucl}$ and the quantity bE begins to exceed a only when $E \geq 1$ TeV.

Note that the calculation of the interaction of cosmic rays with nuclei requires experimental knowledge of hadron interaction details. Unfortunately, the data obtained in collider experiments are not able to give complete information because the acceptance angle of accelerator counter experiments is limited to the milli-radian solid angle relative to the beam direction. So, even at $E_\pi/E_N < 0.15$ the reliable data are absent.

Up-to-date calculation of atmospheric neutrino fluxes allow for all decay processes with branching ratios typically larger than 1%.

The contribution of kaons to the neutrino flux in < 10 GeV range is less than 10%. The muon from π decay prove to be polarized at $\sim 30\%$ in the laboratory system. This makes ν_e from μ^+ to have a higher energy compared to the unpolarized case, and, consequently, it increases $(\nu_e + \overline{\nu}_e)/(\nu_{mu} + \overline{\nu}_\mu)$.

As an illustration, we shall give the basic steps of calculations of neutrino fluxes. If one replaces the decay length and the absorbtion length with their mean values, then the approximate formula for the ν_μ flux from $\pi^+ \to \mu^+\nu_\mu$ will look like [48]:

$$\frac{d^2 F_\nu}{dE_\nu d\Omega} = N \int dh \int dE_\pi \int dy \frac{d\Gamma_\nu(E_\pi, E_\nu)}{dE_\nu} D_\pi(E_\pi) R_\pi(E_\pi, h, y) \frac{d^2 J_\pi(E_\pi, y)}{dE_\pi dy}, \quad (18.12)$$

where $d\Gamma_\nu(E_\pi, E_\nu)/dE_\nu$ is the neutrino spectrum from the decay of pions, $D_\pi(E_\pi) = \tau_\pi^{-1}(m_\pi/E_\pi)$ is the decay rate of pion, $R_\pi(E_\pi, h, y)$ is the probability that a pion of energy E_π produced at height y survives at altitude h where it decays. The pion energy spectrum at altitude y, which enter into Eq. (18.12) is given by the expression

$$\frac{d^2 J_\pi(E_\pi, y)}{dE_\pi dy} = \sigma_{inel} N_L \rho(y) \int_{max[E_\pi, E_{cut}(\theta)]}^\infty \frac{d\Gamma_\pi(E_\pi, E_p)}{dE_\pi} \frac{d^2 J_p(E_p, y, \Omega)}{dE_p dy} dE_p, \quad (18.13)$$

where

$$\frac{d^2 J_p(E_p, y, \Omega)}{dE_p dy}$$

is the primary proton energy spectrum at altitude y and

$$\frac{d\Gamma_\pi(E_\pi, E_p)}{dE_\pi}$$

is the energy spectrum of produced pions that could be approximated by the expression

$$\frac{d\Gamma_\pi(x)}{dx} \approx \frac{(1-x)^3}{x},$$

x is the momenta ratio of pion and proton in the laboratory system. When integrating (18.12) over the solid angle, the geomagnetic effect should be taken into consideration. In the integral (18.13) for the low limit, the geomagnetic cut-off (E_{cut}) has to be taken when $E_{cut} > E_\pi$ or the production threshold (E_π) when $E_{cut} < E_\pi$.

For the neutrino flux (ν_μ and $\overline{\nu}_e$) induced by μ^- decays, we have to fulfill the replacement in Eq. (18.12)

$$\frac{d\Gamma_\nu(E_\pi, E_\nu)}{dE_\nu} \to \int dz \int dE_\mu \frac{d\Gamma_\nu(E_\mu, E_\nu)}{dE_\nu} D_\mu R_\mu(E_\mu, z, h) \frac{d\Gamma_\mu(E'_\mu, E_\pi)}{dE'_\mu}\bigg|_{E'_\mu = E_\mu + \Delta E_\mu(z,h)}, \quad (18.14)$$

where $\Delta E_\mu(z, h) = 2.2$ MeV g cm^{-2} $\rho(z)$ is the ionization energy loss of a muon. The factor $d\Gamma_\nu(E_\mu, E_\nu)/dE_\nu$ must involve the muon polarization effects too.

As we can see, the calculation of atmospheric neutrino fluxes represents the complex problem and involves a series of model assumptions. A detailed list of works where computations of the neutrino fluxes has been fulfilled could be found in the book [47]. Here, we only note that there are considerable deviations in estimations of these fluxes by different authors. However, the ratio of the electron neutrino flux to the muon neutrino flux $\Phi(\nu_e)/\Phi(\nu_\mu)$ is practically the same for the most part of works.

18.3 Atmospheric neutrino detection

If one takes into consideration only neutrinos appearing from pion decays, that is, the reactions (18.4) and (18.8), and assumes that all muons decay when they travel through the Earth's atmosphere,* then the simple counting gives the following relationship between neutrino fluxes:

$$r = \frac{\Phi^{\nu_e + \bar\nu_e}}{\Phi^{\nu_\mu + \bar\nu_\mu}} \equiv \left(\frac{e}{\mu}\right) = \frac{1}{2}. \tag{18.15}$$

Detailed counting has to allow for not only other neutrino sources but energies of created neutrinos as well. Thus, for example, at energies in the interval from 0.2 to 2 GeV, we have $r \approx 0.44$.

The first investigations of atmospheric neutrinos were carried out in underground experiments: (i) in the gold mine East Rend (nearby Johannesburg, South Africa) at the depth of 8000 m of water equivalent [49]; and (ii) in the mine Kolar Gold Fields (India) at the depth of 7500 m of water equivalent [50] (review of experiments with atmospheric neutrinos could be found in Ref. [51]). Initially such experiments were completely aimed at looking for a proton decay that is predicted by the GUT. However, the main source of background to the proton decay were just atmospheric neutrinos. Therefore, atmospheric neutrino investigation was first initiated mainly by determination of this background.

Detection of neutrinos is realized primarily by means of charged (weak) currents (CCs)

$$\nu_e(\nu_\mu) + N \to e^-(\mu^-) + ..., \tag{18.16}$$

where N is the target nucleus. To define the experiment sensitivity it is important to know the detector response on impact of electrons and muons. As a rule, only events that are placed into an interior detector volume are used. In Cherenkov water detectors, the difference between ν_e and ν_μ is displayed in the form of Cherenkov radiation rings a created charged lepton induces. In the process, only events with a single Cherenkov ring are employed. Owing to multiple Compton scattering, the ring from the electron is more degraded than that from the muon. Careful investigation of sharpness of these rings allows one to distinguish muon and electron events. Moreover, the muon decay is detected too. To exclude the dependence on errors of determining the absolute fluxes one introduces the ratio of observed events to theoretical predictions:

$$R = r_{obs}/r_{theor}.$$

Obviously, at the absence of any mechanisms of neutrino conversions, this quantity must be equal to 1. Apart from the above-mentioned quantities, the so-called *up-down asymmetry* of μ and e events is of great interest for experimentalists. Explain what it means. One may

---

*It is true for energies 0.1–2 GeV.

consider that a neutrino flux is produced in the interval of heights at the sea level from 10 to 20 km. If one ignores the influence of a geomagnetic field, then cosmic ray distribution could be considered to be isotropic. Then, the neutrino flux production intensity will be the same in all points of the Earth's atmosphere (up-down symmetry). The existing uncertainties in calculations of the absolute flux of atmospheric neutrinos reach 20% while the uncertainties for the ratio $\Phi^{\nu_\mu}/\Phi^{\nu_e}$ have the order of 5%. However, if one compares two neutrino fluxes that have come from opposite parts of the Earth's atmosphere, then one may manage without knowledge of these quantities. Imagine that we are interested in a muon component of atmospheric neutrino flux. So, two muon neutrino fluxes fall on a underground detector: upper $\Phi_u^{\nu_\mu}$ (which arose above detector at the height h) and down $\Phi_d^{\nu_\mu}$ (which arose in the atmosphere at opposite side of the Earth and travelled the distance $L = 2R_\oplus + h, R_\oplus \approx 6.4 \times 10^3$ km). If one assumes, there are no mechanisms to change a neutrino flux on its way to a detector, then $\Phi_d^{\nu_\mu} = \Phi_u^{\nu_\mu}$. Therefore, the equality of ingoing and outgoing Φ^{ν_μ} fluxes must take place not only for total fluxes, (which have been integrated over all possible directions of a zenith angle θ_z), but for differential fluxes $d\Phi_u^{\nu_\mu}(\theta_z) = d\Phi_d^{\nu_\mu}(\pi - \theta_z)$ too.

In Table 18.1 the results of R measurements in the first experiments with atmospheric neutrinos are presented.* From the results of NUSEX [52] and Frejus [53] Collaborations

TABLE 18.1

List of the atmospheric neutrino experiments.

Experiment name	Detector type	Exposure (kt·year)	$R = (\mu/e)_{obs} : (\mu/e)_{theor}$
NUSEX	Iron Calorimeter	0.74	$0.96^{+0.32}_{-0.28}$
Frejus	Iron Calorimeter	1.56	$1.00 \pm 0.15 \pm 0.08$
Kamiokande	Water	7.7	$0.60^{+0.06}_{-0.05} \pm 0.05$, when E_s
Kamiokande	Cherenkov	8.2	$0.57^{+0.08}_{-0.07} \pm 0.07$, when E_m
IMB-3	Water	7.7	$0.54 \pm 0.05 \pm 0.012$, when E_s
IMB-3	Cherenkov	2.1	$1.4^{+0.4}_{-0.3} \pm 0.3$, when E_m
Soudan-2	Iron Calorimeter	5.1	$0.68 \pm 0.11 \pm 0.06$

it followed that the theoretical and observed ratios μ/e coincide within the limits of experimental error. However, the data obtained by Kamiokande [54], IMB [55], and Soudan-2 [56] testified that the observed value μ/e was much less than the theoretical one. Moreover, the results of Kamiokande showed that the deficit of μ events had the dependence on the zenith angle and the number of the downward-going and upward-going μ-like events did not equal to each other, that is, the up-down symmetry was absent. All of the above-mentioned anomalies are referred to as the *atmospheric neutrino problem.*

After putting into operation the NT of the second generation Super-Kamiokande (Super-K), the results of neutrino fluxes detection become more reliable. Super-K may be used not only for observation of solar (low energy) neutrinos but for the investigation of atmospheric neutrinos as well. The Super-K design makes it possible to detect atmospheric neutrinos

---

*The division of E_s into E_m is defined by a visible energy of an emitted lepton, namely, at $E_{vis} < 1.33$ GeV the case in point is E_s while at $E_{vis} > 1.33$ GeV—E_m.

with energies from 100 MeV to 1000 GeV. Neutrino detection is realized by the Cherenkov radiation registration of relativistic particles when they are moving in a water detector volume. When hitting the Super-K detector atmospheric neutrinos can undergo the following interactions (the symbol NC (CC) over the narrows means that the given process results from neutral (charged) weak currents):

$$\nu_l + e^- \xrightarrow{CC,NC} \nu_l + e^-, \qquad l = e, \mu, \tau, \tag{18.17}$$

$$\nu_l + N \xrightarrow{CC} l + N', \tag{18.18}$$

$$\nu_l + N \xrightarrow{NC} \nu_l + N, \tag{18.19}$$

$$\nu_l + N \xrightarrow{CC} l + N' + \text{meson}, \tag{18.20}$$

$$\nu_l + N \xrightarrow{NC} \nu_l + N' + \text{meson}, \tag{18.21}$$

$$\nu_l + N \xrightarrow{CC} l + N' + \text{hadrons}, \tag{18.22}$$

$$\nu_l + N \xrightarrow{NC} \nu_l + N' + \text{hadrons}, \tag{18.23}$$

$$\nu_l + {}^{16}O \xrightarrow{CC} l + {}^{16}O + \pi, \tag{18.24}$$

$$\nu_l + {}^{16}O \xrightarrow{NC} \nu_l + {}^{16}O + \pi. \tag{18.25}$$

The cross section of the process (18.17) at $E_\nu = 1$ GeV has the order of 10^{-41} cm^2, that is, it proves to be less than all the remaining cross sections by a factor of 10^{-3}. Therefore, it may be surely neglected. In the remaining reactions, besides electrons and muons, other relativistic particles appear. Thus, for example, in the reaction (18.25), the π^0 meson decays into the two γ quanta each of them generates the Cherenkov cone. Hence, the π^0 meson is identified as a two-ring event. Note that the maximum number of simultaneously detected rings at the Super-K equals 5.

The analysis of the Super-K data gave the following values of R [57]:

$$R = 0.651^{+0.019}_{-0.018} \pm 0.05, \qquad \text{at } E_s,$$

$$R = 0.711 \pm 0.036 \pm 0.085 \qquad \text{at } E_m.$$

The measurement results also showed that the zenith angle distribution of the muon neutrino flux significantly differs from calculations while the electron neutrino flux did not display the similar anomaly. Thus, the up-down symmetry was

$$\left[\frac{\Phi_u(-1.0 < \cos\theta_z < -0.2)}{\Phi_d(+0.2 < \cos\theta_z < 1.0)} \right]_{obs} = 0.52 \pm 0.05, \qquad \text{at } E_m. \tag{18.26}$$

The absence of the similar symmetry was a demonstration that some mechanism that leads to changing the neutrino flux on its way to the detector exists. The neutrino oscillation hypothesis could assume the role of such a mechanism. At that, since the atmospheric electron neutrino fluxes crossing the detector are equal, we may be dealing with the $\nu_\mu - \nu_\tau$ mixing only. If the neutrino energy is equal to 1 GeV, then for the downward-going neutrino flux the ratio L/E is approximately equal to 10^4 km/(1 GeV) while for the upward-going neutrino flux it is considerably less $L/E \sim 10$ km/(1 GeV). This suggests that in the former case the oscillations have time do develop while in the latter they do not. In other words, for the Φ_u flux the survival probability of the muon neutrino $\mathcal{P}_{\nu_\mu \nu_\mu}$ must be set equal to 1 whereas for the Φ_d flux $\mathcal{P}_{\nu_\mu \nu_\mu}$ is defined by the expression

$$\mathcal{P}_{\nu_\mu \nu_\mu} = 1 - \sin^2 2\theta_{\mu\tau} \sin^2 \left(\frac{\Delta m_{\mu\tau}^2 L}{4E} \right). \tag{18.27}$$

Note, the atmospheric neutrino experiments measure $\mathcal{P}_{\nu_\mu \nu_\mu} = \mathcal{P}_{\bar{\nu}_\mu \bar{\nu}_\mu}$. So, the Super-K results with atmospheric neutrinos are also evidence in favor of the neutrino oscillation hypothesis.

19

Results and perspectives

> *It was likewise known that the woodpecker was wholly taken up with writing a "History of the Wilderness" on the bark of pine-trees, but this undertaking, too, would not be long-lived, because as the work progressed the bark was being nibbled and reduced to pieces by outlaw ants. Thus, it is evident that the forest folk were quite ignorant of the past and present and never showed any interest in the future. In other words, they simply dallied, wrapped in the obscurity of the times.*
>
> M. E. Saltykov Shchedrin, "Bears in Government"

So, there are the three physical effects that can change the neutrino flavor. The first is the vacuum neutrino oscillations. The second effect is due to the fact that a neutrino may possess an anomalous magnetic dipole moment (AMM), which results in an interaction with an external magnetic field. The third is the Mikheyev–Smirnov–Wolfenstein (MSW) effect.

Electron neutrinos are produced in the Sun's core, then they travel 700,000 km along the solar radius in the medium with variable density and a sufficiently intensive magnetic field. In these conditions at particular values of oscillation parameters (OPs) and AMM, the decrease of electron neutrinos from an initial flux is possible. After leaving from the solar surface the neutrino flux flies 150,000,000 km in a vacuum before it will reach a terrestrial detector. Here, the conversion of the electron neutrinos are also available, thanks to the vacuum oscillations. The solar neutrinos can reach the detector right away during the day or during the night after penetrating the Earth's interior. In the last case, at the particular values of the OPs, the influence of the Earth's matter could change the neutrino flavors ratio in the flux. To date, the information concerning the neutrino sector structure may be divided into three parts:

(i) The neutrino oscillations with the solar frequency that give information about Δm_{21}^2, θ_{12} and estimate θ_{13} as a preliminary. This class includes in its experiments with solar and reactor neutrinos.

(ii) The neutrino oscillations with the atmospheric frequency in which the parameters Δm_{31}^2, θ_{23} and θ_{13} are determined.[*]

(iii) Non-oscillations experiments that give information concerning the absolute values of the neutrino masses.

In Ref. [58] the global fit of the solar and reactor neutrino data in the assumption of oscillations only between active neutrinos ($\nu_{e,\mu,\tau}$) was fulfilled. The results of the following experiments were taken into account: the data released in the summer 2008 by the MINOS (Main Injector Neutrino Oscillation Search) collaboration [59], the data from the neutral current counter phase of the SNO* experiment (SNO-NCD) [60], the latest KamLAND [61]

---

[*]Since the relation

$$\Delta m_{12}^2 + \Delta m_{31}^2 + \Delta m_{23}^2 = 0$$

takes place, then, for the atmospheric neutrino parameters, one may use either Δm_{31}^2 or Δm_{23}^2.

*Sudbury Neutrino Observatory.

and BOREXINO [62] data, as well as the results of a recent re-analysis of the Gallex/GNO solar neutrino data presented at the Neutrino 2008 conference. The results are as follows. Fig. 19.1 illustrates how the determination of the leading solar oscillation parameters θ_{12}

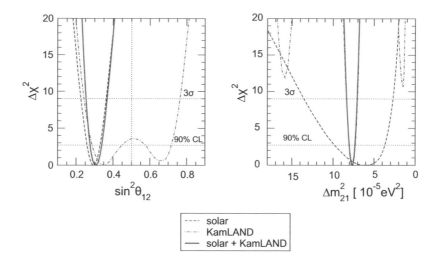

FIGURE 19.1

The constraints on $\sin^2\theta_{12}$ and Δm_{21}^2.

and Δm_{21}^2 emerges from the complementarity of solar and reactor neutrinos. From the global three-flavor analysis it follows (1σ errors)

$$\sin^2\theta_{12} = 0.304^{+0.022}_{-0.016} \qquad \Delta m_{21}^2 = 7.65^{+0.23}_{-0.20} \times 10^{-5} \text{ eV}^2 \qquad (19.1)$$

The numerical χ^2 profiles shown in Fig. 19.1 have very good accuracy in the Gaussian shape $\chi^2 = (x - x_{\text{best}})^2/\sigma^2$, when the different σ for upper and lower branches are used as given in Eq. (19.1). Spectral information from KamLAND data leads to an accurate determination of Δm_{21}^2 with the remarkable precision of 8% at 3σ, defined as $(x^{\text{upper}} - x^{\text{lower}})/(x^{\text{upper}} + x^{\text{lower}})$. It appears that the main limitation for the Δm_{21}^2 measurement comes from the uncertainty on the energy scale in KamLAND of 1.5%. KamLAND data also start to contribute to the lower bound on $\sin^2\theta_{12}$, whereas the upper bound is dominated by solar data, most importantly by the CC/NC solar neutrino rate measured by SNO. The SNO-NCD measurement reduces the 3σ upper bound on $\sin^2\theta_{12}$ from 0.40 to 0.37. In Fig. 19.2 the values of leading atmospheric oscillation parameters θ_{23} and $|\Delta m_{31}^2|$ obtained from the atmospheric and accelerator neutrino data are shown. The following best fit points are (1σ errors):

$$\sin^2\theta_{23} = 0.50^{+0.07}_{-0.06}, \qquad |\Delta m_{31}^2| = 2.40^{+0.12}_{-0.11} \times 10^{-3} \text{ eV}^2. \qquad (19.2)$$

The determination of $|\Delta m_{31}^2|$ is dominated by spectral data from the MINOS long-baseline ν_μ disappearance experiment, where the sign of Δm_{31}^2 (that is, the neutrino mass hierarchy) is undetermined by present data. The measurement of the mixing angle θ_{23} is still largely dominated by atmospheric neutrino data from Super-K with a best fit point at maximal mixing. Small deviations from maximal mixing due to subleading three-flavor effects have been found in [63].

It should be stressed that agreement of the theory with solar neutrino experiments is guaranteed providing that the solar neutrino deficit is caused by the MSW effect in the

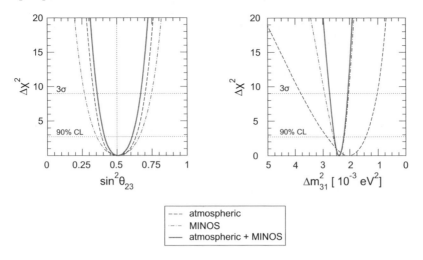

FIGURE 19.2

The leading atmospheric oscillation parameters θ_{23} and $|\Delta m_{31}^2|$.

adiabatic regime. The solution found in this case has large vacuum mixing angles (LAM solution). Note that calculations within any solar model needs the values of the cross sections of nuclear reactions that constitute different branches of the most intensive solar cycle, namely, the pp cycle. Because of the exceptionally low probability of penetration through the Coulomb barrier at corresponding energies (in the region of the so-called *Gamow peak*), these cross sections are so small that they could be measured only recently. LUNA (Laboratory Underground Nuclear Astrophysics) is one of such experiments. From the beginning it was based on an ion accelerator 50 kV placed underground in Gran Sasso National Laboratory (LNGS), which gives a high-intensive beam of $^3He^+$ (500 mA). In the process of the experiment, one managed to determine the total cross section σ of the reaction

$$^3He + {}^3He \rightarrow 2p + {}^4He \qquad (19.3)$$

as a function of the energy. It turns out that $\sigma(E)$ could be presented in the form

$$\sigma(E) = \frac{S(E)}{E} \exp\left(-31.3 z_1 z_2 \sqrt{\frac{M}{E}}\right), \qquad (19.4)$$

where M is the reduced mass of initial nuclei in atom mass units, z_1 and z_2 are the nucleus charges and E is the energy in the center-of-mass frame expressed in keV. The factor $S(E)$ is called an *astrophysical coefficient* and used for extrapolation of the cross section in the low-energy region providing smoothness of the function $\sigma(E)$. Since 2001 the LUNA-2 experiment employing the 400 keV accelerator has been in operation. LUNA-2 investigates the reactions $^{14}N(p,\gamma)^{15}O$, $^3He(^4He,\gamma)^7Be$, and $^7Be(p,\gamma)^8B$.

Now we proceed to the discussion of the experimental status of the third oscillation angle, θ_{13}. Measuring this angle represents subleading effects.* It is necessary to note the following. If the CP violation effect in the lepton sector enters into consideration, then the

---

*Among these effects are also the establishment of the CP violation in neutrino oscillations for a value of the Dirac CP phase $\delta \neq 0, \pi$ and the identification of the type of the neutrino mass hierarchy, which can be normal ($\Delta m_{31}^2 > 0$) or inverted ($\Delta m_{31}^2 < 0$).

neutrino mixing matrix in the SM takes the form

$$U = \begin{pmatrix} c_{12}c_{13} & s_{12}c_{13} & s_{13}e^{-i\delta} \\ -s_{12}c_{23} - c_{12}s_{23}s_{13}e^{i\delta} & c_{12}c_{23} - s_{12}s_{23}s_{13}e^{i\delta} & s_{23}c_{13} \\ s_{12}s_{23} - c_{12}c_{23}s_{13}e^{i\delta} & -c_{12}s_{23} - s_{12}c_{23}s_{13}e^{i\delta} & c_{23}c_{13} \end{pmatrix}. \qquad (19.5)$$

From (19.5) it follows that the CP phase always appears in combination with $\sin\theta_{13}$, that is, the value of $\sin\theta_{13}$ will finally determine the sensitivity in looking for the CP violation. Therefore, the first step in the investigation of the CP parity violation is the exact definition of θ_{13}. At present, there exist two strategies for the definition of this oscillation angle. They are: usage of collider neutrino beams and very exact experiments with reactor antineutrinos.

The first strong limit on θ_{13} was obtained by the CHOOZ Collaboration [64]. The CHOOZ experiment detected antineutrino fluxes from two reactors located at the distance 1 km. It represented the disappearance experiment. Since the values of L/E in the CHOOZ experiment were on the right from the first maximum of atmospheric oscillations, the reached sensitivity was maximal. Not having discovered oscillations, CHOOZ defined the limit on θ_{13}

$$\sin^2 2\theta_{13} = 0 \pm 0.05 \qquad (\theta_{13} < 10^0). \qquad (19.6)$$

Similar to the case of the leading oscillation parameters, also the bound on θ_{13} emerges from an interplay of different data sets. An important contribution to the bound comes, of course, from the CHOOZ reactor experiment combined with the determination of $|\Delta m_{31}^2|$ from atmospheric and long-baseline experiments. Due to a complementarity of low- and high-energy solar neutrino data, as well as solar and KamLAND data, one finds that also solar+KamLAND provide a nontrivial constraint on θ_{13}, see, for example, Ref. [65]. So, at 90% CL (3σ) the following limits are obtained

$$\sin^2\theta_{13} \leq \begin{cases} 0.060 \ (0.089) \ (\text{solar+KamLAND}) \\ 0.027 \ (0.058) \ (\text{CHOOZ+atm+K2K+MINOS}) \\ 0.035 \ (0.056) \ (\text{global data}). \end{cases} \qquad (19.7)$$

As has been noted in Ref. [66] the slight downward shift of the SNO CC/NC ratio due to the SNO-NCD data leads to a *hint* for a nonzero value of θ_{13}. From the combination of solar and KamLAND data we find a best fit value of $\sin^2\theta_{13} = 0.03$ with $\Delta\chi^2 = 2.2$ for $\theta_{13} = 0$, which corresponds to a 1.5σ effect (86% CL). We recall that the survival probability in KamLAND is given by

$$P_{ee}^{\text{KamL}} \approx \cos^4\theta_{13}\left(1 - \sin^2 2\theta_{12}\sin^2\frac{\Delta m_{21}^2 L}{4E}\right) \qquad (19.8)$$

leading to an anticorrelation of $\sin^2\theta_{13}$ and $\sin^2\theta_{12}$ while in contrast, for solar neutrinos one has

$$P_{ee}^{\text{solar}} \approx \begin{cases} \cos^4\theta_{13}\left(1 - \frac{1}{2}\sin^2 2\theta_{12}\right) & \text{low energies} \\ \cos^4\theta_{13}\sin^2\theta_{12} & \text{high energies.} \end{cases} \qquad (19.9)$$

Experiments with a neutrino beam from a remote accelerator are the important program point of investigating the neutrino oscillations. This stage is mandatory because natural neutrino fluxes are not controllable, that is, their characteristics are not defined with a fair degree of accuracy. In contrast, the content and energy spectrum of an experimentally produced beam are controlled. Since 1999 in Japan the K2K experiment (the KEK to Kamioka long-baseline neutrino oscillation experiment) has been in operation: muon neutrino beam with energies 2–3 GeV produced in KEK was directed to the Super-K detector, which is located at the distance 250 km. This experiment was completed in November 2004. As a result, 112 neutrino events are observed in the Super-K while the expected number of events without oscillations is 158. The probability to observe the deficit without neutrino oscillations is estimated to be 0.0015% and this deficit confirms the prediction of the neutrino oscillation.

In FERMILAB a muon neutrino beam is produced to irradiate the MINOS detector that was built in the mine SAUDAN at a distance 735 km away from the accelerator in 2004. The experiment gives the information in the ν_μ disappearance regime. The latest data confirm the energy-dependent disappearance of ν_μ, showing significantly less events than expected in the case of no oscillations in the energy range ≤ 6 GeV, whereas the high-energy part of the spectrum is consistent with the no oscillation expectation.

The CNGS (CERN to Gran Sasso) project on the ν_τ appearance is very perspective too. Its design is based on a previous CERN experiment, the West Area Neutrino Facility (WANF), where the SPS collider was the source of the ν_μ neutrinos and neutrino detection at a distance about 850 m was realized by two detectors, CHORUS and NOMAD (Neutrino Oscillation Magnetic Detector) [67]. If CERN WANF belonged to the class of *short-baseline* experiments, then CNGS belongs to the class of *long-baseline* experiments. The CNGS is a ν_μ beam produced with 400 GeV protons extracted from the SPS complex at CERN. During one year in a mode where the use of the SPS is shared with LHC operation, 4.5×10^{19} protons on target can be delivered, assuming 200 days of operation.* A primary proton beam impinges on a graphite target. This creates a secondary beam containing kaons and pions. Two specific magnets, a *horn* and a *reflector*, focus the secondary beam towards Gran Sasso and select high energy (20–50 GeV) π^+ and K^+. This secondary beam travels through a decay tunnel, a 1-km-long vacuum tube; π^+/K^+ being not stable, they decay into $\mu^+\nu_\mu$. At the end of the decay tunnel is a hadron stop, consisting of 3 m of graphite and 15 m of iron, and designed to stop all remaining protons and π/K. Muons pass through the monitors and are absorbed later on, in the rock. The neutrino beam is a high-energy beam optimized for the ν_τ appearance with a mean neutrino energy of about 17 GeV. We recall that in order to be sensitive in the oscillation parameter region delimited mainly by the Super-K results, the average energy of the ν_μ beam must be ~ 17 GeV. The fractions ν_e/ν_μ, $\nu_{\overline{\mu}}/\nu_\mu$, and ν_τ/ν_μ in the beam are expected to be as low as 0.8%, 2%, and 10^{-7}, respectively. So, the experiment is aimed at detecting the appearance of tau-neutrinos from the oscillation of muon-neutrinos during their 3 millisecond travel from Geneva to Gran Sasso. The ν_τ appearance search is based on the observation of events produced by the CC interaction with the τ decaying in all possible decay modes. Since the expected event rate is small, it is crucial to separate efficiently the ν_τ CC events from all the other flavor neutrino events and to keep the background at a very low level. From the beginning two detectors are scheduled to identify the events by exploiting the τ specific properties characterized by a non-negligible lifetime and the presence of a missing transverse momentum due to the final state ν_τ produced in the τ decays. The two proposed detectors, OPERA (Oscillation Project with Emulsion-tRacking Apparatus) and ICARUS (Imaging Cosmic And Rare Underground Signals), use two different approaches for identifying the events. The choice made by OPERA is to observe the τ decay topology in nuclear emulsions. In OPERA, τ resulting from the interaction of ν_τ is observed in *bricks* of photographic emulsion films interleaved with lead plates. The apparatus contains about 150,000 of such bricks for a total mass of 1300 tons and is complemented by electronic detectors (trackers and spectrometries) and ancillary infrastructure. ICARUS should separate the ν_τ CC events from the background

---

*The SPS complex operates in supercycles, for example, for the 2007 data a supercycle lasted 39.5 s, and consisted of a long fixed target cycle, 3 CNGS cycles, and one machine development cycle. Each CNGS cycle is 6 s long. It includes 2 extractions, lasting 10.5 μs each, and separated by 50 ms. Each detector reading is logged for every extraction.

through kinematical criteria using a large volume TPC (Time Projection Chambers) filled with liquid argon. The ICARUS detector was expected to have an initial effective volume of 3 kt with a 10-year running time. OPERA was expected to have an effective volume of 2 kton and a running time of 5 years. However, to fulfill such a program in full measure proved to be impossible. Then the CHGS Collaboration is limited by the use of the OPERA detector only. The OPERA detector was installed in the Gran Sasso underground laboratory. The 2400 meters of rock above the experimental halls provide a very efficient cosmic ray shielding. If the $\nu_\mu \to \nu_\tau$ oscillation hypothesis is confirmed, the number of τ's produced via the CC interaction at the Gran Sasso will be about 15 /kton/year for $\Delta m^2 = 2.5 \times 10^{-3}$ eV^2 at full mixing. After two short runs in 2006 and 2007, OPERA started significant data taking in 2008 when about 1700 neutrinos interacted in the target bricks. The Collaboration is finishing the extraction of data from the emulsions and their analysis, locating the final sample of vertices for the ν_τ candidate search [68]. Charm-induced events study will be an important tool to improve control of efficiencies and background. A new physics run has started in June 2009; 3.5×10^{19} protons on target could be delivered in about 160 days and $\sim$3600 events collected in the OPERA bricks. About 2 τ's will then be expected in the total sample 2008–2009, opening the way to the direct observation of $\nu_\mu \to \nu_\tau$ oscillations.

There are two collider experiments, T2K (Japan) and NOνA (USA), which are planned to start in the near future. They will search for the appearance of electron neutrinos in a beam of mainly muon neutrinos, from the decay of pions and kaons produced at a proton accelerator. The following project is also discussed: a neutrino detector is located in Gran Sasso and a neutrino source, muon collider producing a neutrino flux with the energy 10–30 GeV, is either in FERMILAB (L =7400 km) or in KEK (L =8800 km).

There are three reactor neutrino oscillation experiments (Double Chooz, Daya Bay, and RENO) currently under construction, which are expected to start data taking soon (see, for review, [69]). Reactor experiments look for the disappearance of $\bar{\nu}_e$, governed by θ_{13}, where the neutrinos are produced in the nuclear fission processes in commercial nuclear power plants. All three above-mentioned experiments use a liquid scintillator doped with Gadolinium in order to exploit the coincidence of a positron and a neutron from the reaction $\bar{\nu}_e + p \to e^+ + n$. The two crucial parameters determining the final sensitivity are the total exposure (proportional to the detector mass times the time-integrated thermal power of the reactor over the lifetime of the experiment) and the systematic uncertainty. All three proposals use the concept of near and far detectors in order to reduce uncertainties on the initial neutrino flux.*

Let us proceed to the discussion of non-oscillation experiments. Among these are: (i) cosmological observations (Large Scale Structures, anisotropies in the Cosmic Microwave Background, and so on), that in good approximation probe the sum of neutrino masses, $m_{cosmo} \equiv \sum_i m_i$; (ii) neutrinoless double-beta decay experiments, that probe the absolute value of the ee entry of the neutrino Majorana mass matrix m, $|m_{eff}^{(\nu_e)}| = |\sum_i U_{ei}^2 m_i|$; and (iii) β-decay experiments, that in good approximation probe $m_{\nu_e}^2 \equiv (m \cdot m^\dagger)_{ee} = \sum_i |U_{ei}^2| m_i^2$.

We shall be limited by the two first directions. Note that cosmological observations do not distinguish whether the neutrino is a Majorana or Dirac particle. A cosmological constraint on the sum of the neutrino masses is primarily a constraint on the relic Big Bang neutrino density Ω_ν. One can relate this density to the sum of the individual mass eigenstates $\sum m_i$ as given by [70],

$$\Omega_\nu = \frac{\sum m_i}{93.14 H_0^2 \text{ eV}}, \tag{19.10}$$

---

*From the expression for the electron antineutrino survival probability (see, for example, (19.7)) it follows that for each reactor two optimal distances for localization of detectors, a near and far basis line, exist.

where H_0 is the Hubble constant. The direct effects of the neutrinos depend on whether they are relativistic, nonrelativistic, and also the scale under consideration. The neutrinos have a large thermal velocity as a result of their low mass and subsequently erase their own perturbations on scales smaller than what is known as the *free streaming length* [71]. This subsequently contributes to a suppression of the statistical clustering of galaxies over small scales and can be observed in a galaxy survey. The abundance of neutrinos in the universe can also have a direct effect on the primary cosmic microwave background anisotropies if nonrelativistic before the time of decoupling (that is, when sufficiently massive). However, one of the most clear effects at this epoch is a displacement in the time of matter-radiation equality. All these cosmological effects can be used to impose the following bounds on the neutrino masses

$$\sum_i m_i < (0.17 - 2.0) \text{ eV}. \tag{19.11}$$

If there are just three ν_i, and their spectrum is either normal ($\Delta m^2_{31} > 0$) or inverted ($\Delta m^2_{31} < 0$), then Eq. (19.11) implies that the mass of the heaviest ν_i (m_{heav}) cannot exceed (0.07–0.7) eV. Moreover, this mass obviously cannot be less than $\sqrt{\Delta m^2_{atm}}$, which in turn is not less than 0.0048 eV. Then we have

$$0.0048 \text{ eV} < m_{heav} < (0.07 - 0.7) \text{ eV}. \tag{19.12}$$

Now we shall consider looking for the $0\nu2\beta$ process. By 2004 the results were presented by two Collaborations, MAINZ and TROITSK. At that, comparable limits were obtained: $m^{(\nu_e)}_{eff} = (-1.2 \pm 2.2 \pm 2.1)$ eV and $m^{(\nu_e)}_{eff} = (-2.3 \pm 2.5 \pm 2.0)$ eV. Combining both results and summing the errors, we get

$$m^{(\nu_e)}_{eff} = -1.7 \pm 2.2 \text{ eV}, \quad \text{that is} \quad m^{(\nu_e)}_{eff} < 2.0 \text{ eV at} \quad 99\% \text{ C.L.} \tag{19.13}$$

To date, five Collaborations, HM (Heidelberg-Moskow), IGEX (International Germanium EXperiment), CUORICINO (Cryogenic Underground Observatory for Rare Events), DAMA/LXe (DArk MAtter) and NEMO3 (Neutrino Ettore Majorana Observatory), offered the data concerning searches of the $0\nu2\beta$ decay (see, for review, [72]). The results are presented in Table 19.1. The factor h enters to the matrix element of the $0\nu2\beta$ decay. It

TABLE 19.1
The results of looking for the neutrinoless double beta decay.

nucleus	Present bound on $|m^{(\nu_e)}_{eff}|/h$ in eV at 90% CL	
^{76}Ge	0.35	HM
^{76}Ge	0.38	IGEX
^{130}Te	0.42	CUORICINO
^{100}Mo	1.7	NEMO3
^{136}Xe	2.2	DAMA/LXe

is calculated within nuclear physics and its form depends on the nucleus kind. The results presented in the literature gave different limits. Thus, for example, for nuclei Ge and Te the following limits

$$0.3 < h(^{76}\text{Ge}) < 2.4, \quad 0.4 < h(^{130}\text{Te}) < 2.7 \tag{19.14}$$

exist. As clear from Table 19.1 the main present experiments are HM, IGEX, and CUORI-CINO. Two last experiments reached very low background ($\sim 0.18/\text{keV} \cdot \text{kg} \cdot \text{year}$), high detection efficiency, 70% and 84%, and sizeable energy resolution—the total width at half-maximum is equal to 4 keV and 7 keV, respectively. These parameters are considerably better than in the HM experiment. In Table 19.2 the data of HM, IGEX, and CUORICINO Collaborations are listed.

TABLE 19.2
The data on $0\nu2\beta$ decay.

Nucleus and experiment	Total events	Background	Predicted signal	Exposure in 10^{25} nuclei·yr (99% C.L)
^{76}Ge HM	21	20.4 ± 1.6	$76\lvert m_{eff}^{(\nu_e)}/\text{eV}\rvert^2/h^2$	25
^{76}Ge IGEX	9.6	17.2 ± 2.0	$23.5\lvert m_{eff}^{(\nu_e)}/\text{eV}\rvert^2/h^2$	7
^{130}Te CUORICINO	24	35.2 ± 4.0	$21.5\lvert m_{eff}^{(\nu_e)}/\text{eV}\rvert^2/h^2$	0.14

The definitive establishment of the neutrino sector structure allows us to solve the problem: "Whether the SM is the true model of the electroweak interaction or a model with the other gauge group must come to take the SM's place." In the SM extension new gauge bosons, new Higgs bosons, or superpartners of the SM particles are usually present. These additional particles interact with leptons. Some of them have masses lying at the electroweak scale. Therefore, when propagating through matter, neutrinos will be subjected to nonstandard interactions, whose intensity may prove to be comparable with the SM interaction. This, in turn, will lead to modification of the potential matter and essentially influence the oscillation pattern. And at the same time, it is obvious that the vacuum neutrino oscillation has the same form both in the SM and in the SM extensions.

The left-right model (LRM) gives the most consistent description of the neutrino behavior and neutrino properties. In this model, as in any the SM extensions, the motion of the neutrino flux in a matter is described within the hybrid scheme, *neutrino oscillations + nonstandard neutrino interactions*. This, in turn, means that the investigations could go in two different directions. The first approach, which is more traditional, uses the current best-fit points for the neutrino oscillation parameters in order to obtain the bounds on nonstandard neutrino interactions (see, for example, [73, 74]). The second way [75, 76, 77] is to set the values of the corrections to the matter potential and find allowed regions of the neutrino oscillation parameters with the help of solar and/or atmospheric neutrino experiments. An explanation of mass smallness of the ordinary neutrino ν_i due to the seesaw mechanism is one of the LRM advantages. Heavy neutrinos N_i ($i = e, \mu, \tau$) are partners of ν_i in the seesaw mechanism. However, it is not excluded that this is not a single function they perform in nature. Thus, at the normal mass hierarchy ($m_{N_1} < m_{N_2} < m_{N_3}$), N_2 and N_3 could prove to be unstable particles and provide the baryon asymmetry of the universe, thanks to the leptogenesis [78]. The N_1 neutrino could happen to be stable and the N_1 gatherings having produced during the first instant of the universe existence must be still present around us. Therefore, the N_1 neutrino may be successfully considered as a candidate on a cold matter in the universe. The reason of non-observation of these neutrinos lies in the smallness of their interaction with ordinary matter, that is, they belong to the class of weak interacting massive particles. However, as it was shown in Ref. [79], in the LRM, the N_l neutrino properties can be defined without their direct detection, that is, using indirect methods.

The first neutrino telescope (NT), Homestake, was put into operation in 1967. This year could be called the birth date of a new branch of astronomy, neutrino astronomy (NA). The NA presents studies of cosmic objects on the basis of incoming neutrino fluxes. Without NA, electromagnetic waves (visible light, IR and UV radiation, short wavelength radio waves) were the only accessible type of radiation falling down on the Earth from outer space. Note that electromagnetic waves are emitted only from the surface layer of celestial bodies. In the process of motion from their source to the terrestrial observer, electromagnetic waves interact with cosmic rays, particles of the Earth's atmosphere, and so on, losing the major part of information about their progenitors. Following the steps of the conventional gamma-astronomy that covers a very wide range of the wavelengths $\lambda = (1 \div 10^{-12})$ m, NA shows a tendency to expansion of its potential to cover the whole energy range of neutrinos occurring in the universe. By their generation sources and energy range these neutrinos may be subdivided into cosmological (relict), stellar and high-energy cosmic neutrinos.

At early stages of the universe expansion neutrinos are in thermal equilibrium with matter. About 1 s after the Big Bang, the temperature has dropped to $\sim 10^{10}$ K. At that, the particle concentration in a cosmic plasma decreased and the free neutrino path increased so much that they went out of the thermal equilibrium with a plasma. Hot neutrino gas involving all the kinds of neutrino and antineutrino separated from matter and, expanding with the universe, became cold as an independent matter substance. Neutrino gas remaining from that epoch is called *relic neutrinos*. The temperature of this gas is presently equal to ~ 1.9 K and the average energy amounts to 5×10^{-4} eV. With the use of modern experimental techniques the detection of relict neutrinos is very difficult due to extremely small cross sections of their interaction with matter.

Two classes of stellar neutrino sources exist. The former relates to the quiescent (static) stars, similar to our Sun, producing neutrinos in the process of nuclear fusion reactions that provide the observable luminosity. The energy of these neutrinos range from a few fractions to several dozens of MeV. The latter is represented by the collapsed stars, that is, those collapsing to a neutron star or to a black hole. At the final stage of stellar cores evolution their density is increased up to 10^{10}–10^{15} g/cm^3, and their temperature—up to 10^{10}–10^{12} K. The principal mechanism responsible for the energy loss in these conditions is emission of neutrinos through the reactions:

$$e^+ + e^- \to \nu_l + \overline{\nu}_l, \qquad e^- + p \to n + \nu_e, \qquad e^+ + n \to p + \overline{\nu}_e.$$

The energy carried away by neutrinos (all types of neutrinos are radiated) may amount to tens of percentages of the star mass. The duration of neutrino radiation comes to 10–20 s, and the average energy is 10–12 MeV. The outbursts of the supernova SN 1972E, SN 1987A, SN 1993J may be taken as an example of such radiation. The stellar neutrinos are successfully detected by modern experimental facilities.

Cosmic neutrinos are defined as those produced by cosmic rays. The energies, especially convenient to search for the local sources of cosmic neutrinos, are tens of GeV and above. The lower limit of this range is determined by the requirement of a smallness of the angle between the momentum directions of incident neutrino and outgoing particle (for example, muon) during the reaction used for neutrino registration. This requirement is essential in determining the direction for the source. As the energy is reduced, the angle in question increases and the background of atmospheric neutrinos within the solid angle in the source direction grows too. Note that the energies of cosmic neutrinos could reach 10^{12} GeV.

The effect of magnetic fields on the neutrino is minor. The cross section of neutrino scattering from the interstellar matter is also small. To illustrate, for the νN-interaction the characteristic cross section in case of high-energy neutrinos ($E_\nu \sim 1 \div 10^3$ TeV) measures $\sim 10^{-35} \div 10^{-33}$ cm^2 (in case of low-energy neutrinos it is still smaller). If one assumes that the matter along the whole neutrino path has the density that is equal to the galactic density (≈ 1 nucleon/cm^3), then the mean free path of neutrino amounts to $\sim 10^{33} \div 10^{35}$ cm, being well in excess of the universe radius. Neutrino radiation is the only radiation type that comes to the terrestrial observer from the extraterrestrial source carrying almost invariable information about the progenitor. Thus, NA is characterized by a number of unique features making it superior to gamma-astronomy. Enumerate some of them.

First, operating in the high- and superhigh-energy region, NA can widen the horizon of the observable universe and deliver information about extremely distant cosmological epochs. γ-astronomy is inefficient in the high-energy region because of a very small free path of the associated γ-quanta due to their scattering from relict radiation in the intergalactic space.

Second, in the universe one can find the objects radiating extremely low γ-fluxes, whereas their neutrino fluxes are very great. Such objects are called the *hidden sources*. Among these objects are young supernova shells, active galactic nuclei, black holes, and so on. Consequently, the following merit of the NA is its effectiveness in the detection of hidden sources. Also, the NA is used to search for bright phases of galaxies and antimatter in the universe.

Third, analysis of the high-energy neutrino spectra from cosmic sources, in principle, enables registration of the relict neutrino background of the universe. Actually the calculations demonstrate that in the case when high-energy neutrinos are scattered from the background neutrinos

$$\nu_l + \overline{\nu}_l \to Z^* \to l^- + l^+,$$

at the energy of $(E_\nu)_r = m_Z^2/(2m_\nu)$ one can observe the resonance associated with a Z-boson. Then, for the source-radiated neutrinos with the energy $E_\nu = (E_\nu)_r$, reducing the neutrino flux will be within the limits from 15 to 50%.

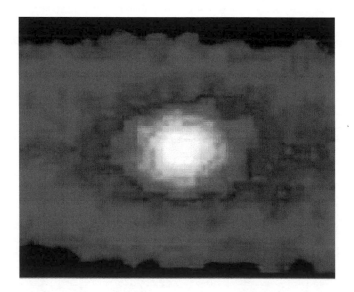

FIGURE 19.3
The Sun's neutrinography.

Fig. 19.3 presents the first neutrino image of the Sun (Sun neutrinography). However, the Sun in the neutrinography is greater in size than in ordinary photography. This stems from the fact that the direction of neutrinos arrival in modern NTs is determined less accurately than the direction of photons. At the same time, the NA is still making the first steps, and its maturity may be attained only upon the definitive establishment of a structure of the neutrino sector and production of high-resolution NT. No doubt that with its help we shall then manage to obtain complete and exact information concerning the Sun's structure. Studying the solar activity will be one of the main sources of receiving this information. Solar flares represent the characteristic manifestation of the Sun's activity. Let us demonstrate that these phenomena could be really investigated using the NA.

At certain conditions the evolution of active regions (ARs) on the Sun may lead to the appearance of solar flares (SFs), which represent the most powerful events of the solar activity. Magnetic energy of sunspots is transformed into kinetic energy of matter emission (at a speed of the order of 10^6 m/s), energy of hard electromagnetic radiation, and fluxes of so-called solar cosmic rays (SCRs). For the most part, the SCRs consist of protons with $E_k \geq 10^6$ eV, nuclei with charges $2 \leq z \leq 28$ and energy within an interval from 0.1 to 100 MeV/nucleon, and electrons with $E_k \geq 30$ keV. Relative content of nuclei with $Z \geq 2$ reflects mainly solar atmosphere composition while proton portion depends on flare power (for big SF power is about of $\sim 10^{29}$ erg $\cdot$ s^{-1}). A pretty popular mechanism of the SF appearance is based on breaking and reconnection of magnetic field strength lines of neighboring spots. In accordance with this model, a change of magnetic field configuration in sunspots which are grouped in pairs of opposite polarity might lead to the appearance of a *limiting strength line* common for the whole group. Through the limiting line the redistribution of magnetic fluxes takes place, which is necessary for the magnetic field to have the minimum energy. The limiting strength line rises from photosphere to corona. From the moment of the line appearance an electric field induced by magnetic field variations causes a current along the line that takes the form of a current layer due to the interaction with a magnetic field. As the current layer prevents the magnetic fluxes redistribution, the process of magnetic energy storage of the current layer begins. The duration of appearance and formation period of the current layer (an initial SF stage) varies from several to dozens of hours. The second stage (an explosion phase of the SF) has a time interval of 1–3 minutes. It begins from the appearance in some part of the current layer of a high-resistance region that results in a current dissipation. Then, due to penetration of the magnetic field through the current layer a strong magnetic field appears perpendicular to it. An arising magnetic force breaks the current layer and throws out a plasma at a great speed. The height where particles acceleration takes place is not the same for different SFs. Acceleration regions may be located either in the chromosphere where plasma particle concentration is $n \sim 10^{13}$ cm^{-3} or in the corona at $n \sim 10^{11}$ cm^{-3}. The particle distribution, according to energies and charges during motion through the interplanetary medium, is defined both by a mechanism of their acceleration at the SF and by peculiarities of the exit from an acceleration region. For high-energy particles $E_k \geq 10^8$ eV the time dependence of the flux intensity near the Earth represents an asymmetrical bell-like curve with the very quick increase (minutes or dozens of minutes) and a slow decrease (from several hours to one day). The produced flux on the Earth's surface for the most powerful SF may reach $\sim 4500\%$ in comparison to the background flux of cosmic particles. The concluding stage (the hot phase of the SF) is characterized by the existence of a high temperature coronal region and can continue for several hours. The heating of dense atmospheric layers leads to an evaporation of a large amount of gas that favors a long-continued existence of a dense hot plasma cloud. One of the characteristic futures of flares is their isomorphism, that is, the repetition in one and the same place with the same field configuration. A small flare may repeat up to 10 times per day while a large one may take place the next day and even several times during the active region lifetime. Besides, the stronger a flare is, the larger magnetic field gradient precedes its appearance.

It turns out, that a neutrino flux passing through the SF region changes its content. For the first time, this correlation was predicted in Ref. [80]. The detailed analysis of the resonance conversions of a neutrino beam at the SF was given in Ref. [23]. Let us list the factors that result in the change of the neutrino beam content:

(i) The change in twisting rate of magnetic field $\dot{\Phi}$ and electric current density j_z in an active zone from pre-flare to post-flare periods may lead to the disappearance or appearance of the neutrino resonance transitions.

(ii) The adiabaticity conditions may be violated on all three SF phases, resulting in the increase of the ν_{eL} flux observed by terrestrial detectors.

(iii) When a neutrino flux crosses a dense plasma cloud evaporated during the SF in the upper atmospheric layers, the conditions for the resonance transitions could be satisfied.

(iv) A flux of particles accelerated up to gigantic energies will also influence the neutrino moving along it (in the simplified form the flux motion could be imagined as a motion with the different speed of three regions, each separately consisting of electrons, nuclei, and protons, respectively).

It should be stressed that the correlation between neutrino flux and solar flares is predicted both within the SM and within any SM extension.

We note one more circumstance connected with the Sun's activity. Neutrinos crossing sunspots, practically all the time in their motion through the Sun and its atmosphere, are in the region of an intensive magnetic field. According to the existing viewpoints the structure of this field is rather complicated

$$\text{rot } \mathbf{B}_z \neq 0, \qquad \dot{\Phi} \neq 0,$$

and the value of the magnetic field B can vary from $\sim 10^8$ Gs in the central region to $\sim 10^3$ Gs in the coronal part of the Sun. A resonance picture for such neutrinos might prove to be richer than for neutrinos, which on their way out, do not face sunspots. Therefore, a comparison of neutrino fluxes crossing and escaping the sunspot regions will serve as a source of information on the electromagnetic field structure of the Sun.

Information incompleteness concerning the Earth's structure needs the development of new methods. Among them is a neutrino geophysical tomography (geotomography) possessing very promising perspectives. All methods of geotomography are based on acquiring the information about the physical properties of elements occurring in the Earth's thickness with the use of the summarized effects measured at its surface. Presently, the data on the Earth's structure are acquired using seismic and gravitational tomography. Seismic tomography is associated with registration of time spent by seismic waves on covering the distance from the internal regions of the Earth to the detectors positioned on the surface at different distances from the source. Seismic waves are produced by earthquakes, the deepest seismic foci of which are allocated at a level of about 700 km. An analysis of the measurement results makes it possible to determine the values of seismic speeds for bulk (transverse and longitudinal) waves, or so-called speed profiles. Nevertheless, the features of the outer Earth shells may be successfully established by seismic tomography, while three inner shells including the outer core, transition core region, and inner core are inaccessible as irradiance of these regions by seismic waves is extremely weak. It is assumed that within the outer core the convection processes governing the magnetic field of the Earth are proceeding, and that it is rotating faster than the solid Earth by $1 - 3°$ per year. Also, the inner core oscillations near the Earth's center are expected. We can only guess about the state of the matter in the inner core. But taking into account that seismic waves are transmitted through the core, we may conclude that the aggregate state of the inner core is a solid. By now this core remains inadequately studied due to the shielding effect of the liquid interlayer (outer core) inaccessible for seismic waves, i.e., for seismic tomography there is no physical effect in this region. One of the latest hypotheses [81] suggests that at the center of the Earth a mixture of uranium and plutonium maintaining the continuous nuclear reaction could be found. This core representing a *giant natural nuclear reactor* is almost 8 km in diameter. Due to the activity of this nuclear core, a high-power magnetic field formed around the Earth protects our planet against hazardous cosmic rays, capable of wiping out all forms of living biological objects in a few seconds. This natural reactor provides energy for continental drift and manifests itself as volcanic eruption.

Thus, applicability of seismic tomography is limited; its measurement accuracy is relatively low and, what is more, it fails to control the initial conditions. The potentialities of gravitational tomography are limited even more, as it is based on measurements of the terrestrial gravitational field by changes in the free fall acceleration values. Neutrino tomography can outperform the seismic and gravitational tomography for accuracy by some orders of magnitude, enabling determination of the Earth's structure at a radically new degree of quality. The development of neutrino tomography will be realized in two directions. The former is based on the use of high-energy collider neutrinos. The scattering cross section of neutrinos from nucleons σ_ν proves to be proportional to the neutrino energy E_ν, namely $\sigma_\nu \simeq 10^{-35} E_\nu$ cm^2. So, the part of neutrinos withdrawn from the initial beam through the interaction with nucleons of the matter nuclei is proportional to the nucleon number N_m at the beam path per unit area. On the other hand, N_m is determined by:

$$N_m = N_A m(L) = N_A < \rho > L,$$

where N_A is an Avogadro number, $m(L)$ is a summarized matter mass in 1 cm^2, and $< \rho >$ is an average matter density along a path L. Then, one can obtain the mass $m(L)$ by measuring the absorption degree of neutrinos in the path L. Detailed information about $m(L)$ enables one to establish the in-depth variation of the matter density without any additional assumptions. This is the great advantage of neutrino tomography compared to the seismic one, where the matter density may be recovered only on the introduction of additional assumptions concerning the values of the elastic modulus for the in-depth regions of the Earth. Since the cross section σ_ν is the same for nucleons of the matter, both at the surface and in-depth of the Earth, its value may be found by the laboratory measurements and hence neutrino tomography is free from the above-mentioned ambiguities. Let us estimate the neutrino energies required for geotomography. Attenuation of neutrino flux J through absorption is exponential in character:

$$J(L) = J(0) \exp\left[-\frac{L}{L_\nu}\right],$$

where $L_\nu = (N_A \rho \sigma_\nu)^{-1}$ is an absorption length in which the flux is lowered by a factor equaling to $e = 2.718281828....$ The total mass in the path of a neutrino beam passing along the Earth's diameter comes to about 1.2×10^{10} g/cm^2. This value is associated with $\rho = 10$ g/cm^3 giving:

$$L_\nu = 1.7 \times 10^5 \text{ km}/E_\nu.$$

Comparison between the obtained absorption length and the Earth's diameter demonstrates that these values correlate at the neutrino energies of the order of TeV. For instance, at $E_\nu = 10$ TeV the Earth will absorb about half of the initial neutrino beam. Registration of the neutrino beams transmitted through the Earth's thickness with energies over the range from a few fractions to tens of TeV enables one to obtain a detailed neutrinography of the Earth. This type of neutrino geotomography makes it possible to have information about the in-depth distribution of nucleons. A detector at the far side of the Earth, serving as a photographic film in X-ray radiography, will be used to record the process of withdrawing the initial neutrinos from the beam.

The second type of neutrinography is based on the MSW effect. In this case there is no need in such huge neutrino energies. The detector provides registration of the events connected with transitions of neutrinos from one flavor to another ($\nu_l \rightarrow \nu_{l'}$). The sources may be both natural neutrinos—the neutrinos coming from the solar and stellar nuclear reactions, and artificial neutrinos—the reactor or collider neutrinos. Note that in this case the resonance neutrino conversions are influenced exclusively by the interaction with the matter electrons N_e. As the probability of resonance transitions is dependent on the energy as well, by the appropriate selection of the neutrino energy one can ensure fulfillment of the resonant conditions for the particular regions of the Earth, thus making the measurements more sensitive to certain values of $< \rho >$ and less sensitive to some others. The principal merit of this method consists in the possibility to measure not only the attenuation of the initial neutrino beam but also the flavor composition of the final neutrino beam. It should be emphasized that, as the first method provides the in-depth nucleon composition and the second method gives the in-depth electron profile, the combination of these two methods makes it possible to obtain a detailed map of the Earth structure by neutrino geotomography. The time is coming when neutrino tomography will be used for the studies of other planets as well. Note that a detection system used in neutrino tomography need not be stationary. In the case of the collider neutrinos the initial beam may have different orientation angles. Aiming the neutrino beam in such a way as to its exit to the surface takes place at the water spaces, we can use floating objects for a system of neutrino detectors. Neutrino detectors may be also mounted on artificial satellites.

Nuclear reactor monitoring in the "on-line" regime is one more a field of activity of applied neutrino physics. The nuclear reactor represents exceptionally clean and powerful source of electron antineutrinos whose spectrum is formed as a result of the β-decay of the four main fissioning isotopes ^{235}U, ^{239}Pu, ^{238}U, and ^{241}Pu that are part of the nuclear fuel. The unique penetrating power of antineutrino allows to avoid the distorting medium effect and to detect antineutrinos that are practically identical to that produced by actinoid fission, independently on a source-to-detector distance. For their detection the inverse β-decay reaction is traditionally used

$$p + \bar{\nu}_e \rightarrow n + e^+. \tag{19.15}$$

The reaction (19.15) possesses two important properties that are at the heart of "on-line" antineutrino diagnostics of reactor core. First, the antineutrino flux produced by the reactor is proportional to the number of fissions taking place in an active zone. Because the definite energy (200 MeV) is released in every fission, then the antineutrino flux is proportional to the reactor thermal power (W_{rtp}). Second, theoretical calculations and experimental data show that the spectra of antineutrinos emitted by different components of nuclear fuel differ from each other. For example, antineutrinos emitted by ^{239}Pu have smaller energy than that emitted by ^{235}U. It means that the antineutrino spectrum is "softened" as ^{235}U burns down in the active zone. As a result, the positronium spectrum becomes softer as well.

Let us find the theoretical expressions determining both the reactor power and the temporal evolution of nuclear fuel isotopic structure. In order to obtain the relation between the intensity of the detected antineutrino events $n_{\bar{\nu}}$ and W_{rtp} we must calculate the total cross section of the reaction (19.15) for monoenergetic antineutrinos. This cross section is defined by the expression

$$\sigma_{\bar{\nu}p}(E_{\bar{\nu}}) = \sigma_0(E_{\bar{\nu}})[1 + \delta_{thr}][1 + \delta_{WM} + \delta_{rec}][1 + \delta_{rad}], \tag{19.16}$$

where δ_{thr} determines the cross section behavior close to the reaction threshold, δ_{rec} determines the recoil, δ_{WM} is associated with weak magnetism*, and δ_{rad} is caused by radiative corrections. The analytical expressions for all corrections and their detailed discussion are given in Refs. [82], while the "naive" cross section $\sigma_0(E_{\overline{\nu}})$ corresponding to the approximation of infinitely heavy nucleons ($E_n \approx m_n$, $E_{e+} \ll m_n$) is as follows [82]

$$\sigma_0(E_{\overline{\nu}}) = \frac{2\pi^2(E_{\overline{\nu}} - \Delta)\ln 2}{m_e^5(f \cdot \tau/2)}[(E_{\overline{\nu}} - \Delta)^2 - m_e^2]^{1/2}, \tag{19.17}$$

where $E_{\overline{\nu}} - (m_n - m_p) = E_{\overline{\nu}} - \Delta = E_{e+}$ is the total positron energy in the reaction (19.15)†, $\Delta = 1.293$ MeV is the threshold of the process (19.15), $(f \cdot \tau_{1/2})$ is the so-called reduced neutron half-life [83], for which the phase space factor of a neutron $f = 1.7146$ is determined to an accuracy of 0.01%, and the half-life is $\tau_{1/2} = \tau \ln 2$ ($\tau = 887.4 + 0.2$ s).

The intensity of the detected antineutrino events $n_{\overline{\nu}}$ is connected with reactor thermal power by the relation [84]

$$n_{\overline{\nu}} = \frac{\gamma \epsilon_0 N_p \Xi_{\overline{\nu}} W_{rtp}}{4\pi R^2 E_f}, \tag{19.18}$$

where

$$\Xi_{\overline{\nu}} = M_{\overline{\nu}} < \sigma_{\overline{\nu}p} >, \qquad M_{\overline{\nu}} = \int_{E_{thr}}^{E_{max}} \rho(E_{\overline{\nu}})dE_{\overline{\nu}}, \qquad < \sigma_{\overline{\nu}p} > = \frac{\int_{E_{thr}}^{E_{max}} \sigma_{\overline{\nu}p}\rho(E_{\overline{\nu}})dE_{\overline{\nu}}}{\int_{E_{thr}}^{E_{max}} \rho(E_{\overline{\nu}})dE_{\overline{\nu}}},$$

E_f is an average energy per fission at the given fuel composition, N_p is the number of hydrogen atoms in the target, R is the effective distance between the source (reactor core) and the detector, ϵ_0 (γ) is the detection efficiency of positrons (neutrons) in the process (19.15), $M_{\overline{\nu}}$ is the number of electron antineutrinos per fission, and $\rho(E_{\overline{\nu}})$ is the energy spectrum of antineutrinos emitted by fission products of all fuel components (it is expressed in MeV^{-1} × fission^{-1}).

In reality, due to the heterogeneity of isotopic composition of active zone, we have to take into account that apart from ^{235}U the contributions into the common number of fissions give other fissioning isotopes too. Let us introduce the quantity a_i that represents the contribution from i-th isotope (the index $i = 5, 8, 9, 1$ extends over the four main fissioning isotopes ^{235}U, ^{238}U, ^{239}Pu, and ^{241}Pu) to total fission cross section and that depends on the method of spectrum $\rho(E_{\overline{\nu}})$ determination. Then, allowing for

$$\sum_i a_i = 1, \tag{19.19}$$

we obtain

$$\Xi_{\overline{\nu}} = \sum_i a_i \Xi_{\overline{\nu}i}, \qquad \rho(E_{\overline{\nu}}) = \sum_i a_i \rho_i(E_{\overline{\nu}}), \qquad E_f = \sum_i a_i E_{fi}. \tag{19.20}$$

Now we can obtain the basic balance equation of reactor antineutrino spectrometry, which describes the contribution a_i from each actinoid to experimentally measured energy spectrum $\eta(E_{\overline{\nu}})$. Take into consideration that

$$n_{\overline{\nu}} = \frac{1}{\Delta t} \int_{E_{thr}}^{E} \eta(E_{\overline{\nu}})dE_{\overline{\nu}}, \tag{19.21}$$

---

*Here, weak magnetism means the inclusion of the magnetic moments and mass difference of nucleons.

†At the energies of antineutrinos produced in the reactor we may neglect the neutron recoil energy.

where Δt is the measured time. Then, substituting (19.18) into (19.21) and differentiating the both sides of the obtained equation over E, we arrive at [85]

$$\eta(E_{\bar{\nu}}) = \frac{\gamma \epsilon_0 N_p}{4\pi R^2} \sum_i \rho_i(E_{\bar{\nu}}) \sigma_{\bar{\nu}p}(E_{\bar{\nu}}) \lambda_i \Delta t, \qquad (19.22)$$

where λ and λ_i are the average and partial fission rates of nuclear fuel, respectively

$$\lambda_i = a_i \lambda, \qquad \lambda = \frac{W_{rtp}}{E_f}. \qquad (19.23)$$

In order to solve Eq. (19.22) relative to λ_i, we must know not only $\sigma_{\bar{\nu}p}(E_{\bar{\nu}})$, but $\eta(E_{\bar{\nu}})$ and $\rho_{\bar{\nu}i}(E_{\bar{\nu}})$ as well. The energy spectrum $\eta(E_{\bar{\nu}})$, the spectrum "on detection place" of $\bar{\nu}_e$, can be determined directly from the positron kinetic energy T_{e+} as it differs from the detected antineutrino energy only on the threshold energy 1.804 MeV. However, due to the partial absorption of annihilation quanta in a useful capacity of the detector, the observable positron energy always will be $\Delta T_{e+} \approx 0.6$ MeV higher than the true kinetic energy, and it is necessary to take it into consideration when the energy spectrum $\eta(E_{\bar{\nu}})$ is obtained [86]

$$E_{\bar{\nu}} = (T_{e+} - \Delta T_{e+}) + 1.804 \text{ MeV} + O\left(\frac{E_{\bar{\nu}}}{m_n}\right), \qquad (19.24)$$

where $O(E_{\bar{\nu}}/m_n)$ mainly accounts for the nuclear recoil. The spectra ρ_i "on birthplace" of $\bar{\nu}_e$ are distinguished not only by "place" but also by the way of their determination. The procedure of obtaining the spectrum ρ_i for given fissionable nucleus is based on the fact that ratio of the $\bar{\nu}$-spectrum (ρ_i) to the β-spectrum (ρ_β) of fission products in secular equilibrium conditions possesses a good stability that does not depend on decay schemes [87]. Although the stability of this ratio does not have any clear physical substantiation, nevertheless, experimentally measured β-spectra of the fission fragments of ^{235}U, ^{239}Pu, and ^{241}Pu coincide with theoretical results [87, 88] with a good precision. The spectra obtained in such a way are called the converted spectra.

The technique of subsequent calculations is as follows (see, for example, Ref. [84]). Suppose we obtain the date collection presented in Table 19.3. To substitute the values of

TABLE 19.3

The experimental values of $\eta(E_{\bar{\nu}})$ and the converted energy spectra $\rho_i(E_{\bar{\nu}})$ in the 2–9 MeV range [86]. Errors are given in % at 90 CL. For the converted spectra ρ_i, the decimal exponent is shown in brackets.

$E_{\bar{\nu}}$,	$\eta(E_{\bar{\nu}})$,	$\rho_i(E_{\bar{\nu}})$, (MeV · fission)$^{-1}$			
MeV	MeV^{-1}	^{235}U	^{239}Pu	^{238}U	^{241}Pu
2.0	21.285+2.21	0.130(+1)+4.2	0.107(+1)+4.5	0.153(+1)	0.124(1)+4.3
3.0	74.143+2.40	0.673(0)+4.2	0.491(0)+4.3	0.835(0)	0.623(0)+4.0
4.0	73.042+3.11	0.283(0)+4.2	0.190(0)+4.4	0.386(0)	0.270(0)+3.9
5.0	53.357+4.73	0.105(0)+4.2	0.576(-1)+5.2	0.152(0)	0.920(-1)+4.4
6.0	25.500+1.91	0.370(-1)+4.3	0.177(-1)+6.8	0.549(-1)	0.267(-1)+5.6
7.0	7.428+1.73	0.105(-1)+4.7	0.468(-2)+11	0.176(-1)	0.683(-2)+7.0
8.0	0.134+0.56	0.136(-2)+6.0	0.500(-3)+35	0.313(-2)	0.890(-3)+11
9.0	-	0.560(-4)+27	0.398(-4)	0.124(-3)	0.470(-4)+90

$\eta(E_{\bar{\nu}})$ and $\rho_i(E_{\bar{\nu}})$ into (19.22) leads to the system of linear equations. It is clear that the common determinant of the system under consideration has many "zeroes" and this system as a whole could become quasi-degenerated. It means that we deal with the ill-conditioned system of linear equations whose solutions may be unstable with respect to small changes of input data.* Let us rewrite Eq. (19.22) in the following form

$$y = \sum_{i=1}^{n} \lambda_i x^{(i)}, \tag{19.25}$$

where

$$y = \eta(E_{\bar{\nu}}), \qquad x^{(i)} = \frac{\gamma \epsilon_0 N_p}{4\pi R^2} \Delta t \rho_i(E_{\bar{\nu}}) \sigma_{\bar{\nu}p}(E_{\bar{\nu}}). \tag{19.26}$$

After the discretization on energy, Eq. (19.25) will look like

$$y_k = \sum_{i=1}^{n} \lambda_i x_k^{(i)}, \qquad k = 1, 2, \dots, N, \tag{19.27}$$

where N is the number of energy values, at which the spectrum was measured. At this stage one usually employs the least-squares method. Demanding the minimum of the expression

$$\chi = \sum_{k=1}^{N} \left(\sum_{i=1}^{n} \lambda_i x_k^{(i)} - y_k \right)^2 \tag{19.28}$$

$(\partial \chi / \partial \lambda_i = 0)$, we arrive at the system of linear algebraic equations relative to λ_i

$$A\lambda - u, \tag{19.29}$$

where

$$A_{ij} = \sum_{k=1}^{N} x_k^{(i)} x_k^{(j)}, \qquad u_j = \sum_{k=1}^{N} x_k^{(j)} y_k, \qquad j = 1, 2, \dots, n.$$

Further, we assume that the matrix A_{ij} and vector u_j are given with the errors represented by the Euclidean norms

$$||A_k - A|| \leq h, \qquad ||u_\delta - u|| \leq \delta.$$

Then, in accordance with the Tikhonov regularization method, in order to find the normal solution of Eq. (19.22) or, equally, the system of Eqs. (19.29), we should determine a minimal norm vector on set of vectors $\lambda = (\lambda_1, \lambda_2, \dots, \lambda_n)$, satisfying the requirement

$$||A\lambda - u|| = 2(h||\lambda|| + \delta). \tag{19.30}$$

Let us use the Lagrange multiplier method, that is, find the vector λ^α minimizing the smoothing functional

$$M^\alpha(\lambda, u, A) = ||A\lambda - u||^2 + \alpha||\lambda||^2. \tag{19.31}$$

Here the parameter α is defined from the condition

$$||A\lambda^\alpha - u|| = 2(h||\lambda^\alpha|| + \delta). \tag{19.32}$$

The probe quantities λ_i are the solution of the system of linear algebraic equations

$$\alpha \lambda_i + \sum_{j=1}^{n} \sum_{k=1}^{n} A_{ki} A_{kj} \lambda_j = \sum_{j=1}^{n} A_{ji} u_j, \qquad i = 1, 2, \dots, n \tag{19.33}$$

---

*In other words, the problem of this type belongs to ill-posed problem class and the regularization method by Tikhonov [89] should be used for its solution.

that are obtained from the minimum conditions of the functional M^α

$$\frac{\partial M^\alpha}{\partial \lambda_j^\alpha} = 0, \qquad j = 1, 2, \ldots, n. \tag{19.34}$$

At a given initial approximation of regularization parameter α, the substitution of the probe sequence $\{\lambda_i\}$, obtained from Eq. (19.33), into the condition (19.32) leads to the equation relative to the parameter α. Solving this equation, we obtain the "true" value of α. Then, the solution of Eq. (19.33) at the found value of the parameter α will be the sought solution of Eq. (19.22) (or, equally, the system of Eqs. (19.29)) relative to λ_i^α.

So, we see that the use of antineutrino detectors opens up possibilities for developing the neutrino on-line technology of temporal evolution of nuclear fuel isotopic structure and reactor power.

The first to exploit antineutrino detection as a tool for reactor monitoring were Russian physicists [90]. Multiple basic and applied experiments were conducted over the course of many years, beginning in 1982, at the Rovno Atomic Energy Station in Kuznetsovsk, Ukraine. The antineutrino source was a Russian VVER-440* loaded with low-enriched uranium (LEU) fuel[†] with a nominal power of 440 MWt. For the antineutrino flux to be detected, the inverse β-decay reaction (19.15) was used. Antineutrinos that was emitted in this reaction interacted with free protons present in the detection medium. The numerical value of the cross section per target proton for this interaction is of order 10^{-43} cm^2, which is large in comparison to most other possible antineutrino interaction mechanisms, and which enables cubic meter scale detectors at tens of meter standoff from standard nuclear reactors.

In the Rovno experiments, two neutrino detectors were in operation. The first was the scintillation spectrometer (ScS), while the second was the water integral neutrino detector (WIND). The ScS represented a modernized version of the organic liquid scintillator detector that was first realized by F. Raines and C. Cowan in 1953. In the ScS both the positron and the neutron are detected in close time coincidence compared to other backgrounds. The positron and its annihilation γ-quanta produce scintillation signals within a few nanoseconds of the antineutrino interaction: this is collectively referred to as the prompt scintillation signal. This is followed by a second, delayed scintillation flash arising from the cascade of γ-quanta, which come from decay of the excited nuclear state of the neutron-absorbing element following neutron capture. The scintillation light is recorded in photomultiplier tubes, with the number of scintillation photons proportional to the deposited positron and neutron-related energies and the time of each deposition recorded. The neutron capture can occur on hydrogen, for which a 2.2 MeV γ-quantum is released. Often, a dopant with a high neutron-absorption cross section is used to improve the robustness of the signal. This brings the dual benefits of reducing the capture time and increasing the energy released in the γ-cascade following capture, compared to 2.2 MeV for hydrogen. For example, a 0.1% concentration of gadolinium that has the highest neutron absorption cross section of any element, reduces the capture time from about 200 microseconds to tens of microseconds relative to hydrogen, and increases to 8 MeV the neutron-related energy available for deposition in the detector (arising from neutron capture de-excitation γ-quanta). The ScS that was used in Rovno experiments [88] consisted of 1050 liters of Gd-doped organic liquid scintillator, viewed through light guides by 84 photomultiplier tubes. A 510 liter central volume was used as the primary target, with a 540 liter surrounding volume, separated by a light reflecting

---

*Water-moderated water-cooled power reactor.

[†]LEU has a lower than 20% concentration of ^{235}U. For use in commercial light water reactors, the most prevalent power reactors in the world, uranium is enriched to 3 to 5% ^{235}U.

surface, employed as shield against external γ-quanta, and as a capture volume for gamma-rays emitted respectively by the positron annihilation and neutron capture. The deployed location of the detector was 18 meters vertically below the reactor core, providing substantial overburden for screening of muons. An active muon rejection system was apparently not used in this deployment. The intrinsic efficiency of this detector was approximately 30%, while the gross average daily antineutrino-like event rate was 909+6 per day, with a reactor-off background rate of 149+4 events per day (i.e., a net antineutrino rate of about 760+7 per day).*

In the WIND, the reaction (19.15) is detected only by produced neutrons with the help of the proportional ^3He-counters. The neutron detection consists of two stages. Neutrons with the energy of about few MeV are slowed down by means of collisions with protons of a scintillator and diffuse into multi-wire ^3He-counter where they are registered through the reaction

$$n + {}^3He \rightarrow p + {}^3H + 765 \text{ keV}. \tag{19.35}$$

Finally, two charged products of this reaction, proton and triton, cause ionization in multi-wire chamber. The ^3He-counters that were used in the Rovno experiments [91] had the internal diameter 31 mm, the length about 1 m and were filled up by ^3He and Ar. Distilled water, contained in a cubic tank with the edge 1300 mm, served both as a target for neutrinos and as a moderator for neutrons. 256 ^3He-counters put into water formed the 16×16 matrix with the step 70 mm.

Even the first Rovno experiments clearly demonstrated sensitivity to both the power and the isotopic content of the core, based on rate and spectral measurements. They revealed the expected variation in the antineutrino count rate due to the burnup effect. The 5–6% change in rate is well matched to a prediction based on a model of the fuel isotopic evolution in the core, combined with the theoretically predicted antineutrino emission spectra for each isotope. The absolute precision on the reconstructed thermal power of the reactor is 2%, with the largest uncertainties arising from imperfect knowledge of the detection efficiency and detector volume. For comparison, direct thermal power measurements by the most accurate methods have precision in the range 0.5–1.5% depending on the method. The change in the antineutrino spectrum over the course of the fuel cycle is also visible in the Rovno data. The variation in spectra is most pronounced at the highest energies, consistent with predictions, and is caused by the net consumption of 521 kg of fissile material (both plutonium and uranium) over the course of the fuel cycle. While not directly quoting an uncertainty on this value derived from the antineutrino measurement, the Rovno group independently estimated fuel consumption from the reactor's thermal power records, and found a value of 525+14 kg, close to the antineutrino derived value. Thus, the Rovno experiments clearly demonstrate many of the expected requirements for antineutrino-based safeguards.

Apart from using the inverse β-decay, there are other mechanisms for detecting reactor antineutrinos. They are radiochemical methods, antineutrino-electron scattering, and coherent antineutrino-nucleus scattering. Radiochemical methods refers to a family of possible inverse beta decay interactions with various nuclei, following the generic process

$$\bar{\nu}_e + (A, Z) + Q_\beta \rightarrow e^+ + (A, Z - 1), \tag{19.36}$$

where Q_β is the energy threshold for the interaction, and A and Z are the mass number and proton number of the target nucleus, respectively. The interaction probabilities decrease with increasing Z, and energy thresholds vary with isotope. Accounting for these effects,

---

*The ScS used in the beginning of Rovno experiments registered approximately useful antineutrino-like events per day.

only a few targets have rates comparable with the inverse beta decay on protons, most notably the transition $^3He \rightarrow {}^3H$, which has a very low (0.018 MeV) energy threshold. Generally, these approaches are undesirable for detection on time scales of days to months, either because the materials are inconvenient or expensive is in the case of He, or because they involve radiochemical recovery of converted isotopes following long exposure times. Concepts of operation involving long acquisition times might still employ detectors of this kind.

Antineutrino-electron scattering involves the exchange of the neutral gauge boson with an electron in a convenient target medium, with water being a common choice. It has an approximately 100-fold lower interaction probability per target than inverse beta decay on proton, requiring correspondingly larger detector sizes, and is, therefore, not as readily applicable for security applications.

Coherent antineutrino-nucleus scattering is collective neutral current interaction that occurs via coherent summing of the neutral gauge exchange amplitudes between the antineutrino and all the nucleons in any nucleus. The coherent addition of nucleon amplitudes makes the interaction probability significantly higher than inverse beta decay on protons, by factors of 10–100. While theoretically very well motivated, the very low energy transfer in the process makes it extremely difficult to detect—indeed no experiment has yet succeeded in measuring coherent neutrino scattering.

At present the antineutrino detectors having about one cubic meter in volume that are based on the reaction (19.15) possess high statistic accuracy 1–2% and could detect a net antineutrino rate of about 4000 per day. The relatively simple design of these detectors, with their low channel counts, readily available raw materials, and low maintenance requirement, demonstrates that the International Atomic Energy Agency's (IAEA) criteria for low cost, simple unattended remote monitoring capabilities can all be met. Twenty-eight years after the Russsian demonstrations at Rovno and six years after the first IAEA experts meeting on this topic, there are now many efforts underway around the world to explore the potential of the antineutrino detectors. The reader could find the evolution of these efforts in *Proceedings of annual "Applied Antineutrino Physics" Workshops*.

Instead of an epilogue

The deep philosophy is hidden in the great book—the Universe, always opened to our inquisitive look. But it's possible to read this book, only having learnt to understand its language, having learnt to read the letters from which it consists.

G. Galilei

Looking at the night firmament, it is difficult to imagine, that all this star magnificence and our Earth were born at the explosion of the fire ball compressed until the sizes of Planck's length (initial singularity). All symmetry and all laws defining further dynamics of the universe, have been programmed in an initial singularity just as DNA molecules predetermine the future of a person. Just as Big Ben tolls at midnight serving as a signal of the beginning of a new day, this explosion has announced the universe's birth. It has occurred simultaneously everywhere, having filled from the very beginning all space existing at that moment that has been closed on itself as a sphere surface. Even now, 14 billion years later, the universe feels the consequences of this explosion: huge islands of stars and galaxies, go different directions with a speed close to the light velocity. More recently (1998 1999) we have learned about the existence of one more component of the cosmological environment, a cosmic vacuum, thanks to which the universe's expansion never will stop and will never be replaced by compression. If the baryon matter and dark matter create the gravitation, the cosmic vacuum creates the antigravitation. In the modern universe the vacuum and antigravitation prevail over the usual matter and gravitation. Therefore, galaxies run up from each other with increasing speeds. So, the universe has a beginning in time, but it has neither beginning nor end in space. It will extend unlimitedly and it will face fading in a boundless cold. However, long before this sad ending, mankind will face problems of space scale. *Nothing is eternal under the moon* and our Sun is not an exception of this savage law.

Approximately in $\sim 10^{10}$ years in the Sun's core all hydrogen fuel will burn out. Therein radiating pressure will fall and the gravitational forces will start to compress the Sun; that, in turn, will lead to temperature increase and to helium burning. As at the transformation of helium into carbon, more amounts of energy will be already released, the luminosity of the Sun should increase. Energy release will lead to increase of radiation pressure at external layers of the Sun that will cause its expansion, that is, the Sun will turn into the red giant. At that moment the Sun first of all will burn the Earth to ashes (because of the increased release of energy), and then as a result of huge expansion will absorb its remains. Before it will occur, our descendants should leave the Earth and search for a haven on other planets of the Milky Way. However, even earlier they will unfortunately face one more problem. The galaxy nearest to us—Andromeda nebula, approaches us with a speed ~ 100 km/s. In five–six billions years, both galaxies will collide. It is obvious that it is better to be the observer of this event, than the direct participant, that is, mankind will be compelled to replace our galaxy with an other.

To so incomprehensible and senseless scenario of the world, we have come, leaning, first of all, on the existing standard model of elementary particle physics and unquestionable results of astronomical and astrophysical researches. Millstones of mathematics and experiment have reduced to dust all others, sometimes even more favorable for us, models of evolution of the universe. As it is sad, but at a such scenario our fine Earth is only a tiny and temporary island in a stunningly hostile ocean of the universe. An island, which all the same should be left to continue a human civilization. Only fundamental science and, first of all, physics can lead this history to a more or less happy end. As in the days of Copernicus, it was clear, that there is no force in the world that could stop scientific progress. Despite a rather wide spectrum of measures, even the inventive fathers of medieval inquisition did not manage to make this. And the numerous official brotherhood, demanding an every second practical exit from any scientific research and, in any way, not wishing to understand that progress is an indissoluble chain of *fundamental science* $\longrightarrow$ *applied science* $\longrightarrow$ *production*, has not managed to stop the development of fundamental science. Although, apparently, it is simple to understand. Actually, take any device or mechanism and track the history of its production back in time. You are convinced, that there was always a source, the certain law discovered in fundamental science. Unfortunately, the time interval between the discovery of this law and the masterpieces of human hands often lasts decades, and even longer. So, created by Maxwell in 1860–1865, the theory of an electromagnetic field was realized in technological discoveries only at the end of the 1890s, and the positron that had been predicted by Dirac in 1928 and was discovered in cosmic rays already in 1932, did not become an energy source at our power stations until now.

For the present time, classical physics as the source of producing new technologies has practically exhausted its possibilities. New directions are based on discoveries in the standard model of the strong and electroweak interactions already constructed. Controlled thermonuclear fusion, the appearance of neutrino astronomy and geotomography, nanotechnologies, and the prospect of nuclear weapons destruction with the help of the collider neutrinos can serve as convincing examples. The time will come and the Gauge Unified Theory will become a source of new technologies and, subsequently, it will be replaced by the Unified Field Theory.

Will it be possible to assert, that the picture of a structure of the Universe constructed for today gives answers to all questions? Unfortunately, for today not all experimental facts find explanations. Thus, for example, within the standard cosmological models (SC) confusing uncertainty concerning the initial singularity until now remains; there is not any theoretical bases for calculation of the modern density of the cosmic vacuum. Moreover, the relict radiation, which registration has played an important role in the formation of the Big Bang model, has seriously puzzled theorists-cosmologists. Recently, data on fluctuations of background microwave radiation by means of the automatic space station MAP (Microwave Anisotropy Probe) have been obtained. The information collected by MAP allowed us to construct the detailed map of small fluctuations of relic radiation temperature in the universe. The mean temperature is now near $T_m = 2.73$ K, distinguished on different sites of the heavenly sphere on parts per million of degrees. It has thus appeared, that the *cold* and *warm* areas discovered by space telescope are located in the heavenly sphere not in a random way, as it should be after the Big Bang, and built in a certain order concerning an axis called the *Harm Axis* (Fig. 20.1). Moreover, within the Galaxy Zoo project it was clarified, that the number of visible spiral galaxies twirled clockwise is not corresponding to that of the visible spiral galaxies twirled in the opposite side.* During the studying of the orientation of 1660 spiral galaxies it was found out, in particular, that they are twirled mainly in a direction of the Harm Axis, that does not keep within the framework of the SC. As follows from Fig. 20.1, the Harm Axis penetrates all the observable universe and divides

---

*From the viewpoint of the terristrial observer, these numbers must be equal in accordance with the Big Bang theory.

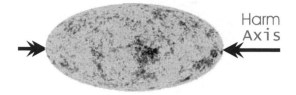

FIGURE 20.1

The map of the relic radiation in the universe (WMAP Science Team, NASA).

it in two parts. As this takes place, at the top and bottom parts of a map of relic radiation it is possible to find regions of equal temperatures that are symmetrized. On the other hand, these symmetric parts containing bipolar temperatures, also are the unified universe. The received map of relict radiation could be considered as one more acknowledgment of a principle of the duality underlying the universe: *All real—bipolar.* For a live matter such poles, for example, are: "Good and Evil," "Life and death," "Science and art." In the sphere of the lifeless nature examples can be: "Corpuscular-wave dualism," "Matter and antimatter," "Gravitation and antigravitation," "Holographic duality", and the same "Harm Axis." Such is the reality, and the unipolar world would appear like the play where there are no dynamics, and the end would be guessed at already right after curtain raising.

A variety of questions in the physics of a microworld is not solved also. Until now the mechanism of spontaneous symmetry breaking is not established and the definitive choice between the existing GUT is not made. Only the first steps in the construction of the Unified Field Theory (UFT) are being made. However, in the near future increasing the energy of accelerators and improving the accuracy of observations in γ- and in neutrino-astronomy will lead to an increase in the number of reliable experiments as in microphysics and the physics of a macrocosmos. Time will show the cost of their explanations within the existing picture of the world. At that, one ought not to forget that the considerable amount of experimental facts have already found solutions both within the SC and within the standard model describing the electroweak and strong interactions. This, in turn, means that the modern theory of the universe construction could appear approximate at worst, but not wrong in any way. As to the SC it is necessary to remember that it will be considered accomplished only after the creation of a successful gravitation theory, that is, after construction of the UFT. So, for today the basic contours of the theory describing the structure and evolution of the universe are established only, but also behind them already even now it is possible to make out a prototype of a true picture of the world. So that is how things are.

References

[1] T. P. Cheng and L. F. Li, Phys. Rev. **D22**, 2860 (1980).

[2] A. Zee, Phys. Lett. **B93**, 389 (1980).

[3] L. Wolfenstein, Nucl. Phys. **B175**, 93 (1980).

[4] T. D. Lee, Phys. Rev. **D8**, 1226 (1973).

[5] S. L. Glashow and S. Weinberg, Phys. Rev. **D15**, 1958 (1977).

[6] J. Velhinho, R. Santos, and A. Barroso, Phys. Lett. **B322**, 213 (1994).

[7] B. W. Lee, R. E. Schrok, Phys. Rev. **D16**, 1444 (1977).

[8] A. Zee, Phys. Lett. **B161**, 141 (1985).

[9] R. N. Mohapatra, Phys. Lett. **B201**, 517 (1988).

[10] M. V. Voloshin, Yad. Phys. **48**, 804 (1988).

[11] R. Barbieri, R. N. Mohapatra, Phys. Lett. **B218**, 225 (1989).

[12] L. Wolfenstein, Phys. Rev. **D17**, 2369 (1978).

[13] S. P. Mikheev, A. Yu. Smirnov, Yader. Fiz. **42**, 1441 (1985).

[14] L. Landau, Phys. Zeits. d. Sowjetunion **2**, 46 (1932).

[15] S. J. Parke, Phys. Rev. Lett. **57**, 1725 (1986); W. C. Haxton, Phys. Rev. Lett. **57**, 1271 (1986).

[16] C. Zener, Proc. Roy. Soc. **A137**, 696 (1932).

[17] P. Pizzochero, Phys. Rev. **D36**, 2293 (1987).

[18] T. K. Kuo and J. Pantaleone, Phys. Rev. **D39**, 1930 (1989).

[19] S. J. Parke S. J., Phys. Rev. Lett. **57**, 1275 (1986); W. C. Haxton, Phys. Rev. **D35** 2532 (1987); S. T. Petcov, Phys. Lett. **B191**, 299 (1987).

[20] S. Toshev, Phys. Lett. **B196**, 170 (1987); T. Kaneko, Progr. Theor. Phys. **78**, 532 (1987).

[21] M. Voloshin, M. Vysotskii, and L. B. Okun, ZETP, **91**, 754 (1986).

[22] O. M. Boyarkin and T. I. Bakanova, Phys. Rev. **D62**, 075008 (2000).

[23] O. M. Boyarkin, Phys. Rev. **D53**, 5298 (1996).

[24] V. P. Meytlis, H. R. Strauss, Solar Physics **145**, 111 (1993).

[25] M. Altman et al., Phys. Lett. **B490**, 16 (2000).

[26] J. N. Bahcall, S. Basu, and M. N. Pinsonneault, Phys. Lett. **B443**, 1 (1998); J. N. Bahcall, S. Basu, and M. N. Pinsonneault, Asrophys. J. **555**, 990 (2001).

[27] G. E. Kocharov, Yu. N. Starbunov, Acta Phys. Acad. Sci. Hung. **29**, 353 (1970).

[28] P. Demarque et al., Nature **246**, 33 (1973).

[29] A. J. Prentice, Mon. Not. R. Astron. Soc. **163**, 331 (1973).

[30] D. D. Clayton, M. J. Newman, R. J. Talbot, Asrophys. J. **201**, 489 (1975).

[31] Y. Fukuda et al. (Super-Kamiokande Collaboration), Phys. Rev. Lett. **81**, 1562 (1998).

[32] Q. R. Ahmad et al. (SNO Collaboration), Phys. Rev. Lett. **89**, 011301 (2002).

[33] K. Egushi et al. (KamLAND Collaboration), Phys. Rev. Lett. **90**, 021802 (2003).

[34] S. Fukuda et al. (Super-Kamiokande Collaboration), Phys. Rev. Lett. **86**, 5651 (2001).

[35] K. Zuber, *Neutrino Physics*, (IOP Publ., Bristol, 2004).

[36] X. Shi and D. N. Schramm, Phys. Lett. **B283**, 305 (1992).

[37] B. Aharmim. et al. (SNO Calloboration), Phys. Rev. **D72**, 055502 (2005).

[38] G. L. Fogli, E. Lisi, Astropart. Phys. **3**, 185 (1995).

[39] D. Muller, Astrophys. J. **374**, 356 (1991).

[40] M. J. Rayan, J. F. Ormes, and V. K. Balasubrahmanyan, Phys. Rev. Lett. **28**, 985 (1972).

[41] R. Belotti et al., Phys. Rev. **D60**, 052002 (1999).

[42] M. Boezio et al., Astrophys. J. **518**, 457 (1999).

[43] T. Sanuki et al., Astrophys. J. **545**, 1135 (2000).

[44] J. Alcaraz et al., Phys. Lett. **B490**, 27 (2000).

[45] E. N. Parker, *Interplanetary Dynamical Processes*, (Interscience, New York, 1963).

[46] C. Stormer, *The Polar Aurora*, (Oxford Clarendon Press, London, 1955).

[47] M. Fukugita, T. Yanagida, *Physics of Neutrinos*, (Springer, 2003).

[48] L.V. Volkova, Yader. Fiz. **31**, 1510 (1980).

[49] F. Reines et al., Phys. Rev. Lett. **15**, 429 (1965).

[50] C. V. Achar et al., Phys. Lett. **18**, 196 (1965).

[51] T. Kajita, Y. Totsuka, Rev. Mod. Phys. **73**, 85 (2001).

[52] M. Aglietta et al., Eur. Phys. Lett. **8**, 611 (1989).

[53] K. Daum et al., Zeits. Phys. **C66**, 417 (1995); C. Berger C. et al., Phys. Lett. **B227**, 489 (1989).

[54] K. S. Hirata et al., Phys. Lett. **B205**, 416 (1988).

[55] R. Becker-Szendy et al., Phys. Rev. **D46**, 3720 (1992); R. Clark, Phys. Rev. Lett. **79**, 491 (1997).

[56] W. A. Mann et al., Nucl. Phys. Proc. Suppl. **91**, 134 (2000).

[57] Y. Fukuda et al., Phys. Lett. **B433**, 9 (1998); Phys. Lett. **B436**, 33 (1998).

[58] T. Schwetz, M. Tortola, and J. W.F. Valle , New J. Phys. **10**, 113011 (2008).

[59] , P. Adamson et al. (MINOS Collaboration), Phys. Rev. Lett. **101**, 131802 (2008).

[60] B. Aharmim et al. (SNO Collaboration), Phys. Rev. Lett. **101**, 111301 (2008).

[61] S. Abe et al. (KamLAND Collaboration) Phys. Rev. Lett. **100**, 221803 (2008).

[62] The Borexino Collaboration, Phys. Rev. Lett. **101**, 091302 (2008).

[63] M. C. Gonzalez-Garcia and M. Maltoni, Phys. Rept. **460**, 1 (2008) [arXiv:0704.1800].

[64] M. Apollonio et al. Phys. Lett. **B466**, 415 (1999).

[65] S. Goswami and A. Yu. Smirnov, Phys. Rev. **D72**, 053011 (2005).

[66] G. L. Fogli et al., Phys. Rev. Lett. **101**, 141801 (2008).

[67] The CHORUS Collaboration, Phys. Lett. **B497**, 8 (2001); The NOMAD Collaboration, Nucl. Instr. and Meth. **A404**, 96 (1998).

[68] R. Acquafredda et al. (OPERA Collaboration), JINST **4**, P04018 (2009).

[69]) P. Huber et al., JHEP **0911**, 044 (2009).

[70] S. Dodelson, *Modern cosmology*, (Academic Press, 2003).

[71] J. Lesgourgues and S. Pastor, Physics Reports **429**, 307 (2006).

[72] A. Strumia, F. Vissani, arXiv:hep-ph/0606054v2.

[73] S. Bergmann et al., Phys. Rev. **D62**, 073001 (2000).

[74] N. Fornengo et al., Phys. Rev. **D65**, 013010 (2002).

[75] O. M. Boyarkin, D. Rein, Phys. Rev. **D53**, 361 (1996).

[76] O. M. Boyarkin, T. I. Bakanova, Phys. Rev. **D62**, 075008 (2000).

[77] O. M. Boyarkin, Yader. Fiz. **64**, 354 (2001); O. M. Boyarkin, G. G. Boyarkina, Phys. Rev. **D67**, 073023 (2003).

[78] W. Buchmuller and M. Plumacher, Int. J. Mod. Phys. **A15**, 5047 (2000).

[79] O. M. Boyarkin, G. G. Boyarkina, and T. I. Bakanova, Phys. Rev. **D70**, 113010 (2004).

[80] O. M. Boyarkin and D. Rein, Zeits. Phys. **C67**, 607 (1995).

[81] J. M. Herndon, Proc. Nat. Acad. USA **100**, 3047 (2003).

[82] P. Vogel, Phys. Rev. **D29**, 1918 (1984); S. A. Fayans, Nucl. Phys. **B42**, 929 (1985).

[83] P. M. Rubtsov et al., Nucl. Phys. **B46**, 1028 (1987).

[84] V. D. Rusov et al., J. Appl. Phys. **96**, 1734 (2004).

[85] V. D. Rusov, V. I. Vysotsky, and T. N. Zelentsova, Nucl. Radiat. Safety **1**, 66 (1998).

[86] V. I. Kopeykin, L. A. Mikaelyan, and V. V. Sinev, Nucl. Phys. **B60**, 230 (1997).

[87] P. Vogel et al., Phys. Rev. **C24**, 1543 (1981).

[88] A. A. Borovoy et al., Nucl. Phys. **B36**, 400 (1982).

[89] A. N. Tikhonov and V. Y. Arsenin, *Methods of Ill-posed Problems Solution*, (Nauka, Moscow, 1979).

[90] L. Mikaelyan, in *Proc. Intern. Conf. "Neutrino 77"*, (Nauka, Moscow) **2**, 383 (1978).

[91] A. A. Kuvshinnikov et al., Yader. Fiz. **52**, 472 (1990).

Index